Meramec Library
St. Louis Community College
11333 Big Bend Road
Kirkwood, MO 63122-5720
314-984-7797

WITHDRAWN

Space Exploration and Humanity

Space Exploration and Humanity

A Historical Encyclopedia

VOLUME 1

History Committee of the American Astronautical Society

Stephen B. Johnson, General Editor

Timothy M. Chamberlin, Michael L. Ciancone, Katherine Scott Sturdevant, and Rick W. Sturdevant, Section Editors

David Leverington, Technical Consultant

Santa Barbara, California • Denver, Colorado • Oxford, England

Copyright 2010 by ABC-CLIO, LLC

All rights reserved. No part of this publication may be reproduced, stored in a retrieval system, or transmitted, in any form or by any means, electronic, mechanical, photocopying, recording, or otherwise, except for the inclusion of brief quotations in a review, without prior permission in writing from the publisher.

Library of Congress Cataloging-in-Publication Data

Space exploration and humanity : a historical encyclopedia / History Committee of the American Astronautical Society ; Stephen B. Johnson, general editor ; Timothy M. Chamberlin . . . [et al.], section editors.

p. cm.

Includes bibliographical references and index.

ISBN 978–1–85109–514–8 (hard copy : alk. paper) — ISBN 978–1–85109–519–3 (ebook)

1. Astronautics—History. 2. Astrophysics—History. 3. Astronautics—United States—History. 4. Outer space—Exploration—History. 5. Astronautics and civilization. 6. Astronautics—Historiography. 7. Astrophysics—Historiography. I. Johnson, Stephen B., 1959– II. American Astronautical Society. History Committee.

TL788.5.S636 2010

629.403—dc22 2010009664

ISBN: 978–1–85109–514–8

EISBN: 978–1–85109–519–3

14 13 12 11 10 1 2 3 4 5

This book is also available on the World Wide Web as an eBook.
Visit www.abc-clio.com for details.

ABC-CLIO, LLC
130 Cremona Drive, P.O. Box 1911
Santa Barbara, California 93116-1911

This book is printed on acid-free paper ∞

Manufactured in the United States of America

Portions of the glossary are reprinted from New Cosmic Horizons: Space Astronomy from the V2 to the Hubble Space Telescope (copyright (c) 2000 David Leverington) and Babylon to Vogager and Beyond (copyright (c) 2003 David Leverington). Both reprinted with the permission of Cambridge University Press.

Contents

VOLUME 1

VOLUME 2

Preface

The American Astronautical Society (AAS) strives to be the premier network of professionals, both technical and non-technical, dedicated solely to the exploration and development of space. Our membership includes engineers, scientists, teachers, lawyers, historians, artists, journalists, and entrepreneurs.

Our dedication to space is expressed as we host pre-eminent conferences on topics as varied as space law and astrodynamics, and as we bestow honors such as the John F. Kennedy Astronautics Award, the Carl Sagan Memorial Award, and the Eugene M. Emme Award for Astronautical Literature.

We are proud to have had members of the AAS History Committee provide distinguished service to this publishing effort both as members of the editorial staff and as authors of various articles. This tradition of excellence through leadership has enabled AAS to flourish.

We are also very proud to include this space history encyclopedia in our family of distinguished publications, which includes the AAS History series. As the encyclopedia articulates the challenges and accomplishments of the past, it lays a vital foundation for the future.

Mark K. Craig
President
American Astronautical Society

Foreword

Since the early 1960s, space has become an important part of the human story. Weather satellites beam images that allow scientists to forecast hurricanes and save lives, while other Earth-observing satellites take the pulse of the planet in many ways. Communications satellites make possible instant messaging, international radio broadcasts, and live television from around the world. Global Positioning System satellites give people their location on the planet to within a few meters. Reconnaissance satellites are essential to national security. Peering outward, space probes have studied the universe, taken humankind back to the beginning of time, and revealed its place in the history of cosmic evolution. Humans have seen Earth from the Apollo spacecraft, witnessed Earthrise from the Moon, and imaged a pale blue dot from a vantage point at the edge of the solar system. People have learned about life in outer space, from the extension of humankind into the cosmos—exploring the space environment, conducting experiments in microgravity, and constructing the space station—and from the search for extraterrestrial life. In the future, space will become more important as commercial possibilities become reality.

This two-volume encyclopedia on *Space Exploration and Humanity* fills a large gap in the literature of space exploration. While many encyclopedias of space have been published, this is the first comprehensive space history encyclopedia. Many of the articles are written by space historians, and all of the articles collectively describe, in a historical way, how the space enterprise has evolved during the past 50 years. The encyclopedia covers the scientific and technical sides of human and robotic spaceflight and also civilian, commercial, and military applications. It is unique in giving attention to societal aspects of the space program, an area overdue for attention. The reader will find entries related to media and popular culture, space education, and the economics of spaceflight.

The entries are international in scope, and show how space exploration has evolved from the purview of two superpowers to become an essential activity of many nations. In this encyclopedia, the reader can find a succinct summary of the Soviet/Russian space program ("Russia") and the Chinese human space missions ("Shenzhou"), read about the development and theory of strategic uses of space ("Military Space Doctrine"), gain an overview of ballistic missiles ("Ballistic Missile"), and find the history and results of many spacecraft that have studied the solar system (for example,

"Voyager") and the universe beyond (for example, "*Hubble Space Telescope*," "*COBE, Cosmic Background Explorer*," and "*WMAP, Wilkinson Microwave Anisotropy Probe*"). Readers can also discover how to have their ashes lofted into space, as space enthusiasts Gene Roddenberry and Gerard K. O'Neil have done ("Burial Services").

Space Exploration and Humanity will provide an entrée into the fascinating and far-reaching issues in the growing field of space history. The American Astronautical Society (see entry) is to be congratulated for sponsoring a volume that will become a standard reference in the field.

Steven J. Dick
NASA Chief Historian
Washington, DC

Acknowledgments

This encyclopedia is the result of the efforts of more than 100 authors and several editors and also the patience and guidance of the project coordinators at ABC-CLIO and AAS. At ABC-CLIO, Steve Danver kicked off this project and deserves much credit for envisioning the need and getting it started and funded. David Tipton was able to see it through; both Steve and David provided encouragement and an occasional kick in the pants as needed! Jim Kirkpatrick at AAS provided consistent support for this work.

Over the course of the seven years from the beginning to end of manuscript development, several people stand out. Mike Ciancone and Rick Sturdevant volunteered at the beginning to be area editors, and of our initial editors they alone were able to see their sections through from start to finish. David Leverington acted as our technical consultant for space science, meteorology, and Earth remote sensing, and also for European topics. His technical expertise and attention to high quality ensured the accuracy of our most technically difficult areas and helped keep us on the straight and narrow way. David also provided his glossary definitions and permission to utilize them from his two books, *New Cosmic Horizons*, and *Babylon to Voyager and Beyond*. I greatly appreciated these, as it saved time and ensured technical accuracy for some of the most difficult technical terms. Joni Wilson was our dedicated and effective copyeditor; her timely work and determination to work through more than 600 items, some several times, were invaluable. Tim Chamberlin, Katie Berryhill, and Kathy Sturdevant all contributed significantly as area editors and as our education editor. Howard Trace served as our ace fact-checker, which is a tedious but crucial task that we desperately needed. Bill Dauphin and Llyn Kaimowitz were able to move two of our sections along before other events pulled them from the project, and I appreciate their efforts. We also had important contributions from several authors who stepped in to research and write difficult articles. David Leverington, John Ruley, and David Hartzell in particular were instrumental in effectively writing some of our toughest and most problematic articles. Tim Chamberlin, Katie Berryhill, and Rick Sturdevant researched and wrote many articles, as did Peter Gorin and Bart Hendrickx, who took on the most of our difficult Soviet and Russian articles.

Finally I thank all of our authors for their contributions in writing one or several articles, in addition to their patience in seeing this effort through to its successful conclusion. Without their support, this project would not have gotten started, let alone completed.

Stephen B. Johnson

About the Editors

Timothy M. Chamberlin is design editor for the *Tulsa World* newspaper in Tulsa, Oklahoma. His work includes coverage of the space shuttle *Columbia* accident and the future of spaceflight for *The State* newspaper in Columbia, South Carolina. He has developed a website for finding relevant information about current U.S. space policy and law, and has written about space advisory committees appointed by the president for *Space Times* magazine. He is an honorary member of the American Astronautical Society (AAS) and serves as editor of the AAS History Committee's newsletter *Explorer*. Tim holds an MS in space studies from the University of North Dakota, North Dakota. He serves as the section editor for the Human Flight and Microgravity Science section of this encyclopedia.

Michael L. Ciancone is an engineer at the NASA Johnson Space Center in Houston, Texas, where he provides technical support on human spaceflight safety to the Constellation Program. He is also a bibliophile who has maintained an active interest in pre-Sputnik rocket societies and spaceflight visionaries, as well as the cultural and social impacts of spaceflight. In connection with these interests, he chairs the American Astronautical Society (AAS) History Committee and coordinates the review panel for the annual Eugene M. Emme Award for astronautical literature. He is a member of the History Committee of the International Academy of Astronautics. Michael has written papers on spaceflight visionaries such as David Lasser and Luigi Gussalli, and has served as curator for an exhibit at the Western Reserve Historical Society on "Cleveland and Outer Space, The Cleveland Rocket Society (1933–37)." He is the author of *The Literary Legacy of the Space Age—An Annotated Bibliography of Pre-1958 Books on Rocketry & Space Travel* (1998) and the volume editor for the papers presented during the International Academy of Astronautics (IAA) History Symposium of the 2002 World Space Congress in Houston, Texas. He serves as the section editor for the Space and Society section of this encyclopedia.

Dr. Stephen B. Johnson is associate research professor at the National Institute for Science, Space, and Security Centers at the University of Colorado at Colorado Springs, Colorado, and is currently assigned to the NASA Marshall Space Flight

Center for the Ares I and Constellation programs. He acquired his doctorate at the University of Minnesota in the history of science and technology, and has written *The Secret of Apollo*, which won the 2002 Eugene M. Emme Award for astronautical literature, *The United States Air Force and the Culture of Innovation 1945–1965*, numerous journal articles in publications such as *Technology and Culture*, *History and Technology*, *Air Power History*, and *Journal of Industrial History*, and essays in books including *Critical Issues in the History of Spaceflight*, *The Societal Impact of Spaceflight*, and *The Business of Systems Integration*, and was the editor of *Quest: The History of Spaceflight Quarterly* from 1998 to 2005. Stephen has contributed to encyclopedias such as *The Readers Guide to the History of Science* and *American National Biography*. He is the general editor for this encyclopedia and the section editor for the Astrophysics and Planetary Science, Civilian and Commercial Space Applications, and Technology and Engineering sections.

Dr. David Leverington, now retired, is a writer on the history of astronomy and space research. Educated at the University of Oxford, UK, with a degree in physics, he was, in the 1970s, design manager of GEOS, Europe's first geosynchronous scientific satellite, and program manager, at the European Space Agency, of *Meteosat*, Europe's first meteorological satellite. Later he was engineering director in BAE Systems' Space Division, responsible, among other things, for *Giotto*, the spacecraft that intercepted Halley's Comet in 1986, the Photon Detector Assembly for the *Hubble Space Telescope*, the Envisat/Polar Platform, the Medium Energy Detector Assembly for *Exosat*, subsystems on the *ISEE-B*, *Ulysses*, and *Hipparcos* satellites, and remote sensing instruments that flew on the American *UARS* and *TIROS N* satellites. He was deputy CEO of British Aerospace Communications and on the management board of the UK's Earth Observation Data Centre. He has written three books: *A History of Astronomy from 1890 to the Present*, *New Cosmic Horizons*; *Space Astronomy from the V2 to the Hubble Space Telescope*; and *Babylon to Voyager and Beyond: A History of Planetary Astronomy.* David has been published in many journals including *Nature* and the *Quarterly Journal of the Royal Astronomical Society*, and is a contributor to Elsevier's *Encyclopedia of the Solar System, 2nd Edition*. He is included in the marquis' *Who's Who in the World 2006* and is a fellow of the UK's Royal Astronomical Society. He serves as technical consultant for this encyclopedia's Astrophysics and Planetary Science section and its meteorological and civilian Earth observation articles.

Katherine Scott Sturdevant, MA and PhD candidate, is professor of history at Pikes Peak Community College in Colorado Springs, Colorado. She has taught a wide range of undergraduate American History courses for 25 years and has been a scholarly historical editor for nearly 30 years. She authored *Organizing and Preserving Your Heirloom Documents* and *Bringing Your Family History to Life through Social History*, received state and national teaching excellence awards, and is active in many forms of curriculum innovation, public history publishing, interpretative training, source collection and preservation, historic preservation, and speaking for general audiences. She acted as the education editor for the early stages of this encyclopedia, ensuring at the outset that the level of writing and editing is appropriate for the target audience of high school seniors, college freshmen, and general readers.

Dr. Rick W. Sturdevant is deputy director of history at Headquarters Air Force Space Command, Peterson Air Force Base in Colorado Springs, Colorado. He acquired his PhD in 1982 from the University of California, Santa Barbara, California, and joined the U.S. Air Force history program in April 1984 as chief historian for Air Force Communication Command's Airlift Communications Division at Scott Air Force Base, Illinois. In 1985, he moved to Peterson Air Force Base to become chief historian for the Space Communications Division, and in 1991 moved to the Air Force Space Command (AFSPC) history office. In addition to producing classified periodic histories and special studies for AFSPC, Rick has published extensively on the subject of military aerospace history in such periodicals as *Space Times*, *High Frontier: The Journal for Space & Missile Professionals*, and *Journal of the British Interplanetary Society*, and essays in a number of books including *Beyond the Ionosphere: Fifty Years of Satellite Communication*, *To Reach the High Frontier: A History of U.S. Launch Vehicles*, and *Harnessing the Heavens: National Defense through Space*. He has also contributed to encyclopedias such as *Air Warfare: An International Encyclopedia* and *Encyclopedia of 20th-Century Technology*. Rick is editor of the International Academy of Astronautics and American Astronautical Society (AAS) history series. A recipient of the Air Force Exemplary Civilian Service Award and the AAS President's Recognition Award, he was elected an AAS Fellow in 2007. He serves as the section editor for the Military Applications section of this encyclopedia.

Encyclopedia Purpose, Structure, and Standards

In the summer of 2002 academic reference publisher ABC-CLIO approached the Smithsonian Institution National Air and Space Museum (NASM) proposing to create the first encyclopedia of space history. NASM personnel recommended that ABC-CLIO discuss the proposal with the History Committee of the American Astronautical Society (AAS), a particularly active group that has managed the editing and publication of the Univelt History Series that contains the proceedings of the History Symposia of the International Academy of Astronautics from the International Astronautical Congresses since 1967. After discussions in the fall of 2002, the History Committee agreed to take on the project, and AAS and ABC-CLIO signed an agreement to develop the product you are now reading. Members of the History Committee formed an editorial team, consisting of a general editor, section editors, and a copyeditor.

Encyclopedia Goals, Scope, and Limitations

From the beginning, the AAS editorial team took an expansive view of what a space history encyclopedia should contain. It should move far beyond the traditional topics of human spaceflight, space technology, and space science to include political, social, cultural, and economic issues, and also commercial, civilian, and military applications.

A key early decision was to create brief, 100-word, sidebar biographies, some with images, to introduce important individuals, rather than preparing full-length articles on individuals. There were two reasons for this decision: the fact that bibliographic encyclopedias already exist, and the inclusion of some people but not others would lead to disputes about who should and should not be included. The editorial team concluded that relevant people would be identified in the relevant articles. We do not mention people merely because they held a prominent position. Rather they had to have done something "historically important," such as make a discovery, institute a key change, or perform some other action worthy of mention. Of course, what good is a rule without exceptions? The committee decided that two individuals, Wernher von Braun and Sergei Korolev, had made such wide-ranging contributions to space history that they warranted specific full-length articles.

The editorial team also needed to establish the academic level to which the encyclopedia should be written. Addressing this issue required accounting for ABC-CLIO's marketing strategy. As a publisher of history encyclopedias primarily for university and high school libraries, the most plausible level for ABC-CLIO was that of the high school senior and university undergraduate. For most topics, this level was relatively easy to achieve, but some topics were virtually impossible to write about at this level without dramatic expansion to explain esoteric topics. The most difficult topics were those on astrophysics and certain aspects of planetary science articles, where the need to explain and the need to be concise could not be jointly met. In these cases we decided that being concise was the more critical factor, with the inclusion of a glossary to explain key terms. A table of acronym definitions is equally important, given the penchant in aerospace for multiple acronyms.

The typical "use case" envisioned by the committee was a student performing research for a history paper, in which space topics were a primary or secondary topic within that paper. We envisioned each encyclopedia article as a source of basic information and as the starting point for further research. As such, we then decided that we would include with each article the top three or four sources to which the student could go to for further information. This also makes the encyclopedia useful for professional historians and other scholars.

Authors, Reviews, and Fact Checks

The credibility of any encyclopedia depends on the quality of the articles, and this in turn depends on the quality of the authors, combined with the thoroughness of the review process. The author-recruiting strategy was straightforward. For topics on which historians have done the research and published, we attempted to recruit that historian to write the article. In these pages you will find many of the leading space historians represented. In cases where we were unsuccessful in getting "the" expert historian, or for topics on which no historical research had been published, we next tried to find a scientific or technical expert who participated in the project or subject in question, or who was a scientific or technical expert in that discipline or topic. For topics in which neither of these approaches was successful, we recruited others, often students in graduate space programs, to research and write the article.

All articles have been reviewed by the area editor and the general editor, and all articles were copyedited, which included extensive fact-checking. In addition, for topics on which the editors did not feel sufficiently competent to review, we searched for expert outside reviewers. In addition we performed a series of random and targeted spot-checks. Despite all these checks, it is certain that some errors remain; we hope that these are few, and minor!

Because we engaged the services of many authors, the role of the editorial staff was particularly important as we worked to ensure consistent depth of detail, use of terminology, and tone from one article to the next and among areas.

Structure

The arrangement of the encyclopedia's articles was another critical decision. The usual trade is a balance between topical and alphabetical arrangement. Given our target audience of high school and undergraduate students, the editorial team decided that a topical structure would be more useful, because it would allow students to find articles on similar subjects grouped together. This would allow them to browse through the pages near the article they first looked up to find other topics that might be useful to them, but of which they might not be aware. Within each level of the topical hierarchy, the articles are arranged alphabetically.

The hierarchical structure affects the contents of the "See also" references listed at the end of each article. Articles that are above, below, or in the same subsection as the article being reviewed do not appear in the See also listing because they are organizationally proximate. The See also lists therefore refer to articles of interest that are organizationally distant. In the Table of Contents, the section, subsection structure is clearly marked by indenture levels. These indenture levels are reflected in the font sizes of the articles in the main text, and also in the page headers. Left page headers show the Level 1 and 2 Table of Contents location of the articles on the page, and Right page headers show the Level 3 and 4 Table of Contents location of the articles on the page. Together, they provide a clear indicator of your location in the encyclopedia, with respect to the entire Table of Contents.

Another critical aspect of the encyclopedia's hierarchical structure was the need to provide historical integration across articles and topics. If we wrote only about a plethora of projects, this would provide a misleading picture of how each project fit into a larger evolution of similar or related projects. The area and subarea overview articles focus on showing the time evolution and interconnections among the subjects within each area. At higher levels, the overviews provide connections to ever-larger topics, such as the relationship among military, civilian, and commercial activities, or the relationship among political, economic, and military strategies of different nations. For each area overview, a timeline of events is provided, which gives a sense of the major events for each major subject area. Preparing timelines also proved to be an effective mechanism to detect errors in and among articles.

Article Contents

The intent of each article is to present as concisely as possible the basic historical facts of the subject, along with its historical connection to other topics. Given the wide range of topics in the encyclopedia, there is no single method that is appropriate for all articles and subjects. However, a few things can be said about the goals and approaches used to provide this information.

Because this is a history encyclopedia, as opposed to a technical encyclopedia, priority is given to the development of the topic over time, as opposed to a pure technical description of "what it is." For a historical encyclopedia, it is critical to identify the people and organizations involved, and when. As an example, in the article on the planet Jupiter, the reader will indeed learn many of the scientific facts about Jupiter and its moons. However, this is done by identifying who discovered each fact, when this occurred, how it was discovered, and how this discovery related to earlier and later observations. We do not have a table showing the specific facts about Jupiter itself. Tables in this encyclopedia generally provide lists of the major projects (and key facts about them, including dates) involved with a topic, and in some complex spacecraft, a list of its instruments and their specifications and capabilities. In the case of instruments, the list of capabilities matters only because many discoveries were based on the capabilities of these instruments, as compared to less capable instruments of previous spacecraft observing the same phenomena.

For articles on science missions, and for many other spacecraft-related articles, we developed an informal template of the information that was most effective. This generally included a brief description of the project and its goals, who (or which organizations) designed, built, managed, and operated the spacecraft, including prime contractors instead of only governments because, even for government-funded systems, contractors developed the systems and thus had critical but frequently overlooked roles. We then described its launch and operational history, and finally the key observations or discoveries made by the spacecraft, with a brief description of why that was important in a broader technical and/or historical sense.

We have also attempted to avoid a common pitfall of writing a history that is heavily influenced by one perspective. To that end, we attempted to ensure that authors avoided adopting a NASA-centric or U.S.-centric view. The encyclopedia is still more U.S.-centric than we would like, but given the state of historical research and the fact that this is written in English, this is difficult to avoid.

Editorial Standards and Quirks

Through the course of developing the encyclopedia, we decided on a number of editorial standards, and resolved a number of issues. A few of the more interesting ones are noted here.

- All acronyms are defined when used for the first time in every article, either in the article or in a table associated with an article, though not necessarily both.
- NASA (National Aeronautics and Space Administration) and U.S. (United States) are the only two acronyms not defined in first usage.
- Acronyms are generally in all capital letters. There are some exceptions, such as Intelsat and Inmarsat, as these are typically not capitalized in common usage.
- Some organizations that are in all capital letters are not actually acronyms. Three examples are RAND, TRW, and EUMETSAT. These are derived from words that were then shortened to be acronyms. However, the organizations use the acronym as the actual name, and they are in capital letters as the organization's name.

- Some projects, such as Tiros, have names derived from acronyms, but are not actually acronyms, and are hence not capitalized.
- Preparing a list of acronyms uncovered a number of instances of differing definitions for the same acronym, such as SPOT. We attempted to determine the "most accurate" definition.
- Russian acronyms are defined in English, but the acronym itself is transliterated from the Russian.
- Some projects and organizations have their acronyms as the most commonly known name. Examples include SPOT, SOHO, and LAGEOS. In these cases the acronym is shown first and its definition later in parentheses.
- Units are not defined in the article. The definitions are only in the units list.
- Spacecraft and launcher numbers are in regular Indo-European numerals, without hyphens, such as *Mariner 2*, as opposed to *Mariner II* or *Mariner-2*. There are many variations in various documents, even for the same project or spacecraft name. There are some cases where Roman numerals are consistently used in the source documentation; in these cases we follow that usage. We keep hyphens in certain cases where they are needed to avoid confusion, as occurs with some Soviet systems.
- Italics are used for specific spacecraft names, but not a series, project, or mission. Therefore we use *Mariner 2* for the specific spacecraft, but Mariner for the series of spacecraft, or Mariner 2 for a clear reference to the mission and not the spacecraft.
- We use the term "crewed" or "piloted" or "human" instead of "manned" to avoid inappropriate gender usage. However, we use "manned" when it is the proper historical usage, such as the "manned Moon missions" for Apollo, where in fact these missions used only men and were referred to as such.
- We use "Soviet Union" always, and never "Union of Soviet Socialist Republics (USSR)."
- "Soviets" is a derogatory term in Russian when used alone, but not when used as an adjective such as "Soviet launcher."
- After considerable research, we concluded that the first Soviet satellite was *Sputnik*, not *Sputnik 1*, although the second satellite was *Sputnik 2*.
- We avoid direct quotations that would require footnotes.
- For non-English words and phrases (including publication titles), we include the English translation in parentheses. This is particularly useful in the case of acronyms that are based on non-English words. For example, RKA is the acronym for the Russian Space Agency, but the acronym is based on the Russian words, not the translation.

Stephen B. Johnson

Using This Encyclopedia

(Professors, please invite students to read this section.)

For You, the Undergraduate Student

Welcome to the universe of space history, where few undergraduates have gone before! The following explanations will help demystify it for you.

Believe it or not, this impressive (perhaps even intimidating) two-volume set, *Space Exploration and Humanity: A Historical Encyclopedia*, was produced primarily for *you*, the undergraduate student. The editors and writers are space historians, engineers, and scientists who want you to be able to access and appreciate space history's importance as a field of study for the future. The editors have sought to be comprehensive and authoritative for the sake of *your* research. They sought the most knowledgeable authorities to compose the articles. They have defined terms, cross-referenced entries, and meticulously checked accuracy so that *you* could rely on this encyclopedia.

The information is often technical, esoteric, and might even seem at times to be stated in a language foreign to some. Space scientists, engineers, and historians are used to speaking and writing to other space scientists, engineers, and historians. While it is helpful that you understand basic science and general history, the editors have tried not to assume familiarity with technical subject matter or language. Therefore there are helpful devices throughout the encyclopedia so that you might

- find a space history topic of interest that might never have occurred to you;
- search for information on that topic in the encyclopedia;
- grasp the unfamiliar, technical vocabulary;
- know how to find the full names of any acronyms that appear;
- use the index and "see also" to find cross-referenced items;
- use the best sources recommended in the bibliographies and conduct additional research beyond this encyclopedia to expand or update its information;
- learn how complex technologies, such as satellites, work;
- find inspiration in the achievements of many remarkable people.

Trusting the Encyclopedia

There are other space encyclopedias, but this is the first comprehensive space *history* encyclopedia. That means it traces and analyzes space science, space technology, space discoveries, space development, space leaders, and space issues across time, in chronological and topically logical progression. It also means that historians, who analyze cause and effect through time, have written the overviews and gathered the work of scientists and engineers, in order to weave small narratives into a larger tapestry. The editors have tried to make this encyclopedia an authoritative source.

Most articles in this encyclopedia have been written by historical, scientific, or engineering experts on the subject. Often the author performed the original historical or technical research on the topic, or otherwise participated in the matters described. All articles have been reviewed by professional space historians. Some entries are virtually primary sources of information, because they are the first historical articles written on the subject.

General Education Courses for which This Encyclopedia Is Useful

- history of western or world civilization, especially modern
- American history surveys, especially modern
- recent, contemporary, or twentieth-century world or U.S. history
- humanities, especially modern
- history of science and technology
- economics
- physics
- astronomy
- world geography
- public speaking (for unusual speech topics)
- English composition (for unusual essay topics)

You might find a topic for a research project to suit any of the above courses or the explanation of the technology behind hundreds of innovations in spaceflight, satellites, and rocketry. Realize, too, that not all space history is from the Cold War era (circa 1945–1985), nor is all of it United States or United States versus Soviet Union. It is truly a *global* topic. Remember that the great scientists and space visionaries were often from other parts of the world beyond the United States. Countries that have their own space or satellite programs discussed in this encyclopedia include Australia, Brazil, Canada, France, Germany, India, Israel, Italy, Japan, and Russia.

Human Interest, Not Just Technical Information

The editors and writers knew that you would find space history more engaging if it held human interest. Therefore watch for topics such as science fiction, popular cultural fads about space, animals in space, or what food or clothing is used for space travel. If you are concerned about the environment, consider what space debris does to it. If you are interested in women's roles, look up women astronauts. Need inspiration for your own career choices? Space careers are a subsection. History is about people and organizations, not just technical or scientific developments. So the writers identified key people and organizations in the articles whenever appropriate.

How to Select a Topic of Interest for Your Research

There are six overviews of the major sections that can guide you to a topic for further reading and research:

- Astrophysics and Planetary Science
- Civilian and Commercial Space Applications
- Human Spaceflight and Microgravity Science
- Military Applications
- Space and Society
- Space Technology and Engineering

The editors suggest that you read or scan these overviews to determine themes and theses in space history that might serve you as research topics. Each overview's author is an expert in that field who also served as that section's editor. Each major section has a timeline and many subsections that indicate the subtopics or subthemes of the major theme. Thus they form an outline for you.

As you read the articles, try following the organizational pattern *they* follow. The articles have a logical order that can model how you might cover your topic. The authors and editors developed each article to be of the greatest use in the smallest space possible. So, for example, each article about a scientific spacecraft would include a brief description of it, its prehistory, when and how it became operational, and its scientific or technical results, this all in chronological order. That is standard operating procedure in history, but most people are not used to historical methods applied to scientific and technological topics. Historical methods can help lay people—non-space science people—understand.

Using the Timelines

Each of the six major sections has a timeline. Note that the first one—Astrophysics and Planetary Science Timeline—begins as far back as 500 BCE. Thus historians trace ideas and inventions back as far as records will allow. Consider using the timeline information integrated with history or science classes to see the processes that led to modern space technology.

Using the Article Bibliography and "See Also"

The editors believe that the most useful part of this encyclopedia for you might be the brief bibliographies at the end of all articles. Think about it: each author supplied three to four best references for learning more about the topic covered in the article. The authors are the experts. So they have filtered for you the most essential, reliable, and helpful sources you need to create your own longer research paper or presentation on that topic. These bibliographies are similar to first footnotes in scholarly journal articles, where the expert author tells you the top few best sources on what he or she has just stated. There *are* no notes. Notes might often bury the most useful sources for you. Instead the authors tell you boldly where to go for more.

At the end of an article, "see also" refers to a cross-listing. Look for the other article in this encyclopedia that is related to the one at hand. The "see also" will not appear for an article that is within the same section, because to understand the articles in that section, you really need to be reading the overview of that section and scanning all its articles. So "see also" means go to that named article in another section to find related information.

Tables, Their Purposes and Layout

The authors and editors have provided tables whenever technical data became too detailed for a standard narrative article, yet was instructive to provide. As you use the tables, realize that their factual information is not duplicated in the articles, plus the tables make it easier to see and compare. So, for example, if a table lists launch vehicles, it would offer the name, country of origin, dates of service, number of launches and failures, and comments. If the table lists science satellites, it would show lists of their instruments and who built them. Thus you have statistical information to support some of your space history research topics.

Understanding the Acronyms

Technical fields, technical organizations, and government bureaucracies are infamous for their arrays of acronyms. In this encyclopedia, therefore, watch for two main devices to help with these. First, the name is fully spelled out for the reader with the acronym following it parenthetically: Intercontinental Ballistic Missile (ICBM). Then the author uses the acronym from that point on, rather than repeat the full name. Second, a list at the back of the encyclopedia shows you the terms, their acronyms, and their multiple meanings. Remember that this list is in this encyclopedia because you might run across the acronyms without definitions in many other sources, including modern and future news stories or reports. Come back to this encyclopedia for enlightenment.

Understanding the Units of Measurement

If scientific or technical units of measurement seem bewildering, there is a table of abbreviations at the back of the encyclopedia that defines each of them. This encyclopedia does not attempt to explain their technical meanings and uses. For that, you will need to investigate further with technical sources appropriate to each measurement.

Is It All Greek to You? Using the Glossary

The glossary at the back of the encyclopedia is a dictionary to help you with the many technical terms that can be overwhelming. The editors made the glossary extremely comprehensive, for *your* sake. They strove to include any term that might stop a lay reader or an undergraduate student but, to keep to a practical length, they did not define terms that should be in any standard dictionary, especially general science terms. The most technically challenging section of the encyclopedia is Astrophysics and Planetary Science. Many of these terms and ideas are definitely not in your normal vocabulary, and the editors found that there was no easy way to simplify them in the short space available for each article. Not surprisingly this section requires by far the most glossary definitions. You will find this glossary both extensive and necessary. It could function for you as a dictionary of space history.

Using the Index for Your Topic

Students too often forget that there are indexes at the back of books. This encyclopedia index is intended to help you find all the other places that mention the topics that you discover through reading the overviews. Topics repeat themselves throughout this huge encyclopedia, but those repetitions might be deep within articles you might not suspect. The index will help you find them. Also glance at the index to see what topics have the most listings. Some times that, too, suggests likely themes for your papers and presentations.

Biographies and Sidebars

There are fascinating individuals who are key to any entry in this encyclopedia. The editors attempted to mention them where most appropriate in each article. Use the index to find them. Then the editors created sidebars that are brief biographies of some major figures. Look at a few of these, such as Arthur C. Clarke. The editors agreed there were two men for whom they felt compelled to offer full-length biographies: Wernher von Braun and Sergei Korolev. These could best be called the "fathers" or the "czars" of opposing space programs. An excellent topic for further research and writing in college courses, however, would be to expand any of the biographies of the many individuals responsible for space exploration and development.

You Can Be the Next Space Historian

There has been tremendous effort to ensure that this encyclopedia is up to date, as of its submission for publication. Yet every day, as the authors wrote and the editors edited, there were new discoveries, developments, and disasters related to outer space. The editors could not change the encyclopedia in every such instance. Thus there is ever-increasing room for you to do additional research. The authors and editors are also students and researchers and hope you will join them and be the next space historians.

Space History

Space history includes the scientific investigation of the objects and environment beyond Earth, the various military, civilian, and commercial applications in space, the political, economic, and cultural motivations and ramifications of spaceflight, and flights of humans into space.

Astronomy began with simple observations of fixed stars, the movements of the Sun, Moon, "wandering stars" (planets), comets, and meteors. Ancient Babylonians and Greeks began to measure and record these movements and recognized regular patterns that enabled prediction of their future motions. The Greeks developed geometric models of these movements, which were propagated and extended by Islamic astronomers during the Dark Ages in Europe. Early modern Europeans recovered the knowledge of ancient astronomy and then moved beyond it, associating physical causes with these motions. In the late seventeenth century Isaac Newton united the understanding of movements on Earth with those in the heavens with the laws of classical mechanics and of gravitation. These physical laws provided the basis for calculating the forces necessary to escape Earth's gravity and for navigating in space.

Separately from these scientific developments, by the eleventh century the technology of rocketry developed in China, and then during the next three centuries migrated west to Europe. Using scientific and technological knowledge of their times, a few writers created fictional stories about traveling into space. In the nineteenth century Jules Verne's fictional accounts of travels to the Moon were particularly compelling in their technical plausibility, inspiring many young people, including most of the pioneers of rocketry and spaceflight visionaries of the early twentieth century.

Three of these visionaries became particularly influential: Konstantin Tsiolkovsky in Russia, Robert Goddard in the United States, and Hermann Oberth, a German-speaking Romanian. All three were keenly interested in spaceflight and calculated the performance of rockets needed to reach space, the propellants needed, and a variety of other aspects of spaceflight. Goddard in particular began the practical development of rockets. In the 1920s–30s the work of these three men, the rapid evolution of aviation, and the growth of science fiction encouraged many other young men (there were few women) to form amateur societies to encourage the spaceflight idea and in some cases to build rockets. A few of their members became leaders in the development of rockets

and spaceflight, including the Soviet Union's Sergei Korolev and Germany's Wernher von Braun.

The amateur societies could experiment with rockets, but from the 1920s through World War II only the military had the need for and the resources to seriously develop rocket technologies. The military in several nations recruited some society members to develop solid- and liquid-propellant rockets for applications, for example, short-range rockets, such as the Soviet Katyusha, rocket-assisted aircraft takeoff (enabling takeoff from short airfields or aircraft carriers), and ballistic missiles. Recruiting the young Wernher von Braun, the German Army successfully developed the V-2 ballistic missile in World War II, though it did not change the course of the war. The Soviet Union imprisoned most of its rocketry leaders during Stalin's purges in the late 1930s, which put most Soviet rocketry efforts on hold until after the end of World War II.

After the war's end, the Allied powers competed to acquire German technologies, including the V-2. Most of the V-2's leading engineers and managers went to the United States, but others were recruited or (in the case of many in the Soviet zone of occupation) coerced into contributing to the rocket and missile programs of the Soviet Union, France, and the United Kingdom. The United States, the Soviet Union, and the United Kingdom began by launching V-2s, and the Soviet Union then manufactured its own V-2 replica, known as the R-1. Later Soviet rockets improved on the R-1 design. Before the end of World War II, the U.S. Army began ballistic missile development at the Jet Propulsion Laboratory (JPL), which led to the Corporal and Sergeant missiles. The U.S. Army also funded von Braun's team to develop the Redstone and Jupiter missiles. The U.S. Air Force (USAF) developed both cruise missiles (Navaho) and ballistic missiles (Atlas, Thor, and Titan). Nuclear-tipped ballistic missiles were formidable and unstoppable weapons. Because U.S. nuclear warheads were smaller than their Soviet counterparts at the time, the early U.S. intercontinental ballistic missiles (ICBMs) were also less powerful than the first Soviet ICBM, the R-7. This would have significant implications for the early space race.

From the 1930s to the early 1950s spaceflight seemed to the general public and many political and military leaders in the United States to be a "Buck Rogers" fantasy, referring to the character who appeared in comic books and movies, and on radio and television, from 1929 to 1951. Popular books of speculative nonfiction on the use of rockets for human spaceflight by David Lasser and P. E. Cleator in the 1930s and Willy Ley and G. Edward Pendray in the 1940s, and Arthur C. Clarke in the 1950s, among others, began to change these perceptions. Several articles describing the possibilities of spaceflight appeared in *Collier's* magazine from 1952 to 1954, and television shows produced by Walt Disney starting in 1955, with astronomers and space leaders, including Wernher von Braun. These articles and shows were supported by realistic space art by Chesley Bonestell, which helped readers and viewers to visualize how the future might appear.

Outside the United States, a similar though not identical evolution occurred, in which perceptions of the possibilities of spaceflight shifted from fantasy to plausibility. In the Soviet Union, amateur societies, public lectures, traveling displays, articles, and films on spaceflight topics were popular in the 1920s. Tsiolkovsky's nonfiction and fiction work became quite influential, as did Aleksei Tolstoy's science fiction work and later

film *Aelita: Zakat Marsa* (*Aelita: Sunset of Mars*—novel 1922–23, film 1924). With the imposition of rigid Stalinism in the 1930s, these utopian ideas were diverted to more practical, technological approaches implemented by engineering leaders whose original exposure to spaceflight came from the utopian space fad of the 1920s. In other nations and languages, speculative nonfiction works on the possibilities of spaceflight began to appear. Hermann Oberth's book *Wege zur Raumschiffahrt* (*Ways into Space*—1929) was the most important popular nonfiction work in the German-speaking world. The film *Frau im Mond* (*Woman in the Moon*—1929) played a similar role in popularizing science fiction spaceflight.

In the 1940s–50s U.S. political and military leaders were desperate to gather information about Soviet military capabilities and intentions. Given the secretive nature of the communist regime, U.S. presidents authorized occasional clandestine reconnaissance aircraft overflights of the Soviet Union to gather this information. As Soviet air defenses improved, the Dwight Eisenhower presidential administration recognized that the Soviet Union would soon be able to shoot these aircraft down. To counter this, in March 1955 the USAF issued requirements to begin the development of a reconnaissance satellite. Because there was no guarantee that the Soviet Union would not also try to destroy an orbiting reconnaissance satellite, Eisenhower also needed a political strategy to prevent this outcome.

The political solution was at hand, thanks to ongoing scientific efforts. In the late 1920s–30s some scientific investigations of the upper atmosphere focused on its electromagnetic properties, spurred by the recognition that shortwave radio signals were sometimes reflected off the upper atmosphere. The U.S. Naval Research Laboratory developed equipment to send and monitor reflected radio pulses to what became known as the ionosphere. Studies of the ionosphere figured prominently in subsequent research, and in 1950 several geophysical scientists proposed a third International Polar Year for the years 1957–58. This followed two earlier international scientific efforts to coordinate geophysical measurements at the poles. In 1954 the International Union of Geodesy and Geophysics accepted and expanded the proposal to be a worldwide effort called the International Geophysical Year (IGY). In May 1955 the U.S. National Security Council authorized the development of a scientific satellite for IGY, with one of the reasons being the need to test the principle of the Freedom of Space to ensure that a later reconnaissance satellite would be able to overfly the Soviet Union. President Eisenhower publicly announced in July 1955 that the United States would develop a satellite for IGY, and the Soviet Union soon followed suit.

The Soviet Union *Sputnik* satellite was the first into space in October 1957, which served the U.S. reconnaissance strategy even better since no countries complained about the Soviet Union's satellite overflight of their nations. The U.S. political strategy paid off, for by 1960 U.S. Corona satellites were imaging the Soviet Union. Soviet leaders publicly objected to these reconnaissance satellite overflights, but after the Soviet Union's Zenit reconnaissance satellites (first orbited 1962) proved valuable, the Soviet Union quietly accepted U.S. space reconnaissance. The Strategic Arms Limitations Talks agreements of 1972 formally legitimized the use of "national technical means" in space, a euphemism for reconnaissance satellites. Space-based reconnaissance was a critical capability that helped keep the Cold War from escalating.

Domestic and international reactions to *Sputnik* caught the Eisenhower administration by surprise. Observers and media around the world viewed *Sputnik* as evidence that the Soviet Union had overtaken the United States in scientific and technical capabilities, and hence that the communist system was superior to its capitalist-democratic adversary. In the United States, critics of the Eisenhower administration used *Sputnik* as a way to highlight the administration's flaws. Soviet leaders were equally surprised by the international reactions and somewhat belatedly began to understand and use the space program as a propaganda mechanism. These political forces pushed both nations into the technological competition known as the space race.

In the first few years of the space race, the United States held an edge in many spacecraft technologies. The Soviet Union's primary advantage was the heavy-lift capability of the R-7 launcher, which was a major asset for human spaceflight and placing a spacecraft near or on the Moon. In important practical applications, such as scientific experiments, communications, weather observation, reconnaissance, navigation, and weapons development, the United States jumped to a significant lead with the first satellites. Since the race to the Moon and human spaceflight yielded the highest media and propaganda value, Soviet Premier Nikita Khrushchev directed the rapid exploitation of these aspects of the space program, along with the continued military development of ICBMs. The United States faced political pressure to catch up with these Soviet successes.

The political pressure on the U.S. space program reached a climax with the Soviet launch of the first man in space, Yuri Gagarin, on 12 April 1961. Already on the defensive from the failed U.S.-backed invasion of Cuba by Cuban exiles at the Bay of Pigs, President John Kennedy sought a way to defeat the Soviet Union in the space race. Based on inputs from NASA and others, in May 1961 Kennedy directed the fledgling space agency to undertake the Apollo program to land a man on the Moon and return him safely before the end of the decade. This new "race" dominated media coverage and public perceptions of space. The Soviet Union secretly began to develop its response.

Landing humans on the Moon required the development of many new technologies. The first was a gigantic launch vehicle. Von Braun's team developed the Saturn V for Apollo, while Korolev's team created the N-1 for its manned lunar landing program. The success of the Saturn V with its liquid-hydrogen upper stages, and the failure of the N-1, largely determined the course of the race to place a man on the Moon. The manned lunar program also required the ability of two spacecraft to rendezvous and dock in space. Recognizing the large discrepancy in capabilities between the Mercury and Apollo programs, NASA created the Gemini program, which in the mid-1960s demonstrated orbital maneuvering and rendezvous capabilities. The Soviet Union's equivalent system was the Soyuz, which Korolev's team began to develop in 1962. While these new projects emerged, the Soviet Union exploited its early advantage with a series of space firsts with its Vostok and Voskhod spacecraft, including the first multicrew mission, the first woman in space, and the first spacewalk. With Gemini, by 1965 the United States performed the first successful docking maneuvers, and it undertook longer missions than the Soviet Union. Both nations developed robotic

spacecraft to fly by, crash land, and then soft land on the Moon. The Soviet Union claimed the first successful robotic landing with *Luna 9* in February 1966.

The United States won the manned Moon race with the *Apollo 11* mission in July 1969. It won for several reasons, including the skills of von Braun's team to develop the Saturn V, a large budget of more than $21 billion in then-year dollars, and a unified team that used and further developed the methods of systems management. Another reason was the tragic loss of three astronauts in a ground test of *Apollo 204* (later called *Apollo 1*) in January 1967. The resulting delay, during which NASA implemented many technical and organizational improvements, helped ensure Apollo's success. By contrast, the Soviet Union's lunar programs suffered from much more severe bureaucratic infighting than did NASA, between Korolev's team and that of Vladimir Chelomey about the methods and roles needed to win the race. Korolev's death in 1966 was a blow from which the Soviet space program never recovered. Among the leaders of the Soviet program, only Korolev managed to navigate among and control the many organizations and technologies required for the complex lunar missions, with sufficient political clout. By comparison with the United States, the Soviet Union's manned lunar program suffered from a lack of funding, which prevented development of the testing equipment needed to test the N-1 first stage. When the first tests of the N-1 (1969) and the *Soyuz 1* (April 1967) failed, in the latter case with the death of a cosmonaut, the Soviet Union's last chances to win the space race also failed. After the success of Apollo, the Soviet Union's leaders proclaimed that they had never intended to land a man on the Moon, arguing that robotic exploration was more cost-effective. The Soviet Union reprioritized its efforts and propaganda to focus instead on human-occupied space stations.

The space race also featured a robotic race into the solar system, starting with the Moon and then to the nearest planets. As in the human spaceflight race, the Soviet Union held the initial edge with its R-7 rocket enabling larger payloads to be sent to the Moon than the rockets of the United States. In January 1959, *Luna 1* was the first spacecraft to reach the Moon, while in October *Luna 3* gathered the first grainy images of the lunar far side. However, the United States countered with *Mariner 2*, which successfully flew by Venus in 1962. Two years later, *Mariner 4* became the first spacecraft to successfully fly by Mars. Though the Soviet Union made many attempts to reach Mars, most of the probes failed, and those that reached Mars provided little scientific data before failing. The United States claimed many successes at Mars, including *Mariner 9*, which in 1972 showed that the planet featured huge extinct volcanoes and massive canyons, and *Viking 1* and *2*, which in 1976 became the first spacecraft to successfully land on Mars. Soviet efforts were rewarded at Venus. From 1962 to 1983, the Venera program featured orbiters and landers, which successfully landed on the planet's surface where they measured temperatures as high as 475°C.

The Soviet Union and the United States also competed in military space. In 1960 one of the claims that U.S. presidential candidate John Kennedy made against Richard Nixon (who had been Eisenhower's vice president) was that the Soviet Union led in the deployment of ICBMs. Both candidates were briefed on satellite reconnaissance, which steadily accumulated evidence that the United States held a substantial lead, though by fall 1960 this evidence was not yet definitive. After the election, the evidence became conclusive, but by that time Kennedy was committed by campaign promises to further expansion of ICBMs. By late 1962 at the time of the Cuban Missile Crisis, Kennedy's knowledge of United States nuclear superiority allowed him to force Khrushchev to withdraw Soviet missiles from Cuba. After the crisis, the Soviet leadership vowed to expand its nuclear force so it would never be in a position of inferiority again. This led to a massive expansion of the Soviet ICBM and nuclear submarine force in the 1960s–70s. The United States was also developing its nuclear submarine force, and the U.S. Navy developed and deployed the Transit navigational satellites to provide the initial positions to Polaris submarines. This provided the positional accuracy needed to launch nuclear strikes against the Soviet Union. In turn the Soviet Union developed the Tsiklon navigational satellites to support its submarine forces. Geodetic satellites that precisely measured the shape of Earth and Earth's gravitational field were also crucial to precise ballistic missile targeting. These were priorities for nations with long-range ballistic missiles, including the United States, the Soviet Union, and France.

Providing early warning, and defending against potential nuclear strikes from these massive arsenals, became a priority for both nations. The United States developed the Defense Support Program, which deployed satellites that detected and relayed the infrared signatures of ballistic missile launches. The Soviet Union countered with its Oko early warning satellites. Both nations also developed antiballistic missile systems in the 1960s. In 1972 the Anti-Ballistic Missile (ABM) Treaty limited the United States and the Soviet Union to the deployment of ABM systems to protect two locations in each nation.

Both nations also developed and tested antisatellite (ASAT) systems in the 1960s. The United States deployed ground-based ASATs in the 1960s, developed an air-launched ASAT in the 1980s, and funded another ground-based system in the 1990s. It also tested a laser system to blind reconnaissance satellites. The Soviet Union developed similar systems and also tested a satellite that could maneuver in orbit to destroy a target. China successfully tested a ground-based direct ascent ASAT in 2007. Antisatellite systems have not been prohibited in any space treaties, leading to concerns about an ASAT arms race.

The possibility of using spacecraft as telecommunications relays was understood as early as 1945 by British engineer and future science fiction author Arthur C. Clarke. As the space age dawned in the late 1950s, the U.S. military, the newly formed NASA, and private industry (in particular, American Telephone and Telegraph Company—AT&T) began development of communications satellites (comsats). In the early 1960s NASA tested a passive aluminized balloon as a comsat, while the military and AT&T successfully tested repeater satellites in medium Earth orbit (MEO) that received, amplified, and rebroadcast radio signals. MEO satellites minimized the communications delay due to light-speed travel time, but required ground stations that could track their movements and a constellation of many MEO satellites to provide continuous coverage. By contrast a geostationary satellite system required only three satellites to provide continuous worldwide coverage, except at the poles. NASA funded the Syncom project, which in 1963 proved the feasibility of geosynchronous satellites. Because of the Soviet Union's far northern location, geostationary satellites (which orbited above the equator) were not viable. In the 1960s it developed its Molniya comsat system to provide both military and civilian communications.

While the U.S. military and the North Atlantic Treaty Organization (NATO) developed and deployed military comsat systems to provide secure communications, civilian and commercial leaders considered how to organize the development of a nonmilitary system. The Kennedy administration had three major goals: to prevent AT&T from expanding its domestic communications monopoly; to ensure a commercial (private, profit-making) implementation for the global system; and to ensure U.S. influence over that global comsat system. Military and NASA comsat funding had created competition to AT&T, and the Kennedy administration expanded this by creating the Communications Satellite (Comsat) Corporation in 1963 to implement the U.S. portions of the global system. The administration also started negotiations with Japan, Canada, and European nations to create the International Telecommunications Satellite (Intelsat) Organization, inaugurated in 1964. Comsat Corporation became the manager of Intelsat, which soon deployed a geosynchronous satellite system. Because most nations used their own government-run postal and telecommunications departments to operate their portions of Intelsat, the U.S. determination to ensure private operation of Intelsat staked out a position for commercial enterprise in space. The United Nations rejected a Soviet proposal to ban private enterprise in space in favor of a proposal that national governments had to supervise the actions of, and were liable for, damages caused by individuals or corporations from their countries. This provision was included in the 1967 Outer Space Treaty.

In the negotiations to create Intelsat, other nations won a key concession from the United States to ensure that the final Intelsat rules would be determined in negotiations that began in 1971. The resulting new regulations stipulated that Intelsat share ownership was determined by usage, so that by the late 1970s U.S. control of Intelsat ended. Another provision was the allowance of regional comsat systems that would not compete with the worldwide Intelsat system. France, Germany, Italy, the United Kingdom, and then most west European nations, through the European Space Research Organisation, began development of comsats in the late 1960s. In the 1970s these efforts came to fruition with successful experimental comsats.

By the 1980s the next generation of comsats in Europe, Japan, and the United States were being deployed into operational domestic and regional systems. Many governments and private corporations purchased comsats to provide communications services. New applications, such as direct broadcast satellite television and radio and satellite Internet services, proliferated, such that by the twenty-first century communications was by far the largest space economic sector, with more than $80 billion in worldwide revenues. The growth of telephony and these new applications spawned new companies, such as Pan American Satellite Corporation, Société Européenne des Satellites, DIRECTV, Echostar, Sirius Satellite Radio, and new government-dominated consortia such as Inmarsat, Eutelsat, and Arabsat. The U.S. and other military organizations sent their most critical communications through military comsats, but they also sent many routine messages through nonmilitary comsat systems to reduce costs. U.S. manufacturers such as Hughes, Ford Aerospace, and General Electric competed with European comsat manufacturers, including Marconi and Alcatel, to sell comsats.

The growing demand from various nations and private companies to launch comsats fed an international competition to launch them. In the 1960s–70s comsat operators had to contract for service to NASA, but the first successful launch of Ariane in 1979 broke this U.S. monopoly. Comsat operators in the early 1980s could choose between Ariane and NASA's Space Shuttle, which first flew in 1981. After the *Challenger* disaster in January 1986, the U.S. government removed the Shuttle from the commercial launch business, but two years previously the President Ronald Reagan administration had authorized private U.S. companies to provide commercial launches licensed through the Department of Transportation. The demise of the Soviet Union signaled the further expansion of competition. By the early 1990s Russian and Ukrainian corporations and government organizations joined China in offering commercial launches. Because these nations could significantly undercut U.S. and European launch prices, the U.S. government negotiated agreements to limit the number of launches it could offer and required that they charge high prices nearly equivalent to Western launch prices. The former Soviet organizations developed joint ventures with Western companies to help overcome political barriers and to aid marketing.

A brief spike in the number of comsats to be launched caused by the development of MEO satellite constellations Iridium, ICO, and Globalstar created a spurt of investment in the development of purely private launchers, including innovative reusable designs. However, these efforts came to naught, and the new companies went bankrupt along with the three corporations that had created the initial demand.

Commercial success was an unexpected outcome of military global navigation systems. By the late 1960s commercial shipping and offshore exploration applications were using signals from the U.S. Navy's transit system, which was sufficient for applications in which relatively low precision (roughly 100 meters) and its slow update rate of several hours were acceptable. In 1973 the U.S. Department of Defense began the tri-service (Army, Navy, and Air Force) Global Positioning System (GPS) program, with experimental satellites orbited starting in 1978. Though a full constellation of 24 MEO satellites was declared fully operational only in 1995, by the time of the First Gulf War in 1991 it was sufficiently functional to support allied operations. In its first major military test, these GPS applications, which included precision weapons targeting and navigation of aircraft, land, and sea units, were spectacularly successful, making GPS a household word. Access to GPS precision signals, which was made permanent in 2000 by the President William Clinton administration, spurred the development of commercial applications implemented through addition of various functions to GPS receivers. These included critical applications, such as timing of Wall Street transactions and commercial aircraft guidance; practical and scientific uses, such as truck tracking and measurement of continental drift; and personal applications, such as pet monitoring and golf shot estimation. The multibillion-dollar commercial success of GPS spurred the development of ground- and space-based regional systems to enhance GPS signals, and the European Union–ESA Galileo project to provide a guaranteed private commercial service.

Another practical application that evolved in the 1950s was weather monitoring from space. Sounding rockets observed major events, such as hurricanes before they were spotted from the ground, and the Tiros satellites launched by NASA starting in

1960 proved their utility. In parallel with Tiros, the USAF developed and launched its Defense Meteorological Satellite Program series to support Corona, as it made no sense for Corona to waste precious film taking pictures of cloud-tops above the Soviet Union. The Weather Bureau used improved Tiros designs for its first operational satellites, while NASA improved satellite and sensor technologies with Nimbus. In the 1980s a Nimbus satellite confirmed the existence of an ozone hole over Antarctica. Following the U.S. lead, the Soviet Union developed and deployed its Meteor series. In the 1970s the United States and major nations around the world (including European countries, India, Japan, China, and Russia) developed and deployed geosynchronous and polar-orbiting weather satellites, which formed a worldwide system organized through the Coordination Group for Meteorological Satellites. Nations operated their weather satellites as publicly provided services, shared data, and even moved and shared satellites when untimely satellite failures occurred.

Observation of the land and sea surface was useful for many applications. Military applications, such as monitoring ship movements, were developed by the United States (White Cloud) and the Soviet Union (EORSAT—Electronic Intelligence Ocean Reconnaissance Satellite, and RORSAT—Radar Ocean Reconnaissance Satellite). Oceans could also be monitored for surface heights, temperatures, currents, and biomass, for which purpose several nations developed specialized spacecraft and instruments. Land remote sensing also had several uses, including scientific assessments of seasonal and long-term land cover changes and their relationship to human and natural activities, government evaluation of urban growth, and commercial uses, such as mineral prospecting and mapmaking. *Landsat 1*, launched in 1972, was the first civilian land remote sensing satellite. Its multiple uses caused difficulties for U.S. policy makers, who debated whether Landsat should be a government-provided public service or a private commercial operation. Congress privatized Landsat in the mid-1980s, but the resulting high imagery costs resulted in user complaints. With the launch of France's SPOT (Satellite Pour l'Observation de la Terre) satellites starting in 1986, the new competition for imagery sales and the existing complaints led Congress in 1992 to reverse policy and revert Landsat to a public service. This 1992 law also enabled U.S. companies to provide high-resolution (1 meter) imagery. The first high-resolution system, Space Imaging's *Ikonos 1*, was launched in 1999. From that time forward, governments, corporations, and individuals could purchase high-resolution imagery of most locations in the world, a capability formerly available only to the two superpowers. By the 1990s Earth-monitoring satellites from several nations (including the European nations, Japan, India, Russia, Israel, Canada, and China) were providing continuous coverage of Earth's atmosphere, oceans, and land surfaces, which enabled research into global climate change and a host of practical uses.

Deep space probes enabled direct, close-up observation of objects in the solar system. After the early missions to Venus and Mars, NASA began planning for missions to more distant planets. *Mariner 10* was the first spacecraft to use the gravity assist technique, swinging by Venus to enable flybys of crater-pocked Mercury in 1974–75. NASA sent *Pioneer 10* to Jupiter in 1973 and *Pioneer 11* past Jupiter (1974) and then to Saturn (1979). JPL's *Voyager 1* and *2* also flew by Jupiter and Saturn from 1979 to 1981, but *Voyager 2* then went farther to fly by Uranus (1986) and Neptune (1989). These probes, along with the later *Galileo* probe to Jupiter and *Cassini* to Saturn, provided more accurate measurements and observations of these planets' dynamic atmospheres, including a vast hurricane at Saturn's southern pole, and of each planet's ring system. Voyager revealed the diverse geology of their many moons, including the active volcanoes of Io, the ice-covered oceans of Europa, the methane rivers and lakes on Titan, water plumes on Enceladus, Miranda's fragmented surface, and active nitrogen geysers on Triton. In the 1990s NASA restarted its exploration of Mars, with several JPL-managed orbiter and rover missions. The Mars Exploration Rovers proved that water had once existed on Mars's surface, spurring speculation that life may have existed on the Mars, or might still exist under the surface. The United States, Japan, ESA, and the Soviet Union all contributed missions to explore comets and asteroids, including a flotilla of spacecraft to Halley's Comet in 1986 and NASA's *Near Earth Asteroid Rendezvous* spacecraft that landed on the surface of 433 Eros in 2000. NASA's *Genesis* spacecraft returned samples of the solar wind to Earth in 2004, and *Stardust* returned samples from comet Wild 2 in 2006.

Spacecraft proved to be important tools for astronomy and astrophysics. Certain wavelengths of the electromagnetic spectrum are blocked by Earth's atmosphere, and observations from space are free of atmospheric distortions and light pollution. A series of solar and astronomical observatories from several nations (primarily the United States, European nations, and Japan) made observations of fundamental importance. Space-based observations from spacecraft, such as the U.S. *Mariner 2*, *Solar Maximum Mission*, and Orbiting Solar Observatories, the NASA-ESA Solar and Heliospheric Observatory, ESA's *Ulysses*, and Japan's *Yohkoh*, confirmed the previously hypothesized existence of the solar wind, coronal mass ejections, and the existence of X-ray emissions from the base and apex of solar flares. Early astronomical observatories focused on high-energy X rays and even higher–energy gamma rays. NASA's *Uhuru* spacecraft, launched in 1970, was the first dedicated to extra-solar X-ray sources. Those sources located in the Milky Way galaxy were later found through ground-based and space-based observations to be mainly in binary star systems, including neutron stars and black holes. More distant X-ray sources were often identified in quasars and active galactic nuclei. Gamma-ray observations frequently occurred in bursts, usually identified with cataclysmic explosions, such as novae and supernovae. These bursts were first observed with the USAF's Vela 5 nuclear monitoring satellites in 1969. Later scientific satellites included ultraviolet and infrared observatories, which detected low energy wavelengths. These discovered a variety of protostars, dust clouds, and the existence of water in gas clouds and deuterium in the interstellar medium. The *Cosmic Background Explorer* (launched in 1989) and *Wilkinson Microwave Anisotropy Probe* (launched in 2001) provided the first observations of the variations of the cosmic background radiation. These provided information that supported estimates of the universe's age (13.7 billion years) and composition (4 percent matter, 23 percent dark matter, and 73 percent dark energy).

After the manned Moon race of the 1960s, the Soviet Union shifted its priorities to crewed space stations, while NASA used remaining Apollo hardware to launch its *Skylab* space station in 1973 and began development of the Space Shuttle in 1972. For both nations, space station development began with military space stations in the 1960s, to be used for reconnaissance. The USAF started the Manned Orbiting Laboratory (MOL) program in 1963 on the cancellation of its Dyna-Soar piloted spaceplane, but MOL was canceled in 1969 in favor of the KH-9 robotic reconnaissance satellite program and *Skylab*. Several of MOL's astronauts and technologies transferred to NASA. The Soviet Union began several space station studies in response to MOL, but settled on Vladimir Chelomey's Almaz design. The Soviet leadership decided to combine the Almaz hull with Soyuz electronics, with the new system called the Long-Duration Station (DOS). Both military Almaz and civilian DOS stations flew under the public Salyut label from 1971 to 1984, with the Almaz missions proving that crewed reconnaissance was ineffective in comparison to robotic reconnaissance. The crew of *Soyuz 11*, returning from *Salyut 1* in 1971, died on reentry, but otherwise the civilian DOS missions were successful, with the Soviet Union learning how to operate long-duration missions and using Salyut as a destination for its Interkosmos program of guest cosmonauts from Soviet allies. The Apollo-Soyuz Test Project, which orbited in July 1975, was a brief demonstration of political détente in which U.S. and Soviet crews rendezvoused in orbit and performed joint science experiments.

As NASA's Space Shuttle flight program was trying to accelerate its flight schedule in the early 1980s, NASA successfully lobbied to start the Space Station Freedom (SSF) program, which U.S. President Ronald Reagan announced in 1984. Over the next eight years NASA struggled to define SSF's mission and design while recruiting international partners (ESA, Canada, and Japan), and responding to congressional budget cuts and changes. The resulting cost overruns and schedule delays nearly led to its cancellation by June 1993. In the meantime discussions had begun one year earlier between U.S. President George H. W. Bush and Russian President Boris Yeltsin regarding possible collaboration in space. The Soviet Union had launched the first component of its modular *Mir* space station in 1986 and had progressively expanded and operated it until the Soviet Union dissolved as a political entity in 1991. The resulting financial chaos made it extremely difficult for the new Russian state to continue funding its space program. The September 1993 agreement between the United States and Russia to create the *International Space Station* (*ISS*) saved the U.S. space station program, by integrating Russia into the community of democratic nations, and provided Russia the funding it needed for its space program to continue. The first *ISS* module, the Russian *Zarya*, was launched in November 1998, and construction continued until the loss of the Space Shuttle *Columbia* on reentry in February 2003. The Shuttle fleet was grounded while problems with the Shuttle's external tank insulation were resolved, and *ISS* construction resumed in 2006. ESA's *Columbus* module and Japan's *Kibo* module were attached in 2008. As of fall 2009, *ISS* construction was planned for completion in 2010 or 2011.

The loss of *Columbia* led the U.S. leadership to reconsider NASA's future. In January 2004 President George W. Bush announced a new Vision for Space Exploration, in which NASA would complete *ISS* construction, retire the Shuttle by 2010, and return to the Moon in the following decade. NASA began development of its Constellation program later that year, but a gap of several years loomed after the retirement of the Shuttle before a new replacement system would became operational.

The President Barack Obama Administration threw these plans into doubt in February 2010, proposing the cancellation of the Constellation program. China launched its first taikonaut in October 2003, becoming only the third nation with a human spaceflight launch capability. U.S. businessman Dennis Tito became the first of several space tourists, with cash-strapped Russia selling him a place on the flight of Soyuz TM-32 to *ISS* in April 2001. In 2004 Scaled Composites Corporation won the X-Prize of $10 million by flying *SpaceShipOne* on two suborbital flights into space within one week, without any government funding. Bigelow Aerospace launched two inflatable modules into space in 2006–7 as experiments leading toward a space hotel design. In 2008 billionaire Richard Branson's Virgin Galactic announced a partnership with Scaled Composites to build a tourist-capable *SpaceShipTwo*.

From the 1940s to the early twenty-first century, rocketry and spaceflight were supported by the development of professional societies and educational programs. The amateur societies of the 1920s–30s had to evolve in a postwar world in which rocketry was a military-funded reality and spaceflight was dominated by governments. The American Rocket Society, founded in the 1930s, merged in 1963 with the Institute of Aeronautical Sciences to form the American Institute of Aeronautics and Astronautics. The International Astronautical Federation was founded in 1951, and the American Astronautical Society in 1954. Universities developed space-related engineering curricula, usually associated with aeronautical engineering programs to become aerospace engineering programs. Space interdisciplinary programs came into being in the late 1980s with the University of North Dakota's Department of Space Studies and the International Space University. Conversely, led by human spaceflight, planetary probes and images from space observatories were used by national governments to inspire young people to enter mathematics, science, and engineering programs. The Soviet Union created Young Cosmonauts groups in the early 1960s for this purpose. The United States created its first space camp in 1982, and these were replicated in several other nations over the next two decades.

By the twenty-first century, space endeavors were integrated into the world's military, economic, and political structures and had permeated deeply into the cultures of many of the world's nations.

Stephen B. Johnson

Bibliography

William E. Burrows, *This New Ocean: The Story of the First Space Age* (1999).
Roger Handberg, *International Space Commerce: Building from Scratch* (2006).
T. A. Heppenheimer, *Countdown: A History of Space Flight* (1997).
Walter A. McDougall, *. . . the Heavens and the Earth: A Political History of the Space Age* (1985).

Space Historiography

Space historiography refers to the techniques and ideas used by historians investigating the history of space endeavors. Historical understanding of space exploration and use has evolved from relatively simple first-person accounts to sophisticated analyses and narratives drawing from many sources and theories.

The first accounts of space exploration, as with most other recent human actions, came from journalists, popularizers, and participants in the events. As space exploration is an unusual human activity, many people in both groups saw themselves as involved in important events and wrote about their roles in these events, in addition to the projects and activities in which they found themselves. Among the early accounts were the histories written by associates and members of Wernher von Braun's rocket team, by both Germans and Americans who worked with von Braun in the United States. *Rockets, Missiles, and Men in Space* (1968) by Willy Ley and *History of Rocketry and Space Travel* (1966) by Wernher von Braun and Frederick Ordway represent two examples. Because of the secretive nature of the Soviet regime, personal accounts remain crucial sources to understand Soviet space history. The most important of these include *Rockets and People* by Boris Chertok, published in several volumes, translated into English and edited by Asif Siddiqi and published by NASA. There are many first-person accounts written by, with, or about astronauts and cosmonauts, which have remained popular over the decades, but provide a narrow view of space history. Of these, Andrew Chaikin's *A Man on the Moon: The Voyages of the Apollo Astronauts* (1994) is a fine example.

Since the 1960s the NASA History Office has sponsored many high-quality research projects, books, and short monographs. The majority of these have been histories of specific projects, though there are also administrative histories, histories of the NASA field centers, personal memoirs, collections of interviews, collections of primary source documents (such as the *Exploring the Unknown* series under the editorial leadership of John Logsdon), and conference proceedings (including the outstanding *History of Rocketry and Astronautics* series published by Univelt and edited by members of the History Committee of the American Astronautical Society that compiles the papers presented during the History Symposia of the International Academy of Astronautics from International Astronautical Congresses since 1967). The European Space Agency (ESA) followed suit with its own historical research program, which published *A History of the European Space Agency 1958–1987* (2000) by John Krige et al., and many shorter research papers on the history of the national space programs of the ESA Member States. The U.S. military also has a long-standing tradition of historical research and publication, and include several project and institutional histories, including an overview of the U.S. Air Force's space history *Beyond Horizons: A Half Century of Air Force Space Leadership* (1997) and collected primary documents, *Orbital Futures: Selected Documents in Air Force Space History* (2004), both edited by David Spires. Separately from these government-published works, the vast majority of space history books are of specific projects and nations, and these remain the bulk of space history publications. Norman Friedman's *Seapower and Space: From the Dawn of the Missile Age to Net-Centric Warfare* was particularly noteworthy for its description of the relationship of space programs to naval warfare and strategy.

With spaceflight's origins and continuing programs with military and intelligence organizations, and with one of the space race's main protagonists, the Soviet Union, being an extremely secretive society, the declassification of military and intelligence records and the fall of the Soviet Union and subsequent opening of its archives were major events in space history. For the U.S. military and intelligence story, the declassification of the Corona program in 1995 opened a window into political strategies and secret projects that were crucial to the history of spaceflight. The proceedings of the conference that announced Corona led to *Eye in the Sky: The Story of the Corona Spy Satellites* (1998) edited by Dwayne Day et al. Many other books and articles on Corona and other military and intelligence programs followed. The fall of the Soviet Union provided scholars with access to many Soviet and Russian archives. Asif Siddiqi was among the first to take advantage of this, and his groundbreaking and authoritative *Challenge to Apollo: The Soviet Union and the Space Race, 1945–1974* (2000) set forth the previously unknown internal history of the Soviet space program. *The Kremlin's Nuclear Sword: The Rise and Fall of Russia's Strategic Nuclear Forces 1945–2000* (2002) by Steven Zaloga provided a similar first look into the Soviet ballistic missile and missile defense programs based on archival material. Other books and articles subsequently appeared on a regular basis regarding various Soviet projects.

Synthetic, overview histories that cross organizational and project boundaries remain rare, but these are extremely valuable. Walter McDougall's 1985 Pulitzer Prize–winning book *. . . the Heavens and the Earth: A Political History of the Space Age* was the first major synthesis of space history, interestingly written by a general political historian as opposed to a specialized space historian. Despite its age, it remains an outstanding political overview through the 1960s space race. Two later overview histories, *Countdown: A History of Space Flight* (1997) by T. A. Heppenheimer and *This New Ocean: The Story of the First Space Age* (1999) by William E. Burrows, provided further technical, military, and scientific depth, with both extending McDougall's Cold War–framework and interpretations. For space science, David Leverington's *New Cosmic Horizons: Space Astronomy from the V2 to the Hubble Space Telescope* provided the first overview of space science observations and discoveries performed from above Earth's surface.

The most recent additions to different types of space history are those that draw from broader currents in general history and social sciences, and which apply these to subjects that cross the usual project and institutional boundaries. An important branch of these works are political studies, exemplified by *Spaceflight and the Myth of Presidential Leadership* (1997), edited by Roger D. Launius and Howard E. McCurdy. Works drawing from social history and theory include William Bainbridge's *The Spaceflight Revolution: A Sociological Study* (1976), Pamela Mack's *Viewing the Earth: The Social Construction of the Landsat Satellite System* (1990), and Donald McKenzie's *Inventing Accuracy: A Historical Sociology of Nuclear Missile Guidance* (1990). Cultural history was a later entrant into space history studies, with major works including *Space and the American Imagination* (1997) by Howard McCurdy, *Right Stuff, Wrong Sex: America's First Women in Space Program* (2004) by Margaret Weitekamp, and *Astrofuturism* (2003) by De Witt Douglas Kilgore. Several works crossed traditional boundaries and drew from several technical, social, and historical approaches. Two good examples of this literature are *The Secret of Apollo: Systems Management in American and European Space Programs* (2002) by Stephen B. Johnson and *Digital Apollo: Human and Machine in Spaceflight* (2008) by David A. Mindell. A fine example of deep historical scholarship on a heavily explored topic is *Von Braun: Dreamer of Space, Engineer of War* (2007) by Michael J. Neufeld.

Despite the many books and articles written on space history, there remain many subjects to be explored. Most obvious are the new projects ongoing at any given time. In most cases, current projects do not have official histories, as it usually requires a number of years to pass before historians begin to investigate. Other topics remain closed due to classification or secretiveness on the part of the nations or organizations involved. These include the military and intelligence services, secretive societies such as Communist China, and private corporations, which often do not wish to publicize their ongoing activities or do not keep detailed records that enable historians to perform research (nor do they generally retain the archives of companies that they take over). Even when the records are open and exist, some subjects are technically more difficult or simply have not caught the attention of space historians. These include subjects such as space business and economics, labor, telecommunications, military command and control, meteorology, the fragmented history of space technology development, and scientific projects studying the near-Earth environment (ionosphere, magnetosphere, etc.).

The only dedicated journal of spaceflight history is *Quest: The History of Spaceflight Quarterly*, which evolved from a semipopular, semiprofessional magazine in the mid-1990s to a professional journal in 1998. Many excellent articles on the history of spaceflight can be found in *Spaceflight*, and in the *Journal of the British Interplanetary Society*. Occasional professional articles in space history can also be found in *Technology and Culture* (the journal of the Society for the History of Technology) and *Isis* (the journal of the History of Science Society).

Space history as a discipline has evolved from simple reminiscences and straightforward project histories to include a diversity of topics and approaches, signaling its maturity as a discipline and drawing from many historical, cultural, and technical approaches.

Stephen B. Johnson

Bibliography

Stephen B. Johnson, "The History and Historiography of National Security Space," in *Critical Issues in the History of Spaceflight*, ed. Steven J. Dick and Roger D. Launius (2006), 481–548.

Roger D. Launius, "American Spaceflight History's Master Narrative and the Meaning of Memory," in *Remembering the Space Age: Proceedings of the 50th Anniversary Conference*, ed. Steven J. Dick (2008), 353–85.

Asif A. Siddiqi, "American Space History: Legacies, Questions, and Opportunities for Future Research," in Dick and Launius, *Critical Issues*, 433–80.

Margaret A. Weitekamp, "Critical Theory as a Toolbox: Suggestions for Space History's Relationship to the History Subdisciplines," in Dick and Launius, *Critical Issues*, 549–72.

Astrophysics and Planetary Science

Astrophysics and planetary science began with the early civilizations of Egypt, Babylon, Greece, China, and others who began observing the Sun, Moon, and stars and recording what they saw. Some civilizations, like the Babylonians, were more interested in predicting the movements of these celestial bodies across the sky, whereas others, such as the Greeks, tried to understand these movements. To do this they produced schematic models of the universe, most of which had Earth at the center.

Europe went through a relative dark age, after the demise of the Greek and Roman empires, that ended in the later Middle Ages with the Renaissance. In 1543 Nicolaus Copernicus rejected the idea of an Earth-centered universe, which was generally accepted at the time, and published his theory with the Sun at the center. Then in the early seventeenth century Johannes Kepler proved that the orbits of the planets around the Sun were ellipses.

A key scientific concept that helped revolutionize seventeenth-century astronomy was that of "force," or "anima motrix" ("moving spirit") as Kepler called it, emanating from the Sun to keep the planets in their orbits. One type of force considered by Kepler was magnetic force, which had been studied by William Gilbert in the sixteenth century. Galileo Galilei undertook a number of experiments to try to understand the laws of motion, but it was Isaac Newton with his universal theory of gravitation, published in 1687, who solved the problem of what caused the planets to orbit the Sun in the way that they did.

In the meantime, telescopes had become available in the early seventeenth century. As a result, Galileo was able to observe the mountains of the Moon, see the phases of Venus, and discover the four large moons of Jupiter. Large sunspots can be seen with the naked eye, but telescopes enabled smaller ones to be studied. As a result Johann Fabricius was able to show that the Sun rotated on its axis. A little later Christiaan Huygens discovered Saturn's ring.

Further details were observed on the Sun, Moon, and planets over the next century. In 1750 Thomas Wright suggested that the Milky Way was a large rotating disc of stars, and in 1775 Immanuel Kant suggested that the "nebulous stars" were systems of stars like the Milky Way. It was not until the twentieth century that both ideas were shown to be correct.

Astronomers of the early nineteenth century thought that they would never be able to discover the composition of the Sun and stars. But in 1859 Gustav Kirchhoff and Robert Bunsen showed that this was possible, in principle, by measuring the positions of the dark lines in their spectra. That same year Kirchhoff discovered gaseous sodium and iron in the Sun, and four years later William Huggins showed that stars are made of the same elements present on Earth.

John Draper took the first photograph of the Moon in 1840, but early astronomical photographs were of little use. However, from 1858 to 1872 a daily record of sunspots was produced at Kew, London, with photographic equipment designed by Warren De la Rue. The photographic process was very complicated at that time, but in 1871 dry photographic plates became available that were sensitive enough to allow photographs to be taken of nebulae and the dimmer stars. In 1886 Paul and Prosper Henry produced a photograph of the Pleiades open star cluster that showed 1,400 stars.

Progress in planetary astronomy virtually stopped at the end of the nineteenth century as, although better telescopes had become available, their visual images were blurred by atmospheric turbulence. Occasional brief periods of very good "seeing" enabled astronomers to see some planetary details, but the exposures required by photographs were far too long to take advantage of these brief periods. It was also difficult to interpret planetary spectra, whether observed visually or by photography, as it was difficult to sort out which spectral lines were due to Earth's atmosphere and which to the planets. It was also difficult to interpret molecular spectra of planetary atmospheres compared with the atomic spectra of stars.

Albert Einstein published his special theory of relativity in 1905 proposing the equivalence of mass and energy. So, at long last, it appeared that the source of the Sun's energy would soon be discovered. But this required a much better understanding of atomic physics than existed at that time. In 1913 Niels Bohr proposed his quantum theory of the atom, which explained hydrogen's atomic spectra, and in 1932 James Chadwick discovered the neutron. During this period the structure and forces present in the atom gradually became clearer, and this enabled Charles Critchfield, Hans Bethe, and Carl von Weizsäcker to propose two alternative schemes for the generation of heat in stars.

In 1916 Karl Schwarzchild published his theory of black holes, based on Einstein's general theory of relativity that had been published the previous year. In 1924 Wolfgang Pauli postulated his exclusion principle, which led to the concept of a degenerate gas at very high pressures. In the following year Walter Adams discovered the first white dwarf star, which was found to have an incredibly high density. Six years later Subrahmanyan Chandrasekhar published his theory of white dwarfs, using Pauli's principle, and shortly afterward Walter Baade and Fritz Zwicky proposed their theory of neutron stars.

The 1920s saw significant progress in understanding the large-scale structure of the universe. First Edwin Hubble proved, for the first time, that the Andromeda and M33 nebulae were galaxies of stars outside the Milky Way. Then in 1929 Hubble and Milton Humason found that the recessional velocity of distant galaxies, as determined by their red shift, varied linearly with distance. This was two years after Georges Lemaître had proposed what is now called the Big Bang theory of the universe based on Einstein's theory of relativity. This theory had a big boost in 1964 when Arno Penzias and Robert Wilson discovered the microwave background radiation at exactly the correct temperature predicted by the Big Bang theory.

Astronomical research up to the start of World War II had been largely undertaken from the surface of Earth using visible light. However, a few balloon experiments had helped to characterize cosmic rays, and in 1932 Karl Jansky had detected radio waves that appeared to be coming from the center of the Milky Way. This was to be the start of an astronomical revolution, using frequencies outside the optical waveband.

During the war the Sun was found to emit radio waves, and after the end of the war a number of former radar engineers, mostly in the United Kingdom, Australia, and the Netherlands, began to find a number of radio sources using relatively simple radio telescopes. At first it was unclear if these sources were stars, but in the late 1940s two of them, Cen A and Vir A, were identified as galaxies. Then in 1963 Maarten Schmidt discovered that the radio source 3C 273 was a galaxy receding at 15 percent of the speed of light. This was the first quasar to be discovered.

After World War II a number of rockets became available in the United States, in particular for scientific research. A captured V-2 rocket launched in October 1946 photographed the Sun's ultraviolet spectrum down to 2,300 Å for the first time. Three years later, Herbert Friedman proved that the Sun emitted X rays using a V-2 experiment. Later experiments showed that the majority of the (soft) X rays in the 8–20 Å region were produced by the solar corona. Then in 1956 Friedman's group found that solar flares produced very high energy X rays. Six years later Riccardo Giacconi's group found the first source of non-solar X rays, Sco X-1, using an Aerobee rocket experiment. This was the first indication of the enormous amount of energy that some cosmic sources were emitting continuously.

So after the war radio astronomy had begun to develop, and X-ray astronomy had begun a little later with experiments mounted on rockets. Another key astronomical resource became available in 1957, initially as part of the International Geophysical Year, when the Soviet Union started the space age with the launch of *Sputnik*. The first major space-age discovery was made the following year of a radiation belt around Earth by James Van Allen using experiments on U.S. Explorer spacecraft. In 1959 the Soviet Union *Luna 1* was the first spacecraft to reach the Moon, *Luna 2* discovered the solar wind, and *Luna 3* produced the first images of the far side of the Moon.

The early years of the space age turned into a competition between the United States and the Soviet Union to impress both their own public but, more particularly, other countries during this difficult period of the Cold War. The Soviet Union won the early rounds of the lunar competition, with the first soft landing on the Moon (*Luna 9*) and the first lunar orbiter (*Luna 10*), both in 1966, but the United States soon overtook Soviet efforts with its Lunar Orbiter and manned Apollo programs. Although all these projects were politically motivated, they nevertheless produced a wealth of geological information about the Moon. This ultimately led to a new consensus about the origin of the Moon as being caused by the impact of a small planetary-sized body with the early proto-Earth.

The United States was the first to reach a planet, with *Mariner 2*'s flyby of Venus in 1962, measuring a surface temperature of 425°C. But the Soviet Union Venera program was a highly successful series of missions, starting with *Venera 4*, which released

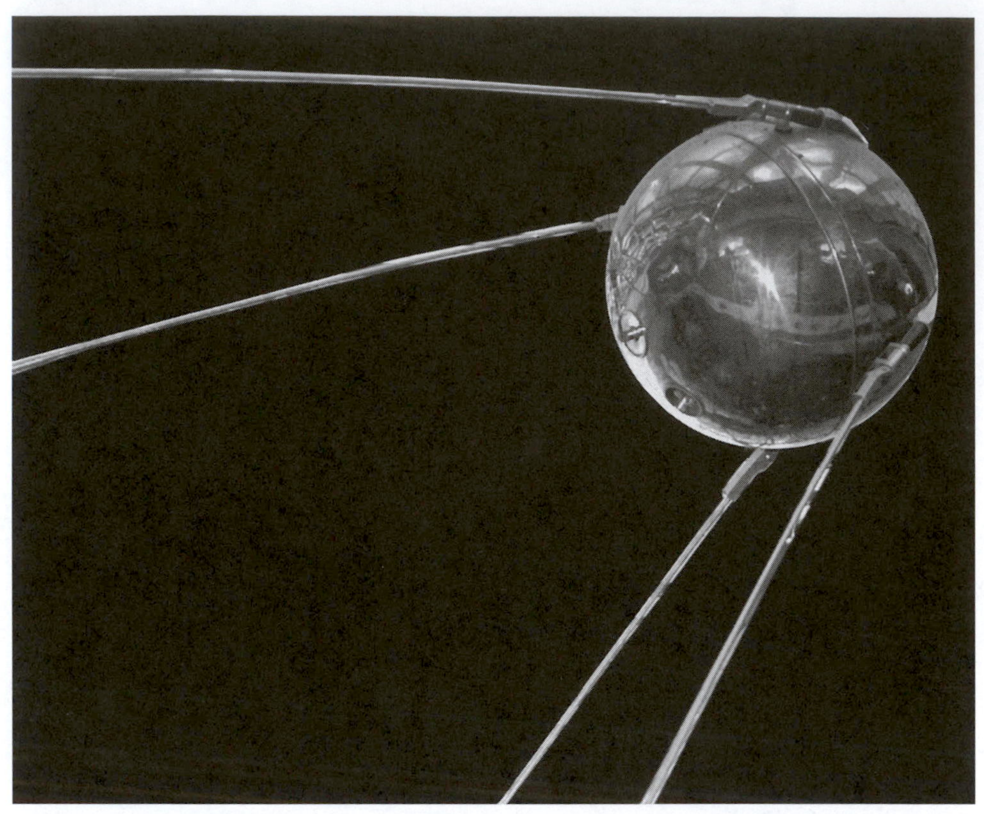

Model of Sputnik. *A Soviet contribution to the scientific collaboration of the International Geophysical Year, it was the first artificial satellite to orbit the Earth. (Courtesy NASA/Goddard Space Flight Center)*

a landing capsule in 1967, and *Venera 7*, which was the first spacecraft to successfully soft-land a capsule on Venus. Both the Soviet Union and the United States imaged the surface of Venus using orbiting radars to see through its dense clouds. The Soviet Union was far less successful in its attempts to reach Mars, whereas the United States had many successful visits to the planet, starting with *Mariner 4* in 1965. The early spacecraft showed a disappointing crater-strewn surface, but *Mariner 9* showed in 1972 that there were giant volcanoes and a large canyon system on the surface. Later spacecraft were to provide evidence for a limited amount of subsurface water. The United States was the only nation during the first phase of planetary exploration to send spacecraft to the outer planets, starting with *Pioneers 10* and *11*, followed by *Voyagers 1* and *2* in the 1970s. These discovered rings around Jupiter and imaged an amazing amount of detail, including spokes and shepherding satellites, in Saturn's rings. They found active volcanoes on Io and imaged all the major planetary satellites.

Over time the U.S. and Soviet planetary programs became more oriented to scientific research and less to political prestige. Collectively these missions have gathered a vast amount of information about the planets, their satellites, and minor solar system bodies, such as comets and asteroids, which could not have been acquired by ground-based or Earth-orbiting observatories.

Since the start of the space age other countries have launched astronomical spacecraft, aided initially by the United States, which in 1959 offered to launch scientific experiments or complete scientific spacecraft of other friendly countries. So an American Scout rocket launched *Ariel 1* in 1962, which included British-built experiments to investigate Earth's ionosphere and solar radiation. In the same year an American Thor Agena launched a Canadian ionospheric spacecraft called *Alouette*. Then in 1964 a Scout launched an Italian spacecraft called *San Marco 1* to investigate Earth's upper atmosphere. In the same year a number of Western European nations joined to form the European Space Research Organisation (ESRO), which in 1975 was subsumed into the European Space Agency (ESA).

In the early days of space research most astronomical work was undertaken by the United States and the Soviet Union, but then Western European nations and Canada started to develop their own astronomical space programs, first as separate countries, then together as part of ESRO and ESA. Japan and China were the next countries to launch scientific satellites in the early 1970s. Initially ESRO had a small research program, concentrating mainly on observations of the Sun and Earth's magnetosphere, while the United States and the Soviet Union were spending vast amounts of money on lunar and planetary probes like the American Surveyors, Mariners, Pioneers, Vikings, and Apollos, and the Soviet Veneras and Mars spacecraft. In addition, the United States launched Earth-orbiting astronomical observatories, including *Uhuru* (launched in 1970, X-ray mission), *Copernicus* (1972, ultraviolet), and *Einstein* (1978, X ray).

Astronomy remains dependent on ground-based telescopes operating mainly in the optical, radio, and microwave wavebands, as they are very much cheaper than space observatories both to build and to operate. By the early twenty-first century there were a number of mountain-based optical telescopes with mirror diameters of 8–10 m, compared with the *Hubble Space Telescope*'s 2.4 m mirror. Adaptive optics, operating mainly in the infrared, is now used with many ground-based telescopes to dramatically reduce problems previously caused by atmospheric turbulence. In the radio waveband there are a number of very large ground-based interferometers, which have enabled resolutions to be produced in the milliarcsec region. In 1997 Japan launched the *HALCA* (*Highly Advanced Laboratory for Communications and Astronomy*) radio observatory spacecraft that formed an interferometer with ground-based radio observatories in the Very Large Baseline Interferometry Space Observing Program. A resolution of about 0.3 milliarcsec was achieved.

Before 1957 astronomy was carried out by a relatively small number of astronomers, and former radio engineers, worldwide using relatively inexpensive equipment. For example the 200 in (5 m) Palomar telescope, the largest optical telescope in the world, only cost about $100 million at 2007 prices. However, large spacecraft and their associated launch, tracking, and operational costs would come to dwarf this.

Before World War II American astronomy was paid for largely by wealthy philanthropists, but the situation changed dramatically with the start of the space age when large sums of government money became involved in funding the space program. At that time the best optical telescopes in the world were in the United States, but that was not the case for radio telescopes, where the best were in the United Kingdom and Australia. Unfortunately, access to the best American optical telescopes was limited to members of various privately funded observatories. As a result in October 1957, fortuitously the same month that *Sputnik* was launched, a number of American universities joined together to found AURA, the Association of Universities for Research in Astronomy, with the aim of building an optical observatory in the United States using state-of-the-art telescopes. Funded by the National Science Foundation, this was to soon become the Kitt Peak National Observatory in Arizona, which by 1964 had both an 84 in (2.1 m) and 36 in (0.9 m) telescope operational, with plans for a 150 in (3.8 m) telescope.

During this period, 1957–64, the United States also began to catch up with Australia and the United Kingdom in the provision of radio telescopes, with the building of the 1,000 ft (300 m) fixed Arecibo dish in Puerto Rico, paid for by the U.S. Department of Defense, and the 300 ft (90 m) transit telescope at the National Radio Astronomy Observatory, Green Bank, West Virginia. But these facilities were built in a piecemeal way and were not part of an overall American astronomical facility strategy. As a result, in the early 1960s the National Academy of Sciences (NAS) asked Lick Observatory Director Albert Whitford to chair a review of the then-current situation and recommend a 10-year program to improve ground-based astronomical facilities. This was to result in the first of a series of decennial reports, with all subsequent ones including both space-based and ground-based facilities.

Shortly after the launch of *Explorer 1* in 1958, NASA, aided by its advisory committees, developed a program of observatory spacecraft, starting with the existing program of small Explorer spacecraft. This was supplemented by a solar program also using small observatory spacecraft, and a deeper space observatory program using large Earth-orbiting Optical Astronomical Observatories (OAOs). Astronomers wanted these OAOs to be proof-of-concept spacecraft, to be followed by larger observatories later. However, NASA wanted to launch large, prestigious observatories immediately to try to upstage the Soviet Union. The result was four, 2-ton spacecraft, two of which were failures. These different perceptions of what was required for scientific spacecraft was not unusual in the early days of the American space program, built, as it was, with two major targets in mind, neither of which was scientific. One was military and the other was to demonstrate to the world the superiority of American technology by, for example, landing men on the Moon and launching spacecraft to the planets.

A similar problem faced by NASA in the 1960s was how to persuade the astronomical community to back the building of a 3.0 m diameter Large Space Telescope (LST), when the money could be used, in principle, to buy many, larger Earth-based telescopes instead. Many astronomers refused to understand or accept that, if the LST was not built, the money saved would not be used to build large numbers of these ground-based telescopes. The NAS set up an ad hoc committee on the LST to review matters, which recommended in 1969 that, not only should the LST be built, but the number of large ground-based optical telescopes should also be increased. This would help to offset astronomers' concerns about the LST, but also avoid them asking for access to the LST when ground-based facilities could do some tasks as well as, or even better, in some cases.

The LST story also showed the importance of political lobbying in getting a program approved. In 1972 the Greenstein Report (the decennial report following the Whitford Report) only endorsed the LST as a long-term goal, despite pleading by two committee members, George Field and Donald Morton, to give it high priority. So, not surprisingly, the House Appropriations Subcommittee refused in 1974 to back the LST program. That same year the NAS Space Science Board reevaluated the LST program on the basis that the approval of the Space Shuttle in 1972 now allowed in-orbit servicing. John Bahcall and Lyman Spitzer used this opportunity to push the LST case and persuaded each of the 23 members of the Greenstein Committee to back the LST as a priority program just two years after they had rejected such an idea. As a result Congress agreed to fund the LST (later called the *Hubble Space Telescope—HST*) phase B study. When reflecting in 1989 on this and similar events, John Bahcall observed that the reason that radio astronomy seemed to have a better record in obtaining federal funding than optical astronomy was because radio astronomers worked as a more coherent community than optical astronomers, who were a more fragmented group. Clearly decisions on funding astronomical facilities either on the ground or in space was not based on pure scientific need. Other factors were at least as important.

How successful were the various ground- and space-based observatories? Some space observatories became well known, such as the *HST*, partly because it took beautiful images, and partly because it was backed by an excellent organization, the Space Telescope Science Institute. Although many studies have shown that the *HST* is the most productive modern astronomical facility in terms of number of papers published, once costs have been taken into account it does not come out too well on a cost-efficiency basis. The *Chandra X-ray Observatory*, ESA's *XMM-Newton* (*X-ray Multi-Mirror Newton*), and the ground-based, 10 m diameter Keck telescope all proved to be far more cost-effective on this basis than the 2.4 m diameter *HST*. Two reasons that drove up the costs of the latter was its long gestation period, with many changes in design, and the plan to have in-orbit servicing by the Space Shuttle, which was extraordinarily expensive.

The vast array of ground- and space-based facilities touched on above have yielded rich discoveries, which it is impossible to do justice to in this article. A scan of the adjacent timeline gives some idea of progress.

For many years many people, not just astronomers, wondered whether there were planets around other stars. In 1991 Alex Wolszczan discovered one around a pulsar, and four years later Michel Mayor and Didier Queloz discovered the first planet orbiting a normal star. Since then some 200 exoplanets have been found (as of early 2009). However, despite many attempts, so far no signs of intelligent life have been found away from Earth anywhere in the universe.

David Leverington

MILESTONES IN THE DEVELOPMENT OF ASTROPHYSICS AND PLANETARY SCIENCE

ca. 500 BCE The Pythagoreans conclude that Earth is spherical.

ca. 410 BCE Hicetas suggests that Earth spins on its axis.

ca. 260 BCE Aristarchus proposes a Sun-centered universe.

ca. 230 BCE	Eratosthenes measures the diameter of Earth correct to better than 1 percent.
ca. 130 BCE	Hipparchus produces his star catalog containing the position of about 850 stars. He also discovers the precession of the equinoxes.
1543	Nicolaus Copernicus publishes his theory of a Sun-centered universe.
1572–74	Tycho Brahe concludes that the supernova of that period is a newly visible star, and is not in the solar system.
1577	Brahe shows that comets are astronomical objects and are not in Earth's atmosphere.
1600	William Gilbert suggests that Earth behaves like a large magnet.
1604	Johannes Kepler shows that the orbit of Mars is an ellipse.
1608–9	Telescopes become available in Europe. The earliest ones appear to have been made in Holland.
1610	Galileo Galilei concludes that there are mountains on the Moon, discovers the four large moons of Jupiter, and the phases of Venus.
1611	Johann Fabricius concludes that the Sun is rotating on its axis by observing the movement of sunspots.
1620	Francis Bacon comments that the shorelines on either side of the Atlantic could fit into each other.
1635	Henry Gellibrand finds that Earth's magnetic field is drifting with time.
1659	Christiaan Huygens discovers Saturn's ring.
1663	Giovanni Domenico Cassini deduces a rapid rotation period for Jupiter.
1668	Isaac Newton makes the first reflecting telescope.
1672	Christiaan Huygens observes Mars's south polar cap.
1675	Giovanni Domenico Cassini discovers that Saturn's ring is divided into two.
1687	Isaac Newton's publishes his universal theory of gravitation.
1691	Giovanni Domenico Cassini observes Jupiter's polar flattening, caused by its rapid rotation.
1693	Edmund Halley discovers the secular acceleration of the Moon.
1705	Giovanni Domenico Cassini suggests that Saturn's rings are composed of numerous small particles.
1717	Halley predicts the return of the comet of 1698, now called Comet Halley.
1733	Chester Moore Hall makes the first achromatic telescope, greatly reducing chromatic aberration.
1750	Thomas Wright suggests that the Milky Way is a large, rotating disc of stars.
1755	Immanuel Kant suggests that the solar system condensed out of a cloud of gas that started to rotate as it contracted.
1761	Mikhail Lomonsov observes that Venus has an extensive atmosphere.
1775	Immanuel Kant suggests that "nebulous stars" are systems of stars like the Milky Way.
1778	The Compte du Buffon suggests that Jupiter has not completely cooled down since it was formed.

1781 William Herschel discovers Uranus.

1782 Edward Pigott and John Goodricke conclude that the intensity variations of Algol are due to it being periodically eclipsed by a companion star.

1783 John Michell suggests that if stars are heavy enough, light would be prevented by gravity from leaving the surface.

1789 William Herschel completes his 40 ft Newtonian reflector telescope of 48 in (122 cm) diameter.

1796 Pierre Laplace proposes his nebula hypothesis of the origin of the solar system.

1801 Giuseppe Piazzi discovers the first asteroid, Ceres.

1802 William Wollaston observes that the solar spectrum has dark lines, later called Fraunhofer lines.

1834 Friedrich Bessel concludes that the density of the Moon's atmosphere is less than 0.2 percent that of Earth.

1838 Friedrich Bessel makes the first reliable distance measurement of a star, 61 Cygni.

1840 John William Draper takes the first photograph of the Moon.

1843 Heinrich Schwabe discovers the solar sunspot cycle.

1846 Johann Galle and Heinrich d'Arrest discover Neptune, following calculation of its expected position by Urbain Le Verrier.

1852 Edward Sabine, Rudolf Wolf, and Alfred Gautier independently conclude that there is a correlation between sunspots and disturbances in Earth's magnetic field.

1857 James Clerk Maxwell proves mathematically that Saturn's rings must be composed of numerous small particles.

1858 Richard Carrington discovers that the latitude of sunspots changes over a solar cycle.

1859 Richard Carrington discovers the differential rotation of the Sun

Gustav Kirchhoff and Robert Bunsen explain the presence of dark and bright lines in laboratory spectra.

Gustav Kirchhoff concludes that Fraunhofer lines show that there is sodium and iron on the Sun.

1863 William Huggins shows that stars are made of the same elements as seen on Earth and on the Sun.

1866 Daniel Kirkwood discovers gaps in the orbital periods of asteroids caused by Jupiter's gravity.

Giovanni Schiaparelli shows that the Perseid meteors are caused by Earth passing through a ring of particles orbiting the Sun, produced by Comet Swift-Tuttle.

1877 Giovanni Schiaparelli produces a map of Mars showing linear features called canali, which are mistranslated into English as canals, indicating an artificial origin.

1878 George Darwin suggests that the Moon spun off the proto-Earth when it was still molten.

1888 Karl Küstner finds that the latitude of the Berlin Observatory is not constant.

1892 Seth Chandler explains the motion of Earth's poles, with respect to its surface, as being the superimposition of two effects of 14- and 12-month periods.

1893 Edward Maunder discovers that there were very few sunspots from about 1645 to 1715, now called the Maunder Minimum. This was later found to be a period of exceptionally cold winters in Europe and North America.

1897 Emil Wiechert suggests that Earth has a dense metallic core made mostly of iron.

The 40 in (102 cm) diameter Yerkes refractor is completed.

1905 Albert Einstein publishes his special theory of relativity proposing, among other things, the equivalence of mass and energy.

1908 The 60 in (1.5 m)-diameter Mount Wilson refractor is completed.

George Ellery Hale discovers large magnetic fields in sunspots.

Frank Taylor suggests that the mid-Atlantic ridge is the line of rifting between Africa and South America, which are gradually moving apart.

1909 Bernard Brunhes finds evidence of Earth's magnetic field reversal.

Andrija Mohorovičić discovers the boundary between Earth's crust and mantle.

1910 Joel Stebbins proves that Algol is an eclipsing binary by measuring its luminosity with an early selenium photometer.

1911 Victor Hess discovers cosmic rays.

1912 Henrietta Leavitt shows that the periods of Cepheid variables correlate with their absolute intensities. This enabled their distances to be deduced from their periods and apparent intensities.

Alfred Wegener proposes his theory of continental drift.

1913 Henry Norris Russell shows how the absolute intensity of stars correlates with their spectral type by publishing what was later called the first Hertzsprung-Russell diagram.

Niels Bohr proposes his quantum theory of the atom.

1914 Beno Gutenberg detects the interface between Earth's core and mantle about 3,500 km from the center of Earth.

1915 Einstein publishes his general theory of relativity, which explains, among other things, the precession of Mercury's perihelion.

1916 Karl Schwarzschild publishes his theory of black holes.

1919 Creation of the International Astronomical Union to foster international cooperation in astronomy.

1920 The famous, inconclusive debate between Heber Curtis and Harlow Shapley about the nature and distance of spiral galaxies.

1922 Alexander Friedman publishes his equations describing an expanding universe.

1924 Ejnar Hertzsprung discovers the first flare star, DH Carinae.

	Arthur Eddington shows that for most stars, their absolute luminosity is a function of mass.
	Edwin Hubble identifies Cepheid variables in the Andromeda and M33 nebulae, enabling their distances to be calculated. This proved that they were galaxies of stars outside the Milky Way.
1925	Cecelia Payne analyzes stellar spectra and concludes that hydrogen and helium are the most abundant elements in stellar atmospheres.
	Walter S. Adams shows that Sirius B is very dense, as predicted theoretically by Arthur Eddington 10 years earlier. It is the first proven white dwarf star.
1927	Jan Oort explains how the Milky Way rotates.
1929	Edwin Hubble and Milton Humason find that the recession velocity of distant galaxies varies linearly with distance.
	Arthur Holmes suggests that Earth's continents are floating on convective currents in the mantle.
1930	Clyde Tombaugh discovers Pluto.
1931	Subrahmanyan Chandrasekhar publishes his theory of white dwarfs.
1932	Karl Jansky is the first to unambiguously detect radio emissions from the cosmos, which appear to come from the center of the Milky Way.
1934	Walter Baade and Fritz Zwicky propose their theory of neutron stars.
1938	Charles Critchfield and Hans Bethe describe the proton-proton chain reaction to explain the generation of heat in stars.
	Hans Bethe and Carl von Weizsäcker independently describe the carbon-nitrogen-oxygen cycle to explain the generation of heat in heavy stars.
1941	Bengt Edlén concludes that the temperature of the solar corona is at least 2 million K.
1942	James Hey and George Southworth independently discover that the Sun is a radio source.
1943	Kenneth Edgeworth suggests that there are a large number of small bodies in the solar system outside the orbit of Pluto. Eight years later Gerard Kuiper makes the same suggestion.
	Carl Seyfert investigates galaxies, now called Seyfert galaxies, which have intense cores and unusual emission lines. They were the first active galaxies to be identified.
1944	Hendrick van de Hulst predicts that interstellar hydrogen should emit 21-cm radio waves.
	Gerard Kuiper detects methane in Titan's atmosphere.
1949	Herbert Friedman finds that the Sun emits X rays.
	Harold Urey proposes his cold accretion theory of Earth and Moon.
	The 200 in (5.1 m) diameter Palomar reflector sees first light.
1950	Jan Oort suggests that long period comets originate in a cloud about 100,000 AU from the Sun.
1950–51	Fred Whipple proposes his dirty snowball theory of a comet's nucleus.

1951 Harold Ewen and Edward Purcell discover hydrogen's 21 cm radiation.

1952 Walter Baade discovers that there are two types of Cepheid variables, causing the estimated age of the universe to be doubled, making it more than the estimated age of Earth for the first time.

1954 Scott Forbush shows that the number of cosmic rays received on Earth varies with the solar cycle.

1956 Cornell Mayer, Russell Sloanaker, and T. P. McCullough deduce a surface temperature for Venus of about 300°C based on its thermal radio emissions.

Herbert Friedman shows that solar flares emit X rays.

1957 Geoffrey Burbidge, Margaret Burbidge, Willy Fowler, and Fred Hoyle show how heavy elements can be produced in stars.

Eugene Parker publishes his theory of the solar wind.

Max Waldmeier discovers coronal holes.

Foundation in the United States of AURA, the Association of Universities for Research in Astronomy.

Launch of *Sputnik* on 4 October starts the space age.

1957–58 The International Geophysical Year runs from July 1957 to December 1958.

1958 Launch of *Explorers 1*, *3*, and *4*. James Van Allen's experiments discover a radiation belt around Earth.

Vanguard 1 shows that Earth is pear shaped.

Jan Oort, Gart Westerhout, and F. J. Kerr produce the first map of the overall structure of the Milky Way using radio telescopes.

NASA is created.

1959 Launch of *Luna 1*, the first spacecraft to reach the Moon. It finds that the Moon has no measurable magnetic field.

Luna 2 discovers the solar wind.

Luna 3 provides first images of the far side of the Moon.

1960 Herbert Friedman and colleagues produce the first image of the Sun in X rays.

Robert Leighton, Robert Noyes, and George Simon discover five-minute oscillations on the surface of the Sun.

Launch of *Tiros 1*, the first weather satellite to provide routine imagery.

Harry Hess produces his theory of sea floor spreading.

1961 Launch of *Explorer 11*, the first gamma-ray spacecraft. During its five-month lifetime, it detected just 22 gamma rays.

1962 Launch of *Mariner 2*, the first spacecraft to reach Venus. It shows that Venus's surface temperature is about 425°C.

Riccardo Giacconi and colleagues discover Sco X-1, the first non-solar source of X rays.

The 63 in (1.6 m) diameter McMath Pierce solar telescope is completed on Kitt Peak.

1963 Richard Goldstein and Roland Carpenter unambiguously resolve Venus's axial rotation period using radar.

Maarten Schmidt discovers the first quasar, 3C 273. It had a redshift of 0.16.

IMP 1 discovers Earth's bow shock.

1964 Arno Penzias and Robert Wilson discover the microwave background radiation, evidence for the Big Bang theory of the universe.

Launch of *Mariner 4*, the first successful spacecraft to reach Mars. It measures a surface atmospheric pressure of about 5 millibar and shows a surface covered in craters.

1965 Gordon Pettengill and Rolf Dyce unambiguously determine Mercury's axial rotation period using radar.

1966 *Luna 9* makes first soft landing on the Moon, discovering that the rocks appear to be volcanic.

Frederick Vine and Drummond Matthews find conclusive evidence for sea floor spreading by measuring magnetic reversals on the ocean floor.

Thomas Brock finds thermophiles in Yellowstone National Park geysers that can tolerate temperatures in excess of 60°C.

Launch of *ATS 1* provides cloud images of Earth from geosynchronous orbit.

1967 The *Venera 4* lander shows that the atmosphere of Venus is 96 percent carbon dioxide.

Mariner 5 measures the surface atmospheric pressure on Venus to be about 75–100 b.

OSO 3 measures the spectrum of a solar flare, indicating a temperature of 30 million K.

Jocelyn Bell and Antony Hewish discover the first pulsar.

1968 Launch of *OAO 2*, the first dedicated ultraviolet spacecraft.

Raymond Davis and colleagues find that the number of neutrinos detected from the Sun is inconsistent with the theory of solar energy production.

David Staelin and Edward Reifenstein discover a pulsar in the Crab Nebula. A month later John Comela finds that its pulsation rate is slowing down.

1969 Charge coupled devices are developed by William S. Boyle and George E. Smith.

Neil Armstrong and Buzz Aldrin are the first men to land on the Moon with *Apollo 11*. The rock samples returned to Earth confirm that the maria are volcanic, and give an age for the Tranquility lavas of 3.65 billion years.

Gerry Neugebauer and Bob Leighton publish their Two Micron (infrared) catalog.

Ray Klebesadel, Ian Strong, and Roy Olson discover gamma-ray bursts using the military *Vela 5A* and *B* spacecraft (announcement delayed until 1973).

1970 Launch of *Uhuru*, the first spacecraft dedicated to non-solar, X-ray astronomy.

Experiments placed on the Moon by the Apollo astronauts show that the Moon has a large, hot core.

1971 Paul Murdin, Louise Webster, and Tom Bolton conclude that the X-ray source, Cyg X-1, contains a black hole.

David Scott and James Irwin of *Apollo 15* bring back a rock that proves to be 4.5 billion years old, or just 100 million years younger than the Moon.

OSO 7 takes the first image of a coronal mass ejection.

1972 *Mariner 9* images massive volcanoes, an enormous canyon system, and sinuous channels on Mars, suggesting that water may once have flowed on the surface.

Launch of *OAO 3* (*Copernicus*), which detects deuterium in the interstellar medium and finds that massive, hot stars emit powerful stellar winds.

Evans, Belian, and Conner discover X-ray bursts using the military *Vela 5A* and *B* spacecraft (announcement delayed until 1976).

Launch of *Pioneer 10*, the first spacecraft to visit Jupiter.

OSO 7 discovers gamma-ray emission from a solar flare.

1973 Godfrey Sill, Andrew Young, and Louise Young suggest that Venus's clouds are composed mainly of sulfuric acid.

Robert Dicke shows that the five-minute oscillations represent the vibration of the Sun as a whole.

Skylab images coronal holes in X rays.

Pioneer 10 detects helium on Jupiter and confirms that Jupiter emits about twice as much energy as it receives from the Sun.

Launch of *Pioneer 11*, the first spacecraft to visit Saturn.

Launch of *Mariner 10*, the first spacecraft to visit Mercury. It is found to be cratered, with a relatively strong magnetic field.

1974 John Grindlay and John Heise detect the first X-ray emission from a flare star using the *ANS* spacecraft.

Joe Taylor and Russell Hulse discover the first binary pulsar, PSR 1913+16. The orbital precession rate of 4.2°/year is in good agreement with Einstein's theory of general relativity.

1976 Landers from the *Viking 1* and *2* spacecraft fail to find any clear evidence of life, past or present, on Mars.

Mario Acuña and Norman Ness suggest that the reduction of high energy particles near Jupiter measured by *Pioneer 11* is either due to an undiscovered ring or satellite.

1977 Launch of *Voyager 2*, the first spacecraft to visit Uranus and Neptune.

Astronomers discover the rings of Uranus.

John Corliss, Robert Ballard and colleagues discover hydrothermal vents at the bottom of the ocean that support life.

1978 Dale Cruikshank and Peter Silvaggio detect methane on Triton.

The *Pioneer Venus Multiprobe* spacecraft shows that there are sulfuric acid clouds on Venus.

Launch of *Seasat*, which provides the first global views of Earth's ocean floor topography.

1979 Stanton Peale, Patrick Cassen, and Ray Reynolds suggest that there is widespread volcanism on Io.

Voyager 1 discovers Jupiter's rings and active volcanoes on Io. Europa's icy surface is thought to cover a possible liquid water ocean.

Voyager 2 shows that there are hot spots on Io at temperatures of up to 200°C. The volcanoes, or geysers, seem to be based on either sulfur dioxide or sulfur.

Peter Goldreich and Scott Tremaine suggest that there may be shepherding satellites on either side of Uranus's thin rings.

1980 Richard Willson and colleagues show, using *Solar Max*, that the solar constant decreases when sunspots cross the solar meridian.

Andy Collins and Richard Terrile discover the first shepherding satellites, Prometheus and Pandora, in the solar system, on either side of Saturn's narrow F ring.

Voyager 1 shows that Saturn's ring system consists of about 1,000 narrow rings.

Richard Terrile discovers spokes on Saturn's B ring in *Voyager 1* images.

Voyager 1 measures the temperature of Titan's surface to be very close to methane's triple point, where it can exist in solid, liquid, or gaseous form.

1981 Completion of the Very Large Array radio telescope in New Mexico.

Alan Guth proposes his inflationary model of the very early universe, in which it expands exponentially, thus solving the horizon problem.

1982 Shrivinas Kulkani and colleagues discover the first millisecond pulsar, PSR 1937+214.

1983 Launch of *IRAS*, the first infrared spacecraft.

1984 Astronomers find the first unambiguous evidence of a partial ring around Neptune.

1985 Richard Willson and colleagues show, using *Solar Max*, that the solar constant reduces toward solar minimum.

Joseph Farman and colleagues discover the ozone hole above the Antarctic.

1986 *Voyager 2* discovers Uranus's highly inclined, off-axis magnetic field.

The *Vega* and *Giotto* spacecraft take the first image of a comet's nucleus when they image the nucleus of Comet Halley.

1987 Supernova 1987A, discovered independently by Ian Shelton, Oscar Duhalde, and Albert Jones, is the first supernova to be detected in the Milky Way since the invention of the telescope.

1989 *Voyager 2* finds that Neptune has a significant internal heat source. Neptune also has very high wind speeds and a highly inclined, off-axis magnetic field.

	Voyager 2 detects eruptions in progress on the surface of Triton.
1990	Launch of *Ulysses*, the first spacecraft to fly out of the ecliptic and view the poles of the Sun.
	Launch of the *Hubble Space Telescope*, the first of NASA's Great Observatories.
1991	Launch of the *Compton Gamma-Ray Observatory*, the second NASA Great Observatory.
	Adaptive optics declassified. They compensate for atmospheric turbulence for ground-based telescopes.
	The *Galileo* spacecraft produces the first closeup image of an asteroid, Gaspra.
	Alex Wolszczan discovers the first exoplanet.
1992	David Jewitt and Jane Luu discover the first Kuiper Belt object.
	The 10 m diameter Keck 1 telescope completed on Mauna Kea.
	George Smoot and colleagues find an anisotropy in the microwave background radiation using the *COBE* spacecraft.
	EUVE discovers the first extreme ultraviolet source outside the Milky Way.
1993	The *Galileo* spacecraft discovers Dactyl, a companion to the asteroid Ida. It is the first asteroid companion to be found.
1994	Remnants from Comet Shoemaker-Levy 9 collide with Jupiter leaving short-lived visible scars. This is the first time that a comet has been seen to collide with Jupiter.
1995	The *Galileo* spacecraft is the first to be put into orbit around Jupiter. It included a probe that was the first to descend through Jupiter's outer atmosphere.
	Launch of *SOHO*, which was to image hundreds of sun-grazing comets.
1996	David Jewitt and Jane Luu discover the first scattered disc object.
1997	*Mars Global Surveyor* finds abundant evidence that liquid water once flowed on Mars.
	The *BeppoSAX* spacecraft discovers a gamma-ray burst, GRB 970508, which is found to be at a cosmological distance of at least 4 billion light years.
	Data from the *Galileo* spacecraft indicates that Callisto may have a subsurface ocean of liquid water.
	Launch of the *HALCA* radio observatory. It forms an interferometer with ground-based radio telescopes to achieve a resolution of 0.3 milliarcsec.
	The *SOHO* spacecraft discovers the Sun's magnetic carpet transferring energy into the corona.
1998	Two teams headed by Adam Reiss and Saul Perlmutter independently find evidence that the cosmological constant is non-zero, reigniting the concept of dark energy.
1999	Launch of the *Chandra X-Ray Observatory*, the third NASA Great Observatory.
2000	Data from the *Galileo* spacecraft indicates that Ganymede may have a subsurface ocean of liquid water.

2001	Launch of the *Wilkinson Microwave Anisotropy Probe* that detects the composition of the early universe.
	NEAR-Shoemaker lands on the asteroid Eros.
2002	Completion of the four 8.2-m diameter telescopes of the European Southern Observatory's Very Large Telescope on Cerro Paranal.
2003	Launch of the *Spitzer Space Telescope*, the fourth and final NASA Great Observatory.
2004	The *Cassini-Huygens* spacecraft starts to orbit Saturn, the first spacecraft to do so.
	The Mars Exploration Rovers, *Spirit* and *Opportunity*, find convincing evidence that water once existed on the surface of Mars.
2004–5	The *Cassini* spacecraft finds that Enceladus is emitting plumes of water and water ice.
2005	*Cassini-Huygens* discovers what appear to be lakes of liquid methane on Titan.
	A probe from the *Deep Impact* spacecraft impacts Comet Tempel 1.
	The *Huygens* probe lands on Titan. It images drainage channels during its descent and lands on a soft methane surface, where it images cobbles of water ice.
2006	Launch of *New Horizons*, the first probe to Pluto.
	The International Astronomical Union reclassifies Pluto as a dwarf planet, together with the scattered disc object Eris and the asteroid Ceres.

David Leverington

See also: Economics of Space Science, Nations, Space Science Policy

Bibliography

David Leverington, *Babylon to Voyager and Beyond: A History of Planetary Astronomy* (2007).

———, *New Cosmic Horizons: Space Astronomy from the V2 to the Hubble Space Telescope* (2000).

Malcolm Longair, *The Cosmic Century: A History of Astrophysics and Cosmology* (2006).

Jean-Louis Tassoul and Monique Tassoul, *A Concise History of Solar and Stellar Physics* (2004).

ASTROBIOLOGY

Astrobiology is the study of the origins, evolution, distribution, and future of life in the universe. The term astrobiology came into widespread use beginning in 1995, when NASA named its Ames Research Center the lead center for a revived and transformed program stemming from NASA's exobiology program. Exobiology began to congeal as a scientific discipline in the mid-1960s and in turn drew on a tradition of discussion about the possibility of life in the universe. Today the terms astrobiology and exobiology are sometimes used interchangeably, but astrobiology is much broader than exobiology as originally conceived. Astrobiology placed life in the context of its planetary history, encompassing the search for planetary systems, the study of biosignatures, and the past, present, and future of life. Astrobiology science also added new techniques and concepts to exobiology's repertoire.

Geneticist Joshua Lederberg coined the term exobiology in 1960, when he realized that spaceflight held the potential to address the problems of the origin of life and its existence throughout the universe. Four subjects, themselves not yet fully formed disciplines, fed into the new field of exobiology: planetary science, planetary systems science, origins of life, and the Search for Extraterrestrial Intelligence (SETI). By 1965 declarations of the new discipline had been made in the United States and in the Soviet Union.

The most fully formed of these subjects was planetary science. If scientists were to find life beyond Earth, it would most likely be on a planet's surface. There was a tradition of observations of the Moon and planets, and particularly Mars. Percival Lowell's late-nineteenth-century claim of artificially constructed canals on Mars was largely shown spurious by Eugène Antoniadi's detailed observations of the planet in 1909. However, claims of vegetation, bolstered by ground-based observations of dark areas on the planet that grew and shrank with the seasonal waning and waxing of the polar caps, persisted into the 1960s. In the late 1950s spectroscopic evidence of Martian vegetation was believed to be in hand, but it also proved spurious. Nonetheless these observations and others were enough to make the search for life on Mars the prime goal of NASA's space biology program. This goal led to the Mariner and Viking missions to Mars. These did not find life, but the *Mars Odyssey* and Mars Exploration Rovers found evidence of past water on Mars, thus making it possible that life existed in the past.

By contrast, the search for planets beyond the solar system was in a rudimentary state in the 1960s. Astronomer Peter van de Kamp announced in 1963 the existence of a possible planet around Barnard's star, using delicate astrometric methods. Though this claim proved spurious, it took decades to disprove, and meanwhile the possibility of extrasolar planets fed the intense excitement of the embryonic field of exobiology. Three more decades passed before definitive evidence was in hand for planets around other solar-type stars.

The field of origins of life studies was given impetus with the Urey-Miller experiment in 1953. Working under chemist Harold Urey at the University of Chicago, Stanley Miller showed how complex organic molecules could be produced under conditions presumed to exist in the primitive Earth atmosphere. This set off a chain of work on prebiotic experiments throughout the next two decades. By the 1980s, however, further studies cast into doubt the reducing (hydrogen-rich) atmosphere that Miller had presumed, and with it the question of the relevance of such experiments.

The final component of exobiology was SETI. Frank Drake undertook the first search using radio sources in 1960, and during the next few decades many more searches were undertaken in the United States, the Soviet Union, and elsewhere, using increasingly sophisticated equipment. Congress terminated NASA's SETI program in 1993, but SETI continued with private funding.

HAROLD CLAYTON UREY
(1893–1981)

(Courtesy Library of Congress)

Harold Urey was a physical chemist whose interests extended into the fields of astronomy, astrophysics, biology, and geology. He received the 1934 Nobel Prize in Chemistry for his discovery of deuterium. In collaboration with his student, Stanley Miller of the University of Chicago, he performed the classic 1953 Urey-Miller experiment that created amino acids under laboratory conditions that were believed to simulate early Earth. He was also a pioneer in the use of nuclear isotope studies of asteroids and a theorist of planetary evolution. As a founding member of the Space Studies Board of the National Academy of Sciences, Urey was both a strong promoter of space exploration and one of NASA's harshest critics.

Dave Medek

Following the failure to detect life on Mars with the Viking landers in 1976, the field of exobiology might have declined. However, while no spacecraft returned to Mars for two decades, ground-based experiments on prebiotic synthesis continued. NASA's exobiology program funded a wide array of groundbreaking work, including the research of Carl Woese on three domains of life (Archaea, Eubacteria, and Eukarya); the mass extinction work of Luis Alvarez, David Raup, and Joseph Sepkoski Jr.; and the work of Carl Sagan and Chris Chyba on delivery of organics from outer space.

During the 1990s a number of extraordinary events fueled the new discipline of astrobiology. In 1995 Michel Mayor and Didier Queloz found a planet around the solar-type star known as 51 Pegasi. During the next decade astronomers, in particular the team led by Geoff Marcy and Paul Butler, discovered more than 130 other extrasolar planets. Although all were gas giants, these discoveries hinted that the more difficult detections of Earth-size planets would come in the near future. In 1993 NASA researchers claimed that a meteorite found in the Antarctic, dubbed ALH84001 (the first found in 1984 in the Allan Hills region), had its origin on Mars. Detailed analysis of the rock showed what NASA scientists and their colleagues interpreted as nanofossils—extremely small, fossilized remnants of life that once existed on Mars. Although years of further study have convinced many scientists that these artifacts are not biogenic, they caused immense excitement and gave a boost to astrobiology's fortunes and funding. Almost simultaneously, the *Galileo* spacecraft provided firmer evidence of a previous discovery of the *Voyager* probes: that the Jovian moon Europa had a surface of cracked ice, beneath which, in all likelihood, was an ocean of water.

Where there was water, scientists reasoned, there could be life, especially since tidal friction between Jupiter and its satellites provided thermal energy far from the Sun.

CARL SAGAN (1934–1996)

Carl Sagan, in addition to his research studies of Venus and Mars, perhaps did more than any other scientist to popularize astronomy in the United States. His initial efforts, including articles in the weekly newspaper supplement *Parade*, met with skepticism and condescension by his colleagues, but Sagan persisted, writing and producing the enormously successful Public Broadcasting Service series *Cosmos*. In 1980 he cofounded and became the first president of the Planetary Society. Sagan was also a pioneering researcher and promoter of astrobiology and the Search for Extraterrestrial Intelligence.

Amy Sisson

Meanwhile at Ames Research Center, where NASA's Life Sciences laboratories had been located since the early 1960s, exobiology was transformed into astrobiology. In the face of threatened closure in 1995, Ames, which had been criticized for having a fragmented mission in life science, space science, and Earth science, argued that this was actually a strength when it came to interdisciplinary work necessary for the search for life beyond Earth. Convinced by this argument, NASA made Ames the lead center for astrobiology, and the word astrobiology was first published in its modern context in the NASA Strategic Plan for 1996. By 1998 the first astrobiology roadmap had been created, and 11 U.S. institutions became charter members of the Astrobiology Institute. The biennial Astrobiology Science Conferences drew about 700 participants.

Origins of life studies also fed into the new optimism about extraterrestrial life. Scientists found life at extreme pressures and temperatures around deep-sea hydrothermal vents, fueled by energy and nutrients seeping from Earth's crust. More generally, life was found in a variety of extreme environments, including caves, inside deep rock, and in highly acidic and salty conditions. These discoveries showed that life was much more adaptable than previously thought. At the same time, the discovery of complex organics in molecular clouds in space, at the level of amino acids, gave credence to the idea that life could be ubiquitous because its building blocks were common in outer space.

Not everyone shared in the optimism. Geologist Peter Ward and astronomer Donald Brownlee argued that the conditions for advanced life are so stringent that Earth is a "Rare Earth" in terms of harboring intelligence. The universe may be filled with microbes, they concluded, but not intelligence. A universe full of microbes would in itself arguably constitute the greatest discovery in the history of science, making possible a "universal biology" rather than just a natural history of life on Earth. A universe full of intelligence would be even more interesting, and this was the province of SETI, which continued its search programs with increasingly sophisticated equipment.

With the demise of NASA's SETI program in the wake of congressional termination of funding, SETI was no longer a programmatic part of NASA's astrobiology effort. It became the domain of private SETI programs, but remained an intellectual part of astrobiology.

The guiding principle for astrobiology is the concept of cosmic evolution, the connected evolution of the constituent parts of the cosmos. Physical evolution of material objects such as planets, stars, and galaxies has been accepted for much of the twentieth century, especially after 1950, giving rise to our understanding of the physical universe that is now well known. Biological evolution throughout the universe is the domain of the extraterrestrial life debate. Although it remains unproven, the belief in otherworldly life has the status of a worldview, with implications for many areas of human endeavor. The worldview, in which cosmic evolution commonly ends in life, mind, and intelligence, has been called the "biological universe." Astrobiology is the science that will decide between these two worldviews. Scientists have also postulated a third worldview, the postbiological universe in which biology has been replaced by artificial intelligence. Entering the twenty-first century, humans are in a position similar to that of 400 years ago, when natural philosophers were weighing the relative merits of the heliocentric and the geocentric worldviews. The proof of a biological (or postbiological) universe would have implications even more profound than the acceptance of the Copernican theory.

Steven J. Dick

See also: Exoplanets, Mars, Viking Biology

Bibliography

David J. Darling, *Life Everywhere: The Maverick Science of Astrobiology* (2001).

Steven J. Dick, *The Biological Universe: The Twentieth-Century Extraterrestrial Life Debate and the Limits of Science* (1996).

Steven J. Dick and James E. Strick, *The Living Universe: NASA and the Development of Astrobiology* (2004).

Donald Goldsmith and Tobias Owen, *The Search for Life in the Universe* (1980).

Peter Ward and Donald Brownlee, *Rare Earth: Why Complex Life is Uncommon in the Universe* (2000).

Exobiology. *See* Astrobiology.

Planetary Protection

Planetary protection is the practice of preventing the biological contamination of solar system bodies by Earth life (known as forward contamination) and preventing the contamination of Earth by possible life forms that may be returned from other solar system bodies (back contamination). Planetary protection prevents contamination that would interfere with scientific efforts to find evidence of life elsewhere or to understand chemistry on lifeless worlds in their natural states, which may help in understanding life on Earth. It is also necessary to protect the Earth's biosphere if extraterrestrial life exists.

The idea of planetary protection dates to the beginning of the space age, with scientists interested in the possibility of extraterrestrial life raising questions about forward and back contamination. The international scientific community organized early to express its concerns about planetary protection and has remained involved in the global dialogue about the subject ever since. At the 1956 congress of the International Astronautical Federation, scientists identified a need for coordinating international efforts to prevent interplanetary contamination with the commencement of spaceflight. In 1957 the U.S. National Academy of Sciences examined the issue and recommended that the United States adopt noncontaminating spaceflight practices. In 1958 the International Council of Scientific Unions introduced planetary quarantine standards for solar system exploration missions and formed an interdisciplinary Committee on Space Research (COSPAR), which served as the focal point of international activities relating to planetary protection. By 1960, NASA had codified lunar and planetary spacecraft decontamination and sterilization procedures, which were applied to the Ranger lunar exploration project. NASA established its first official planetary quarantine program in 1963, and COSPAR established its first formal interplanetary contamination probability standards in 1964. In 1967 the United Nations Treaty on Principles Governing the Activities of States in the Exploration and Use of Outer Space was ratified. Its Article IX requires signatories to avoid contamination in the conduct of solar system exploration. NASA's Apollo program established planetary quarantine requirements for both astronauts and lunar materials. Some members of the space community viewed these measures as necessary, while others viewed them as burdensome. The planetary quarantine program for astronauts was ended after the *Apollo 14* mission, as scientists concluded there was no life on the Moon. NASA required its Viking landers, sent to Mars in the mid-1970s with life-detection experiments, to be sterilized before launch. The *Galileo* spacecraft was not sterilized. However, when scientists discovered the possibility of a warm-water ocean on Europa, they decided to target the spacecraft at Jupiter itself before its propellant ran out, to prevent any future impact on Europa.

Cleaning and sometimes sterilization are required for spacecraft intended to explore planetary environments of particular interest in the search for evidence of extraterrestrial life, such as the surface and subsurface of Mars. Limits on the allowable biological contamination carried by a spacecraft are established if it is intended to contact such an environment. Planetary protection methods include the use of clean rooms and microbial barriers in spacecraft construction, assembly, and testing; sterilization of spacecraft components; and microbial assays of spacecraft surfaces. Dry-heat sterilization, while the only certified sterilization method, caused problems for electronic components and some other materials, as demonstrated on the Ranger program, which endured several failures in part to excessive heat sterilization. Other methods have been tested. COSPAR and NASA changed the policy for Mars landers in the mid-1990s. Landers without life-detection experiments now have less stringent cleaning requirements.

The U.S. Apollo and Soviet Union Luna missions of the 1960s–70s established a precedent of returning extraterrestrial materials to Earth, a practice that continued in twenty-first-century comet and asteroid sample-return missions. COSPAR policies for future Mars sample-return missions included hermetically sealed sample containers, breaking the chain of contact between the return vehicle and the Martian surface, and a comprehensive quarantine protocol.

Linda Billings and John D. Rummel

See also: Outer Space Treaty, Ranger, Viking

Bibliography

Charles Phillips, *The Planetary Quarantine Program: Origins and Achievements, 1956–1973* (1974).

Task Group on Issues in Sample Return; Space Studies Board; Commission on Physical Sciences, Mathematics, and Applications; U.S. National Research Council, *Mars Sample Return: Issues and Recommendations* (1997).

Task Group on the Forward Contamination of Europa; Space Studies Board; Commission on Physical Sciences, Mathematics, and Applications; U.S. National Research Council, *Preventing the Forward Contamination of Europa* (2000).

Search for Extraterrestrial Intelligence

Search for Extraterrestrial Intelligence (SETI) is a term most commonly used in connection with searches for artificial radio signals from intelligence beyond Earth using radio telescopes. Although early radio pioneers, including Nikola Tesla and Guglielmo Marconi, believed they had detected such signals, the modern scientific search began with Frank Drake's Project Ozma in April 1960. SETI became a small NASA research and development program during the 1970s–80s and was briefly operational for one year before Congress terminated it in 1993. The search continued in the twenty-first century in privately funded programs around the world.

Interest in SETI was fueled by rising belief in the possibility of extraterrestrial intelligence at the dawn of the space age. In 1959 Cornell physicists Giuseppe Cocconi and Philip Morrison published a seminal article "Searching for Interstellar Communications" in *Nature*. On technical grounds they suggested that such communications would most likely take place in the radio portion of the electromagnetic spectrum. They suggested further that searches should take place at 21 cm wavelengths, the 1420 MHz radio emission of neutral hydrogen, the most abundant element in the universe. This was the first of several "magic frequencies" that became prominent in the SETI literature of the twentieth century.

Frank Drake's first search, with an 85 ft radio telescope at the National Radio Astronomy Observatory in Green Bank, West Virginia, employed the 21 cm strategy. It was unsuccessful but precipitated a conference on the subject the following year at Green Bank, the first of many such conferences held throughout the following decades. In the Soviet Union, radio astronomer I. S. Shklovskii became the champion of SETI in the 1960s, and a group of Soviet astronomers began their own series of observations. Shklovskii and Carl Sagan's book, *Intelligent Life in the Universe* published in 1966, became the foundational book for SETI enthusiasts for the remainder of the century.

SETI as a research program took a major step forward with NASA's increasing interest in the 1970s. Led by John Billingham at NASA's Ames Research Center in

California, NASA sponsored various studies on the problem, most notably those led by Bernard M. Oliver in 1971 and by Philip Morrison in 1975–76. Oliver and Billingham's *Project Cyclops* study laid much of the technical groundwork for SETI, while the landmark volume *The Search for Extraterrestrial Intelligence*, edited by Morrison et al., gave the field its name and, more important, provided the rationale for SETI as a NASA program.

With funding at the level of about $1.5 million per year, in 1983 NASA Ames Research Center and Jet Propulsion Laboratory (JPL) embarked on a program to build the instrumentation necessary for a systematic SETI effort. Ames concentrated on a "targeted search" of about 1,000 solar-type stars within 100 light-years, using the giant radio telescope at Arecibo, Puerto Rico, while JPL employed an "all-sky survey" approach, using the 34 m antennas of NASA's Deep Space Network. Ames proposed to cover the 1–3 GHz range of frequencies for its search, while JPL would cover 1–10 GHz. The central effort in searching these billions of channels was the construction of "multi-channel spectrum analyzers," capable of covering millions of channels simultaneously and analyzing them for unusual signals. The NASA program became operational on 12 October 1992, symbolically the quincentennial of Columbus's landfall in the New World. Observations using both strategies continued for only one year, when Congress terminated the entire NASA SETI effort as part of budget cuts. NASA excluded SETI from its efforts in astrobiology.

The targeted search continued under the privately funded SETI Institute. Several other observational programs continued around the world, amid debate on the chances of success. Critics point out that no definitive signal has been received from the proposed extraterrestrials after 45 years of searching. Optimists insist that only a small portion of Earth's galaxy has been searched, and that a successful detection would be the greatest discovery in the history of science.

SETI has gone through several crises that have raised questions about the validity of its program and, beyond that, of humankind's place in the universe. Most notably is the so-called Fermi paradox, which points out that if extraterrestrial civilizations are widespread, and given the time scales of millions of years during which they could have undertaken interstellar travel, they should be here. They are not here (Unidentified Flying Object reports notwithstanding), therefore they do not exist. There are many answers to this paradox, but it is an argument that must be taken seriously. On such arguments hinge humankind's place and destiny in the cosmos, with implications for philosophy, religion, and other human endeavors.

Steven J. Dick

See also: Astrobiology

Bibliography

Steven J. Dick, *The Biological Universe: The Twentieth-Century Extraterrestrial Life Debate and the Limits of Science* (1996).

Steven J. Dick and James E. Strick, *The Living Universe: NASA and the Development of Astrobiology* (2004).

Philip Morrison et al., *The Search for Extraterrestrial Intelligence* (1977).

I. S. Shklovskii and Carl Sagan, *Intelligent Life in the Universe* (1966).
Stephen Webb, *Where Is Everybody? Fifty Solutions to the Fermi Paradox and the Problem of Extraterrestrial Life* (2002).

ASTROPHYSICS

Astrophysics is the study of physical and chemical processes in astronomy. Astronomy had been limited to the study of the appearance, position, and motion of astronomical objects until the nineteenth century. At that time it was thought that it would never be possible to discover the physical and chemical processes occurring in and around these astronomical objects. But the situation changed completely in 1859, when Gustav Kirchhoff and Robert Bunsen found that absorption lines in the spectra of luminous objects in the laboratory showed which elements they were composed of. Kirchhoff immediately extended this work to the solar spectrum, in which he first identified sodium and iron. This was the birth of astrophysics.

In 1863 William Huggins found that absorption lines in the spectra of stars were caused by hydrogen, iron, sodium, calcium, and magnesium. This showed, for the first time, that stars consisted of the same elements as those found on Earth and on the Sun. Work by various astronomers and physicists in the late nineteenth and early twentieth centuries showed that the temperature, pressure, and magnetic fields present in astronomical objects could be deduced from a detailed analysis of their spectra. This worked particularly well for the Sun, but there were problems in interpreting planetary spectra because it was unclear which lines in their spectra were due to the planets and which were due to the intervening Earth's atmosphere. Sounding rockets and Earth-orbiting spacecraft overcame this problem.

The first major step in understanding the source of energy in the Sun and stars was taken in 1905, when Albert Einstein published his special theory of relativity. In this he proposed the equivalence of mass and energy, according to the relationship Energy (E) = mass (m) × velocity of light squared (c^2). However, it took another 30 years or so before the nuclear processes present in the interior of the Sun and stars were basically understood. It then became evident that the temperature in the interior of the Sun and stars was of the order of tens of million degrees K. This compared with surface temperatures in the range of 1,000–100,000 K. But in 1941 Bengt Edlén concluded, from the solar corona's spectrum, that the corona had a temperature of at least 2 million K. It was unclear, and still was in 2007, how such a high temperature could be generated above a solar photosphere with a temperature of only about 6,000 K. This was the first direct evidence that very high temperatures are present in the universe outside the interior of stars.

X rays, which are produced by very energetic processes, are absorbed by Earth's atmosphere, so extraterrestrial X-ray sources could not be detected from the ground. Gamma rays, which are produced by even more energetic processes, are also absorbed by Earth's atmosphere.

Edward Hulbert and Lars Vegard had independently suggested in 1938 that some of the ionization regions in Earth's atmosphere may be caused by solar X rays. In 1949

Herbert Friedman showed, using an experiment on a V-2 rocket, that the Sun was an emitter of X rays, which, he showed three years later, were sufficiently intense to sustain Earth's ionospheric E region.

It was clear that, with the instruments available in the late 1950s, it would be impossible to detect X rays from Sun-like stars because of their distance. But a number of astronomers suggested that other types of stars may produce detectable X rays. In particular, Riccardo Giacconi and colleagues suggested that certain exceptional objects, such as very hot stars, rapidly spinning stars with large magnetic fields, flare stars, and supernova remnants, may emit measurable X rays.

In 1962 Giacconi and his team, using a sounding rocket experiment, detected the first source, now called Sco X-1, of non-solar X rays. This was the first indication of the enormous amount of energy that could be produced in astrophysical processes, other than in supernova and in the interior of stars. It was later found that the ratio of X-ray to visible intensity was 10^3 for Sco X-1, compared with only 10^{-6} for the Sun. Giacconi and colleagues also found a diffuse source of X rays across the whole sky that they scanned.

The early 1960s was an important period in the development of astrophysics. In 1963 Maarten Schmidt discovered the first quasar, when he found that the spectrum of a blue star, which was a part of the radio source 3C 273, had, for then, a very large red shift of 0.16, indicating a distance of some two billion light-years. So it must be intrinsically very intense. Then in 1967 Jocelyn Bell and Antony Hewish, using a radio telescope, discovered the first pulsar, which was pulsing with a frequency of just 1.34 seconds. Thomas Gold and Franco Pacini suggested that this pulsar was a rapidly rotating neutron star, with a typical diameter of just 10–20 km.

NASA launched *Uhuru* in 1970, the first spacecraft dedicated to non-solar X-ray astronomy. Unlike stars at optical wavelengths, X-ray sources tended to be highly variable in their X-ray emissions. About 300 new sources were found, including binary X-ray sources, globular clusters, spiral galaxies, active galaxies, clusters of galaxies, and quasars. Later work showed that most X-ray sources in the Milky Way are binary stars consisting of a rapidly rotating, highly magnetic neutron star and a normal star. In the case of Cyg X-1, however, the collapsed object in the binary was not a neutron star but, most likely, a stellar mass black hole. So by the mid 1970s the universe was seen no longer as the relatively predictable place of the early twentieth century, with just one, or more, solar systems, stars, and galaxies, and the occasional nova or supernova, but one with exotic objects including pulsars, neutron stars, and black holes.

Uhuru, which had no imaging capability, could only locate X-ray sources to within about 1°. The *Einstein* observatory spacecraft, launched in 1978, was the first to be able to image X-ray sources. It found that not only did violent objects emit X rays, but so did cool, optically dim, red dwarf stars. *Einstein* also undertook two deep sky surveys to try to resolve into discrete objects the X-ray background radiation discovered in 1962. Astronomers found that most of the resolved objects in these surveys were quasars or active galactic nuclei, but a significant amount of the X-ray background radiation still remained unresolved.

The Uhuru *satellite during preflight tests at NASA's Goddard Space Flight Center. (Courtesy NASA/Goddard Space Flight Center)*

During the subsequent 25 years or so a number of X-ray spacecraft observatories were launched (see table on page 28), which have been able to resolve fainter and fainter sources and analyze their spectra. This has resulted in many different exotic sources being detected, including X-ray bursters, cataclysmic variables (consisting of a binary of a white dwarf surrounded by an accretion disk of material stripped from a normal star, and a normal star), polars (cataclysmic variables in which the white dwarf is highly magnetic, thus preventing the formation of the accretion disk; the spin rate of the white dwarf being synchronized to the orbital period of the binary), intermediate polars (like polars, but the white dwarf's spin rate is not synchronized), quasiperiodic oscillations (probably connected with black holes, neutron stars, and white dwarfs), and very small high energy sources (possibly supermassive black holes) at the center of active galaxies.

The first gamma-ray spacecraft, *Explorer 11*, was launched in 1961. Throughout five months it detected just 22 gamma rays. Six years later *OSO 3* detected about 600 gamma rays. Although *OSO 3* could not detect individual sources, it showed that the galactic equator was a clear source, with the galactic center the most intense part. In 1971 a balloon-borne experiment identified the Crab pulsar as a gamma-ray source.

Since the 1960s a number of spacecraft have been used to identify an increasing number of gamma-ray sources, probably the most interesting of which are gamma-ray bursts, which were first detected by the *Vela 5 A* and *B* spacecraft in 1969. They can last between a few milliseconds and a few minutes. The long-duration bursts, which last more than two seconds, are thought to be caused by the core collapse of a Wolf-Rayet star to form a black hole, but the cause of the short-duration bursts was, in 2007, unclear.

Selected Space Observatory Missions (Part 1 of 2)

(A) X Ray and Gamma Ray			
Years	**Mission**	**Meaning of Acronym**	**Energies Detectable (keV)**
1961–1961	*Explorer 11*		≥ 50,000
1962	Sco X-1 rocket		2–10
1967–1969	*OSO 3*	Orbiting Solar Observatory	≥ 50,000
1969–1979	*Vela 5A* and *5B*		3–750
1970–1973	*Uhuru* (*SAS 1*)	Small Astronomy Satellite	2–20
1972–1973	*SAS 2*		20,000–1,000,000
1974–1976	*ANS*	Astronomical Netherlands Satellite	0.1–30
1974–1980	*Ariel 5*		0.3–40
1975–1982	*COS B*	Cosmic Ray Satellite	30,000–5,000,000
1977–1979	*HEAO 1*	High Energy Astronomy Observatory	0.1–10,000
1978–1981	*Einstein* (*HEAO 2*)		0.2–20
1979–1981	*HEAO 3*		50–10,000
1983–1986	*EXOSAT*	European X-ray Observatory Satellite	0.04–50
1987–1991	*Ginga*		1–500
1989–1998	*Granat*		2–100,000
1990–1999	*ROSAT*	Röntgen Satellite	0.1–2.4
1991–2000	*Compton GRO*	Gamma Ray Observatory	20–30,000,000
1993–2000	*ASCA*	Advanced Satellite for Cosmology and Astrophysics	0.4–10
1995–	*RossiXTE*	X-ray Timing Explorer	2–250
1996–2002	*BeppoSAX*	Satellite per Astronomia X	0.1–300
1999–	*Chandra*		0.1–10
1999–	*XMM-Newton*	X-ray Multi-Mirror	0.1–15
2000–	*HETE 2*	High Energy Transient Explorer	0.5–400
2002–	*INTEGRAL*	International Gamma-Ray Astrophysics Laboratory	3–10,000
2004–	*Swift*		0.2–500

Selected Space Observatory Missions (Part 2 of 2)

(B) Ultraviolet			
Years	**Mission**	**Meaning of Acronym**	**Waveband (nm)**
1968–1973	*OAO 2*	Orbiting Astronomical Observatory	100–430
1972–1980	*Copernicus* (*OAO 3*)		70–330
1978–1996	*IUE*	International Ultraviolet Explorer	115–320
1992–2002	*EUVE*	Extreme Ultraviolet Explorer	7–76
1999–2007	*FUSE*	Far Ultraviolet Spectroscopic Explorer	90–120
2003–	*GALEX*	Galaxy Evolution Explorer	135–280

(C) Infrared			
Years	**Mission**	**Meaning of Acronym**	**Waveband (μm)**
1983–1983	*IRAS*	Infrared Astronomical Satellite	12–100
1995–1998	*ISO*	Infrared Space Observatory	2–240
2003–	*Spitzer*		3–180

(D) Microwave			
Years	**Mission**	**Meaning of Acronym**	**Waveband**
1989–1993	*COBE*	Cosmic Background Explorer	1 μm–1 cm
2001–	*WMAP*	Wilkinson Microwave Anisotropy Probe	22–90 GHz

The next wavebands to be investigated by spacecraft were the ultraviolet and infrared. Earth's atmosphere is transparent mainly from about 310–900 nm (that is, the visible spectrum, plus a very small part of the near ultraviolet and near infrared) plus radio wavelengths. There are also some atmospheric windows in parts of the infrared, which allow limited work to be undertaken from Earth's surface. But the only way to detect most ultraviolet and infrared energy is from above Earth's atmosphere.

Hot objects with temperatures between 10,000–300,000 K have their peak intensity in the ultraviolet, whereas those with temperatures of below 4,000 K have their peak intensity in the infrared. So ultraviolet objects are basically hot, and infrared objects relatively cool.

The first dedicated ultraviolet observatory was *OAO 2*, launched in 1968. It detected ultraviolet emission from more than 5,000 stars and found a huge cloud of hydrogen around Comet Tago-Sato-Kosaka. Four years later *OAO 3*, also known as *Copernicus*, detected deuterium in the interstellar medium. This was the first time that deuterium had been detected anywhere in the universe, except on Earth. Its observed concentration, of 15 parts per million by mass, is an important constraint on whether the universe is open and will go on expanding forever, or closed, in which case the expansion will eventually cease and go into reverse.

Copernicus showed that massive, hot stars, such as Wolf-Rayet stars, emit powerful stellar winds much more powerful than the solar wind. *IUE* showed that the velocity of these stellar winds was in the region of 1,500–3,500 km/s. At this rate, these stars would lose one solar mass of material in about 25,000 years. The Wolf-Rayet star HD 192163 was known to be surrounded by a ring nebula (NGC 6888). Analysis of this nebula's spectrum showed that it was being heated up to 60,000 K by the shock wave caused by the interaction of the Wolf-Rayet star's stellar wind and the interstellar gas.

In the 1970s it gradually became clear that the solar system is almost at the center of a Local Bubble, about 250 light-years in diameter, in which the density of interstellar gas is much lower than elsewhere in the local region of the Milky Way. As a result, it should be possible to detect extreme ultraviolet sources within the bubble, but possibly not outside, because of the attenuation of the extreme ultraviolet in the surrounding denser regions of the Milky Way. It was something of a surprise, therefore, when in 1992 *EUVE* detected the BL Lac object PKS 2155304, which was the first extreme ultraviolet source found outside the Milky Way. Since then a number of spacecraft have charted the structure of the Local Bubble and of other inhomogeneities in the distribution of interstellar gas in the Milky Way.

The first infrared sky survey was carried out in the 1960s by Gerry Neugebauer and Bob Leighton, using the 2.2 μm window to detect more than 5,000 stars with a ground-based telescope. Their work was extended up to 27 μm, using sounding rockets, in the 1970s, when objects and interstellar dust were detected with temperatures of only about 100 K. Then in 1983 the first infrared spacecraft, *IRAS* was launched. It detected protostars (stars in the early stages of development) in interstellar dust clouds, stars still surrounded by remnants of their original dust clouds, and dust shells emitted by stars near the end of their lives. *IRAS* also found ultraluminous infrared galaxies, one of which was emitting about 10,000 times as much energy as the Milky Way.

The next infrared spacecraft, *ISO*, found that water molecules are common in the Milky Way. It also found star-forming regions in both normal and colliding galaxies. Since its launch in 2003, the *Spitzer* infrared spacecraft was the first to detect light emitted by an exoplanet. It also imaged possible planet-forming gas clouds around stars and participated, along with the ground-based James Clerk Maxwell Telescope, in observing all prominent star-forming regions within about 1,500 light-years of the solar system, in the so-called Gould's Belt Survey.

In 1964 Arno Penzias and Robert Wilson accidentally discovered the remnants of the Big Bang, when they detected microwave radiation all over the sky with a black body temperature of about 3.5 K. Twelve years later, David Wilkinson and colleagues found an anisotropy in this radiation, using a balloon-borne experiment, due to the motion of the Milky Way relative to the microwave background.

If the microwave background radiation is a true indication of the early universe, it should have structure, as otherwise it would be difficult to see how the galaxies could have formed. George Smoot and colleagues found such an anisotropy, of just one part in 100,000, in 1992 using the *COBE* microwave spacecraft. In the following decade, the *WMAP* spacecraft was able, for the first time, to produce an image showing the variability of this background radiation throughout the whole sky. *WMAP* also showed that the universe was about 13.7 billion years old, and that it consisted of just 4 percent normal matter, 23 percent dark matter, and 73 percent dark energy.

David Leverington

See also: European Space Agency, National Aeronautics and Space Administration

Bibliography

Philip Charles and Frederick Seward, *Exploring the X-ray Universe* (1995).
David Leverington, *New Cosmic Horizons: Space Astronomy from the V2 to the Hubble Space Telescope* (2000).
Malcolm Longair, *The Cosmic Century: A History of Astrophysics and Cosmology* (2006).

High-Energy Astrophysics

High-energy astrophysics is the branch of astronomy that seeks to explain high-energy processes in the universe. It is concerned with X-ray and gamma-ray radiation and cosmic rays, which are the nuclei of atoms that have been stripped of their electrons and accelerated to ultra high speeds. X rays, gamma rays, and cosmic rays are spawned by some of the most massive and powerful objects in the universe, including active galaxies, black holes, supernovae, and pulsars.

Although German physicist Wilhelm Röntgen and French physicist Paul Villard discovered X rays and gamma rays in 1895 and 1900, respectively, the field of high-energy astrophysics is relatively new. Because Earth's atmosphere efficiently absorbs these energetic types of radiation, scientists did not observe cosmic sources of X rays or gamma rays until after World War II. Additionally, even though Austrian physicist Victor Hess discovered cosmic rays during a series of balloon flights between 1911 and 1914, interactions between these high-speed extraterrestrial particles and Earth's atmosphere likewise limited scientific studies for years.

Herbert Friedman of the U.S. Naval Research Laboratory showed, in 1949, using an experiment on a V-2 sounding rocket, that the Sun emitted X rays. Seven years later Friedman's group found that solar flares emitted very high energy X rays. But the total emitted X-ray energy flux from the Sun was very low, and many astronomers doubted that it would be possible to detect X rays emitted from remote stellar objects. Others disagreed, however, and in 1958 Riccardo Giacconi, George W. Clark, and

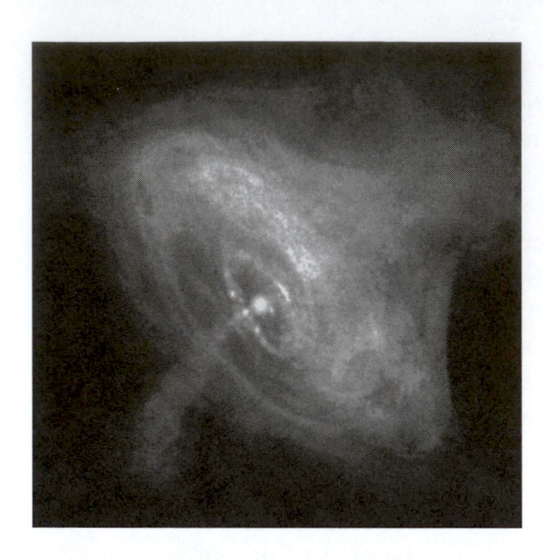

An image of the dynamic rings, wisps, and jets around the pulsar in the Crab nebula, observed in X-ray wavelengths by Chandra *in 2001. (AP/Wide World Photos)*

Bruno Rossi of American Science and Engineering (AS&E) wrote a seminal paper in which they suggested that there are a number of different types of astronomical object that may be X-ray emitters.

The first X-ray source outside the solar system, Sco X-1, was discovered by Giacconi's group using an Aerobee sounding rocket in 1962. In the following year they also discovered a diffuse X-ray background radiation. Then in 1964 Friedman's group used the occultation of the Crab nebula, a well-known supernova remnant, by the Moon to show that the Crab's X-ray source was not a point source, and so could not be a neutron star, as some astronomers had thought. Finally in 1964 the AS&E group measured the X-ray spectrum of Sco X-1 and showed that it was not a neutron star either.

The first U.S. satellite, *Explorer 1*, carried James Van Allen's cosmic ray detector in 1958. It detected an unexpected reduction in cosmic ray intensity with altitude above about 800 km. But it was difficult to interpret the data, and it was not until the launch of *Explorer 3* that Van Allen proposed that charged particles trapped in Earth's magnetic field had affected the instrument. Three years later *Explorer 11* opened the field of gamma-ray astronomy when it detected 22 discrete gamma-ray sources distributed randomly throughout the sky.

In 1967 the U.S. Department of Defense began launching its advanced Vela series of satellites. Although not intended for astrophysical studies, these Vela satellites were the first to detect short-lived, intense gamma-ray bursts. The *Compton Gamma Ray Observatory* (*CGRO*), *BeppoSAX* (*Satellite per Astronomia X*), and *Swift* satellites have more recently produced much data on many thousand of gamma-ray bursts, and seem to have shown that they originate outside the Milky Way. But the exact mechanism of their production is still unclear more than 30 years since their discovery.

These early discoveries showed that the universe was far more dynamic than previously thought. As a result, there have been many high energy space missions since these early days. For example, in December 1970 NASA launched *Uhuru*, the first satellite dedicated to X-ray astronomy. Two years later NASA launched *SAS 2* (*Small Astronomy Satellite 2*), which produced the first gamma-ray survey of the sky, showing that the galactic center was a strong source of gamma rays. In 1974 *ANS* (*Astronomical Netherlands Satellite*), a joint Netherlands/ U.S. satellite, detected X-ray emission from flare stars, and in the same year *Ariel 5*, a United Kingdom satellite, detected X rays from the center of a type of active galaxy called Seyfert galaxies. In the following year the European Space Agency (ESA) placed *COS-B* (*Cosmic Ray Satellite, Option B*) in orbit. It found a number of point sources, one of which, 3C273, was a quasar. Additionally, a NASA series of Orbiting Solar Observatories provided new data on celestial X-ray sources, even though its primary objective was to study the Sun.

The international community launched larger satellites in the late 1970s–80s to better study the X-ray and gamma-ray sky. Between 1977 and 1979 NASA placed a series of High Energy Astronomy Observatories (HEAO) into orbit. *HEAO 2*, later renamed *Einstein*, carried the first imaging X-ray telescope into space, designed by a Harvard-Smithsonian group led by Giacconi. This highly successful spacecraft imaged supernova remnants and detected more than 1,000 new sources. *HEAO 3* detected gamma-ray emission from the center of the Milky Way caused by electron-positron annihilation. The year 1979 also saw the launch of Japan's first X-ray satellite, *Hakucho* (formerly *Corsa B*). *Hakucho*'s success encouraged Japan to launch *Tenma* (formerly *Astro B*) in 1983 and *Ginga* (formerly *Astro C*) in 1987. *Tenma* provided measurements of iron in active galactic nuclei. ESA also launched *Exosat* (*European X-ray Observatory Satellite*) in1983, and three European nations—France, Bulgaria, and Denmark—collaborated with the Soviet Union to launch *Granat* (formerly *Astron 2*) in 1989.

The frontiers of high-energy astrophysics advanced steadily in the 1990s and early 21st century, as improved instrumentation and increased computing power enabled spacefaring nations to place ever more sophisticated observatories into orbit. This period saw the *ROSAT* (*Röntgen Satellite*), the Broad Band X-ray Telescope, and Diffuse X-Ray Spectrometer flown on the Space Shuttle, *CGRO*, *ASCA* (*Advanced Satellite for Cosmology and Astrophysics*), and *BeppoSAX*. Missions still returning scientific data in 2005 included the *RXTE* (*Rossi X-ray Timing Explorer*), the *Chandra X-ray Observatory*, ESA's *XMM-Newton* (*X-ray Multi-Mirror Newton*), the *HETE 2* (*High Energy Transient Explorer 2*), the *INTEGRAL* (*International Gamma-Ray Astrophysics Laboratory*), and *Swift*. All these missions have returned valuable data and led to new insights about dynamic processes in the universe.

David Leverington and David A. Medek

See also: European Space Agency, Japan, National Aeronautics and Space Administration

Bibliography

John D. Fix, *Astronomy: Journey to the Cosmic Frontier* (1995).

David Leverington, *New Cosmic Horizons: Space Astronomy from the V2 to the Hubble Space Telescope* (2000).

NASA's High-Energy Astrophysics Science Archive Research Center. http://heasarc.gsfc.nasa.gov/.

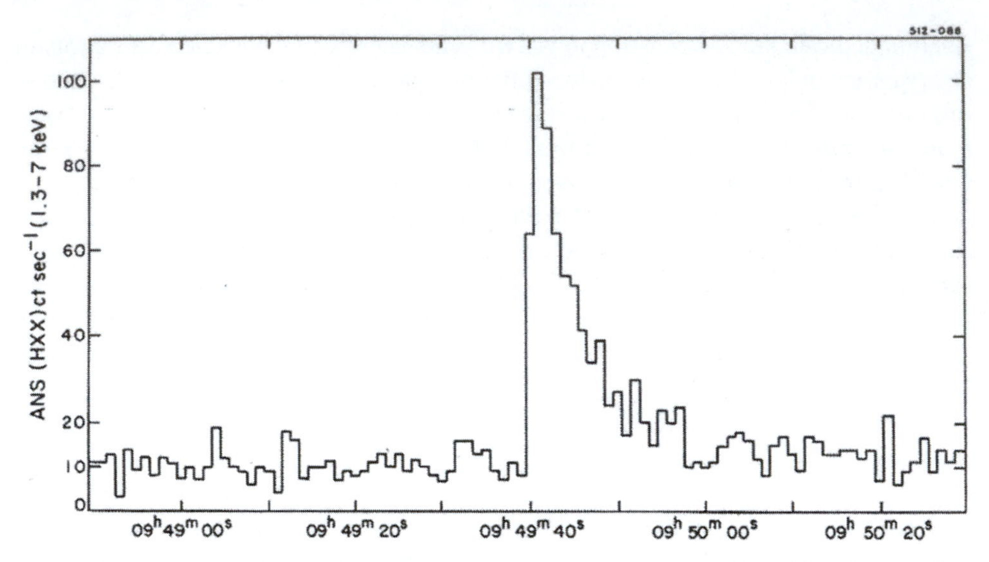

X-ray burster discovered in the source 3U 1820–1830 by ANS in September 1975. This was the first X-ray burster discovery to be announced. (Courtesy J. Grindlay et al, ApJ 205 (1976), p L128, reproduced by permission of the AAS)

ANS, Astronomical Netherlands Satellite. *ANS, Astronomical Netherlands Satellite* (*Astronomische Nederlandse Satelliet*) was a joint Netherlands/U.S. program to observe the universe in both the ultraviolet and X-ray wavebands. Built by an industrial consortium of Fokker-VFW and Philips, the *ANS* spacecraft was three-axis stabilized with a pointing accuracy of 1 arcmin. Both the ultraviolet spectrophotometer (150–330 nm) and the soft X-ray experiment (0.16–7 keV) were provided by the Dutch, while the hard X-ray experiment (1.5–30 keV) was an American contribution.

The 120 kg spacecraft was launched on 30 August 1974 by an American Scout rocket from Vandenberg Air Force Base, California, into a Sun-synchronous, polar orbit. Unfortunately, a first-stage guidance failure injected *ANS* into a highly elliptical orbit instead of the planned circular orbit. This complicated observational scheduling and caused initial problems with background radiation. Nevertheless, *ANS* made some important discoveries, including X-ray emissions from the flare stars UV Ceti and YZ Canis Minoris and X-ray bursts from the X-ray source 3U 1820-30, which appeared to be in a globular cluster. This was thought to be the first X-ray burster to be discovered, although, unknown to the astronomical community, they had previously been detected by the American military Vela spacecraft.

David Leverington

See also: Scout, Vela 5

Bibliography

David Leverington, *New Cosmic Horizons: Space Astronomy from the V2 to the Hubble Space Telescope* (2000).

Joost van Kasteren, *An Overview of Space Activities in the Netherlands* (2002).

Ariel 5. *Ariel 5* was the fifth spacecraft to be launched for the United Kingdom (UK) by NASA. It followed an offer made by the Americans in 1959 to launch scientific equipment from other countries free of charge, subject to certain conditions. *Ariel 5*, built by Marconi Space and Defence Systems, was the first UK spacecraft to be devoted to X-ray research and was the first to be controlled in orbit by UK engineers.

Ariel 5 was first proposed as an X-ray research spacecraft in 1968, some two years before the launch of the first such American spacecraft, *Uhuru*. Researchers wanted their spacecraft to be placed into an equatorial orbit, so like *Uhuru* it was launched from the Italian San Marco platform off Kenya.

Ariel 5, which was launched on 15 October 1974, contained six experiments: five British and one American. They were to measure the position, spectrum, and polarization of X-ray sources.

During more than five years, the spacecraft conducted long-term monitoring of many X-ray sources. Numerous X-ray transients were discovered, one or two of which may contain black holes. The X-ray intensity of some Seyfert galaxies was found to vary in about a day, indicating that their X-ray sources were only about one light day across. Highly ionized iron line emissions were detected in some supernova remnants and galaxy clusters, implying extremely high plasma temperatures.

David Leverington

See also: United Kingdom

Bibliography

Harrie Massey and M. O. Robins, *History of British Space Science* (1986).

ASCA, Advanced Satellite for Cosmology and Astrophysics. *ASCA, Advanced Satellite for Cosmology and Astrophysics*, was a Japanese mission that combined X-ray imaging and spectroscopy in the 0.5–10 keV range to study a variety of stellar and galactic objects. The Institute for Space and Astronautical Science (ISAS) built the 420 kg spacecraft and launched *ASCA* using an ISAS M-3S-II rocket on 20 February 1993 into a nearly circular 550 km orbit. Originally named *Astro D*, it was renamed after launch to *ASCA* for both the Japanese word for flying bird (asuka) and the NASA *ASCA* acronym. The spacecraft was operated from ISAS and the Kagoshima Space Center launch facility with download telemetry support from NASA's Deep Space Network.

The ASCA instrumentation consisted of four grazing incidence X-ray telescopes (XRT), each with a different focal plane detector. Two of them contained Gas Imaging Spectrometers (GIS) and the other two contained Solid-State Imaging Spectrometers (SIS). The latter used X-ray Charge-Coupled Device detectors, the first such sensors successfully placed in orbit. The XRTs and SIS were jointly produced by Japanese and American institutions, while the GIS was a Japanese-only instrument.

The mission began with a six-month calibration and performance verification period. After this, all observing time was open to peer-reviewed proposals, with time allocated 50 percent to Japan, 15 percent to the United States, 25 percent to

Japan–United States collaboration, and 10 percent to Japan-Europe. All data was made publicly available worldwide.

ASCA observed more than 2,000 targets, with typical observations of a half day or more per target (with some shorter and several monitoring campaigns). The *ASCA* operations team actively worked with other space-based telescopes to produce simultaneous and coordinated observations, including *ROSAT* (*Röntgen Satellite*), *RXTE* (*Rossi X-ray Timing Explorer*), *EUVE* (*Extreme Ultraviolet Explorer*), *IUE* (*International Ultraviolet Explorer*), and *Hubble Space Telescope*. *ASCA* provided the first X-ray look at an ongoing supernova event (SN1993J), observed bright early-type galaxies, and examined accretion disk emission in X-ray binaries in addition to observing active galactic nuclei and other galaxies and coronal stellar sources.

Designed for a nominal two-year prime mission life, attitude control was lost on 14 July 2000 during a geomagnetic storm, and no further experiments were performed after that date. *ASCA* reentered the atmosphere on 2 March 2001 after an eight-year life.

Alex Antunes

See also: Goddard Space Flight Center, Japan

Bibliography

Goddard Space Flight Center, High Energy Astrophysics Science Archive Research Center. http://heasarc.gsfc.nasa.gov/.

Institute of Space and Astronautical Science, Advanced Satellite for Cosmology and Astrophysics. http://www.isas.jaxa.jp/e/enterp/missions/mm/ascaproj/index.shtml.

F. Makino and T. Ohashi, eds., *New Horizon of X-Ray Astronomy—First Results from ASCA* (1994).

Astro D. *See ASCA, Advanced Satellite for Cosmology and Astrophysics.*

BeppoSAX, Satellite per Astronomia X. *BeppoSAX*, *Satellite per Astronomia X*, was an X-ray satellite project of the Italian Space Agency (ASI—Agenzia Spaziale Italiano) with the participation of the Netherlands Agency for Aerospace Programs. It was originally SAX and was renamed in honor of Italian physicist and high-energy astronomy pioneer Giuseppe (Beppo) Occhialini. The spacecraft was built by Alenia Spazio, and the mission managed by the ASI Science Data Center in Frascati, Italy.

An Atlas-G Centaur rocket placed the satellite directly into a 600 km orbit from Cape Canaveral, Florida, on 30 April 1996. *BeppoSAX* carried the first X-ray payload capable of a wide spectral coverage from 0.1–300 keV with good energy resolution and imaging capabilities. A Phoswich (phosphor sandwich) Detector System could monitor gamma-ray bursts (GRB) from 60–600 keV at 1 msec temporal resolution. The spacecraft also carried two Wide Field Cameras to study long-term variability of sources and X-ray transient phenomena at 5 arcminutes resolution in the 2–30 keV range. *BeppoSAX* supported identification of the positions of these transients within 4–8 hours, allowing other systems to observe the object in other wavebands before their intensity faded below detection thresholds. The spacecraft was employed to study a wide range of high-energy phenomena making nearly 1,500 observations of active galactic nuclei, compact galactic sources, galaxies and galaxy clusters, supernovae remnants, stellar coronas, and GRBs.

BeppoSAX made history on 28 February 1997 when it discovered the first fading X-ray afterglow of a GRB. It discovered 50 more during the lifetime of its mission, determining their positions with unprecedented precision and permitting researchers to demonstrate the extragalactic nature of these objects. After nearly six years of operation, well beyond original expectations, *BeppoSAX* ended its operational phase on 30 April 2002 and reentered Earth's atmosphere a year later.

Stephen L. Rider

See also: Italy

Bibliography

BeppoSAX Mission. http://heasarc.gsfc.nasa.gov/docs/sax/saxgof.html.
Johan A. M. Bleeker et al., eds., *The Century of Space Science* (2002).
Daphne Burleson, *Space Programs Outside the United States: All Exploration and Research Efforts, Country by Country* (2005).
Govert Schilling, *Flash! The Hunt for the Biggest Explosions in the Universe* (2002).

Chandra X-ray Observatory. The *Chandra X-ray Observatory* was the third in a series of Great Observatories launched into Earth orbit by NASA. *Chandra* complemented the *Hubble Space Telescope* (*HST*), the *Compton Gamma Ray Observatory*, and the *Spitzer Space Telescope* by observing the universe across the X-ray portion of the electromagnetic spectrum. *Chandra* was largely the brainchild of Riccardo Giacconi, who discovered the first cosmic X-ray sources in 1962, and was a successor to the *Einstein X-ray Observatory*, which was launched in 1978. *Chandra*'s partner in space was the European Space Agency's X-ray spacecraft, *XMM-Newton* (*X-ray Multi-Mirror Newton*).

The initial proposal for *Chandra* was submitted to NASA in 1976, but cost overruns associated with *HST* and arguments within NASA and the astronomical community delayed its approval as a new program for 12 years. Even then, approval by Congress was conditional on the production of a satisfactory set of X-ray mirrors. This was because Congress did not want to repeat the mistake of *HST* by giving *Chandra*, then called the Advanced X-ray Astrophysics Facility (AXAF), blanket program approval. NASA selected TRW as prime contractor for AXAF, with Perkin-Elmer and Eastman Kodak responsible for the mirrors. NASA's Marshall Space Flight Center managed the program.

The X-ray mirrors passed their test in 1991, but in the following year there was another budget crisis at NASA, mainly caused by large cost overruns on the *International Space Station*. This resulted in severe financial pressure on the AXAF budget, which eventually led to the deletion of the originally foreseen in-orbit servicing by the Space Shuttle and splitting the mission in two. One spacecraft would primarily provide imaging and the other, with lower resolution mirrors, would provide spectroscopy. In 1993 the latter was canceled. Five years later, NASA renamed the remaining spacecraft the *Chandra X-ray Observatory*, in honor of astrophysicist and Nobel Laureate Subrahmanyan Chandrasekhar.

The crew of the Space Shuttle *Columbia* deployed the *Chandra* observatory into orbit on 23 July 1999 as the largest civilian satellite ever launched by the Shuttle. The Smithsonian Astrophysical Observatory controlled *Chandra*'s science and flight operations.

Like *XMM-Newton*, *Chandra*'s orbit was highly elliptical and reached more than one-third of the way to the Moon, thereby permitting long-duration exposures of cosmic X-ray sources. *Chandra*'s mirrors were the smoothest and most precisely shaped mirrors ever made as of 2006, providing a resolution of about 0.5 arc seconds, 10 times better than *XMM-Newton*. However, while *Chandra* provided sharper images than *XMM-Newton*, the latter could detect fainter objects as it had a larger collecting area.

Chandra exceeded its specified five-year operational lifetime. It provided observations of young stars in dust and gas clouds, even finding evidence of star formation close to the black hole at the center of the Milky Way. *Chandra* appeared to have found the clearest evidence for dark matter. It also found a neutron star with a mass of at least 40 solar masses, which is a surprise, as such massive stars are normally thought to end their lives as black holes rather than neutron stars. *Chandra* detected X-ray emission from Jupiter's auroral zones and measured the size of the X-ray absorbing part of Titan's atmosphere. It found direct evidence of the dark energy accelerating the expansion of the universe, by observing galaxy clusters. However, the biggest accomplishment of *Chandra* was its resolution of the enigmatic X-ray background radiation into individual X-ray sources.

David Leverington and David A. Medek

See also: Marshall Space Flight Center, TRW Corporation

Bibliography

Chandra X-ray Observatory Center. http://chandra.harvard.edu/.

John M. Logsdon et al., eds., *Exploring the Unknown: Selected Documents in the History of the U.S. Civil Space Program, Vol. V: Exploring the Cosmos* (2001).

Wallace H. Tucker and Karen Tucker, *Revealing the Universe: The Making of the Chandra X-ray Observatory* (2001).

Compton Gamma Ray Observatory. *The Compton Gamma Ray Observatory* (CGRO) was the second of NASA's "Great Observatories" to be launched into space. It was named after Arthur Holly Compton, the Nobel Prize–winning physicist, who carried out cutting-edge research in high-energy physics in general and cosmic rays in particular.

In 1976 the Space Studies Board of the National Academy of Sciences endorsed NASA's concept for a large Earth-orbiting gamma-ray observatory. The design work on this new Gamma Ray Observatory or GRO, as it was then called, was started in 1977, and in the following year NASA selected its complement of five instruments. GRO was to be launched by a Space Shuttle into a 300 km altitude orbit, from which it would climb to its operational altitude using an orbital maneuvering engine. Like the *Hubble Space Telescope*, it was designed for on-orbit servicing, at which time the Shuttle would also replenish its propellant supply.

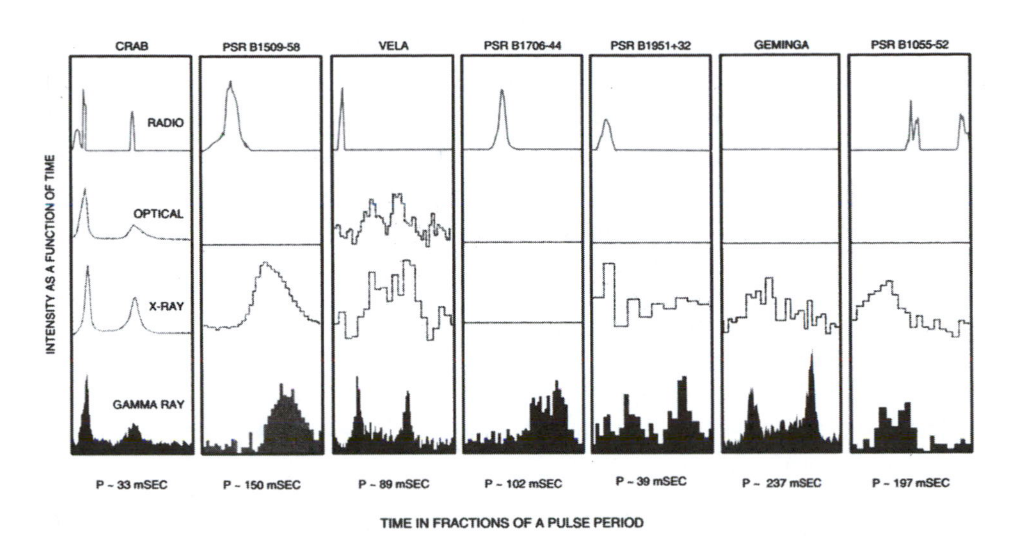

Gamma-ray pulsars as observed by the CGRO, and as detected in other wavebands. Before the CGRO the Crab and Vela pulsars were the only gamma-ray pulsars known. (Courtesy NASA)

The program soon ran into trouble because of its high cost. However, President Jimmy Carter endorsed the project, and NASA cut costs in other parts of the science budget to help. In 1980 Congress approved the program for a 1981 start. The on-orbit servicing requirement was canceled to save money, as was one of the five instruments. Launch was scheduled for 1988, but the *Challenger* disaster in January 1986 caused an inevitable program delay, as no other launch vehicle was powerful enough to launch the 16-ton spacecraft.

Managed by the Goddard Space Flight Center (GSFC), with TRW Corporation as prime contractor, *CGRO* carried four instruments: the Burst and Transient Source Experiment (BATSE), the Oriented Scintillation Spectrometer Experiment (OSSE), the imaging Compton Telescope (COMPTEL), and the Energetic Gamma Ray Telescope (EGRET).

NASA's Marshall Space Flight Center developed BATSE, which operated over the energy range from about 20 keV to 2 MeV. It was designed to quickly locate the position of short duration gamma-ray bursts and enable other observatories to locate them and hence determine their nature. When *CGRO* was launched, the nature of these gamma-ray burst sources was a complete mystery. OSSE was designed to observe discrete sources over the range 0.1–10 MeV. It was built by Ball Aerospace for the Naval Research Laboratory. COMPTEL was built by a consortium led by the Max Planck Institute for Extraterrestrial Physics in Garching, Germany and operated over the range 1–30 MeV. EGRET was developed by a group led by GSFC. The instrument's main function was to complement COMPTEL and map the sky over the energy range from 20 MeV to 30 GeV.

The *CGRO* was deployed from the Space Shuttle *Atlantis* on 5 April 1991, orbiting Earth at the relatively low altitude of 450 km to avoid the Van Allen radiation belts, which would have compromised the experiments. This low altitude would cause the orbit to decay relatively quickly, however, so the onboard propellant was to be used to boost it back to its normal orbital altitude every year or so, over its nominal lifetime of four years.

At the time of launch, there were only two gamma-ray pulsars known, *CGRO* discovered 10; one gamma-ray quasar was known, *CGRO* discovered 70; and 500 gamma-ray bursts had been detected, *CGRO* detected 2,500. Importantly, the source of these gamma-ray bursts seemed to be evenly distributed across the sky, indicating that they came from beyond the Milky Way. *CGRO* also detected gamma-ray emission from supernova remnants, black hole candidates, and even Earth above thunderstorms.

One of *CGRO*'s gyroscopes failed in December 1999, which made the spacecraft highly vulnerable to another gyroscope failure. This could have resulted in a loss of control and an uncontrolled reentry with a possible loss of life on the ground. NASA ruled out the possibility of a Space Shuttle repair or retrieval mission and decided to end the project. So on 4 June 2000, NASA undertook a controlled reentry of the spacecraft into the Pacific Ocean.

David Medek and David Leverington

See also: Goddard Space Flight Center, TRW Corporation

Bibliography

John D. Fix, *Astronomy: Journey to the Cosmic Frontier* (2005).
David M. Harland, *The Space Shuttle: Roles, Missions and Accomplishments* (1998).
John M. Logsdon et al., eds., *Exploring the Unknown: Selected Documents in the History of the U.S. Civil Space Program: Volume V: Exploring the Cosmos* (2001).
Govert Schilling, *Flash! The Hunt for the Biggest Explosions in the Universe* (2002).

COS B. *COS B* (Cosmic Ray Satellite, Option B), a European gamma-ray astronomy satellite operational from 1975–82, was a major step forward in high-energy astronomy observations. In the mid-1960s, the European Space Research Organisation (ESRO) had decided on, and had begun building, its first generation of scientific satellites and began considering its second generation. The cosmic ray and trapped radiation working group (COS) of the Launching Programme Advisory Committee investigated two projects: a joint X-ray–gamma-ray satellite, *COS A*, and a dedicated gamma-ray satellite, *COS B*, eventually deciding on the latter in 1969, as it would provide a more capable instrument to compete with proposed U.S. missions.

A consortium of scientific institutions, known as the Caravane Collaboration, in France, West Germany, the Netherlands, and Italy built the components of the gamma-ray detector, while Messerschmitt-Bölkow-Blohm was the satellite prime contractor. An American Delta 2913 launcher placed the satellite into a highly elliptical orbit in August 1975, to ensure that it spent most of its time above the Van Allen radiation belts.

COS B significantly expanded the observations of gamma-ray events (photons) from around 8,000, before its launch, to more than 200,000 in the 30 MeV to 5 GeV energy range. It observed roughly 50 percent of the sky, providing gamma-ray maps in three energy bands, along with a catalog of some 25 discrete candidate sources at energy ranges of more than 100 MeV. One of *COS B*'s goals was to identify in gamma-ray wavelengths sources observed in other wavelengths. Its ability to do this was limited, since its angular resolution was only 1–2 degrees, and it detected on average only about five photons per hour, making it difficult to precisely locate any but the most intense sources. *COS B* data allowed scientists to create the first complete gamma-ray map of the Milky Way galaxy.

Stephen B. Johnson

See also: European Space Research Organisation

Bibliography

European Space Agency. http://www.esa.int/esaCP/index.html.

A. Russo, "The *COS B* Satellite: A Case Study in ESRO's Selection Procedures," in *A History of the European Space Agency 1958–1987*, vol. 1, *The Story of ESRO and ELDO 1958–1973*, ed. J. Krige and A. Russo (2000).

Einstein Observatory. *See* High Energy Astronomical Observatory Program.

EXOSAT, European X-ray Observatory Satellite. *EXOSAT, European X-ray Observatory Satellite*, was a medium-sized, X-ray spacecraft of the European Space Agency (ESA) built by the COSMOS consortium led by Messerschmitt-Bölkow-Blohm as the prime contractor, and launched by an American Delta rocket on 26 May 1983. Its mission, when originally conceived in the late 1960s, was to measure the position of X-ray sources by lunar occultation. But by the time that the program was approved in 1973, its original mission was obsolete. As a result, ESA decided to produce a more sophisticated spacecraft with a mass three times that originally envisaged.

EXOSAT's payload consisted of a pair of low-energy imaging telescopes, a medium-energy detector array, and a gas-scintillation proportional counter. They were designed to measure the precise location and time variability of X-ray sources, the mapping of extended sources, and broadband spectroscopy. The experiments covered the range 0.04–50 keV and had spatial and time resolutions of up to 10 arc seconds and 10 μs, respectively. There were numerous problems with the low-energy system in orbit, but the remainder of the spacecraft operated satisfactorily past its two-year design lifetime.

A key element of *EXOSAT* was its high-eccentricity Earth orbit, which allowed it to undertake uninterrupted observations of up to 76 hours. This was crucial to the understanding of many sources. During its lifetime, *EXOSAT* observed active galactic nuclei, stellar coronae, white dwarfs, cataclysmic variables, X-ray binaries, galactic bulge X-ray sources, and supernova remnants. In particular, it was the first spacecraft to observe quasi-periodic oscillations in low-mass X-ray binaries including Scorpius X-1, Cygnus X-2, and X-ray pulsars. In addition, *EXOSAT* clearly showed for the first time that not all flickering, X-ray sources were black holes.

David Leverington

See also: European Space Agency

Bibliography

G. Altmann et al., *ESA Bulletin* (August 1982): 6–33.

Philip A. Charles and Frederick D. Seward, *Exploring the X-ray Universe* (1995).
David Leverington, *New Cosmic Horizons* (2000).

Fermi Gamma-ray Space Telescope. The *Fermi Gamma-ray Space Telescope* was launched by NASA on 11 June 2008 on a Delta 2 rocket into a 565 km circular orbit. Originally called the *Gamma-ray Large Area Space Telescope* (*GLAST*), the spacecraft was renamed *Fermi*, after the Nobel Prize–winning physicist Enrico Fermi, following an on-orbit checkout phase. *Fermi* was 2.9 m high, spanned 15 m with its solar arrays deployed, and weighed 4,303 kg. Managed and operated by Goddard Space Flight Center, the prime contractor for the spacecraft was General Dynamics Advanced Information Systems.

Fermi was designed with two instruments: the Large Area Telescope (LAT), its primary instrument, and the Gamma-ray Burst Monitor (GBM). The LAT was a pair-conversion telescope in which the incoming gamma rays are converted to a positron/electron pair, which leave detectable electrical traces. The GBM contained twelve lower-energy and two higher-energy scintillation detectors. Together the LAT and GBM spanned over seven orders of magnitude in energy: from 10 keV to 30 MeV for the GBM, and from 20 MeV to more than 300 GeV for the LAT. The LAT covered about 20 percent of the sky at a time, which enabled it to scan the entire sky in roughly three hours. This rapid full-sky coverage was important to detect high energy transients.

The *Fermi* mission was designed to explore the extremes of density, temperature, and magnetic fields. *Fermi* was expected to detect thousands of blazars—supermassive black holes in the cores of galaxies that emit jets of particles at near light speed. Other targets included gamma-ray pulsars, gamma-ray bursts, and supernova remnants, where shocks accelerate charged particles to create gamma-ray emissions. *Fermi* was also to conduct a search for gamma rays from annihilating dark matter particles, which might be concentrated in the Milky Way galaxy. *Fermi*'s nominal mission length was slated for five years, with a goal of ten years.

By January 2009 *Fermi*'s GBM had detected more than 100 gamma-ray bursts, with three of these bursts also detected by the LAT at energies greater than 1 GeV. *Fermi* discovered a new class of pulsars, which emit pulses only in gamma radiation. It also detected gamma rays from more than a half-dozen millisecond radio pulsars, none of which were previously known to be gamma emitters.

Lynn Cominsky

See also: Goddard Space Flight Center

Bibliography

W. Atwood et al., "Window on the Extreme Universe," *Scientific American*, November 2007.
R. Cowen, "Gammas from Heaven." *Science News*, 3 November 2007.
NASA *Fermi*. http://fermi.gsfc.nasa.gov/.

Ginga. *See* Japanese High-Energy Satellites.

Granat. *Granat* was a Soviet Union satellite with major French contributions, whose primary purpose was to study galactic and extra-galactic X-ray and gamma-ray sources from a high apogee orbit. Built by Lavochkin Scientific Production Association and managed by the Space Research Institute (IKI), the 4.4 metric ton *Granat* was launched 1 December 1989 aboard a Proton rocket into a highly elliptical initial orbit to allow long-term monitoring of sources, with an apogee of 200,000 km and perigee of 2,000 km. By 1994 lunar and solar gravitational perturbations increased the perigee to 59,000 km and decreased its apogee to 144,500 km. When the Soviet Union collapsed in the early 1990s, CNES (French Space Agency) funded *Granat* operations.

Granat was designed to observe the universe using seven instruments at energies ranging from X ray to gamma ray (2 keV–100 MeV). It included three French instruments: a coded mask telescope, spectrometer, and gamma-ray burst detector. Denmark contributed an X-ray all-sky burst detector. The remaining instruments were developed in the Soviet Union: an X-ray telescope, a coded mask imaging telescope, and an X-ray spectrometer.

During its first four years of service, the satellite focused on deep imaging and spectroscopy of the Milky Way's center and broadband observations of probable black holes and X-ray novae. In September 1994 the orbiting observatory was placed in a survey mode because its attitude control fuel was depleted. *Granat*'s survey work continued until November 1998, when it stopped transmitting data.

Major discoveries included deep imaging of the galactic center; discovery of electron-positron annihilation lines from a galactic microquasar, a binary system jetting material at nearly the speed of light and from Nova Muscae, an X-ray binary believed to consist of a low-mass object orbiting a massive object, possibly a black hole; and the study of spectra and time variability of black-hole candidates.

David Takemoto-Weerts

See also: Russia (Formerly the Soviet Union)

Bibliography

Centre d'Etude Spatiale des Rayonnements *Granat*. http://www.cesr.fr.
High Energy Astrophysics Science Archive Research Center. http://heasarc.gsfc.nasa.gov/.

Hakucho. *See* Japanese High-Energy Satellites.

HETE 2, High Energy Transient Explorer 2. *HETE 2, High Energy Transient Explorer 2*, was a small U.S. high energy spacecraft launched by a Pegasus rocket into 620 km circular equatorial orbit on 9 October 2000. The Pegasus rocket was released from an L-1011 aircraft that had taken off from the Kwajalein Missile Range in the South Pacific. The spacecraft bus was built by the Massachusetts Institute of Technology (MIT), with instruments provided by U.S., Japanese, and French institutions. *HETE 2*'s main mission was to detect and locate gamma-ray bursts (GRBs) to better than 10 arcseconds. The coordinates were then sent to ground-based optical, infrared, and radio observatories within seconds or minutes to allow them to find the GRB counterparts and their afterglows. A secondary mission objective was to undertake an X-ray sky survey, covering about 60 percent of the celestial sphere, and detect X-ray bursts and bursts from soft gamma-ray repeaters.

The launch of *HETE 1*, *HETE 2*'s predecessor, had failed in 1996. Shortly afterward, NASA agreed to the launch of a replacement spacecraft, *HETE 2*, using flight spare hardware from the first spacecraft. In the meantime, experience with *BeppoSAX* caused the HETE team to replace the ultraviolet cameras in *HETE 1*'s design with soft X-ray and optical cameras. The optical cameras were used as star trackers.

The scientific payload of *HETE 2* consisted of a French Gamma-Ray Telescope (FREGAT), a Japanese Wide Field X-ray Monitor (WXM), and a U.S. Soft X-ray Camera (SXC). FREGAT, which operated over the range 6–500 keV, detected the GRB, and one or both of the X-ray monitors/cameras were used to define its exact position in celestial coordinates using onboard processing. The SXC defined the source positions to an order of magnitude better accuracy than the WXM.

In the first two and a half years of operation *HETE 2* detected about 250 GRBs, of which it localized 43. Twenty-one of these led to the detection of X-ray, optical, or radio afterglows, of which eleven had measurable redshifts. The mission confirmed the connection between long duration GRBs and Type Ic core collapse supernovae.*HETE 2* also observed numerous X-ray bursts, and bursts from soft gamma-ray repeaters, and a limited number of X-ray flashes.

In 2005 *HETE 2*'s data allowed the *Chandra X-ray Observatory* and the *Hubble Space Telescope* to identify the X-ray afterglow and, for the first time, the optical afterglow of a short duration GRB (GRB 050709). This showed the cosmological origin of this short GRB, and indicated that it was caused by the merging of two compact binaries.

David Leverington

See also: Pegasus

Bibliography

HETE 2 MIT. http://space.mit.edu/HETE/.

D. Q. Lamb et al., "Scientific Highlights of the *HETE-2* Mission," *New Astronomy Reviews* (2004).

High Energy Astronomy Observatory Program. The High Energy Astronomy Observatory program (HEAO), managed by NASA's Marshall Space Flight Center, consisted originally of four 11-ton spacecraft, but in 1972 the fourth spacecraft was canceled to save money. Shortly afterward, NASA suggested that the mass of each spacecraft should be reduced to 3 tons to enable them to be launched by an Atlas-Centaur. The first two spacecraft were dedicated to X-ray astronomy. *HEAO 1* was a survey spacecraft, while *HEAO 2* was the first X-ray imaging astronomical spacecraft. *HEAO 3* was dedicated to gamma- and cosmic-ray astronomy. All three spacecraft were built by TRW and launched into almost circular, low Earth orbits.

HEAO 1 was launched in August 1977. The spacecraft rotated approximately once every 30 minutes about the Sun–Earth line, allowing the experiments to slowly scan the sky and cover the whole celestial sphere in six months. It surveyed the whole sky almost three times over the 0.2 keV to 10 MeV energy band and produced a number of pointed observations of selected sources. *HEAO 1* measured the X-ray background radiation from 3–50 keV, in addition to the variability of a variety of sources, including active galactic nuclei (AGN), X-ray binaries, and cataclysmic variables. It discovered the Cygnus Superbubble and a number of strong, soft X-ray sources.

HEAO 2, otherwise called the *Einstein Observatory*, was launched in November 1978. Its X-ray telescope was able to image X rays in the 0.15–3 keV energy band with an angular resolution of a few arc seconds. Three spectrometers were also included in the payload. *Einstein* imaged supernova remnants for the first time, many radio galaxies, and detected X-ray jets from the radio galaxies Centaurus A and M87. It found that ordinary red dwarfs, O- and B-type stars, and flare stars in their quiescent state were strong X-ray emitters. Additionally, it measured many X-ray transients and undertook medium and deep X-ray surveys of portions of the celestial sphere, detecting about 1,000 new sources, including many quasars, AGNs, clusters of galaxies, and BL Lacertae objects (a special class of active galaxies).

HEAO 3, which was launched in September 1979, was a survey spacecraft like *HEAO 1* but operating in the hard X-ray and gamma-ray band from 50 keV to 10 MeV. Its results were much more modest than those of *HEAO 1* and *2* because it was more difficult to detect and locate gamma-ray sources. *HEAO 3* undertook a sky survey of narrow line gamma-ray emission. It detected 511 keV gamma rays (produced by electron-positron annihilation) from the central region of the Milky Way. Further, it observed the black hole candidate Cygnus X-1 for 170 days and detected gamma rays and flickering, hard X-ray emission. The gamma-ray radiation gradually decreased in intensity as the hard X-rays increased.

David Leverington

See also: Goddard Space Flight Center

Bibliography

Herbert Friedman and Kent S. Wood, "X-ray Astronomy with HEAO 1," *Sky and Telescope* (December 1978): 490–94.

Dennis Overbye, "The X-ray Eyes of Einstein," *Sky and Telescope* (June 1979): 527–34.

Wallace Tucker and Karen Tucker, *The Cosmic Inquirers: Modern Telescopes and Their Makers* (1986).

Hinode. *See* Japanese High-Energy Satellites.

Hinotori. *See* Japanese High-Energy Satellites.

INTEGRAL, International Gamma-Ray Astrophysics Laboratory. *INTEGRAL, International Gamma-Ray Astrophysics Laboratory*, was launched by the Russian Space Agency on 17 October 2002 on a four-stage Proton launcher from Baikonur Cosmodrome. Russia waived the cost of the launcher in exchange for observing time. *INTEGRAL* was selected in 1993 as the second medium mission within the European Space Agency's (ESA) Horizon 2000, a long-term plan for space science, which mapped out roughly a dozen medium- and large-sized missions.

The 4-ton spacecraft, built by an industrial consortium led by Alenia Aerospazio of Italy, was the first spacecraft capable of observing simultaneously in gamma-ray, X-ray, and visible wavelengths. It was designed to conduct high-resolution spectroscopy and fine source imaging of high-energy sources with energies up to 10 MeV, including gamma-ray bursts, neutron stars, active galactic nuclei, and regions surrounding black hole candidates. *INTEGRAL*'s highly elliptical, three-day orbit kept it mostly outside Earth's radiation belts, which would interfere with gamma-ray detection. More than 90 percent of the orbit could be used for scientific observations.

The payload module weighed 2 tons, in part due to the shielding necessary to protect the sensors from background radiation. The four experiment packages comprised a gamma-ray spectrometer, a gamma-ray imager, an X-ray monitor, and an optical camera. The latter two helped identify gamma-ray sources. *INTEGRAL* used the same service module design as *XMM-Newton* (*X-ray Multi Mirror*), a savings made possible by the missions' similar requirements.

All ESA member states, in addition to the United States, Russia, Poland, and the Czech Republic participated in the mission. Its expected lifetime was just over two years, but was extended and was still operating as of early 2006.

INTEGRAL was part of the Gamma-Ray Bursts Coordination Network, detecting about 10 gamma-ray bursts per year and relaying information to the network for follow-up observations. Major scientific results include resolving the diffuse gamma-ray glow in the center of the Milky Way galaxy into a small number of discrete sources, estimating a rate of one supernova per 50 years in the Milky Way by mapping gamma-ray emission from radioactive aluminum, and revealing that the black hole at the center of Earth's galaxy was once more active.

Katie J. Berryhill

See also: European Space Agency

Bibliography

ESA *INTEGRAL*. http://www.esa.int/esaMI/Integral/.
C. Winkler et al., "The *INTEGRAL* Mission," *Astronomy and Astrophysics* (2003).

Japanese High-energy Satellites. Japanese high-energy satellites included nine spacecraft successfully launched by the Institute for Space and Astronautical Science (ISAS) as of mid-2007. High-energy observations included spectroscopic and imaging data, often with precise timing of the X-ray photon events. ISAS assigned prelaunch names to the spacecraft and missions (for example, *Astro E2*) and then renamed successful launches with more evocative Japanese names (for example, *Suzaku*). The four solar missions were *Taiyo*, *Hinotori*, *Yohkoh*, and *Hinode*. These looked at soft– and hard–X-ray emissions from the Sun. The five astrophysical missions were *Hakucho*; *Tenma*; *Ginga* (*Astro C*); *Asuka* (*ASCA*, *Advanced Satellite for Cosmology and Astrophysics*; *Astro D*); and *Suzaku* (*Astro E2*). They observed soft and hard X rays of specific sources in addition to the cosmic X-ray background.

ISAS high-energy satellites used new and established technology, typically with an upgrade on one or more proven instruments plus one new concept. As primary instruments, *Tenma*, *Ginga*, and *ASCA* used gas scintillator proportional counters (*ASCA* with imaging capability). *ASCA* and *Suzaku* both had X-ray Charge-Coupled Device detectors.

ISAS launched its satellites from Kagoshima Space Center (KSC) on Japanese launch vehicles, such as the M-5. The *Astro E* and *Corsa A* spacecraft failed to

Japanese High-Energy Satellites (Part 1 of 2)

Name	Mission Dates	Type	Instruments	Comments
Taiyo (*SRATS*)	1975–80	solar	Two ultraviolet (UV) radiometers, four UV photon counters, X-ray detector	Soft X-ray and UV observations of Sun
(*Corsa A*)	1976	astrophysics	See *Hakucho*	Launch failure
Hakucho (*Corsa B*)	1979–85	astrophysics	Eleven X-ray proportional counters	Investigated cosmic X-ray bursts and X-ray stars
Hintori (*Astro A*)	1981–91	solar	Imaging collimator telescope with scintillation proportional counters	Hard X-ray images of solar flares
Tenma (*Astro B*)	1983–89	astrophysics	Gas Scintillation Proportional Counter, X-Ray Focusing Collectoring, Hadamard X-ray Telescope, Transient Source Monitor, Radiation Belt Monitor/GRB Detector	X-ray spectra plus timing of sources and bursts
Ginga *Astro C*)	1987–91	astrophysics	Large area proportional counter, all sky monitor, gamma-ray burst detector	Discovered many new X-ray sources
Yohkoh (*Solar A*)	1991–2005	solar	Hard X-ray telescope (HXT) and soft X-ray telescope (SXT) with Charge-Coupled Device (CCD), wide-band spectrometer, Bragg crystal spectrometers	Defined Sun as an active star, established magnetic reconnection as a likely driver of solar flares and coronal mass ejection
Asuka/*ASCA* (*Astro D*)	1993–2001	astrophysics	X-ray telescope with two gas imaging spectrometer proportional counters and two solid-state imaging spectrometer CCD cameras	Black holes, supernova remnants, cosmic X-ray background
(*Astro E*)	2000	astrophysics	See *Suzaku*	Launch failure
Suzaku (*Astro E2*)	2005–	astrophysics	Five SXT, one HXT, four X-ray imaging CCD spectrometers, one X-ray spectrometer microcalorimeter, hard X-ray detector scintillator	First X-ray calorimeter flown in space, but it failed after one month and was not used

(Continued)

Japanese High-Energy Satellites (Part 2 of 2)

Name	Mission Dates	Type	Instruments	Comments
Hinode (*Solar B*)	2006–	solar	Solar optical telescope with magnetograph and polarimeter at five minute resolution, X-ray telescope, extreme UV imaging spectrometer	High spatial and time resolution investigations of flares, especially at very high or low latitudes

achieve orbit. Using a combination of new parts and existing flight spares, ISAS built replacements and launched them as *Suzaku* and *Hakucho*. Due to Kagoshima's location, most launches were to a 31 degree inclination at approximately 600 km height, with nearly circular orbits. The satellites were operated from KSC by ISAS. Researchers and graduate students at ISAS took turns serving as touban, or duty scientists, to operate the satellite and provide quick-look data. Data was downloaded to KSC with additional support from NASA's Deep Space Network.

ISAS engaged in much international collaboration. *Tenma* involved the European Space Agency (ESA), which with the United Kingdom (UK), also participated in *Hinode*. *ASCA* and *Suzaku* were joint Japan–United States missions, while the United States and the UK participated in *Ginga* and *Yohkoh*. Data older than one year was made publicly available worldwide through Goddard Space Flight Center and other gateways. Multi-satellite coordinated observations were common. *Ginga* observations were coordinated with those of the *IUE* (*International Ultraviolet Explorer*), and *Hubble Space Telescope*; *ASCA* observations with the *ROSAT* (*Röntgen Satellite*) and the *RXTE* (*Rossi X-ray Timing Explorer*); *Suzaku* observations with *Chandra X-ray Observatory*.

The most notable findings of the Japanese high-energy satellite missions were in the areas of compact objects (such as black holes, pulsars, and active galactic nuclei) and in solar physics. *Ginga* discovered a number of transient black hole candidates, and *ASCA* probed the detailed structure and metal abundances of compact objects. *Suzaku* continued this, particularly with the black holes in the center of active galaxies. Before *Yohkoh*, the solar surface and corona—apart from discrete noticeable active features such as flares and coronal mass ejections—was assumed to vary mildly and on timescales of days; *Yohkoh* observed activity corona-wide varying over minutes and indicating the future missions would require a higher cadence, or rate of data frames. *Hinode* returned extremely high resolution images of solar plasma caught in coronal loops, with a time resolution as high as two seconds.

Alex Antunes

See also: European Space Agency, Goddard Space Flight Center, Japan, National Aeronautics and Space Administration

Bibliography

ISAS English-language. http://www.isas.jaxa.jp/e/index.shtml.

ROSAT, Röntgen Satellite. *ROSAT, Röntgen Satellite,* was an X-ray and extreme-ultraviolet (EUV) observatory named after Wilhelm Conrad Röntgen, the German physicist who discovered X rays. During its eight-year mission, *ROSAT* conducted an extensive all-sky survey and provided detailed information about selected X-ray and EUV sources by permitting long duration observations (up to three hours).

The Max-Planck-Institute for Extraterrestrial Physics proposed *ROSAT,* and the observatory was built by Dornier and operated by the German Space Operation Center. It primarily consisted of two telescopes, the X-ray telescope built by Germany and the wide field camera built by the United Kingdom, which extended into the EUV. NASA launched *ROSAT* on a Delta II rocket on 1 June 1990 from Cape Canaveral and provided the high-resolution imager, part of the X-ray telescope. Originally NASA planned to provide a free launch on a Space Shuttle, but this was changed following the *Challenger* accident. The high-resolution imager was an upgraded version of the imager from the earlier U.S. *Einstein* satellite.

Although *ROSAT* was not the first all-sky survey performed in X rays, it greatly increased the number of known X-ray objects from 6,000 to more than 150,000. It also significantly increased the number of extreme ultraviolet sources, detecting more than 700. The findings from *ROSAT* were quite varied. They ranged from finding variable X-ray emissions from galaxies with suspected black holes at their centers to discovering that comets emit X rays. Using *ROSAT* data, astrophysicists discovered "middleweight" black holes, a new class of black holes between 100 and 10,000 times the Sun's mass not predicted from existing theories. *ROSAT* also observed Jupiter when Comet Shoemaker-Levy 9 collided with it.

Z Cameo Johnson-Cramer

See also: Delta, Germany

Bibliography

B. Aschenbach et al., *The Invisible Sky: ROSAT and the Age of X-Ray Astronomy* (1998).

High Energy Astrophysics Science Archive Research Center. http://heasarc.gsfc.nasa.gov/.

Max Planck Institute for Extraterrestrial Physics: X-ray Astronomy. http://www.mpe.mpg.de/main.html.

Roentgen Satellite ROSAT. http://heasarc.gsfc.nasa.gov/docs/rosat/rosgof.html.

ROSAT Guest Observer Facility. http://heasarc.gsfc.nasa.gov/docs/rosat/rosgof_ov.html.

"X-ray Astronomy," *MPE Annual Report 1998* (1999).

RXTE, Rossi X-ray Timing Explorer. *RXTE, Rossi X-ray Timing Explorer*, was a satellite launched into low Earth orbit on 30 December 1995 on a Delta 2 rocket. The spacecraft was named in honor of Dr. Bruno Rossi, whose team was credited with detecting the first non-solar source of X rays from a sounding rocket flight in 1962. *RXTE* was designed, built, and managed by NASA's Goddard Space Flight Center.

The mission's primary purpose was the study of temporal ("timing") and broadband spectral phenomena associated with stellar and galactic systems containing compact and extremely energetic objects. Such objects include rapidly spinning pulsars, black holes, active galactic nuclei (AGNs), and gamma-ray (and X-ray) burst events. The timing of these events reveals the physics underlying the environment around the objects. The scientific instruments on *RXTE* span an energy range of 2–220 keV. The highly resolved timescales of the objects studied range from microseconds to several months.

RXTE carried three science instruments, two of which were pointed. The Proportional Counter Array (PCA) was comprised of five large proportional counters. The PCA was sensitive from 2–60 keV and had a field of view of 1°. Sources as faint as 1/1000th of the Crab Nebula could be detected in a few seconds. The High Energy X-ray Timing Experiment (HEXTE) consisted of two rocking clusters of detectors that spanned an energy range of 15–250 keV. HEXTE's field of view was coaligned with the PCA. The only non-pointed instrument on *RXTE* was the All-Sky Monitor (ASM). The ASM was sensitive to energies from 2–10 keV. This instrument alerted astronomers to high-energy events (such as flares) and also changes of state in X-ray sources. Other space-based observatories could, in turn, be alerted to observe these high-energy phenomena. The instrument had three rotating shadow cameras that scan about 80 percent of the sky every 90 minutes (about one orbit of Earth).

The *RXTE* made numerous discoveries. It observed a three-hour long superburst from the compact low mass X-ray binary 4U 1820-30, timed the variability of X rays in AGNs, and detected X-ray flashes from millisecond pulsars. As of 2009 the RXTE mission continued to operate.

Rob Landis

See also: Goddard Space Flight Center

Bibliography

Goddard Space Flight Center *RXTE*. http://heasarc.gsfc.nasa.gov/docs/xte/xte_1st.html.

***SAS 2* and *3*, *Small Astronomy Satellites 2* and *3*.** *SAS 2* and *3*, *Small Astronomy Satellites 2* and *3* were NASA's successors to the X-ray observing spacecraft *SAS 1*, also known as *Uhuru*. Both missions, orbited by Scout launchers, were managed by Goddard Space Flight Center, which also built the *SAS 2* spacecraft and gamma-ray telescope. The Johns Hopkins University Applied Physics Laboratory built the *SAS 3* spacecraft, with its payload built by the Massachusetts Institute of Technology.

SAS 2, which studied gamma rays in the energy range above 35 MeV, was launched in November 1972 from the Italian San Marco launch facility off the Kenyan coast, as were the other SAS satellites. *SAS 2* made 27 directed observations covering approximately 55 percent of the sky, including most of the galactic plane. Observations were curtailed when its low-voltage power supply failed in June 1973. The satellite provided the strongest evidence that gamma radiation in the Milky Way was directly associated with certain features in the galactic plane. It also demonstrated that certain celestial objects such as the Crab and Vela pulsars were the sources of powerful gamma-ray emissions.

SAS 3 launched in May 1975 and operated for four years. It was devoted to the search for and study of X-ray sources. Specifically, the spacecraft was designed to locate bright X-ray sources to a high degree of accuracy; study specific emission sources within a defined energy range; and maintain an ongoing search for novae, flares, and other short-term events producing X-ray emissions. *SAS 3* had a slow, gyroscopically controlled spin rate of one rotation every 95 minutes that could be stopped to allow the satellite to observe a fixed point for approximately 30 minutes. This feature allowed detailed data gathering from sources such as pulsars, bursters, or transients. The four instruments on board could be pointed at selected targets.

The project's major scientific findings were the discovery of the Rapid Burster, an unusual source among 12 X-ray burst emitters; the first detection of X rays from a highly magnetic white dwarf binary system; the accurate location of dozens of X-ray sources; the first location of a quasar by means of its X radiation; and the identification of X-ray sources within the centers of globular clusters.

David Takemoto-Weerts

See also: Goddard Space Flight Center, Johns Hopkins University Applied Physics Laboratory

Bibliography

Goddard Space Flight Center High Energy Astrophysics Science Archive Research Center SAS 2 and SAS 3. http://heasarc.gsfc.nasa.gov/docs/sas2/sas2.html and http://heasarc.gsfc.nasa.gov/docs/sas3/sas3.html.

Small Astronomy Satellite 1. *See Uhuru.*

Suzaku. *See* Japanese High-Energy Satellites.

Swift. The *Swift* gamma-ray burst satellite launched into low Earth orbit on a Delta 7320 on 20 November 2004. Spectrum Astro built the spacecraft, and Goddard Space Flight Center managed the project, which included an international team of scientists from the United States, United Kingdom, and Italy. There were three telescopes on board *Swift*: the Burst Alert Telescope (BAT), the X-ray Telescope (XRT), and the Ultraviolet-Optical Telescope (UVOT).

Swift's main scientific objective was to study the gigantic cosmic explosions known as gamma-ray bursts. First discovered in 1967 data from the Vela satellites, the origins of these bursts remained mysterious until the first clues emerged in 1997 with the discovery of X-ray and optical afterglows. By studying these afterglows, scientists deduced that at least the longer gamma-ray bursts seemed to be associated with the supernovae of massive stars in distant galaxies. Evidence indicated that the shorter gamma-ray bursts are caused by the merger of two neutron stars or a neutron star and a black hole.

Swift was designed to first detect gamma-ray bursts using the BAT, then "swiftly" turn (in about one minute) to focus the narrower fields of view of the XRT and UVOT onto the target area. Scientists thus obtained high-quality simultaneous data across a wide energy range. They correlated these data with observations from ground-based telescopes in a follow-up network that spanned radio, infrared, and visible light energies.

Swift detected its first gamma-ray burst on 17 December 2004 and subsequently detected more than 200 more in the first two years of the mission. In addition to studying gamma-ray bursts, *Swift* recorded data from a record-breaking gamma-ray flare from a magnetar. This giant flare produced more light at Earth than any celestial object except for the Sun and was attributed to a starquake on the surface of a highly magnetized neutron star in Earth's galaxy at a distance of more than 20,000 light-years.

Lynn R. Cominsky

See also: Goddard Space Flight Center

Bibliography

Neil Gehrels et al., "The Brightest Explosions in the Universe," *Scientific American* (December 2002):84–92.

GSFC *Swift*. http://swift.gsfc.nasa.gov/docs/swift/swiftsc.html.

Robert Naeye, "Dissecting the Bursts of Doom," *Sky and Telescope* (August 2006):30–37.

Taiyo. *See* Japanese High-Energy Satellites.

Tenma. *See* Japanese High-Energy Satellites.

Uhuru. *Uhuru*, also called *SAS 1* (*Small Astronomy Satellite 1*), was the first satellite dedicated to X-ray astronomy. Its aim was to produce a catalog of sources, including their positions, intensities, and spectra. Riccardo Giacconi of American Science and Engineering Corporation proposed such a spacecraft in 1963, just after the discovery of the first non-solar X-ray sources. The spacecraft was built by the Applied Physics Laboratory of Johns Hopkins University.

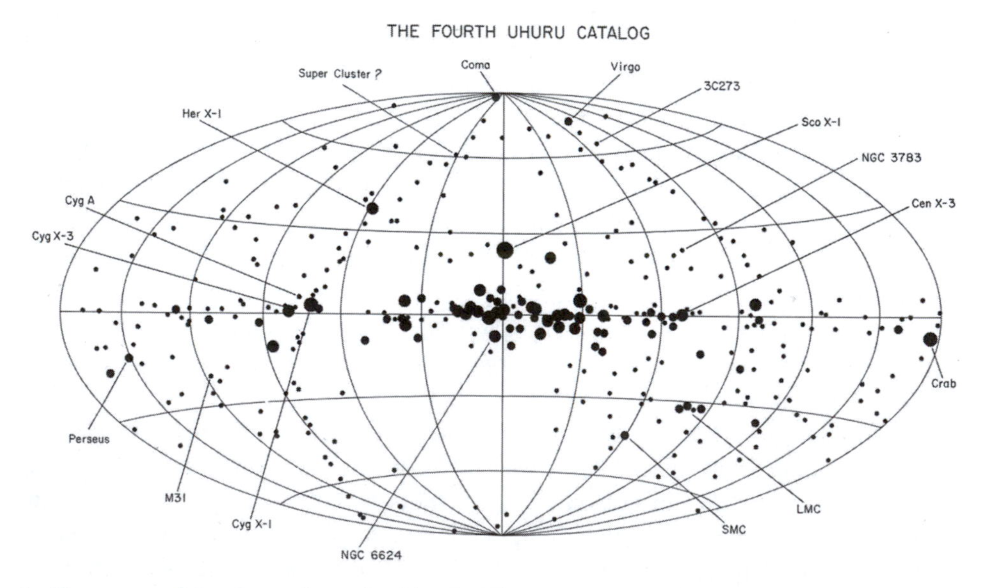

An X-ray map of the sky as determined by the Uhuru *spacecraft. The plane of the Milky Way is along the equator, and the center of the Milky Way in the center of the plot. (Courtesy Harvard-Smithsonian Center for Astrophysics)*

RICCARDO GIACCONI (1931–)

(Courtesy Getty Images)

Riccardo Giacconi was an American Science and Engineering Company physicist who in 1962 led a team that discovered Sco X-1, the first X-ray source outside the solar system, using an Aerobee sounding rocket. He pioneered the use of grazing incidence X-ray telescopes, later used on most large X-ray spacecraft. Giacconi was the driving force behind *Uhuru*, NASA's first dedicated X-ray spacecraft, and its highly successful follow-on, the *Einstein Observatory*, which was the first spacecraft to image X-ray sources. In 1981 he became the first director of the Space Telescope Science Institute and in 1993 the director of the European Southern Observatory. In 2002 he was one of the recipients of the Nobel Prize in Physics for his work in X-ray astronomy.

David Leverington

The 140 kg *Uhuru* (Swahili for freedom) was launched on 12 December 1970 (Kenyan Independence Day) by a U.S. Scout rocket from the Italian San Marco platform off the equatorial coast of Kenya. This location enabled NASA to use the smallest possible launcher to put the spacecraft into a near-equatorial, circular orbit.

The 65 kg payload of *Uhuru* consisted of two X-ray telescopes covering the range 2–20 keV that scanned the sky as the spacecraft spun about its axis. Strong X-ray sources were located to within about 1 arc minute and weak ones to about 15 arc minutes. Data was transmitted in both real time and, after recording by a tape recorder, during a data dump once per orbit to a single ground station. However, the tape recorder failed six weeks after launch, so NASA had to use additional ground stations to receive data in real time.

During its lifetime of just more than two years, *Uhuru* detected 339 discrete sources, 100 of which had accurate-enough locations to enable their counterparts to be identified at other wavelengths. *Uhuru* found that X-ray sources tended to be highly variable in their X-ray intensity.

Uhuru helped to show that Centaurus X-3 and Hercules X-1 were both pulsars in a binary system, and Cygnus X-1 was a black hole, also in a binary system. In addition, *Uhuru* measured X-ray emission from supernova remnants, globular clusters, spiral and Seyfert galaxies, and clusters of galaxies.

David Leverington

See also: Johns Hopkins University Applied Physics Laboratory

Bibliography
J. Leonard Culhane and Peter W. Sanford, *X-ray Astronomy* (1981).
Richard F. Hirsh, *Glimpsing an Invisible Universe: The Emergence of X-ray Astronomy* (1983).
"X-ray Results from 'Uhuru,' " *Sky and Telescope* (June 1971): 341.

Vela 5. *Vela 5* was the fifth pair of satellites in the Vela series. These classified satellites were developed to monitor Nuclear Test Ban Treaty compliance; however, the Vela satellites also provided useful scientific data. The Vela program was run jointly by the Advanced Research Projects Agency of the U.S. Department of Defense and the U.S. Atomic Energy Commission, and managed by the U.S. Air Force. All six pairs of Vela satellites were built by TRW, with each pair placed in roughly 118,000 km orbits by a single launcher, each pair separated by 180 degrees.

They were spin-stabilized with the spin axis pointed at Earth, which allowed coverage of the entire celestial sphere twice per orbit of roughly 112 hours. *Vela 5B* contributed the most to science, with its 10 year lifetime between 23 May 1969 and 19 June 1979. *Vela 5A* started having significant equipment failure on 24 July 1970. Vela 5 satellites were equipped with two X-ray detectors (0.2–12 keV) and six gamma ray detectors (150–750 keV). Since nuclear detonations and high-energy cosmic phenomena produce energies in these same wavebands, these military spacecraft made many useful scientific observations, though these were not released for a number of years.

Vela 5's cosmic X-ray detectors performed an all-sky survey, while the other X-ray detectors searched for solar X rays. Though Vela 5 was the first to observe (with the *Astronomical Netherlands Satellite*) X-ray bursts, these results were kept secret until after observations of X-ray bursts from the *Astronomical Netherlands Satellite* were published in 1975. Vela 5 satellites were the first to detect gamma-ray bursts with observations in 1969, but these were not declassified and announced until 1973. Due to *Vela 5B*'s long lifetime, it was uniquely able to observe the long-term variability of X-ray sources.

Z Cameo Johnson-Cramer

See also: Nuclear Detection, TRW Corporation

Bibliography
High Energy Astrophysics Science Archive Research Center. http://heasarc.gsfc.nasa.gov/.
National Space Science Data Center Master Catalog, Spacecraft, *Vela 5B*. http://nssdc.gsfc.nasa.gov/nmc/spacecraftDisplay.do?id=1969-046E.
Vela 5B Satellite. http://heasarc.gsfc.nasa.gov/docs/vela5b/vela5b.html.
Harry Waldron, "SMC Remembers the Vela Program," *SMC Astro News*, 7 February 1997.

XMM-Newton, X-ray Multi-Mirror Newton. *XMM-Newton*, X-ray Multi-Mirror Newton, was launched by the European Space Agency (ESA) on 10 December 1999 on an Ariane 5G rocket into an unusual 48-hour elliptical orbit, which took the spacecraft nearly one-third of the distance to the Moon, allowing for long observations that were not interrupted by occultations by Earth. *XMM-Newton* was the second cornerstone of ESA's Horizon 2000 Science Program and the biggest scientific satellite ever

Artist's impression of the XMM-Newton spacecraft in orbit. (Courtesy European Space Agency/ D. Ducros)

built in Europe. The total length of *XMM-Newton* was 10 m, and with its solar arrays deployed, the satellite spanned 16 m. The prime contractor for the spacecraft was Dornier Satellitensysteme, which led an industrial consortium involving 46 companies from 14 European countries and one company in the United States.

The instruments on board *XMM-Newton* included three EPIC (European Photon Imaging Camera) X-ray Charge-Coupled Device cameras, two Reflection Grating Spectrometers, and the Optical Monitor (OM). All the X-ray cameras and spectrometers gathered light from three X-ray telescopes that each employed 58 nested, concentric, gold-plated nickel mirrors to focus the high-energy light. The OM, a 30 cm aperture Ritchey-Chretien, was coaligned with the main X-ray telescope to provide a multi-wavelength capacity, by viewing light in the 170–600 nm spectral range.

XMM-Newton was initially called XMM, X-ray Multi-Mirror, because of the design of the mirrors. To honor one of history's scientists, ESA attached the name of Isaac Newton to the XMM mission. In addition to imaging sources of X-ray light, *XMM-Newton* also obtained extremely high-quality X-ray spectra. These spectra could be useful in helping scientists understand how an object like a black hole, neutron star, or galaxy produced light, how fast it moved, and even of what elements it was made.*XMM-Newton*'s observing capabilities were complementary to those of NASA's *Chandra X-ray Observatory*. While *Chandra*'s mirrors were capable of producing higher-resolution images, *XMM-Newton* had a larger collecting area, which produced higher-quality spectra.

In its first six years after launch, more than 1,300 refereed publications resulted from *XMM-Newton* data, involving more than 4,000 observations of objects as varied as galaxies, stars, black holes, comets, and planets. Some of *XMM-Newton*'s discoveries included an X-ray map of the most distant, massive structure in the universe (a cluster of galaxies formed when the universe was one-third its present age), X rays from iron in matter swirling around super-massive black holes in distant galaxies that have the properties predicted by Albert Einstein's theory of general relativity, spectral evidence in support of a new class of intermediate-mass black holes, individual hot spots on rotating neutron stars, reflected solar X rays on Jupiter that can be used to monitor flare activity on the far side of the Sun, X rays emitted from a collapsing cloud that signaled the birth of a star, evidence that the well-studied supernova remnant RCW86 originated in 185 CE (making it much younger than previously thought), the discovery of the most distant (10 billion light-years) cluster of galaxies, the discovery of a comet-like ball of gas more than one thousand million times the mass of the Sun hurling through a distant galaxy cluster at more than 750 km per second, the discovery of the first black hole found inside a globular cluster, evidence for a new class of "prompt" supernovae, and contributions to the first three-dimensional map of dark matter as inferred by weak lensing studies. *XMM-Newton* was still operating as of August 2008.

Lynn Cominsky

See also: European Space Agency, Germany

Bibliography

Ron Cowen, "X-rays Unveil Secret Lives of Black Holes," *Science News Online*, 6 January 2001.

Giuseppina Fabbiano, "The Hunt for Intermediate-Mass Black Holes," *Science* 307 (28 January 2005): 533–34.

Ilana Harrus, "Fine-Tuning Astronomy," *Scientific American*, 17 June 2002.

XMM-Newton Education and Public Outreach. http://teachspacescience.org/cgi-bin/search.plex?catid=10000566&mode=full.

Low-Energy Astrophysics

Low-energy astrophysics concerns itself with energies below that of X rays. From lowest to highest energy, this includes the radio, microwave, infrared, visible, and ultraviolet portions of the electromagnetic spectrum, each portion of which reveals different information. Since the atmosphere degrades or completely blocks ground observations of most wavelengths with "windows" allowing radio and visible light to reach Earth's surface, the limitations of observing from within Earth's atmosphere must be overcome using spacecraft dedicated to observing light from blocked portions of the spectrum.

In addition to observation, space has proved a useful platform for studying fundamental physics. In 1976, NASA's *Gravity Probe A* verified General Relativity's prediction of how gravity affects the passage of time. In 2004 *Gravity Probe B* carried an experiment to test the curvature of space-time near Earth.

Most ultraviolet light is absorbed by atmospheric ozone, which led to early experiments with sounding rockets to detect the Sun's "Lyman alpha" line, detected in 1952, with images of the Sun in the Lyman alpha light in 1956. The first large three-axis stabilized satellites with ultraviolet detection capability were NASA's series of four Orbiting Astronomical Observatory (OAO) satellites in the late 1960s and early 1970s, two of which were successful. Operating from 1968 to 1973, *OAO 2* gave astronomers their first ultraviolet observations of many objects, including the discovery of hydrogen clouds around comets. *OAO 3*, also known as *Copernicus*, took high-resolution spectra of hundreds of stars from 1972 to 1981.

A joint project of NASA, the European Space Agency (ESA), and the United Kingdom, the *IUE* (*International Ultraviolet Explorer*) was launched in 1978 and gathered ultraviolet observations until 1996. NASA launched the *EUVE* (*Extreme Ultraviolet Explorer*) in 1992 to observe in shorter wavelength ultraviolet than *IUE*. It performed an all-sky survey and followed up with spectra of bright sources. NASA's *Far Ultraviolet Spectroscopic Explorer*, launched in 1999, observed at ultraviolet wavelengths between that of *IUE* and *EUVE*. In 2003, NASA launched two ultraviolet satellites with narrower scientific goals. *Cosmic Hot Interstellar Plasma Spectrometer* studied the million-degree plasma of the local bubble, while *Galaxy Evolution Explorer* focused on galaxies.

While visible light passes through the atmosphere, the resolution of ground-based observations up to the late twentieth century was limited by atmospheric turbulence to about .5 arcsec. This has led to most large observatories being built at high altitudes and finally to space-based observatories. The *Hubble Space Telescope* (*HST*), launched in April 1990, was designed to provide .1 arcsec resolution at visible wavelengths, and to observe the sky in infrared, visible, and ultraviolet with a variety of imagers and spectroscopes. These were made possible in part by the development of charge-coupled devices, which were initially developed for reconnaissance satellites in the 1970s. Among its accomplishments was to measure the distances to galaxies so that the Hubble Constant, linked to the rate of expansion of the universe, could be determined. Built to be serviced by astronauts from the Space Shuttle, NASA updated and replaced *HST*'s instruments over time, improving its capabilities, though at significant cost for each mission. Other visible light space observatories had less all-encompassing missions. ESA launched *Hipparcos* in 1989 with the goal of high-precision astrometry (star location measurement) in addition to measuring proper motion and two-color photometry. Results were published as the *Millennium Star Atlas*, which contains data for one million stars down to magnitude 11. The Canadian *Microvariability and Oscillation of Stars* satellite was launched in 2003 and measured small brightness variations in stars down to magnitude 6. By the early twenty-first century ground-based system capabilities had advanced significantly due to adaptive optics, such that ground-based 8–10 m telescopes could achieve .03–.05 arcsec resolution.

The atmosphere blocks most infrared light, forcing infrared observations to be done from high altitudes or space. Infrared telescopes, however, posed difficulties in that their detection devices generally require cryogenic cooling. This in turn required the spacecraft to include cryogens such as liquid helium, which boil off, thus limiting the useful life of the spacecraft. Gerry Neugebauer and Bob Leighton made the ground-based infrared survey beginning in the late 1960s, which was followed with a series of eight rocket flights by the U.S. Air Force in the early 1970s. An all-sky survey was performed by the *Infrared Astronomical Satellite*, launched in 1983 as a joint project of the United States, the United Kingdom, and the Netherlands. The spacecraft's observations, which continued for 10 months, led to discoveries of a dust cloud around the star Vega, and an "infrared cirrus" cloud of what were believed to be carbon or hydrocarbon molecules in the Milky Way. These initial discoveries were followed up by ESA's *Infrared Space Observatory* in 1995 and NASA's *Spitzer* in 2003. The two missions showed many new star-forming regions in the Milky Way and other galaxies, that water was common throughout the Milky Way, and that dust obscures many large black holes.

NASA launched the *Submillimeter Wave Astronomy Satellite* in 1998 to study interstellar clouds. It was dedicated to studying specific molecules in these clouds that emit light with submillimeter wavelengths, such as a star-forming region in the Orion Nebula. It also observed water in Comet Tempel 1 to support the *Deep Impact* mission.

Microwave observations allow the study of the cosmic background radiation (CBR), believed to be the leftover radiation from the big bang. Observations from NASA's *COBE* (*Cosmic Background Explorer*) satellite, launched in 1989, led to two discoveries that support the Big Bang theory. *COBE* results showed that the CBR spectrum matched that of a black body and contained small variations believed to result from differences in density in the early universe. These variations led to large-scale structures, such as galaxy clusters. These observations were followed by the *Wilkinson Microwave Anisotropy Probe* in 2001, which made more precise measurements of CBR fluctuations that provided a basis for more precise estimates of cosmological parameters, such as the universe's age, estimated at 13.7 billion years, and the percentages of normal matter (4.4 ± .3 percent), dark matter (21.4 ± 2.7 percent), and dark energy (74.2 ± 3.0 percent) in the universe.

Radio waves can penetrate the atmosphere to reach the surface, and ground-based radio telescopes have been successful. Radio astronomy has benefited greatly from interferometry, which attains high resolution by combining the light from several telescopes. Higher resolution can be obtained by increasing the distance between telescopes, which led to global arrays of radio telescopes. In 1997 Japan launched a radio telescope, *Highly Advanced Laboratory for Communications and Astronomy*, as the space portion of an international network of ground- and space-based radio telescopes from the United States, Europe, Australia, Canada, and Japan. Its elliptical orbit allowed high resolution beyond what is possible with Earth-based telescopes alone.

Spacecraft have allowed astronomers to observe the universe in ways not possible from Earth. Combining the results from each mission gives a broader, more complete picture than is possible by viewing in a single wavelength.

Dean Smith and Stephen B. Johnson

See also: European Space Agency, Japan, National Aeronautics and Space Administration

Bibliography

David Leverington, *New Cosmic Horizons* (2000).

CHIPS, Cosmic Hot Interstellar Plasma Spectrometer. *CHIPS, Cosmic Hot Interstellar Plasma Spectrometer*, was the first NASA University-Class Explorer (a NASA program to launch low-cost science missions and train university students) mission to reach orbit, conducting an all-sky spectroscopic survey of the local interstellar medium (ISM) in the extreme ultraviolet (EUV) waveband from 90 to 265 Å to determine its electron temperature, ionization conditions, and cooling mechanisms. Prior studies of the ISM, based on observations of the diffuse X-ray background and sodium abundance, showed that the solar system lies inside a local bubble of roughly 300 light-years extent in which the ISM is much thinner than typical for the Milky Way galaxy as a whole. Astrophysicists built collisional ionization equilibrium (CIE) models to predict other properties of the ISM, showing that highly ionized iron lines should be prominent in EUV. Managed by Goddard Space Flight Center, the University of California, Berkeley, was the spacecraft project manager for the $18 million mission, with SpaceDev, Inc., building the spacecraft.

The three-axis stabilized spacecraft weighed 60 kg and was the size of a large suitcase. *CHIPS* was launched on 12 January 2003 as a secondary payload on a Boeing Delta 2 from Vandenberg Air Force Base, California, into a 600 km orbit at 94° inclination.

CHIPS observations of highly ionized iron were much fainter than predicted, which posed a puzzle regarding the validity of the CIE models of the local ISM. *CHIPS* also observed comets NEAT-Q4, LINEAR-T7, and Comet Tempel 1 (*Deep Impact*'s interception in 2005) in conjunction with other spacecraft, providing upper limits on EUV emissions from these comets. The CHIPS mission evolved to provide high resolution spectral data on the Sun, and Earth atmospheric spectra by observing the Sun at it set through Earth's atmosphere. The spacecraft was shut down in April 2008.

Stephen B. Johnson

See also: Goddard Space Flight Center, Universities

Bibliography

CHIPS, University of California, Berkeley. http://chips.ssl.berkeley.edu/.

COBE, Cosmic Background Explorer. The *COBE, Cosmic Background Explorer* spacecraft was the first spacecraft dedicated to cosmology. Managed by the Goddard Space Flight Center, it performed an all-sky survey at microwave and infrared wavelengths to measure the structure of the Cosmic Background Radiation (CBR), and so measure the structure of the early universe.

In 1974, NASA turned down three proposals to study the CBR during a competition for Explorer-class missions, approving the *Infrared Astronomical Satellite* (*IRAS*) instead. Two years later, however, NASA suggested that the three CBR teams get together and propose a joint mission. This resulted in the *COBE* spacecraft, to be launched into a polar orbit by either a Delta rocket or the Space Shuttle, carrying three instruments. The Differential Microwave Radiometer (DMR) was designed to detect

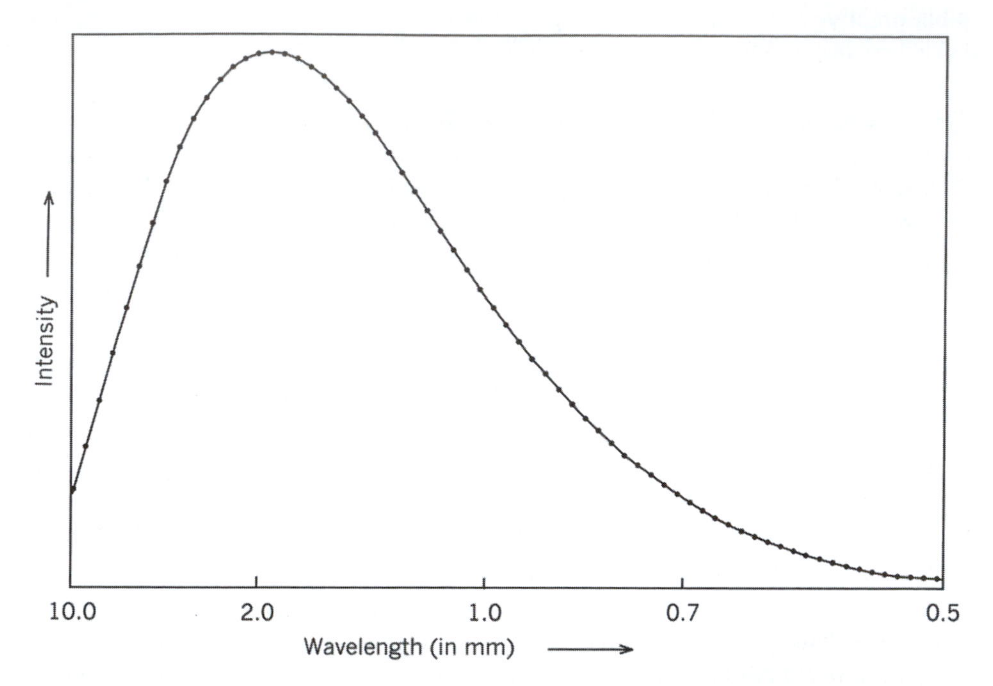

COBE's FIRAS data is shown as small circles. The curve is the theoretical black body curve for a temperature of 2.73 K. The agreement is excellent. (Courtesy David Leverington)

the anisotropy or spatial variation of the CBR, showing the overall structure of the early universe about 300,000 years after the Big Bang. The Far Infrared Absolute Spectrophotometer (FIRAS) was to measure the spectrum of the CBR. If it was found to be that of a black body, this would give added credibility to the Big Bang theory. Finally, the Diffuse Infrared Background Experiment (DIRBE) would detect early infrared galaxies.

NASA accepted the joint proposal, and decided on a Space Shuttle launch from Vandenberg Air Force Base, California. But cost overruns on the *IRAS* program caused NASA to delay *COBE*'s detailed design phase. Cost constraints also resulted in *COBE* using similar designs to *IRAS* of infrared detectors, and of the liquid helium Dewar to cool the detectors. So the success of *IRAS* in 1983 was a great relief to the COBE team. But the explosion of the Space Shuttle *Challenger* on 28 January 1986 caused yet another holdup to the program. Shortly afterward the United States decided to cancel all Space Shuttle launches from Vandenberg, and NASA decided, as a result, to launch *COBE* on a Delta. Unfortunately, this required a major *COBE* redesign, to reduce the spacecraft size and mass by one-half from the Shuttle configuration, so that it would fit on board the new launcher.

NASA launched *COBE* on 18 November 1989. The FIRAS instrument ran out of liquid helium, as expected, 10 months later. This also degraded the performance of the DIRBE, but the DMR continued working normally until NASA switched it off in December 1993.

John Mather, the FIRAS Principal Investigator (PI), announced in January 1990 that the CBR radiation was that of a black body to a very high accuracy, so supporting the Big Bang theory. Then two years later George Smoot, the DMR PI, announced that the CBR was very slightly anisotropic. This showed, for the first time, that the universe was not perfectly smooth shortly after the Big Bang, thus providing the seeds for galaxy formation. The DIRBE measured both the infrared radiation from the early universe and that from more recent galactic sources. In addition, it showed that the plane of the Milky Way is slightly warped and measured the distribution of dust in the solar system. In 2006, Mather and Smoot received the Nobel Prize in Physics for the discovery of the anisotropy and black-body form of the CBR.

David Leverington and David Medek

See also: Goddard Space Flight Center

Bibliography

John C. Mather and John Boslough, The Very First Light: The True Inside Story of the Scientific Journey Back to the Dawn of the Universe (1996).

George Smoot and Keay Davidson, Wrinkles in Time: The Imprint of Creation (1993).

Copernicus. *Copernicus* was the fourth and most successful of NASA's Orbiting Astronomical Observatories. It was managed by Goddard Space Flight Center.

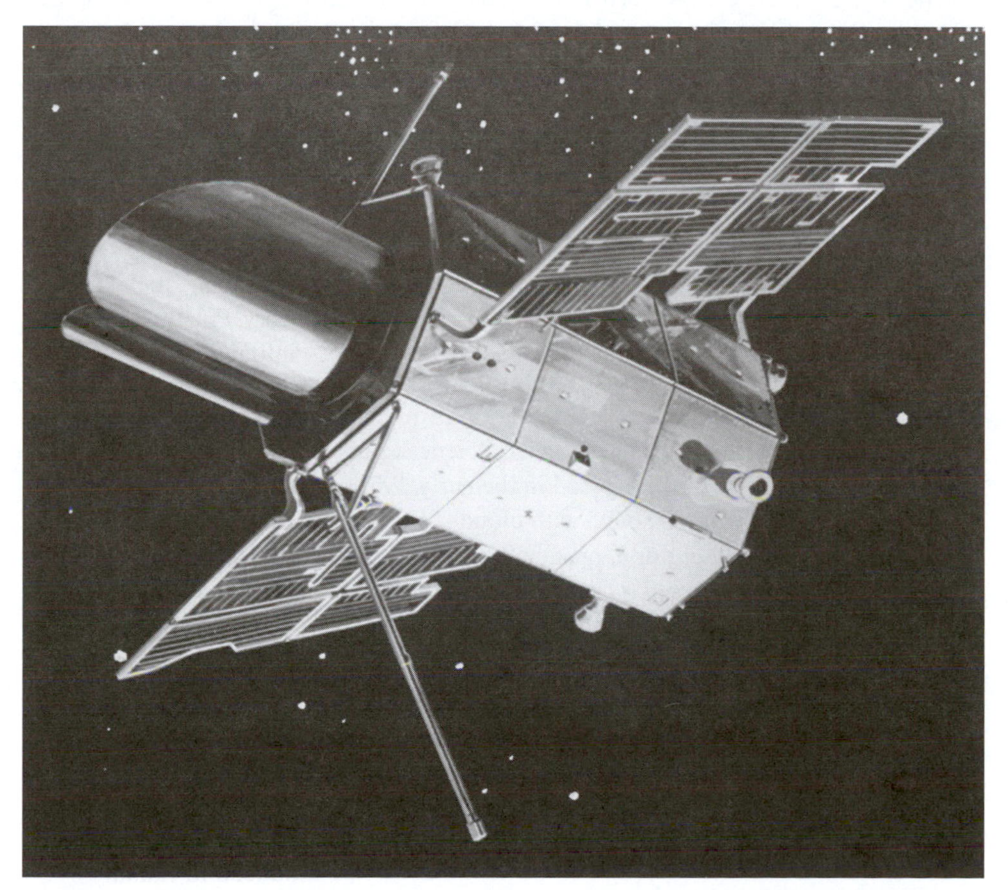

Artist's impression of the Copernicus *spacecraft in orbit. (Courtesy NASA/National Space Science Data Center)*

It was launched into a low Earth orbit by an Atlas-Centaur on 21 August 1972. With a launch mass of 2,220 kg, *Copernicus* was the heaviest robotic spacecraft so far orbited by NASA.

The main instrument on *Copernicus* was an 80 cm diameter ultraviolet telescope with a scanning spectrometer built by Princeton University. It was designed to measure absorption lines in the interstellar medium and of bright stars. It covered the range 70–330 nm, and had a pointing accuracy of 0.1 arc seconds. *Copernicus* also included an X-ray experiment developed by University College London to observe already-known X-ray sources at lower energies than had previously been possible.

The spacecraft made the first detection of deuterium in the interstellar medium as predicted by the Big Bang theory. *Copernicus* also found that the abundance of some elements in the interstellar gas was orders of magnitude lower than that observed in newly formed stars and in the Sun. The spacecraft also measured thin, highly ionized gas at a temperature of about 200,000 K in the regions between interstellar hydrogen clouds. In the Orion region, fast-moving clouds were detected having a temperature of about 10,000 K. These were thought to have been produced either by old supernovae or by shock fronts created by high-velocity stellar winds from the hottest stars.

The X-ray experiment provided valuable data on a number of known X-ray sources, including Cygnus X-1, Cygnus X-3, and the radio galaxy Centaurus A.

David Leverington

See also: Goddard Space Flight Center

Bibliography

David Leverington, *New Cosmic Horizons: Space Astronomy from the V2 to the Hubble Space Telescope* (2000).

EUVE, Extreme Ultraviolet Explorer. *EUVE, Extreme Ultraviolet Explorer*, was the 67th mission in the successful Explorer series and the first dedicated extreme-ultraviolet (EUV) observatory. EUV radiation is quickly blocked by the gases of the interstellar medium. Stuart Bowyer, Space Sciences Laboratory (SSL), University of California, Berkeley, (UCB) proposed in the early 1970s that the interstellar medium might be more like "wonton soup" with clouds of gas and clear channels in between rather than the relatively uniform, opaque "pea soup" that was the prevailing theory. In 1972 the *Copernicus* spacecraft (*Orbiting Astronomical Observatory 3*) observations provided evidence that the interstellar medium contained hot, less dense regions. In a series of sounding rocket flights, and in 1975 with a telescope on board the Apollo-Soyuz Test Project, Bowyer's team proved the wonton soup model was correct, and therefore that many more stars could be studied in the EUV waveband than had been previously thought. Data from the *ROSAT* (*Röntgen Satellite*), launched in 1990, confirmed this.

In 1984 NASA approved *EUVE*, with Goddard Space Flight Center managing the project and performing system integration. Goddard funded Bowyer's SSL team to perform a survey of the sky in the 70–760 Ångstrom EUV wavelengths, for which it designed and built three telescopes for the sky survey and a single Deep Survey/Spectrometer telescope to observe faint objects and perform spectroscopic analysis.

Fairchild Aerospace built the spacecraft bus: the Explorer Platform (EP) based on the Multi-Mission Modular Spacecraft (MMS) architecture, which weighed more than 3 tons at launch. EP was to have been a reusable platform with a new science instrument carried by the Space Shuttle. After the loss of the Space Shuttle *Challenger*, *EUVE* was modified to fly on board a Boeing Delta II 6920-10 rocket.

EUVE launched from Cape Canaveral Air Force Station on 7 June 1992 and received two mission extensions. The most significant was to move control of the spacecraft from NASA's Goddard Space Flight Center to UCB's Center for EUV Astrophysics. This was one of the first missions to be completely transferred from NASA control to a university. *EUVE* also pioneered "lights out" automated operations of spacecraft. UCB implemented systems for automatic control of the spacecraft that allowed significant cost savings by removing the need for staff to be at the control facilities on a continuous basis.

Besides the survey of the sky, *EUVE* detected EUV objects outside Earth's galaxy for the first time. It also studied comets and watched Comet Shoemaker-Levy 9 crash into Jupiter. *EUVE* observed more than 1,000 stars, supernova remnants, quasi-stellar objects, and other objects. NASA turned off *EUVE* on 2 February 2001. It reentered Earth's atmosphere a year later.

Stephen B. Johnson

See also: Apollo-Soyuz Test Project; Explorer; Goddard Space Flight Center; *ROSAT, Röntgen Satellite*, Universities

Bibliography

Martin A. Barstow and Jay B. Holberg, *Extreme Ultraviolet Astronomy* (2003), 115–45.

Stuart Bowyer, "Extreme Ultraviolet Astronomy," *Scientific American*, August 1994.

NASA Headquarters Press Release, 17 November 2000. http://www.gsfc.nasa.gov/news-release/releases/2000/h00-181.htm.

Space Telescope Science Institute EUVE Archive. http://archive.stsci.edu/euve/.

University of California, Berkeley, Space Sciences Laboratory EUVE Archive. http://www.ssl.berkeley.edu/euve/.

FUSE, Far Ultraviolet Spectroscopic Explorer. *FUSE, Far Ultraviolet Spectroscopic Explorer*, was a mission that was part of NASA's Origins program. The Johns Hopkins University developed *FUSE* in collaboration with the Canadian Space Agency and CNES (French Space Agency). *FUSE* was launched 24 June 1999 into low Earth orbit. The spacecraft, built by Orbital Sciences Corporation, was operated by the Johns Hopkins University Center for Astrophysical Sciences and participated in many joint observations with telescopes operating in other parts of the electromagnetic spectrum.

FUSE's unique use of mirror segments rather than a single mirror optimized its capabilities over the ultraviolet (UV) range, as two segments were coated to reflect the shortest wavelengths and the other two to reflect the longer UV wavelengths. *FUSE* used a high-resolution far ultraviolet spectrograph to study stars, the interstellar medium, and distant galaxies. The primary mission lasted three and one-half years. One primary mission objective was to study the abundance of deuterium, a heavy hydrogen isotope formed only in the Big Bang, found in the near interstellar medium, the far Milky Way, and intergalactic clouds. Another primary objective was to study the distribution of atoms and ions in hot interstellar gas to determine its composition, how well it is mixed, and what types of processes can heat it to more than one million degrees. The *FUSE* spectrograph had such a high sensitivity that it was used to study interstellar matter not only in the halo and disk of the Milky Way galaxy, but also in two neighboring galaxies, the Large and Small Magellanic Clouds. One of the most important results from *FUSE* was abundance measurements of deuterium towards stars of varying distances from the Sun. Understanding the distribution of deuterium is an important part of understanding the chemical evolution of the Milky Way galaxy. Other *FUSE* accomplishments included: the first detection of molecular nitrogen in interstellar space; studying the debris disks around young stars; and discovering molecular hydrogen in Mars' atmosphere, supporting evidence of early Martian oceans.

FUSE spent two months in safe mode after two of the four reaction wheels used to maintain the satellite's attitude failed in late 2001. A third reaction wheel failed in December 2004. Science operations resumed in November 2005 after operations teams were able to develop, test, and uplink a revised control software package. *FUSE* operated well beyond its three-year design life, playing a unique role in multiwavelength observations of many objects, and collecting information that will lead to a better understanding of the early universe. Operators shut it down in October 2007.

Virginia D. Makepeace

See also: Explorer, Johns Hopkins University Applied Physics Laboratory

Bibliography

Far Ultraviolet Spectroscopic Explorer. http://archive.stsci.edu/fuse/.

GALEX, Galaxy Evolution Explorer. *GALEX*, *Galaxy Evolution Explorer*, a NASA ultraviolet astronomy satellite, launched 28 April 2003, conducted the first wide area imaging survey of the sky in ultraviolet. By the late 1990s the extragalactic sky had been imaged in essentially every portion of the electromagnetic spectrum, except ultraviolet. An international science team led by the California Institute of Technology proposed the mission specifically to study how galaxies in the universe evolved through time, to determine when the stars that make up galaxies originally formed, and to discover which mechanisms were responsible for triggering star formation. As part of the Explorer program managed by Goddard Space Flight Center, the GALEX project was managed by Jet Propulsion Laboratory (JPL).

By observing galaxies in ultraviolet light, the population of massive stars (hot enough to shine in ultraviolet) is quantified. Because these stars have comparatively short lifetimes, their numbers can help determine their galaxy's star formation rate. By surveying large numbers of galaxies at varying distances from Earth (and therefore varying ages going back toward the Big Bang), the history of star formation across 80 percent of the age of the universe is revealed. To efficiently survey large areas of the sky, *GALEX* combined a wide field of view, 0.5 m aperture telescope (1.25 degrees) with large format detectors to simultaneously image in two ultraviolet bands,

1,350–1,750 Å and 1,750–2,800 Å. A grism (a prism with a ruled grating on one surface) was used to disperse the light for low resolution spectroscopy.

NASA selected *GALEX* in the 1997 Small Explorer competition, a program intended to provide frequent opportunities to launch low-cost space astrophysics and space physics missions. Orbital Sciences Corporation was the spacecraft contractor. The instrument was built at JPL, incorporating detectors developed at the University of California, Berkeley. France contributed special ultraviolet optical elements, and South Korea provided assistance with the ground data system. A Pegasus rocket placed the 280 kg *GALEX* in a 690 km altitude circular Earth orbit.

The *GALEX* surveys, containing millions of galaxies, stars, and more exotic objects, provided astronomers an ultraviolet window on the universe and enabled a wide range of scientific investigations.

James Fanson

See also: Orbital Sciences Corporation, Pegasus

Bibliography

D. Christopher Martin et al., "The Galaxy Evolution Explorer: A Space Ultraviolet Survey Mission," Part 2, *Astrophysical Journal Letters* 619 (20 January 2005): 1–6.

***Gravity Probes A* and *B*.** *Gravity Probes A* (GP A) and *B* (GP B) were experiments that served to validate Albert Einstein's theory of general relativity by providing precise measurements of space and time in Earth's vicinity and shedding new light on how gravity affects the large-scale structure and evolution of the universe.

NASA launched *GP A* on 18 June 1976 from Wallops Flight Center. A Scout D rocket carried a highly accurate atomic clock, developed by Robert Vessot and Martin Levine of the Smithsonian Astrophysical Observatory, on a suborbital trajectory with an altitude of 6,200 miles. The experiment confirmed that gravity affects the flow of time, because *GP A*'s clock ran faster than an identical clock on the ground as the gravitational field weakened with increasing altitude.

Leonard Schiff, William Fairbank, and Robert Cannon envisioned *GP B* in 1959, joined at Stanford University by Francis Everitt in 1962. NASA began funding the project in 1964, although state-of-the-art technology could not produce the high-quality gyroscopes and other equipment necessary. A NASA review in the early 1970s concluded that nine new technologies were required. NASA attempted to cancel the project seven times, but congressional support ensured the project's longevity for four decades. On 20 April 2004, NASA launched *GP B* aboard a Delta II rocket from Vandenberg Air Force Base. The spacecraft collected scientific data for approximately 16 months to confirm two predictions of general relativity: the "geodetic effect," in which Earth's mass warps space and time near the planet, and the "frame dragging effect," in which Earth's rotation twists and drags space and time around the planet. As of mid-2006 the team has not released the final results.

GP B was the first space science mission with a university as prime contractor. Stanford University built the science instruments and subcontracted spacecraft construction to Lockheed Martin. NASA's Marshall Space Flight Center managed the program.

David A. Medek

See also: Universities, Lockheed Martin Corporation, Marshall Space Flight Center

Bibliography

John M. Logsdon et al., eds., *Exploring the Unknown: Selected Documents in the History of the U.S. Civil Space Program, Vol. 5, Exploring the Cosmos* (2001).
Stanford University Gravity Probe B. http://einstein.stanford.edu/.
John Archibald Wheeler, *A Journey into Gravity and Spacetime* (1990).

HALCA, Highly Advanced Laboratory for Communications and Astronomy. *HALCA*, the *Highly Advanced Laboratory for Communications and Astronomy*, was a Japanese radio astronomy satellite and the core element of the multinational VLBI (Very Long Baseline Interferometry) Space Observatory Programme (VSOP). The VLBI project incorporated ground-based radio telescopes from the United States, Canada, Europe, Australia, and Japan. VSOP added spacecraft to the network. While the initial demonstrations of VLBI in combination with a spacecraft occurred in the 1980s, the system using *HALCA* was the first operational version.

Originally known as *Mu Space Engineering Spacecraft* (*MUSES*) *B*, the second satellite in the Institute of Space and Astronautical Science's MUSES series, it was renamed *HALCA* after being launched into a highly elliptical orbit from Kagoshima, Japan, on 12 February 1997, using an M-5 launch vehicle. Before its last VSOP observation in October 2003, its 8 m dish, operating at 1.6 and 5 GHz, used interferometry techniques, in tandem with ground-based radio telescopes, to form a virtual telescope with a diameter more than twice the size of Earth, based on its orbital perigee of roughly 21,000 km. In the process, it achieved the highest angular resolution (0.3 milliarcsec), roughly three times that of ground-only arrays, ever obtained at these or any other wavebands. *HALCA* was also originally designed to operate from 22.0–22.3 GHz, but this capability was damaged, probably due to vibration at launch.

Extragalactic targets that *HALCA*/VSOP studied were the accretion disks of active galactic nuclei, quasars, radio galaxies, and BL Lacertae objects. OH masers in star-forming regions and pulsar emission areas were also frequently observed. *HALCA* was finally commanded to power down on 30 November 2005.

Stephen L. Rider

See also: Japan

Bibliography

HALCA Mission. http://www.isas.jaxa.jp/e/enterp/missions/halca/index.shtml.
H. Hirabayashi et al., "The VLBI Space Observatory Programme and the Radio-Astronomical Satellite *HALCA*," *Publications of the Astronomical Society of Japan* 52 (2000): 955–65.
VLBI Space Observatory Programme. http://www.vsop.isas.jaxa.jp/vsop2/.

***Hipparcos*.** *Hipparcos*, the High Precision Parallax Collecting Satellite (officially "HiPParCoS"), was a European Space Agency spacecraft designed to measure the position, parallax, proper motion, and intensity of stars to an unprecedented accuracy. This first astrometry spacecraft was named after Hipparchus of Nicaea, who published a catalog of the positions of 1,080 stars in 129 BCE.

The one-ton *Hipparcos* spacecraft was designed and built by a European consortium with Matra as prime contractor. At its core was a telescope with mirrors accurate to 1/60 of the wavelength of light. The spacecraft was designed to allow a much better understanding of the movement of stars near the Sun, the physics of stars, and the age of the universe.

Hipparcos was successfully launched by an Ariane 44LP from Kourou, French Guiana, into a geosynchronous transfer orbit on 8 August 1989. But the spacecraft's apogee motor failed to fire, leaving it in its transfer orbit of 220 × 35,600 km. This would have subjected *Hipparcos* to unacceptable air drag and to particle radiation in the Van Allen belts, which would have significantly degraded the spacecraft. But shortly after launch, engineers were able to use *Hipparcos*'s onboard thrusters to increase its perigee to about 530 km, and so significantly reduced these effects. However, this new compromise orbit required the use of three ground stations, instead of the one originally envisaged, and significantly complicated the calculation of stellar positions to the required accuracy. The spacecraft operated until 15 August 1993.

It had been originally expected that *Hipparcos* would have measured the position of about 120,000 stars to an accuracy of 0.002 arcsec. The final Hipparcos Catalogue that was published in 1997 listed their positions to an accuracy of 0.001 arcsec and intensities to within 0.002 magnitude. In addition, the star mappers were used to produce the Tycho Catalogue, which contained more than 1 million stars with positions and intensities an order of magnitude less. Additional data reduction combined with data from existing catalogues resulted in the Tycho-2 Catalogue, containing positions, proper motions, and magnitudes for 2.5 million stars.

David Leverington

See also: European Space Agency

Bibliography

John K. Davies, *Astronomy from Space: The Design and Operation of Orbiting Observatories* (1997).

Catherine Turon, "From Hipparchus to Hipparcos; Measuring the Universe, One Star at a Time," *Sky and Telescope* (July 1997): 28–34.

***Hubble Space Telescope*.** The *Hubble Space Telescope* (*HST*) was an enormously successful spacecraft operating in the near ultraviolet through infrared wavelengths. Its history dates back to ideas advanced by Lyman Spitzer Jr. in 1946 about the potential of large telescopes in space for scientific discoveries by getting above the distorting and blocking effects of Earth's atmosphere. In the 1960s and early 1970s, space astronomy in optical and ultraviolet wavelengths centered on the development of the Orbiting Astronomical Observatories (OAOs). Some of the astronomers who worked on the OAOs, including its chief advocate Spitzer, were also crucial for

In 1995, the majestic spiral galaxy NGC 4414 was imaged by the Hubble Space Telescope *as part of the HST Key Project on the Extragalactic Distance Scale, measuring the periods of Cepheid variables found in this and similar galaxies. (Courtesy NASA/The Hubble Heritage Team/Space Telescope Science Institute/Aura)*

the development of what was generally known by the late 1960s as the Large Space Telescope (LST).

The LST was intended to be an automated reflecting telescope controlled from the ground, whose primary light collecting mirror was to be 3 m in diameter. By the early 1970s many contractors and astronomers were at work on the LST, which was being designed to be launched onboard and serviced by the Space Shuttle. LST's designers also benefited greatly from experience with reconnaissance satellites. LST nevertheless ran into serious political opposition in 1974 because of its high cost. Although eventually approved by the White House and Congress in 1977, by this time the LST had become the more modestly named Space Telescope (renamed *HST* in 1983), with a smaller 2.4 m mirror. It would carry a reduced complement of scientific instruments with NASA inviting the European Space Agency (ESA) as a junior partner.

NASA's Marshall Space Flight Center managed the design and development project, which involved hundreds of companies, government labs, and universities across the United States and Europe. Lockheed Missile and Space Company was the prime contractor, and Perkin-Elmer Corporation designed and manufactured the optical system and its support structures. A new institution, the Space Telescope Science Institute, based at Johns Hopkins University, was founded in 1981 expressly to manage the scientific operations of the *HST* in addition to play a central role in general design issues.

The initial complement of scientific instruments contained two cameras (the Wide Field/Planetary Camera and the ESA-funded Faint Object Camera), two spectrographs (the Faint Object Spectrograph and the Goddard High Resolution Spectrograph), and a High Speed Photometer, designed to measure brightness fluctuations of astronomical objects during even very brief time periods. All these instruments were designed to be replaceable in orbit so that Shuttle astronauts could later switch them for different instruments. In this way *HST*'s designers intended to keep the telescope close to the state of the art. *HST*'s Fine Guidance Sensors, the primary role of which was to aid in pointing the telescope to its astronomical targets for long periods, also served as a sixth scientific instrument as they could measure precisely the angular separation of astronomical objects.

The building of *HST* was beset by many technical and management problems, delays, and cost overruns, raising its design and development costs to more than $2 billion (1990 dollars). Despite these difficulties, the 12.5-ton *HST* was launched into orbit on 24 April 1990 onboard the Space Shuttle *Discovery*. Shortly after, tests of the telescope's optics revealed that its primary mirror suffered from an optical flaw known as spherical aberration resulting in images not nearly as sharp as anticipated. The problem was quickly traced to a misshapen primary mirror, a consequence of the incorrect assembly of a device used to check the mirror's surface during its polishing. That the misassembly was not found before launch was the product of technical and management failings exacerbated by intense cost and schedule pressures to finish the mirror.*HST* became a symbol of scientific and technological failure.

Its public and scientific reputation was restored by an extremely successful Space Shuttle mission in December 1993. Astronauts performed a series of spacewalks to replace several telescope components. A newly installed camera and a specially designed device known as COSTAR (Corrective Optics Space Telescope Axial Replacement) corrected for the spherical aberration, such that *HST*'s optical system performed as well as had been expected before launch. COSTAR's insertion, however, meant that one of the initial instruments, the High Speed Photometer, had to be removed. The cost of the repair mission, at around $700 million, was also high. There were, then, significant scientific and dollar costs to be paid for the blunder in polishing the primary mirror.

In addition to gathering a wealth of scientific data, *HST* generated spectacular (and to some degree carefully constructed) images that regularly made news. By the mid-1990s *HST* had become widely seen as a uniquely powerful, cutting-edge scientific tool.

The initial plans to regularly upgrade the telescope by swapping in orbit new scientific instruments for older and less powerful ones has worked well. In 1993 the Wide Field/Planetary Camera 2 replaced the Wide Field/Planetary Camera 1, in 1997 the Space Telescope Imaging Spectrograph and the Near Infrared Camera and Multi-Object Spectrometer replaced the two initial spectrographs, and in 2002 the Advanced Camera for Surveys was switched with the Faint Object Camera.

With the aid of these instruments, *HST* has made massive contributions to many areas of astronomy, from planetary science to the study of the most distant regions of the universe. Among the most important results are: long exposure views of a small area of sky that produced the "Hubble Ultra Deep Field," then (2003) the deepest image of the universe ever secured and a crucial tool for the study of the evolution of galaxies; helping to establish the existence of "dark energy" and the fact that galaxies move apart faster with time, and the discovery of numerous proto-solar systems that are detected as dark, flattened disks against bright backgrounds of nebular gas.

Sending a spacecraft beyond the atmosphere is not the only way to try to combat its harmful effects. Another is to use adaptive optics in which the optics adapt to changing conditions in the atmosphere. The first adaptive optics systems were proposed in the early 1950s. But it was only in the early 1990s that, following advances in computer technology, practical adaptive optics systems were developed for astronomy, although they were in use by the Department of Defense before this time. Adaptive optics works over a limited field of view, however, and so the *HST*'s ability to observe at high resolution over relatively wide fields is still unique.

In early 2004 as a result of safety concerns following the 2003 Space Shuttle Columbia accident, NASA announced plans to cancel the fifth planned Shuttle servicing mission to the telescope, in effect sentencing the telescope to death within a few years as failure of various critical components would inevitably follow unless they were replaced. But following intense congressional pressure and the installation of a new administrator, NASA announced in 2006 that it would fly one final Shuttle servicing mission. HST may go down as the most famous and productive telescope ever fashioned, in addition to one of the most successful science missions launched to date by NASA. The cost to operate HST in orbit has been around $250 million per year to which have to be added the costs of the initial launch, new instruments and equipment installed during servicing missions, and the Shuttle costs associated with those missions. These hefty costs, therefore, have meant HST's success came in part at the expense of delays in other space science programs.

Robert W. Smith

See also: Lockheed Corporation, Marshall Space Flight Center, Space Shuttle

Bibliography

Mario Livio et al., eds., *A Decade of Hubble Space Telescope Science* (2003).

Carolyn Collins Petersen and John C. Brandt, *Hubble Vision: Further Adventures with the Hubble Space Telescope* (2003).

Robert W. Smith, *The Space Telescope: A Study of NASA, Science, Technology, and Politics* (1993).

IRAS, Infrared Astronomical Satellite. The *IRAS*, *Infrared Astronomical Satellite*, was a Dutch-inspired spacecraft that, because of Dutch funding difficulties, became a NASA/Netherlands/United Kingdom (UK) program. NASA provided the infrared telescope, cooling system, and launcher; the Netherlands the spacecraft bus; and the UK the ground station and a "quick-look" data-analysis facility.

There were many development problems with the detectors, primary mirror, and cooling system. At one stage the experimenters considered recommending that

IRAS be stripped and completely repaired, but NASA ultimately decided to launch as is.

IRAS was designed to provide an all-sky survey in four wavebands, from 12–100 μm with a resolution of 4 arcmin, and to make detailed spectroscopic and photometric observations of selected sources. The payload consisted of a 57 cm diameter, f/9.6-aperture reflecting telescope, which was cooled to a temperature of less than 10 K by superfluid helium to reduce its own infrared emission. The focal plane detectors were cooled to 2.5 K by the same system.

Unlike most scientific Delta launches (which usually launched from Cape Canaveral), the 1-ton *IRAS* spacecraft was launched in January 1983 from the Western Test Range in California into a Sun-synchronous, polar orbit. *IRAS* operated autonomously on a program uplinked from its control center every 12 hours. Likewise, data was stored on a tape recorder and transmitted to Earth every 12 hours.

The helium coolant was finally exhausted after 300 days, when the mission was terminated. By then *IRAS* had observed 95 percent of the sky at least four times, and had made thousands of point observations. The main *IRAS* catalogue, which was published in 1984, contained about 245,000 sources.

IRAS detected protostars, or stars in their early stages of development, in interstellar dust clouds, plus stars like Vega and β-Pictoris still surrounded by remnants of their original dust clouds, and dust shells emitted by stars like Betelgeuse near the end of their lives. It also observed infrared patterns radiating at 35 K all over the sky, which looked similar in structure to atmospheric cirrus clouds, and discovered a number of ultraluminous galaxies, which emit most of their energy in the infrared.

David Leverington

See also: Ball Aerospace and Technologies Corporation, Goddard Space Flight Center, United Kingdom

Bibliography

John K. Davies, *Astronomy from Space: The Design and Operation of Orbiting Observatories* (1997).

Wallace Tucker and Karen Tucker, *The Cosmic Inquirers: Modern Telescopes and Their Makers* (1986).

Joost van Kasteren, *An Overview of Space Activities in the Netherlands* (2002).

***ISO*, Infrared Space Observatory.** *ISO*, *Infrared Space Observatory*, was a European Space Agency (ESA) spacecraft designed as a follow-on to the *IRAS* (*Infrared Astronomical Satellite*) spacecraft, which had detected some 245,000 infrared sources in 1983. *IRAS*'s mission had been to find new sources, but that of *ISO* was to observe specific infrared sources continuously.

ISO, which was designed and manufactured by a European industrial consortium led by Aérospatiale of France, was launched by an Ariane 4 rocket on 17 November 1995. Its initial orbit was then changed to its operational 24-hour orbit of 1,000 × 70,600 km using the spacecraft's hydrazine thrusters. Instrument observations were limited to about 17 hours per day when the spacecraft was above the Van Allen radiation belts, as these generally degraded instrument performance. Two ground stations were used during the operational phase. One was ESA's ground station at Villafranca, Spain, and the other was NASA's at Goldstone, California, which was funded by the United States and Japan in return for guaranteed access to *ISO*.

The *ISO* payload consisted of a 60 cm diameter, reflecting telescope enclosed in a cryostat, which was cooled with 2,300 liters of superfluid helium at a temperature of 1.8 K. The telescope's field of view was shared among four instruments: an imager, a photopolarimeter, and two spectrometers. Routine operation started on 4 February 1996 and continued until 8 April 1998 when *ISO* ran out of coolant. This period was some 30 percent longer than originally anticipated.

Astronomers anticipated that *ISO* would probably find water molecules in the gaseous envelopes surrounding newly formed, massive stars. But to most astronomers' surprise, *ISO* also found water molecules in stars near the end of their lives. In fact, it found that water molecules are common in the Milky Way.

As gas clouds collapse, they gradually heat to produce stars. *ISO* detected a number of these star-forming regions in both normal and colliding galaxies, where star formation had been triggered by the collision shock wave. The earliest phase of cloud collapse was found in a two-solar mass dark cloud that had a temperature of just 13 K.

David Leverington

See also: Aérospatiale, European Space Agency

Bibliography

M. F. Kessler et al., "ISO's Astronomical Harvest Continues," *ESA Bulletin* 99 (September 1999): 6–15.

———., "Looking Back at ISO Operations," *ESA Bulletin* 95 (August 1998): 87–97.

David Leverington, *New Cosmic Horizons: Space Astronomy from the V2 to the Hubble Space Telescope* (2000).

IUE, International Ultraviolet Explorer. The *IUE*, *International Ultraviolet Explorer*, was originally the *Large Astronomical Satellite* (*LAS*) of the European Space Research Organisation (ESRO), but escalating costs forced its cancellation in 1967. British astronomer Robert Wilson then suggested to NASA that a modified version of *LAS* could fill the gap between the last *Orbiting Astronomical Observatory* spacecraft and what became the *Hubble Space Telescope*. NASA agreed in principle, but increased its planned orbit from low Earth orbit to geosynchronous, so that Earth obscured less of the spacecraft's sky and allowed continuous communications with a given ground station.

The 4.2 m long spacecraft was built around its 45 cm diameter telescope. This fed a spectroscope operating in the ultraviolet from 115–320 nm. A high-resolution spectrum was produced for bright objects or a low-resolution one for dim objects. There was no imaging capability.

NASA Goddard Space Flight Center managed the program and built the spacecraft bus, while the United Kingdom provided the detectors. The European Space Agency (ESRO's successor) provided the solar arrays and one of the two ground control stations. As *IUE* was always in view of one of these ground stations, it allowed the spacecraft to be used like a ground-based observatory, which was a novel idea at the time, with astronomers directing their observations from the ground stations. In its 18-year lifetime, 15 years longer than planned, *IUE* produced more than 100,000 spectra and provided data to more than 3,000 astronomers.

IUE, which was launched by a Thor-Delta on 26 January 1978, was used to study many different types of object from comets to stars to galaxies. In particular, it provided valuable new data on stellar winds from hot, massive stars, such as Wolf-Rayet stars, indicating that they were losing mass at the rate of one solar mass per 25,000 years. *IUE* also detected absorption lines in nebulae surrounding Wolf-Rayet stars. In addition, *IUE* provided valuable data on the chromosphere of cool stars and on their stellar winds. It also gave valuable insights into the structure of active galaxies like Seyfert galaxies.

David Leverington

See also: European Space Agency, Goddard Space Flight Center

Bibliography

John K. Davies, *Astronomy from Space: The Design and Operation of Orbiting Observatories* (1997).

Y. Kondo, ed., *Exploring the Universe with the IUE Satellite* (1987).

J. Krige et al., *A History of the European Space Agency 1958–1987* (2000).

MOST, Microvariability and Oscillation of Stars. *MOST, Microvariability and Oscillation of Stars* (Microvariabilité et Oscillations STellaire) was a Canadian Space Agency satellite, Canada's first space telescope, the first microsatellite capable of pointing to accuracies of about 1 arcsec, and the first satellite dedicated to ultra-precise astronomical photometry. It could measure brightness oscillations in stars to about 1 part per million, equivalent to the change produced by looking at a streetlight 1 km away and moving a human eye back and forth by 0.5 mm. This allowed *MOST* to study subtle oscillations in stars to probe their internal structure through asteroseismology and the variations in light associated with giant exoplanets in orbit around Sun-like stars.

MOST was launched 30 June 2003 by a Russian three-stage Rockot from the Plesetsk Cosmodrome. Its low Earth polar orbit was Sun-synchronous, allowing it to monitor selected stars for up to 60 days without interruption. The 54 kg, suitcase-sized spacecraft housed a single optical telescope with a 15 cm collecting mirror, feeding a CCD (Charge Coupled Device) camera with twin detectors: one for science, the other for tracking.

Early scientific results from *MOST* included the first direct measurement of differential rotation in a star other than the Sun; seismic determination of the age of eta Boötis to an accuracy of about 1 percent, better than for any star other than the Sun; setting a stringent limit on the reflectivity of the giant planet in the system HD 209458 and constraining the nature of its atmosphere and cloud cover; proof that the close-in planet around tau Boötis has forced the outer layers of the star to rotate in synchronization with its orbit; and the controversial result showing no oscillations in the bright, well-studied star Procyon. Scientists thought Procyon was a good candidate for asteroseismology. Not detecting these waves in Procyon prompted new theoretical modeling of convection in the outer layers of evolved stars.

Jaymie Mark Matthews

See also: Canada

Bibliography

University of British Columbia's *MOST.* http://www.astro.ubc.ca/MOST/index.html.

Gordon Walker et al., "The *MOST* Asteroseismology Mission: Ultraprecise Photometry from Space," *Publications of the Astronomical Society of the Pacific* (September 2003): 1023–35.

SIRTF, Space Infrared Telescope Facility. *See Spitzer Space Telescope.*

Spitzer Space Telescope. *Spitzer Space Telescope*, named after astrophysicist Lyman Spitzer Jr., was a space-based infrared observatory. Launched by a Delta rocket on 25 August 2003 with a specified two-and-a-half year operational life and five-year target life, it used a 0.85 m telescope and included three cryogenically cooled science instruments: the Infrared Array Camera for large-area surveys, the Infrared Spectrograph for moderate-resolution spectroscopy, and the Multiband Imaging Photometer for imaging photometry. NASA's Jet Propulsion Laboratory (JPL) was responsible for project management and operation of the spacecraft. Lockheed Martin Missiles and Space designed, integrated, and tested the spacecraft. The California Institute of Technology's Spitzer Science Center managed science

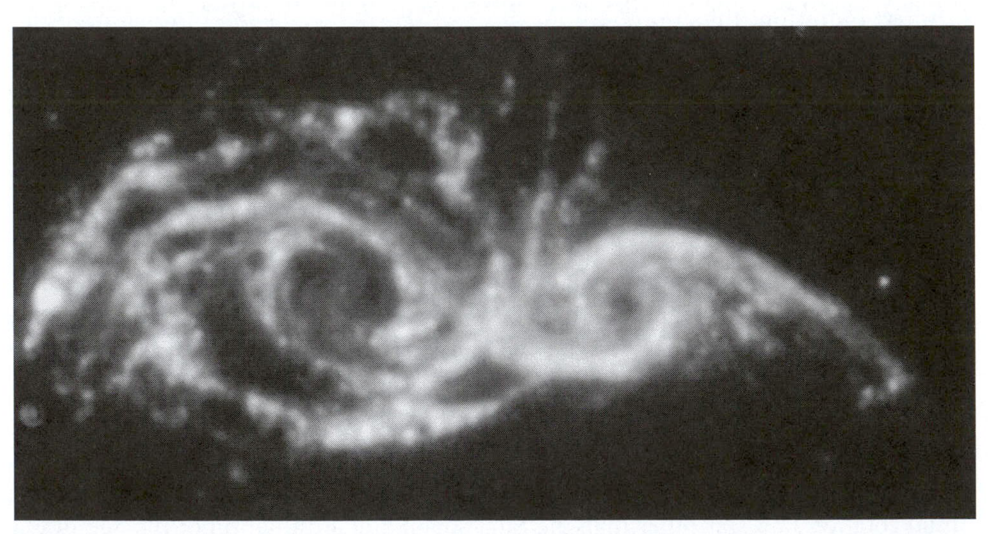

Composite image by the Spitzer Space Telescope *and the* Hubble Space Telescope *showing shape-shifting galaxies that have taken on the form of a giant mask. The "eyes" are actually the cores of two merging galaxies, called NGC 2207 and IC 2163, and the mask is their spiral arms. The image consists of infrared data from NASA's* Spitzer Space Telescope *(the mask) and visible data from NASA's* Hubble Space Telescope *(the eyes). This cosmic interaction will come to an end when the galaxies meld into one. The infrared data from* Spitzer *highlights the galaxies' dusty regions, while the visible data from* Hubble *indicates starlight. (Courtesy NASA/European Space Agency/Jet Propulsion Laboratory/Space Telescope Science Institute/D. Elmegreen (Vassar))*

observations by evaluating observation requests and creating the schedule that JPL implemented.

Three decades of development went into the *Spitzer* from proposal to launch. In the early 1970s the project was first proposed as a Space Shuttle–based observatory named the Shuttle InfraRed Observatory. The name was changed to Shuttle Infrared Telescope Facility (SIRTF) in 1979. At the time, the Shuttle was projected to fly weekly flights of up to 30 days duration, which was an advantage for the cryogenic system, in which maintaining the cold helium's 1.5 K temperature was a major problem. After *Infrared Astronomical Satellite* demonstrated the improved observational capability of a free-flying infrared observatory over one attached to the heat-radiating Space Shuttle, in 1983 NASA switched to the free-flying Space Infrared Telescope Facility. NASA invited the European Space Agency (ESA) to collaborate on this project, but after postponements and cancellations related to the International Solar Polar Mission (later *Ulysses*), ESA declined. The increasing importance of SIRTF was evident in the National Research Council–commissioned Bahcall report, a two-year survey of the astronomical community that provided recommendations for ground- and space-based observatories for the 1990s.

However, NASA's budget outlook changed dramatically in the early 1990s and required a redesign of SIRTF, reducing its scientific capabilities, but bringing the development cost from $2.2 billion to less than $500 million and the telescope weight from 5,700 kg down to 850 kg. Modifying its orbit to an Earth-trailing heliocentric orbit allowed its cryogenically cooled instrument to be launched warm and then cooled by the low temperatures in space. It also eliminated the interference of Earth's infrared radiation, thus reducing the amount of coolant needed to maintain the required temperature of 5.5 K. To reduce weight, engineers switched to using lightweight beryllium for the mirrors and decided to cool only the infrared detectors instead of the entire telescope. The onboard computer's size and cost were reduced by designing it to download all collected data one to two times each day and allowing only one science instrument to operate at a time. NASA successfully used an experimental management technique involving the industrial contractors who would potentially build the spacecraft and instruments in the preliminary design process, rather than having them bid after the design and requirements were completed.

Spitzer's scientific mission was to study star and planet formation, galaxy formation and evolution, the origins of energetic galaxies and quasars, and the distribution of matter and galaxies. It helped astronomers to research star and planet formation by detecting a warm disk of material around the sun-like star HD69830, interpreted to indicate a possible asteroid belt. *Spitzer* has provided spectra and structure data on protoplanetary nebulae and infrared data about HD189733b, a Jupiter-like planet orbiting a nearby star. The spacecraft also detected infrared light echoes (reflected light from the supernova remnant) in dust surrounding supernova remnant Cassiopeia A. Some astronomers interpreted these echoes, which are not seen in X-ray or radio emissions, to mean that the supernova is a rare magnetar, a neutron star with an extraordinarily strong magnetic field.

Spitzer made progress in discovering the origins of active galaxies and quasars by imaging distant active galaxies and quasars that were undetectable in visible light due to surrounding dust. Quasars are thought to be super-massive black holes surrounded by gas and dust that light up as the black hole sucks in surrounding material. Astronomers suspected that these "invisible" quasars existed because the amount of X-ray emission detected in the universe, a key predictor of quasars, was much greater than the number of visible quasars observed. This discovery indicates that dust obscures most super-massive black hole growth.

Kathleen Mandt

See also: Jet Propulsion Laboratory, Lockheed Martin Corporation

Bibliography

George H. Rieke, *The Last of the Great Observatories* (2006).
Spitzer Science Center. http://ssc.spitzer.caltech.edu/.

SWAS, Submillimeter Wave Astronomy Satellite. *SWAS*, *Submillimeter Wave Astronomy Satellite*, was a radio observatory whose mission was to study the composition of interstellar gas clouds and determine how they cool as they collapse to form stars and planets. One important goal was searching for and determining the abundances of water vapor, molecular oxygen, neutral carbon, isotopic carbon monoxide, and isotopic water—molecules key to life on Earth.

NASA approved *SWAS* as a Small Explorer Project in 1989, and it was launched on 5 December 1998 from Vandenberg Air Force Base, aboard a Pegasus-XL rocket into a 600 km orbit at 70 degree inclination. Built and managed by Goddard Space Flight Center, the spacecraft contained a single instrument, an elliptical off-axis Cassegrain telescope that used two submillimeter wave radiometers as detectors for signals from 487 to 556 GHz.

Through May 2005, *SWAS* studied more than 300 interstellar clouds within the Milky Way galaxy. Among its findings, *SWAS* observed a region of active star formation in the Orion Nebula, in which the chemical formation of water is occurring rapidly. Toward the star CW Leonis, *SWAS* detected large amounts of water vapor, which astronomers hypothesize was released by the mass vaporization of billions of comets following a surge in heat output by the dying star. However, for most clouds, *SWAS* detected less water vapor and molecular oxygen than had been predicted. Astronomers believed that water was present in these clouds, but predominantly in the form of small particles of solid water–ice rather than in the form of water vapor. Thus they thought that large amounts of water–ice became entrained in the cloud collapse to form stars and planets. *SWAS* made its final observations and ended its scientific odyssey in 2005 when it joined a large number of other telescopes in observing Comet Tempel-1 as the *Deep Impact* spacecraft launched an impactor into the nucleus of this comet.

Gary Melnick

See also: Goddard Space Flight Center

Bibliography

Gary J. Melnick et al., "The *Submillimeter Wave Astronomy Satellite*: Science Objectives and Mission Description," *Astrophysical Journal* 539 (2000).
SWAS. http://www.cfa.harvard.edu/swas/.

WIRE, Wide Field Infrared Explorer. *WIRE, Wide Field Infrared Explorer*, the fifth Small Explorer mission, was designed to study the evolution of starburst galaxies. Goddard Space Flight Center built, managed, and operated the *WIRE* spacecraft. In addition to orbiting the smallest hydrogen cryostat to date, which cooled its infrared detectors to 6.8 K, the 259 kg spacecraft utilized the first all-bonded graphite-composite spacecraft structure.

On 4 March 1999, 30 minutes after launch on a Pegasus XL rocket, a subtle design flaw created an electronic transient, prematurely releasing the instrument cover and exposing the telescope's internals to the heat of Earth and the Sun. This caused a rapid sublimation of the frozen hydrogen into vapor; all the hydrogen was lost within 36 hours, ending the primary mission before it started. After recovering the spacecraft from a 60 RPM spin, NASA had a working spacecraft with a dead instrument.

Derek Buzasi of the University of California at Berkeley soon proposed using the star tracker to gather asteroseismology data. *WIRE* is credited with discovering the first multimodal oscillations on a cool star other than Earth's Sun. These oscillations in brightness are caused by seismic waves that propagate through the star. Measurement of the frequencies of these waves enables astronomers to bound fundamental parameters such as the star's mass.

David Everett

See also: Goddard Space Flight Center

Bibliography

D. Buzasi et al., "The Detection of Multimodal Oscillations on Ursae Majoris," *The Astrophysical Journal Letters* 532 (1 April 2000): L133–36.

David F. Everett et al., "Recovery of the Wide-Field Infrared Explorer Spacecraft," *Proceedings of the 14th Annual American Institute of Aeronautics and Astronautics/Utah State University Conference on Small Satellites* (August 2000).

Goddard Space Flight Center *WIRE*. http://sunland.gsfc.nasa.gov/smex/wire/.

WMAP, Wilkinson Microwave Anisotropy Probe. The *WMAP, Wilkinson Microwave Anisotropy Probe*, was launched on 30 June 2001 to characterize the tiny temperature differences in the cosmic microwave background radiation. The Big Bang theory, developed in 1948 by Ralph Alpher, Robert Herman, and George Gamow, predicted that the universe should be bathed in an afterglow of cosmic microwave background radiation. This is the cool remnant of an unimaginably hot early epoch of the universe. Arno Penzias and Robert Wilson detected the uniform sky glow in 1965, and NASA's *Cosmic Background Explorer* satellite determined the precise temperature of the microwave background in 1990 to be 2.73° above absolute zero and first detected tiny temperature differences across the sky (anisotropy) in 1992. David T. Wilkinson of Princeton University and Charles L. Bennett of the Goddard Space Flight Center (GSFC) formed a partnership that led to a 1995 proposal to NASA to obtain detailed full sky maps of the anisotropy to reveal a wealth of information about the properties of the universe. Competitively selected for study in 1996, the mission was confirmed for development in 1997.

Built by Princeton University in partnership with GSFC, the *Microwave Anisotropy Probe* was launched aboard a Delta 2 rocket. Following a lunar gravity-assist boost, the probe was placed in a Lagrange point (L2) orbit of the Sun-Earth system, 1.5 million km beyond Earth. From this vantage point, it surveyed the whole sky over the course of a year, and in February 2003 the first detailed full sky maps of the cosmic background radiation were announced. NASA named the mission in honor of Wilkinson. Cosmological theories make specific predictions about the observable temperature patterns in the cosmic background radiation. Using the precise new measurements, *WMAP* definitively answered many questions about the universe, as scientists compared the observed patterns with those predicted by various theories. The sky maps revealed the content of the universe to be 4 percent regular matter, 23 percent cold dark matter, and 73 percent exotic dark energy, which was not yet understood in 2008. *WMAP* data also showed the age of the universe is 13.7 billion years (± about 100 million years) and when the first stars formed (200 million years after the Big Bang, which is earlier than expected). Specific versions of inflation theories (theories about the first fraction of a second of the universe) were ruled out. New questions arose: What caused the early reionization of the universe? What is the nature of dark energy? Does it change with time, and if so, how and why? What is dark matter? What are the limits of what we can learn about the very first moments of the universe? Is the universe really isotropic and homogeneous on the largest scales?

Charles L. Bennett

See also: Goddard Space Flight Center

Bibliography

Michael D. Lemonick, *Echo of the Big Bang* (2003).
WMAP. http://map.gsfc.nasa.gov/.

PLANETARY SCIENCE

Planetary science includes the study of all objects orbiting the Sun—that is, planets, asteroids, and comets—and of planets orbiting other stars.

Priests of many early civilizations noticed that not all heavenly bodies appeared fixed relative to one another. Not only did the Moon move relative to the fixed stars, but so did the wandering stars or planets, as the ancient Greeks called them. Occasionally a comet would appear with its apparently threatening tail.

Many early civilizations studied the position and visibility of the planets for religious or astrological purposes. The Babylonians of the first millennium BCE used mathematics to try to predict the positions of the Moon and planets known to them (Mercury, Venus, Mars, Jupiter, and Saturn) in the sky, while the ancient Greeks tried to understand what orbits these objects described in space. Much of this early Greek work was lost to Europe when the Roman Empire collapsed in the fifth century, but it survived in the Middle East and reappeared in Islamic Spain in the twelfth century in Arabic translations.

Using an early telescope, in 1609 Galileo Galilei observed that the Moon's terminator had an irregular shape and concluded that this was because the Moon had mountains and valleys. This was quite unlike the pure spherical body of Aristotle's cosmology, which was taught in universities of the day. Two years later, Johann Fabricius concluded that sunspots were on the surface of the Sun, which also contradicted Aristotle's teaching that the Sun was a perfect body. Fabricius concluded that their movement showed that the Sun was rotating.

In 1610 Galileo observed that Jupiter has four moons that were orbiting the planet. He also found that Venus exhibits a full set of phases like the Moon. This showed that Ptolemy's Earth-centered model of the universe was incorrect, overturning yet another ancient theory. Galileo then entered a long argument with the Vatican about the interpretation of his observations and his contention that Nicolaus Copernicus was correct in the latter's 1543 book *De Revolutionibus Orbium Caelestium* (*On the Revolution of the Heavenly Spheres*) that described a Sun-centered universe.

Galileo believed, like Copernicus, that the planets orbited the Sun in a circle, using epicycles to modify their orbits. However, in 1605 Johannes Kepler had shown that the orbit of Mars was an ellipse, which he proved for Venus in 1614 and for Mercury one year later. By 1618 Kepler had developed his three laws of planetary motion, but his ideas of what caused the planets to move in the way he proposed were not as successful. This was left to Isaac Newton to solve by his universal theory of gravitation in *Philosophiae Naturalis Principia Mathematica* (*Mathematical Principles of Natural Philosophy*) of 1687.

The observed characteristics of the Moon and planets gradually became clearer as telescopes improved during the next two centuries, although it was unclear as to whether the patterns observed on the planets were true surface features or clouds. Mercury, in particular, was difficult to observe, as it is small and close to the Sun in the sky, so it was impossible to see any clear features. Some astronomers thought that they could see an atmosphere, but others disagreed.

In 1761 Mikhail Lomonsov observed that Venus was surrounded by a luminous ring when it had just started its transit across the Sun. As a result he concluded that Venus has an extensive atmosphere. In 1667 Giovanni Domenico Cassini had deduced that Venus's rotation period was about 24 hours, and this had been generally accepted until the latter part of the nineteenth century, when Giovanni Schiaparelli and others concluded that it has a synchronous rotation period of 224.7 days. In the early part of the twentieth century, various periods were proposed.

In 1666 Cassini used marks on the surface of Mars to deduce a rotation period of 24 h 40 min, which is within three minutes of its correct value. A few years later, Christiaan Huygens first unambiguously detected Mars's white south polar cap. In 1719 Giacomo Maraldi observed that the south polar cap disappeared, only to return again later, indicating its nonpermanent nature. William Herschel suggested that this was because it consisted of ice and snow that melted in the southern summer.

Jupiter is the largest planet in the solar system. Cassini used markings on the planet in 1663 to deduce a rotation period of 9 h 56 min, and two years later he observed a prominent spot that may have been an early appearance of the Great Red Spot. Cassini later concluded that Jupiter's angular rotation rate depended on latitude and observed its polar flattening due to its rapid rotation rate. In 1778 the Compte de Buffon suggested that Jupiter had not cooled down fully since its formation, a popular idea among astronomers in the second half of the nineteenth century. Some astronomers even suggested that Jupiter was partly self-luminous, like a miniature sun.

Galileo observed what he thought were two moons of Saturn in 1610, but two years later he found that they had disappeared. Then they reappeared the following year. It was not until 1659 that Huygens explained that Galileo had not seen two moons but a thin, flat, solid ring around the planet, which was inclined to the ecliptic and so changed its appearance with time. A few years later Cassini observed that the ring was divided into two by a dark line, now called the Cassini Division. He also speculated that the two rings were not solid, but composed of swarms of small satellites.

Uranus, the first planet to be discovered since antiquity, was discovered by William Herschel in 1781. The first asteroid, Ceres, was discovered by Giuseppe Piazzi in 1801. Then Neptune was discovered by Johann Galle and Heinrich d'Arrest in 1846, following the calculation of its expected position by Urbain Le Verrier.

More details were seen on Mars, Jupiter, and Saturn in the nineteenth and early twentieth centuries, as telescopes continued to improve, but turbulence of Earth's atmosphere was a real limitation. Nevertheless, Pluto and numerous planetary satellites were discovered.

Gustav Kirchoff and Robert Bunsen's work on spectroscopy, which they began in 1859, held out the hope that the constituents of the atmospheres of the various planets could be deduced. This proved more difficult than expected, however, owing to the problem of extracting the effects of Earth's atmosphere from the observations. As a result, planetary work slowed in the first half of the twentieth century, until the start of the space age. Then it became possible to visit the Moon and planets and to observe them from Earth-orbiting spacecraft. A summary of our knowledge just before the start of the space age in 1957 is shown in the table on page 81.

A synchronous axial rotation period of 88 days had gradually been accepted for Mercury in the early twentieth century. But in 1965 Gordon Pettengill and Rolf Dyce found a period of 59 ± 5 days from Doppler shifts of radar signals transmitted using the Arecibo radio telescope. Almost immediately Giuseppe Colombo pointed out this was about two thirds of the synchronous rotation rate. Later work showed that the two-thirds "lock" was correct, with a rotation period of 58.65 days.

The only spacecraft to visit Mercury in the twentieth century was *Mariner 10*, which flew by in 1974–75, imaging about half the surface. It found that it was much like the Moon, with a large number of impact craters. However, there were no lunar-like maria. *Mariner 10* also found that Mercury, surprisingly for such a small planet, has a significant magnetic field. Then in 1991 Earth-based radar studies indicated that Mercury's permanently shadowed polar craters may contain water ice.

In 1957 Charles Boyer discovered V-shaped clouds on Venus that had a rotation rate of about four days retrograde. Five years later Richard Goldstein and Roland Carpenter then solved the mystery of Venus's axial rotation rate using the Jet Propulsion

State of Knowledge Just Before Start of the Space Age, October 1957 (Part 1 of 2)

Mercury	*Axial rotation period 88 days.* Little or no atmosphere.
Venus	Axial rotation period unknown. Has an atmosphere. Vague clouds observed from time to time. Carbon dioxide detected in atmosphere. Other constituents uncertain. Atmospheric pressure uncertain but thought to be higher than on Earth. Surface temperature uncertain. Estimates ranged from 80–300°C.
Earth	Shape non-symmetrical owing to non-symmetrical internal mass distribution. Spin axis moves slightly with respect to surface with period of just more than one year. Crust thickness varies from 5 km under deep oceans to about 70 km under parts of continents. Dense metallic core, mostly of iron. Basic structure of ionosphere known. Link known among magnetic storms, auroral displays, radio interference, and sunspots.
Moon	No detectable atmosphere. Unclear as to how much of surface has been modified by impacts and how much by volcanic activity. Depth of surface dust unknown. Figures up to a few meters proposed. Possibly elemental life-forms, like bacteria, present on surface.
Mars	*Surface atmospheric pressure thought to be in the region of 50–120 mb.* Carbon dioxide discovered in atmosphere, but *main constituent, at something like 98 percent, thought to be nitrogen.* Surface temperature varies from about –80°C at poles in winter to +20°C at equator in summer. Thin polar icecaps of water ice or hoar frost observed, which largely disappear in local summer. Some form of life, possibly simple vegetation, in the form of lichens or moss, present on the surface. Planetwide dust storms observed occasionally. Two small moons, Phobos and Deimos.
Jupiter	Methane and ammonia detected in atmosphere. Core of metallic hydrogen. Great Red Spot and small white and dark spots observed to continuously change their appearance with time. 400 km/h equatorial current. Emits radio waves. Four Galilean and eight other satellites. Four Galilean satellites have synchronous rotation.

(Continued)

State of Knowledge Just Before Start of the Space Age, October 1957 (Part 2 of 2)

Saturn	Methane and ammonia detected in atmosphere. White spots observed occasionally. 1,400 km/h equatorial current. Three rings (A, B, and C), with Cassini Division between B and A, and with Enke Division in the A ring. Nine satellites. Titan, the largest, has methane detected in its atmosphere.
Uranus	*Axial rotation period thought to be about 10h, 45m.* Spin axis inclined at about 98° to Uranus's orbit around the Sun. (So spin direction is retrograde.) Methane detected in atmosphere. Five satellites.
Neptune	Axial rotation period thought to be about 16h prograde. Methane detected in atmosphere. Two satellites, including Triton, the larger, whose orbit is retrograde.
Pluto	Size and mass unclear. Axial rotation period 6d, 9h. No known satellites.
Asteroids	Ceres is largest, with diameter of about 800 km. Kirkwood gaps known. Asteroid families identified. First asteroid (Apollo) found whose orbit crosses that of Earth. Kuiper-Edgeworth belt hypothesised beyond the orbit of Neptune.
Comets	Various carbon, nitrogen, and hydrogen compounds found in the heads of comets, and ionized molecules found in their tails. Reason why comet's tails point directly away from Sun unclear, as radiation pressure from Sun thought not to be powerful enough. Biermann suggested that cause of this effect could be elementary particles emitted by the Sun. *Typical nucleus size thought to be about a few hundred kilometers.* Oort cloud hypothesised at great distance from Sun as source of long period comets.

The main "facts" that have since been found to be incorrect are shown in italics.

Laboratory's radar. In the following year they announced that it was about 240 days retrograde. This was a big surprise, as the only other planet known to have a retrograde axial rotation was Uranus, and that was really rotating on its side.

Many spacecraft have visited Venus, and a number had probes that landed on the surface. The surface temperature was found to be about 440°C at its mean surface elevation, and the surface atmospheric pressure about 93 bar. The surface temperature varies by only about 1°C throughout the course of a year, owing to the insulation provided by the dense, cloudy atmosphere. The atmosphere was found to be about 96.5 percent carbon dioxide, which, via its greenhouse effect, is the main reason for the very high surface temperature. The clouds were found to be about 75 percent sulfuric acid.

Plate tectonics do not operate on Venus, but the surface is relatively young, like Earth, with both volcanic and tectonic features, such as the coronae, which are unique to Venus. The lack of water is thought to have resulted in a stiff outer shell to the planet, which has prevented plate tectonics from taking place, while enabling the surface to support high features such as Ishtar Terra, which includes the 11,500 m high Maxwell Montes. Although Venus has some highland regions, about 80 percent of the surface, compared with about 30 percent for Earth, is within ± 1 km of the average elevation.

As far as Earth is concerned, it had been generally appreciated for some time that the shorelines on both sides of the Atlantic Ocean appeared as though they could fit into each other. In 1912 Alfred Wegener suggested that the land masses on both sides had once been part of the same landmass called Pangaea, which had broken apart about 100–200 million years ago. This idea was largely dismissed until new evidence in the 1950s–60s caused his ideas to be resurrected, modified, and expanded into the modern theory of plate tectonics.

The first discovery of the spacecraft age was Van Allen's discovery in 1958 of the radiation belts, which now bear his name, of protons and electrons around Earth. The detailed structure of these belts was found to vary with solar activity. In 1959 the Soviet *Luna 2* spacecraft detected charged particles emitted by the Sun in the so-called solar wind. Then in 1963 the American *IMP 1* (*Interplanetary Monitoring Platform*) spacecraft detected the bow shock, where Earth's magnetic field meets the solar wind.

The *Luna 3* spacecraft took the first photographs of the far side of the Moon in 1959, showing that it was broadly similar to the visible side, except that it had fewer maria. Later spacecraft showed that maria covered only 2.5 percent of the far side, compared with 16 percent for the Moon as a whole. *Lunar Orbiter 1* showed that there are mass concentrations or mascons under the maria, and other spacecraft showed that the dust on the Moon's surface was relatively thin.

Following the visits of the Apollo astronauts (1969–72), it became apparent that the Moon's crust of feldspar had formed within about 100 million years of the formation of the Moon, which was about 4.6 billion years ago. The regolith, which consists of fine dust to boulders, has been produced by numerous meteorite impacts. It covers the crust and is a few meters thick over the maria but about 10–15 m thick over the highlands. The crust is, on average, about 60 km thick, and below that is

Key Solar System Exploration Spacecraft (Part 1 of 4)

Spacecraft	Country	Launch	Target	Mission	Comments
Luna 1	Soviet Union	2 Jan 1959	Moon	Impact	First to flyby Moon.
Luna 2	Soviet Union	12 Sep 1959	Moon	Impact	First to impact Moon.
Luna 3	Soviet Union	4 Oct 1959	Moon	Flyby	First to photograph Moon's far side.
Mariner 2	USA	27 Aug 1962	Venus	Flyby	First to fly by Venus. Detected high surface temperature.
Ranger 7	USA	28 Jul 1964	Moon	Impact	First to photograph Moon during impact sequence.
Mariner 4	USA	28 Nov 1964	Mars	Flyby	First to fly by Mars. Imaged craters.
Luna 9	Soviet Union	31 Jan 1966	Moon	Lander	First Moon soft landing. Transmitted images from lunar surface.
Luna 10	Soviet Union	31 Mar 1966	Moon	Orbiter	First to orbit Moon.
Surveyor 1	USA	30 May 1966	Moon	Lander	Designed to find safe landing areas for Apollo manned spacecraft.
Luna Orbiter 1	USA	10 Aug 1966	Moon	Orbiter	Imaged Moon to find landing sites for Apollo astronauts.
Venera 4	Soviet Union	12 Jun 1967	Venus	Lander	First to descend through Venus's atmosphere.
Mariner 5	USA	14 Jun 1967	Venus	Flyby	Detected high surface atmospheric pressure.
Apollo 8	USA	21 Dec 1968	Moon	Manned	First manned spacecraft to orbit Moon.
Apollo 11	USA	16 Jul 1969	Moon	Manned	First manned lunar landing.
Venera 7	Soviet Union	17 Aug 1970	Venus	Lander	First successful planetary lander.
Lunar 16	Soviet Union	12 Sep 1970	Moon	Sample return	First automatic sample return.
Lunar 17	Soviet Union	10 Nov 1970	Moon	Rover	First automatic lunar rover.
Mariner 9	USA	30 May 1971	Mars	Orbiter	First Mars orbiter. Imaged volcanoes and valleys.
Apollo 15	USA	26 Jul 1971	Moon	Manned	First manned lunar rover.

Key Solar System Exploration Spacecraft (Part 2 of 4)

Spacecraft	Country	Launch	Target	Mission	Comments
Pioneer 10	USA	3 Mar 1972	Jupiter	Flyby	First Jupiter flyby.
Apollo 17	USA	7 Dec 1972	Moon	Manned	First geologist on the Moon.
Pioneer 11	USA	6 Apr 1973	Jupiter Saturn	Flyby	First Saturn flyby.
Mariner 10	USA	3 Nov 1973	Venus Mercury	Flyby	First Mercury flyby.
Venera 9	Soviet Union	8 Jun 1975	Venus	Lander and orbiter	First photograph from the surface of a planet.
Viking 1	USA	20 Aug 1975	Mars	Lander and orbiter	First photographs from surface of Mars. Life laboratory—results generally thought to be negative.
Viking 2	USA	9 Sep 1975	Mars	Lander and Orbiter	Spacecraft twin of *Viking 1*.
Voyager 2	USA	20 Aug 1977	Jupiter Saturn Uranus Neptune	Flyby	First flybys of Uranus and Neptune.
Voyager 1	USA	5 Sep 1977	Jupiter Saturn Titan	Flyby	Spacecraft twin of *Voyager 2*.
Pioneer Venus Orbiter	USA	20 May 1978	Venus	Orbiter	First radar maps of Venus.
Pioneer Venus Multiprobe	USA	8 Aug 1978	Venus	Probes	Four atmospheric probes. Found clouds are mainly of sulphuric acid.
International Cometary Explorer (ICE)	USA	12 Aug 1978	Comets	Flyby	Renamed from *ISEE 3 (International Sun-Earth Explorer)*. Flew by comets Giacobini-Zinner and Halley.

(Continued)

Key Solar System Exploration Spacecraft (Part 3 of 4)

Spacecraft	Country	Launch	Target	Mission	Comments
Venera 15	Soviet Union	2 Jun 1983	Venus	Orbiter	Radar mapper.
Vega 1	Soviet Union	15 Dec 1984	Venus Venus Halley	Lander Balloon Flyby	An identical *Vega 2* launched 6 days later had same mission.
Sakigake	Japan	8 Jan 1985	Halley	Flyby	
Giotto	ESA (European Space Agency)	2 Jul 1985	Halley	Flyby	Closest spacecraft approach to Comet Halley's nucleus.
Suisei	Japan	18 Aug 1985	Halley	Flyby	
Magellan	USA	5 May 1989	Venus	Orbiter	Radar mapper.
Galileo	USA	18 Oct 1989	Jupiter	Orbiter and probe	Flew by asteroids Gaspra and Ida. First spacecraft to orbit Jupiter and release probe to descend through Jupiter's outer atmosphere.
Clementine	USA	25 Jan 1994	Moon	Orbiter	Produced first compositional and topographical map.
Near-Earth Asteroid Rendezvous (NEAR)	USA	17 Feb 1996	Eros	Flyby and lander	Flew by asteroid Mathilde and landed on asteroid Eros.
Mars Global Surveyor	USA	7 Nov 1996	Mars	Orbiter	Surface Geology Mission.
Mars Pathfinder	USA	2 Dec 1996	Mars	Lander	Deployed rover Sojourner.
Cassini	USA	15 Oct 1997	Saturn	Orbiter	Launched with Huygens.
Huygens	ESA	15 Oct 1997	Titan	Lander	Launched with Cassini.
Lunar Prospector	USA	6 Jan 1998	Moon	Orbiter	Impacted polar region of Moon at end of life to try to detect water ice. None found.
Deep Space 1	USA	24 Oct 1998	Braille	Flyby	Flew by asteroid Braille and comet Borrelly.

Key Solar System Exploration Spacecraft (Part 4 of 4)

Spacecraft	Country	Launch	Target	Mission	Comments
Stardust	USA	6 Feb 1999	Wild 2	Sample return	Flew by comet Wild 2 and returned sample to Earth.
Mars Odyssey	USA	7 Apr 2001	Mars	Orbiter	Mineralogical and Geological Survey. Detected evidence for water in topsoil.
Hayabusa	Japan	9 May 2003	Itokawa	Sample return	Flew by asteroid Itokawa, sample on its way back to Earth (2007).
Mars Express	ESA	2 Jun 2003	Mars	Orbiter	Detected atmospheric methane.
Spirit	USA	10 Jun 2003	Mars	Rover	Detected evidence of earlier presence of water on surface.
Opportunity	USA	7 Jul 2003	Mars	Rover	Detected evidence of earlier presence of water on surface.
SMART 1	ESA	27 Sep 2003	Moon	Orbiter	Impacted Moon at end of life.
Deep Impact	USA	12 Jan 2005	Tempel 1	Comet flyby and impactor	
Mars Reconnaissance Orbiter	USA	12 Aug 2005	Mars	Orbiter	
Venus Express	ESA	9 Nov 2005	Venus	Orbiter	Global atmospheric measurements.

Note: Only missions that successfully reached their target by mid-2007 are included. Excludes those spacecraft in Earth orbit and those exploring the Sun.

The Planets

	Mean Distance from Sun (AU)	Orbital Period (Years/ Days)	Orbital Eccentricity	Inclination of Orbit to Ecliptic	Equatorial Diameter (km)	Mass (Earth = 1)	Mean Density (g/cm^3)	Spin Period	Inclination of Equator to Orbit	No. of Satellites Known (mid 2007)
Mercury	0.39	0.24/88	0.21	7.0°	4,880	0.06	5.44	58.7 days	0°	0
Venus	0.72	0.62/225	0.01	3.4°	12,100	0.81	5.25	243.0 days	177°	0
Earth	1.00	1.00/365	0.02	0°	12,760	1.00	5.52	23.9 hours	23°	1
Mars	1.52	1.88	0.09	1.9°	6,790	0.11	3.94	24.6 hours	25°	2
Jupiter	5.2	11.86	0.05	1.3°	143,000	318	1.33	9.93 hours	3°	63
Saturn	9.5	29.5	0.05	2.5°	120,500	95	0.69	10.66 hours	27°	60
Uranus	19.2	84.0	0.05	0.8°	51,100	14.5	1.30	17.24 hours	98°	27
Neptune	30.1	164.8	0.01	1.8°	49,530	17.1	1.76	16.11 hours	29°	13

the mantle which, about four billion years ago, was the source of the maria basalts when it was still fluid. The measured heat flow to the surface indicated that the temperature of the Moon's core is at least 1,000°C. There was no evidence of any life-forms, water, or organic material in any of the rock and dust samples brought to Earth.

In 1994 the *Clementine* spacecraft showed that the Moon's crust ranged in thickness from about 100 km for a part of the far side to only 10 km for the maria on the near side. This partly explains why the center of mass of the Moon is about 2 km from the geometrical center in the direction of Earth. A few years later some scientists, using data from the *Lunar Prospector* spacecraft, concluded that, as for Mercury, there appeared to be evidence of water ice deposits in permanently shadowed craters at the poles. This conclusion has been hotly contested.

The first great surprise in spacecraft observations of Mars was the discovery, in 1965, of numerous craters by *Mariner 4*. The spacecraft also measured a ground-level atmospheric pressure of only 4–7 millibars, implying that at least 50 percent of the atmosphere was carbon dioxide, rather than the few percent previously thought. This thin atmosphere would not shield the surface from solar ultraviolet radiation, which would have killed any life long ago. There was also no evidence that water had ever existed on Mars. These results were a big disappointment to those who had expected to see evidence of some form of life, but *Mariner 4* had imaged only 1 percent of the surface. In 1969 *Mariner 6* and *7* found that the atmosphere was almost 100 percent carbon dioxide, with only minute amounts of water vapor. In addition, the surface of the polar ice caps appeared to be covered with frozen carbon dioxide, rather than the expected frozen water.

Three years later, *Mariner 9* imaged a completely different Mars from that seen by the previous three Mariners in their brief flybys, with massive volcanoes and a giant canyon system, called Valles Marineris. Evidently Mars varied dramatically from one region to another. *Mariner 9* also detected water vapor over the south polar cap, where it was summer. So maybe there was a substantial amount of frozen water there.

During the late twentieth and early twenty-first century, numerous spacecraft have been sent to Mars, many with landers designed to look for signs of water or elementary forms of life that may exist under rocks, or just under the surface, protected from ultraviolet radiation. Collectively they found that there is water present on Mars in the form of ice, particularly at both poles, and hydrated minerals just beneath the surface. But no evidence has been found of liquid water on the surface, although there is considerable evidence of liquid water flows in the past. So far no evidence of life, even in fossilized form, has been found on the surface.

In the early 1970s, *Pioneers 10* and *11*, the first spacecraft to fly by Jupiter, characterized Jupiter's extensive magnetosphere and found that its bow shock moved significantly on a daily basis in response to varying pressure of the solar wind. Jupiter's intense magnetic field (see table on page 85) was found to consist of a dipole field near the planet plus a non-dipolar field farther out. Intense radiation belts were also detected, which were most intense inside the 20 R_J (Jupiter radii) boundary of the dipole field. The spacecraft detected helium in Jupiter's atmosphere for the first time, showed that

Magnetic Fields of the Outer Planets Compared to Earth

	Equatorial Radius (km)	Inclination of Equator to Orbit	Angle between Magnetic and Spin Axes	Dipole Offset (km)	Dipole Offset (radii)	Magnetic Field at Equator at Surface (gauss)	Dipole Moment (Earth = 1)	Average Sunward Distance of Magnetopause (planetary radii)
Jupiter	71,490	3.1°	9.6°	7,000	0.10	4.28	19,000	70
Saturn	60,270	26.7°	0.0°	2,400	0.04	0.22	540	22
Uranus	25,550	97.9°	59°	7,700	0.30	0.23	50	20
Neptune	24,765	28.8°	47°	14,000	0.55	0.14	28	27
Earth	6,378	23.4°	10.8°	460	0.07	0.31	1	10.4

The second of the volcanic plumes on Io detected by Voyager 1 *rises up about 200 kilometers above the surface. (Courtesy NASA/National Space Science Data Center)*

the helium-to-hydrogen ratio was similar to that of the Sun, and measured a brightness temperature that strongly indicated that Jupiter had an internal source of heat.

The two Voyager spacecraft, which flew past Jupiter in 1979 and 1980, showed the structure of its cloud movements in great detail. The spin rate of Jupiter's core was found to be 9 h 55 min 30 s by measuring the variability of its radio emissions. The maximum wind velocity, relative to this core, was found to be about 500 km/h around the equator. To most astronomers' surprise, the Voyagers also discovered a ring system around the planet.

It was generally thought, prior to the *Voyager 1* intercept, that the surface of Io, the innermost of the four Galilean moons, would be old and cratered. But three days before *Voyager 1*'s closest approach, Stanton Peale and colleagues suggested that there would be widespread volcanism, because of the effect of Jupiter's gravity on its surface, causing flexing. The volcanoes were found, but much to everyone's surprise, a number of them were found to be active during both Voyager intercepts. The other three Galilean moons were all found to be covered with ice, with varying amounts of cratering. Callisto has the most heavily cratered surface, and so is the oldest, and Europa the least. The surface of Europa was extremely flat and covered with stripes some tens of kilometers wide, which some astronomers thought were cracks in the icy surface that was floating on a water substrate.

The *Galileo* spacecraft, which went into orbit around Jupiter in 1995, included a probe that was released and descended into Jupiter's atmosphere. The wind velocity was seen to increase as the probe descended, consistent with the idea that Jupiter had an internal source of heat. The mother spacecraft found that Ganymede has a magnetic field, that Callisto appears to be covered by dark material several meters thick, and that Europa, Ganymede, and Callisto all appear to have water oceans under their icy surfaces.

Pioneer 11 discovered that Saturn has a magnetic field and bow shock, which it crossed some 24 R_S from Saturn. Radiation belts were found to be prevented from

A Voyager image of Saturn's spokes, taken from a distance of 4 million kilometers. They changed their appearance over periods of less than one hour. (Courtesy NASA/Jet Propulsion Laboratory)

forming in the inner magnetosphere by Saturn's rings. *Voyager 1* found that Saturn's radio emission varied with a period of 10 h 39.4 min, which was taken to be the rotation period of Saturn's core. Relative to this, wind speeds reached up to 1,700 km/h at the equator. *Voyager 1* showed that, like Jupiter, Saturn emits about 1.8 times as much energy as it receives from the Sun.

In 1980–81, the two Voyager spacecraft detected a number of new rings around Saturn. In addition, Richard Terrile discovered radial "spokes" on Saturn's B ring, which changed their appearances during a timescale of the order of 20 minutes or so, and which seemed to be due to some sort of electromagnetic effect. Many new satellites were discovered, two of which, Prometheus and Pandora, were on either side of the narrow F ring, which had been first detected by *Pioneer 11*. These two satellites seemed to be shepherding or stabilizing the ring. A little later a new satellite called Atlas was found apparently restricting the outward expansion of the A ring. Then in 1990 Mark Showalter discovered a 20 km satellite called Pan, in the Voyager images, responsible for producing the Encke gap in the A ring.

Titan is Saturn's largest satellite. In 1979 John Caldwell and colleagues measured Titan's surface temperature as about 87 K using the Very Large Array radio telescope. From this Donald Hunten concluded theoretically that the surface atmospheric pressure would be about 2.0 bar. In the following year, *Voyager 1*'s radio occultation experiment produced a value of 1.5 bar at 94 K. The atmosphere appeared to be about 90 percent nitrogen, with methane as a minor but important constituent. It was thought that methane would exist like water on Earth in solid, liquid, and gaseous form.

The *Cassini-Huygens* spacecraft that arrived at Saturn in 2004 appeared to have found large lakes of liquid methane on Titan. It also discovered plumes of water and water ice being ejected from cracks in the icy surface of Enceladus into Saturn's E ring. In 2007 Don Gurnett and colleagues concluded that the mass of these water particles was slowing the rotation rate of Saturn's plasma disk, which, rather than Saturn's core, is the source of Saturn's variable radio emission.

Uranus was found to have a set of narrow rings during stellar occultation observations from Earth in 1977. These were the first planetary rings to be discovered since those of Saturn in the seventeenth century. They were followed by the detection of rings around Jupiter in 1979 and rings or ring arcs around Neptune in the following decade. So all four of the major planets are known to have complete or partial rings.

Voyager 2, the only spacecraft to visit Uranus and Neptune as of 2007, detected Uranus's magnetic field during its flyby in 1986. Uranus, which essentially spins on its side, has a magnetic axis tilted at 59° to its spin axis and a magnetic center offset 0.3 R_U from its geometric center. Timing of the radio emissions produced a core rotation rate of 17 h 14 min, much longer than generally expected. *Voyager 2* confirmed pre-Voyager observations that Uranus's internal heat source, if it existed, was very weak.

Unlike the case of Uranus, pre-Voyager observations of Neptune had indicated that it had a more normal spin axis orientation of 29°, and that it emitted more heat than it received from the Sun. But *Voyager 2* showed that Neptune's magnetic axis was, like Uranus's, inclined at a large angle to its spin axis, with a magnetic center a large distance from its geometric center. Neptune was found to have much greater wind speeds than Uranus, however, presumably driven by its internal heat source, but Neptune's magnetic field, whose magnitude should be linked to internal planetary motion, was less intense than on Uranus.

Prior to *Voyager 2*, Neptune's large moon Triton was known to orbit Neptune in a retrograde sense, which indicated that it had been captured rather than having been formed around Neptune. *Voyager 2* found that Triton's size and density was almost identical to those of Pluto, which added to this captured idea. It also found that Triton's surface was highly reflective, so its temperature was only 38 K, and most of its atmosphere was frozen out onto the surface. Nevertheless, Voyager found that there were what appeared to be geyser-like eruptions in progress on its surface. Triton's sparse atmosphere was found to be mostly nitrogen, with nitrogen ice on the surface being contaminated by trace amounts of methane and other compounds.

No spacecraft has reached Pluto as of 2007, so information about the planet has been largely produced by Earth-based observatories. In 1976 Dale Cruikshank and colleagues detected methane ice on Pluto's surface, indicating that Pluto has a much higher reflectivity and so was somewhat smaller than previously thought. Then two years later James Christy found evidence that Pluto has a moon, now called Charon. Later work showed that the center-to-center distance of Charon from Pluto was only 20,000 km, that Charon's diameter was about half that of Pluto, and that Pluto and Charon's axial spin rates are the same as Charon's orbital period. So Pluto and Charon are really a binary planet, rather than a planet and satellite. In 2006 the International Astronomical Union downgraded Pluto from a planet to a new category of dwarf planet, following the discovery of other bodies in the outer solar system of similar size.

Kenneth Edgeworth and Gerard Kuiper had independently suggested in the mid-twentieth century that there should be a large number of planetesimals or asteroids outside the orbit of Neptune, left over from the formation of the solar system. The technology of the time was not sufficient to find such small bodies, but in 1992 David Jewitt and Jane Luu found the first such member of this so-called Kuiper-Edgeworth (K-E) belt, when they imaged a 240 km diameter object, which orbited the Sun beyond Neptune's orbit. So far more than 1,000 of such K-E belt objects have been found, the largest of which, called Eris, has a diameter about the same as Pluto. It is now, like Pluto, listed as a dwarf planet.

A number of spacecraft have imaged a few of the near-Earth and main belt asteroids. The *Galileo* spacecraft imaged Gaspra in 1991 and Ida in 1993 on its way to Jupiter. The 18 × 11 × 9 km Gaspra was found to be irregular in shape, with crater counts indicating an age of a few hundred million years. Ida, on the other hand, was 56 km long and had a very small companion, Dactyl. Since then Earth-based observations have shown that more than 100 asteroids have companions.

In 1997, the *NEAR-Shoemaker* (*Near Earth Asteroid Rendezvous*) spacecraft imaged the 59 × 47 km Mathilde, which was seen to have at least four 20–30 km diameter craters. Mathilde's density of 1.3 g/cm^3 indicated that it was highly porous, which is probably why it was able to survive the large impacts that created the craters. Then in 2000 *NEAR-Shoemaker* orbited Eros and, one year later, landed on the asteroid. Finally, in 2005 Hayabusa collected samples from the asteroid Itokawa, which, in 2007, were on their way back to Earth for analysis.

The first coordinated attempt at cometary exploration was provided by the small fleet of spacecraft that intercepted Halley's Comet in 1986. They found that Halley had an extremely dark, irregular 16 × 8 × 8 km nucleus partly covered with what appeared to be impact craters. The Vega and *Giotto* spacecraft found that the inner coma consisted mainly of water, and that water vapor and dust production rates, although of the order of 10 tn/s, were low enough to allow Halley about 1,000 orbits of the Sun.

In 2004 the *Stardust* spacecraft collected dust samples from Comet Wild 2 and returned them to Earth. Then in 2005 the *Deep Impact* spacecraft released a probe that deliberately impacted the surface of Comet Tempel I. The resulting debris cloud was analyzed by the flyby spacecraft, and Earth-orbiting and ground-based telescopes. It was found to contain, inter alia, water, water ice, carbon, and hydrocarbons.

The first detection of planets, or exoplanets, around other stars was made by Alex Wolszczan in 1991 using the Arecibo radio telescope. Surprisingly, the parent star was a millisecond pulsar, as it was thought that the formation of pulsars would have ejected any planets. Four years later the first discovery of a planet around a "normal" star was made by Michel Mayor and Didier Queloz. Since then about 200 exoplanets have been found.

Clearly it is easier to detect the largest exoplanets, and many of the early discoveries were of planets even larger than Jupiter. Now planets about the size of Neptune can be detected. *COROT*, the first spacecraft dedicated to finding exoplanets was launched in December 2006, and its better than expected performance has given hope that it will be able to find exoplanets not much larger than Earth.

David Leverington

See also: Space Science Organizations, Space Science Policy, Spacecraft

Bibliography

David Leverington, *Babylon to Voyager and Beyond: A History of Planetary Astronomy* (2007).
Lucy-Ann McFadden, et al., eds., *Encyclopedia of the Solar System*, 2nd ed. (2007).

Asteroids

Asteroids are small rocky bodies, smaller than dwarf planets, which orbit the Sun. Some authorities include Kuiper Belt Objects (KBO), which orbit the Sun generally beyond the neighborhood of Neptune, in their definition of asteroids, but others exclude them. KBOs are covered in the "Pluto and Kuiper Belts" article. Many asteroids reside in the main belt between Mars and Jupiter, and some, known as near-Earth asteroids, approach or cross Earth's orbit.

Spacecraft Close-up Asteroid Observations

Name	Country	Asteroid	Encounter Date	Closest Distance (km)	Comments
Galileo	United States (US)	951 Gaspra	29 Oct 1991	1,600	19 × 12 × 11 km. S-type. Metal-rich, magnetized.
Galileo	US	243 Ida	28 Aug 1993	2,390	52 × 24 × 15 km. S-type. Primarily olivine. Density 2–3. 1 g/cm^3. First moon of asteroid found, Dactyl.
Galileo	US	Dactyl	28 Aug 1993	2,390	1.2 × 1.4 × 1.6 km. 27 hr. orbit. Olivine, orthopyroxene, clinopyroxene.
Near Earth Asteroid Rendezvous (NEAR) Shoemaker	US	353 Mathilde	27 Jun 1997	1,212	66 × 48 × 46 km. C-type. Density 1.3 g/cm^3. Albedo 0.03.
Deep Space 1	US	9969 Braille	28 Jul 1999	26	2.2 × 1 km. V-type? Basalt?
Cassini	US	2685 Masursky	23 Jan 2000	1,600,000	15 × 20 km. S-type.
NEAR Shoemaker	US	433 Eros	14 Feb 2000 orbit; 12 Feb 2001 landing.	0	34 × 11 × 11 km. S-type. Density 2.7 g/cm^3. Pyroxene, olivine. Solid body.
Hayabusa	Japan	25143 Itokawa	25 Nov 2005 landing.	0	535 × 294 × 209 m. S-type. Olivine. Sample return mission. Uncertain if sample gathered. Planned return to Earth in 2010.
Rosetta	Europe	2867 Steins	5 Sep 2008	800	2 × 5 km. E-type. Albedo 0.1.

Giuseppe Piazzi, a member of the Celestial Police (a group organized by Franz Xaver von Zach to search for a "missing planet" believed to be located between Mars and Jupiter), accidently discovered the first asteroid, 1 Ceres, in 1801. When Heinrich Olbers discovered 2 Pallas in 1802, astronomers realized these objects were much smaller than planets. Later that year, William Herschel coined the term asteroid, Latin for starlike, to refer to these objects. Prior to the application of photography to asteroid detection in 1891 by German astronomer Max Wolf, astronomers had discovered some 300 asteroids. By 1866 U.S. astronomer Daniel Kirkwood noticed gaps in the distribution of their orbital periods, soon correlated to whole number ratios of their orbital periods to that of Jupiter, whose orbital resonance moved them out of these orbits or prevented their initial formation in these orbits. The use of photography sharply increased the number of asteroids discovered each year, many by amateurs. In 1918 Kiyotsugu Hirayama identified families of asteroids with similar orbits, postulating that these groups resulted from the fracturing of parent asteroids.

By 1940 nearly 2,000 asteroids were known, some of which passed close to Earth. These posed a risk of collision with Earth, as highlighted by Luis Alvarez and Walter Alvarez's 1980 theory that an asteroid impact likely precipitated the extinction of the dinosaurs. These ideas along with the collision of comet Levy-Shoemaker with Jupiter in 1994 contributed to the development of asteroid and comet detection programs. Dedicated instruments for detection programs began in 1980 with the University of Arizona Spacewatch. Asteroid detection and tracking received a major boost when U.S. Air Force (USAF) space surveillance equipment was modified for the purpose in the mid-to-late 1990s. By early 2009, USAF equipment had identified more than 123,000 asteroids and Spacewatch more than 21,000; all combined search programs had identified more than 440,000 asteroids.

As possible primordial remnants from the early solar system, the physical characteristics of asteroids were of great scientific interest to assess theories of solar system origins. Scientists determined asteroid physical characteristics primarily through ground-based observations, supplemented by spacecraft observations beginning in 1991 when the *Galileo* probe took close-up photographs of 951 Gaspra and in 1993 observed Ida and its moon, Dactyl, the first observed satellite of an asteroid. *NEAR Shoemaker*, the first dedicated asteroid probe, orbited 433 Eros in 2000 and landed on its surface the next year. In 2005 the Japanese *Hayabusa* spacecraft landed on 25143 Itokawa to acquire and return a sample to Earth, scheduled for 2010. These probes found that asteroids are heavily cratered, with many being "rubble piles" of loose rock. Some, like Eros, are solid bodies. The discovery of small bodies in the outer solar system, starting with Chiron in 1977, started a debate about whether some of these distant small bodies were comets, asteroids, or something else altogether. In September 2007 NASA launched the *Dawn* spacecraft to rendezvous with asteroids 4 Vesta (August 2011) and subsequently with 1 Ceres (February 2015). In early 2009 the European, Japanese, and U.S. space agencies were planning further asteroid missions.

Stephen B. Johnson and Amy Sisson

See also: Space Surveillance

Bibliography

Ronald E. Doel, *Solar System Astronomy in America: Communities, Patronage, and Interdisciplinary Research, 1920–1960* (1996).
David Leverington, *From Babylon to Voyager: A History of Planetary Astronomy* (2003).
Minor Planet Center. http://www.minorplanetcenter.org/iau/mpc.html.
Curtis Peebles, *Asteroids: A History* (2000).

Hayabusa. *Hayabusa* was a Japan Aerospace Exploration Agency (JAXA) mission to explore and return samples from an asteroid and to demonstrate several new technologies, including a xenon ion engine and an autonomous navigation system to approach the target without human interaction. NASA was originally to supply a rover, but its cancellation in November 2000 forced JAXA's Institute of Space and Astronautical Science (ISAS) to rapidly develop its own lander, called *MINERVA* (*Micro/Nano Experimental Robot Vehicle for Asteroid*). It was designed to "hop" on the surface of near-Earth asteroid 25143 Itokawa to collect soil samples. The spacecraft carried a multispectral telescopic imager, a laser altimeter, a near-infrared spectrometer, and an X-ray fluorescence spectrometer. ISAS built and operated the spacecraft.

The spacecraft, originally called *MUSES C* (*Mu Space Engineering Spacecraft*), was renamed *Hayabusa* (falcon in Japanese) after its launch on 9 May 2003 on an M-V-5 launcher from Kagoshima Space Center. While en route, a large solar flare damaged the solar panels, reducing its power output and hence reducing the power of *Hayabusa*'s ion engines. This delayed the spacecraft's arrival at Itokawa to September 2005, three months later than initially planned, and shortening its stay at the asteroid since the return to Earth had to begin in November. After initially surveying the surface from a distance of 20 km, on 12 November, *MINERVA* was released when *Hayabusa* was at too high an altitude and escaped into space. The first attempt to land on 19 November failed to collect samples and controllers commanded an emergency ascent. A second attempt occurred six days later, though it was uncertain if the spacecraft successfully acquired samples, since the pellet firing needed to collect samples probably did not occur. On 9 December communications was lost when a propellant leak triggered a safe hold mode. Communications was regained in February 2006, and in April 2007 flight controllers used its ion engines to start the return journey to Earth. *Hayabusa* was expected to return to Earth three years later than initially planned in 2010, where ISAS would analyze any samples.

Hayabusa's observations show that Itokawa is roughly 535 × 294 × 209 m in size, with relatively large albedo and color variations. It features regions of rough terrain with large boulders and smooth gravelly terrain. The asteroid appears to be a chemically homogeneous, olivine-rich, Class S "rubble pile." If *Hayabusa* successfully returns samples, scientists expect them to shed more light on the early origins of the inner solar system.

Stephen B. Johnson

See also: Japan

Bibliography

J. Kelly Beatty, "The Falcon's Wild Flight," *Sky and Telescope* (September 2006): 34–38.
Science, June 2006.
JAXA *Hayabusa*. http://www.jaxa.jp/projects/sat/muses_c/index_e.html.

***MUSES C*, Mu Space Engineering Spacecraft.** *See Hayabusa.*

***NEAR Shoemaker*, Near Earth Asteroid Rendezvous.** The *NEAR Shoemaker, Near Earth Asteroid Rendezvous* spacecraft was the first space probe to achieve orbit around an asteroid (433 Eros) and land on the surface. Designed and built by the Johns Hopkins University Applied Physics Laboratory, *NEAR Shoemaker* was the first launch of NASA's Discovery program, which aimed to fund innovative, low-cost missions. The project's total cost was $220.5 million, including $43 million for the launcher and $60.8 million for operations.

The 805 kg spacecraft was launched on a Delta 2 launcher from Cape Canaveral on 17 February 1996. *NEAR* flew within 1,212 km of the medium-sized (66 × 48 × 46 km) carbonaceous (C-type) asteroid 253 Mathilde on 27 June 1997, measuring a density of 1.34 g/cm^3, and detecting five craters more than 20 km in diameter. *NEAR Shoemaker*'s initial rendezvous maneuver with Eros was aborted on 20 December 1998, resulting in a flyby of the silicaceous (S-type) asteroid, which allowed a density estimate of 2.7 g/cm^3. After one orbit around the Sun, the spacecraft rendezvoused with Eros again and achieved orbit around Eros on 14 February 2000.

NEAR Shoemaker's multispectral imager revealed a cratered surface strewn with boulders and covered with fine regolith, perhaps tens of meters deep, and a several-kilometer-wide valley between the two peaks that formed the ends of the 34 × 11 × 11 km asteroid. Larger craters, 500–1,000 m in diameter, were found in expected abundance, but smaller craters, such as those less than 30 m in diameter, were less common than expected in comparison to the Moon. Perhaps a large impact spread ejecta across the surface or caused seismic shaking. As controllers continually refined the spacecraft's orbit, they constructed maps of the asteroid's complex gravity field so they could maintain increasingly complicated orbital trajectories at lower altitudes. Scientists used onboard X-ray/gamma-ray and near infrared spectrometers to detect the composition of surface materials. The data indicated that Eros is more mechanically coherent compared to other S-type asteroids, such as Gaspra and Ida, which the *Galileo* spacecraft found to be rubble piles loosely bound by gravity. The spacecraft also carried a laser rangefinder to build high-resolution topographic profiles of the asteroid and a magnetometer that detected no magnetic field.

Controllers commanded *NEAR Shoemaker* to descend to the surface on 12 February 2001 for the first landing on an asteroid, an accomplishment for which it was not designed. Although no images were transmitted after landing, controllers received signals from the spacecraft indicating that it survived the low-speed impact.

Eric Reynolds

See also: Johns Hopkins University Applied Physics Laboratory

Bibliography

Johns Hopkins University Applied Physics Laboratory NEAR. http://near.jhuapl.edu/.

Joseph Veverka et al., "NEAR at Eros: Imaging and Spectral Results," *Science* (22 September 2000): 2088–97.

Comets

Comets are icy bodies in the solar system that become visible as they slowly vaporize during relatively close approaches to the Sun. Their appearance was often seen as a harbinger of impending doom.

Greek philosopher Aristotle formularized in his *Meteorologica* (340 BCE) that comets occurred when the Sun or planets warmed and ignited parts of Earth's upper atmosphere. Danish astronomer Tycho Brahe accurately observed the comet of 1577, proving that it was located beyond the Moon and hence that Aristotle was incorrect. In 1705 English astronomer Edmund Halley used Newton's theory of gravitation to predict the reappearance of the comet that later bore his name.

The first photographs of a comet were taken in 1858, and spectroscopic observations started in 1864. Photography offered detailed imagery of a comet's structure, and spectroscopy provided insight into its chemical composition. By 1867 astronomers clarified the relationship between comets and meteor showers, showing that meteor showers were the remnants of comet tails. In the early 1950s, U.S. astronomer Fred Whipple developed a model in which comet nuclei formed into "dirty iceballs" from ice and dust as part of the solar system, but far from the Sun. German astronomer Ludwig Biermann hypothesized in 1951 how what was later called the solar wind interacted with comets to create ion tails in addition to dust tails. In 1950 Dutch astronomer Jan Oort proposed that the source of long-period comets—later called the Oort cloud—was a region of space about one light-year from the Sun. U.S. astronomer Gerald Kuiper proposed in 1951 a ring-shaped region past the orbit of Neptune as the source for short-period comets, later termed the Kuiper (or Edgeworth-Kuiper) Belt.

Since 1985 robotic spacecraft have been used to more closely study comets. The *International Cometary Explorer* (*ICE*) spacecraft (formerly *ISEE 3–International Sun-Earth Explorer*) was the first to fly by a comet, Giacobini-Zinner in September 1985, finding water ions, but no bow shock. *ICE* then contributed to the international flotilla of spacecraft to study Halley's Comet in March 1986. This also included Japan's *Suisei* and *Sakigake*, the Soviet Union's *Vega 1* and *2*, and the European Space Agency's (ESA) *Giotto*, which made the closest approach at 600 km from the comet's nucleus. Collectively these spacecraft measured the chemical composition of the coma (mainly water, but other simple organic compounds), a dark albedo of 4 percent, the extent of the bow shock and ionopause, and Halley's size. *Giotto* also flew by Comet Grigg-Skjellerup in July 1992, finding a bow shock much closer to the comet's nucleus, and much less dust, but unusual magnetic waves near the nucleus. NASA's *Deep Space 1* (*DS1*), *Stardust*, and *Deep Impact* probes were the next comet visitors, with *DS1* flying by Comet Borrelly in September 2001, *Stardust* collecting samples from the coma of Wild 2 in January 2004, and *Deep Impact* deliberately crashing an impactor into comet Tempel 1 in July 2005. *Stardust* surprisingly found olivine in Wild 2's dust samples, indicating that its dust formed much closer to the Sun than expected.*Deep Impact* revealed a relatively weak comet surface and new compounds, including sulfides

Space Missions with Significant Comet Observations (Part 1 of 2)

Name	Country	Comet	Encounter Date	Closest Distance (km)	Comments
ISEE 3/ICE	United States (US)	Giacobini-Zinner	11 Sep 1985	7,800	Detected water ions. No bow shock measured.
ISEE 3/ICE	US	Halley	25 Mar 1986	28,100,000	
Suisei	Japan	Halley	8 Mar 1986	151,000	Detected 10 million km radius neutral hydrogen sphere. Solar wind speed decreased from 400 km/s to 200 km/s upon crossing bow shock, 54 km/s at closest approach.
Sakigake	Japan	Halley	11 Mar 1986	6,990,000	Plasma wave intensity increased up to closest approach.
Vega 1	Soviet Union	Halley	6 Mar 1986	8,890	Measured dust production 10 tons/s and water vapor production 40 tons/s. Bow shock at 1.1 million km. Surface temperature 330 K.
Vega 2	Soviet Union	Halley	9 Mar 1986	8,030	Measured dust production 5 tn/s. Different nucleus image than *Vega 1*, indicating rotation.
Giotto	Europe	Halley	14 Mar 1986	596	Measured 12,000 dust impacts, dust production 3 tons/s, water vapor production 15 tons/s. Ionopause at 4,700 km, neutral molecules and cold ions inside. Size $16 \times 8 \times 8$ km. Bow shock at 1.3 million km.
Giotto	Europe	Grigg-Skjellerup	10 Jul 1992	200	Bow shock at 17,000 km. Measured 3 dust impacts, magnetic waves at 25,000 km. Camera inoperative.

Space Missions with Significant Comet Observations (Part 2 of 2)

Name	Country	Comet	Encounter Date	Closest Distance (km)	Comments
Deep Space 1	US	Borrelly	22 Sep 2001	2,200	Albedo .04. 60 km long dust jets. Size 8 × 4 × 4 km. Gas clouds offset of nucleus by 2,000 km.
Contour	US	Encke and Schwassman-Wachmann 3			Launched 3 Jul 2002, contact lost 15 Aug 2002.
Stardust	US	Wild 2	2 Jan 2004	240	Collected samples from coma, returned to Earth January 2006. Found olivine and polycyclic aromatic hydrocarbons.
Deep Impact	US	Tempel 1	4 Jul 2005	500 for flyby; 0 for impactor	Impactor plus flyby. Surface structure weaker than expected. Numerous compounds detected from impact, including sulfides, silicates, carbonates.
Deep Impact	US	Hartley 2	Planned Oct 2010		DIXI—Deep Impact Extended Investigation mission.
Stardust	US	Tempel 1	Planned Feb 2011		Stardust NeXT—New Exploration of Tempel 1, after *Deep Impact* excavation.
Rosetta	Europe	67P / Churyumov-Gerasimenko	Planned May 2014		First attempted soft landing on comet nucleus.

and silicates. As of 2008 *Stardust* was being retargeted to Tempel 1 to explore the *Deep Impact* excavation, *Deep Impact* was aimed at Comet Hartley 2, and ESA's *Rosetta* was headed for comet 67P/Churyumov-Gerasimenko to perform a soft landing on its nucleus. In general, robotic probes have confirmed Whipple's hypothesis of a "dirty iceball," although they might be better considered as "icy dirtballs."

The impact of comet Shoemaker-Levy 9 on Jupiter in July 1994 emphasized the danger that comets pose to planets. Scientists used evidence of past cometary impacts, and the apparent discovery of polycyclic aromatic hydrocarbons in *Stardust*'s samples, to argue that comets were a mechanism of panspermia, an idea developed in the nineteenth century that the origin of life on Earth came from space.

Tim Chamberlin and Stephen B. Johnson

See also: European Space Agency, National Aeronautics and Space Administration

Bibliography

David Leverington, *Babylon to Voyager and Beyond: A History of Planetary Astronomy* (2003).

David H. Levy, *The Quest for Comets: An Explosive Trail of Beauty and Danger* (1994).

Donald K. Yeomans, *Comets. A Chronological History of Observation, Science, Myth, and Folklore* (1991).

Deep Impact. *Deep Impact*'s main mission was the better understanding of active cometary nuclei by sending an impactor to collide with Comet Tempel 1 and observing the resulting crater and its ejecta plume with a flyby spacecraft and other Earth and space-based observatories. Consisting of the impactor and flyby spacecraft, *Deep Impact* was built by Ball Aerospace Technologies, managed by Jet Propulsion Laboratory, and launched 12 January 2005 on a Delta II rocket.

The flyby craft's design incorporated both a medium- and high-resolution imager and an infrared spectrometer. A duplicate medium-resolution imager on the impactor allowed both navigation and close-up photos of the nucleus. The impactor was a unique spacecraft with a total mass of 372 kg, which included a 128 kg, almost-pure spherical copper cap that acted as the cratering mass, and 7–8 kg of unused fuel. Mission goals included identification of primordial ices, investigation of cometary evolution, determination of nucleus composition, and study of impact history.

Twenty-four hours before impact, the impactor separated from the flyby spacecraft. The collision with Tempel 1 on 4 July 2005 was analogous to the detonation of about 4.5 tons of TNT and excavated an estimated 100–200 m (330–660 ft) crater, although ejecta prevented direct observation. At the time of impact, the spacecraft was still 10,000 km distant from the comet, but approaching rapidly. Thirteen minutes after the collision, the main craft deployed its shield for 46 minutes to prevent damage to its instruments as it flew by at a closest approach of roughly 500 km.

More than two dozen professional ground-based optical and radio telescopes, in addition to the *Hubble*, *Spitzer*, and *Chandra* space telescopes, studied the 10.3 km/s collision in infrared, visible, ultraviolet, and radio wavelengths. Science results included detection of both gaseous and solid water, sulfides, silicates, amorphous carbon, carbonates, and phyllosilicates. Both before and after impact, the Keck 2 telescope detected methanol, ethane, water, and hydrogen cyanide, while the *Spitzer Space Telescope* captured spectra of amorphous carbon, carbonates, phyllosilicates, sulfides, gaseous and solid water, and both crystalline and amorphous silicates. The ejecta contained those gases plus carbon monoxide, formaldehyde, acetylene, and methane. Differences in pre- and post-impact abundances helped better assess the nucleus's composition. Impact craters and water ice were observed on its surface. After impact a larger than expected cloud of fine dust 1–100 μm in size indicated a weaker and more powdery surface than expected. Steam and carbon dioxide were superheated to more than 1,000 degrees K.

After the Tempel 1 encounter, *Deep Impact* was renamed EPOXI, an acronym of two missions that NASA assigned to the spacecraft: EPOCh—Extrasolar Planet Observation and Characterization, and DIXI—Deep Impact Extended Investigation. *Deep Impact*'s role in EPOCh would be to characterize extrasolar planets by observing stars they were known to transit. In September 2007 *Deep Impact* was retargeted to fly by comet Hartley 2 for the DIXI mission.

Durand Johnson

See also: Ball Aerospace and Technologies Corporation, Jet Propulsion Laboratory

Bibliography

Ron Cowen, "A Grand Slam," *Science News*, 9 July 2005.
JPL *Deep Impact*. http://solarsystem.nasa.gov/deepimpact/index.cfm.
NASA *Deep Impact*. http://www.nasa.gov/mission_pages/deepimpact/main/index.html.

Giotto. *See* Halley's Comet Exploration.

Halley's Comet Exploration. Halley's Comet exploration, a series of six spacecraft flew by Halley's Comet in 1986: two from the Soviet Union (*Vega 1* and *2*), two from Japan (*Suisei* and *Sakigake*), one from the European Space Agency (ESA) (*Giotto*), and one from the United States (*International Cometary Explorer* or *ICE*). The fly-by distances varied from 596 km for *Giotto* to 28.1 million km for *ICE*.

The orbit of Halley's Comet is inclined 18° to the ecliptic, and the most convenient places to intercept it are the two points where that orbit intercepts the ecliptic. That occurred on 27 November 1985 and on 10 March 1986, when the comet was 1.85 Astronomical Units (AU) and 0.89 AU from the Sun, respectively. The comet would be most active when it was closest to the Sun, and so the international agencies decided to aim for an encounter around the March date.

Halley's Comet orbits the Sun with a period of 76 years in a retrograde direction. As a result, a spacecraft launched from Earth, would intercept the comet at a very high velocity; in this case about 260,000 km/h. This would then be the velocity that any cometary dust particles would strike the spacecraft. *Giotto* was the most vulnerable spacecraft as it was to pass closest to Halley's nucleus, so a two-layer bumper shield protected the spacecraft as much as possible from dust impacts.

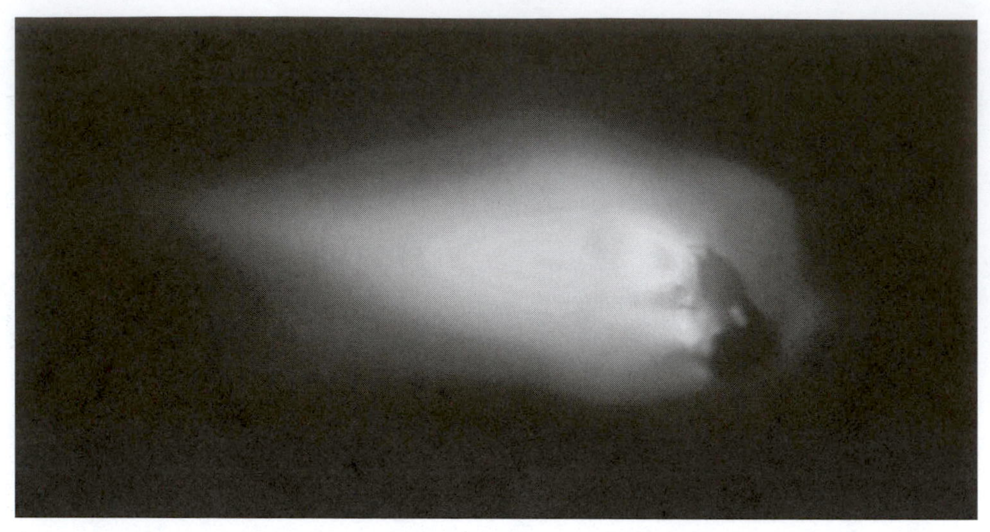

A composite image of the nucleus of Halley's Comet as imaged by Giotto. *(Courtesy European Space Agency/Max Planck Institute for Solar System Research, 1986, 1996)*

The key to the successful Halley intercepts was international collaboration, which was informally organized through the Inter-Agency Consultative Group to solve common problems. For example, it was expected that by the time that *Giotto* was launched, the position of Halley's nucleus in space would only be known to about 1,000 km or so, and yet ESA hoped to send *Giotto* to within about 1,000 km of the nucleus. Fortunately, the two Vega spacecraft were due to fly about 10,000 km from Halley a few days before *Giotto*. So, a rapid feedback of results from the Soviet Union of the nucleus's exact position should allow ESA to fine tune *Giotto*'s orbit, provided the position of the Vega spacecraft themselves was known accurately enough. The international team needed NASA's Deep Space Network to do this.

The Soviet Vega spacecraft had been launched to Halley's Comet via Venus, where they had released planetary probes. They had then swung by Venus en route to Halley. Early on 6 March 1986 Moscow time, *Vega 1* crossed the comet's bow shock about 1.1 million km from the nucleus. Then at a distance of 320,000 km, *Vega 1* detected its first impact by dust particles. It produced a number of images of the nucleus that appeared to have a surface temperature of about 330 K. This is about 100 K greater than the sublimation temperature of water ice in space, so any water ice in the cometary nucleus must sublimate below this dark insulating surface. The shape of the nucleus imaged by *Vega 2* looked quite different from that imaged by *Vega 1*, indicating that it had rotated between the two intercepts.

The Japanese *Suisei* spacecraft passed within about 150,000 km of Halley 53 hours after *Vega 1* and before *Vega 2*. For some months, its ultraviolet telescope had been observing the enormous hydrogen corona that extended some 10 million km from the comet and found that it tended to brighten at a frequency of 52.9 hours. This was similar to the rotation period of the nucleus deduced from ground-based observations.

Suisei also showed that the 400 km/s solar wind of interplanetary space was reduced by about 50 percent on crossing the bow shock, reaching a minimum of 54 km/s at the spacecraft's closest approach.

Sakigake passed some 7 million km from Halley three days after *Suisei*, but even at this enormous distance, it felt the effect of the comet. Not only was *Sakigake* within the hydrogen corona, but the plasma wave intensity increased up to the period of closest approach and then decreased.

On 13 March, *Giotto* crossed Halley's bow shock, 1.3 million km from the nucleus. About one minute before closest approach, *Giotto* crossed the ionopause 4,700 km from the nucleus (the only spacecraft to do so) where the interplanetary field fell to zero. Then 7.6 seconds before closest approach, the spacecraft signal suddenly stopped, because a relatively large dust particle had set the spacecraft nutating. Half an hour later, however, the onboard nutation dampers had stabilized the spacecraft sufficiently for its signals to be picked up again on Earth.

Giotto found that inside the ionopause there was a stream of neutral molecules and cold ions flowing away from the nucleus at about 1 km/s. These neutral molecules were the agent that dragged dust particles away with them to form the coma or head of the comet. The cometary nucleus was seen to be an irregular potato-shaped mass covered with bumps and hollows, one of which clearly looked like a crater. Bright jets could be seen streaming toward the Sun in both the Vega and *Giotto* images. At 16 × 8 × 8 km, the nucleus was somewhat larger than some had expected, but the Vegas and *Giotto* found that it was also darker than predicted with an albedo of only about 4 percent.

The Vegas and *Giotto* found Halley's inner coma to consist mainly of water, with smaller amounts of carbon monoxide, carbon dioxide, methane, ammonia, and polymerized formaldehyde. It was thought that the polymerized formaldehyde might be the reason why the nucleus was so dark. The three spacecraft also found that the dust particles consisted of carbon, hydrogen, oxygen, nitrogen, and simple compounds of these elements, or of mineral-forming elements such as silicon, calcium, iron, and sodium. Water vapor and dust production rates were about 20–40 tons/s and 5–10 tons/s, respectively, but even at these rates there is still enough material in the cometary nucleus to allow about 1,000 more orbits of the Sun.

David Leverington

See also: Deep Space Network, European Space Agency, Japan, Science Policy

Bibliography

Nigel Calder, *Giotto to the Comets* (1992).

David Leverington, *New Cosmic Horizons: Space Astronomy from the V2 to the Hubble Space Telescope* (2001).

Rüdeger Reinhard, "The Halley Encounters," in *The New Solar System*, ed. J. Kelly Beatty and Andrew Chaikin, 3rd ed. (1990).

International Cometary Explorer. *See* Halley's Comet Exploration.

Rosetta. *Rosetta*, the European Space Agency (ESA) comet probe, consisted of an orbiter and a lander. The orbiter was built by an international consortium led by Astrium, Germany, and the lander by another consortium led by the German Aerospace Research Institute (DLR). Arianespace had planned to launch *Rosetta* in January 2003 with an Ariane 5 rocket, but an earlier Ariane failure caused the launch to be postponed until March 2004. As a result, the ESA Science Programme Committee decided to change *Rosetta*'s target comet from 46P/Wirtanen to 67P/Churyumov-Gerasimenko.

To reach its target, *Rosetta* planned to use gravity assists by Earth (in 2005, 2007, and 2009) and Mars (in 2007). It would also pass through the asteroid belt twice, with a consequent risk of damage, before crossing the orbit of Jupiter. In the process, it flew by asteroid 2867 Steins on 5 September 2008.

If the mission goes as planned, *Rosetta* will be the first spacecraft to orbit a comet's nucleus, the first to fly alongside a comet as it heads toward the Sun, and the first to land a soft-lander onto a comet's nucleus. As a result, 1,670 kg of the total spacecraft mass of about 3,000 kg consisted of propellant, while the lander weighed just 100 kg. Rendezvous with the comet is expected in May 2014, and the mission is due to end in December 2015.

David Leverington

See also: European Aeronautic Defence and Space Company (EADS), European Space Agency

Bibliography

Claude Berner et al., "Rosetta: ESA's Comet Chaser," *ESA Bulletin* No. 112 (November 2002): 10–37.

John Ellwood et al., "Rosetta's New Target Awaits," *ESA Bulletin* No. 117 (February 2004): 5–13.

Sakigake. *See* Halley's Comet Exploration.

Stardust. *Stardust* was the fourth mission of NASA's Discovery Program of small, inexpensive missions. The $210 million Stardust mission was the first to capture cometary material and return it to Earth for analysis. The mission's concept, developed by Peter Tsou (who created techniques for relatively benign intact capture of hypervelocity particles using very low density aerogel) of the Jet Propulsion Laboratory (JPL), was subsequently selected in late 1995. Managed by JPL, the 385 kg *Stardust* spacecraft was built by Lockheed Martin Astronautics. It incorporated a Navigation Camera, a real-time Comet and Interstellar Dust Analyzer, and a Dust Flux Monitor.

NASA launched *Stardust* from Cape Canaveral on 7 February 1999 aboard a Delta II rocket. Five years later, it encountered and took high-resolution images of Comet Wild 2 at a distance of 240 km and fulfilled its primary mission of collecting dust samples from the comet's coma using the silica-based substance, aerogel. Two years later, on 15 January 2006, the parachuting reentry capsule returned to Earth with extraterrestrial dust samples. *Stardust* also collected dust grains suspected of being of interstellar origin and set a new distance record for a solar-powered spacecraft by operating at a distance of 2.72 AU from the Sun.

Much to their surprise, scientists found olivine in Wild 2 dust. Since olivine can only be created at high temperatures, this indicates that the dust in Wild 2 formed relatively close to the young Sun. Comet Wild 2 contains minerals more typical of chondritic (stony, not differentiated by a parent body) asteroids than what scientists expected of comets. The comet was also deficient in silicate materials often found in comet interplanetary dust particles, as would be expected from an object that spent most of its life in the Kuiper Belt. Scientists were also surprised by the apparent measurement of polycyclic aromatic hydrocarbons, which play important roles in terrestrial biochemistry.

In July 2007, NASA decided to send the *Stardust* spacecraft to fly by comet Tempel 1 (expected 14 February 2011) in the Stardust-NExT (New Exploration of Tempel 1) mission so as to attempt to image the hemisphere of the comet that had been impacted by *Deep Impact* in 2005.

Stephen L. Rider and Stephen B. Johnson

See also: Jet Propulsion Laboratory, Lockheed-Martin Corporation

Bibliography

Don E. Brownlee et al., "The *Stardust* Mission: Returning Comet Samples to Earth," *Meteoritics and Planetary Science* 32 (1997): A22.

Stardust Mission. http://stardust.jpl.nasa.gov/home/index.html.

Suisei. *See* Halley's Comet Exploration.

Vega. *See* Halley's Comet Exploration.

Earth Magnetosphere and Ionosphere

The Earth magnetosphere and ionosphere shield Earth from charged particles in the solar wind. The first indicator of what would eventually be recognized as Earth's magnetism was when the Chinese discovered, sometime in the first millennium CE, that when a lodestone or a magnetized needle is placed on a float it always pointed north-south. It then gradually became clear over the centuries that a freely pivoted magnetic needle did not point due north, and that the force on the needle was not horizontal, as its north-pointing end always slanted downward in the northern hemisphere. William Gilbert undertook many experiments in magnetism in the late sixteenth century and, in his *De Magnete* published in 1600, concluded that Earth was like a large magnet.

George Graham noticed in 1724 that a magnetic compass needle was sometimes unstable, indicating a geomagnetic disturbance. Then on 5 April 1741, Graham in London and Anders Celsius and Olof Hiorter in Sweden observed that such a geomagnetic disturbance occurred at the same time as a polar aurora. Later that century Henry Cavendish used triangulation to estimate the height of aurorae as between about 85–115 km.

In 1852 Edward Sabine, Rudolf Wolf, and Alfred Gautier independently identified a correlation between sunspots and disturbances in Earth's magnetic field. On 1 September 1859 Richard Carrington observed two white light solar flares moving over the surface of a large sunspot, followed about 18 hours later by the start of a

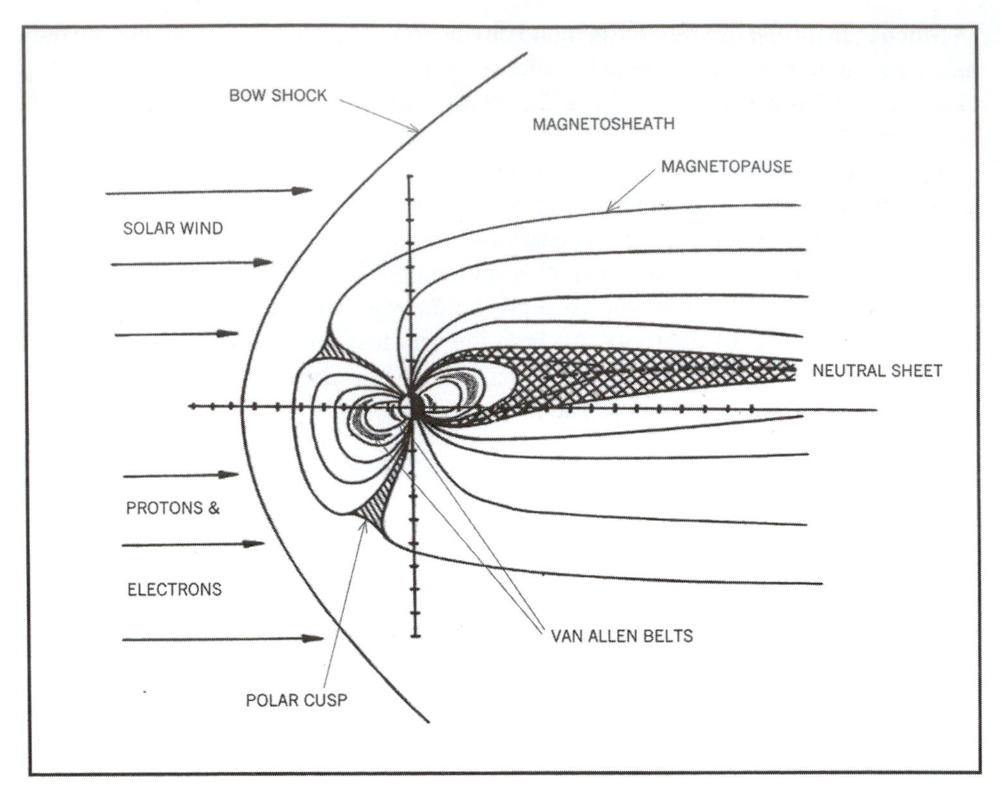

The effect of the solar wind on the near-Earth environment, producing the bow shock, magnetosheath, magnetopause, and neutral sheet in the geomagnetic tail. (Courtesy David Leverington)

spectacular auroral display, which was seen as far south as the West Indies, peaking about 36 hours after the white light flares had been seen. The white light solar flares had lasted just five minutes, but at exactly the same time Kew Observatory had recorded a similar very brief magnetic disturbance, followed later by a prolonged disturbance coincident with the aurorae. The simultaneous occurrence of the flare and the magnetic disturbance on Earth was a big surprise, as it implied that the disturbance had been caused by something that had been emitted by the flare at the speed of light.

Kristian Birkeland researched the auroral phenomena at the turn of the nineteenth century. He suggested that the aurora was caused by electrons from the Sun guided by Earth's magnetic field lines toward Earth's magnetic poles. Shortly afterwards Carl Størmer undertook a theoretical analysis of the motion of charged particles in Earth's magnetic field and showed that such a particle from the Sun can reach Earth only in two narrow zones centered near Earth's magnetic poles. Charge particles would be trapped in these zones, traveling from one polar region to the other, and back again. In the 1930s Sydney Chapman and Vincent Ferraro developed a theory in which the Sun emitted a huge cloud of plasma at the time of a solar flare, which initially compressed Earth's magnetic field.

Ludwig Biermann suggested in 1951 that the Sun continuously emitted charged particles, not just during solar storms. Then in 1957–58 Eugene Parker proposed his solar wind theory, in which charged particles emitted continuously by the Sun drew magnetic field lines in the corona out into the solar system.

In a parallel line of research Guglielmo Marconi found in 1902 that he could transmit radio waves across the Atlantic. This was a surprise as it was thought, at the time, that radio waves were transmitted in straight lines, except for possibly minor deviations caused by differences in air density. But Oliver Heaviside and Arthur Kennelly independently explained the effect, by suggesting that the radio waves had been reflected off an electrically conducting layer, in what was later called the ionosphere, about 100 km above Earth's surface.

In 1924 Edward Appleton discovered that there was not one layer in the ionosphere but two. One, later called the E or Heaviside layer, at an altitude of about 100 km, allowed long distance radio transmission at night, and a second, at about 250 km, later called the F_2 or Appleton layer, allowed long distance transmission during the day.

Further work over the next decade or so by numerous researchers revealed that there are actually three layers in the ionosphere, the D, E, and F layers centered at about 80, 120, and 300 km altitude, all of whose reflectivity varied during 24 hours. During daytime, the Sun apparently created the D and E layers, and changed the configuration of the F layer into an F_1 and F_2 layer. In the 1930s these changes were generally attributed to ultraviolet solar radiation. But Lars Vegard and E. O. Hulburt independently speculated in 1938 that the source could be solar X rays.

Howard Dellinger showed in the 1930s that problems with the reception of shortwave radio waves were often, but not always, associated with solar flares seen on the Sun a short time before. Then in 1939 Thomas Johnson and Serge Korff found that the ionization of Earth's atmosphere was quickly disrupted at the time of solar flares, and suggested that the cause of the disruption was solar X rays.

Investigations of electromagnetic phenomena benefitted greatly from the ability to place instruments above Earth's atmosphere. For example, Herbert Friedman found, using a V-2 experiment in 1949, that the Sun emitted X rays. Three years later he was able to show that the X-ray intensity was sufficient to sustain the E region. At about the same time Richard Tousey and Friedman independently showed that the intensity of the Lyman- ultraviolet solar radiation was sufficient to maintain the D region. In 1956 Friedman's group found that solar flares were the source of very high energy X rays that disrupted the ionosphere.

On 1 May 1958 James Van Allen announced that he had discovered a belt of elementary particles, later called the inner Van Allen belt, around Earth above an altitude of about 1,000 km, from his experiments on the *Explorer 1* and *3* spacecraft. Later that year he discovered a second radiation belt at an altitude of about 15,000 km with an experiment on *Pioneer 3*. Shortly afterwards photographic emulsions carried on a Thor-Able rocket proved, for the first time, that most of the particle radiation in the inner Van Allen belt was due to protons.

Luna 2 in 1959 was the first spacecraft to detect the solar wind, which Parker had predicted, outside the influence of Earth's magnetic field. Subsequent spacecraft showed that there is a region, called the magnetosphere, around Earth that is dominated by Earth's magnetic field. Inside the magnetosphere, the geomagnetic field is relatively well ordered.

HERBERT FRIEDMAN (1916–2000)

Herbert Friedman was a Naval Research Laboratory physicist whose sounding rocket experiments made enormous strides in astronomy and solar physics. Among many achievements, he used a Geiger counter on a V-2 to find the source of the upper atmosphere's ionization in 1949, and in 1963 confirmed the existence of both the diffuse X-ray background and discrete sources of X rays in space. He also used a sounding rocket to observe an occultation of the Crab Nebula by the Moon in 1964, proving that the nebula is a source of X rays.

Katie Berryhill

The magnetosphere is bordered by the magnetopause, and in the magnetosheath, between the magnetopause and the bow shock, the magnetic field is variable.

Explorer 10 provided the first indication of the tail-like structure of the magnetopause around local midnight in March 1961. Later that year, *Explorer 12* was the first spacecraft to clearly detect the magnetopause near the noon meridian. Then in 1963 John W. Freeman, Van Allen, and Laurence Cahill announced the discovery of the magnetosheath using *Explorer 12*. The following year Freeman found that the distance of the magnetopause from Earth at local noon varied from 8 R_E (Earth radii) to beyond *Explorer 12*'s perigee of 13 R_E, depending on the intensity of the solar wind.

In 1963, *Interplanetary Monitoring Platform 1* (*IMP*) provided the first clear-cut evidence of the bow shock, which it crossed many times between about 13–25 R_E from Earth, depending on the bow shock's position relative to the Sun-Earth line. On Earth's night side, *IMP 1* detected a geomagnetic tail that appeared to extend well beyond the spacecraft's apogee of 32 R_E. *IMP 1* also discovered a very narrow region around local midnight, where the geomagnetic field fell almost to zero. This so-called neutral sheet was sandwiched between a region to its north, where the geomagnetic field was pointing toward Earth, and a region to its south, where the field lines pointed in the opposite direction. In 1966 Samuel J. Bame et al. found, using particle data from the Vela spacecraft, that the neutral sheet was embedded in a plasma sheet about 6 R_E thick.

The *Orbiting Solar Observatory 7* spacecraft detected clouds of protons leaving the Sun in coronal mass ejections in 1971. These are similar to the clouds envisaged by Chapman and Ferraro in the 1930s, but their clouds had an equal amount of positive and negative charge making them electrically neutral.

The *International Sun–Earth Explorer 3* spacecraft began a number of passes through the geomagnetic tail in 1982, the furthest of which was about 240 R_E from Earth. It found that, while the near-Earth plasma flow was mostly Earthward, there was a clear outflow in the distant plasma sheet, which appeared to accelerate as it got beyond about 80 R_E from Earth. This acceleration was thought to have been caused by the reconnection of magnetic field lines near the neutral line.

In 1998 *Geotail* provided evidence of magnetic reconnection in the geomagnetic tail when it detected intense bursts of energetic particles at about 13 R_E downwind in the magnetosphere. Then, in the following year, the *Wind* spacecraft became the first to detect magnetic reconnection, when it observed a change in direction of 180° in jets of plasma about 60 R_E downwind in the geomagnetic tail. In 2000 the *Polar* spacecraft detected Alfvén waves, propagating along magnetic field lines from the geomagnetic tail toward Earth, which had enough energy to create aurorae during magnetospheric substorms.

Three consecutive substorms were observed simultaneously by the four Cluster and two Double Star spacecraft on 26 September 2005. They showed that magnetic reconnection coincided in space and time with that of the current disruption process, which is characterized by large amplitude and turbulent magnetic field fluctuations. A few tens of seconds later, the *IMAGE* (*Imager for Magnetopause-to-Aurora Global Exploration*) spacecraft imaged aurorae. So it appeared as though magnetic reconnection in the geomagnetic tail was a cause of aurorae.

On 23 March 2007 the THEMIS (Time History of Events and Macroscale Interactions during Substorms) constellation of five spacecraft measured the rapid expansion of the plasma sheet, and the simultaneous westward expansion of the visible aurora at about 15° longitude/minute. This was the first time that such observations had been made simultaneously by any observatories. In the following year the THEMIS spacecraft detected magnetic reconnection about 20 R_E from Earth in the geomagnetic tail, followed 90 seconds later by the brightening of the aurora observed by a ground station network, and 90 seconds after that by near-Earth current disruption.

The key to recent work in this area has been the simultaneous observations of a number of spacecraft and ground stations.

David Leverington

See also: Earth, Sun

Bibliography

Wilmot N. Hess, *The Radiation Belt and Magnetosphere* (1968).

David P. Stern, "A Brief History of Magnetospheric Physics before the Spaceflight Era," *Reviews of Geophysics* 27 (1989).

———, "A Brief History of Magnetospheric Physics during the Space Age," *Reviews of Geophysics* 34 (1996).

James A. Van Allen, *Origins of Magnetospheric Physics* (2004).

Akebono. *See EXOS.*

Alouette. *See* Canada.

Cluster. Cluster was a European Space Agency (ESA) mission to study Earth's magnetosphere, the region where Earth's magnetic field is constrained by the solar wind. Regions of particular interest were the polar cusp, where solar wind particles can ride down Earth's field lines to its magnetic polar regions, and the magnetotail, where the magnetosphere tails away from Earth, blown by the solar wind.

Cluster consisted of four identical spacecraft to characterize the magnetosphere in three dimensions, and to allow them to differentiate between spatial and temporal

An artist's impression of the Cluster quartet. (Courtesy European Space Agency)

features. The mission was approved in 1986 as a joint program with the *SOHO* (*Solar and Heliospheric Observatory*) spacecraft, on condition that the total cost of both missions was reduced by 50 percent. To save money, ESA decided to launch all four spacecraft on an Ariane 5 qualification flight, which the Ariane program was willing to accept provided Cluster paid the additional costs.

The Cluster spacecraft were designed by an international industrial consortium led by Dornier in Germany. Each spacecraft carried 11 experiments built in Europe and the United States to study charged particles and electric and magnetic fields. In the launch configuration, the four 1.3 m high, 2.9 m diameter cylindrical spacecraft sat one on top of another under the Ariane 5 shroud. Unfortunately, the launch from French Guiana in June 1996 was a failure, with the launcher breaking up 37 seconds into the flight because of high dynamic loads.

After the failure, ESA investigated building four more identical spacecraft, but paying for a production Ariane 5 launcher was impossible. Instead, ESA decided to launch the four new spacecraft (*Salsa*, *Samba*, *Rumba*, and *Tango*) of the Cluster 2 program in pairs on two Russian Soyuz launchers. These launches took place successfully in July and August 2000 from Baikonur in Kazakhstan, putting the spacecraft into a parking orbit. From there they achieved their operational 20,000 × 120,000 km polar orbits, with a period of 57 hours, using onboard propellant. The spacecraft were flown in a tetrahedral formation with spacecraft to spacecraft distances that could be varied from 600–20,000 km throughout their mission. To do this, about 50 percent of the spacecraft's mass was propellant. As the orbits were inertially fixed, they bisected Earth's magnetotail at apogee in about August annually, and passed through the northern cusp about six months later.

In December 2003 and July 2004, the Chinese National Space Administration launched its Double Star spacecraft carrying seven instruments identical to those on the Cluster 2, plus eight instruments of its own. These were China's first magnetospheric research spacecraft and were designed to operate with Cluster 2 in an extended constellation. The *TC 1* (*Explorer 1*) equatorial spacecraft was launched into a 570 × 79,000 km orbit inclined at 29° to the equator, and the *TC 2* polar spacecraft into a 700 × 39,000 km polar orbit.

The Cluster spacecraft have determined the characteristics of the magnetotail current sheet, which separates the northern and southern parts of the magnetotail, for the first time. They have determined the bow shock thickness and have discovered vortices of solar wind particles that appeared to be tunneling their way into the magnetosphere. Magnetic reconnection in the cusp region and elsewhere also seemed to allow solar wind particles to enter Earth's magnetosphere. Magnetic reconnection was observed in the turbulent plasma of the magnetopause. This was the first time magnetic reconnection had been observed in any turbulent plasma; the results had far reaching significance in the fields of astronomy and plasma physics in general.

On 27 December 2004, radiation from a starquake on a neutron star was recorded by Double Star and Cluster satellites, giving the first observational evidence of cracks in a neutron star's crust.

David Leverington

See also: China, European Space Agency

Bibliography

J. Credland and R. Schmidt, "The Resurrection of the Cluster Scientific Mission," *ESA Bulletin* 91 (August 1997): 5–10.

ESA. http://www.esa.int/esaCP/index.html.

J. Krige et al., *A History of the European Space Agency, 1958–1987*, vol. 2, *The Story of ESA, 1973–1987* (2000).

Double Star. Double Star was China's first spacecraft program to study Earth's magnetosphere. It consisted of two 660 kg spacecraft, *TC* (Tan Ce or Explorer) *1* and *2*, both designed and launched by the Chinese National Space Administration (CNSA). Half of the experiments on board were provided by Chinese and half by European research establishments. All but one of the European experiments was identical to those flown on the European Space Agency's (ESA) Cluster spacecraft. The two Double Star spacecraft operated simultaneously, in an extended constellation, with the four Cluster spacecraft that had been launched a few years earlier.

In 1992 the Chinese Academy of Sciences had agreed to collaborate with ESA on its Cluster mission. Then, six years later, CNSA invited ESA to collaborate on its complementary Double Star program. ESA finance was approved in 2001, which enabled the first ESA-funded experiments to be launched on a Chinese spacecraft.

TC 1, the equatorial spacecraft, was launched in December 2003 by a Long March (Chang Zheng—CZ) –2C from Xichang in south Sichuan province into a 570 × 79,000 km orbit inclined at 28.5° to the equator. This enabled *TC 1* to study Earth's magnetotail, where particles are accelerated toward Earth's magnetic poles by magnetic reconnection processes. *TC 2*, the polar spacecraft, was launched seven months later by a CZ–2C from Taiyuan in Shanxi into a 700 × 39,000 km polar orbit. This enabled it to study the physical processes taking place above the magnetic poles.

On 26 September 2005 the Double Star and Cluster spacecraft detected three consecutive substorms during a period of just two hours. Magnetic reconnection was found to occur in the magnetotail much closer to Earth than usual, and was almost colocated with the current disruption process. In turn auroral brightenings occurred just a few tens of seconds later. These, and similar observations, have provided a much better understanding of the role of magnetic reconnection and current disruption in causing magnetospheric substorms and their associated auroral effects.

David Leverington

See also: China, European Space Agency

Bibliography

ESA Double Star. http://sci.esa.int/science-e/www/area/index.cfm?fareaid=70.

Early Explorer Spacecraft. Early Explorer spacecraft include those Explorer spacecraft (numbered 1–16) launched between 1958 and 1962, inclusive. They were small, Earth-orbiting scientific spacecraft weighing between 14 kg (*Explorer 1*) and 108 kg (*Explorer 16*). Although relatively simple, they were generally successful in researching the novel space environment.

The Explorer program had its origins in the U.S. Army Ballistic Missile Agency (ABMA) 1954 proposal, led by Wernher von Braun, to launch a 2.3 kg scientific spacecraft into Earth orbit using a modified Redstone rocket. ABMA proposed that this Minimum Satellite Vehicle or Project Orbiter could be part of the U.S. contribution to the International Geophysical Year. In 1955 the proposal was turned down in favor of the Naval Research Laboratory Project Vanguard.

Project Vanguard soon ran into technical trouble, and it was not ready to launch when the Soviet Union launched *Sputnik* and *Sputnik 2* in late 1957. Public pressure

JAMES VAN ALLEN (1914–2006)

James Van Allen was an atmospheric physicist and pioneer in the use of rockets and satellites as scientific tools. After World War II, he developed radiation counters to place on captured V-2 rockets, and in 1952 he launched the first "rockoons," which used balloons to carry sounding rockets to high altitudes for launch. He followed this accomplishment by developing radiation counters to go into Vanguard and Explorer satellites. These devices discovered regions of high-energy particles around Earth—the Van Allen belts. Van Allen continued studies of Earth and other planets, winning numerous scientific awards.

Matt Bille and Erika Lishock

persuaded President Dwight D. Eisenhower to authorize ABMA to launch its scientific spacecraft as soon as possible, using a modified Jupiter C Intermediate Range Ballistic Missile with a modified Redstone as first stage. The spacecraft was *Explorer 1*, built by the Jet Propulsion Laboratory of the California Institute of Technology and launched successfully from Cape Canaveral on 31 January 1958 into a 360 × 2,550 km orbit inclined at 33° to the equator. The instruments included a cosmic ray detector, developed by James Van Allen of Iowa State University. There was no onboard data storage, so data could only be obtained when the spacecraft was within visibility of a ground station. This made it difficult to separate altitude effects from longitude/latitude effect, but the particle counts appeared to saturate at about 800 km before reducing to zero at higher altitudes.

From left to right, William Pickering, James Van Allen, and Wernher von Braun pose with a model of Explorer 1, *the first U.S. satellite successfully placed into orbit in 1958. The spacecraft is in the right portion with the striped markings. The non-striped section is the final launcher stage. (Courtesy NASA/Jet Propulsion Laboratory)*

The launch of *Explorer 2* failed, but the payload on *Explorer 3* was almost identical to that on *Explorer 1*, although it also included a tape recorder. This enabled Van Allen to make sense of his results from both *Explorers 1* and *3*. On 1 May 1958 he announced the discovery of a belt of elementary particles around Earth, soon called the inner Van Allen belt. *Explorer 4*, launched on 26 July, mapped the belt for 2.5 months.

Explorer 6, built by Space Technology Laboratories of what later became TRW Corporation, launched in August 1959, was designed to study elementary particles around Earth, galactic cosmic rays, and geomagnetism. It also transmitted the first experimental television picture of Earth's cloud cover and was a target for an early antisatellite test. *Explorer 7* produced heat balance data by measuring the solar radiation incident on Earth and that radiated back to space.

In March 1961 *Explorer 10*, which had been launched into a highly eccentric orbit with an apogee of 180,000 km, obtained the first indications of the tail-like structure of the magnetopause around local midnight. The spacecraft had no solar cells, so transmission ceased after 60 hours when the spacecraft's batteries were discharged. *Explorer 11* was a gamma-ray observatory spacecraft launched in April 1961. Although operational for five months, it only detected 22 gamma rays. Other early Explorer spacecraft undertook atmospheric density research (*Explorer 9*), magnetospheric research (*Explorers 12*, *14*, and *15*), and measurements of micrometeoroids (*Explorers 13* and *16*).

David Leverington

See also: International Geophysical Year, Jet Propulsion Laboratory, United States Army, Vanguard

Bibliography

William E. Burrows, *This New Ocean: The Story of the First Space Age* (1998).

John M. Logsdon, *Exploring the Unknown: Selected Documents in the History of the U.S. Civil Space Program, Vol. I* (1995).

James A. Van Allen, *Origins of Magnetospheric Physics* (1983).

European Space Research Organisation Scientific Satellite Program. The European Space Research Organisation (ESRO) scientific satellite program was first described in the 1961 *Blue Book* produced by the Scientific and Technical Working Group (STWG) set up to advise potential ESRO Member States. It divided the eight-year program into short-, medium-, and long-term projects. The short-term projects were mainly based on sounding rocket investigations of the upper atmosphere, including the aurora, from a launch range in northern Sweden. Some astronomy payloads were also envisaged to be launched by sounding rockets from lower latitudes, using existing national ranges in Europe. Medium-term projects consisted of small Earth-orbiting spacecraft to study near-Earth space and to observe the Sun, while long-term projects included large Earth-orbiting astronomical observatories and lunar probes. The eight-year program proved too ambitious. While ESRO launched 168 sounding rockets between 1964 and 1972, only eight spacecraft were launched in the same period.

ESRO Scientific Spacecraft (Part 1 of 2)

(A) Spacecraft designed and launched under ESRO management (before 1975)						
Spacecraft	**Launch Date**	**Launcher**	**Spacecraft Mass (kg)**	**Orbit**	**Mission Study**	**Comment**
ESRO 2A	30 May 1967	Scout	74		Solar astronomy and cosmic rays	Launcher failed
ESRO 2B (*Iris*)	17 May 1968	Scout	80	1,090 × 330 97°	Solar astronomy and cosmic rays	
ESRO 1 (*Aurorae*)	3 October 1968	Scout	85	1,540 × 260 94°	Polar ionosphere	
HEOS A1	5 December 1968	Thor Delta	105	223,000 × 400 28°	Interplanetary environment	
ESRO 1B (*Boreas*)	1 October 1969	Scout	85	390 × 290 85°	Polar ionosphere	Launch problem limited spacecraft life to 52 days
HEOS A2	31 January 1972	Thor Delta	110	240,000 × 400 90°	Interplanetary environment	
TD 1A	12 March 1972	Thor Delta	470	550 × 525 98°	Ultraviolet and high energy astronomy	
ESRO 4	22 November 1972	Scout	130	1,170 × 245 91°	Ionosphere and magnetosphere	

(Continued)

ESRO Scientific Spacecraft Part (2 of 2)

(B) Programs approved under ESRO management, but spacecraft launched by the European Space Agency						
Spacecraft	**Program Approved**	**Launch Date**	**On-orbit Space-craft Mass (kg)**	**Orbit Used**	**Mission Study**	**Comments**
COS B	2 July 1969	9 August 1975	275	99,000 × 440 90°	Gamma ray observatory	
GEOS 1	2 July 1969	20 April 1977	275	38,400 × 2,100 26°	Magnetosphere	Thor Delta launcher malfunction
GEOS 2	2 July 1969	14 July 1978	275	35,600 × 35,400 26°	Magnetosphere	Flight of flight spare spacecraft
SAS D, to be called *IUE*	14 July 1971	26 January 1978	430	46,000 × 26,000 29°	Ultraviolet observatory	Joint NASA/ United Kingdom/ ESRO program
IMP D, to be called *ISEE B*	12 April 1973	22 October 1977	160	138,000 × 340 29°	Magnetosphere	Joint program with NASA's *ISEE A* and *C*
HELOS, to be called *EXOSAT*	12 April 1973	26 May 1983	510	192,000 × 360 72°	X-ray observatory	

In 1963 the Launching Programme Sub-Committee (LPSC) of the STWG recommended that one of the first two ESRO spacecraft (*ESRO 1*) should investigate the ionosphere in polar regions, while the other (*ESRO 2*) should undertake solar astronomy and cosmic ray studies. They would be launched by an U.S. Scout rocket. That same year four Thor-Delta (TD)–launched spacecraft were added to the program. In 1964 the LPSC recommended that ESRO's first highly eccentric orbit satellite (*HEOS A*) should be devoted to cosmic ray studies, and the second HEOS, to be launched one year later, should investigate the interplanetary medium. So when ESRO formally came into being in 1964, its initial program had been largely defined.

During the next several years, difficulties in designing and building these spacecraft within the agreed budget caused a number of problems. As a result, ESRO agreed to launch *ESRO 2* before *ESRO 1* and reduce the number of TD spacecraft. The *Large Astronomical Satellite* (*LAS*), carrying a high-performance ultraviolet telescope, would be ESRO's first long-term project. Subsequently, the number of TD spacecraft was reduced again, to counter Italian objections to the high price. The *LAS* was also canceled.

ESRO 2 (later called *ESRO 2A*), ESRO's first satellite was launched by a Scout rocket from Vandenberg Air Force Base in California on 30 May 1967. Unfortunately, the Scout malfunctioned and the spacecraft did not enter orbit. However, the launch of the second ESRO spacecraft, called *ESRO 2B*, also by a Scout, was successful in the following year.

ESRO was subsumed into the European Space Agency in 1975. By that time it had a number of spacecraft in orbit, mostly devoted to investigating the solar wind and its interaction with Earth's magnetosphere and atmosphere. ESRO had also entered the field of high energy astronomy with the launch of *TD 1A*. In addition, ESRO Member States had approved a further six spacecraft, including *COS B* that was to produce the first gamma-ray map of the Milky Way.

David Leverington

See also: European Space Research Organisation

Bibliography

John Krige and Arturo Russo, *Europe in Space 1960–1973* (1994).

J. Krige and A. Russo, *A History of the European Space Agency, 1958–1987*, vol. 1, *The Story of ESRO and ELDO, 1958–1973* (2000).

Harrie Massey and M. O. Robins, *History of British Space Science* (1986).

EXOS. *EXOS*, or exospheric satellite, was a series of scientific satellites designed and launched by the Institute of Space and Astronautical Science (ISAS) to study Earth's upper atmosphere. The first two, *EXOS A* and *B*, were renamed *Kyokko* (aurora) and *Jikiken* (magnetosphere) after their February and September 1978 launches on board M-3H boosters from Kagoshima Space Center (KSC). Both were in highly elliptical orbits. Together they were Japan's contribution to the International Magnetospheric Study, organized by the Scientific Committee on Solar-Terrestrial Physics (SCOSTEP). *Kyokko* took ultraviolet auroral images every two minutes.

ISAS launched *EXOS C* from KSC on an M-3S booster in February 1984 and renamed the spacecraft *Ohzora* (sky). *Ohzora*'s mission included studying low-energy particles and plasma waves in addition to the composition and structure of the upper atmosphere as part of SCOSTEP's Middle Atmosphere Program.

EXOS D, renamed *Akebono* (dawn), was launched from KSC on an M-3SII booster in February 1989 to investigate phenomena related to the aurora and plasma. While not officially part of the International Solar Terrestrial Physics (ISTP) program, it did conduct coordinated observations with the ISTP satellites. The spacecraft was still operating in late 2008, having operated through more than a complete solar cycle.

Kyokko obtained the first ultraviolet image of the aurora from space, showing that the magnetosphere plasma is disturbed when an aurora begins. *Jikiken* observed the magnetosphere in kilometer radio waves, showing wave-particle interactions. *Ohzora* observed absorption spectra of sunlight by minor constituents of the middle atmosphere, and high-energy particles above the poles. *Akebono* measurements led to the discovery of a reduction in plasma density during magnetic storms, long-term variations of radiation belt particles, and acceleration of particles by electric fields parallel to the magnetic field lines.

Katie J. Berryhill

See also: Japan

Bibliography

Japan Aerospace Exploration Agency. http://www.jaxa.jp.

FAST, Fast Auroral Snapshot Explorer. *FAST, Fast Auroral Snapshot Explorer*, was the second mission in NASA's Small Explorer Satellite program. It was designed to carry out in situ measurements of acceleration physics and related plasma processes that create Earth's aurora. It measured electric and magnetic fields and fluxes of electrons and ions in high time and space resolution during auroral crossings.

FAST was managed by NASA Goddard Space Flight Center and built at GSFC with instruments provided by the University of California at Berkeley and at Los Angeles. It was launched on 21 August 1996 by a Pegasus-XL rocket into an elliptical 4,175 km × 350 km orbit with 83° inclination. The satellite crossed the auroral region four times every 133 minutes in orbit. The orbit reached high into the atmosphere, where the particles that create the aurora are accelerated.

The spacecraft instruments included electric field detector and Langmuir probes, flux-gate and search coil magnetometers, and ion mass spectrograph and electrostatic analyzers to detect energetic electrons and ions. *FAST* measurements revealed that parallel electric fields are created in a funnel-shaped region of the dayside geospace (called cusp) that accelerate electrons downward, thus creating aurorae, and ions upward, preventing them from reaching the upper atmosphere. *FAST* results showed how auroral kilometric radiation is created by a resonance of electric fields above the aurora. In situ observations in the auroral acceleration region on the nightside of Earth revealed its composition of hot, low density plasma devoid of cold plasma, which was contrary to earlier results from Sweden's 1986–87 Viking mission. Scientists showed that ion injections into the ring current are the primary source for magnetic perturbations during magnetic storms. Wave measurements demonstrated how the magnetotail couples to the lower ionosphere through electromagnetic waves that travel along magnetic field lines.

As of October 2008, the FAST spacecraft was still operating and collecting data.

Harald Frey

See also: Goddard Space Flight Center, Pegasus

Bibliography

"An Overview of the Fast Auroral SnapshoT (FAST) Mission," *Space Sciences Reviews* 98, no. 1–2 (2001).

G. Paschmann et al., eds., *Auroral Plasma Physics* (2002).

Robert F. Pfaff, ed., *The FAST Mission* (2001).

IMAGE, Imager for Magnetopause-to-Aurora Global Exploration.

IMAGE, Imager for Magnetopause-to-Aurora Global Exploration, was the first mission in NASA's Medium-Class Explorer program. It was also the first satellite mission dedicated to remotely sensing and imaging Earth's magnetosphere. Lockheed Martin built *IMAGE*, which had 1,600-ft antennae, and the Southwest Research Institute managed the project. A Delta II rocket launched the satellite on 25 March 2000 from Vandenberg Air Force Base into an elliptical, polar orbit (1,000 km × 46,000 km). *IMAGE* observed the northern hemisphere at the beginning of its mission, but the fast orbital precession provided a much better view of the southern hemisphere after four years. The spacecraft stopped working on 18 December 2005.

The satellite carried six scientific instruments. The Extreme Ultraviolet Imager for the first time imaged the plasmasphere and showed its erosion during magnetic storms and its interaction with high-energy particles of the radiation belts. The Far Ultraviolet Imager, along with its Spectrographic Imager sub-instrument, provided the first global images of proton aurorae. Three instruments, the Low-Energy, Medium Energy, and High-Energy Neutral Atom imagers, detected energetic neutral atoms from 10 eV to 500 keV. These neutral atoms come primarily from aurorae, the solar wind, and the radiation belts, respectively. The Radio Plasma Imager was the first active radio plasma instrument to sound the plasma density distribution in the magnetosphere.

Together, these instruments provided the first comprehensive images of the plasma in the inner magnetosphere and how different plasma regions interact with one another and respond to solar wind forcing. Japan's Communications Research Laboratory and the National Oceanic and Atmospheric Administration's Space Environment Center used *IMAGE* data to help prepare space weather forecasts.

Harald Frey

See also: Lockheed Martin Corporation

Bibliography
James L. Burch, ed., *The IMAGE Mission* (2000).
———, *Magnetospheric Imaging: The IMAGE Prime Mission* (2003).

Interball. Interball was a Russian project that consisted of two pairs of spacecraft, which gathered data to better understand solar-terrestrial physics. Each satellite pair consisted of a large, Russian Prognoz M spacecraft manufactured by Lavochkin design bureau and a smaller Magion subsatellite manufactured by the Geophysical Institute in the Czech Republic. Interball was part of a larger multinational effort coordinated by the Interagency Consultative Group for Space Science, which included the Russian Space Agency, NASA, the European Space Agency, and the Japan Institute of Space and Aeronautics Sciences. To enable comprehensive and time-coordinated views of the magnetosphere and solar wind, data from Interball was coordinated with observations from other spacecraft, including *Geotail*, *Wind*, *Polar*, *SOHO* (*Solar and Heliospheric Observatory*), and the German *Equator-S* as part of the International Solar Terrestrial Physics Science initiative.

The Interball project followed from earlier Soviet Prognoz solar observation spacecraft launched between 1972 and 1983. One pair, known as the *Tail Probe*, was launched in 1995 into an elongated elliptical orbit with a 63° inclination and a 200,000 km apogee to study Earth's magnetotail and operated until late 2000. The other couplet, orbited in 1996 and dubbed the *Auroral Probe*, was placed into the same inclination at an apogee of 20,000 km above the North Pole and its associated auroral zone and operated until early 1999.

Interball examined the transmission of solar wind energy to the magnetosphere, how it is stored there, and how it is dispersed in the atmosphere, ionosphere, and magnetosphere during magnetospheric substorms, which are phenomena associated with auroral and magnetic disturbances. Observations from Interball showed more details of the dynamic relationship between the magnetosphere and the solar wind, showing, for example, that changes in the boundary between the magnetosphere and solar wind regime sometimes occur quite rapidly, on the order of a minute; that plasma convection along the boundary region is directed toward the magnetotail equatorial plane; and that at high latitudes, the boundary region is relatively unstable.

Knowledge gained from the Interball project has practical applications also, especially for nations at high latitudes such as Russia, Canada, and Scandinavian countries, where magnetospheric substorms can seriously disrupt radio-wave transmissions, telephone communications, radar, and global positioning systems.

David Takemoto-Weerts and Stephen B. Johnson

See also: Soviet Union

Bibliography
Centre National d'Etudes Spatiales Interball. http://internet1-ci.cst.cnes.fr:8010/guide/interball/.
Goddard Space Flight Center Interball. http://nssdc.gsfc.nasa.gov/nmc/spacecraftDisplay.do?id=1996-050C.
Russian Space Research Institute (IKI) Interball. http://www.iki.rssi.ru/interball/.

International Solar-Terrestrial Physics Science Initiative. The International Solar-Terrestrial Physics (ISTP) Science Initiative was an international scientific collaboration started in the 1980s to develop a comprehensive understanding of the generation and flow of energy from the Sun through Earth's space environment (geospace). The main goal was to define the relationships among physical processes that link the different regions of this dynamic system. NASA, the European Space Agency (ESA), and the Japanese Institute of Space and Astronautical Science (ISAS) created ISTP as the union of the NASA Global Geospace Spacecraft (GGS) and the ESA/NASA Solar Terrestrial Programme (STP), with the goal to carry out several solar-terrestrial missions during the 1990s. It combined the international resources and science communities, using coordinated space missions and complementary ground facilities and theoretical efforts, to obtain simultaneous investigations of the Sun-Earth environment during an extended time period.

The ISTP initiative combined measurements from the *Geotail*, *Wind*, *Polar*, and *SOHO* (*Solar and Heliospheric Observatory*) spacecraft. The original concept also included NASA's *Equator* and the European *Cluster* spacecraft. Budget and schedule limitations led NASA to delete *Equator*, but the German *Equator-S* satellite later recovered part of its science. Additionally, the catastrophic failure of the first flight of the Ariane 5 rocket with the four *Cluster* spacecraft on board in June 1996 compromised these efforts. Only after ESA decided to rebuild the *Cluster* spacecraft and launch them on two separate Soyuz-Fregat launches in July and August 2000 was the complement of space missions complete. The *Geotail* satellite was launched on 24 July 1992 and was a collaboration between ISAS and NASA to study the dynamics of the magnetotail. The NASA spacecraft *Wind* was launched on 1 November 1994 and spent a long time around the Lagrangian L1 point to study the properties of the solar wind. The NASA *Polar* satellite, launched in February 1996, measured the entry of plasma into the magnetosphere, measured the flow of plasma to and from the ionosphere, and imaged the aurora in multiple wavelengths. The *SOHO* mission was a joint venture between ESA and NASA, launched on 2 December 1995, to provide an uninterrupted view of the Sun from the Lagrangian L1 point.

Results obtained during the main ISTP phase greatly enhanced understanding of the interaction between the Sun and Earth's geospace. *SOHO*'s continuous observation of the Sun allowed the early detection of solar flares and coronal mass ejections that impact Earth's environment with large fluxes of energetic particles. *Wind* provided information about the particle and electric and magnetic field properties of the plasma that is continuously impinging on Earth's magnetosphere as the solar wind. The *Polar* spacecraft observed the impact of these energetic particles on the magnetosphere, how the impacting energy was transferred and distributed, and how the ionosphere reacted to such energy input with bright auroras and the outflow of ions. *Geotail* observed how the energy was stored in the magnetotail and how this energy was periodically released in geomagnetic disturbance events called substorms. The four *Cluster* spacecraft, while flying in formation allowing them to differentiate between spatial structures and temporal features, solved many mysteries of the plasma interaction, such as the transport of solar wind particles across the magnetopause through the merging or reconnection between magnetic fields of opposite polarity, the acceleration of particles inside the magnetosphere, the coupling between regions of the magnetosphere through plasma waves, the creation of auroras, and the response of different plasma regions to the energy input from outside.

Harald Frey

See also: European Space Agency, Goddard Space Flight Center, Japan

Bibliography

Mario Acuña et al., "The Global Geospace Program and Its Investigations," *Space Science Reviews* 71 (1995): 5–21.

Phillipe Escoubet et al., "The Cluster Mission," *Annales Geophysicae* 19 (2001): 1197–1200.

Atsuhiro Nishida, "The GEOTAIL Mission," *Geophysical Research Letters* 21 (15 December 1994): 2871–74.

Keith Ogilvie and Michael Desch, "The WIND Spacecraft and Its Early Scientific Results," *Advances in Space Research* 20 (1997): 559–68.

International Sun-Earth Explorer. International Sun-Earth Explorer (ISEE) was an international science mission to study the Sun and its influence on the near-Earth space environment. The mission featured a successful collaboration between NASA and the European Space Agency (ESA) in support of the International Magnetospheric Study (IMS) program, which was begun in 1970 under the auspices of the Special Committee on Solar-Terrestrial Physics to gather coordinated data on the near-Earth plasma environment from instruments on the ground, on balloons, sounding rockets, and spacecraft in the period 1976–79.

In the late 1960s both NASA and the European Space Research Organisation (ESRO) assessed new magnetosphere study missions. At a NASA-ESRO meeting in February 1971, the two organizations discussed a new dedicated joint magnetosphere research mission, which became ISEE. ESRO approved its portion of the project in April 1973 and designated it, along with ESRO's *GEOS* (*Geostationary Satellite*—not to be confused with NASA's GEOS) spacecraft, as part of the orbiting segment of the IMS. A Memorandum of Understanding between NASA and ESA in March 1975 officially assigned two American and one European spacecraft to the ISEE mission. This was the first NASA-ESA mission, with both organizations contributing full spacecraft on equal terms.

ISEE was a joint mission involving three spacecraft: a mother-daughter pair (*ISEE 1*, also known as *Explorer 56*, and *ISEE 2*) and *ISEE 3* (*Explorer 59*). Engineers at NASA's Goddard Space Flight Center built the two larger spacecraft (*ISEE 1* weighed 329 kg and *ISEE 3* 479 kg). ESA provided the 158 kg *ISEE 2*, built by Dornier Systems of Germany.

The mission was designed to investigate (1) solar-terrestrial relationships at the boundaries of Earth's magnetosphere, (2) the solar wind near Earth and the shock wave that forms where the wind meets Earth's magnetosphere, (3) the structure and

NASA engineers evaluate the International Sun-Earth Explorer C, later called ISEE-3, inside Goddard's dynamic test chamber, 1978. (Courtesy NASA/Goddard Space Flight Center)

movement of the plasma sheets, and (4) cosmic rays and solar flare emissions in the Sun-Earth region. To this end, *ISEE 1* carried 14 experiments, *ISEE 2* carried 8, and *ISEE 3* carried 15.

ISEE 1 and *2* were launched together on 22 October 1977 into a highly eccentric orbit with an apogee of 23 Earth radii. The launch was accomplished using a Thor-Delta booster from Cape Canaveral. *ISEE 3* was launched by the same type vehicle from Cape Canaveral on 12 August 1978. To provide long-term observation of the Sun-Earth system while minimizing the need for station-keeping fuel, *ISEE 3* was placed in a "halo" orbit around the Earth-Sun Lagrange 1 point, roughly perpendicular to the Earth-Sun line, approximately 235 Earth radii toward the Sun.

ISEE 1 and *2* reentered Earth's atmosphere in 1987, while *ISEE 3* found a second life in a mission conceived by its NASA flight director, Robert Farquhar. In 1982, *ISEE 3* was renamed the *International Cometary Explorer* (*ICE*). A series of thruster firings and lunar gravity-assist flybys nudged it into a heliocentric orbit designed to intercept Comet Giacobini-Zinner. *ICE* flew through the comet's tail on 11 September 1985. *ICE* also studied Halley's Comet from a distance of 28 million km, then continued cosmic ray studies and measurements of the ionic composition of the solar wind. *ICE* observations were coordinated with those made by another joint ESA-NASA mission, the solar probe *Ulysses*, when the two were on the same solar radial line. *ICE* was shut down in 1997.

Matt Bille

See also: Comets, Explorer

Bibliography

A. C. Durney and K. W. Ogilvie, "Introduction to the ISEE Missions," *Space Science Reviews* 22, no. 6 (1978).

NASA Goddard Project Information, "Spacecraft Sketch" for each spacecraft (n.d.). http://heasarc.gsfc.nasa.gov/docs/heasarc/missions/isee3.html.

National Space Science Data Center, "International Sun-Earth Explorers (ISEE) Project Information" (2003). http://nssdc.gsfc.nasa.gov/nmc/spacecraftDisplay.do?id=1978-079A.

Space Studies Board Commission on Physical Sciences, Mathematics, and Applications (CPSMA), *U.S.-Europe Collaboration in Space Science* (1998).

ISIS, International Satellite for Ionospheric Studies. *See* Canada.

***Jikiken*.** *See EXOS.*

***Kyokko*.** *See EXOS.*

***Ohzora*.** *See EXOS.*

***Ørsted*.** *Ørsted* was the first Danish satellite. Named after the discoverer of electromagnetism, it was an auxiliary payload on a Delta booster, and delays with the main payload held the launch back for three years. After a record-breaking 10 scrubbed launch attempts, the satellite was launched from Vandenberg Air Force Base, California, on 23 February 1999 into a low Earth, near-polar orbit. The 62 kg satellite was built by Terma A/S and managed by the Danish Meteorological Institute. It was the first satellite to participate in the International Decade of Geopotential Research.

The spacecraft carried five instruments: a fluxgate magnetometer, an Overhauser magnetometer, a star imager, a solid-state charged particle detector, and a Global Positioning System precision receiver. This was the second satellite, after *Magsat*, to precisely map the global magnetic field, so it provided the first accurate global survey of field changes. Compared to *Magsat* data, *Ørsted* data showed a decline in field strength that led scientists to wonder whether Earth's magnetic field was heading toward a pole flip, as it had done often in the planet's history. Since *Ørsted* survived beyond its one-year design life, it was also able to operate in conjunction with *CHAMP* (*Challenging Minisatellite Payload*). It was still operational, except for attitude determination and control, as of mid-2007.

Katie J. Berryhill

Bibliography

Danish Meteorological Institute. http://web.dmi.dk/projects/oersted/.

Peter Stauning, "Ørsted Satellite—The Danish Miracle in Space," *Nordic Space* 15 (February 2007): 4–8.

***Rumba*.** *See* Cluster.

***Salsa*.** *See* Cluster.

Samba. *See* Cluster.

SAMPEX, Solar Anomalous and Magnetospheric Particle Explorer. *SAMPEX, Solar Anomalous and Magnetospheric Particle Explorer*, was the first of the Small Explorer missions. It carried three U.S. instruments and one German instrument designed to measure from low Earth orbit the composition and quantity of energetic solar particles, anomalous cosmic rays (ACRs), and magnetospheric trapped ions, at energies from about 0.1 to several hundred MeV. Energetic solar particles interact with Earth's magnetic field, impact satellite operations, and sometimes disrupt power distribution on the ground. ACRs, named for their unusual energy spectrum that differs from galactic cosmic rays, are a class of ions from outside the solar system that appear to come from the heliopause, where the solar wind meets the interstellar medium. Magnetospheric trapped ions, especially protons, impact satellite operations, and precipitating electrons affect the chemistry and dynamics of Earth's upper atmosphere. SAMPEX was one of the first NASA missions to use fiber-optic communications and gallium arsenide solar arrays.

Proposed by the University of Maryland, the 157 kg *SAMPEX* was designed, built, and operated by Goddard Space Flight Center. *SAMPEX* was launched on 3 July 1992 from Vandenberg Air Force Base, California, on a Scout rocket into a 550 × 675 km orbit at an 82° inclination.

SAMPEX's identification of the elements and isotopes of anomalous cosmic rays and measurement of their energies have improved understanding of heliopause dynamics. While mapping magnetosphere particles, the mission discovered a new radiation belt around Earth inside the lower Van Allen belt. *SAMPEX*'s measurements of precipitating electrons have led to a better understanding of the chemical dynamics associated with the ozone layer.

David Everett

See also: Goddard Space Flight Center, Systems Management

Bibliography

D. N. Baker et al., "An Overview of the Solar, Anomalous, and Magnetospheric Particle Explorer (SAMPEX) Mission," *IEEE Transactions on Geosciences and Remote Sensing* 31 (1993): 531–41.

J. R. Cummings et al., "New Evidence for Anomalous Cosmic Rays Trapped in the Magnetosphere," *Proceedings of the 23rd International Cosmic Ray Conference* 3 (1993): 428–31.

GSFC SAMPEX. http://sunland.gsfc.nasa.gov/smex/sampex/.

San Marco 1. *See* Italy.

Soviet Magnetosphere and Ionosphere Missions. Soviet magnetosphere and ionosphere missions began with *Sputnik 2* and *Sputnik 3* in late 1957 and early 1958. Both satellites detected particles trapped in Earth's radiation belts, although the distinction of recognizing the belts as such fell to James Van Allen when analyzing the results from the early U.S. Explorer satellites. However, *Sputnik 3* did show that the inner Van Allen belt was mostly composed of protons, rather than the low energy electrons hypothesized by Van Allen. The *Luna 1* and *Luna 2* Moon probes provided data on Earth's magnetic field on their way to the Moon in 1959, with *Luna 2* showing time variations in the electron flux and energy spectrum within the outer belt. In 1964 the Soviet Union launched two pairs of Elektron satellites to continue studies of the inner and outer radiation belts.

After that, geophysical studies were performed by three categories of satellites. The first of these was the DS (Dnepropetrovskiy sputnik or Dnepropetrovsk satellite) series of satellites, developed by Mikhail Yangel's OKB-586 (Special Design Bureau, renamed KB Yuzhnoe in 1966) in the Ukrainian city of Dnepropetrovsk. DS satellites used common buses that could serve as platforms for a wide variety of scientific, military, and applications missions. They weighed roughly 300 kg and consisted of two semispherical compartments connected by a cylindrical section. They were launched in the 1960s–70s by Kosmos boosters (the 63S1 and 11K63 versions) from Plesetsk and Kapustin Yar.

One type of DS satellite in the first-generation series (DS-MG) was outfitted to study Earth's magnetic field. The onboard instruments mapped the spatial distribution of Earth's magnetic field, although coverage was limited to about 75 percent of Earth's surface due to the satellites' relatively low 49° inclination.

The second-generation DS buses came in three slightly different types: DS-U1 (battery-powered), DS-U2, and DS-U3 (both with solar panels). The first two types saw missions for ionospheric and magnetospheric studies, many flown with instruments developed by Eastern Bloc countries under the Interkosmos program.

One subclass of DS-U1 satellite (DS-U1-1K) studied the structure and changes in the ionosphere, focusing on such things as the concentration of positively charged ions and the concentration and temperature of ionospheric electrons. The battery-powered DS bus was selected for these satellites to make sure that electric charges produced by solar panels would not disturb the surrounding ionospheric plasma.

The DS-U2 series had seven subclasses for geophysical observations. DS-U2-I satellites studied the propagation of low-frequency radio waves through the ionosphere. DS-U2-GK satellites observed several parameters associated with auroras. A modified version of these satellites (DS-U2-GKA) was built in cooperation with France and launched under the name Aureole (spelled Oreol in Russian) to study charged particles responsible for auroras before they interacted with the ionosphere. DS-U2-MG satellites continued the studies of Earth's magnetic field begun by the DS-MG satellites, increasing coverage to 94 percent of Earth's surface, thanks to higher inclinations of 71° and 82°. A single DS-U2-IP satellite probed the ionosphere up to an altitude of about 2,000 km and one DS-U2-K satellite focused on charged particle flows and the relation of their intensity with the brightness of auroras. Several DS-U2-IK satellites studied the ionosphere, magnetosphere, and radiation belts.

In the early 1970s the Yangel bureau began work on a new satellite bus called Automatic Universal Orbital Station (AUOS) with two subtypes: AUOS-Z with an Earth-pointing attitude control system, and AUOS-SM with a Sun-pointing attitude control system. The bulk of the AUOS-Z series was devoted to ionospheric and

Soviet and Russian Magnetosphere and Ionosphere Missions

DS Series Type	Launched Between	Number of Launches	Comments
DS MG	18 Mar 1964–24 Oct 1964	2	Officially announced as *Kosmos 26* and *Kosmos 49*.
DS U1 IK	25 Dec 1969–30 Nov 1972	2	Officially announced as *Interkosmos 2* and *Interkosmos 8*.
DS U2 I	24 May 1966–14 Dec 1968	3	Officially announced as *Kosmos 119*, *Kosmos 142*, and *Kosmos 259*.
DS U2 GK	19 Dec 1968–13 Jun 1970	2	Officially announced as *Kosmos 261* and *Kosmos 348*.
DS U2 GKA	27 Dec 1971–26 Dec 1973	2	Officially announced as *Aureole* and *Aureole 2*.
DS U2 MG	20 Jan 1970–10 Aug 1970	2	Officially announced as *Kosmos 321* and *Kosmos 356*.
DS U2 IP	17 Nov 1970	1	Officially announced as *Kosmos 378*.
DS U2 K	4 Jun 1971	1	Officially announced as *Kosmos 426*.
DS U2 IK	7 Aug 1970–1 Dec 1975	7	Officially announced as *Interkosmos 3*, *Interkosmos 5*, *Interkosmos 9*, *Interkosmos 10*, *Interkosmos 12*, *Interkosmos 13*, and *Interkosmos 14*.

AUOS Series Type	Launch Date	Number of Launches	Comments
AUOS-Z-R-O	29 Mar 1977	1	Officially announced as *Kosmos 900*.
AUOS-Z-M-IK	24 Oct 1978	1	Officially announced as *Interkosmos 18*. Launched with *Magion* subsatellite.
AUOS-Z-I-IK	27 Feb 1979	1	Officially announced as *Interkosmos 19*.
AUOS-Z-M-A-IK	21 Sep 1981	1	Officially announced as *Aureole 3*.
AUOS-Z-I-E	18 Dec 1986	1	Officially announced as *Kosmos 1809*.
AUOS-Z-AV-IK	28 Sep 1989	1	Officially announced as *Interkosmos 24*. Launched with *Magion 2* subsatellite under "Aktivnyy" research program.
AUOS-Z-AP-IK	18 Dec 1991	1	Officially announced as *Interkosmos 25*. Launched with *Magion 3* subsatellite.

Prognoz Series Type	Launched Between	Number of Launches	Comments
Prognoz	14 Apr 1972–26 Apr 1985	10	Officially announced as Prognoz.
Prognoz-M2	2 Aug 1995–29 Aug 1996	2	Officially announced as *Interball 1* and *Interball 2*. Launched with *Magion 4* and *Magion 5* subsatellites.

magnetospheric studies. The AUOS-Z satellites had an 800 kg standard service module with eight solar panels and a special boom for gravitational stabilization. Mounted on top of the service module was a payload module with a maximum mass of 400 kg. Launches were with Kosmos-3M and Tsiklon-3 rockets from the Plesetsk cosmodrome.

Ten AUOS-Z satellites were orbited between 1976 and 1991, and seven were dedicated to particles and fields studies. Four (*AUOS-Z-M-IK*, *AUOS-Z-I-IK*, *AUOS-Z-AV-IK*, *AUOS-Z-AP-IK*) were flown under the Interkosmos program, one (*AUOS-Z-M-A-IK*) in cooperation with France, and two (*AUOS-Z-R-O* and *AUOS-Z-I-E*) with Kosmos labels. Three of the Interkosmos missions carried Czechoslovakian subsatellites called Magion that were released from the primary spacecraft for joint studies of the magnetosphere and ionosphere. Two of the satellites (*Interkosmos 18* and *19*) contributed data to the International Magnetospheric Study (IMS) conducted between 1976 and 1979.

The third category was Prognoz satellites developed by the NPO (Scientific-Production Association) Lavochkin design bureau. These satellites were mainly intended to study solar wind and solar flares but also focused on their interaction with Earth's plasma mantle and magnetosphere. The satellites consisted of a pressurized cylinder with four triangular solar panels deployed at right angles to the main body. They were spin-stabilized and had a variety of antennas and other scientific equipment attached to the hemispherical ends of the main body and to the solar panels. Weighing about 1 ton, they were launched into highly elliptical orbits extending out to 200,000 km by the four-stage version of the Soyuz rocket known as the Molniya rocket. Ten Prognoz satellites were launched from Baikonur between 1972 and 1985 and four of them (*Prognoz 4*, *5*, *6*, and *7*) were involved in the IMS program.

Two modified versions of the satellite (Prognoz-M2) were built under an Interkosmos project known as Interball, designed to investigate the magnetospheric tail and auroral zones. Interball consisted of two pairs of spacecraft, a Prognoz-M2 and a Czechoslovakian Magion subsatellite. One pair had orbits of 500 km × 200,000 km (tail probes) and the other had orbits of 500 km × 20,000 km (auroral probes). The satellites carried a variety of plasma and charged particle detectors and were launched by Molniya rockets from Plesetsk in 1995 and 1996.

Bart Hendrickx

See also: Elektron, Interkosmos, Luna, Soviet Union, Sputnik

Bibliography

D. Hart, *The Encyclopedia of Soviet Spacecraft* (1987), 79–81.

S. N. Konyukhov et al., *Rakety i kosmicheskie apparaty konstruktorskogo byuro Yuzhnoe* (2000), 109–51, 157–75.

Tango. *See* Cluster.

THEMIS, Time History of Events and Macroscale Interactions during Substorms. THEMIS, Time History of Events and Macroscale Interactions during Substorms, was the fifth of NASA's Medium Class Explorer missions. It was designed to answer fundamental questions on the nature of substorm instabilities that explosively release energy stored in Earth's magnetotail. To do this, THEMIS consisted of five identical spacecraft placed in various Earth orbital configurations over time, plus a network of ground stations in auroral regions in the northern United States and Canada. These ground stations, some of which were in schools, included specially installed magnetometers and cameras.

The five small THEMIS spacecraft, each weighing 126 kg, were launched together on a Delta II on 17 February 2007. Almost 40 percent of the mass of each spacecraft was fuel to allow for orbital maneuvers. The program was managed by the Goddard Space Flight Center, with the spacecraft bus built by Swales Aerospace and the scientific instruments integrated by the University of California, Berkeley, which also operated the spacecraft. These instruments included magnetometers, electric field instruments, electrostatic analyzers, and particle detectors.

Initially the five THEMIS spacecraft followed one after another in a "string of pearls" configuration in the coast phase, with hundreds of kilometers to 2 R_E (Earth radii) along-track separations. In September 2007 mission operators moved them to more distant orbits with apogees on the dawn side of the magnetosphere. Then in December 2007 the first tail science phase began with apogees in the magnetotail.

The THEMIS spacecraft detected their first substorm on 23 March 2007. They measured the rapid expansion of the plasma sheet at a speed consistent with the simultaneous expansion of the visible aurora. This was the first ever simultaneous observation of the rapid westward expansion of such a disturbance in space and nearer the ground. During the two-hour event the disturbance moved westward at about 15° longitude/minute, or about 10 km/s at auroral altitudes, much faster than expected.

On 26 February 2008 the THEMIS spacecraft constellation and associated ground stations were the first observatories to track the effects of a solar substorm in detail. They detected magnetic reconnection about 20 R_E from Earth in the magnetic tail, followed 90 seconds later by the brightening of the aurora.

David Leverington

See also: Goddard Space Flight Center

Bibliography

V. Angelopoulos, "The THEMIS Mission," *Space Science Reviews Online* (April 2008).

Vassilis Angelopoulos et al., "Tail Reconnection Triggering Substorm Onset," *Science*, 15 August 2008.

THEMIS. http://www.nasa.gov/mission_pages/themis/main/index.html.

TIMED, Thermosphere Ionosphere Mesosphere Energetics and Dynamics. *TIMED, Thermosphere Ionosphere Mesosphere Energetics and Dynamics*, was the initial mission in NASA's Solar Terrestrial Probes (STP) program. Others included Japan's *Solar B* (with instruments provided by the STP program) and

STEREO (*Solar Terrestrial Relations Observatory*). STP's goal was to investigate the dynamic nature of the Sun, and the physics that links the Sun, Earth, and the rest of the solar system. TIMED studied the influences of the Sun on the mesosphere and lower thermosphere/ionosphere (MLTI), the least understood region of Earth's atmosphere, focusing on a region about 60–180 km above Earth's surface. While other satellites had studied sections of the MLTI, it is a difficult region to study because it is too high for balloons and aircraft, and too low for direct satellite measurements. Advances in remote sensing technology allowed TIMED to obtain the first global view of the MLTI. The dynamic MLTI can have significant effects on communications and spacecraft.

NASA launched the 587-kg spacecraft into a 625 km, high-inclination circular orbit on 7 December 2001 on a Delta II booster from Vandenberg Air Force Base, California. The Johns Hopkins University Applied Physics Laboratory built and operated the spacecraft, while Goddard Space Flight Center managed the program.

TIMED carried a spatial scanning, far-ultraviolet spectrograph to measure the temperature and composition of the MLTI, in addition to energy inputs from aurora, a multichannel infrared radiometer to sense heat emitted by the atmosphere; a spectrometer and photometers to study the variations and effect of solar extreme ultraviolet radiation; and an interferometer to monitor wind and temperature within the MLTI.

In 2003 some of the largest geomagnetic storms on record occurred, and TIMED measured the atmosphere's response. Along with other satellites, it seemed to detect penetration into the atmosphere of solar particles normally deflected by Earth's magnetosphere. Coordinated observations with other satellites and ground-based data improved the accuracy of TIMED observations. Designed for a two-year mission, TIMED was still operating in 2007, with an extended mission approved through 2010. The continuous data acquired from TIMED provided a baseline for future study.

Katie J. Berryhill

See also: Goddard Space Flight Center, Johns Hopkins University Applied Physics Laboratory

Bibliography

Applied Physics Laboratory TIMED. http://www.timed.jhuapl.edu/WWW/index.php.

Dave Kusnierkiewicz, "A Description of the TIMED Spacecraft," *American Institute of Physics Conference Proceedings* 387 (January 1997):115–22.

Earth Science

Earth science is a special case of planetary science, focusing on the study of Earth and its major components, including the lithosphere (crust and upper mantle), hydrosphere (water in all its manifestations, most prominently including the oceans), atmosphere, and biosphere (the global sum of ecosystems). This article will not discuss magnetosphere or ionospheric research.

Before the space age, Earth science as such did not exist. Rather there were separate but related studies of geology, geography, geodesy and geophysics, oceanography, meteorology, ecology, and related disciplines.

In 1796 Pierre Laplace proposed his so-called "nebular hypothesis" in which the solar system was formed by the gravitational collapse of the original solar nebula. Physicists realized in the nineteenth century that the Earth would heat up as it initially contracted and would then gradually cool. Continents were assumed to be fixed in position, formed through the crinkling of Earth's crust as the planet cooled.

The alternate idea that the continents might not be fixed grew initially from the recognition of the matching pattern of the continental landforms on either side of the Atlantic, first made by English philosopher Francis Bacon in 1620. German naturalist Friedrich Alexander von Humboldt suggested around 1800 that they had once been connected. In 1912 geologist Alfred Wegener postulated that the continents moved, but his colleagues did not accept this idea, partly due to his inability to identify a mechanism that could create this movement. In the 1950s Harold Urey challenged another aspect of the prevailing views by postulating a theory of cold accretion, in which Earth had formed cool but then heated through radioactive processes. At the dawn of the space age in the late 1950s, geodesists and geographers had mapped Earth's surface, but there was significant uncertainty in measuring the precise positions of the continents with respect to one another. Oceanographers had mapped the major ocean currents, such as the Gulf Stream, and understood much about the chemical constituents of the oceans. In the 1930s weather balloons had provided information about the constituents at higher levels of the atmosphere. Experiments with computer-based predictions of weather began in the 1950s, but were hampered by the lack of computing power and by the lack of global measurements to feed into the atmospheric models. Studies of the oceans and the atmosphere used measurements at particular points around the globe, leaving huge gaps where no measurements existed.

One of the primary benefits of observations from space was to fill in these gaps with truly global measurements. Space observations were of limited utility by themselves. They had to be compared with concurrent observations and measurements on the ground so as to understand what the space observations were actually measuring. This article emphasizes observations made from space and how these supported Earth science, but in the bigger picture, space observations were simply one of many techniques used in the course of the development of these sciences. For example, among the most prominent of these developments was the theory of global warming, for which satellite measurements provided many pieces of evidence, including a slight imbalance in Earth's radiation budget, glacial melting, and daily global temperature measurements. However, there were numerous other ground-based observations that were of equal or greater importance than space observations, both of which fed sophisticated computer models of Earth's climate.

Observations of Earth from space found their origins in photography and scientific experiments from suborbital sounding rockets, starting with V-2 rockets launched from White Sands, New Mexico, starting in 1946. In the early 1950s, U.S. and British scientists conceived of an International Geophysical Year for 1957–58, which was intended as a program of coordinated observations of geophysical phenomena around the globe by scientists from many nations during the next active portion of the solar cycle. Both the United States and the Soviet Union announced that they would launch satellites to aid with these studies. Tracking and analysis of the orbits of *Sputnik* (4 October 1957) and *Explorer 1* (31 January 1958) led to geodetic proof of Earth's slight "pear shape."

One of the first uses of Earth observing satellites was to aid weather forecasting. The Tiros program (first launch 1960) demonstrated the value of satellites for weather forecasting, leading to the first operational satellite systems in the United States and elsewhere. The first geosynchronous satellite to provide weather observations was *Applications Technology Satellite 1* (*ATS 1*), launched in 1966. Geosynchronous satellites proved successful, so that by the twenty-first century, India, China, Japan, the United States, and Europe operated geosynchronous weather satellites. By 1969, NASA's *Nimbus 3* meteorological satellite provided vertical profiles of temperature, water vapor, and ozone concentrations using "sounding techniques" that involved inverting multispectral infrared emission spectra. Later Nimbus spacecraft provided global maps of ozone distribution and began sea ice and snow mapping. *Nimbus 7* (1978) initiated the first measurements of ocean biological productivity through observation and interpretation of ocean color, the first global total ozone content measurements, and a scanning radiometer that greatly enhanced sea ice concentration measurement capability. It was particularly important in confirming the existence of the ozone hole above Antarctica, which had been hypothesized by British scientists in 1985 using ground-based measurements. By the early twenty-first century, spacecraft observations provided a significant proportion of the measurements necessary for computer simulations of the atmosphere for scientific purposes and for weather forecasting.

In October 1959 the *Explorer 7* spacecraft launched with a variety of instruments, including an experiment to measure Earth's radiation budget (comparing the amount of solar energy absorbed by Earth to the energy radiated away from Earth). Radiation budget studies advanced significantly with the launch of the *Solar Maximum Mission* (*SMM*) in 1980, which was the first of many satellites to carry active cavity radiometers. *SMM* demonstrated that total solar irradiance was not constant, but varied by about 0.1 percent throughout the 11-year solar cycle, with a peak at solar maximum and a low at solar minimum. NASA's *ERBS* (*Earth Radiation Budget Satellite*), launched in October 1984, measured the total solar radiation reflected by Earth and the total thermal emission leaving Earth. It also carried an instrument for measuring the vertical distribution of aerosol particles and ozone. Although these satellites carried high-precision instruments, their absolute calibration contained significant uncertainties, such that it was necessary to have overlapping missions to gain a history of total solar irradiance and its variation. In January 2003 the *Solar Radiation and Climate Experiment* carried a precise and accurate Total Irradiance Monitor and also the first instrument for measuring the spectral variation of solar radiation at the top of Earth's atmosphere.

The dynamics of the atmosphere depend critically on the oceans. NASA's *Tiros N* satellite (1978) carried a vertical sounder that included a multispectral infrared sounder and the first microwave sounding instrument, plus a high-resolution radiometer that began a long history of sea surface temperature measurements from space in addition to measurements of terrestrial vegetation and its seasonal variation due

Selected Earth Science Spacecraft, Missions, and Accomplishments (Part 1 of 5)

Satellite	Launched	Accomplishment
Tiros 1	1960	First experimental polar-orbiting meteorology satellite, which used two television (TV) cameras. First ingest of satellite data into weather forecasting models.
Kosmos 4	1962	First Soviet satellite with cloud cover TV system, comparable to Tiros.
Tiros 7	1963	Carried low-resolution energy budget radiometer (the first orbital climate instrument) that failed after three months.
ATS 1, 3	1966, 1967	*ATS 1* provided the first geostationary weather observations of the tropics and mid-latitudes; first full-Earth cloud cover images using a Spin Scan Camera. *ATS 1* was located above the eastern Pacific and *ATS 3* above the western Atlantic.
Meteor 1	1969	First dedicated Soviet weather satellite, offering day and night coverage of clouds, snow, and ice cover.
Nimbus 3, 4	1969, 1970	*Nimbus 3* provided the first U.S. weather satellite to make day and night global measurements from space of temperature, water vapor, and ozone at varying levels in the atmosphere. *Nimbus 4* provided the first global maps of ozone distribution.
Landsat 1–3	1972–1978	First moderate-high resolution visible and infrared images of the land surface for land cover change detection. It used a multispectral scanner (MSS) and a series of video cameras.
SMS 1, 2	1974, 1975	Synchronous Meteorological Satellite, operational prototype geosynchronous satellites designed to make atmospheric observations in the tropics and subtropics.
Nimbus 5, 6	1972, 1975	*Nimbus 5* mapped and measured sea ice for the first time. *Nimbus 6* generated data on Earth's radiation budget.
GOES 1–11	1975–present (2008)	Geostationary Operational Environmental Satellite, weather observations in the tropics and subtropics; cloud tracked winds, hurricane monitoring and tracking, severe storm warnings, fire monitoring (American sector).
Meteosat	1977–present (2008)	Weather observations in the tropics and subtropics; cloud tracked winds, severe storm warnings, and dust storms (European sector).
Tiros N	1978	First microwave sounding of the atmosphere, allowing temperature measurement through clouds. Advanced Very High Resolution Radiometer (AVHRR) provided four-band multispectral imaging with 1 km resolution, enabling ocean surface temperature and vegetation health to be monitored.
Nimbus 7	1978	First satellite to measure natural and human-made atmospheric pollution. Total Ozone Mapping Spectrometer (TOMS) confirmed existence of Antarctic ozone hole in 1985. Coastal Zone Color Scanner (CZCS) made first global observations of ocean biological activity using ocean. Scanning Multichannel Microwave Radiometer (SMMR) greatly enhanced observations of sea ice concentration in polar regions begun by *Nimbus 5*.

(Continued)

Selected Earth Science Spacecraft, Missions, and Accomplishments (Part 2 of 5)

Satellite	Launched	Accomplishment
NOAA 6–18	1978–present (2008)	National Oceanic and Atmospheric Administration operational polar-orbiting satellite system, carrying AVHRR and a vertical temperature and moisture sounder system (TOVS).
Seasat	1978	Used imaging synthetic aperture radar and other instruments to measure ocean surface properties with unprecedented precision. Seasat operated for only 105 days, but inspired use of imaging radar and scatterometers on later spacecraft.
AEM 2	1979	Stratospheric Aerosol and Gas Experiment (SAGE) measured the vertical distribution of atmospheric aerosols and ozone and was used to track plumes from volcanic eruptions in conjunction with instruments on board *Nimbus 7*.
SMM	1980	Measured total solar irradiance for nine years, correlating decrease in solar irradiance with solar activity (in conjunction with *Nimbus 7*).
Meteor 31	1981	Also identified as Meteor-Priroda. Soviet atmospheric research satellite that carried a wide range of instrumentation to measure cloud cover, precipitation, atmospheric water exchange, surface temperature, reflected sunlight, and radiant heat in the atmosphere.
Landsat 4, 5	1982, 1985	Improved land cover change detection and famine early warning using a Thematic Mapper (TM) and MSS.
ERBS	1984	Earth Radiation Budget Satellite carried the Earth Radiation Budget Experiment (ERBE) scanner and nonscanner instruments, beginning a continuous measurement of total solar irradiance from the Sun and reflected solar energy and emitted thermal energy from Earth. The nonscanner instrument continued operation until October 1999. It also carried the SAGE II that measured the vertical distribution of atmospheric aerosols and ozone. ERBE instruments were also carried on *NOAA 9* (1985) and *NOAA 10* (1986)
UARS	1991	The first satellite dedicated to studying stratospheric science, especially the processes that led to ozone depletion. Measured stratospheric ozone and the role of chlorine monoxide in regulating its concentration (especially in the Antarctic ozone hole). *UARS* also measured winds and temperatures in the stratosphere and the energy input from the Sun.
ERS 1, 2	1991, 1995	*ERS 1*, developed by ESA, used synthetic aperture radar to provide all-weather imaging of land and oceans. *ERS 2* added the Global Ozone Monitoring Experiment to measure the concentration of ozone and other trace gases in the atmosphere. First measurements of tropospheric nitrogen dioxide from space. Both spacecraft measured ocean surface wind speed and direction using the scatterometer technique first developed for *Seasat*.
TOPEX/Poseidon	1992	Joint U.S.-French mission to obtain global measurements of sea-surface topography using radar altimetry to an accuracy of 4.2 cm, contributing to improved forecasting of

Selected Earth Science Spacecraft, Missions, and Accomplishments (Part 3 of 5)

Satellite	Launched	Accomplishment
		the 1997–98 El Niño. Patterns of ocean circulation (including eddies) and tides established with unprecedented accuracy.
Radarsat 1	1995	Canadian synthetic aperture radar satellite provided high-resolution, all-weather imagery. Detected ice breakers, oil spills, and routine surveillance of the entire Arctic region in all weather conditions.
SOHO	1995	International partnership between NASA and ESA to study the sun from its deep core to the outer corona and the solar wind. Solar and Heliospheric Observatory carried an instrument to make precise measurements of the total solar irradiance reaching Earth.
TOMS-EP	1996	Long-term monitoring of Earth's ozone layer in conjunction with other TOMS sensors on board *Nimbus 7* and *Meteor 3* spacecraft. Confirmed the existence of large ozone holes above Antarctica in 2000 and 2003.
Midori I, II	1996, 2002	Midori was developed by the Japan Aerospace Exploration Agency with contributions from NASA and CNES. *Midori I* measured ocean vector winds, total ozone, ocean color, aerosol concentration, and trace gas constituents of the atmosphere, but failed after 10 months. *Midori II* was a successor mission that carried NASA's SeaWinds scatterometer for ocean vector winds, but also contained a microwave imager and global imager. *Midori II* also failed after 10 months.
OrbView 2	1997	Sea-Viewing Wide Field-of-View Sensor (SeaWiFS) measured global ocean bio-optical properties arising from photosynthetic activity of marine phytoplankton, most notably chlorophyll-a, using the technique of ocean color.
TRMM	1997	*TRMM*, jointly operated by NASA and the Japan Aerospace Exploration Agency, measured rainfall between 35° N and 35° S using precipitation radar. Other instruments on the spacecraft allowed global lightning detection, sea surface temperature, and continued measurement of Earth's radiation budget.
Landsat 7	1999	Repetitive acquisition of moderate resolution observations of Earth's land mass, coastal boundaries, and coral reefs using the Enhanced Thematic Mapper Plus. Together with *Landsat 5*, monitors land cover change (deforestation, urban expansion), vegetation stress (famine early warning), and glaciers and seasonal snow pack.
QuikSCAT	1999	Acquired accurate, high-resolution, continuous all-weather measurements of global radar cross-sections and near-surface vector winds over the ice-free global oceans using a wide-swath scatterometer. Complements *ERS 1*, *ERS 2*, and *Envisat* scatterometers. Same instrument that was on Japan's *Midori II* spacecraft, which lasted only 10 months on orbit.
Terra	1999	Earth Observing System (EOS) spacecraft that provided global data on the state of the atmosphere, land, and oceans, and their interactions with solar radiation and with one another. It measures ocean color, sea surface

(Continued)

Selected Earth Science Spacecraft, Missions, and Accomplishments (Part 4 of 5)

Satellite	Launched	Accomplishment
		temperature, fires, Earth radiation budget, aerosol and cloud properties, carbon monoxide, and vegetation health and evolution.
Acrimsat	1999	Active Cavity Radiometer Irradiance Monitor III instrument was (in 2008) extending solar irradiance measurement with much higher absolute accuracy and stability (better than 0.1 percent) to provide a continuous record of solar irradiance since 1975, based on measurements from earlier spacecraft.
Jason 1	2001	EOS mission that was a follow-on to *TOPEX/Poseidon*, measuring sea-level anomalies, significant wave height, currents, and ocean eddies.
Meteor 3M	2001	A partnership with the Russian Aviation and Space Agency, the SAGE III on board the spacecraft measured the vertical distribution of atmospheric aerosols and ozone using both solar and lunar occultation at high northern and southern latitudes.
GRACE	2002	A joint NASA and German mission, GRACE enabled the most precise measurements of Earth's mean and time-variable gravity field ever obtained, thereby providing continental and regional water balance through monitoring of changes in aquifers.
Envisat 1	2002	ESA environmental remote sensing spacecraft, with ten sensors to monitor ocean surface topography, atmospheric chemical constituents, ocean vector winds, ocean biological productivity, desertification, and fires. Produced global map of NO_2 pollution in 2004.
Aqua	2002	EOS spacecraft that provided global data on the state of the atmosphere, land, and oceans, with a particular emphasis on the water cycle. It complemented Terra observations on two instruments, but also included a hyperspectral atmospheric sounder for vertical profiles of temperature and water vapor, and provided a precise scanning microwave radiometer that provided measurements on sea ice concentration, ocean wind speed, surface temperature, cloud water content, and surface soil moisture.
ICESat	2003	Ice, Cloud, and land Elevation Satellite was an EOS mission that provided ice sheet altimetry and sea ice thickness, using satellite lidar for the first time. First global measurements of vertical cloud properties and multilayer clouds in addition to vegetation elevation and structure.
SORCE	2003	Solar Radiation and Climate Experiment, EOS spacecraft that provided the first measurements of spectral solar irradiance that complemented a precise total solar irradiance measurement.
Aura	2004	EOS spacecraft that measured the atmosphere's chemistry and dynamics, investigated questions about ozone trends, air-quality changes, and their linkages to climate change. Extended satellite total ozone record begun by TOMS in 1978, and enhanced observations of tropospheric ozone, nitrogen dioxide, and sulfur dioxide.

Selected Earth Science Spacecraft, Missions, and Accomplishments (Part 5 of 5)

Satellite	Launched	Accomplishment
PARASOL	2004	Polarization and Anisotropy of Reflectances for Atmospheric Sciences coupled with Observations from a Lidar, a French-built satellite that carried the Polarization and Directionality of the Earth's Reflectances (POLDER) instrument for monitoring the global distribution of cloud and aerosol properties using polarization of reflected sunlight.
CALIPSO	2006	Cloud-Aerosol Lidar and Infrared Pathfinder Satellite Observations, a joint NASA and CNES mission that used space-borne lidar to probe the vertical distribution of aerosol and cloud properties in the atmosphere.
CloudSat	2006	A joint NASA and Canadian Space Agency mission that used spaceborne radar to probe the vertical distribution of cloud systems and measure their liquid water and ice-water content.
MetOp 1	2006	First operational polar-orbiting satellite system of the EUMETSAT (Europe) that carried AVHRR and TOVS provided by the United States (NOAA) and an Infrared Atmospheric Sounding Interferometer for improved temperature and water vapor sounding provided by Europe. It also carried the Global Ozone Monitoring Experiment 2 (GOME 2), a follow-on to GOME that flew on *ERS 2* that measures atmospheric chemical constituents of the atmosphere. It also carried a scatterometer for measuring vector winds above the global ocean.
OSTM	2008	Ocean Surface Topography Mission, *Jason 2*, follow-on to *Jason 1*, measured sea-level anomalies, significant wave height, currents, and ocean eddies.

to the growth and senescence of vegetation associated with photosynthesis of green vegetation. That same year, ocean remote sensing made other major strides with the launch of the Coastal Zone Color Scanner (CZCS) on *Nimbus 7*, which made the first measurements of chlorophyll and ocean biology, and *Seasat*, which carried the first synthetic aperture radar and scatterometers. *Seasat* lasted only 105 days, but showed the value of scatterometers to measure wind speed and direction above the ice-free global ocean.

NASA followed during the next decade with a number of Earth probes with specific environmental objectives. The *UARS* (*Upper Atmosphere Research Satellite*), 1991, carried instruments for studying stratospheric chemistry leading to ozone depletion, stratospheric temperature and winds, and total solar irradiance, as well as spectral solar irradiance in the far ultraviolet. *TOPEX/Poseidon* (1992) was the start of a fruitful collaboration between NASA and CNES (French Space Agency) to make precision radar altimetry measurements of the global ocean. This started observations of ocean circulation, eddies, tides, currents, and sea level variations, and played a vital role in studies of ocean circulation in the Pacific associated with El Niño and La Niña events, and of the thermohaline "global conveyor," in which warm surface waters are driven northward, only to cool near the poles and sink, flowing along the ocean floors and eventually upwelling in other locations. In both cases, scientists had pieced together evidence of these global circulation movements by the 1970s and 1980s; by the 1980s and 1990s satellites provided precise global data, such as sea surface temperatures, phytoplankton mass, and sea surface heights to support their theories and models.

By the 1990s other nations joined the constellation of atmospheric and ocean observation satellites. In 1991 the European Space Agency (ESA) launched *European Remote Sensing Satellite 1* (*ERS 1*), followed in 1995 by *ERS 2*. Both carried radar altimeter instruments similar to *TOPEX/Poseidon* but in different orbits, a scatterometer for ocean vector winds, synthetic aperture radar, and a new along-track scanning radiometer for precise sea surface temperature measurements. *ERS 2* added an ozone-monitoring experiment. In 1996 Japan launched *Midori 1*, which carried several instruments, including an interferometer for studying greenhouse gases, a limb atmospheric spectrometer, a NASA scatterometer, and a CNES instrument for measuring atmospheric aerosols and clouds. When it failed the next year, NASA quickly developed and launched *QuikSCAT* (1999), which carried the SeaWinds scatterometer to replace *Midori 1*'s lost capability. *Midori 2* (2002) carried five instruments, including SeaWinds. In 1997, NASA, again partnering with Japan, launched the *TRMM* (*Tropical Rainfall Measuring Mission*). It carried instruments that could measure the three-dimensional distribution of rainfall in the tropics, precipitation, and sea surface temperature beneath tropical storms (such as hurricanes and typhoons), Earth's radiation budget, and the occurrence of lightning. In conjunction with CNES, NASA launched the radar altimeter missions *Jason 1* (2001) and *Jason 2* (2008), which measured sea surface height to an accuracy of 3 cm. By the early twenty-first century, India (Indian Remote Sensing series) and China (Feng Yun and Haiyang series) deployed their own satellites with atmosphere- and ocean-observation capabilities.

By the 1990s scientists recognized that oceanic biological organisms played significant roles in climate studies. To enhance observations of primary production, defined as the mass of carbon-based material produced by organisms in the global oceans, NASA worked with Orbital Sciences Corporation's *OrbView 2* satellite (1997) that carried the Sea-viewing Wide Field-of-view Sensor, or SeaWiFS, a follow-on to *Nimbus 7*'s CZCS. NASA processed and distributed worldwide the ocean color data on chlorophyll-a concentration.

Monitoring physical, biological, and human phenomena on land was another major area of spacecraft development. Growing from the success of Earth photographs taken from Mercury and Gemini, in 1965 the U.S. Geological Survey proposed development of a land resource remote sensing satellite. Despite some opposition from the U.S. military and intelligence communities, which did not want to draw attention to their secret satellite reconnaissance missions, and a number of foreign nations, which objected to U.S monitoring of their natural resources, this led to the Landsat program. *Landsat 1–3*, launched between 1972 and 1978, each carried a video camera and a multispectral scanner that provided the first moderate-high resolution, visible, and infrared images of the land surface. *Landsat 4* and *5*, launched in 1982 and 1985, added a Thematic Mapper with higher spatial resolution. With Landsat's lengthening record, it became possible to examine land cover change over time, including growth

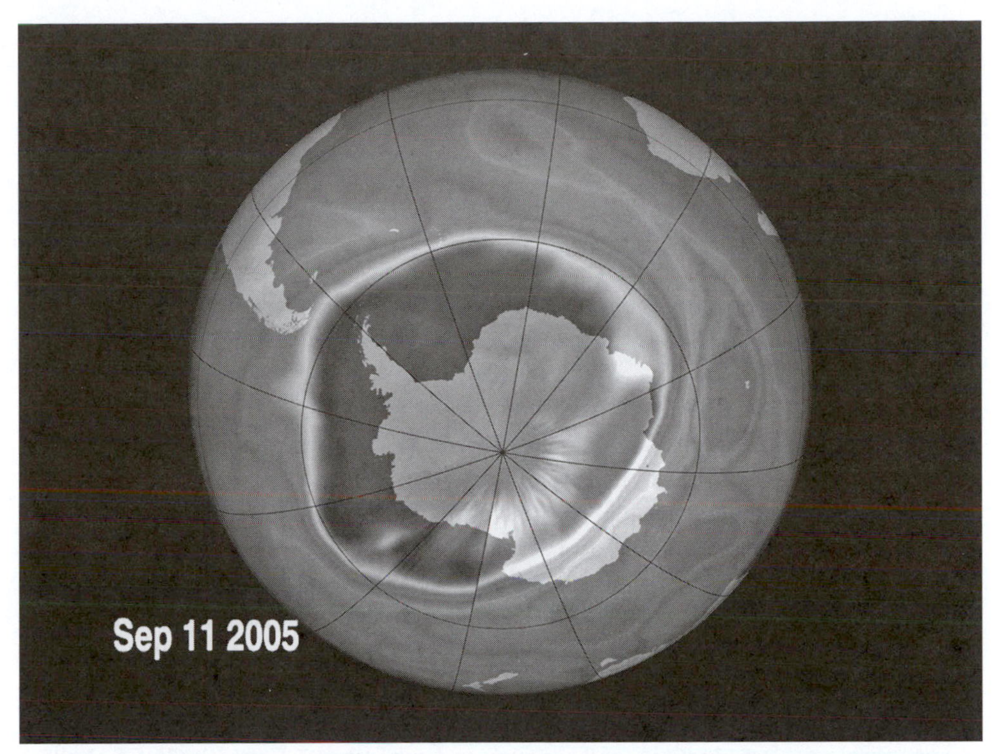

An image based on data acquired by the Aura *satellite, showing the "ozone hole" over Antarctica during the Antarctic winter of 2005. Each year the hole, actually a region of exceptionally depleted ozone in the stratosphere, expands during the Antarctic winter. (Courtesy NASA/Goddard Space Flight Center Scientific Visualization Studio)*

of urban areas, land cover transformation from biomass burning and agricultural clearing, and glacier and snow pack evolution. These data were soon adopted by the United Nations and used in its Famine Early Warning System. *Landsat 7*, launched in 1999, provided an enhanced thematic mapper with a panchromatic (broadband visible) band at 15 m spatial resolution that, together with the multispectral bands, sharpened land features in the imagery. Its enhanced data handling capability enabled extensive mapping of Antarctica, coral reefs, lava lakes, and volcanoes worldwide.

The Soviet Union, India, France, and others soon developed their own land resource satellites. The Soviet Union's first series of land remote sensing satellites used for civilian purposes were Meteor Priroda (later called Resurs-O), first orbited in 1974. It transmitted low-resolution data from orbit via radio, while the Resurs-F series took higher-resolution imagery and sent the images to Earth via film capsules. India emphasized land remote sensing satellites to aid national development and launched its first experimental remote sensing satellite, *Bhaskara 1*, in 1979. The operational system, the Indian Remote Sensing series, first launched in 1988. France placed its first SPOT (Satellite Pour l'Observation de la Terre) satellite in orbit in 1986. In the 1990s and 2000s, other nations developed their own land remote sensing systems for scientific, commercial, educational, and military purposes.

Space observations revolutionized studies of Earth's ice features, by providing the data needed to measure ice depth and dynamics. Landsat provided data on the extent of snow and ice coverage, showing decreasing glacier sizes around the world. Cryosphere studies further developed with the launch of Canada's *Radarsat 1* (1995) and *2* (2007). NASA's *Ice, Cloud, and Land Elevation Satellite* (2003) carried the first laser altimeter (lidar) into low Earth orbit from which ice sheet mass balance, cloud and aerosol heights, and land topography and vegetation characteristics were measured. By 2006 data from these satellites showed that the world's major ice sheets in Antarctica and Greenland were losing ice to the oceans at their peripheries, though there was continued debate as to whether this is balanced by growth in ice thickness at the center of the sheets.

By the late 1980s the importance of integrated measurements of land, ocean, and atmospheric phenomena became more apparent. NASA began the development and deployment of its Earth Observing System (EOS) of multiple Earth-orbiting satellites. These satellites included multiple instrument platforms and focused single-instrument missions. It also included international partnerships with France, Japan, the United Kingdom, Brazil, Russia, Canada, and others. The first three multi-instrument EOS platforms were *Terra* (1999), *Aqua* (2002), and *Aura* (2004). Each had special instruments for land, sea, and atmosphere, respectively. *Terra* provided global data on the state of the atmosphere (clouds, aerosols, water vapor, carbon monoxide), land (surface temperature, reflectance, vegetation canopy and biophysical properties, fire), oceans (biological productivity and sea surface temperature), Earth's radiation budget and the effects of clouds on reflected solar and emitted thermal radiation. *Aqua* carried Earth radiation budget instruments and added an infrared sounder and microwave imagers to derive the vertical profile of temperature and water vapor in the troposphere and stratosphere. Its microwave scanning radiometer enabled measurements of above-ocean water vapor, precipitation, soil moisture, and sea ice concentration, contributing to hydrological cycle studies.*Aura* provided instruments to measure troposphere and lower stratosphere chemical constituents, helping to unravel human contributions to atmospheric chemical changes. Its Dutch-Finnish Ozone Monitoring Instrument extended the record of total ozone begun by NASA's *Nimbus 7*. *Aura*'s Microwave Limb Sounder helped identify anthropogenic chlorine's role in the destruction of ozone above Antarctica.

Other nations also orbited multi-instrument missions to add to the pool of combined data from many ground- and space-based sources. In 2001 Russia launched *Meteor 3M* carrying NASA's Stratospheric Aerosol and Gas Experiment III solar and lunar occultation instrument, which measured the vertical distribution of atmospheric aerosol particles, ozone, and water vapor from the upper troposphere through the stratosphere. In 2002, ESA launched *Envisat 1*, which contained sophisticated sensors for measurements of ocean surface topography, atmospheric chemical constituents, and primary productivity, primarily in coastal waters.

Geodetic and navigation satellites provided an important means of studying classical geology and geophysics. Military and intelligence organizations needed to know the precise locations of their targets and assets around the world, in particular for ballistic missile launches. They funded the bulk of the geodetic and navigation satellites and systems starting in the early 1960s. Satellite orbit variations held clues to the underlying geology at the points of those variations. Some geodetic satellites were simple multiple mirror designs to allow ground-based laser ranging (bouncing laser beams from different locations on Earth off the satellite to determine its precise location). Once the satellite orbits were known, more sophisticated navigational satellites, such as the U.S. Global Positioning System (GPS) and the Soviet Union's Global Navigation Satellite System, with precise timing signals, could be used to precisely determine locations on Earth.

Analysis of variations in satellite orbits yielded scientific discoveries, such as the slowly increasing altitude of the Canadian north, which is rebounding from the removed weight of the last ice age ice sheet. Geophysicists revised their estimates of deep Earth viscosity, since larger viscosities were necessary to account for the continuing rebound. In the 1960s geologists using deep ocean floor topographic mapping information and magnetic studies from various ship- and submarine-based surveys discovered that the mid-ocean ridges were spreading over time, formed from the upwelling of magma. This led to the acceptance of plate tectonics by the late 1960s and early 1970s. By the 1990s, GPS satellites provided the primary mechanism to be able to measure the slow millimeters per year movement of the various plates with respect to one another. GPS satellites also enabled precise studies of local ground movements due to volcanism and earthquakes. Other specialized satellites such as *Magsat* (1979) enabled global mapping of Earth's magnetic field variations. Measurements from the Gravity Recovery and Climate Experiment (2002) supported studies of gravity variations over time due to surface and deep currents in the ocean, ground water storage on land masses, and exchanges between ice sheets or glaciers and the ocean.

By the end of the first decade of the twenty-first century, considerable progress had been made in quantifying the spatial and temporal distribution of atmospheric constituents, ocean properties, solar radiation and variations on the Sun, volcanic eruptions and their effect on climate, and land cover and land use change. The view of Earth is greatly enhanced by continuous monitoring provided by space agencies throughout the world, coupled with advances in computer processing and workstation visualization capability, Internet distribution of results, and rapid availability of data to observe natural and human-made hazards.

Michael D. King and Stephen B. Johnson

See also: Civilian Remote Sensing, Commercial Remote Sensing, European Space Agency, India, Japan, National Aeronautics and Space Administration

Bibliography

Michael D. King et al., eds., *Our Changing Planet: The View from Space* (2007).

Paul Lowman, *Exploring Space, Exploring Earth: New Understanding of the Earth from Space Research* (2002).

National Research Council, *Earth Observations from Space: The First 50 Years of Scientific Achievements* (2008).

World Climate Research Program, *A Short History of a Long Success Story* (2006).

ALOS, Advanced Land Observation Satellite. *See* Japanese Land Observation Missions.

ATLAS, Atmospheric Laboratory for Applications and Science. ATLAS, Atmospheric Laboratory for Applications and Science, was flown on a series of three Space Shuttle missions. Occurring during three years, ATLAS was funded as part of NASA's Mission to Planet Earth (later known as Earth Science Enterprise) to simultaneously study the chemistry of Earth's middle atmosphere and the energy output of the Sun, and how they affect global levels of ozone. Additionally, since the instruments could be precisely calibrated before and after flight, they allowed calibration of similar instruments onboard satellites, in particular the *Upper Atmosphere Research Satellite*, which was launched six months before the ATLAS 1 and had direct counterparts to two ATLAS instruments. ATLAS was an international project, with investigators from the United States, Belgium, Germany, France, the Netherlands, and Switzerland. Marshall Space Flight Center managed the ATLAS program.

The Space Shuttle *Atlantis* flew ATLAS 1 and ATLAS 3 in March 1992 (Space Transportation System—STS-45) and November 1994 (STS-66), respectively, while the Space Shuttle *Discovery* flew ATLAS 2 in April 1993 (STS-56). All used European *Spacelab* pallets to house the instruments. Seven instruments flew on all three flights, allowing scientists to study temporal changes. These included two instruments to study the composition of the atmosphere, a spectrometer to measure ozone concentration, two instruments measuring the total energy from the Sun (the solar constant), and two studying the distribution of solar energy and the ultraviolet component, in particular.

ATLAS Instruments (Part 1 of 2)

Instrument	Purpose	Flown on ATLAS Missions	Contractor
Millimeter-wave Atmospheric Sounder (MAS)	Water vapor, ozone, chlorine monoxide in mesosphere and stratosphere; temperature and pressure	1, 2, 3	European Aeronautic Defence and Space Astrium GmbH
Atmospheric Trace Molecule Spectroscopy (ATMOS)	Distribution by altitude of different stratospheric gases	1, 2, 3	NASA/Jet Propulsion Laboratory
Shuttle Solar Backscatter Ultraviolet (SSBUV)	Ozone concentrations	1, 2, 3	NASA/Goddard Space Flight Center
Solar Spectrum Measurement (SOLSPEC)	Solar energy distribution by wavelength (infrared through ultraviolet)	1, 2, 3	Centre National de la Recherche Scientifique (CNRS), France

ATLAS Instruments (Part 2 of 2)

Instrument	Purpose	Flown on ATLAS Missions	Contractor
Solar Ultraviolet Irradiance Monitor (SUSIM)	Solar ultraviolet radiation	1, 2, 3	U.S. Naval Research Laboratory
Active Cavity Radiometer Irradiance Monitor 2 (ACRIM 2)	Solar constant	1, 2, 3	NASA/Jet Propulsion Laboratory
Solar Constant (SOLCON)	Solar constant	1, 2, 3	Royal Meteorological Institute of Belgium (IRMB), Belgium
Atmospheric Lyman-Alpha Emissions (ALAE)	Abundances of hydrogen and deuterium	1	CNRS, France
GRILLE Spectrometer	Mapping trace atmospheric molecules	1	France/Belgium
Atmospheric Emissions Photometric Imaging (AEPI)	Charged particle and plasma environment	1	Lockheed Palo Alto Research Laboratory
Space Experiments with Particle Accelerators (SEPAC)	Charged particle and plasma environment; Produced first artificial auroras in upper atmosphere	1	Southwest Research Institute
Imaging Spectrometric Observatory (ISO)	Composition of the atmosphere (down to parts per trillion)	1	NASA/Marshall Space Flight Center
Cryogenic Infrared Spectrometers and Telescopes for the Atmosphere-Shuttle Pallet Satellite (CRISTA-SPAS)	Disturbances in trace gases in middle atmosphere	3 (free-flying, deployed and retrieved during mission/reflown on STS-85)	German Space Agency—DLR (Deutsches Zentrum für Luft- und Raumfahrt)
Experiment of the Sun for Complementing the Atlas Payload and for Education 2 (ESCAPE 2)	Effect of solar extreme ultraviolet radiation on temperature and composition of upper atmosphere	3 (student designed and built)	University of Colorado, Boulder
Far Ultraviolet Space Telescope (FAUST)	Wide-field images in the far ultraviolet of large-scale phenomena	1 (and *Spacelab 1*/ STS-9)	University of California, Berkeley

Among the many results of these missions, ATLAS 1 measured changes in middle atmosphere chemistry due to the Mount Pinatubo eruption nine months previously. The seasonal difference in launch dates specifically enabled scientists to study processes involved in the atmospheric shifts that take place as the northern hemisphere moves from summer to winter, in addition to the Antarctic ozone hole, which usually peaks in the southern spring. Data from all three missions was the continuation of systematic data collection to determine benchmarks for measurements of atmospheric processes and solar variability.

Katie J. Berryhill

See also: Space Shuttle, *Spacelab*

Bibliography

Jack A. Kaye and Timothy L. Miller, "The ATLAS Series of Shuttle Missions," *Geophysical Research Letters* 23, no. 17 (1996): 2285–88.

Atmospheric Science and Climate Research. Atmospheric science and climate research benefited significantly from the use of space-based instrumentation in the late twentieth and early twenty-first centuries. Prior to World War II, study of the atmosphere was largely performed using traditional instruments at weather stations on the surface, augmented by automatic instruments carried aboard balloons. Measurement of ocean properties was performed using sensors aboard ships. During the war a German Army rocket research team under Wernher von Braun commissioned balloon researcher Erich Regener to develop an automatic instrument package to replace the warhead on a V-2 rocket. Von Braun's interest was driven by the need to better understand the extreme upper atmosphere, in order to design a projected two-stage intercontinental missile that could reach the United States from bases in Nazi-occupied Europe.

While Regener's instrument package never flew, its discovery by U.S. scientists after the war inspired a program of sounding rocket research that began in 1946. Upper atmosphere research using balloons and sounding rockets received significant military support after World War II, as it was crucial for the development of ballistic missiles and reentry vehicles. By 1954 rocket-based pressure and temperature measurements were used in a major revision of the U.S. standard atmosphere tables. Similar research programs were also carried out in the Soviet Union, France, the United Kingdom (UK), and (on a limited scale) Japan in the late 1940s and 1950s.

A major limitation of rocket-based measurements was their extremely limited duration, which caused some scientists to oppose the entire program. An artificial Earth satellite would obviate this restriction and was one of the scientific justifications for President Dwight Eisenhower's 1955 announcement of plans for what became Project Vanguard, to be launched during the International Geophysical Year of 1957–58. The Vanguard satellites' spherical design was selected to simplify computation of atmospheric density.

Many early atmospheric science experiments aboard satellites focused on weather measurements. *Explorer 6* broadcast the first television pictures taken from orbit in 1959, showing cloud formations. It was followed by the *Tiros 1* weather satellite in 1960.

Selected Atmospheric Science and Climate Research Spacecraft, Missions, and Instruments (Part 1 of 4)

Experiment	Launched	Comments
Sounding rockets	1946–present (2008)	Wide variety of rockets launched by the United States, Soviet Union/Russia, European countries, and Japan providing brief measurement of atmospheric temperature, pressure, composition, and density. The first widely accepted scientific result was a revised U.S. standard atmosphere table published in 1954.
Explorer 6	1959	First television (TV) pictures from space, showing cloud formations.
Vanguard 2	1959	First measurement of cloud cover distribution from orbit, with poor results due to unsatisfactory spin axis orientation.
Tiros 1, 2	1960	First series of dedicated meteorology satellites, using TV cameras.
Kosmos 4	1962	First Soviet satellite with cloud cover TV system, comparable to Tiros.
Tiros 7	1963	Carried low-resolution heat budget radiometer (possibly the first orbital climate instrument) that failed after three months.
Nimbus 1, 2	1964, 1966	*Nimbus 1* provided the first global high-resolution images of cloud systems. *Nimbus 2* included an additional infrared instrument to study the effect of water vapor and ozone on Earth's heat balance and ocean temperature measurement.
Meteor 1	1969	First dedicated Soviet weather satellite, offering day and night coverage of clouds, snow, and ice cover.
Nimbus 3, 4	1969, 1970	*Nimbus 3* (1969), carried the first space-based sounding instruments provided vertical profile of atmospheric temperature, water vapor, and ozone concentration; also located and interrogated remote sensor platforms. *Nimbus 4* (1970) provided the first global maps of ozone distribution.
Nimbus 5, 6	1975	*Nimbus 5* mapped and measured sea ice. *Nimbus 6* generated data on Earth's radiation budget.
Explorer 58	1978	Heat Capacity Mapping Mission instrument provided high spatial resolution (500 m^2) maps of Earth's surface temperature.
Global Atmospheric Research Program (GARP), First GARP Global Experiment (FGGE)	1978	Joint program of the United Nations World Meteorological Organization and International Council of Scientific Unions. Provided continuous weather observations from five geostationary satellites (two from the United States and one each from Japan, Europe, and the Soviet Union) spaced approximately equally around Earth's equator for one year, paving the way for later

(Continued)

Selected Atmospheric Science and Climate Research Spacecraft, Missions, and Instruments (Part 2 of 4)

Experiment	Launched	Comments
		internationalprograms in global weather observation and climate research.
Tiros N	1978	First microwave sounding of the atmosphere, allowing temperature measurement through clouds. French-provided Argos system allowed the Tiros N series to collect, store, and transmit data from balloons and buoys. Advanced Very High Resolution Radiometer provided four-band multispectral imaging with 1 km resolution, enabling ocean surface temperature measurement.
Nimbus 7	1978	First satellite to measure natural and human-made atmospheric pollution. Total Ozone Mapping Spectrometer (TOMS) instrument confirmed existence of Antarctic ozone hole in 1985. Coastal Zone Color Scanner instrument revealed previously unsuspected complex turbulent mixing.
Seasat	1978	Used imaging synthetic aperture radar and other instruments to measure ocean surface properties with unprecedented precision. Seasat operated for only 105 days, but inspired use of imaging radar on later spacecraft.
Explorer 60	1979	Stratospheric Aerosol and Gas Experiment measured the effect of changes in solar radiation on atmospheric aerosols and ozone and was used to track ash plumes from volcanic eruptions in conjunction with instruments on board *Nimbus 7*.
Solar Maximum Mission	1980	Measured total solar irradiance for nine years, correlating decrease in solar irradiance with sunspot activity (in conjunction with *Nimbus 7*).
Meteor 31	1981	Also identified as *Meteor-Priroda*. Soviet atmospheric research satellite that carried a wide range of instrumentation to measure cloud cover, precipitation, atmospheric water exchange, surface temperature, reflected sunlight, and radiant heat in the atmosphere.
Earth Radiation Budget Experiment (ERBE)	1984, 1985, 1986	ERBE, carried by *Earth Radiation Budget Satellite* (1984) and also by *NOAA 9* (National Oceanic and Atmospheric Administration, 1985) and *NOAA 10* (1986), began continuous measurement of total incident solar radiation and reflected energy from Earth using scanning and non-scanning instruments. The nonscanning instruments continue in operation in 2008.
European Remote Sensing (ERS) 1, 2	1991, 1995	*ERS 1* (1991) used synthetic aperture radar to provide all-weather imaging of land and oceans. *ERS 2* (1995) added Global Ozone Monitoring Experiment to measure the concentration of ozone and other trace gases in the atmosphere.

Selected Atmospheric Science and Climate Research Spacecraft, Missions, and Instruments (Part 3 of 4)

Experiment	Launched	Comments
Atmospheric Laboratory for Applications and Science 1, 2, 3	1992–94	Instrument platform carried on Space Transportation System (STS)-45, STS-56, and STS-66. Measured changes in middle atmosphere chemistry due to the Mount Pinatubo volcano eruption, continued systematic data collection to determine benchmarks for measurements of atmospheric processes and solar variability.
TOPEX/Poseidon (*Topography Experiment*)	1992	Joint U.S.-French mission to obtain global measurements of sea-surface heights using radar altimetry to an accuracy of 4.2 cm, enabling mission scientists to forecast the 1997–98 El Niño.
Radarsat 1	1995	Canadian observation satellite using synthetic aperture radar to provide high-resolution, all-weather imagery. Detected collapse of Ross Ice Shelf in 2002.
TOMS-EP (Total Ozone Mapping Spectrometer-Earth Probe)	1996	Long-term monitoring of Earth's ozone layer in conjunction with other TOMS sensors on board *Nimbus 7* and *Meteor 3* spacecraft. Confirmed the existence of large ozone holes above Antarctica in 2000 and 2003.
Tropical Rainfall Measuring Mission (*TRMM*)	1997	*TRMM*, jointly operated by NASA and National Space Development Agency (Japan), measured rainfall between 35 degrees N and 35 degrees S using precipitation radar. Other instruments on the spacecraft allowed global lightning detection and continued measurement of Earth's radiation budget.
ACRIMSAT	1999	Active Cavity Radiometer Irradiance Monitor III instrument is currently (2008) extending solar irradiance measurement with much higher absolute accuracy and stability (better than 0.1 percent) to provide a continuous record of solar irradiance since 1975, based on measurements from earlier spacecraft.
Living Planet Program	1999–present (2008)	European Space Agency (ESA) program including science-driven Earth Explorer missions, operationally oriented Earth Watch missions, and Global Monitoring for Environment and Security missions in collaboration with the European Commission.
Earth Observing System (EOS)	2000–present (2008)	The EOS constellation supports NASA's Mission to Planet Earth and includes *Terra* (2000), *Aqua* (2002), *Aura* (2004), and other spacecraft that function as a sensor web in conjunction with ground- and sea-based sensors. In 2005, the web detected the eruption of an apparently dormant volcano on Sumatra and autonomously commanded *Terra* to begin observing before most scientists realized the eruption had occurred.

(Continued)

Selected Atmospheric Science and Climate Research Spacecraft, Missions, and Instruments (Part 4 of 4)

Experiment	Launched	Comments
Thermosphere Ionosphere Mesosphere Energetics and Dynamics (TIMED)	2001	Observed interactions between the Sun and upper layers of Earth's atmosphere, including what appeared to be penetration of solar particles deeper than expected during 2003 geomagnetic storms. Observations continued as of 2007 to provide a baseline for future research.
Envisat 1	2002	ESA environmental remote sensing spacecraft, with ten sensors to monitor global warming, ozone hole, and desertification. Produced global map of NO_2 pollution in 2004.
Jason 1	2002	Follow-on to *TOPEX/Poseidon*, measuring monitor global climate interactions between the sea and the atmosphere, including sea-level anomalies, significant wave height, and ocean wind speed.

Intended as an experimental series, Tiros results proved so valuable that the satellite and its successors quickly became an operational system. The Soviet Union began its comparable Meteor system of weather satellites in 1964.

In an odd reversal of plans, NASA's sophisticated Nimbus design for an operational weather system proved too expensive for the Weather Bureau; instead it became a platform to develop instrumentation for remote sensing. Nimbus produced a wide range of significant results, including infrared measurements related to the effect of water vapor and ozone on Earth's heat balance and ocean temperature by *Nimbus 2* in 1966; space-based sounding (vertical profile) of atmospheric temperature, water vapor, and ozone concentration by *Nimbus 3* in 1969; and sea ice mapping by *Nimbus 5* in 1975.

Despite such early efforts, space-based instruments contributed little to the understanding of ocean circulation until 1978, when the Advanced Very High Resolution Radiometer instrument carried onboard *Tiros N* made possible sea surface temperature measurements covering entire oceans. That same year the Coastal Zone Color Scanner instrument on *Nimbus 7* provided high-resolution maps of chlorophyll concentration revealing complex turbulent mixing features not suspected before, and a wide range of instruments, including imaging synthetic aperture radar onboard the Jet Propulsion Laboratory (JPL) *Seasat* spacecraft, measured ocean surface properties, including wind speed and direction, topography, and polar ice. These observations have since been combined with in situ water column measurements to provide evidence of anthropogenic global warming in the oceans.

Until late in the twentieth century space-based instruments were limited to detecting surface properties of the ocean. Deep ocean circulation features, such as the thermohaline circulation (or "great ocean conveyor"), were identified by in situ methods (including current flow and radioisotope measurement) and did not benefit from space-based instrumentation until the development of satellite altimetry in the 1970s. By the 1990s centimeter-accuracy altimetry instruments carried aboard spacecraft, including the *European Remote Sensing 1* satellite and the joint U.S.-French *TOPEX/Poseidon*, provided evidence of internal tides, a previously unsuspected major feature of circulation in the deep ocean. These measurements assisted mission scientists in forecasting the 1997–98 El Niño event.

Space-based sensors measured global pollution, including the *TOMS-EP*, first carried onboard *Nimbus 7* launched in 1978. A 1974 laboratory study had suggested that the ozone layer could be depleted by interaction with chlorofluorocarbons (CFCs). In 1984 ground-based data from the United Kingdom's Antarctic survey indicated that ozone concentrations were heavily depleted above the South Pole in the spring, producing what became known as a seasonal ozone hole. This was confirmed by a research team, under Richard McPeters at NASA's Goddard Space Flight Center, which had already observed, in TOMS data, ozone levels below previously observed limits. By 1987 multiple observations that the hole above the Antarctic was forming seasonally—and increasing in size each season—led to the signing of the Montreal Protocol, under which signatory nations agreed to phase out CFCs, allowing atmospheric ozone levels to recover. While CFCs have not been completely phased out, in 2007 a slight decline was observed in the Antarctic hole's maximum size, and the minimum concentration of ozone in the hole was rising slowly since 1998.

Concern about anthropogenic effects on the global climate increased in the 1970s, driven in part by observation of cooling trends in previous decades, which led some scientists to suspect Earth might be entering a new ice age. Further examination of data from a wide range of surface-based sensors eventually refuted this notion. Interestingly, while most of the early measurements came from ground-based weather stations, a major step forward in the analysis came when a research team under NASA's James Hansen analyzed the available data using methods developed for planetary research. Hansen's team discovered that while a cooling trend had indeed occurred, throughout the previous three decades in the northern hemisphere, it was more than compensated for by a warming trend in the southern hemisphere. A research group at the UK's University of East Anglia confirmed Hansen's results and extended them into the nineteenth century by using historical records. By 1988 a consensus developed that Earth was warming, though not everyone (then or since) was convinced. A variety of space-based sensors was developed to measure radiation from the Sun incident on Earth, and radiation emitted from Earth into space. Early measurements in the 1960s showed that Earth was warmer and darker than expected, indicating that some mechanism must exist to transport energy from the equator to the poles. By 1984 the NASA ERBE measured the Sun's total energy incident on Earth and total thermal energy emitted by Earth with sufficient precision to show that clouds effectively doubled Earth's albedo, reducing the sunlight reaching the surface but also trapping heat that would otherwise be radiated to space. By the end of the twentieth century, a wide consensus developed that Earth was warming, leading to negotiation of the Kyoto Protocol, which attempted (with little success) to reduce emission of carbon dioxide and other greenhouse gases into the atmosphere.

Some of the most dramatic evidence of global warming came from synthetic aperture radar images taken by the Canadian *Radarsat*, which revealed an extensive decline in the Greenland and Antarctic ice sheets. Beyond the visual impact of the images, precise measurement of the ice sheet flow showed complex networks of ice streams moving at variable velocity that appeared to be increasing as a result of climate change. The ability of satellites to continuously monitor areas of interest paid off when *Radarsat* observed the collapse of the Ross Ice Shelf in 2002.

Other platforms dedicated to environmental and climate research included the European Space Agency's *Envisat 1*, launched in 2002, which carried 10 sensors to monitor global warming, stratospheric ozone, and desertification. The first global map of nitrogen dioxide pollution in 2004 was based on *Envisat 1* data.

Another major twentieth-century development was international cooperation in atmospheric science and climate research. As early as 1961, a provision of United Nations (UN) resolution 1721 called for the use of satellites and other space-based assets "to advance the state of atmospheric science and technology so as to provide greater knowledge of basic physical forces affecting climate and the possibility of large-scale weather modification." This led, in 1967, to the establishment of GARP, jointly sponsored by the World Meteorological Organization (WMO) and the International Council of Scientific Unions.

GARP coordinated a series of observational programs to characterize particular features of the atmospheric circulation, culminating in the First GARP Global Experiment (FGGE) of 1978–79, which provided continuous global weather observations for one year from five geostationary satellites and a wide range of instruments aboard lower orbiting satellites, in addition to ground-based instruments. These measurements provided the basis for a vastly improved global atmospheric circulation model including previously unquantifiable features, such as global water vapor flux and distribution of kinetic energy in ocean eddy currents. FGGE's success led to the establishment of the WMO's World Climate Research Program (WCRP) in 1980. WCRP-supported projects have since investigated coupling between ocean surface temperature and atmospheric circulation, interaction between clouds and the surface hydrological cycle, stratospheric processes and the Arctic climate, and data from a wide range of space- and ground-based sensors to develop predictive numerical models. Among the results was what WCRP called "major breakthroughs in operational seasonal forecasting" that could account for features such as El Niño events in the Pacific.

A wide range of bilateral and multilateral atmospheric research programs have also been operated using satellites. Examples include the 1997 *Tropical Rainfall Measuring Mission*, jointly operated by NASA and the Japan Aerospace Exploration Agency, which measured rainfall between 35 degrees N and 35 degrees S using precipitation radar. Other instruments on the spacecraft allowed global lightning detection and continued measurement of Earth's radiation budget. A surprising discovery from this mission was the extent to which smoke from forest fires in the Amazon inhibited rainfall.

Yet another development was the fusion of data from multiple sensors, whether space-, air-, sea-, or ground-based. Examples include *Nimbus 3*, which located and interrogated a variety of other sensor platforms beginning in 1969, and the French-provided Argos system, which allowed Tiros N series satellites to collect, store, and transmit data from balloons and buoys beginning in 1978. By 1999, with the launch of its *Terra* spacecraft, NASA began explicit use of sensor webs involving multiple spacecraft and ground-based instruments in its EOS. In 2005 the EOS web autonomously detected an unexpected volcanic eruption and automatically commanded *Terra* to begin recording measurements before most human observers were even aware of the event.

By the first decade of the twenty-first century, extensive atmosphere and climate databases had been compiled by a variety of satellites throughout a period of decades. NASA Active Cavity Radiometer Irradiance Monitor 3 instrument extended solar irradiance measurement to provide a continuous record of solar irradiance since 1975, based on measurements from earlier spacecraft. Satellite data was key to the WMO Global Climate Observing System, which supported the UN Framework Convention on Climate Change and the Kyoto Protocol.

John D. Ruley

See also: Civilian Meteorological Spacecraft, Civilian Remote Sensing

Bibliography

David H. DeVorkin, *Science with a Vengeance: How the Military Created the U.S. Space Sciences after World War II* (1992).
Janice Hill, *Weather from Above* (1991).
National Research Council, *Earth Observations from Space: The First 50 Years of Scientific Achievements* (2008).
P. Krishna Rao et al., *Weather Satellites: Systems, Data, and Environmental Applications* (1990).
World Climate Research Program, *A Short History of a Long Success Story* (2006).

CALIPSO, Cloud-Aerosol Lidar and Infrared Pathfinder Satellite Observations. *See* Earth System Science Pathfinder.

CHAMP, Challenging Minisatellite Payload. *CHAMP, Challenging Minisatellite Payload*, was designed to study Earth's gravity field, magnetic field, and atmosphere. The 500 kg spacecraft was launched on 15 July 2000 from Plesetsk, Russia, on a Kosmos-3M launch vehicle into a 454 km, non-Sun-synchronous, near-polar orbit. This orbit provided global coverage and allowed the satellite to collect data at all local times, to enable separation of the effects of time-related phenomena, such as tides. The German Aerospace Research Center (DLR) funded the mission with GeoForschungsZentrum Potsdam managing the project. Jena-Optronik GmbH, from the former East Germany, was chosen to design and build the satellite in collaboration with former West German partners, as part of a DLR initiative to encourage development of the space industry in that part of the newly reunited Germany. NASA, CNES (French Space Agency), and the U.S. Air Force Research Laboratories each provided an instrument: a Global Positioning System (GPS) Blackjack receiver, a precision accelerometer, and a digital ion drift meter (DIDM), respectively.

CHAMP's laser retroreflector allowed a laser from Earth to be reflected off the satellite to track the satellite's position. This was used in combination with the GPS receiver to characterize Earth's medium- to long-wavelength gravity field. An accelerometer measured orbital perturbations not due to gravity. Using a fluxgate magnetometer and an Overhauser magnetometer, *CHAMP* measured Earth's magnetic field with an order of magnitude better accuracy over *Magsat*, the first dedicated magnetic field mission. The GPS receiver, which had an accuracy of 2–3 cm, also had an altimetry mode, to experiment with the possibility of detecting reflected GPS signals off the ocean's surface, and an occultation mode to view the signals from setting GPS satellites through the atmosphere. The DIDM measured the ion distribution within the ionosphere.

Magnetic measurements from *CHAMP* made it possible to map an ocean's tidal flow. Because sea water is highly conductive, the movement of the ocean through Earth's magnetic field induces electric fields and currents, which in turn generate magnetic fields seen in the *CHAMP* data. *CHAMP* was still operating in 2007, two years past its design life.

Katie J. Berryhill

See also: Germany

Bibliography

GeoForschungsZentrum *CHAMP.* http://op.gfz-potsdam.de/champ/.
Christoph Reigber et al., *Earth Observation with* CHAMP*: Results from Three Years in Orbit* (2004).

Cloudsat. *See* Earth System Science Pathfinder.

Daichi. *See* Japanese Land Observation Missions.

Earth Observing System (EOS). Earth Observing System (EOS) was a NASA-based initiative of the U.S. Global Change Research Program. It consisted of a series of Earth observation spacecraft and their associated ground-based data processing and distribution system.

The EOS program was started in the late 1980s by NASA in an attempt to understand the behavior of Earth's atmosphere, oceans, land, and cryosphere as a single interactive system. This would enable scientists to better understand the magnitude and causes of climate change and make recommendations to mitigate its effects. The program was descoped in 1992 and 1994, reducing the number of instruments to be flown from 30 to 17.

Terra (Latin for "earth"), the first satellite in the EOS program, consisted of a bus built by Lockheed Martin and a suite of five instruments (see table on page 155), three American and one each from Japan and Canada. They were designed to provide simultaneous information on clouds, water vapor, aerosols, trace gases, the land and oceans, and their mutual interactions, together with changes in Earth's radiation energy budget. *Terra*'s instruments produced about 850 gigabytes of information per day.

The 5.2-ton spacecraft, which had a design lifetime of six years, was launched by an Atlas launcher in December 1999 from Vandenberg Air Force Base into a 705 km Sun-synchronous, near-polar orbit. Its descending orbit crossed the equator at 10:30 a.m. local time (hence *Terra*'s original name EOS AM 1) to enable the tropics to be viewed before the afternoon clouds increasingly obscured the ground.

NASA's Goddard Space Flight Center operated *Terra* via the Tracking and Data Relay Satellite System (TDRSS). Goddard also received, processed, and disseminated science data through the Earth Observing System Data and Information System (EOSDIS). Science data recorded onboard were transmitted at Ku band to TDRSS at 150 Mbps. In addition, real time MODIS and ASTER experiment data were transmitted directly to users at X band.

Earth Observing System (EOS) Instruments (Part 1 of 3)

(A) *Terra* Spacecraft			
Instrument	**Wavebands (μm)**	**Measurements and Main Applications**	**Comments**
Advanced Spaceborne Thermal Emission and Reflection Radiometer (ASTER)	Visible, 0.52–0.86 Short Infrared, 1.6–2.4 Thermal Infrared, 8.1–11.7	Provision of high resolution (15 m in visible, 90 m in thermal infrared) images of Earth's surface in 14 different wavebands. Studies of land surface temperature and elevation, volcanic processes, plant evaporation, vegetation, soil, and cloud characteristics.	Instrument designed and built in Japan. Data collection limited to 8 minutes per orbit.
Clouds and the Earth's Radiant Energy System (CERES)	Visible and Infrared, 0.3–5.0 Thermal Infrared, 8.0–12.0 Visible to far Infrared, 0.35–125	Long term measurement of Earth's radiation budget from top of atmosphere to surface for climate change analysis. Limb to limb coverage, altitude, liquid water content, and optical depth of clouds.	CERES was of ERBE (Earth Radiation Budget Experiment) heritage. Similar instrument flown on *Aqua* and *TRMM* (*Tropical Rainfall Measuring Mission*). Cross track measurements continue (*ERBS*, *Earth Radiation Budget Satellite*) mission.
Multi-angle Imaging Spectro-Radiometer (MISR)	Visible	Global and local surface albedo, vegetation and aerosol properties. Helped to distinguish between different types of clouds.	Nine push-broom cameras, eight operating at various angles off-nadir.
Moderate resolution Imaging Spectroradiometer (MODIS)	Visible and Infrared, 0.4–14.5.	Low resolution imagery (250–1,000 m) in 36 spectral bands. Sea surface and land temperature measurements to 0.2 and 1.0 K, respectively. Ocean color measurements. Information on vegetation, biomass burning, snow cover, cloud, and aerosol properties.	Instrument heritage AVHRR and HIRS (*POES*, *Polar Operational Environmental Satellite*), TM (Landsat), CZCS (*Nimbus 7*). Same instrument flown on *Aqua*.
Measurements of Pollution in the Troposphere (MOPITT)	Infrared, 2.3–4.7	Carbon monoxide concentration profile and total column methane from ground up to 13 km. (Carbon monoxide is a byproduct of biomass burning and methane is an important greenhouse gas). Studied how these gases interact with surface, ocean, and biomass systems.	Instrument designed and built in Canada. Self calibrating instrument. Problems with height resolution after 18 months of operation.

(Continued)

Earth Observing System (EOS) Instruments (Part 2 of 3)

(B) *Aqua* Spacecraft			
Instrument	**Wavebands**	**Measurements and Main Applications**	**Comments**
Atmospheric Infrared Sounder (AIRS)	Visible, 0.4–0.9 μm Infrared, 3.7–15.4 μm	Measured vertical water vapor and temperature atmospheric profiles, cloud properties and trace greenhouse gases such as ozone, carbon monoxide, and methane.	Had 2,378 spectral channels, giving an unprecedented spectral resolution.
Advanced Microwave Scanning Radiometer for EOS (AMSR E)	Microwave, 7–89 GHz	Sea surface wind speed, sea surface temperature, cloud water and water vapor, rainfall, sea ice concentration and snow cover, soil moisture.	Instrument designed and built in Japan. Modified version of the AMSR flown on *Midori II*.
Advanced Microwave Sounding Unit A (AMSU A)	Microwave, 23–89 GHz	Vertical atmospheric temperature and moisture profiles from 45 km to Earth's surface.	First flown on *NOAA 15* in 1998.
CERES, as on *Terra*			
Humidity Sounder for Brazil (HSB)	Microwave, 150 and 183 GHz	Atmospheric humidity profiles. Detected precipitation beneath the clouds.	Brazilian contribution. Design based on AMSU-B flown on *NOAA 15*. Designed and built in United Kingdom. Failed after nine months.
MODIS, as on *Terra*			

(C) *ICESat* Spacecraft			
Instrument	**Wavebands (μm)**	**Measurements and Main Applications**	**Comments**
Geoscience Laser Altimeter System (GLAS)	Visible, 0.53 Short Infrared, 1.1	Measured ice sheet and land topography, provided information on the height and thickness of clouds and vegetation characteristics.	Laser ranging system.
GPS Blackjack Receiver		Precise orbit determination, timing, and geolocation.	

Earth Observing System (EOS) Instruments (Part 3 of 3)

(D) *Aura* Spacecraft			
Instrument	**Wavebands**	**Measurements and Main Applications**	**Comments**
High Resolution Dynamics Limb Sounder (HIRDLS)	Infrared, 6–18 μm	Measured atmospheric concentrations of water vapor, ozone, methane, and other trace gases. Measured temperature of upper atmosphere and location of polar stratospheric clouds.	Joint U.S./United Kingdom instrument. Heritage: LRIR (*Nimbus 6*), LIMS and SAMS (*Nimbus 7*), ISAMS and CLAES (*UARS*). 80 percent of aperture accidentally blocked.
Microwave limb Sounder (MLS)	Microwave 120–2500 GHz	Measured chemical species involved in the destruction of stratospheric ozone. Unique ability to measure trace gases in presence of ice clouds and volcanic aerosols.	Greatly enhanced version of the MLS on *UARS*.
Ozone Monitoring Instrument (OMI)	Ultraviolet, 0.27–0.37 μm Visible, 0.37–0.50 μm	Study of ozone chemistry. Distinguished between smoke, dust, and sulfate aerosol particles and measured cloud pressures.	Joint Netherlands, Finland, and U.S. project. Heritage: TOMS (*Nimbus 7*, etc.), and GOME (*ERS*, European Remote Sensing) and SCIAMACHY (*Envisat*).
Tropospheric Emission Spectrometer (TES)	Infrared, 3.2–15.4 μm	Measured ozone and other trace gas concentrations down to ground level; the only Aura instrument to do so. Helped to understand tropospheric-stratospheric exchange.	

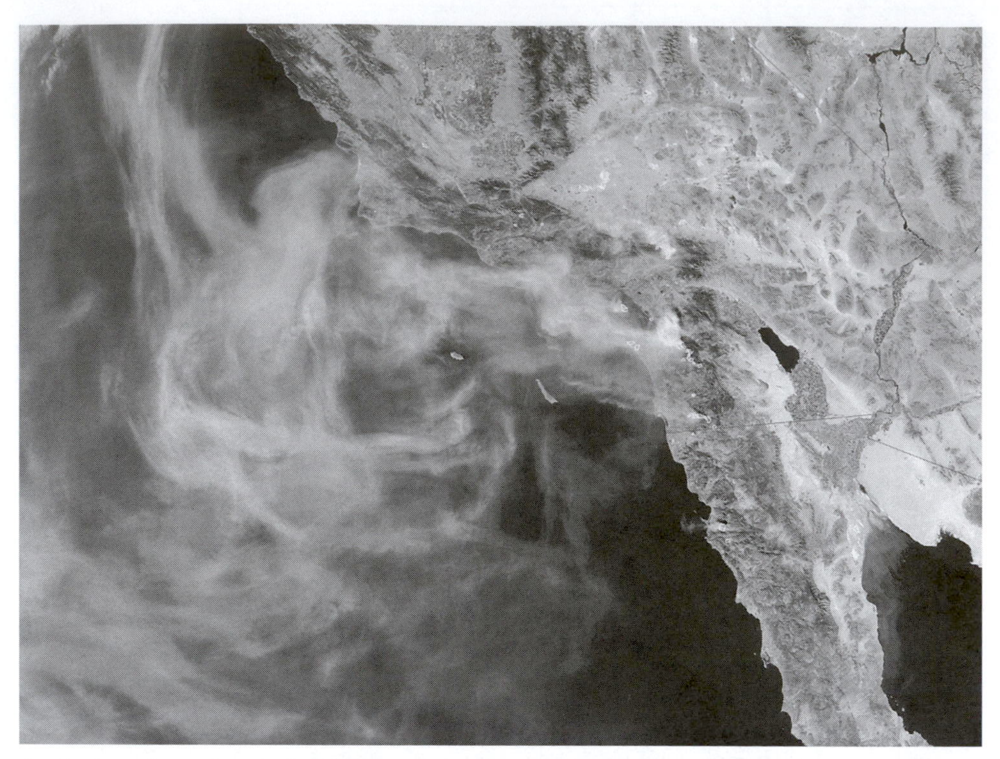

Smoke from forest fires in California on 24 October 2007, taken by the Terra *spacecraft. (Courtesy MODIS Rapid Response Team/NASA/Goddard Space Flight Center)*

For some time in 2001, *Terra* flew in formation with *Landsat 7*, *EO 1* (*Earth Observing 1*), and the Argentine *SAC C* (*Satélite de Aplicaciones Científicos C*) in the so-called "morning constellation," so that similar data could be compared between the different spacecraft.

Terra was followed by *Aqua*, which was launched in May 2002 by a Delta II from Vandenberg into a 705 km Sun-synchronous, near-polar orbit. The 3.1-ton *Aqua* (Latin for water) was built by TRW with a design life of six years. Its ascending orbit crossed the equator at 1:30 p.m. local time, hence *Aqua*'s original name EOS PM 1.

Aqua provided a large amount of information on Earth's water cycle. This included data on atmospheric water vapor, clouds and precipitation, sea and land ice and snow, and soil moisture. It also provided information on aerosols, on trace atmospheric gases, such as ozone, carbon monoxide and methane, on Earth's radiation balance, and on the temperature of the air, land, and sea. This resulted in *Aqua* being used by the National Oceanic and Atmospheric Administration (NOAA) and the European Centre for Medium-Range Weather Forecasts to improve weather forecasts.

ICESat (*Ice, Cloud, and Land Elevation Satellite*), the next of the original EOS satellites, was launched from Vandenberg by a Delta 7000 in January 2003. Originally called EOS LAM, for Laser Altimetry Mission, the 1-ton spacecraft was built by Ball

Aerospace using its Commercial Platform 2000 as the spacecraft bus. Unlike its two EOS predecessors, it was not launched into a Sun-synchronous orbit, but into a 590 km circular, near-polar orbit whose ground track precessed westward. The spacecraft's design lifetime was three years, with a goal of five years. It was controlled in orbit by a team at the University of Colorado. The spacecraft mission was to provide data on ice sheet mass balance, cloud and aerosol heights, land topography and vegetation characteristics using a laser ranging or lidar system.

Aura (Latin for breeze, previously EOS Chem 1) was the EOS atmospheric mission designed to monitor the complex interactions of atmospheric constituents produced by natural and human-made sources. *Aura* was a follow-on to the *Upper Atmosphere Research Satellite* (*UARS*). It was built by Northrop Grumman, weighed 3.0 tons, and had a design lifetime of six years.

Aura was launched in July 2004 from Vandenberg into the same altitude, Sun synchronous, near polar orbit of Aqua by a Delta II. It was a member of the so-called A-train constellation in which *CloudSat*, *CALIPSO* (Cloud-Aerosol Lidar and Infrared Pathfinder Satellite Observations) and the French *PARASOL* (Polarization and Anisotropy of Reflectances for Atmospheric Science coupled with Observations from a Lidar) spacecraft followed *Aqua* in the same orbit. *CloudSat* followed 30–60 seconds behind *Aqua*, *CALIPSO* 15 seconds behind *CloudSat*, and so on, thus enabling scientists to make a comparison with the results of similar instruments on different spacecraft.

One of *Aura*'s main missions was to provide a better understanding of ozone chemistry, and so enable a better assessment to be made of the causes and magnitude of stratospheric ozone depletion. *Aura*'s instruments also enabled scientists to monitor air pollution on a daily basis. The U.S. Environmental Protection Agency had identified carbon monoxide, nitrogen dioxide, sulfur dioxide, ozone, aerosols, and lead as crucial pollutants. *Aura* measured all but the latter.

Carbon dioxide is by no means the only greenhouse gas. Methane, water vapor, tropospheric ozone, and aerosols are also important to any study of climate change. *Aura* measured these and other trace gases in the upper troposphere and lower stratosphere, which are thought to be crucial in understanding climate change.

David Leverington

See also: Ball Aerospace and Technologies Corporation; Goddard Space Flight Center; Lockheed Martin Corporation; National Oceanic and Atmospheric Administration; Northrop Grumman Corporation; TDRSS, Tracking and Data Relay Satellite System

Bibliography

Goddard Space Flight Center. http://www.gsfc.nasa.gov/.
Jet Propulsion Laboratory. http://www.jpl.nasa.gov/.
NASA Headquarters. http://www.gsfc.nasa.gov/.

Earth System Science Pathfinder. Earth System Science Pathfinder (ESSP) was a series of low- to moderate-cost, small- to medium-sized spacecraft examining unique areas of Earth science research. The program complemented the larger satellites of the

Earth Observing System (EOS) by rapid development of global observation missions. As of 2007, ESSP consisted of four active satellites: two GRACE (Gravity Recovery and Climate Experiment) spacecraft, *CloudSat*, and *CALIPSO* (*Cloud-Aerosol Lidar and Infrared Pathfinder Satellite Observations*).

The first ESSP mission was GRACE, a joint project of NASA and the Deutsches Zentrum für Luft- und Raumfahrt (DLR), developed and managed by the Jet Propulsion Laboratory. Launched 17 March 2002 on a Rockot booster from the Plesetsk Cosmodrome, the twin GRACE satellites, designed and built by European Aeronautic Defence and Space Company Astrium GmbH, were placed in co-planar, near-polar orbits approximately 220 km apart. GRACE mapped Earth's gravity field by precisely measuring the separation between the satellites, using the Global Positioning System and a K-band microwave ranging system. Passing over a greater mass concentration causes the lead satellite to pull away from the trailing satellite, followed by the trailing satellite catching up as it passes the same area. The GRACE maps have significance for global climate change research because gravity field variations occur due to relatively fixed features, such as mountains and ocean trenches, and variable factors, such as changes in water content. Previous measurements of Earth's gravity field measured large-scale features. With GRACE, scientists gathered global data from a single source every 30 days

ESSP Satellites and Instruments

Satellite	Instruments	Notes
GRACE	Twin satellites, each had K-Band Ranging Assembly, Office National d'Études et de Recherches Aérospatiales SuperSTAR Accelerometer, two Technical University of Denmark Star Camera Assemblies, and Jet Propulsion Laboratory BlackJack Global Positioning System Receiver	Satellites were identical except for the communications frequencies used.
CloudSat	Cloud Profiling Radar, 94 GHz	Design was almost identical to the DC-8 based Airborne Cloud Radar.
CALIPSO	Cloud-Aerosol Lidar with Orthogonal Polarization, one 1,064 nm and two 532 nm receiver channels with a 1 m receiver telescope; Wide-Field Camera (visible); Infrared Imager Radiometer, three channels in thermal infrared at 8.7 mm, 10.5 μm, and 12.0 μm	Light Detection and Ranging (LIDAR) was follow-on to the Lidar In-space Technology Experiment flown on the Space Shuttle *Discovery* in 1994 and the Geoscience Lawer Altimeter System on-board *ICESat*.
PARASOL	Polarization and Directionality of the Earth's Reflectances (POLDER) instrument working in nine spectral bands with 15 filters and polarizers in the visible and near infrared	POLDER instrument was slightly changed from that flown on *Midori I* and *II*, with changes in some of the channels, including addition of a 1020-nm band to complement LIDAR data from *CALIPSO*.

with a precision several orders of magnitude better than previous measurements, revealing the finer-scale and time-variable features of the geoid (a mathematical model of Earth's gravity field). This improved the accuracy of satellite altimetry, synthetic aperture radar interferometry, and digital terrain models. GRACE was a follow-on to the CHAMP (Challenging Minisatellite Payload) mission, which measured the gravity field at an order of magnitude better accuracy than previous measurements. GRACE improved on that by a factor of 10 to 50.

The next two ESSP missions launched together on 28 April 2006 on a Delta II from Vandenberg Air Force Base, California. *CloudSat* and *CALIPSO* provided coordinated observations from a circular, Sun-synchronous, 705 km orbit, and were designed to investigate the role of clouds and aerosols, small particles suspended in the air, in Earth's climate and air quality.

CloudSat was a partnership between the United States and Canada, managed by Jet Propulsion Laboratory. Ball Aerospace designed and built the spacecraft. *CloudSat* flew the first space-based millimeter-wave radar, which was able to detect the vertical structure of clouds, and image clouds and rainfall simultaneously, both unprecedented space-based capabilities. Early results showed the radar was able to see through all but the heaviest precipitation.

CALIPSO, one of the Proteus series of satellites, was a joint mission between NASA and CNES (French Space Agency), managed by Langley Research Center and built by Alcatel Alenia Space. Its primary instrument was a polarization lidar (light detection and ranging) instrument that used a laser to actively find aerosols and clouds in the atmosphere, different from previous satellites that used passive observations. The lidar operated in two wavelengths and provided high vertical resolution of clouds and aerosols. It carried a Wide-Field Camera operating in push-broom mode with 125 m resolution.

As part of the A-Train constellation of Earth science satellites, *CloudSat* and *CALIPSO* flew in formation with *Aura*, *Aqua*, and *PARASOL* (*Polarization and Anisotropy of Reflectances for Atmospheric Science coupled with Observations from a Lidar*), to be joined in 2008 by NASA's *Orbiting Carbon Observatory* (the last and smallest of the constellation). The entire A-train of satellites provided the first full suite of cloud- and aerosol-observing instruments, with each of the six satellites crossing the equator a few minutes apart at approximately 1:30 p.m. local time (giving rise to the nickname "afternoon constellation" or A-Train). *Aura* and *Aqua* were part of the EOS.

PARASOL, built by CNES, was the second in the Myriade series of microsatellites, launched from Kourou, French Guiana, as an auxiliary payload on an Ariane 5G+ booster on 18 December 2004. The spacecraft was named for the parasol effect that results in the reflection or scattering of about one-third of the Sun's radiant energy by suspended particles in the atmosphere, or aerosols. Using instruments based on the POLDER instrument flown on the Japanese *Midori 1* and *2* satellites, *PARASOL* measured the polarization and direction of scattered light, particularly in the regions covered by *CALIPSO*'s lidar. Among other things, this allowed scientists to better distinguish between natural and human-made aerosols.

Katie J. Berryhill

See also: France, Germany, Goddard Space Flight Center, Jet Propulsion Laboratory

Bibliography

CNES *PARASOL*. http://smsc.cnes.fr/PARASOL/.
NASA *CALIPSO*. http://www.nasa.gov/mission_pages/calipso/main/index.html.
NASA *CloudSat*. http://www.nasa.gov/mission_pages/calipso/main/index.html.
University of Texas GRACE. http://www.csr.utexas.edu/grace/.

Elektron. Elektron was a series of four satellites launched in pairs by the Soviet Union in 1964 and designed to study Earth's radiation belts discovered by U.S. Explorer satellites in 1958. The OKB-1 (Experimental Design Bureau) of Sergei Korolev built the satellites.

The first pair (*Elektron 1* and *Elektron 2*) was launched on 30 January 1964, followed by the second pair (*Elektron 3* and *Elektron 4*) on 10 July 1964. The first satellite of each pair flew in a 400 × 7,000 km orbit, focusing primarily on the inner radiation belt. Each had a mass of 355 kg and was equipped with six small solar panels. The second satellite of each pair was placed into a 460 × 68,000 km orbit and mainly studied the outer belt. Each of these satellites weighed 465 kg and had solar cells mounted on the body. Orbital inclination for all satellites was about 61°, making it possible to study the belts at higher latitudes than earlier U.S. satellites. The different orbits for each pair were achieved by releasing the first satellite when the upper stage of the Vostok launch vehicle was still firing. Additional objectives were to study micrometeorites, solar radiation, and cosmic rays. Originally the satellites were supposed to study artificial radiation belts caused by high-altitude nuclear explosions, but the Partial Test Ban Treaty signed in August 1963 made this task redundant. Data obtained by the Elektrons made it possible to assess the danger that the radiation belts posed to human and robotic space missions.

Bart Hendrickx

See also: Rocket and Space Corporation Energia

Bibliography

Bart Hendrickx, "Elektron: The Soviet Response to Explorer," *Quest* 8, no. 1 (2000): 37–45.

EO 1, Earth Observing 1. *EO 1, Earth Observing 1* was the first Earth-observing mission in NASA's New Millennium Program. The satellite was launched 21 November 2000 from Vandenberg Air Force Base on board a Delta 7320 booster for a one-year mission to validate 13 new technologies to reduce the cost and enhance the performance of future Earth science missions. Goddard Space Flight Center (GSFC) built and managed the spacecraft.

There were three primary instruments on board the 528 kg spacecraft, which flew in a 705 km Sun-synchronous orbit in formation with *Landsat 7*. The Advanced Land Imager produced images directly comparable with *Landsat 7*'s Enhanced Thematic Mapper, with a multispectral spatial resolution of 30 m and a 10 m panchromatic band, demonstrating the ability to establish data continuity between past and future Landsat imagers, as required by the Land Remote Sensing Act of 1992. The Hyperion instrument was a hyperspectral imager, providing 30-m spatial resolution in 220 spectral bands. The Atmospheric Corrector tested technology for correcting the errors in apparent surface reflectances caused by atmospheric absorption and scattering.

Some of the spacecraft bus technologies tested on *EO 1* were an X-Band Phased Array Antenna; a Lightweight Flexible Solar Array; a Pulsed Plasma Thruster, designed to provide low-mass, low-cost electromagnetic propulsion for precision attitude control; and the ability to autonomously fly in formation with other satellites. *EO 1* flew in formation with *Landsat 7* until September 2005, when engineers began lowering its orbit in preparation for eventual reentry.

Following the successful primary mission, NASA and the U.S. Geological Survey (USGS) entered a partnership for an extended mission. The USGS Center for Earth Resources Observation and Science processed and distributed the data to users. In *EO 1*'s extended mission, NASA researchers used it as part of a network of sensors working together. The software controlling this sensor web included artificial intelligence to allow various sensors to autonomously image science events, such as volcanic eruptions, wildfires, floods, and ice breakup. In April 2005 the sensor web detected the eruption of an apparently dormant volcano on Sumatra and autonomously commanded *EO 1* to begin observing before many scientists had realized that the eruption had occurred. *EO 1* could also be triggered by other sensors, such as those on the *Terra* and *Aqua* satellites. This technology could be used in the future for sensors at Mars or those monitoring solar activity.

Katie J. Berryhill

See also: Goddard Space Flight Center

Bibliography

GSFC *EO 1*. http://eo1.gsfc.nasa.gov/.
USGS *EO 1*. http://eo1.usgs.gov/.

ERBS, Earth Radiation Budget Satellite. *ERBS*, *Earth Radiation Budget Satellite*, was one of the first satellites designed specifically for launch by a Space Shuttle. Launched by *Challenger* on 5 October 1984 into a 610 km, 57° inclination orbit, the 226 kg *ERBS* was the first of three satellites involved in the Earth Radiation Budget Experiment (ERBE), the others being *Nimbus 6* and *7*. The radiation budget, or the balance between incoming energy from the Sun and the energy that Earth radiates, is the primary indicator of global climate change. The first satellite to measure Earth's radiation budget was *Explorer 7* in 1959, carrying instruments developed by Verner Suomi, a pioneer in satellite meteorology. *ERBS*'s mission was to accurately measure incoming solar energy and both the longwave and shortwave radiation emitted back into space by Earth.

Built by Goddard Space Flight Center and managed by Langley Research Center, the satellite carried ERBE instruments, including three scanning detectors and five nonscanning detectors. Data from *ERBS* in combination with other satellites allowed scientists to create a 24-plus-year record of the total solar irradiance that indicated a possible .037 percent per decade increase in the amount of energy received by Earth from the Sun. If correct and sustained, this trend could have significant climate effects.

In addition to the ERBE instruments, *ERBS* carried the Stratospheric Aerosol Gas Experiment 2, which provided ozone measurements throughout the spacecraft's operation. Scientists used this data along with that from other spacecraft to study the mechanisms leading to the ozone hole above the Antarctic and the effects on the ozone layer of volcanic eruptions, fossil fuels, and the use of chlorofluorocarbons (CFC). This was key information during the creation of the Montreal Protocol, which led to almost complete elimination of CFC use in the industrialized world.

ERBS had a planned two-year operational life. The scanner portion of the instrument package failed in February 1990. The nonscanner unit was shut down after 21 years of operation in October 2005.

Katie J. Berryhill

See also: Goddard Space Flight Center, Langley Research Center, Nimbus

Bibliography

Langley Research Center's Earth Radiation Budget Experiment. http://asd-www.larc.nasa.gov/erbe/ASDerbe.html.

Explorer. The Explorer program, an ongoing series of diverse spacecraft, dates back to the first U.S. satellite, *Explorer 1*, built by the Jet Propulsion Laboratory and launched on 31 January 1958 by the Army Ballistic Missile Agency. Since then, numerous spacecraft have used the Explorer name, most managed by NASA's Goddard Space Flight Center.

These spacecraft carried out a myriad of scientific investigations, including atmospheric science, solar monitoring, high-energy particle physics, radio astronomy, micrometeroid detection, geodesy, and more. In general the Explorer spacecraft have been small, relatively simple, and inexpensive missions. However, the use of the Explorer name has not been consistent throughout its history. Some missions were Explorer class but were not called Explorer, while others were given the name without a mission number within the series.

With the advent of the Space Shuttle, the United States was focusing on larger, multipurpose satellites. NASA phased out expendable booster production to encourage use of the Shuttle, a policy shown to be flawed by the *Challenger* disaster. This affected many science missions, including the Explorer program. Only one Explorer mission was launched between late 1981 and late 1989. The push to get instruments on less-frequent launches led to increasing costs. This provided the impetus to resume small satellite launches.

NASA restructured the Explorer program to provide frequent, low-cost launch opportunities. Missions in the program launched within three years of selection. The return to small satellites provided part of the basis for NASA's "faster, better, cheaper" policy of the 1990s, a policy that was abandoned after several failures of small missions. The new Explorer program began with the Small Explorers, missions costing NASA less than $120 million. Then came the Medium-class Explorer program, raising the

Explorer Spacecraft (Part 1 of 6)

Explorer Spacecraft Number	Other Name	Meaning of Acronym	Launch (F = failed during launch or shortly afterward)	Spacecraft mass (kg)	Main Mission (not stated if mission failure)
1			1 Feb 1958	14	Energetic particle studies
2			5 Mar 1958 F	5	
3			26 Mar 1958	14	Energetic particle studies
4			26 Jul 1958	17	Energetic particle studies
5			24 Aug 1958 F	17	
6	*S-2*		7 Aug 1959	64	Magnetospheric research
7	*S-1a*		13 Oct 1959	42	Energetic particle studies
8	*S-30*		3 Nov 1960	41	Ionospheric research
9	*S-56a*		16 Feb 1961	36	Upper atmosphere research. (Spacecraft was a balloon)
10	*P-14*		25 Mar 1961	35	Measured interplanetary magnetic field and particle radiation near Earth
11	*S-15*		27 Apr 1961	37	Gamma ray astronomy
12	*EPE A*	Energetic Particle Explorer	16 Aug 1961	38	Magnetospheric research
13	*S-55a*		25 Aug 1961	86	Technology satellite, micrometeoroids research
14	*EPE B*		2 Oct 1962	40	Energetic particle studies
15	*EPE C*		27 Oct 1962	45	Energetic particle studies. Spacecraft failed to de-spin
16	*S-55b*		16 Dec 1962	108	Technology satellite, micrometeoroid research

(Continued)

Explorer Spacecraft (Part 2 of 6)

Explorer Spacecraft Number	Other Name	Meaning of Acronym	Launch (F = failed during launch or shortly afterward)	Spacecraft mass (kg)	Main Mission (not stated if mission failure)
17	*AE A*	Atmospheric Explorer	3 Apr 1963	185	Atmospheric research
18	*IMP A* or *1*	Interplanetary Monitoring Platform	27 Nov 1963	62	Magnetospheric research
19	*AD A*[1]	Atmospheric Density	19 Dec 1963	8	Atmospheric air density measurements
20	*BE A*	Beacon Explorer	25 Aug 1964	45	Ionospheric research
21	*IMP B* or *2*		4 Oct 1964	62	Magnetospheric research
22	*BE B*		10 Oct 1964	53	Ionospheric and geodetic research
23			6 Nov 1964	134	Micrometeoroid research
24	*AD B*		21 Nov 1964	9	Atmospheric air density measurements
25	*Injun 4*[2]		21 Nov 1964	40	Ionospheric research
26	*EPE D*		21 Dec 1964	46	Energetic particle studies
27	*BE C*		29 Apr 1965	60	Geodetic research
28	*IMP C* or *3*		29 May 1965	58	Magnetospheric research
29	*GEOS*[3] *A* or *1*	Geodetic Earth Observing Satellite	6 Nov 1965	387	Geodetic research
30	*Solrad 8*[4]	Solar Radiation	19 Nov 1965	57	Solar radiation monitoring
31	*DME A*	Direct Measurement Explorer	29 Nov 1965	98	Ionospheric research
32	*AE B*		25 May 1966	225	Atmospheric research
33	*IMP D* or *AIMP 1*	Anchored IMP	1 Jul 1966	93	Magnetospheric research
34	*IMP F* or *4*		24 May 1967	163	Magnetospheric research
35	*IMP E* or *AIMP 2*		19 Jul 1967	104	Magnetospheric research

Explorer Spacecraft (Part 3 of 6)

Explorer Spacecraft Number	Other Name	Meaning of Acronym	Launch (F = failed during launch or shortly afterward)	Spacecraft mass (kg)	Main Mission (not stated if mission failure)
36	*GEOS B* or *2*		11 Jan 1968	469	Geodetic research
37	*Solrad 9*		5 Mar 1968	198	Solar radiation monitoring
38	*RAE A*	Radio Astronomy Explorer	4 Jul 1968	193	Radio astronomy
39	*AD C*		8 Aug 1968	9	Atmospheric research
40	*Injun 5*		8 Aug 1968	71	Magnetospheric research
41	*IMP G* or *5*		21 Jun 1969	174	Magnetospheric research
42	*SAS A* or *Uhuru*	Small Astronomy Satellite	12 Dec 1970	140	X-ray astronomy
43	*IMP I* or *6*		13 Mar 1971	288	Magnetospheric research
44	*Solrad 10*		8 Jul 1971	260	Solar radiation monitoring
45	*SSS A*	Small Scientific Satellite	15 Nov 1971	52	Magnetospheric research
46	*MTS A*	Meteoroid Technology Satellite	13 Aug 1972	90	Micrometeoroid research
47	*IMP H* or *7*		23 Sep 1972	376	Magnetospheric research
48	*SAS B* or *2*		15 Nov 1972	186	Gamma-ray astronomy
49	*RAE B*		10 Jun 1973	328	Radio astronomy
50	*IMP J* or *8*		26 Oct 1973	371	Magnetospheric research
51	*AE C*		16 Dec 1973	658	Atmospheric research
52	*Injun 6*		3 Jun 1974	27	Magnetospheric research
53	*SAS C* or *3*		7 May 1975	197	X-ray astronomy
54	*AE D*		6 Oct 1975	676	Atmospheric research
55	*AE E*		20 Nov 1975	721	Atmospheric research
56	*ISEE 1* or *A*[5]		22 Oct 1977	340	Magnetospheric research

(Continued)

Explorer Spacecraft (Part 4 of 6)

Explorer Spacecraft Number	Other Name	Meaning of Acronym	Launch (F = failed during launch or shortly afterward)	Spacecraft mass (kg)	Main Mission (not stated if mission failure)
		International Sun Earth Explorer			
57	*SAS D* or *IUE*	International Ultraviolet Explorer	26 Jan 1978	672	Ultraviolet astronomy
58	*AEM 1* or *HCMM*	Applications Explorer Mission or Heat Capacity Mapping Mission	26 Apr 1978	134	Thermal Earth mapping
59	*ISEE 3* or *C* or *ICE*	International Cometary Explorer	12 Aug 1978	479	Magnetospheric research
60	*AEM 2* or *SAGE*	Stratospheric Aerosol and Gas Experiment	18 Feb 1979	147	Stratospheric, aerosol and ozone research
61	*AEM 3* or *Magsat*	Magnetic Field Satellite	30 Oct 1979	181	Mapping of near-Earth magnetic field
62	*DE 1*	Dynamics Explorer	3 Aug 1981	400	Magnetospheric research
63	*DE 2*[6]		3 Aug 1981	400	Magnetospheric research
64	*SME*	Solar Mesospheric Explorer	6 Oct 1981	145	Atmospheric research
65	*CCE*[7]	Charge Composition Explorer	16 Aug 1984	240	Magnetospheric research
66	*COBE*	Cosmic Background Explorer	18 Nov 1989	2,250	Study of microwave background radiation
67	*EUVE*	Extreme Ultraviolet Explorer	7 Jun 1992	3,256	Extreme ultraviolet astronomy
68	*SMEX 1* or *SAMPEX*	Small Explorer or Solar Anomalous and Magnetospheric Particle Explorer	3 Jul 1992	158	Magnetospheric research
69	*RXTE*	Rossi X-ray Timing Explorer	30 Dec 1995	3,035	X-ray astronomy

Explorer Spacecraft (Part 5 of 6)

Explorer Spacecraft Number	Other Name	Meaning of Acronym	Launch (F = failed during launch or shortly afterward)	Spacecraft mass (kg)	Main Mission (not stated if mission failure)
70	*SMEX 2* or *FAST*	Fast Auroral Snapshot Explorer	21 Aug 1996	191	Auroral studies
71	*ACE*	Advanced Composition Explorer	25 Aug 1997	785	Particle research
72	*STEDI 1* or *SNOE*	Student Explorer Demonstration Initiative or Student Nitric Oxide Explorer	26 Feb 1998	115	Atmospheric research
73	*SMEX 4* or *TRACE*	Transition Region and Corona Explorer	2 Apr 1998	213	Solar observatory
74	*SMEX 3* or *SWAS*	Submillimeter Wave Astronomy Satellite	6 Dec 1998	288	Submillimeter astronomy
75	*SMEX 5* or *WIRE*	Wide-field Infrared Explorer	5 Mar 1999	259	Main mission lost. Stellar seismology
76	*STEDI 2* or *TERRIERS*		18 May 1999 F	125	Ionospheric research
77	*MIDEX 0* or *FUSE*	Mid-sized Explorer or Far Ultraviolet Spectroscopic Explorer	24 Jun 1999	1,334	Ultraviolet astronomy
78	*MIDEX 1* or *IMAGE*	Imager for Magnetopause-to-Aurora Global Exploration	25 Mar 2000	494	Magnetospheric research
79	*HETE*[8] *2*	High Energy Transient Explorer	9 Oct 2000		Detection and location of gamma ray bursts
80	*MIDEX 2* or *WMAP*	Wilkinson Microwave Anisotropy Probe	30 Jun 2001	840	Microwave astronomy

(Continued)

Explorer Spacecraft (Part 6 of 6)

Explorer Spacecraft Number	Other Name	Meaning of Acronym	Launch (F = failed during launch or shortly afterward)	Spacecraft mass (kg)	Main Mission (not stated if mission failure)
81	*SMEX 6* or *RHESSI*	Reuven Ramaty High Energy Solar Spectroscopic Imager	5 Feb 2002	304	Solar flare research
82	*UNEX 1* or *CHIPSat*	University-class Explorer or Cosmic Hot Interstellar Plasma Spectrometer satellite	12 Jan 2003	60	Ultraviolet astronomy
83	*SMEX 7* or *GALEX*	Galaxy Evolution Explorer	28 Apr 2003	280	Ultraviolet astronomy
84	*MIDEX 3* or *Swift*		20 Nov 2004	1 × 331	Gamma ray burst astronomy
85–89	*MIDEX 5A–5D*[9] or *THEMIS*	Time History of Events and Macroscale Interactions during Substorms	17 Feb 2007	5 × 128	Auroral studies
90	*AIM*	Aeronomy of Ice in the Mesosphere	25 Apr 2007	197	Noctilucent cloud observation

All dates in UTC/Greenwich Time

Table produced by David Leverington.

[1]Second balloon spacecraft in the series. First spacecraft was *Explorer 9*.

[2]*Injun 1* and *3* (*2* was a failure) launched in 1961 and 1962, were not given Explorer numbers. They undertook magnetospheric research.

[3]Not to be confused with the magnetospheric research satellite *GEOS 1* launched in 1977, called in the United States *ESA* or *ESRO GEOS 1*.

[4]Earlier Solrads were generally sponsored by the military and so were not given Explorer numbers. The 181-kg *Solrad 11A* and *11B*, otherwise called *P74 1c* and *1d*, launched in 1976 were also not given Explorer numbers.

[5]*ISEE B* was an European Space Agency spacecraft, launched on the same launcher as *ISEE A*. It was not given an Explorer number.

[6]*DE 1* and *DE 2* were launched on the same launcher.

[7]Also launched with the German IRM and the British UKS satellites as part of the AMPTE or Active Magnetospheric Particle Tracer Explorer experiment.

[8]The launch of *HETE 1* failed in 1996.

[9]*MIDEX 4* or *FAME* was canceled.

total cost limit to $180 million. There was also the University-class Explorer/Student Explorer Demonstration Initiative program, the lowest-cost missions, and Missions of Opportunity, which flew on non-NASA projects. The Explorer series was the longest-running series of U.S. spacecraft, encompassing 87 successful spacecraft by the end of 2007.

Katie J. Berryhill

See also: Goddard Space Flight Center

Bibliography

Goddard Space Flight Center Explorer Program. http://nssdc.gsfc.nasa.gov/multi/explorer.html.
National Space Science Data Center. http://nssdc.gsfc.nasa.gov/.

Fuyo. *See* Japanese Land Observation Missions.

Geodetic and Earth Orbiting Satellite 3 (GEOS 3). *See* Space Geodesy.

GOCE, Gravity Field and Steady-State Ocean Circulation Explorer. *See* Space Geodesy.

GRACE, Gravity Recovery and Climate Experiment. *See* Earth System Science Pathfinder.

International Geophysical Year. The International Geophysical Year (IGY) was a catalyst for the first U.S. and Soviet Union scientific satellite programs. Its origin dates to 5 April 1950, when U.S. atmospheric physicist James Van Allen hosted a dinner party for five scientist friends including leading British geophysicist Sydney Chapman. At this gathering, American physicist/engineer Lloyd Berkner asked Chapman about the possibility of another International Polar Year (IPY). IPYs, periods devoted to international cooperation in the study of the polar regions, had been organized in 1882–83 and 1932–33. Chapman enthusiastically endorsed the suggestion and became its major international promoter.

Soon the idea of a third IPY was broadened and renamed IGY. The IGY proposal was endorsed by the International Council of Scientific Unions, which appointed an IGY oversight committee in 1952. The organization was formally designated the Special Committee for the IGY (CSAGI in French), headed by Chapman. The IGY was to be a time when scientists from around the globe could perform and coordinate experiments and observations into all areas of geophysics, from weather to ocean currents to glacial movement. Forty-six nations agreed to join the effort, a number that eventually expanded to 67.

From the start, the IGY planned to include a global program of sounding rocket observations of the upper atmosphere. Soon the CSAGI took up the scientifically popular idea of Earth satellites, passing a resolution on 4 October 1954 urging member countries to consider "the launching of small satellite vehicles" as part of IGY. The U.S. and Soviet IGY commissions began working with the scientific and military proponents of satellites, resulting in formal satellite program announcements by both nations. American support for a satellite program within IGY hinged partly on the Dwight Eisenhower administration's strategy to use a U.S. scientific satellite as a pathfinder to set a precedent for satellite overflight of the Soviet Union, so that later U.S. reconnaissance satellites could safely observe the communist nation from space.

The IGY, which ran from July 1957 to December 1958, was a great success. In addition to thousands of other experiments and observations, it eventually included the launch of the world's first satellites. The Soviet Union successfully launched three Sputnik satellites, each bigger and more sophisticated than the last, and the United States orbited one Vanguard and three Explorer scientific satellites before the end of the IGY. The Explorers contributed the most famous scientific result of the entire IGY, the discovery of the Van Allen radiation belts.

Matt Bille and Erika Lishock

See also: Explorer, Sputnik, Vanguard

Bibliography

Matt Bille and Erika Lishock, *The First Space Race* (2004).

Allan Needell, *Science, Cold War, and the American State: Lloyd V. Berkner and the Balance of Professional Ideas* (2000).

James Van Allen, *Origins of Magnetospheric Physics* (1983).

Japanese Land Observation Missions. Japanese land observation missions consisted of *Fuyo* (or *JERS* for *Japanese Earth Resource Satellite*) and *Daichi* (or *ALOS* for *Advanced Land Observation Satellite*). *Fuyo* (Japanese for lotus) was a National Space Development Agency (NASDA) spacecraft built by Mitsubishi, and *Daichi* (Japanese for land) was a Japanese Aerospace Exploration Agency (JAXA) spacecraft built by Nippon Electric Company.

Japan decided in 1978 to build its own Earth observation spacecraft and in 1987 launched its first in the series, *Marine Observation Satellite 1A*, devoted to oceanography. The first land observation spacecraft, the 1.3-ton *Fuyo*, was launched in February 1992 by a Japanese H-1 launcher from Tanegashima. It was placed into a circular 570 km, Sun-synchronous, near-polar orbit, with a repeat cycle of 44 days. *Fuyo* carried a Synthetic Aperture Radar (SAR) for all-weather imaging and a visible/near infrared radiometer. The latter partially failed in 1993. *Fuyo* downloaded real time data and data that had been stored using a 72 Gbit solid state recorder. NASDA's Earth Observation Center (EOC) at Hatoyama received and processed the data.

Fuyo was built to test the onboard instruments and ground system and to provide Earth observation data. It provided global, all-weather imaging for geology, agriculture, forestry, land use, fisheries, environmental protection, disaster prevention and monitoring, and sea ice and coastal surveillance. *Fuyo* showed land deformations caused by earthquakes, using SAR data from two different passes. The spacecraft exceeded its planned two-year lifetime by four years.

Fuyo was followed by the 4-ton *Daichi* spacecraft in January 2006. *Daichi* was launched by an H-2A launcher from Tanegashima into a circular 700 km, Sun-synchronous, near-polar orbit, with a repeat cycle of 46 days. It carried a stereomapping panchromatic radiometer, an improved visible/near infrared radiometer and an

Japanese Land Observation Missions and Measurements

Fuyo			
Instrument	**Wavebands (μ) or Frequency**	**Measurements and Main Applications**	**Comments**
Synthetic Aperture Radar (SAR)	1.28 GHz (L band)	All-weather, high-resolution, high-contrast imaging with accurate imaging of topographical features. Differential interferometric imaging using data from two orbit passes for land deformation measurements.	Ground resolution 18 m.
Optical Sensor (OPS)	7 visible/near infrared bands (0.56–2.34)	High resolution imaging. The infrared bands were used for hydrothermal, vegetation, and soil moisture mapping.	Some stereo capability. Ground resolution 18 m.

Daichi			
Instrument	**Wavebands (μ) or Frequency**	**Measurements and Main Applications**	**Comments**
Panchromatic Remote-Sensing Instrument for Stereo Mapping (PRISM)	Visible (0.52–0.77)	Digital elevation mapping for cartography.	Three independent optical systems viewing forward, backward, and nadir. Ground resolution 2.5 m at nadir.
Advanced Visible and Near Infrared Radiometer- 2 (AVNIR 2)	3 visible (0.42–0.69) 1 near infrared (0.76–0.89)	Imaged land and coastal zones at high spatial resolution.	Successor to AVNIR on *Midori 1* spacecraft. Ground resolution 10 m at nadir. Cross-track pointing for timely views of disaster areas.
Phased Array L-band Synthetic Aperture Radar (PALSAR)	1.28 GHz (L band)	All-weather land observation, as per SAR above.	Higher-performance SAR than on *Fuyo*. PALSAR has many different operating modes and better ground resolution. Best resolution 7 m, depending on mode.

improved SAR. Data could be transmitted to EOC via both the *Japanese Data Relay Test Satellite* and as a direct downlink. *Daichi* had a similar Earth observation mission to *Fuyo*. Because of *Daichi*'s superior resolution, however, it was intended to use it to update 1:25,000-scale maps of Japan.

David Leverington

See also: Japan

Bibliography

JAXA. http://www.jaxa.jp/index_e.html.

Jason 1. *See* Ocean Remote Sensing.

JERS, Japanese Earth Resources Satellite. *See* Japanese Land Observation Missions.

***LAGEOS 1* and *2*, *Laser Geodynamics Satellites*.** *LAGEOS 1* and *2*, laser geodynamics satellites, were passive satellites designed as laser ranging targets. The mission's science goals were to determine Earth's shape and rotational wobble, the rate of continental drift, and to provide further evidence in support of theoretical physicist Albert Einstein's general theory of relativity. Design compromises included the need to be massive enough to minimize nongravitational effects on the orbit, light enough to be launched into a high orbit (about 5,900 km), small enough to lessen the effects of solar radiation pressure, and large enough to fit as many reflectors as possible. The resulting spherical satellites were 411 kg (400 kg for *LAGEOS 2*), 60 cm diameter, with brass cores and covered with 426 retroreflectors, four of which were made of germanium, the rest made of fused silica glass. The stability of the spacecraft's orbits allowed scientists to measure movement of Earth's crust at better than 1 in accuracy.

LAGEOS 1 was designed and built by NASA Marshall Space Flight Center (MSFC), with Bendix Corporation assisting with system integration, and launched by a Delta booster on 4 May 1976 into a 110 degree inclination orbit to enable viewing by a worldwide network of tracking stations. The mission management was shared by MSFC and Goddard Space Flight Center (GSFC). While *LAGEOS 1* was the first spacecraft exclusively designed for laser ranging, retroreflector arrays had been launched in 1965 on *Beacon Explorer C* (*BE C*, *Explorer 27*), in 1975 on NASA's *Geodetic and Earth Orbiting Satellite 3* (*GEOS 3*), and in 1975 on the French *STARLETTE* (Satellite de Taille Adaptée avec Réflecteurs Laser pour les Études de la Terre) satellite. Additionally, retroreflectors were left on the lunar surface during the Apollo program. Numerous spacecraft since *LAGEOS 1* have included retroreflector arrays, providing targets for a global network of satellite laser ranging stations, supported by the International Laser Ranging Service.

LAGEOS 2, a joint NASA/Italian Space Agency (ASI) mission, was built identically to *LAGEOS 1* by Aeritalia Space Systems Group under contract to ASI and deployed by the Space Shuttle *Columbia* on 23 October 1992. The mission was managed by GSFC. Its 53 degree-inclination orbit was chosen for greater coverage of seismically active areas, including California and the Mediterranean, to provide data to possibly understand irregularities seen in *LAGEOS 1*'s orbit.

Using an improved gravitational model based on *GRACE* (*Gravity Recovery and Climate Experiment*) data, scientists reported in 2004 that the plane of the LAGEOS orbits were shifted about 2 m per year in the direction of Earth's orbit. This was the most accurate direct measurement to date of the frame dragging effect predicted by Einstein.

Katie J. Berryhill

See also: *Gravity Probes A* and *B*

Bibliography

Goddard Space Flight Center *LAGEOS*. http://ilrs.gsfc.nasa.gov/satellite_missions/list_of_satellites/lag1_general.html.

International Laser Ranging Service. http://ilrs.gsfc.nasa.gov/.

***Midori I* and *II* and *QuikSCAT*.** *Midori I* and *II* and *QuikSCAT* were three Earth observation spacecraft launched around the turn of the century. The Midori spacecraft were Japanese (Midori means green in Japanese) and had the alternative name of *ADEOS* (*Advanced Earth Observing Satellite*). They were designed to provide data on a wide range of land, ocean, and atmospheric processes.

The 3.5-ton *Midori I*, built by Mitsubishi, was the largest spacecraft built to that date by Japan. It was launched in August 1996 by a Japanese H-2 launcher from Tanegashima into an 800 km circular, Sun-synchronous, near-polar orbit, which crossed Earth's equator at about 10:30 a.m. local solar time every day. It carried eight instruments (see table on page 176), two of which were provided by the United States, and one by France. Data were transmitted directly to ground and via the Japanese *Engineering Test Satellite 6* (*ETS 6*) communications spacecraft. *Midori I* was declared operational in November 1996, but on 30 June 1997 it failed, due to a problem with its solar array.

In its short period of operation, the OCTS and POLDER instruments provided the first routine global observations of ocean color, enabling estimates to be made of phytoplankton pigment concentrations, and TOMS provided contiguous monitoring of ozone and volcanic eruption clouds. In addition, NSCAT provided data on sea surface winds, which were used by weather forecasters, climatologists, and ship owners to route their ships. The loss of their NSCAT scatterometer on *Midori I* was a significant blow to the Americans. They were not as concerned about the loss of their TOMS data, as there was another TOMS on the *TOMS-EP* spacecraft, although it was at a different altitude.

The Americans were in the process of building an improved version of NSCAT, called SeaWinds, when *Midori I* failed. NASA had intended to launch SeaWinds on *Midori II*, but that was delayed while the Japanese investigated the *Midori I* failure. So NASA decided to launch SeaWinds on a "quick-recovery," 1-ton satellite called *QuikSCAT*, where it was the only instrument. Built in just twelve months by Ball Aerospace under the management of Goddard Space Flight Center, *QuikSCAT* used existing hardware, with a design based on the Ball Commercial Platform 2000. It was launched into a similar orbit to *Midori I* by a Titan II from Vandenberg Air Force Base in June 1999. *QuikSCAT* provided high quality wind speed and direction information over 90 percent of Earth's surface every day. It was still operational in June 2007.

Midori I and _II_ and _QuikSCAT_ Instruments (Part 1 of 2)

(A) *Midori I* (Japanese Spacecraft)			
Instrument	**Wavebands (μm) or Frequency**	**Measurements and Main Applications**	**Comments**
Advanced Visible and Near Infrared Radiometer (AVNIR)	3 visible (0.40–0.72) 1 near infrared (IR) (0.82–0.92) 1 Panchromatic (0.52–0.72)	Imaged land and coastal zones at high spatial resolution.	Resolution 16 m for the four spectral bands and 8 m for the panchromatic band.
Improved Limb Atmospheric Spectrometer (ILAS)	1 visible (0.75–0.78) 2 IR (6.0–11.8)	Measured variability of ozone and other trace gases in the stratosphere.	Vertical resolution 2 km over the range 10–60 km.
Interferometric Monitor for Greenhouse Gases (IMG)	3.3–14	Measured the horizontal distribution of greenhouse gases and vertical distribution of atmospheric temperature and water vapor.	
NASA Scatterometer (NSCAT)	13.99 GHz (Ku band)	Measured ocean surface wind velocity and direction, and provided data on sea–air interactions.	Provided by United States. Follow-on to *Seasat* scatterometer. NSCAT wind speed accuracy 2 m/s.
Ocean Color and Temperature Scanner (OCTS)	6 visible (0.40–0.68) 6 IR (0.75–12.5)	Measured global ocean color, sea surface temperature, distribution of phytoplankton and sedimentary material.	
Polarization and Directionality of the Earth's Reflectances (POLDER)	8 visible and near IR (0.44–0.95)	Measured light scattered by aerosols, clouds, oceans and land surfaces.	Provided by France.
Retroreflector in Space (RIS)	Passive	Distribution of ozone and other trace gases.	Corner cube retroreflector to reflect ground-based laser beam back to ground.
Total Ozone Mapping Spectrometer (TOMS)	6 Ultraviolet (0.30–0.36)	Measured total column atmospheric ozone, and global distribution of sulfur dioxide and aerosols produced by volcanic eruptions.	Provided by United States. A TOMS instrument was first flown on *Nimbus 7*.

***Midori I* and *II* and *QuikSCAT* Instruments (Part 2 of 2)**

(B) *QuikSCAT* (U.S. spacecraft)			
Instrument	**Frequency**	**Measurements and Main Applications**	**Comments**
SeaWinds	13.4 GHz (Ku band)	Measured wind speed and direction over ocean. Measured sea ice extent and tracked icebergs.	

(C) *Midori II* (Japanese spacecraft)			
Instrument	**Wavebands (μm) or Frequency**	**Measurements and Main Applications**	**Comments**
Advanced Microwave Scanning Radiometer (AMSR)	8 bands from 6.9–89 GHz	Measured water vapor, precipitation rate, sea surface temperature, sea surface wind speed, sea ice type and extent.	
Global Imager (GLI)	36 bands from visible to far IR (0.38–12.5)	Measured solar radiation reflected from land, ocean or clouds. Produced chlorophyll and vegetation index, surface temperature, snow and ice cover, and cloud distribution.	
Improved Limb Atmospheric Spectrometer 2 (ILAS 2)	1 visible (0.75–0.78) 3 IR (3.0–12.9)	Measured variability of ozone and other trace gases in the stratosphere.	Improved version of ILAS flown on *Midori I*. Vertical resolution 1 km over the range 10–60 km.
POLDER 2	8 visible and near IR (0.44–0.91)	Measured light scattered by aerosols, clouds, oceans and land surfaces.	Provided by France. Improved version of POLDER flown on *Midori I*.
SeaWinds, as on *QuikSCAT*			Provided by United States.

The 3.7-ton *Midori II* spacecraft was launched by a Japanese H-2A launcher from Tanegashima into a similar orbit to *Midori I* in December 2002. It used the same spacecraft bus as *Midori I*, and carried five instruments (see table on page 177), one of which was provided by the United States, and one by France. *Midori II* failed on 25 October 2003, apparently due to a power supply problem. During its short life, it provided data on Earth's water, energy, and carbon cycles.

David Leverington

See also: Goddard Space Flight Center, Japan

Bibliography

Goddard Space Flight Center. http://www.gsfc.nasa.gov/.
Japan Aerospace Exploration Agency. http://www.jaxa.jp/index_e.html.
Jet Propulsion Laboratory. http://www.jpl.nasa.gov/.
NASA Earth Science Reference Handbook (2006).

Momo 1. *Momo 1* (Japanese for peach blossom) was the name of Japan's first satellites dedicated to Earth observation. Originally called Marine Observation Satellites (MOS) until they were renamed after launch, their mission was to study the oceans and sea/land interface areas to improve environmental protection and natural resource utilization, significant because of Japan's dependence on natural resources from its sea and land surroundings. In 1978 a committee of Japan's Science and Technology Agency recommended a long-range plan for Earth observation research, including Landsat ground stations and a domestic satellite program, beginning with MOS. Nippon Electric Corporation built the 740 kg, three-axis stabilized spacecraft, which were managed by the National Space Development Agency of Japan. The spacecraft, which had a design life of two years, operated for eight and six years, respectively. The satellites were launched from Tanegashima Space Center into 909 km, Sun-synchronous orbits: *Momo 1* on an N-II booster on 19 February 1987 and the upgraded engineering model *Momo 1b* as one of three payloads on an H-1 vehicle on 7 February 1990.

Both of the satellites carried three sensors. A multispectral radiometer monitored ice and snow distribution, ocean chlorophyll, and land use. A visible and thermal infrared radiometer also observed snow and ice coverage in addition to monitoring cloud coverage and sea-surface temperatures, while a microwave radiometer measured water vapor content in the atmosphere and precipitation. They also carried a transponder to collect data from ocean buoys.

Data from these spacecraft was used in conjunction with ground-truth data to analyze environmental and land-use changes. Examples include sea depth analysis around the Yap Islands and evidence of the environmental impact of shifting cultivation in the Laotian Nam Kahne watershed. Designed for two-year lifespans, *Momo 1* was operational until 29 November 1995, *Momo 1b* until 17 April 1996. Japan has since launched or participated in several more Earth observation missions.

Katie J. Berryhill

See also: Japan

Bibliography

Goddard Space Flight Center Global Change Master Directory. http://gcmd.nasa.gov/.

Japan Aerospace Exploration Agency Momo. http://www.jaxa.jp/projects/sat/mos1/index_e.html.

Ocean Remote Sensing. Ocean remote sensing (ORS) monitors the state and conditions of Earth's oceans. Understanding the hydrosphere and ocean processes is crucial to understanding the processes of the entire Earth climate system, for both short-term and long-term periods. Oceans act as a major influence to the global climate, retaining and redistributing solar energy from the equator and polar regions. In addition to naturally varying processes, human activities impact ocean conditions, which warranted monitoring from space. ORS is often the only method for data gathering that satisfies scientific needs of timeliness and completeness. Before spaceborne ORS, data collection of ocean phenomenon and human activities in and on the oceans occurred, but typically from sea and airborne platforms.

Several countries launched ORS space missions, starting in the late 1970s. Both civilian and military ORS missions included monitoring of the sea surface temperatures, ocean topography, sea ice changes, water flow, sea-surface wind velocity, wave activity, and biological production. Examples of human activities that warranted monitoring (both civilian and military) included collection of naval fleet surveillance, tracking of commercial shipping activities, search-and-rescue efforts, and detection of environmental threats, such as oil slick monitoring.

Following the establishment of weather monitoring satellites in the 1960s–70s, the world scientific community became more aware of the linkages among the various Earth processes. Earth systems research analyzed linkages among Earth's land, ocean, atmosphere, and biology and attempted to build models via an interdisciplinary approach for understanding and predicting global cycles. Development of computational models required numerical data to understand Earth as a system, and spaceborne missions often provided the best methods for gathering data globally with frequent repeat coverage.

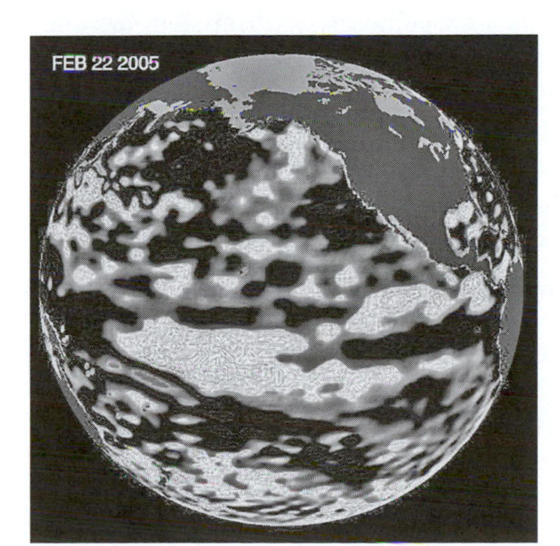

Image from Jason 1 *showing sea surface heights. The El Niño warming occurs in the eastern Pacific as shown here. (Courtesy NASA/Jet Propulsion Laboratory Ocean Surface Topography Team)*

Early weather-monitoring satellites evolved to carry sensors capable of monitoring ocean phenomenon, also atmospheric and land phenomena. In addition to the primary purpose of collecting meteorological data, infrared sensors on board National Oceanic and Atmospheric Administration (NOAA) polar-orbiting satellites provided the ability to monitor sea surface temperature. Experience with early radiometers on board NOAA satellites resulted in the Advanced Very High Resolution Radiometer (AVHRR), which first operated on board *Tiros N*. In 1978 AVHRR provided data on global sea surface temperatures, which was critical to understanding current flows, regional climates, and climate change.

Jet Propulsion Laboratory's *Seasat*, launched in 1978, was the first dedicated civilian ocean-monitoring satellite. *Seasat* carried an array of sensors capable of globally monitoring ocean parameters, such as sea-surface temperatures, sea-surface winds, wave heights, ocean topography, and ice cover. The mission lasted only 110 days due to a power failure, but it provided a wealth of oceanographic and land information and proved the value of new sensor technology, including microwave remote-sensing techniques (such as synthetic aperture radar, radiometers, scatterometers, and near real-time data delivery.

Synthetic aperture radar (SAR) became widely used following *Seasat* and Shuttle-based imaging radar missions, which further proved the usefulness of SAR for oceanographic and geographic observation. The *European Remote Sensing Satellite 1*, launched in 1991, and the *Japanese Earth Resources 1*, launched in 1992, provided operational SAR data to the scientific community for oceanographic monitoring. Canada launched *Radarsat 1* in 1997 to provide SAR data for ice, geologic, and maritime uses, and the Soviet Union and Russia operated the Okean series of satellites, with a primary goal of monitoring sea-ice conditions using microwave radar. SAR data from these missions, and their follow-on missions, provided insight into a variety of topics, including sea ice and polar ice shelf activity, shipping, ocean wave activity, and coastal geography. SAR can directly image large sea vessels, providing government authorities with information on maritime shipping activities and the environmental impacts of those activities, such as oil slick formation. Sea ice monitoring is critical for maritime navigation and climatology purposes. SAR missions provided large amounts of data concerning the polar regions, due to microwave radar's ability to penetrate cloud cover.

Monitoring of ocean topography provided data on the slightly varying height of sea level, caused by changes in temperature, wind, currents, gravity, and upwelling. Missions producing ocean topographic data included the joint NASA and CNES (French Space Agency) *TOPEX/Poseidon* (1992–2001) and *Jasons 1* and *2* (operated from 2001), which provided repeat coverage utilizing microwave radar altimeters. These missions provided insight into ocean circulation and heat transfer, which are critical inputs into global weather models, enhancing the information about weather phenomenon like El Niño and La Niña.

Ocean biological processes played a critical role in Earth's ecosystem. NASA and NOAA launched the Coastal Zone Color Scanner (CZCS) on board *Nimbus 7* in 1978, a proof-of-concept mission that provided multispectral data on biologic activity in the oceans, often near coastal areas. The Sea-viewing Wide Field-of-View Sensor (SeaWiFS), launched in 1997 by NASA in conjunction with Orbital Sciences Corporation, and India's Ocean Color Monitor on *Oceansat 1* also provided ocean biologic activity data.

Scatterometers provided data regarding sea surface wind velocity. Japan's *Midori I* and *II* and NASA's *QuickSCAT* satellites hosted scatterometers, useful for weather forecasting and providing critical planning data to maritime users. NASA's *Skylab* and *Seasat* hosted the first microwave radar scatterometers. Scatterometers provided insight into the heat and momentum changes between the ocean and the atmosphere, which increased the accuracy of weather forecasting models.

NASA's Earth Observation System program launched various satellites in the 1990s and early 2000s for monitoring long-term global conditions of the land, sea, and air. The Moderate Resolution Imaging Spectroradiometer (MODIS) on board the *Terra* (launched 1999) and *Aqua* (launched 2004) satellites provided complete Earth coverage every two days. MODIS was used for various ocean monitoring activities, including sea-surface temperatures and ocean biologic activity. *Aqua* hosted advanced radiometers and infrared sensors for measuring sea-surface temperatures.

Space-based ORS significantly altered and improved understanding of Earth as a system and provided a critical record of measurements and observations regarding global climate change. Global sea-surface temperature records from AVHRR, dating back to 1978, have provided scientists with data regarding Earth's heat redistribution process, also known as the "global conveyor belt," a concept critical to understanding global climate variation. Data from the CZCS and related missions provided new perspectives on the interrelationship between land mass and the oceans and allowed for estimation of ocean chlorophyll and plant biomass, which are critical components of the global carbon cycle. ORS brought new awareness of the impact of global tidal patterns and ocean dynamics, using satellite-based altimeters, capable of providing 2–4 cm of altitude precision. Better information on tidal patterns allowed scientists to build better ocean circulation and thermohaline mixing models, permitting higher accuracy models of global climate change over long periods of Earth's history.

In addition to scientific, meteorological, and maritime users, various militaries operated ORS programs for defense purposes. The U.S. Army's Tiros program, which transferred to NASA in the early 1960s, became the basis of the NOAA meteorological program, which eventually evolved to include ocean monitoring capabilities. Similarly the U.S. Defense Meteorological Satellite Program, evolved to monitor land and ocean characteristics in addition to weather data. The Department of Defense operated an ocean electronic intelligence (ELINT) gathering program, known as White Cloud, to determine the position of other military's naval assets. The Soviet Union operated a space-based radar system Radar Ocean Reconnaissance Satellite and ELINT Ocean Reconnaissance Satellite for monitoring naval assets.

David Hartzell

See also: Civilian Remote Sensing, Reconnaissance and Surveillance

Bibliography

Arthur P. Cracknell, *The Advanced Very High Resolution Radiometer* (1997).

NASA Earth Observation System. http://eospso.gsfc.nasa.gov/.

National Academy of Sciences, *Earth Observations from Space: The First 50 Years of Scientific Achievements* (2008).

NOAA. http://www.noaa.gov/.

Abraham Schnapf, *Monitoring Earth's Ocean, Land, and Atmosphere from Space—Sensors, Systems, and Applications* (1985).

Okean. Okean was a series of oceanographic satellites launched by the Soviet Union and Russia since the late 1970s. One of the primary tasks of the satellites has been to monitor ice conditions in the Arctic to ensure the safety of navigation along the Northern Sea Route, which is of vital importance to the Russian economy. The satellites have also been used to measure water temperature, to observe water pollution and flooded regions, and to track storms and cyclones.

Approval for Okean came in a government decree of 5 May 1977, which also gave the go-ahead for the Resurs-O and Resurs-F remote sensing satellites. The development of Okean was assigned to the Yuzhnoe design bureau in Dnepropetrovsk (Ukraine). Based on the design of the Tselina-D electronic reconnaissance satellites, the Okean satellites weighed roughly 2 tons and were vertically oriented. The standard 550 kg instrument suite featured a side-looking radar with a resolution of about 2 km, low- and high-resolution multispectral scanners, and a microwave scanning radiometer. Also on board was a system to collect data from remote stations on land, water, or ice. Okean satellites were launched from the Plesetsk cosmodrome by the Tsiklon-3 rocket into circular 650 km orbits inclined 82.5° to the equator.

The program began with two experimental satellites (Okean-E), launched as *Kosmos 1076* and *Kosmos 1151* in 1979 and 1980. Two slightly improved experimental satellites (Okean-OE) flew as *Kosmos 1500* and *Kosmos 1602* in 1983 and 1984 and introduced the side-looking radar for all-weather observations. The operational version of the satellite (Okean-O1) was introduced in 1986 and eight of these had been orbited as of 2005 under various names (*Kosmos*, *Okean*, *Okean-O1*, *Sich-1*, and *Sich-1M*). Sich is considered a Ukrainian satellite and aside from traditional tasks has been used more specifically for ecological monitoring of Ukrainian territory.

Okean Satellite Series and Launch Dates

Type	Launched	Number of Launches	Comments
Okean-E	12 Feb 1979–23 Feb 1980	2	Officially announced as Kosmos.
Okean-OE	23 Sep 1983–28 Sep 1984	2	Officially announced as Kosmos.
Okean-O1	28 Jul 1986–24 Dec 2004	9 (including one launch failure)	Officially announced as Kosmos, Okean, Okean-O1, Sich-1, and Sich-1M.
Okean-O	17 Jul 1999	1	Officially announced as Okean-O.

In the late 1980s, Yuzhnoe started the development of two types of advanced Okean satellites carrying a pair of side-looking radars and multispectral scanners with more channels and greater resolution. One type (Okean-O2 or Okean-M) was a 2.5-ton satellite using the same bus as the Tselina 2 electronic intelligence satellites, but it was never flown. The other (Okean-O) was a totally new, horizontally oriented vehicle weighing 6 tn and equipped with 1.5 tons of remote-sensing equipment. The first and only *Okean-O* was launched into a Sun-synchronous orbit from the Baikonur cosmodrome by a Zenit rocket in 1999.

Bart Hendrickx

See also: Russia (Formerly the Soviet Union)

Bibliography

Nicholas Johnson and David Rodvold, *Europe and Asia in Space, 1993–1994* (1995).

S. N. Konyukhov et al., *Rakety i kosmicheskie apparaty konstruktorskogo byuro Yuzhnoe* (2000).

PAGEOS, Passive Geodetic Earth Orbiting Satellite. *See* Space Geodesy.

PARASOL, Polarization and Anisotropy of Reflectances for Atmospheric Science coupled with Observations from a Lidar. *See* Earth System Science Pathfinder.

Proba, Project for On-Board Autonomy. *Proba*, *Project for On-Board Autonomy*, was a European Space Agency (ESA) technology demonstration satellite designed to test autonomous functions on board an Earth observation satellite. *Proba* was launched 22 October 2001 from Sriharikota, India, on India's Polar Satellite Launch Vehicle. A consortium from several European countries and Canada, led by Belgian prime contractor Verhaert Design and Development, built the satellite.

The 94 kg microsatellite was in a Sun-synchronous orbit at 600 km altitude. The satellite navigated itself and adjusted and stabilized its attitude along three axes by means of a star-tracking camera, Global Positioning System attitude sensor, three-axis magnetometer, reaction wheels, and magneto-torquers.

The satellite's scientific payloads were the Compact High Resolution Imaging Spectrometer (CHRIS), a Space Radiation Environment Monitor, a Debris In-Orbit Evaluator, and two other cameras. CHRIS imaged with resolution down to 17 m, the satellite rolling itself to provide up to five viewing angles at one pass. CHRIS operated from near infrared to blue. Out of an available 63 spectral bands for viewing, CHRIS could use up to 19 bands at once in any combination. CHRIS could be programmed, even over the Internet, with only target latitude, longitude, and altitude. *Proba* handled the details autonomously.

Proba was used to monitor environmental conditions on Earth, including studying forests, crops, and solid landfill sites; detecting ancient impact craters; and surveying disaster response. Other operations included wide-spectrum photographic Earth mapping and detection of radiation and space debris. At least 60 Earth-observation programs from various nations used *Proba* for environmental studies. As of 2007 the spacecraft continued to function well, and two follow-on craft, *Probas 2* and *3*, were in development.

Kevin Fitzgerald

See also: European Space Agency

Bibliography

ESA Proba. http://www.esa.int/SPECIALS/Proba/index.html.

QuikSCAT. *See Midori I* and *II* and *QuikSCAT.*

Seasat. *See* Ocean Remote Sensing.

Shuttle Radar Topography Mission. The Shuttle Radar Topography Mission (SRTM) yielded the most complete global topographic map of Earth and was the primary mission objective of the Space Transportation System-99 flight. Launched within the payload bay of Space Shuttle *Endeavour* in February 2000, SRTM was a follow-on to two Shuttle missions in 1994, which flew similar instruments. SRTM was a collaboration between NASA and the National Imagery and Mapping Agency

Image of Mount Saint Helens, which was created using elevation data produced by the Shuttle Radar Topography Mission. Elevation data used in this image were acquired aboard the Space Shuttle Endeavour, *launched on 11 February 2000. (Courtesy NASA/Jet Propulsion Laboratory/ National Geospatial-Intelligence Agency)*

(later renamed the National Geospatial-Intelligence Agency), which funded the mission. The instruments were supplied by NASA, the German Aerospace Center (DLR) and the Italian Space Agency.

The heart of the payload consisted of two synthetic aperture radar antennae. The main antenna was mounted inside *Endeavour*'s payload bay. The other, in the major innovation of the mission, was deployed at the end of a 60 m mast, the largest rigid structure to have flown in space. Each antenna contained both C-band (5.6 cm) and X-band (3 cm) receivers, while only the main antenna transmitted the radar signals. Working in tandem, using a principle known as interferometry, these antennae provided global three-dimensional elevation maps of Earth's surface regions.

The SRTM mission had an orbit inclination of 57°, allowing it to map terrestrial land areas between 56° south and 60° north latitude, approximately 80 percent of Earth's land surface. The radar produced images at a resolution of 90 and 30 m while achieving absolute height accuracy of around 10 m. During the 11 day mission, the SRTM payload operated for 10 days. The United States Geological Survey's Earth Resources Observation System Data Center distributed the C-band data. The German space agency provided ground processing and distribution of the data from the European-supplied X-band radar. The SRTM mast was placed on exhibit at the National Air and Space Udvar-Hazy Museum in Virginia.

Kenneth Peek

See also: Earth Resources Observation Data Center, Space Shuttle

Bibliography

DLR. http://www.dlr.de/en/.
Jet Propulsion Laboratory SRTM. http://www2.jpl.nasa.gov/srtm/.
U.S. Geological Survey SRTM. http://srtm.usgs.gov/.

Space Geodesy. Space geodesy is a subdivision of geodesy, the science that determines the positional and gravitational vector fields of Earth and other planetary bodies. Space geodesy can be characterized by (1) observation of the positions and movements of extraterrestrial objects, and (2) three-dimensional positioning (plus time variations). Under this broad definition traditional astronomic geodesy is excluded, since it generates only orientation, but the employment of lunar occultations and solar eclipses would qualify. However, only those techniques developed for Earth geodesy since the start of the artificial satellite era in 1957 are emphasized in this article. Before then, positioning over intercontinental distances was known to 1 part in 5,000, Earth's flattening to 1 part in 300, and the shape of the geoid (the sea-level surface) practically unknown over the oceans.

The first artificial satellites immediately provided a geodetic breakthrough: an increase in knowledge of Earth's flattening to 1 part in 3,000. These satellites (*Sputnik*, *Explorer 1*, and *Vanguard 1*) were photographed when Sun-illuminated, and the observed change in orbital position caused by the flattening yielded the sharper result. From this beginning, refined observing systems and computing methods during the past 50 years have produced knowledge of the geoid to the 1 m level and positioning on the topographic surface to better than 1 cm, thus solving practically all problems of classical geodesy. By the end of the twentieth century, geodesy shifted its attention to the change of these quantities over time.

These results have vastly improved scientific understanding of Earth. But the chief motivation for space geodetic programs sprang from the tensions of the Cold War. The accuracy of intercontinental ballistic missile trajectories depended on knowledge of distance and direction to the target and the effect of gravity on the missile in flight. Thus the military establishments in both the United States and the Soviet Union provided the major support in funds and resources at the start of the space age and have continued to maintain many subsequent operations, particularly the satellite constellations designed for navigation and precise positioning. As their ballistic missile and space programs developed, other nations, such as France and China, launched geodetic satellites for military purposes.

The 50 year span (1957–2007) of space geodesy can be divided into three eras, each characterized by increasingly sophisticated observing systems. During the first 15 years, optical photography and electronic Doppler observations predominated. These were supplanted in the 1970s by laser ranging, very long baseline interferometry (VLBI), radar altimetry, and space navigation networks, which in the early twenty-first century remained the most prevalent and useful methods operationally, especially for positioning. By the turn of the twenty-first century, new systems for improved information on the gravity field, including satellite-to-satellite tracking and satellite gravity gradiometry, were developed.

Baker-Nunn photographic cameras employed by the Smithsonian Astrophysical Observatory comprised the foremost observing system during the first decade of the satellite era. A landmark article by William Kaula in 1963 using this data provided the first global combined solution for worldwide positioning and the low-degree harmonics of the gravitational field. A purely geometric solution for an intercontinental set of station positions was carried out in the 1960s by camera observations on balloon satellites *Echo* and *PAGEOS*. Optical observations were hampered by weather dependent visibility. Doppler measurements overcame this limitation and the Transit satellites during the 1960s yielded station positions to the 1 m level and a global geoid to the 5 m level.

Around 1970 satellites equipped with mirrors to reflect laser ranging ground signals supplanted optical observations on passive satellites and thereafter remained a vital means of acquiring geodetic data. A most spectacular initial application was *LAGEOS 1*, which was able to exhibit crustal motion between ground stations to the 1 cm level. Reflector arrays left on the surface of the Moon, starting with the first successful Moon landing in 1969, yielded the Earth-to-Moon distance at the 1 cm level.

Beginning in 1973 with *Skylab*, radar altimeters installed in satellites transformed them into active rather than passive vehicles for acquiring data. Dense coverage of the ocean surface provided the means for distinguishing the various components of sea level: the geoid, steady-state circulation, and time-varying phenomena such as tides and seasonal changes. Laser ranging and radar altimetry have continued as standard operational tools on geodetic satellites.

Selected Geodetic Methods and Spacecraft (Part 1 of 2)

Satellite	Country	Timeline	Principal Observing Method	Application
Sputnik	Soviet Union	1957–58	optical	low harmonic gravitational field
Explorer 1	United States (US)	1958	optical	low harmonic gravitational field
Vanguard 1	US	1958	optical	low harmonic gravitational field
Transit	US	1962–96	Doppler	positioning, low harmonic gravitational field
Echo 1, 2	US	1960	optical	positioning
ANNA 1B (Army, Navy, NASA, Air Force)	US	1962	electronic ranging	positioning
PAGEOS (*Passive Geodetic Earth Orbiting Satellite*)	US	1966–97	optical	positioning
Beacon Explorer B	US	1964	laser ranging	first satellite to carry retroreflectors
GEOS 1, 2 (*Geodetic Earth Orbiting Satellite*)	US	1965, 1968	electronic ranging	positioning, gravitational field
Skylab	US	1973	altimeter	sea level determination
Starlette	France	1975	laser ranging	gravitational field, positioning, Earth dynamics
GEOS 3 (*Geodynamics Experimental Ocean Satellite*)	US	1975	altimeter, laser ranging, Doppler	sea level determination, positioning
LAGEOS 1, 2 (Laser Geodynamics Satellite)	US/Italy	1976, 1992	laser ranging	crustal motion, gravitational field, Earth dynamics
Seasat 1	US	1978	altimeter	sea level determination
Ajisai	Japan	1980	laser ranging	positioning, gravitational field
TDRSS (Tracking and Data Relay Satellite System)	US	1983	electronic ranging	positioning
GPS (Global Positioning System)	US	1983	electronic ranging	positioning

(Continued)

Selected Geodetic Methods and Spacecraft (Part 2 of 2)

Satellite	Country	Timeline	Principal Observing Method	Application
GEOSAT (*Geodetic Satellite*)	US	1985	altimeter	sea surface, gravitational field
Etalon 1, 2	Soviet Union	1989	laser ranging	positioning, gravitational field
SPOT 2, 3, 4, 5 (Satellite pour l'Observation de la Terre)	France	1990–2002	Doris (reverse Doppler)	positioning
ERS 1, 2 (European Remote Sensing)	European Space Agency (ESA)	1991, 1995	altimeter, laser ranging	sea surface
TOPEX (*Ocean Topography Experiment*)/ *Poseidon*	US/France	1992	altimeter	sea surface
Stella	France	1993	laser ranging	gravitational field, Earth dynamics
GLONASS (Global Navigation Satellite System)	Russia	1993	electronic ranging	positioning
GFZ 1 (*GeoforschungsZentrum*)	Germany	1995–99	laser ranging	gravitational field, Earth dynamics
GFO (*Geosat Follow-on*)	US	1998	altimeter	gravitational field, sea surface
CHAMP (*Challenging Minisatellite Payload*)	Germany	2000	laser ranging, satellite-to-satellite tracking	gravitational field
Jason	US/France	2001	altimeter, Doris, laser ranging	sea surface
ENVISAT (*Environmental Satellite*)	ESA	2002	altimeter, Doris	sea surface, Earth dynamics
GRACE (*Gravity Recovery and Climate Experiment*)	US/Germany	2002	satellite-to-satellite tracking, laser ranging	gravitational field
Galileo	ESA	2005	electronic ranging	positioning
GOCE (*Gravity Field and Steady-State Ocean Circulation Explorer*)	ESA	2008	satellite-to-satellite tracking, gradiometer	gravitational field

Upon its deployment in the mid-1980s, the dominant operational observing system for geodesy became the U.S. Global Positioning System (GPS), whose full configuration consisted of 24 satellites that transmitte electronic ranging information to ground receivers. GPS almost completely supplanted conventional ground surveying for horizontal and vertical positioning because of its flexibility in station placement, all-weather capability, instant availability, improved accuracy, and cost advantages. A similar array, GLONASS, developed by the Soviet Union and Russia, started operation in the early 1990s. A third constellation, Galileo, a joint initiative of the European Space Agency and the European Commission, was under development in 2008.

Whereas laser ranging, radar altimetry, and GPS successfully handled geodetic questions regarding positioning, sea level, and crustal motion, later satellite observing systems, such as *CHAMP*, GRACE, and *GOCE*, were developed to determine the gravitational field's fine structure. *CHAMP* involved analyzing the ranging between a high satellite (GPS) and a low one (*CHAMP*) whose orbit was steadily decreasing. GRACE consisted of tracking between a satellite pair at the same low altitude. *GOCE* planned to utilize a gravity gradiometer to attempt to attain a geoidal accuracy to the 1 cm level.

The increasing precision of satellite-observing systems enabled extraction of dynamical properties of Earth to a much higher accuracy. Among the most important results were the measurements of movements of the tectonic plates that comprise Earth's crust, including both intraplate deformation and with respect to each other. The rates of change, of the order of 1–10 cm per year, were determined via VLBI, lunar laser ranging, satellite laser ranging, and GPS to better than 1 cm. These observing techniques also provided the means to vastly improve the accuracy of knowledge of polar motion and Earth rotation. The former was calculated to within 5 mm and the latter to within .05 ms per day. Solid Earth tides, which range up to 50 cm, were obtained to an error of 1 percent.

The greater accuracy and density of information, both spatially and over time, flowing from the methods and programs of space geodesy have transformed geodetic practice and affected comprehension in other geophysical areas. Seismology, tectonophysics, oceanography, and planetary exploration utilized data and methodology supplied by space geodesy. Altimeters were employed in the exploration of Mars and Venus, and in determining Earth's sea level change. Variations in the gravity field contributed by low orbit satellites yielded information on subsurface structure and density of Earth, Moon, Mars, and Venus. Precise electronic positioning revolutionized land and sea navigation. Laser ranging furnished strain measurements that influenced earthquake prediction and analysis.

Bernard Chovitz

See also: Early Explorer Spacecraft, Military Geodetic Satellites, Navigation, Sputnik

Bibliography

Günter Seeber, *Satellite Geodesy*, 2nd ed. (2003).

TOMS-EP, Total Ozone Mapping Spectrometer-Earth Probe. *TOMS-EP, Total Ozone Mapping Spectrometer-Earth Probe*, was the third satellite in NASA's TOMS program and the first in a series of Earth Probe missions. TRW Corporation built the spacecraft, and Goddard Space Flight Center managed the mission. Earlier TOMS instruments flew on *Nimbus 7* in 1978 and the Soviet Union's *Meteor 3* in 1991. The TOMS program carried out long-term monitoring of Earth's ozone layer and the release of sulfur dioxide from volcanic eruptions. This latter data was studied by the Federal Aviation Administration to learn how to detect volcanic ash clouds. TOMS acquired ozone data indirectly by comparing incoming solar ultraviolet light with that backscattered by Earth's atmosphere.

Originally scheduled to launch in 1994 on a Pegasus XL booster, *TOMS-EP* was delayed because of failures of the first two Pegasus launches. Its eventual launch from Vandenberg Air Force Base was 2 July 1996, six weeks before the scheduled launch of the Japanese *Midori I*, which carried an identical TOMS instrument, originally intended to be *TOMS-EP*'s successor. As a result, the satellite's mission was modified to augment *Midori I* by observing from a lower orbit, which would have allowed scientists to see the difference between stratospheric and tropospheric ozone. However, when *Midori I* failed the following year, NASA boosted *TOMS-EP*'s orbit to 740 km, not quite to its initially planned 950 km orbit.

After five times the original design life of two years (in part because of the higher orbit), the transmitter on board *TOMS-EP* failed on 2 December 2006. With no way to return the data, spacecraft controllers turned off the spacecraft in mid-2007.

In 2000 and 2003, *TOMS-EP* recorded the two largest Antarctic ozone holes to date. The holes are caused by chemical reactions, but *TOMS-EP* data also detects changes in total ozone related to climate. For example, in 1999 it detected the lowest ozone levels recorded in the northern hemisphere (determined to have been caused by a combination of weather systems); while in 2002, the Antarctic ozone hole was significantly smaller than in previous years, which scientists attributed to above average temperatures in the stratosphere.

Katie J. Berryhill

See also: Meteor, *Midori I* and *II* and *QuikSCAT*, Nimbus, TRW Corporation

Bibliography

Goddard Space Flight Center's *TOMS-EP*. http://disc.sci.gsfc.nasa.gov/gesNews/toms-ep-near-real-time-data.

Space Studies Board, *The Role of Small Satellites in NASA and NOAA Earth Observation Programs* (2000).

TOPEX/Poseidon. *See* Ocean Remote Sensing.

TRMM, Tropical Rainfall Measuring Mission. The *TRMM*, *Tropical Rainfall Measuring Mission*, a joint project of NASA and the Japan Aerospace Exploration Agency, was the first spacecraft devoted to measuring tropical rainfall. Launched 27 November 1997 from Tanegashima Space Center on an H-II booster for a planned three-year mission, the satellite was still operational in late 2006, with an extended mission scheduled through 2009. The mission was managed by Goddard Space Flight Center (GSFC).

The spacecraft, designed and manufactured by GSFC, was initially in a circular 350 km orbit inclined to 35 degrees to maximize coverage of the tropics. The orbit was raised to 402 km in 2001 to extend the satellite's lifespan.

TRMM had five instruments, four of which were developed by NASA and one by Japan, the Precipitation Radar. The TRMM Microwave Imager was based on the microwave radiometer flown on five Defense Meteorological Satellite Program satellites. The Visible and Infrared Scanner descended from the National Oceanic and Atmospheric Administration's Advanced Very High Resolution Radiometer allowed *TRMM* measurements to tie in with those from other meteorological satellites. The Clouds and the Earth's Radiant Energy System instrument was also used on the *Aqua* and *Terra* satellites, but it failed eight months after launch. The Lightning Imaging Sensor evolved from the Optical Transient Detector instrument flown on the *Microlab 1* satellite. These instruments could operate separately or together. To calibrate *TRMM* measurements, ground stations were established in Australia, Brazil, Israel, Republic of the Marshall Islands, Taiwan, and Thailand.

TRMM was the first Earth Science Enterprise mission to provide data publicly via the Internet within one to two days of collection.*TRMM* data provided much more accurate estimates of local and global rainfall in the tropics than previously possible and provided detailed measurements of hurricane rainfall and dynamics.

Katie J. Berryhill and Stephen B. Johnson

See also: Defense Meteorological Satellite Program, Goddard Space Flight Center, Japan

Bibliography

GSFC *TRMM*. http://trmm.gsfc.nasa.gov/.

National Research Council, "Assessment of the Benefits of Extending the Tropical Rainfall Measuring Mission" (2006).

UARS, Upper Atmosphere Research Satellite. *UARS*, *Upper Atmosphere Research Satellite*, was the first and largest of NASA's Mission to Planet Earth spacecraft. It was deployed from Space Shuttle *Discovery* on 15 September 1991, measuring 4.6 m in diameter and 9.8 m long, with a mass of 6,795 kg. The *UARS* mission was to study the chemical and physical processes that occur in Earth's atmosphere between 15 km and 100 km altitude (including the ozone layer), an essential part of understanding the role of the upper atmosphere in global climate and climate change. *UARS* carried 10 instruments from the United States, the United Kingdom, and Canada to measure chemical composition, wind speed, temperature, charged particle energy, solar irradiance, and charged particles injected into the upper atmosphere. Chemical constituents detected include: nitrogen and chlorine species, water vapor, methane, carbon dioxide, carbon monoxide, hydrofluoric and hydrochloric acid, ozone, and sulfur dioxide. The spacecraft, designed and built by General Electric Astro Space, was in a 585 km circular orbit inclined at 57°, an orbit that provided global coverage of the stratosphere and mesosphere every 36 days. The project was managed by Goddard Space Flight Center.

Many of the scientific achievements made by *UARS* relate to seasonal and global movement of various chemical compounds through the layers of the atmosphere. For example, *UARS* instruments monitored ozone depletion over the poles, showed there is a direct relationship between ozone depletion and chlorine, and demonstrated that most of the chlorine in the upper atmosphere is from human-made chlorofluorocarbons. *UARS* tracked the sulfuric acid aerosol formed in the upper atmosphere by the June 1991 Mount Pinatubo eruption, mapping volcanic aerosols for the first time. *UARS* instruments made the first direct measurements of upper atmosphere winds from space.

Although the design lifetime was only three years, *UARS* collected more than 14 years of data—enough for one solar cycle, with 6 of the 10 instruments remaining in operation until the end of the mission. It operated in reduced-power mode throughout most of that time, due to problems with its solar array drive. Operation of *UARS* ended 14 December 2005 at NASA's direction. The Earth Observing System *Aura* mission, launched in July 2004, is the follow-on mission to *UARS*.

Virginia D. Makepeace

See also: Goddard Space Flight Center

Bibliography

JPL Mission and Spacecraft Library UARS Quicklook. http://msl.jpl.nasa.gov/QuickLooks/uarsQL.html.

UARS Science. http://umpgal.gsfc.nasa.gov/.

***V-2* Experiments.** *V-2* experiments were conducted by the U.S. Army from 1946 to 1952, taking advantage of the V-2 rockets and parts captured in Germany in the spring of 1945. Under Project Hermes, a U.S. Army project to develop a tactical ballistic missile, General Electric Corporation contracted to reconstruct as many V-2s as possible, with the Army launching them from White Sands Proving Grounds, New Mexico. Recognizing the value of knowledge about the high atmosphere to the improvement of long-range ballistics, Army Ordnance invited scientists in military laboratories to form a V-2 Panel in January 1946 to coordinate the scientific use of the V-2 flights. Led first by staff from the Naval Research Laboratory (NRL) and later from the Applied Physics Laboratory, this civilian panel coordinated the scientific payloads to be flown on the V-2s and later on sounding rockets of American manufacture.

The first launch occurred in April 1946, and the first successful flight took place the following month. Simple scientific instrumentation was added to early V-2s, including a Geiger counter and undeveloped 35 mm film to detect cosmic radiation. Despite attempts to strengthen the payload compartments, they disintegrated on impact, so little if any data was obtained. The Naval Gun Factory in Washington, DC, designed new payload sections that could be explosively separated from the rocket so they could be recovered, and instruments requiring physical recovery were placed in the lower fin sections. This system's first complete success was V-2 12, launched in October 1946. An NRL solar spectrograph returned ultraviolet measurements, which could not be obtained on the ground because these rays are absorbed by Earth's atmosphere.

The V-2 program included a number of biological experiments. Fungus spores were placed onboard V-2 17, launched 17 December 1946. It reached an altitude of 116 miles, the highest of any Project Hermes launch. Corn seeds and fruit flies were launched

> **HOMER E. NEWELL (1915–1983)**
>
> Homer Newell, a mathematician turned physicist, was deeply involved in early U.S. space efforts. Newell's career and interests spanned both robotic and crewed missions. Newell joined the Naval Research Laboratory in 1944, working on sounding rocket research and then project Vanguard. He joined NASA in 1958, was Associate Administrator for Space Science and Applications from 1963 to 1967, and Associate Administrator of NASA from 1967 to 1973. Newell played a key role in persuading the Astronaut Office to assign geologist Harrison Schmitt to the last Apollo lunar landing and was also involved in a wide range of projects from communications satellites to the lunar receiving laboratory.
>
> *John Ruley*

onboard Hermes V-2s. Other launches were to test the effects of weightlessness and cosmic rays on primates. Three launches of V-2s with Rhesus monkey passengers failed. In summer 1950 a mouse was flown while a camera captured the effects of acceleration and weightlessness.

A major finding of the V-2 program was that the Sun emits X rays. Richard Tousey and Herbert Friedman, both of NRL, showed that radiation in the Lyman alpha range was sufficient to sustain the ionosphere's D-region. Friedman's photon counter experiment, flown on 29 September 1949, proved that X rays detected on earlier flights were from the Sun.

The Army fired 74 V-2 and modified V-2 rockets from White Sands. In addition to the advances in rocketry, scientific instrumentation, and scientific data that the V-2 program supported, the program was crucial in providing experience for many key figures in the nascent U.S. space program, including James Van Allen, who discovered the radiation belts that now bear his name; Homer Newell, who later headed NASA's science programs; Richard Tousey, solar physicist; and Milton Rosen, technical director of the Vanguard program.

William Hartel

See also: United States Army, V-2

Bibliography

David H. DeVorkin, *Science with a Vengeance* (1992).
Gregory P. Kennedy.
Vengeance Weapon 2 (1983).
Frederick I. Ordway III and Mitchell Sharpe, *The Rocket Team* (1979).

World Geodetic System. World Geodetic System (WGS 84) is a global datum and reference frame originally developed in the 1950s by the U.S. Department of Defense (DoD), but in 2007 maintained and updated in collaboration with other international institutions. It is the primary standard global reference system and the basis for the Global Positioning System (GPS) and other space-related systems. The origin point of the system is the seasonally averaged center of Earth's mass (Earth's mass shifts slightly north and south due to seasonal vegetation changes), and the system characterizes both the positional and gravitational fields of Earth.

In 1924 the international ellipsoid reference, developed by geodesist John Fillmore Hayford, served as the basis for the Figure of the Earth (the planet's size and shape) and regionally based datums that described the positions of specific points on Earth's terrestrial surface. In the 1940s the emerging era of long-range bombers and the anticipation of intercontinental ballistic missiles (ICBM) required a system that would link the continents across ocean basins, expedite positioning at sea, and characterize Earth's gravity field to precisely guide ICBMs to their targets. The new era of missiles and satellites required a global datum; satellites enabled it to be created.

The WGS that eventually evolved is based on a geodetic reference system of four parameters describing Earth's mass, rotation, size, and shape. The determined positions of a large number of specific points and the determined gravity field were defined using satellite systems.

The evolution of the system may be characterized by three eras of positional accuracies. In the first era (circa 1957–70) ground-based systems observed satellites visually, or by observing the Doppler shift from satellite signals, as in the U.S. Navy's pioneering Transit system (the program started in 1958). Positional accuracies were around 10 m, but the collective satellite observation systems in use soon provided data used to revise the flattening of Earth (its ellipsoidal shape) in 1958 from its pre-space value, the first triumph of satellite geodesy.

In the second era (circa 1970–90) new technologies allowed relative positional accuracies to improve by a factor of 10, down to near 1-m accuracies. These included laser-ranging systems, using ground-based lasers and a small satellite (*LAGEOS—Laser Geodynamics Satellite*) covered with 400 laser reflectors; Very Long Baseline Interferometry, which derives positional differences by phase changes in signals broadcast from distant stellar sources; positioning by phase comparison systems using satellite signals, particularly GPS; and satellite radar altimetry, which is the inverse of laser-ranging because the active sensor is stationed on the satellite rather than Earth. Radar altimetry was first tested on the crewed Skylab missions. Pioneering satellite altimetry systems included *GEOS-3* (*Geodynamics Experimental Ocean Satellite*, 1975), a NASA system to better define the structure of Earth's irregular gravitational field, and *Seasat* (1978), which was the first satellite designed for remote sensing of Earth's oceans, and *TOPEX-Poseidon* (1992). *TOPEX-Poseidon* was a joint venture between NASA and CNES, the French Space Agency. The mission's paired radar altimeter systems (only one of which was operational at any one time) acquired data on sea surface topography until 2005, when a system malfunctioned. *Geosat* (*Geodetic Satellite*, 1985–89) provided detailed marine gravity maps for the Trident submarine program, and provided the first detailed maps of ocean tides and currents. With the convergence of these increased positional accuracies, dynamic processes can be increasingly differentiated from static conditions, so that, for instance, observations of Earth's sea surface allow instantaneous dynamic sea level heights to be differentiated from the mean.

The third era, which began around 1990, features an array of systems providing positional accuracies to 1 cm or less so that movements of continents may be observed. New satellite systems also include pairs of satellites, such as *GRACE* (*Gravity Recovery and Climate Experiment*, 1992), to characterize large-scale geodetic features of hydrology, polar ice, and climate. *GRACE* is a joint venture between NASA and Deutsche Forschungsanstalt für Luft und Raumfahrt in Germany. *TOPEX-Poseidon* was succeeded by *Jason 1* (2001), which overlapped the earlier system for three years and allowed tandem coverage of sea surfaces, which allowed smaller sea features to be discerned and studies. This was later succeeded by the Ocean Surface Topography Mission (OSTM), orbited on *Jason 2* (2008). OSTM-*Jason 2* is a joint venture between the National Oceanic and Atmospheric Administration, NASA, CNES, and EUMETSAT.

WGS has continuously evolved. A formal presentation of the system, in both higher-accuracy classified and lower-accuracy unclassified versions, was released by DoD in 1972 as WGS 72. Major geodetic revisions led to the release of WGS 84 in 1984. Since then the decision was made to incorporate revisions occasionally but retain the datum name, so WGS 84 refers to the contemporary global datum as observed and maintained by DoD in collaboration with a complex nexus of international scientific institutions, which converge principally in the International Earth Rotation and Reference Systems Service (IERS), established in its present form in 1987, by the International Astronomical Union and the International Union of Geodesy and Geophysics. The IERS is the body responsible for maintaining global time and reference frame standards, upon which is developed and maintained WGS 84.

John Cloud

See also: Ballistic Missiles, EUMETSAT, Global Positioning System, Transit

Bibliography

Bernard Chovitz, "Modern Geodetic Earth Reference Models," *EOS* 62, no. 7 (17 February 1981): 65–67.

Department of Defense World Geodetic System 1984 (2000).

National Geo-spatial Intelligence Agency, *World Geodetic System 1984* (2005).

Günter Seeber, *Satellite Geodesy*, 2nd completely revised and extended edition (2003).

Exoplanets

Exoplanets (or extrasolar planets) are planets orbiting stars other than the Sun. Until their discovery in the late twentieth century, their existence had been a subject of centuries-long astronomical debate. Exoplanet detection is difficult, since they usually cannot be seen directly due to their close proximity to their parent star, as seen from Earth. The most successful detection methods, as of early 2009, were measurements of Doppler shifts in the radial velocity of a star, due to the gravitational tug by an orbiting body; and detection of a transit, in which scientists measure the minute dip in a star's brightness as a planet passes in front of it. Some were also detected during microlensing events and direct imaging.

After several unconfirmed detections in previous decades, the first confirmed identification of planets detected outside the solar system was two large planets found around the millisecond pulsar PSR B1257, announced in 1991 and confirmed in 1993. The first exoplanet orbiting a Sun-like star, 51 Pegasi, was announced in October 1995. The first discovery of a transit of an exoplanet in front of its star (HD 209458) occurred in 1999. Two years later, the discoverers confirmed that the planet's atmosphere contained sodium. By early 2009 astronomers had detected more than 340 exoplanets, mostly by the radial velocity method.

Imaging an exoplanet was the next challenge. Ground-based telescopes were limited to using this method with planets around faint stars. In 2004–5, ground-based and space-based telescopes began to take the first images of extrasolar planets. As these were infrared detections, these planets are relatively hot, compared to the planets in Earth's solar system. In 2008 infrared imaging by the Keck and Gemini North telescopes combined to show the first image of a planetary system: three planets orbiting HR 8799. The first confirmed visible light image of an exoplanet orbiting its star, taken by the *Hubble Space Telescope* and announced in 2008, showed a planet orbiting Fomalhaut.

The first spacecraft dedicated to the exoplanet search was *Convection, Rotation, and Planetary Transits* (*COROT*), launched on a Soyuz booster from Baikonur into an 896 km polar orbit in December 2006. The mission was managed by CNES (French Space Agency) with participation from the European Space Agency, Germany, Spain, Belgium, Austria, and Brazil. The 630 kg spacecraft, designed by Thales Alenia Space, carried a 30-cm telescope with two off-axis mirrors and a two-part camera. Each half of the wide-field camera was designed for one of the satellite's missions: planetary transit detection and asteroseismology.

About four months after launch, *COROT* detected its first exoplanet, observing the transit of a planet less than twice the radius and 1.3 times the mass of Jupiter, orbiting a Sun-like star every 1.5 days. The accuracy of its early data promised the possibility of detecting exoplanets similar to the rocky planets of Earth's solar system.

Katie J. Berryhill

See also: Astrobiology, France

Bibliography

California and Carnegie Planet Search. http://exoplanets.org/.
CNES *COROT.* http://smsc.cnes.fr/COROT/.
Steven J. Dick, *Life on Other Worlds* (1998).
Gregory L. Matloff, *Deep Space Probes: To the Outer Solar System and Beyond* (2005).

Jupiter

Jupiter is the largest planet in the solar system. As viewed in Earth's skies, it is also usually the second brightest planet, exceeded only by Venus. Naked-eye observations from antiquity had established the duration of its circuit through the background stars as 11.86 Earth years. When Galileo turned his telescope toward Jupiter in January 1610, he first realized that Jupiter had four large moons (henceforth called the "Galilean moons"), a discovery that contributed to the acceptance of the Copernican view of the heavens and to the determination of the speed of light. Application of Newton's 1687 laws of motion helped astronomers to realize that Jupiter was a giant, but low-density planet. By then, improved telescopes enabled astronomers to discern latitudinally confined light and dark bands in the atmosphere of Jupiter, along with an enormous oval storm, possibly the same storm that later became known as the Great Red Spot. Before *Pioneer 10*'s encounter, Earth-based observations in ultraviolet to radio wavelengths revealed the main atmospheric gas to be hydrogen, with as-yet-undetected helium suspected to be the second most abundant gas. These spectroscopic observations led to the conclusion that the visible "surface" of Jupiter consisted of clouds of ammonia ice crystals, colored by unknown impurities. Minute quantities of methane and ammonia gas were also detected. Before the early 1972 launch of *Pioneer 10*, Jupiter's moon count had risen to 12, with the 1892 discovery of Amalthea by Edward E. Barnard and seven others by various observers between 1904 and 1951.

Pioneers 10 and *11* encountered Jupiter in December 1973 and 1974, respectively. They detected helium in the atmosphere, measured the intense radiation belts and magnetic field of the giant planet, and provided the best images of the planet up to that time. They also determined that Jupiter's interior was primarily liquid and provided improved masses for the four Galilean moons. Using measurements of mass, volume, rotation, gravitational field, and electromagnetic radiation, mathematical models provided assessments of Jupiter's interior conditions. There is no liquid or solid surface beneath the cloud-tops, although the high pressures and temperatures inside Jupiter compress the materials into liquid-like layers. Below about halfway to the center, hydrogen is compressed into a metallic liquid, somewhat like mercury, and electrical

Voyager 2 image of the Great Red Spot at the top edge, a large white spot and surrounding cloud structure of Jupiter. (Courtesy NASA/Jet Propulsion Laboratory)

currents flowing in this conductive layer are the likely source for the planet's strong magnetic field. At Jupiter's center there may be an Earth-sized core of molten rocky material. In 1975 measurements from the Kuiper Airborne Observatory detected water vapor in Jupiter's atmosphere.

The *Voyager 1* encounter with Jupiter occurred in March 1979; the *Voyager 2* encounter followed four months later. They were the first to view Jupiter's ring system and the three small moons (in addition to Amalthea) that were later found to be the source of its material. They verified the internal rotation period of Jupiter as 9.924 hours, consistent with periodic radio bursts measured at Earth. Images of the planet during approach enabled mapping of the atmospheric wind speeds relative to the rotation rate, disclosing both prograde and retrograde winds, some with speeds up to 150 m/s. Helium was found to constitute 23 percent of the mass of the atmosphere above the cloud tops, which was slightly less than that of the Sun; hydrogen occupied 76 percent of the mass, and all other gases combined were less than 1 percent. Io was seen to have volcanoes on its surface, triggered by gravitational interactions with Europa and Jupiter. Europa's crust is almost devoid of impact craters, but is covered with cracks, a probable indication that it harbors an ocean beneath a thick crust of water ice. Ganymede, the largest moon in the solar system, showed heavily cratered dark regions and lighter regions with grooves and troughs. Callisto, the second largest Galilean satellite, is heavily cratered and may also have a layer of liquid water at great depth.

The first orbital mission to Jupiter was *Galileo*, which was inserted into orbit in December 1995 and ended its life by plunging into the Jupiter atmosphere in early September 2003. The *Galileo Probe* made the first in situ measurements of the changing atmospheric pressure, temperature, and composition with depth below the cloud tops. The orbiter also studied the structure and dynamics of the magnetosphere of Jupiter, and completed high-resolution mapping of the surfaces of all four Galilean satellites, discovering nearly 100 active volcanic regions on the surface of Io. *Galileo* also discovered that the Gossamer Ring was divided into an inner (Amalthea) and outer (Thebe) portion and better characterized the Main and Halo Rings of Jupiter. Detection of small intrinsic magnetic fields on Ganymede and Callisto provided evidence of liquid water deep within their interiors.

During its December 2000 encounter with Jupiter, the *Cassini Orbiter*, whose primary mission was to study the Saturn system, skirted along the outer boundary of Jupiter's magnetic field, greatly increasing the understanding of conditions near that boundary. Cassini also obtained the highest resolution images ever of Jupiter's atmosphere and also studied the cloud motions within that atmosphere for more than four months. It obtained images of Himalia, the only disk-resolved image of any of the myriad of outer satellites of Jupiter known to exist. Due to a combination of spacecraft encounters and much-improved ground-based imaging, by 2007 Jupiter was known to possess at least 63 moons, most of which are captured asteroids in distant inclined and elliptical orbits.

Ellis Miner

See also: *Cassini-Huygens*, *Pioneer 10* and *11*, Voyager

Bibliography

Fran Bagenal et al., eds., *Jupiter: The Planet, Satellites and Magnetosphere* (2004).
Reta Beebe, *Jupiter: The Giant Planet* (1994).
Henry C. Dethloff and Ronald A. Schorn, *Voyager's Grand Tour* (2003).
Tom Gehrels and Mildred Shapley Matthews, eds., *Jupiter* (1976).

Galileo. The *Galileo* mission included an orbiter and probe to provide extended scientific study of Jupiter. The orbiter was designed, built, and operated by NASA's Jet Propulsion Laboratory (JPL), while Ames Research Center managed the probe, built by Hughes Aircraft Corporation. The orbiter propulsion system was provided by West Germany.

Galileo's odyssey began in a series of studies carried out in the early 1970s. These were intended to define the scientific goals of planetary science after the (then as yet unflown) *Voyager* flyby missions to the outer planets. The mission concept became known as JOP, Jupiter Orbiter with Probe, which was finally approved in 1977. JOP gained the name *Galileo* in 1978, in commemoration of the discoverer of Jupiter's four largest moons.

NASA originally slated *Galileo* for launch from a Space Shuttle in 1982, using the Inertial Upper Stage (IUS) to boost in into its planetary trajectory. When the IUS program was canceled in 1981, NASA switched to the Centaur liquid-hydrogen fuel upper stage, with a launch in 1985. After further delays, the *Challenger* accident in January 1986 forced a switch back to a resurrected IUS, and delayed the launch to 1989. The many changes to launch date and upper stage forced JPL to redesign several times the trajectory needed to get *Galileo* out to Jupiter. The final trajectory using the less-powerful IUS included gravitational assists from a Venus flyby and two Earth flybys.

On 18 October 1989, the crew of Shuttle *Atlantis* launched *Galileo*. In April 1991 the *Galileo* team tried to deploy the spacecraft's high gain antenna, necessary to transmit data at high rates back from Jupiter. The antenna only partially deployed. Use of a new data compression scheme and management of the spacecraft's onboard data storage allowed the mission managers to salvage the vast majority of the spacecraft's science capability. In the meantime, *Galileo* flew by the asteroids Gaspra in October 1991 and Ida in August 1993. It returned the first close-up images ever taken of asteroids, demonstrating that they were irregular, cratered, and rocky. Ida even had its own tiny satellite, Dactyl.

As the spacecraft approached Jupiter, JPL controllers released the atmosphere entry probe on 13 July 1995; it entered Jupiter's atmosphere at a speed of 170,000 km/hr the same day *Galileo* achieved orbit, 7 December 1995. The probe transmitted its data to *Galileo* for 61.4 minutes, providing details on the chemical composition of the atmosphere and its temperature and winds. It failed when it reached atmospheric pressures nearly 24 times Earth's sea level pressure, more than twice its design goal. The probe's data indicated that Jupiter's atmosphere had less water, but more helium, than expected based on planetary formation theory and Voyager data.

A high-resolution image of Jupiter's moon Europa taken from Galileo, *showing crustal plates ranging up to 13 kilometers across that have been broken apart and "rafted" into new positions. This resembles pack-ice disruptions on Earth's polar seas, suggesting that these features are enabled by liquid water or soft ice movements. (Courtesy NASA/Jet Propulsion Laboratory)*

Before its fiery end, *Galileo* completed 34 orbits of Jupiter. It also executed multiple flybys of all four Galilean moons. It returned images taken while en route to the planet of Comet Shoemaker-Levy 9 fragments hitting the planet in July 1994, of Jupiter itself and of many of its moons. It provided more detail of the Jovian rings and found organic compounds (based on carbon molecules) on the Galilean moons. It discovered nearly 100 active volcanic regions and witnessed extreme volcanic events on Io, which contains the solar system's largest active volcano. Its observations of Io indicated that the moon's volcanic processes generate both sulfur-rich and silica-rich magmas and that Io has not experienced the differentiation in silicate composition that is expected of ancient, still-active rocky bodies. Io also appears to have a very large, still-molten iron core. Galileo provided strong evidence to support the theory, developed after the Voyager encounters, that Europa has a liquid water ocean that is many times deeper than Earth's oceans, under a global ice sheet. Magnetic field data suggested that Callisto and Ganymede might also have subsurface liquid water oceans. The possible presence of liquid water on these moons has led some scientists to argue that they could support primitive forms of life.

Galileo orbited Jupiter until 21 September 2003, when JPL deliberately plunged it into Jupiter's atmosphere. This was done to protect the Galilean moons from possible contamination from Earth bacteria that could have survived on the orbiter.

Erik M. Conway

See also: Ames Research Center, Asteroids, Inertial Upper Stage, Jet Propulsion Laboratory

Bibliography

Galileo Legacy. http://solarsystem.nasa.gov/galileo/?CFID=35693881&CFTOKEN=30601131.
Bruce Murray, *Journey into Space: The First Thirty Years of Space Exploration* (1989).
David M. Harland, *Jupiter Odyssey: The Story of the Galileo Mission* (2000).
R. M. C. Lopes and D. A. Williams, "Io after Galileo," *Reports on Progress in Physics* 68, no. 2 (February 2005): 303–40.
Michael Meltzer, *Mission to Jupiter: A History of the Galileo Project* (2007).

Kuiper Belt Objects. *See* Pluto and Kuiper Belt Objects.

Lunar Science

Lunar science began with investigation of the Moon's physical nature. Around 300 BCE Aristarchus of Samos deduced the approximate size and distance of the Moon. In 1609 Galileo Galilei pointed the newly invented telescope at the Moon and discovered mountains, depressions, and plains, dispelling Aristotle's notion that the Moon was an unblemished sphere. He believed the mountains were up to four times higher than those on Earth, but eventually measurements proved that this was not so. The use of the Latin word mare, or sea, for the smooth dark plains led many observers and philosophers to consider that the Moon had large Earth-like bodies of water, with a thin atmosphere and perhaps even living creatures.

Lunar maps from Galileo's time to the early 1900s showed ever more detail, but the positions and sizes of small features remained uncertain. Mapping details at the limit of telescopic resolution led to infamous cases of apparent changes and reoccurring controversy about the possibility of life on the Moon. The main scientific gain was in showing the variety of lunar landforms.

With better observations and maps, natural philosophers (later scientists) began to speculate about the nature of their observations. In the 1660s British natural philosopher Robert Hooke experimented to produce crater-like features by dropping objects onto powders and boiling a clay-water mix. Because he did not know of any objects in space that could hit the Moon, he favored a volcanic origin. After the discovery of asteroids in the early 1800s, several astronomers resurrected the impact theory. However, most astronomers continued to favor the volcanic theory because impact craters were unknown on Earth, and the lunar ones were circular despite the fact that most asteroids would hit the surface at an angle, which astronomers then wrongly believed would make the craters elliptically shaped. In the 1890s, noting the morphological differences between volcanoes on Earth and lunar craters, U.S. geologist Grove Karl Gilbert proposed that the craters also were of impact origin. He correctly ascribed central peaks to rebound of rocks compressed by the energy of impact and interpreted lunar rays as splashes of material ejected at impact. However, his ideas were not widely known to astronomers and were forgotten for more than 50 years.

World War I's massive artillery bombardments and resultant cratering led some geologists to note the resemblance between artillery craters and lunar craters. Similar observations in World War II inspired U.S. astronomer and businessman Ralph Baldwin. He studied bomb cratering and experimented with small explosions, discovering that the ratio of depth to diameter of small explosion craters, bomb craters, and lunar craters was similar. In 1949 he published his observations in *The Face of the Moon*. This convinced the majority of U.S. astronomers, but few outside the United States, that lunar craters formed by high-speed impact collisions. Baldwin also recognized that the major circular maria occupied large impact basins and proposed that the maria were volcanic lavas.

In the 1950s and early 1960s there was a vigorous debate between advocates of the hot Moon and cold Moon theories of lunar origin. Astronomer Gerard Kuiper was a leading advocate of the hot Moon theory, which postulated that the Moon originated through the collision of dust and rocks in the early solar system and heated up as it formed, with a hot internal core. Geochemist Harold Urey was the primary advocate of the cold Moon theory, with the Moon again being formed by collisions, but primarily remaining cool during the process. Radioactive processes later generated internal heat. A major discriminator between the theories would be the presence (hot Moon) or absence (cold Moon) of basaltic rocks indicating the presence of lava flows, which were found in abundance during the Apollo expeditions.

The physics of the impact cratering process was investigated further in the 1950s by U.S. Geological Survey scientist Eugene Shoemaker. Shoemaker mapped Meteor Crater in Arizona and nuclear test craters, discovering that crater rims were overturned flaps of near-surface rocks, and he explained in detail the formation of crater rays. With Robert Hackman, in 1961 Shoemaker recognized that the formation of the large impact basins like Imbrium created widespread ejecta, material thrown out during an impact. Over time, ejecta from different impacts formed layers, with more recent formations on top of earlier ones. This stratigraphic mapping concept was systematically applied to the entire Moon and later all other solid planets in the solar system.

In 1959 the Soviet *Luna 3* spacecraft flew past the Moon, obtaining the first coarse images of the lunar farside. They showed that maria are far less common on the farside than the nearside, but more important that space exploration would revolutionize understanding of the Moon and beyond. The 1960s marked a transition from traditional telescopic research to direct lunar exploration. Gerard Kuiper compiled major atlases of the best lunar telescopic photographs. Cataloging lunar craters quantified what was already known—there are many more smaller craters than larger ones. Following an already existing strategy, Kuiper's graduate student William Hartmann used crater counts to estimate that the maria are about 3–3.5 billion years old. Hartmann and Kuiper also created new photographic methods to recognize large-scale features, such as multi-ring basins. U.S. astronomer Tom McCord and his colleagues developed telescopic instruments to measure surface brightness at differing wavelengths, which led to mapping of lunar surface composition.

In the space race's early years, the Moon was the most accessible target beyond Earth orbit for the United States and the Soviet Union. Soviet technology was never sophisticated and reliable enough to provide scientific bonanzas, despite technically successful Luna orbiter, lander, and rover (Lunokhod) missions from 1958 to 1976, some of which returned a small amount of lunar samples that mainly confirmed Apollo results. Nearly all early scientific advancement came from the U.S. missions Ranger, Surveyor, Lunar Orbiter, and Apollo. The Ranger spacecraft (1961–65) were hard landers that obtained increasingly higher resolution images until the moment of collision, demonstrating that ever smaller craters existed all the way to the surface.

Landing humans on the Moon required development of soft-landing technology to determine if the surface would support a lander. In the 1920s Bernard Lyot's polarization studies led to his theory that the lunar surface was covered with volcanic dust. In the 1950s Thomas Gold suggested that this dust (parodied by critics as "Gold dust") was so thick that landers would sink deeply into it. The robotic landers Surveyor (1966–68) and the Soviet *Luna 9* (1966) successfully answered those engineering questions and also showed that the surface of the Moon was relatively smooth, with impact craters of every scale the main topography. The surface was granular and the only visible rocks were crater ejecta. Surveyor made the first chemical analyses of lunar material, finding that maria had compositions consistent with basaltic lava and that the highlands near Tycho were considerably different, being richer in aluminum. Surveyor data cast doubt on Urey's cold Moon theory.

High-resolution images from the Lunar Orbiter spacecraft (1966–67) identified safe landing sites for Apollo and provided details of nearly the entire lunar surface,

The interior of the crater Copernicus from Lunar Orbiter 2 when it was 46 kilometers above the Moon. (Courtesy NASA/Marshall Space Flight Center)

imaging landforms with two to three orders of magnitude better resolution than obtainable from Earth. Scientifically these images revealed details of the morphology of lunar features, but also showed that most surfaces were covered by a deposit that smoothed over the finest details—this was the regolith and mega-regolith, layers of impact ejecta later found to be 10 m to 2 km thick that covered everything but the steepest slopes. The high resolution allowed the physics of ejecta and rays to be understood, as well as the structure of impact craters and basins. Perhaps the most important revelation was the discovery of dozens of large impact basins and the depiction of the bull's-eye pattern of the young, and hence least-modified, Orientale basin.

Apollo 11 and the subsequent Apollo missions (1969–72) transformed lunar science from inferences and speculations into a rigorous discipline based on field study and laboratory analysis. Apollo proved that the Moon was ancient, had no atmosphere or life, and revealed the composition of lunar materials and the processes that formed and modified the Moon. One of the most surprising discoveries was that lunar rocks completely lack water. Widespread fragmented rocks (breccia) demonstrated that impact was a major process. *Apollo 11* samples of tiny aluminum-rich rocks from the highlands led U.S. planetary scientist John Wood and his colleague to hypothesize that the Moon must have totally melted during its formation (a hot Moon theory). At that time low-density, light-colored plagioclase minerals floated to the surface and solidified, forming the highlands crust. Bombardment by giant asteroids, comets, and other debris excavated impact basins that originally looked like Orientale. Millions of years later radioactive decay melted parts of the lunar mantle (a modification of Urey's theory), leading to voluminous eruptions of mare basalts onto the floors and even overflowing the giant impact basins.

Laboratory analyses of Apollo samples continued for decades, but by 1975 the samples and other data led to a new theory of the Moon's origin. William Hartmann and Don Davis, and independently Al Cameron and William Ward, proposed that the Moon came into existence as an accident of the late stages of the formation of Earth. They theorized that about 4.55 billion years ago a planet-sized projectile collided with early Earth, throwing into space a cloud of Earth mantle rocks and bits of the projectile. Rocks within this debris cloud collided and grew to form the Moon. The initial cataclysmic collision and the following giant impacts of accretion meant that the Moon formed at high temperature; any water boiled away into space.

After Apollo and the Soviet Luna missions, the Moon fell from the consciousness of the public and most planetary scientists for more than a decade. Ground-based observers perfected the ability to map lunar surface compositions using multispectral imaging during this period. The next two missions, Japan's *Hiten* (1990) and the U.S. military's *Clementine* (1994), were technology development missions with science as secondary goals. *Clementine*, which the U.S. military developed to test sensor technologies, successfully collected multispectral data of nearly the entire lunar surface and monitoring of its orbital changes led to better definition of mass concentrations under the lunar surface. This data led to improved estimates of the Moon's crustal thickness, with wide variations from 10–120 km in different locations. In the 1990s it was newly recognized that that early phases of mare volcanism were concealed by ejecta from basins and nearby craters, meaning that volcanism began earlier than inferred from Apollo samples. In 1998 *Lunar Prospector* completed a geophysical mapping of the Moon, with results indicating the possibility of abundant hydrogen at the lunar poles. The most tantalizing interpretation was that the hydrogen is the signature of ice deposited by comet impacts and preserved deep on crater floors where the Sun never shines, but 2006 Arecibo radar observations cast doubt on this interpretation.

The twenty-first century saw significantly increased interest in the Moon from many nations. The European Space Agency launched its *SMART 1* (*Small Missions for Advanced Research in Technology*) technology demonstrator in 2003, which detected calcium on the lunar surface and studied lunar librations. Japan launched *Kaguya* in 2007, which improved on the *Hiten* mission with two subsatellites, *Okina* and *Ouna*, to measure gravitational variations. It detected new gravitational anomalies, which indicated differences in crustal structure of impact basins on the near and far sides. China's *Chang'e 1* also reached the Moon in 2007, with instruments to map lunar composition, particularly helium-3. India launched *Chandrayaan 1* in 2008, which included a Moon Impact Probe that impacted Shackleton Crater to release subsurface debris in a search for possible water. NASA's Constellation program, started in 2004, planned to return humans to the Moon. This, along with the intriguing scientific results, resulted in a variety of plans for renewed scientific lunar exploration in several nations, including the United States, Europe, Japan, China, and India.

Charles A. Wood and Stephen B. Johnson

See also: Apollo, National Aeronautics and Space Administration

Bibliography

Donald A. Beattie, *Taking Science to the Moon: Lunar Experiments and the Apollo Program* (2001).
Ronald E. Doel, *Solar System Astronomy in America: Communities, Patronage, and Interdisciplinary Research 1920–1960* (1996).
David Leverington, *Babylon to Voyager and Beyond: A History of Planetary Astronomy* (2003).
William Sheehan and Thomas A. Dobbins, *Epic Moon: A History of Lunar Exploration in the Age of the Telescope* (2001).

Apollo Science. Apollo science was crucial in answering age-old questions about the Moon's origin. When in 1961 President John F. Kennedy announced that the United States would send men to the Moon, he made no mention of science goals the astronauts might achieve there. Just getting to the Moon and back was considered sufficient justification in view of the ongoing space competition with the Soviet Union. To determine science goals, scientists and engineers from NASA and academe worked together in an intense effort to select and develop the experiments that eventually were carried to the Moon on the Apollo missions.

Many in the scientific community greeted the Apollo program as an unprecedented opportunity to find answers to fundamental questions. These dealt with the early history of the solar system, including how the Earth-Moon system formed, when it formed, and whether the Earth and Moon were similar or different in composition. Scientists believed that by conducting different types of experiments and returning samples from the Moon's surface and subsurface, they might answer these questions.

A recommended list of experiments to be carried on the Apollo missions was first compiled in the spring of 1962. NASA selected principal investigators (PIs) who would be responsible for each experiment's design and, eventually, the interpretation of the data collected. In a similar fashion, PIs were selected to analyze the lunar samples. NASA built a biologically secure Lunar Receiving Laboratory (LRL) at the Manned Spacecraft Center near Houston, Texas, to assure that if any deadly pathogen existed in the samples, it could not escape into Earth's environment. When the LRL tested the returned samples, it found no evidence of any life forms, and it released samples to laboratories in the United States and elsewhere. Eventually Apollo brought some 382 kg of lunar material to Earth, much of it still conserved in 2008 for future study.

Fifty-three experiments went to the Moon on Apollo's six successful landing missions. To assure that the astronauts were familiar with each experiment, NASA conducted extensive training and simulations with all crews. Of most interest were the geology traverses during which astronauts would collect samples. Training for the geological traverses was elaborate and time-consuming and was conducted at many locations around the world. NASA included simulations using a small vehicle for the last three missions, when *Apollo 15*, *16*, and *17* carried lunar rovers that permitted the astronauts to travel tens of kilometers from their landing sites.

Astronauts deployed many experiments in conjunction with a sophisticated geophysical station, the Apollo Lunar Science Experiments Package. Although every experiment provided exciting new data, scientists considered three to be of highest importance. The passive seismic experiment that would monitor for "moonquakes" and take gravity measurements, a magnetometer to measure the Moon's magnetic field, and heat flow probes to record the heat radiating from the Moon's interior that would provide clues to the Moon's interior structure and composition. One experiment on the lunar surface, the Laser Ranging Retroreflector, was still in use in 2007.

In addition to the experiments taken to the lunar surface, astronauts conducted 15 different experiments in lunar orbit. In general, these experiments, including a subsatellite and various cameras, made measurements that scientists could compare to surface "ground truth" for extrapolation over wide areas.

As a result of the experiments conducted during Apollo, many age-old questions have been answered. Age dating of some of the returned samples showed extremely old ages, some as high as 4.4–4.5 billion years old. Thus, based on estimates that the Earth is of similar age, it is believed that the Earth and Moon were born together early in the evolution of the solar system. To account for the Moon's Earth-like, but somewhat different mineralogical composition, most scientists now agree that the Moon formed as the result of a collision between a still-forming Earth and another large body of different composition. The remnants of that collision were the original constituents of the present Moon. Over the next billions of years, the Moon experienced intense periods of impacts of large, leftover, fragments of the solar nebula. Those impacts created features such as Mare

Imbrium that were then filled by lava during a billion years of widespread volcanism giving rise to the maria on the Moon's Earth-facing side.

Donald Beattie

See also: Apollo

Bibliography

Donald A. Beattie, *Taking Science to the Moon: Lunar Experiments and the Apollo Program* (2001).

William David Compton, *Where No Man Has Gone Before: A History of Apollo Lunar Exploration Missions* (1989).

Don E. Wilhelms, *To a Rocky Moon: A Geologist's History of Lunar Exploration* (1993).

Chang'e 1. *Chang'e 1* was the first spacecraft in a planned three-phase Chinese lunar exploration program. Designed and manufactured by the China Academy of Space Technology, *Chang'e 1* was launched into a highly elliptical Earth orbit by the China National Space Administration on board a Long March 3A booster from Xichang Satellite Launch Center on 24 October 2007. The spacecraft entered lunar orbit on 5 November 2007, after spending seven days in Earth orbit to check out systems. Two days later, it reached its operational, high-inclination circular orbit 200 km above the Moon.

Named for the mythical Chinese goddess who flew to the Moon, the 2,350 kg spacecraft was based on the Dong Fang Hong 3 communications satellite bus. The payload consisted of eight instruments: a stereo camera to image visible wavelengths, a laser altimeter, a multispectral interferometer spectrometer imager, gamma- and X-ray spectrometers, a microwave radiometer, a high energy particle detector, and solar wind monitors. The spacecraft also carried 30 songs, chosen by popular vote, which it played back to Earth on important Chinese holidays.

Designed for a planned mission life of one year, *Chang'e 1* was to create three-dimensional images and study the composition of the Moon's surface, including detecting abundances of elements and helium-3. *Chang'e 1* tested technology for future Chinese lunar missions. By early 2008 the spacecraft had successfully transmitted images and information from the Moon, including its polar regions.

Katie J. Berryhill

See also: China

Bibliography

Barry E. DiGregorio, "Chinese Satellite Arrives at Moon", *IEEE Spectrum* (November 2007).

S. Huixian et al., "Scientific Objectives and Payloads of *Chang'e 1* Lunar Satellite," *Journal of Earth System Science* 114, no. 6 (December 2005): 789–94.

Clementine. *Clementine* was a technology demonstration mission jointly sponsored by the Ballistic Missile Defense Organization and NASA, built and operated by the Naval Research Laboratory. Launched on 25 January 1994 from Vandenberg Air Force Base, its principal objective was to use the Moon and a near-Earth asteroid as passive targets to demonstrate military sensor performance beyond Earth orbit, and hence immune from potential Antiballistic Missile Treaty restrictions. *Clementine* was one of the first missions to demonstrate the "faster, better, cheaper" concept of space missions: from approval until launch, the Clementine project took only 22 months (far less than usual for missions of comparable complexity) and cost $80 million (Fiscal Year 1993 dollars), including launch vehicle. Its low cost and technical success made it an exemplar for many later small spacecraft missions, such as NASA's Discovery series.

Clementine mapped the Moon for 71 days, returning nearly two million multi-color images in addition to gravity field, topographical, and other measurements. These data allowed geologists to create the first global compositional map of the Moon, illustrating the distribution of rock types. From laser ranging, *Clementine* made the first global topographic map, revealing the enormous South Pole–Aitken basin, an impact crater 2,600 km on the far side of the Moon. Using the spacecraft radio in an improvised experiment conceived during the mapping operations, *Clementine* discovered hydrogen, and hence the possibility of water ice in permanently shaded craters near the south pole of the Moon. On 4 May 1994 *Clementine* left the Moon for the asteroid Geographos. Three days later, an onboard computer failure caused one of the thrusters to fire continuously until the fuel was depleted, causing the spacecraft to spin up while depleting its attitude propellant, terminating the mission. The data returned by *Clementine* greatly improved knowledge of the Moon's topography and mineralogy, on which later missions built.

Paul D. Spudis

See also: Ballistic Missiles and Defenses, United States Navy

Bibliography

Ben Bussey and Paul D. Spudis, *The Clementine Atlas of the Moon* (2004).
Clementine Science Team, various articles in *Science* 226 (1994): 1777–1916.
Paul D. Spudis, "The Moon," in J. Kelly Beatty et al., *The New Solar System* (1999).

Hiten. *See* Lunar Science.

Kaguya. *Kaguya*, also known as SELENE (Selenological and Engineering Explorer), was a Japanese lunar mission launched in 2007 to further scientific understanding of the Moon's origins and geophysical properties. The Japanese Aerospace Exploration Agency (JAXA) managed Kaguya, which was Japan's second lunar mission. Built by the Japanese Institute of Space and Aeronautical Science (ISAS), ISAS and the National Space Development Agency (NASDA) initiated the Kaguya project, prior to their merger with the National Aerospace Laboratory to form JAXA. As part of a JAXA public relations campaign, the Japanese public selected the name "Kaguya," a lunar princess from Japanese folklore.

The mission consisted of a main orbiting platform, *Kaguya*, weighing approximately 3,000 kg at launch, and two 50 kg spacecraft, *Okina* and *Ouna*. *Kaguya* was launched on 14 September 2007 from Tanegashima Space Center on a HII-A launch vehicle. On 4 October 2007 *Kaguya*, which carried the two smaller spacecraft, achieved lunar orbit.

Several days after orbital insertion, *Okina* and *Ouna* separated from the main bus. *Okina* entered an elliptical polar orbit of 100 km × 800 km, and *Ouna* a polar orbit of 100 km × 2,400 km. On 18 October 2007 *Kaguya* began the operational phase with a 100 km, circular, polar orbit for lunar observations. The mission used three different orbits to better measure the Moon's gravitational field variations. On 21 December 2007 the mission entered a 10 month nominal observation period.

The orbit configuration of *Kaguya* and the small satellites permitted very long baseline interferometry measurements, in conjunction with Earth-based measurements, to improve the lunar gravity models. Instruments on board *Kaguya*'s main bus included X-ray and gamma-ray spectrometers, visible and near-infrared imagers and spectrometers, a radar sounder, a laser altimeter, magnetometers, plasma sensors, and radio-science instruments. *Kaguya* carried a high-definition television camera to produce images for public relations. The Terrain Camera acquired images of the *Apollo 15* and *17* landing sites with 10 m resolution.

As of late 2008 the mission was in the primary data collecting phase, with full results expected in late 2009. It had already produced the highest resolution topographic and gravitational anomaly maps to date, which revealed undiscovered basins on both the lunar near and far sides. Future Japanese lunar exploration missions were expected to rely on data collected from *Kaguya* for site selection.

David Hartzell

See also: Japan

Bibliography

JAXA. http://www.jaxa.jp/index_e.html.

Luna. Luna, or Lunik, was a series of spacecraft launched to the Moon by the Soviet Union between 1958 and 1976. The first three launches failed, so those spacecraft were not numbered, while the first successful launch, on 2 January 1959, resulted in the 360 kg *Luna 1* becoming the first spacecraft to reach the Moon. At that time, the Soviet Union never admitted to any launch failures nor provided advance publicity of launches, and so the success of what was apparently its first attempt at a Moon shot was a surprise, especially as the attempt followed four American failures. The lunar spacecraft were initially the responsibility of Sergei Korolev's OKB-1 (Experimental Design Bureau), but in 1965 this responsibility was transferred to the Lavochkin bureau.

Luna 1 flew by the Moon at a distance of about 6,000 km, 34 hours after launch. It detected a zone of intense radiation about 30,000 km from Earth and then went into orbit around the Sun, becoming the first artificial planet. Eight months later *Luna 2* was the first spacecraft to impact the Moon, followed three weeks later by *Luna 3*, which became the first spacecraft to photograph the Moon's far side. Although poor by later standards, these photographs were a sensation, as people had often speculated what the other side of Earth's nearest neighbor would look like.

Luna Satellites and Missions (Part 1 of 2)

Spacecraft	Launched[1]	Mass (kg)	Mission	Comments
Unnamed	23 Sep 1958		Lunar impact	Launcher failure.
Unnamed	12 Oct 1958		Lunar impact	Launcher failure.
Unnamed	4 Dec 1958		Lunar impact	Launcher failure.
Luna 1 or *Mechta*	2 Jan 1959	360	Lunar impact	First spacecraft to reach the Moon. Flew by at a distance of 6,000 km.
Unnamed	18 Jun 1959		Lunar impact	Launcher blown up by range safety due to second stage guidance problem.
Luna 2	12 Sep 1959	390	Lunar impact	First spacecraft to impact the Moon.
Luna 3	4 Oct 1959	279	Photograph rear side of Moon	Took first photograph of Moon's far side. First scientific spacecraft with an attitude control system.
Unnamed	12 Apr 1960	300	Photograph rear side of Moon	Did not reach Moon.
Unnamed	15 Apr 1960	300	Photograph rear side of Moon	Launcher failure.
Sputnik 25	4 Jan 1963		Soft landing of 100 kg capsule	Did not leave Earth orbit.
Unnamed	2 Feb 1963		Soft landing of 100 kg capsule	Launcher failure.
Unnamed	3 Mar 1963		Soft landing of 100 kg capsule	
Luna 4	2 Apr 1963	1,422	Soft landing of 100 kg capsule	Trajectory correction failed. Spacecraft missed the Moon by about 8,500 km.
Cosmos 21	11 Nov 1963		Fly-by	Did not leave Earth orbit.
Unnamed	21 Mar 1964		Soft landing of 100 kg capsule	Launcher failure.
Unnamed	20 Apr 1964		Soft landing of 100 kg capsule	Launcher failure.
Unnamed	4 Jun 1964		Fly-by	Launcher failure.
Cosmos 60	12 Mar 1965		Soft landing of 100 kg capsule	Failed to leave Earth orbit.
Luna 5	9 May 1965	1,476	Soft landing of 100 kg capsule	Retrorocket failed during attempted landing.
Luna 6	8 Jun 1965	1,442	Soft landing of 100 kg capsule	Problem with mid-course correction motor. Spacecraft missed Moon by 160,000 km.

Luna Satellites and Missions (Part 2 of 2)

Spacecraft	Launched[1]	Mass (kg)	Mission	Comments
Luna 7	4 Oct 1965	1,506	Soft landing of 100 kg capsule	Retro-rocket fired too early. Spacecraft crashes onto Moon.
Luna 8	3 Dec 1965	1,552	Soft landing of 100 kg capsule	Retro-rocket fires too late. Spacecraft crashes onto Moon.
Luna 9	31 Jan 1966	1,583	Soft landing of 100 kg capsule	First spacecraft to soft-land on the Moon.

[1]Spacecraft up to and including those launched in 1960 were launched by a SL-3/Luna/R-7 launcher. All subsequent spacecraft were launched by an SL-6/Molniya.

These first-generation Luna spacecraft were designed to fly directly to the Moon from Earth. Nine were launched, but six failed because of problems with their Luna (R-7 based, short-shroud Vostok) launchers. The Soviet Union then radically changed the design of its Luna spacecraft, launching these much larger, second-generation craft into an Earth parking orbit before ejecting them to the Moon. This second generation, each weighing about 1,400 kg, also included a 100-kg capsule, complete with camera, to soft-land on the Moon.

A Molniya (R-7A) rocket launched the first of this second generation, *Sputnik 25*, on 4 January 1963, but the Soviet Union was unable to eject it from its parking orbit. Three months and two more failures later, *Luna 4* successfully left Earth orbit, but a problem with its trajectory-correction maneuver caused it to miss the Moon by about 8,500 km. A number of subsequent attempts failed to achieve or leave Earth orbit until *Luna 5* hit the lunar surface when its retrorocket failed in May 1965. *Lunas 6*, *7*, and *8* also failed, but on 3 February 1966, the landing capsule of *Luna 9* successfully landed on the Moon. Early the next day it became the first spacecraft to transmit images from the lunar surface, showing a relatively dust-free surface with enough rock-free areas to allow a safe landing by a manned spacecraft.

Two months after the successful landing of *Luna 9*, *Luna 10* became the first spacecraft to orbit the Moon. It detected a weak magnetic field, about 0.1 percent that of Earth's, that did not change with altitude, indicating it was probably measuring the interplanetary magnetic field, rather than that of the Moon itself. The Soviet Union then completed a successful 1966 with *Luna 13*, which landed on the Moon with an explosively driven probe. This found that the Moon's soil had the load-bearing characteristics of terrestrial soil and so could support a manned lander.

In spite of these successes, it became clear as the 1960s progressed that the United States had overtaken the Soviet Union in its general space capabilities. Nevertheless, the Soviet Union was gradually improving its launcher and spacecraft designs, with the first of the third-generation Lunas being launched by a Proton rocket as an automatic sample-return mission in January 1969. Unfortunately the launcher failed, and it was not until September 1970 that a Soviet automatic sample-return mission, *Luna 16*, was successful, returning 101 g of material from Mare Fecunditatis. By then the United States had beaten the Soviet Union to put a man on the Moon, with the *Apollo 11* astronauts returning with some 22 kg of lunar rocks the previous year.

Two months after *Luna 16*, *Luna 17* landed a bizarre-looking wheeled vehicle on the Moon. Called Lunokhod, this eight-wheeled machine traveled over the lunar surface, controlled by a team of drivers on Earth. Over the course of 10 months, it traveled a total of 10.5 km in Mare Imbrium, taking thousands of images and performing numerous soil analyses. These confirmed previous indications that the lunar surface, at least in the mare regions, was composed mostly of basalt.

During 1971–76, the Soviet Union launched eight more spacecraft to complete its Luna series with *Luna 24*; one was destroyed during launch, two were orbiters, one contained an updated Lunokhod, and four were sample-return missions. Only two of the latter, namely *Lunas 20* and *24*, returned with samples, however. The *Luna 20* samples, which were returned from a highland region, had a high concentration of anorthosite material like those returned by *Apollo 16* and *17*, which had also landed in a highland region. *Luna 24*, on the other hand, which had landed in Mare Crisium, retrieved core samples from depths of up to 2.3 m that contained basalt of about 3.3 billion years old.

David Leverington

See also: Rocket and Space Corporation Energia, Russian Launch Vehicles

Bibliography

James Harford, *Korolev: How One Man Masterminded the Soviet Drive to Beat America to the Moon* (1997).

Robert Reeves, *The Superpower Space Race: An Explosive Rivalry through the Solar System* (1994).

Lunar Orbiter. Lunar Orbiter was a spacecraft program designed to image possible landing sites for the manned Apollo spacecraft, and monitor micrometeorites and particle radiation in the Moon's vicinity to show that the environment was safe. The program was managed by NASA Langley and the spacecraft was designed by Boeing. NASA launched five Lunar Orbiters between August 1966 and August 1967; all of them operated successfully in lunar orbit.

The key to the Lunar Orbiters was high resolution, and so photographic film was used instead of television imaging. The 70 mm film was processed and scanned on board at a resolution of 300 lines/mm. The camera design was based on that developed for the U.S. Air Force Samos program.

The motion-compensation system on *Orbiter 1* failed, so it was only able to produce medium-resolution images. However, although Lunar Orbiter was not a scientific program, slight deviations in the orbit of *Orbiter 1* indicated that there were mass concentrations or mascons beneath some of the Moon's maria.

The main area of interest on the Moon for Apollo was an equatorial strip extending ±45° from the center of the visible disk, and so *Orbiters 1*, *2*, and *3* were put into equatorial orbits. Their performance was so good, however, that NASA had, by April 1967, settled on eight potential landing sites for Apollo. As a result, NASA re-designated

Orbiters 4 and *5* as scientific spacecraft, launching both into polar orbits. This resulted in almost complete image coverage of both sides of the Moon to at least medium resolution. The micrometeorite and radiation environment was found to be relatively benign for Apollo.

Early analysis showed large differences in crater counts between different parts of the Moon. This indicated that either these were areas of radically different age, or that many craters were of volcanic origin. A few fresh-looking craters were found indicating that, whatever the cause, the surface of the Moon was changing, albeit slowly.

Lunar Orbiter was so successful that the images were still being used in studies in the early twenty-first century.

David Leverington

See also: Apollo, Boeing Company, Langley Research Center, Samos

Bibliography

Bruce K. Byers, *Destination Moon: A History of the Lunar Orbiter Program* (1977).
James R. Hansen, *Spaceflight Revolution: NASA Langley Research Center from Sputnik to Apollo* (1995).
Oran W. Nicks, *Far Travelers: The Exploring Machines* (1985).

Lunar Prospector. The *Lunar Prospector* mission began in late 1988 as a private effort to demonstrate that lunar exploration missions could be conducted quickly and inexpensively outside the NASA bureaucracy. The small, privately funded volunteer team defined the program and designed a simple spin-stabilized spacecraft to provide global maps of the Moon during a one-year, polar-orbiting mapping mission.

The scientific payload consisted of a Gamma Ray Spectrometer (GRS) to map various elements at 150 km resolution, a Neutron Spectrometer (NS) to map hydrogen to search for suspected water ice deposits in the polar regions, an Alpha Particle Spectrometer to map gas release events and determine their frequency, a Magnetometer and an Electron Reflectometer (MAG/ER) to map the lunar magnetic fields at orbital altitudes and at the surface, and the Doppler Gravity Experiment (DGE) to map the lunar gravity field.

Exhausting all efforts to obtain private funding for the mission, after six years principal investigator Alan Binder asked Lockheed Missiles and Space Company to partner in a $63 million proposal to the new NASA Discovery program. In February 1995, NASA selected *Lunar Prospector* to become that program's first peer-reviewed, competitively selected mission. Construction of the spacecraft and instruments began in October 1995 and was completed in 22 months for $24 million.

Lunar Prospector was launched on 6 January 1998 and achieved its nominal 100 km altitude, polar-mapping orbit on 15 January 1998. After nearly one year of mapping from this orbit, *Lunar Prospector*'s altitude was decreased to 30 km for a seven-month extended mission of low-altitude, high-resolution mapping. After

570 days, the spacecraft was intentionally deorbited and impacted in a south polar crater as a final attempt to determine if water ice exists there.

The GRS data showed that the trace element–rich material KREEP (potassium [K], rare Earth elements [REE], and phosphorus [P]) was excavated from the crust/mantle boundary by the Imbrium Basin–forming impact and distributed around the Moon, that post-Imbrium mare volcanism and ejects from younger impacts partially buried the KREEP-rich ejecta, and that the KREEP has been re-excavated by even younger impacts.

The NS data show that hydrogen is concentrated in permanently shadowed craters near the poles. These data are consistent with water ice deposits and/or enhanced deposits of solar wind hydrogen.

The MAG/ER data show that relatively strong magnetic fields are concentrated at the antipodes (180 degrees away) of the youngest of the mare basins. This confirmed that, as the ejecta and plasma from a basin-forming impact circumnavigates the Moon, the plasma sweeps up and amplifies the ambient magnetic fields. At the antipode, where the ejecta, plasma, and trapped magnetic fields congregate and collide, the amplified field strengths magnetize the shocked, hot ejecta rocks.

The DGE data produced the first complete gravity map of the Moon and were used to identify several new mass concentrations in the limbs and on the lunar far side. The gravity data also increased the precision of moment of inertia estimates, and when combined with magnetic data, showed that the iron core radius is roughly 330 km.

Alan Binder

See also: Lockheed Martin Corporation

Bibliography

Alan B. Binder, *Against All Odds* (2005).
Lunar Research Institute. http://www.lunar-research-institute.org/.

Moon. *See* Lunar Science.

Ranger. Ranger was a series of spacecraft launched to the Moon by NASA between 1961 and 1965. Initially the program consisted of five spacecraft, as part of the Jet Propulsion Laboratory's (JPL) attempt to develop a basic spacecraft design suitable for planetary exploration. Two Rangers, the so-called Block I's, were designed to loop around the Moon and return Earthward, and three Rangers, the Block IIs, included a landing capsule. However, before any of these could be launched, in May 1961 President John F. Kennedy committed the United States to a manned lunar program. As a result, the Ranger spacecraft that had, until the president's decision, been designed mostly with science in mind, were redesignated as precursors to the manned program.

The Ranger program started badly when all five spacecraft failed between August 1961 and October 1962. The Block I Rangers failed because of problems with the Atlas-Agena launcher, and the Block IIs failed because of spacecraft problems caused by excessive sterilization temperatures and a lack of adequate onboard redundancy. A board of inquiry into the Block II failures also recommended major changes in the roles of project and system management at JPL. In the meantime, a Block III

Ranger program had been approved; the spacecraft being designed to take high-resolution images of the Moon as they crashed onto its surface.

The approval of these and other spacecraft, and the redesign of the Block IIIs caused by the problems with their predecessors, caused overload problems at JPL, which was managing the Ranger program. These problems were considerably eased when several JPL spacecraft were canceled in 1963 to save money.

The first Block III spacecraft, designated *Ranger 6*, was finally launched a year late in January 1964. Problems with the television cameras resulted in the spacecraft impacting the lunar surface with them switched off. The technical review boards set up at JPL and NASA produced conflicting results, while a congressional inquiry into the Ranger program found management problems at the key interface between NASA and JPL, and at JPL itself.

The next spacecraft, *Ranger 7*, launched just six months after *Ranger 6*, was a resounding success, impacting the Moon just 15 km from its target in the Mare Nubium. *Rangers 8* and *9* were also successful, landing in the Mare Tranquillitatis and crater Alphonsus, respectively.

Images from the Rangers showed that the lunar surface appeared to be covered with a thin layer of dust, indicating there should be no serious problem in landing a spacecraft on the Moon, at least as far as the dust was concerned. However, the bearing strength of the surface could not be determined directly from the Ranger images, and so Apollo mission designers could not be certain that the surface was strong enough to carry a manned lander.

The Ranger images showed craters of all sizes down to the minimum resolved of about 1 m. The edges of most of the smaller craters were rounded, indicating that even on the Moon there was some form of weathering, probably due to the impact of meteorites. But, some of the smaller craters were sharp and fresh looking, indicating that the surface was still changing, albeit slowly. Some craters were clearly secondary craters, not caused by meteorite impact, but caused by the impact of material ejected when other, larger craters were formed.

David Leverington

See also: Apollo, Jet Propulsion Laboratory

Bibliography

R. Cargill Hall, *Lunar Impact: A History of Project Ranger* (1977).

Gerard Peter Kuiper, "Lunar Results from Rangers 7 and 9," *Sky and Telescope* 29 (May 1965): 293–308.

SELENE, Selenological and Engineering Explorer. *See Kaguya.*

Surveyor. Surveyor was a series of spacecraft designed to soft-land on the Moon and find safe landing areas for the manned Apollo spacecraft. The program was managed by the Jet Propulsion Laboratory with Hughes Aircraft as the spacecraft contractor. In total, NASA launched seven spacecraft between 1966 and 1968, of which just two were failures, *Surveyor 2* and *Surveyor 4*.

Alan Bean standing next to the Surveyor 3 *spacecraft on the Moon. The* Apollo 12 *lunar excursion module that brought him to the Moon is in the background. (Courtesy NASA/National Space Science Data Center)*

Initially, when Surveyor, then called Surveyor Lunar Orbiter, was conceived in 1959 as a purely scientific program, it consisted of a main spacecraft to orbit the Moon and a smaller spacecraft probe that would soft-land. In the following year NASA decided that the soft-lander and lunar orbiting spacecraft should be autonomous spacecraft, based on a common bus design, and launched separately. As NASA added Apollo requirements, it became clear that this common bus concept was limiting the design of both spacecraft, and eventually the orbiting spacecraft was canceled and a separate Lunar Orbiter program initiated.

In 1961 a total of 157 kg had been allowed for experiments out of a total spacecraft launch mass of about 1,100 kg for the Surveyor soft-lander. But problems with the design of the Atlas-Centaur launch vehicle and the Surveyor spacecraft eventually meant that when NASA launched *Surveyor 1*, there was only one experiment on board, weighing just 9 kg out of a total spacecraft mass of 995 kg. This experiment was a television system designed to take panoramic images from the Moon's surface.

By the time *Surveyor 1* was launched on 30 May 1966, the program was three years late and vastly over budget. Fortunately, however, the spacecraft performed almost flawlessly. Images from the surface showed that it had landed on a dark, level area of the Oceanus Procellarum, which was littered with craters ranging from a few centimeters to a few hundred meters in diameter.

Surveyor 3 had a soil scoop added to the payload, which showed that the lunar soil in the Oceanus Procellarum where it landed was soft and clumpy like course damp sand. *Surveyor 5* and *6* had an alpha-particle scattering experiment, which found that the chemical composition of the lunar soil in their landing areas of the Mare Tranquillitatis and the Sinus Medii was similar to terrestrial basalt, produced by volcanic activity.

Spacecraft images and onboard data indicated that the surface in the relatively flat areas of the Moon so far visited by Surveyor could support the weight of the Apollo lander. So NASA decided to make *Surveyor 7* a scientific spacecraft, landing it in the lunar highlands. It had a soil scoop that weighed rocks to determine density and an alpha scattering experiment. The latter found that the iron content of the soil, which appeared to have come from deep within the Moon, was lower than measured in the maria regions.

David Leverington

See also: Apollo, Jet Propulsion Laboratory

Bibliography

William F. Mellberg, *Moon Missions: Mankind's First Voyages to Another World* (1997).
Robert Reeves, *The Superpower Space Race: An Explosive Rivalry through the Solar System* (1994).

SMART 1, Small Missions for Advanced Research in Technology 1. *SMART 1, Small Missions for Advanced Research in Technology 1*, was launched 27 September 2003 as an auxiliary payload on an Ariane 5. It was the European Space Agency's (ESA) first lunar probe and was the first of a series of small, relatively low-cost missions designed to test technology for future planetary exploration. The complete mission cost €120 million.

Managed by the European Space Research and Technology Centre with the Swedish Space Corporation as prime contractor, *SMART 1*'s primary objective was to flight test a solar electric primary propulsion system. Developed by Snecma Moteurs of France, it was a stationary plasma thruster, using xenon gas as a propellant, and powered by the spacecraft's solar arrays.

The spacecraft's orbit gradually spiraled out from Earth and was captured by the Moon's gravity on 15 November 2004. From a polar elliptical orbit, it carried out scientific observations using a suite of experimental mini-instruments weighing 19 kg, including a camera, a near-infrared spectrometer, an X-ray spectrometer, a communications experiment, and a radio-science experiment. *SMART 1* science goals included investigations into lunar origin and evolution, crustal composition, searching for water ice at the south pole, and mapping potential lunar resources. Other instruments studied solar variability, the Moon's "wake" in the solar wind, and the Moon's libration. The latter experiment was designed to determine the accuracy with which both longitudinal and latitudinal lunar libration can be detected. Science results included the first remote detection of all the main elements in lunar minerals, including the first detection of calcium on the lunar surface. The camera returned thousands of images of the Moon in unprecedented detail (up to 40 m resolution). After a one-year mission extension, *SMART 1* was maneuvered to a controlled impact onto the Moon on 3 September 2006.

Katie J. Berryhill

See also: European Space Agency

Bibliography

ESA *SMART 1*. http://esa-mm.esa.int/SPECIALS/SMART-1/index.html.

Giuseppe Racca et al., "A Solar-Powered Visit to the Moon: The *SMART 1* Mission," *ESA Bulletin* 113 (February 2003): 14–27.

Mars

Mars has always been the foremost objective of spaceflight, after the Moon, which enjoyed the advantage of proximity. Johannes Kepler's study of Mars's orbit led to his discovery that Mars orbited the Sun in an ellipse, published in his *Astronomia Nova* of 1609. For the next three centuries, Mars remained an elusive object of telescopic study, with few obvious features visible beyond periodic polar caps and a few major features that sometimes disappeared and changed colors. In 1877 U.S. astronomer Asaph Hall discovered two small moons, Phobos and Deimos. That same year, Italian astronomer Giovanni Schiaparelli observed fine linear features that he called "canali" (in Italian, "channels"). When in 1894 U.S. astronomer Percival Lowell also observed these features, he proposed that the inhabitants had built a system of irrigation channels to stave off extinction on their dying desert world. This thrust Mars into the forefront of the public imagination—and became the leading subject of science fiction writers following the publication of H. G. Wells's *War of the Worlds* in 1898.

By the 1950s Mars was regarded as quite possibly an abode of life; seasonal changes in its surface features were reported; there was a thin atmosphere—the best estimates suggested a surface pressure of 80 millibars—and water-ice at the poles. Though the canals had been largely abandoned by astronomers, they continued to feature in science fiction stories and paintings by visionary space artists like Chesley Bonestell.

The first attempts to send a spacecraft to Mars were made by the Soviet Union from Tyuratam in October 1960, both failed before reaching Earth orbit. *Mars 1*, another Soviet mission, was launched in November 1962, but contact was lost four months later, when the spacecraft had reached a record distance across which radio communications had been maintained. In all, the Soviet Union launched five unsuccessful Mars missions between then and the arrival of *Mars 2* in 1971.

In the United States, two Mariner flyby missions were launched toward Mars in November 1964. The first, *Mariner 3*, failed, while *Mariner 4* reached Mars in July 1965, passing within roughly 9,900 km of the surface. It sent back images that, despite being fuzzy, surprisingly revealed a surface thickly peppered with craters. A radio-occultation experiment as the spacecraft slipped behind the planet established that the atmosphere of Mars had a surface pressure of only about 5 mb, and the magnetometer detected no magnetic field. *Mariner 4* conveyed the impression that Mars was a crater-pocked lunar landscape; it produced a mood of depression among mission planners and planetary scientists. While only 1 percent of the planet's surface was imaged, Lowell's dreams of a living world seemed to be shattered. *Mariners 6* and *7*, in July and August 1969, increased the coverage to 10 percent of the Martian surface, but did little to change the impression of a Moonlike world. Based on measures of the temperatures of the south polar cap, it was announced that rather than being a water-ice deposit as long believed, the cap consisted of frozen carbon dioxide.

As the two flyby Mariners were arriving at Mars, Wernher von Braun, who had been designing Mars missions since his pioneering *Marsprojekt* of the late 1940s, presented a bold vision to the U.S. Space Task Force for a manned mission to Mars. To be launched in 1981, the mission would build on Saturn V rocket technology used for the Apollo lunar landings and require development of space shuttles and a nuclear-powered vehicle, assembled in Earth-orbit, to carry the astronaut crews to Mars. At that moment, the Vietnam War and economic issues were beginning to eclipse the space program as a priority, and the Mariner flybys of Mars at that point were not promising. NASA eventually committed itself to a Space Shuttle instead.

In 1971 this gloomy picture began to change. American plans called for *Mariners 8* and *9* to orbit Mars at different inclinations. The Soviet Union readied three ambitious missions, some with both orbiter and lander components. After *Mariner 8* scuttled in the Atlantic Ocean, *Mariner 9*'s flight plan was revised so that it would enter an orbit inclined 65° to the Martian orbit to allow topographic mapping and albedo studies. It was successfully launched in May 1971. Unfortunately, *Mariner 9* arrived in mid-November to find the planet swathed in dust, except for the polar cap and the summits of the Tharsis volcanoes, and was immediately commanded to shut down its television cameras until the dust cleared. The Soviet *Mars 2* and *Mars 3* orbiters, which could not be reprogrammed, sent back a series of mostly blank and uninformative images, though the *Mars 2* capsule crashed onto the surface. The *Mars 3* capsule soft-landed, but could do no more than switch on its television cameras when transmission was unexpectedly lost.

Mariner 9 was a success, whose images were called "the most exciting ever obtained in planetary exploration." They allowed the first detailed global topographic map of Mars to be created and also provided the first close-up views of the satellites, Phobos and Deimos. Mars was revealed as a majestic world of volcanoes, including the 78,000 ft high shield of Olympus Mons, an enormous canyon system (Valles Marineris) stretching along the equator one-fourth of the way around the Martian globe, and what looked like dry riverbeds fashioned during warmer, wetter periods of Martian geologic history by running water.

While another Soviet Mars mission failed in 1973, the United States planned its Viking missions, which combined orbiter and lander components. After a two-year delay largely due to the great complexity of the spacecraft, they were launched in 1975. The *Viking 1 Orbiter* entered Martian orbit in June 1976. After determining that the original landing site was too risky, a smoother site in western Chryse was chosen for the landing, which was successfully achieved on 20 July 1976. The lander began transmitting the first images obtained from the Martian surface, showing rust-colored rock-strewn plains under a salmon-pink sky. *Viking 2* arrived at Utopia, in the far north of Mars, in September 1976. A prime goal of the missions had been the search for microbial life; a labeled release, gas exchange, and pyrolitic release experiment all found evidence for exotic chemistry on Mars—notably, the presence of highly reactive superoxides in the Martian soil. However, most scientists agreed that the results were not suggestive of the presence of life-forms; the mass spectrometer had found no evidence of organic compounds.

After Viking, there was a hiatus until 1988 when the Soviet Union launched two vehicles designed to encounter Mars and its larger moon, Phobos. Radio contact with *Phobos 1* was lost before it arrived at Mars, while *Phobos 2*, though it did achieve orbit and obtained some images of Mars and Phobos, was lost prematurely.

Several Mars missions failed in the 1990s. The costly U.S. *Mars Observer* lost contact three days before orbit-insertion in August 1993. Others were *Mars 96*, a Russian spacecraft which fell to Earth shortly after launch in November 1996, and the Japanese *Nozomi*, which lost power before it reached Mars in 2003. Two U.S. missions of 1999 also ended in failure: *Mars Climate Orbiter*, which failed to achieve Mars orbit; *Mars Polar Lander* (*MPL*), which crashed in the south polar region of Mars; and the *Deep Space 2* probes, attached to the *MPL*, which were supposed to penetrate the Martian surface near the south pole and extract ice cores.

The successes, however, were inspiring. *Mars Pathfinder*, which cost only 1/10 as much as the Viking project, was cushioned by airbags when it reached Ares Vallis on 4 July 1997. It deployed the cigar box–sized *Sojourner* rover, which generated enormous public excitement. Scientists found that the rocks in the area appeared to have been laid along the flowlines of a moving river or flood, evidence that Ares Vallis had been fashioned by running water.

Mars Global Surveyor (*MGS*), which was designed to fulfill some of the mission objectives of the ill-fated *Mars Observer*, achieved Mars orbit in 1997. It carried a sophisticated camera able to image the surface to a resolution of up to 0.5 m, a thermal emission spectrometer used to map the mineralogy of Mars, and a laser altimeter for doing geological prospecting from Martian orbit, in addition to a magnetometer. *MGS* found abundant evidence of the presence of liquid water on the planet during the first billion years or so in its history, since which time Mars has been in a deep freeze, except for local environments in which volcanic thermal processes have produced local irruptions of water onto the surface.

In October 2001, *MGS* was joined by *Mars Odyssey*, which began a comprehensive mineralogical and geological survey of the planet's surface using a thermal imaging system to map the Martian surface in visible and infrared light and a gamma-ray spectrometer to search for hydrogen, which provided strong circumstantial evidence of the existence of abundant subsurface water ice in a broad "water ice" zone stretching from the poles into the Martian temperate zones. Much of the Martian regolith down to a depth of 2 m or so may consist of water-ice.

The European Space Agency's *Mars Express* entered Mars orbit on 25 December 2003, mapping the locations of minerals that formed in water on Mars, though its mostly privately funded *Beagle* lander failed. The camera design combined high-resolution, stereo, and color features to produce images that came close to what the eye would see. One of the most widely discussed findings was the detection of atmospheric methane by the spacecraft spectrometer. Though at first it was suspected that it might be produced by living organisms on the planet, degradation of the inorganic mineral olivine is now deemed a more likely explanation. Another U.S. orbiter, *Mars Reconnaissance Orbiter*, arrived at Mars in 2006 to continue the search for water. It carried the largest camera flown to date on a planetary mission.

In 2003 NASA launched two Mars Exploration Rovers. The first, *Spirit*, landed near Gusev Crater on 3 January 2004; it was joined by *Opportunity*, which arrived on 25 January 2004 in the hematite-rich Meridiani Planum region. Their main purpose was to search for evidence of water on Mars. *Opportunity* was fortunate; it explored a site close to an outcrop of bedrock that revealed stratigraphic features and hematite "blueberries" that could only have been formed in water. Mission controllers expected that the cold Martian winter or dust deposited on the solar panels would doom them, but they survived, and winds cleared the solar panels of dust. As of late 2007, three orbiters—*Mars Odyssey*, *Mars Express*, and *Mars Reconnaissance Orbiter*—in addition to the rovers, were still going strong on Mars. Probably the most important result of Mars exploration, as of 2008, has been the finding that abundant water once occupied the planet's surface.

William Sheehan

See also: Astrobiology, Jet Propulsion Laboratory, Landing Craft, Rovers, Science Fiction

Bibliography

Edward C. Ezell and Linda N. Ezell, *On Mars: Exploration of the Red Planet 1958–1978* (1984).
Robert Godwin, ed., *Mars: The NASA Mission Reports* (2000).
Paul Raeburn, *Mars: Uncovering the Secrets of the Red Planet* (1998).
Michael William Sheehan and Stephen James O'Meara, *Mars: The Lure of the Red Planet* (2001).

Beagle 2. *See Mars Express.*

Early Mariners to Mars. Early Mariners to Mars, comprised *Mariners 3*, *4*, *6*, and *7*, all designed and operated by NASA's Jet Propulsion Laboratory (JPL). The twin 260 kg *Mariner 3* and *4* spacecraft, which were based on JPL's Ranger design, were the first U.S. spacecraft designed to visit Mars. Their main payload was a camera that would take images as the spacecraft flew past Mars at an altitude of about 10,000 km. An onboard tape recorder would save the images for transmission to Earth after encounter. Scientists also planned to measure the density profile of Mars's atmosphere by measuring the effect on Mariner's radio signals as the spacecraft swung behind Mars as seen from Earth.

Mariner 3 failed in November 1964 due to a problem with the Atlas-Agena payload shroud. This required the redesign and manufacture of a completely new shroud before the launch window closed for *Mariner 4* about a month later. NASA successfully launched *Mariner 4* just 23 days after *Mariner 3* on 28 November 1964.

For many years, astronomers knew that Mars's white polar caps reduced in size as local summer approached, which indicated the presence of thin water ice. In addition, there were dark areas on Mars that also changed their appearance with season, indicating that they might be covered with some form of vegetation. Scientists thought Mars's atmosphere could have a surface pressure of about 10–80 millibar and be mostly composed of nitrogen (as on Earth) or argon, with carbon dioxide as a minor constituent. So, at the start of the spacecraft observations, hopes were high that evidence of some form of elementary life would be found.

Mariner 4, the first spacecraft to reach Mars, dramatically changed this view of the red planet. The first photographs showed a dead-looking, crater-covered surface, much like the Moon, with a surface atmospheric pressure of just 4–7 millibar. This low pressure implied, based on Earth-based spectroscopic observations, that the atmosphere was mostly composed of carbon dioxide and that the ice caps were probably frozen carbon dioxide. The spacecraft showed no evidence that liquid water had ever existed on the Martian surface, and it could certainly not exist there now with such a tenuous atmosphere. In addition, solar ultraviolet radiation would have killed any microorganisms long ago in such a thin atmosphere. So conditions appeared to be highly unfavorable for life. But *Mariner 4* had only imaged about 1 percent of the surface at a resolution of 3 km, so astronomers could not be sure.

Mariner 4 found no measurable magnetic field on Mars, indicating that it had no molten magnetic core. The apparent lack of volcanoes also indicated that the layer of rock near the surface had not been molten for a long time. Interestingly, the lack of a significant magnetic field would also allow charged particles emitted by the Sun to impact the surface, which would also be a threat to any life.

Mariners 6 and *7* had much better cameras than *Mariner 4*, and data transmission rates some 2,000 times higher. In addition, the new Mariners both had spectrometers and radiometers to study the atmosphere and polar caps. *Mariner 6* was mainly to observe Mars's equatorial regions and *Mariner 7* the south polar cap.

The 410 kg *Mariner 6*, launched by an Atlas-Centaur on 24 February 1969, was the first planetary spacecraft to be launched direct to its destination, without entering an Earth parking orbit. Just one month later, *Mariner 7* was also on its way to Mars.

Mariner 6 flew over the Martian equator in July 1969 at an altitude of 3,400 km. It measured a surface temperature that varied typically from about 15°C during the day to –75°C at night, suggesting that the surface was covered in a highly insulating material such as dust. In addition to craters, *Mariner 6* showed an almost craterless area that had short jumbled ridges and depressions, unlike anything seen previously. It detected an almost 100 percent carbon dioxide atmosphere.

Mariner 7 generally confirmed the results of *Mariner 6*, although it also found that the Hellas plain was apparently featureless. It detected minute amounts of water vapor in the atmosphere and dust and carbon dioxide crystals in the clouds. The surface of the south polar ice cap appeared to be covered in frozen carbon dioxide in the form of ice or snow.

David Leverington

See also: Jet Propulsion Laboratory, Mars, Ranger

Bibliography

Stewart A. Collins, *The Mariner 6 and 7 Pictures of Mars* (1971).

Edward C. Ezell and Linda N. Ezell, *On Mars: Exploration of the Red Planet 1958–1978* (1984).
Robert Godwin, ed., *Mars: The NASA Mission Reports* (2000).

***Mariner 8* and *9*.** *Mariners 8* and *9* were the first U.S. spacecraft designed to orbit Mars. The Jet Propulsion Laboratory (JPL), which designed and operated the spacecraft, based its design on that of *Mariners 6* and *7*, but added a large retrorocket for orbit insertion. Unfortunately, a fault with the Atlas-Centaur upper stage caused the loss of the *Mariner 8* spacecraft during launch, but the launch of *Mariner 9* three weeks later was a success.

Originally, *Mariner 8* was to map the Martian surface from a polar orbit, while *Mariner 9* observed temporal changes in Mars's low latitude regions from an orbit inclined about 50°. However, with the loss of *Mariner 8*, NASA combined these two missions into a modified *Mariner 9* mission. As a result, its orbital inclination was increased to 65°.

A planetwide dust storm was raging when *Mariner 9* arrived at Mars on 14 November 1971. Nevertheless, it was put into orbit around Mars while waiting for the dust storm to blow itself out. After two days, this orbit was trimmed, as planned, to a 1,400 × 17,100 km orbit inclined at about 65°. This allowed repeat imaging of the same piece of ground every 17 days, to enable the study of transient features, such as seasonal color changes and the seasonal advance and retreat of the polar caps.

Because of the dust storm, the initial surface images received from *Mariner 9* showed limited surface detail, and so the spacecraft was used to take the first detailed images of the two small Martian satellites, Phobos and Deimos. Both were seen to be irregularly shaped and covered in craters.

When routine imaging of Mars started on 2 January it revealed a radically different Mars from that previously observed. Instead of the disappointing, crater-strewn planet previously imaged, Mars was seen to have massive volcanoes and a giant canyon system. The largest volcano, Olympus Mons, was about 25,000 m high, while the canyon system, now called Valles Marineris, was about 4,000 km long, and up to 6,000 m deep. In addition, the Hellas plain was seen to be the floor of a 2,300 km impact basin some 5,000 m deep.

The discovery of large volcanoes indicated that there has been no general plate tectonics on Mars. With no plate movement, hot spots in the underlying mantle would have continued to eject lava through the same place in the crust year after year to build up these enormous volcanoes. In addition, geologists thought that Valles Marineris could possibly be where two plates started to move apart, but stopped as Mars cooled quickly, shortly after the crust was formed.

Other images showed sinuous channels that looked very much like riverbeds. However, because of the low atmospheric pressure, it was clear that water could not exist in liquid form on the surface of Mars today. On the other hand, it could well have existed in the past when the atmosphere was probably denser.

From Earth, astronomers had seen parts of Mars change color with the season, indicating that there may be plant life on its surface. *Mariner 9* showed that the color change was due to nothing more than seasonal winds, with velocities of up to 180 km/h, alternately covering and uncovering the darker substrate with lighter dust.

Mariner 9 found that the water vapor content over the south polar region varied from about 15 precipitable μm during its summer to zero in winter, whereas over the north polar region the content varied from about 25 precipitable μm in its summer to zero in winter. So there appeared to be water ice at both poles that partially evaporated during the local summers. Alternatively, maybe the water vapor content of the atmosphere stayed approximately constant, and the concentrations simply moved from one hemisphere to the other with the seasons.

David Leverington

See also: Jet Propulsion Laboratory

Bibliography

Edward C. Ezell and Linda N. Ezell, *On Mars: Exploration of the Red Planet 1958–1978* (1984).
Robert Godwin, ed., *Mars: The NASA Mission Reports* (2000).
William K. Hartmann and Odell Raper, *The New Mars: The Discoveries of Mariner 9* (1974).
William Sheehan and Stephen James O'Meara, *Mars: The Lure of the Red Planet* (2001).

Mars 1–3. *See* Soviet Mars Program.

Mars 5. *See* Soviet Mars Program.

Mars Exploration Rovers. The Mars Exploration Rovers (MER) were two identical spacecraft launched by NASA using Delta II 7925 and 7925H rockets on 10 June 2003 and 7 July 2003. The first to arrive, named *Spirit*, landed near Gusev Crater on 3 January 2004. Its twin, *Opportunity*, landed 25 January 2004 in the Meridiani Planum region. Both were built and operated by the Jet Propulsion Laboratory (JPL).

The MER mission grew out of two independent proposals for landed Mars missions. The first, a science proposal to fly a geological rover payload called Athena, came from a team led by a Cornell University astronomer. The payload consisted of instruments to determine the mineralogical composition of Martian rocks and soil in addition to high-resolution imagers. It had originally been built for the 2001 Mars lander mission that had been canceled after the 1998 Polar Lander failure.

After the Polar Lander loss, JPL *Mars Pathfinder* veterans proposed flying a mission like *Mars Pathfinder* but replacing the lander and small *Sojourner* rover with a larger, autonomous geological rover. This rover would carry its communications gear with it, signaling Earth directly or via relay through the *Mars Global Surveyor* (*MGS*) or 2001 *Mars Odyssey* orbiters. NASA accepted the JPL rover proposal for the 2003 launch window, but instructed JPL to include the Athena payload in the rover design.

NASA did not formally approve MER until July 2000. With a June 2003 launch window dictated by planetary motion, this gave MER the shortest development period of any Mars mission to date. NASA asked for two identical vehicles to reduce the risk of failure. JPL's engineers also rapidly determined that the design had been highly unrealistic. By mid-2001 they had to undertake an almost-complete redesign to keep the rover mass within the maximum capability of their Delta 2 launch vehicles.

As built, the rovers were 185 kg, more than ten times the mass of *Sojourner*. The project ultimately cost $800 million.

Spirit's chosen landing spot was inside Gusev Crater, which appeared from *MGS* imagery to have been a place where lake water may have ponded long ago, while *Opportunity*'s was inside the large region of hematite spotted by *MGS*. Despite its outward resemblance to a lakebed, evidence of water proved difficult to find at *Spirit*'s landing site, and *Spirit* did not find confirmation until 11 weeks after completing its three-month primary mission. *Opportunity*, however, found such evidence in an outcrop of bedrock within a few weeks. Both vehicles found minerals that on Earth only form in water, leading the mission scientists to conclude that on Mars the minerals had formed in the presence of groundwater more than three billion years ago. Both vehicles far outlived their 90-day primary missions, after which time JPL uploaded new software that allowed them greater autonomy and speed.

While the mission team had expected the rovers to die eventually from lack of power as dust coated their solar arrays, Martian winds kept cleaning them off. Both unexpectedly survived their first Martian winter. In early May 2005 *Opportunity* got stuck in a sand dune, but using a duplicate on Earth, JPL's engineers figured out how to get *Opportunity* to free itself. Each vehicle was still operating in late 2008, after surviving a series of planetary-scale duststorms that had left them nearly powerless. They had traveled a combined total of about 19 km since landing.

Erik M. Conway

See also: Jet Propulsion Laboratory

Bibliography

Athena. http://athena.cornell.edu/.
Joy Crisp et al., "Mars Exploration Rover Mission," *Journal of Geophysical Research* 108 (2003).
Mars Rovers. http://marsrover.nasa.gov/home/.
Steve Squyres, *Roving Mars: Spirit, Opportunity, and the Exploration of the Red Planet* (2005).

Mars Express. *Mars Express*, the first European Space Agency probe sent to Mars, was launched by Starsem aboard a Russian Soyuz-Fregat launcher on 2 June 2003 from Baikonur, Kazakhstan. Built by a consortium led by Astrium Toulouse, *Mars Express* included an orbiter and a lander, called *Beagle 2* (named after the ship in which Charles Darwin sailed). The lander was released from *Mars Express* six days before it entered a highly inclined, elliptical orbit around the Red Planet on 25 December 2003.

Beagle 2 descended to the Martian surface on 25 December 2003, but contact was never established after the planned landing. By February *Beagle 2* was presumed lost, perhaps crashed or situated within unfavorable terrain. One possible reason was that the low atmospheric pressure that day above the landing area provided insufficient drag for the parachutes to slow the lander down.

The orbiter, however, performed well and contributed a wealth of data. Its instruments complemented the work of *Mars Odyssey* (which detected hydrogen ions in the topsoil at the poles) and *Mars Global Surveyor* in inventorying the amount of water on the planet. The Analyzer of Space Plasma and Energetic Atoms 3 measured

A high-resolution stereo camera image of Echus Chasma from ESA's Mars Express. *Echus Chasma is the source region of Kasei Valles which extends 3,000 kilometers to the north, and thought to be one of the largest water source regions on the planet. (Courtesy European Space Agency)*

the penetration depth into the Martian ionosphere of the solar wind and accelerated ionospheric ions. *Mars Express* included visible and near infrared hyperspectral experiments (simultaneous imaging in hundreds of spectral bands) by the Observatoire pour la Minéralogie, l'Eau, les Glaces, et l'Activité (OMEGA) designed to provide high-resolution images of the surface. The OMEGA spectrometer and two other instruments on board were originally developed and built for the failed Russian *Mars 96* mission. It found diverse mineralogy on the surface in addition to the composition of the water and carbon dioxide ices at the polar caps. The presence of kieserite, gypsum, and polyhydrated sulfates indicated the presence of past water on Mars. Additional support for water included the detection of calcium-rich sulfates, showing that water played a role in forming minerals found near the poles.

Other instruments aboard *Mars Express* included the High Resolution Stereo Camera, used to map the terrain and detect surface ice. Data gathered from the Mars Advanced Radar Subsurface and Ionosphere Sounding radar in June and July 2005 showed the existence of a dozen buried basins in the northern hemisphere, indicating that the northern latitudes are much older than previously thought. The results indicated that the age dichotomy between the northern and southern hemisphere crusts dates back to the Early Noachian epoch from the planet's formation to about four billion years ago. *Mars Express* also acted as a communications relay, successfully transmitting images from NASA's *Opportunity* rover to Jet Propulsion Laboratory in August 2005.

Eric T. Reynolds

See also: European Aeronautic Defence and Space Company (EADS), European Space Agency

Bibliography

ESA *Mars Express*. http://www.esa.int/SPECIALS/Mars_Express/index.html.

Agustin Chicarro et al., "Unravelling the Scientific Mysteries of the Red Planet," *ESA Bulletin* 115 (August 2003).

Aline Gendrin et al., "Sulfates in Martian Layered Terrains: The OMEGA/Mars Express View," *Science* 307 (11 March 2005): 1587–91.

R. Lundin et al., "Solar Wind-Induced Atmospheric Erosion at Mars: First Results from ASPERA-3 on Mars Express," *Science* 305 (24 September 2004): 1933–36.

Mars Global Surveyor. *Mars Global Surveyor* (*MGS*) was the first of NASA's new Mars Surveyor program that envisioned sending an orbiter and a lander to Mars every 26 months, when Earth and Mars were in suitable alignment. NASA launched *MGS* on 7 November 1996 on board a Delta 7925 launcher. Surveyor's sister spacecraft was *Mars Pathfinder*. The *Global Surveyor* was a reflight of five of the seven experiments on the ill-fated *Mars Observer*, which disappeared a day before it was supposed to enter Martian orbit in 1993. The Jet Propulsion Laboratory managed the project, and Lockheed Martin Astronautics built the spacecraft.

Unlike earlier Mars missions, *MGS* used aerobraking along with retrorockets to achieve a Sun-synchronous mapping orbit. This substantially reduced the amount of fuel the spacecraft had to carry, allowing use of a smaller, less expensive launch vehicle. After launch, the spacecraft operations team detected a defect in one solar array that limited the amount of pressure that could be put on it, delaying the achievement of the desired orbit until February 1999, almost two years later than planned. However, the spacecraft operated well beyond its design life of two years. A further innovation to decrease propulsion use was the angular momentum management plan, which used reaction wheels to counterbalance the gravitational effects from the uneven terrain of the planet.

Mars Global Surveyor's primary purpose was to define Mars's surface geology globally through remote sensing. The Thermal Emission Spectrometer produced global maps of the mineralogy of Mars, finding, for example, a large region containing hematite, a mineral known only to form in water. The Mars Orbiter Camera, with both wide- and narrow-angle modes, provided imagery of up to 0.5 m resolution, eventually photographing the *Pathfinder* and *Mars Exploration Rover* landers after their arrival. The Mars Orbiter Laser Altimeter provided high precision topographic information. A magnetometer revealed the existence of magnetic striping similar to Earth's in the surface material.

MGS carried a communications relay system designed to receive, store, and retransmit data from landers to Earth. This reduced the cost and complexity of the landers, which could rely on lower-power, and lower-weight, short-range communications gear.

Global Surveyor was the first mission in a systematic effort by NASA to fully characterize the Martian environment and to understand Mars's evolution. It revealed that the Martian surface was evolving slowly, that it had changed more rapidly in the remote past, that it might once have had crustal motion similar to Earth's, that many sites have geologically modern gullies, and that ancient Mars had persistent flows of liquid. Finally its data was used to select geologically interesting landing sites for the Mars Exploration Rovers launched in 2003. Contact with *Global Surveyor* was lost in November 2006, just before the 10th anniversary of its launch.

Erik M. Conway

See also: Jet Propulsion Laboratory, Lockheed Martin Corporation

Bibliography

Arden Albee et al., "Overview of the *Mars Global Surveyor* Mission," *Journal of Geophysical Research* (2001): 23, 291–316.

Mars Global Surveyor. http://mars8.jpl.nasa.gov/mgs/.

Howard McCurdy, *Faster, Better, Cheaper: Low-Cost Innovation in the U.S. Space Program* (2001).

Mars Odyssey. *Mars Odyssey* was placed into orbit around Mars on 24 October 2001 after a 7 April 2001 launch. Its name derived from Stanley Kubrick and Arthur C. Clarke's film *2001: A Space Odyssey*. The Jet Propulsion Laboratory (JPL) managed the project, and Lockheed Martin Astronautics built the spacecraft. Launched on board a Delta II rocket, *Odyssey* was about one-third the dry mass of *Mars Global Surveyor*.

After the 1997 success of *Mars Pathfinder*, NASA and JPL had sent an orbiter and lander to Mars in the 1998 launch window. *Mars Climate Orbiter* was lost to a navigation error, while *Mars Polar Lander*, which had used a Viking-like retrorocket-based landing system, failed during its landing sequence. For these losses, the two organizations received intense public ridicule. The lander for the 2001 mission was initially based on the *Polar Lander*, but NASA canceled it as part of a Mars program overhaul incorporating lessons learned from these missions. NASA redirected the remaining lander funds to the *2001 Mars Odyssey* to ensure the mission's success.

Odyssey carried three scientific instruments, primarily aimed at surface geology. The Thermal Emission Imaging System, an enhanced version of the Thermal Emission Spectrometer instrument on *Mars Global Surveyor*, was designed to map the distribution of water-altered minerals. The Gamma Ray Spectrometer, reflown from the *Mars Observer* mission, was designed to measure water and certain elements in the top meter of Martian surface. The third instrument measured charged-particle intensity in Mars orbit, failing during an unexpectedly powerful solar storm in October 2003. *Odyssey* was also equipped with a communications relay system for future landers.

Odyssey's mapping mission began in February 2002 after aerobraking to lower its orbit. *Odyssey* found particle radiation levels two to three times those in low Earth orbit, with bursts many times that, meaning that human voyagers to Mars may need shielding or storm shelters against solar flares. But *Odyssey*'s principal scientific finding was that based on measurements of neutrons as a proxy for hydrogen, water ice may exist throughout the top meter of Martian surface from about 50 degrees latitude toward each pole.

Erik M. Conway

See also: Jet Propulsion Laboratory, Lockheed Martin Corporation

Bibliography

2001 Mars Odyssey. http://marsprogram.jpl.nasa.gov/odyssey/.

Howard McCurdy, *Faster, Better, Cheaper: Low-Cost Innovation in the U.S. Space Program* (2001).

Stephen Saunders et al., "2001 Mars Odyssey Mission Summary," *Space Science Reviews* 110 (2004): 1–36.

Mars Pathfinder. *Mars Pathfinder* landed on Mars 4 July 1997, the first successful mission to the Red Planet since the 1976 Vikings. *Pathfinder*, along with Near Earth Asteroid Rendezvous, was one of the first two missions assigned to NASA's Discovery program (inaugurated in 1992), which was intended to carry out scientific missions "faster, better, cheaper" than prior missions. At a comparatively low cost of $265 million including its rover, its success legitimized the low-cost approach.

Designed, built, and operated by the Jet Propulsion Laboratory (JPL), *Mars Pathfinder* originated in the late 1980s with the Ames Research Center's Mars Environmental Survey (MESUR) project, which envisaged 16 identical landers equipped with meteorological instruments. The large number of landers was very expensive, and Ames engineers conceived a less expensive Entry, Descent, and Landing system using airbags to reduce costs. These allowed a higher velocity impact with the surface, and allowed a wider range of potential landing sites. MESUR still exceeded NASA's available budget and was replaced in 1991 by a single technology demonstration mission assigned to JPL, *Mars Pathfinder.*

The new mission's primary goal was to demonstrate the ability to land on Mars much more cheaply than the Viking landers had, though *Pathfinder*'s scientific goals were far less demanding. Pathfinder included a lander that used an airbag-based terminal descent system, and a separately funded 62 cm long, 10.5 kg microrover named *Sojourner. Sojourner* communicated with Earth through the communications system on the lander, restricting its range to a few meters. The lander contained a stereo imager for both scientific purposes and rover navigation, while the rover had a "camera mode" and carried an alpha proton X-ray spectrometer donated by Germany's Max Planck Institute for Chemistry. This could measure chemical composition when held against rocks or soil. The lander also carried the meteorology package designed for MESUR.

Seven months after its launch on a Delta 2 rocket from Cape Canaveral on 4 December 1996, *Pathfinder*'s arrival at the Ares Vallis region of Mars produced a media sensation. The project's extremely popular (by mission's end, it counted half a billion hits) and frequently updated website allowed the public to follow the mission closely. *Sojourner* left the lander 6 July. On 8 July the *Sojourner* science team reported its first chemical analysis of a Mars rock named Barnacle Bill, and the rover continued to analyze the soil and rocks near the lander until 24 September 1997, when the lander sent its last image. The lander's demise ended communications with the microrover, which was still operating.

The principal scientific result from the mission was additional indirect evidence that Mars had hosted liquid water sometime during its first billion years. The rover's results also showed that some of the measured rocks were similar in composition to andesites on Earth, unexpectedly rich in silica and suggesting multiple episodes of melting. The region had a complex geologic history. More importantly for NASA and JPL, however, *Pathfinder* showed that by using new technologies, low-cost planetary missions were possible, as long as NASA was willing to accept the higher risks, and that the public loved robotic Mars exploration. Both lessons boosted NASA's Mars exploration program, though a few years later a series of failures of low-cost missions caused NASA to reevaluate the risks it was willing to take.

Erik M. Conway

See also: Jet Propulsion Laboratory, Systems Management, World Wide Web

Bibliography

Mars Pathfinder Legacy. http://nssdc.gsfc.nasa.gov/planetary/mesur.html.
Howard McCurdy, *Faster, Better, Cheaper: Low-Cost Innovation in the U.S. Space Program* (2001).
Andrew Mishkin, *Sojourner: An Insider's View of the Mars Pathfinder Mission* (2003).
Donna Shirley with Danelle Morton, *Managing Martians* (1998).

Mars Reconnaissance Orbiter. *Mars Reconnaissance Orbiter* (*MRO*) was a multipurpose spacecraft designed to study the atmosphere, surface, and subsurface of Mars. Managed by Jet Propulsion Laboratory and built by Lockheed Martin, the spacecraft's primary goal was to understand more about the history of water on Mars. Additionally it was used as a communications relay for lander missions and as a scout for landing sites.

NASA launched the 2,180 kg spacecraft from Cape Canaveral on an Atlas V-401 booster on 12 August 2005. After a seven-month cruise phase, it entered Martian orbit on 10 March 2006. A six-month aerobraking period lowered *MRO* into a 255 × 320 km Sun-synchronous orbit. The nominal scientific mission was scheduled for November 2006 through November 2008, approximately one Martian year, after which the spacecraft was planned for use as a communications orbiter until December 2010.

MRO carried the highest resolution camera flown to another planet to date, the High Resolution Imaging Science Experiment, showing details 1 m across over about 1 percent of the surface during the primary mission. Additionally the spacecraft carried the Compact Infrared Imaging Spectrometer for Mars (CRISM), a wide-swath camera, a shallow subsurface sounding radar, and a 6 m resolution black-and-white camera that had covered around 30 percent of the surface by mid-2008. The Mars Climate Sounder was a compact version of an infrared radiometer flown on the failed *Mars Observer* and *Mars Climate Orbiter* missions. Its objective was to measure variations of water vapor, dust, and temperature in the atmosphere. Another instrument reflown from *Mars Climate Orbiter* was the Mars Color Imager, used to measure variations in ozone and monitor clouds and dust storms. The Gravity Investigation mapped the planet's gravity field.

MRO's discoveries included imaging active avalanches, widespread evidence from CRISM of phyllosilicates, clay-like minerals that form only in the presence of water, surprising findings that the water persisted for thousands of years, and evidence that the Borealis basin, covering 40 percent of the Martian surface, was created by an enormous impact early in Mars' history.

Katie J. Berryhill

See also: Jet Propulsion Laboratory

Bibliography

NASA *MRO*. http://mars.jpl.nasa.gov/mro/.

R. W. Zurek and S. E. Smrekar, "An Overview of the *Mars Reconnaissance Orbiter* (*MRO*) Science Mission," *Journal of Geophysical Research* 112 (2007).

***Opportunity*.** *See* Mars Exploration Rovers.

***Phobos 2*.** *See* Soviet Mars Program.

***Phoenix* Mars Lander.** *Phoenix* Mars Lander, managed by the Jet Propulsion Laboratory (JPL), was the first Scout mission in NASA's Mars Exploration Program. Designed and constructed by Lockheed Martin Astronautics, the lander originally was scheduled to fly as part of the 2001 *Mars Surveyor* mission but was dropped and mothballed after NASA lost contact with the *Mars Polar Lander* in December 1999. Resurrected as *Phoenix*, a Delta 2 rocket launched the lander from Cape Canaveral Air Force Station on 4 August 2007 and completed the first successful landing in a Martian polar region when it touched down in the Green Valley of Vastitas Borealis at 68° north latitude on 25 May 2008 for a planned 90 day mission during the Martian summer.

Phoenix collected data to help answer at least three important questions: the Martian arctic's ability to support life, the history of water at the landing site, and the effect of polar dynamics on the planet's climate. The science team employed a variety of instruments on board the spacecraft, built by U.S., Canadian, and European universities, corporations, and government organizations. A Surface Stereo Imager took high-resolution photographs of the Martian arctic landscape. A robotic arm dug through the surface soil to expose the water-ice layer and deliver samples, via a wheel with 69 small buckets, to the Microscopy, Electrochemistry, and Conductivity Analyzer (MECA). The MECA incorporated a variety of different instruments: an optical microscope; an atomic force microscope containing four separate cells, each with 26 chemical sensors and a temperature sensor; and a Thermal and Electrical Conductivity Probe to measure regolith and atmospheric temperature, humidity, conductivity, dielectric permittivity, and wind speed.

The lander also carried a Thermal and Evolved Gas Analyzer (TEGA) and a Meteorological Station (MET). The TEGA combined eight tiny, high-temperature ovens and a mass spectrometer to measure water ice, how much water vapor and carbon dioxide was given off during heating, the presence of minerals that might have been formed in a wetter and warmer past climate, and organic volatiles such as methane. The MET had wind, pressure, and temperature sensors, plus a LIDAR (light detection and ranging) device for sampling airborne dust particles. The MET and LIDAR instrumentation marked the first significant Canadian involvement in a Mars mission.

Phoenix's nominal mission ended on 2 November 2008 when JPL received its last signal. By that time, *Phoenix* had sent images of dust and sand particles, with the greatest resolution ever returned from another planet. After digging a shallow trench, the lander instruments confirmed the presence of water ice near the surface. Completing the first wet-chemical analysis ever done on any planet other than Earth, it identified high alkalinity but a reasonable number of nutrients needed for life. Evidence accumulated for past interaction between minerals and liquid water, and the lander detected snow falling from Martian clouds but vaporizing before it reached the ground.

Rick W. Sturdevant

See also: Jet Propulsion Laboratory, Lockheed Martin Corporation

Bibliography

J. Kelly Beatty, "Polar Prospector," Sky and Telescope (September 2008): 22–24.
Brent J. Bos, "Red Planet Resolve," Sky and Telescope (September 2008): 96.
NASA *Phoenix*. http://www.nasa.gov/mission_pages/phoenix/main/index.html.
The Planetary Society. http://www.planetary.org/home/.

Sojourner. *See Mars Pathfinder.*

Soviet Mars Program. The Soviet Mars program began with the launch of two spacecraft to Mars in October 1960, designed by OKB-1 (Experimental Design Bureau) under Sergei Korolev, four years before the United States. Both Soviet spacecraft were lost when the third stage of their Molniya launcher failed. The Soviet Union then launched three spacecraft to Mars during the next launch window in 1962, but two of these were lost when their launch vehicles exploded. Contact with the third, called *Mars 1*, was lost when it was 106 million km from Earth, still some three months away from Mars. Before its demise, however, *Mars 1* encountered the Taurid meteor stream and made a series of successful measurements of the interplanetary environment. In 1965 Korolev handed the design of Mars spacecraft to the Lavochkin Design Bureau. But more Soviet failures to reach Mars followed, until in 1971 *Mars 2* and *3* succeeded, just as a major, planetwide dust storm was covering the planet.

Mars 2 and *3* both consisted of an orbiter and lander. Unfortunately, their automated design made it impossible to wait for the dust storm to blow itself out before attempting the landings. The *Mars 2* lander crashed on landing, while the *Mars 3* lander successfully landed, but failed a few seconds later. The orbiters, on the other hand, were relatively successful, measuring a very weak magnetic field and detecting small amounts of water vapor in the atmosphere. They also measured the surface temperature, showing that it had cooled by about 25°C during the dust storm, while the atmospheric temperature had increased.

Unlike their successful Venus program, the Soviet Mars program was to suffer even more failures, with only *Mars 5* in 1973 and *Phobos 2* in 1989 being limited successes. However, these were completely overshadowed by the successful American program.

David Leverington

See also: Mars, Rovers

Bibliography

V. G. Perminov, *The Difficult Road to Mars: A Brief History of Mars Exploration in the Soviet Union* (1999).

Robert Reeves, *The Superpower Space Race: An Explosive Rivalry through the Solar System* (1994).

Spirit. *See* Mars Exploration Rovers.

Viking. The Viking program, managed by NASA's Langley Research Center, included two spacecraft, each consisting of an orbiter and a lander, intended to search for life on Mars. Viking, which eventually cost about $4 billion in 2005 dollars, was among the most expensive and difficult planetary missions ever attempted. The program originated with Jet Propulsion Laboratory (JPL) studies in 1960–67 for a large spacecraft called Voyager that could be sent to Mars or Venus. This was ultimately canceled due to cost problems. A scaled-back mission to Mars, first called Titan Mars 1973, was proposed in November 1967 under Langley's management, and became the basis for Viking.

Each orbiter, built by JPL, included a camera system, a water detector, and an infrared instrument to map the surface temperature. Initially NASA would use the cameras and water detectors to try to locate a smooth, moist surface on which to land the landing module.

Each lander, built by Martin Marietta, was contained in a double-skinned, lens-shaped cover attached to the underside of the orbiter. The cover's outer skin was a bioshield, designed to protect the lander from biological contamination while it was on the launcher, as NASA wanted to avoid contaminating Mars. The inner skin was an aeroshell to protect the lander from the high temperatures caused by aerodynamic braking in the Martian atmosphere. Each lander included a camera system, soil scoop, gas chromatograph mass spectrometer, X-ray fluorescence spectrometer, seismometer, and detectors to measure atmospheric temperature, pressure, and wind velocity. The most sophisticated instrument package on the lander was, however, a miniature biological laboratory to test the soil for signs of life.

Using *Mariner 9* images, Viking managers chose provisional landing sites, where they thought there might be water. The provisional site for *Viking 1* was in the Chryse region, where *Mariner 9* discovered sinuous channels. *Viking 2*'s provisional site was in the Cydonia region, near the southern fringe of the north polar hood, where scientists hoped to find evidence of liquid water.

The Viking *spacecraft with the lens-shaped bioshield at the bottom that contained the lander. (Courtesy NASA/Jet Propulsion Laboratory)*

The Viking spacecraft had initially been scheduled for launch in 1973, but national budgetary problems in December 1969 forced a postponement, resulting in a two-year delay. *Viking 1* and *2* were both successfully launched to Mars in 1975 by Titan 3E-Centaur launch vehicles. *Viking 1* was put into orbit around Mars on 19 June 1976. Unfortunately, the images of the planned Chryse landing site showed that it was on the floor of what looked like a river channel that, although scientifically interesting, would be a risky place to land. Instead a new landing site was chosen in western Chryse.

On 20 July 1976 the *Viking 1* lander touched down just 28 km from its target. Twenty-five seconds after landing, the lander began transmitting the first ever image from the surface of Mars. During the next few months, numerous images of the red, rock-strewn, sandy surface and pink sky were received and analyzed on Earth. It was early summer when the *Viking 1* lander touched down, and on the first day it detected that the atmospheric temperature varied from –86°C at dawn to –33°C in the afternoon. The atmospheric pressure was 7.6 mb, and the wind velocity gusted up to 52 km/h.

The soil was found to be iron-rich clay, and the atmosphere to consist of 95.3 percent carbon dioxide, 2.7 percent nitrogen, 1.6 percent argon, and 0.13 percent oxygen. Interestingly, the mass spectrometer found no evidence of organic matter on the surface. Both the soil and the very fine dust were found to be magnetic.

Viking 2 arrived at Mars on 7 August, but imaging of its proposed landing site showed it was too rough for a safe landing. Eventually, a new site was found in Utopia Planitia about 4° farther north than originally planned. Scientists chose this more northerly position hoping its soil would have more moisture as it was nearer the north polar cap. The *Viking 2* lander touched down only 10 km from this chosen site, which its camera showed had a similar red, rock-strewn surface to that of *Viking 1*.

The *Viking 2* lander measured a minimum temperature of –87°C during the northern summer, but during winter the minimum fell to –118°C. Frost was first observed on the surface in September 1977 when the temperature was –97°C.

Throughout their lifetimes, the *Viking 1* and *2* orbiters together imaged virtually all the Martian surface to at least 300 m resolution, with 2 percent being imaged at about 25 m resolution. Later in the mission, NASA reduced the periapsis of both Viking orbiters, allowing imaging of selected areas at a resolution of 8 m.

The Viking images showed clear signs of flash floods in the Chryse region and many landslides, some of which seem to have occurred in saturated soil. Images also showed dendritic or branching drainage features resembling terrestrial river systems. Temperature and water vapor sensors on the orbiters showed that the north polar cap is made of water ice in the northern summer at a temperature of about –65°C. It is then covered by carbon dioxide ice in the northern winter. In the case of the southern cap, however, the carbon dioxide ice does not completely melt in the southern summer.

Mariner 6 had shown that the south polar cap temperature was about –125°C, which is the equilibrium temperature of carbon dioxide at the south polar atmospheric pressure of about 6 millibars. The infrared thermal mapper on the *Viking* orbiter indicated a south polar temperature as low as –139°C, however, which was difficult to explain unless there was a higher concentration of nitrogen and argon at the south pole than elsewhere on Mars, which could provide the extra cooling.

The water vapor in the Martian atmosphere was found to be highest in low-lying regions, and more water vapor was found during the summer than during the winter, when it was presumably frozen out of the atmosphere. In regions of rough terrain, there were marked daily fluctuations in the amount of water vapor, possibly due to changing wind patters. The greatest concentration of atmospheric water vapor occurs near the edge of the north polar cap in midsummer.

David Leverington

See also: Astrobiology, Langley Research Center, Martin Marietta Corporation

Bibliography

Edward C. Ezell and Linda N. Ezell, *On Mars: Exploration of the Red Planet 1958–1978* (1984).
Robert Godwin, ed., *Mars: The NASA Mission Reports* (2000).
Paul Raeburn, *Mars: Uncovering the Secrets of the Red Planet* (1998).

Viking Biology. Viking Biology, a set of life-detection experiments aboard the two 1976 NASA Viking landers, designed to determine whether microbial life exists on Mars. The biology package (one in each lander) weighed about 15 kg, fit in a volume of 1 cu ft, and cost about $100 million in then-year dollars, including a separate molecular analysis experiment that was relevant to the question of life because of its ability to detect organic molecules.

The Viking project began in 1968. Ames Research Center led the Viking biology science team, and TRW built the biology instrument package. As the first Mars lander, it represented a landmark in the search for life on Mars, an important objective of the U.S. space program since the National Academy of Sciences named the search for extraterrestrial life the "prime goal" of space biology in 1962.

The diverse ideas about the nature of Martian life led to three different biology experiments aboard Viking, each representing a different approach to the problem of life. The experiments tested for life using different philosophies, environmental conditions, and detectors. The "labeled release" experiment, developed by Gilbert Levin, assumed that any Martian microorganisms, like those on Earth, would eat simple organic compounds, decompose them, and produce gases such as carbon dioxide, methane, or hydrogen as end products. Thus a dilute solution of seven organic compounds, radioactively labeled for detection purposes, was added to an incubation chamber containing the Mars soil sample. The experiment tested for the expected labeled release of the gas produced as organisms ate the organics and breathed out the decomposition products.

Vance Oyama of NASA Ames developed the second biology test, the "gas exchange" experiment. In one mode the experiment added slight water moisture and, assuming that any Martian organism in the dry Martian environment would be stimulated to metabolic activity, looked for a gas that would be given off and detected by chromatography, a technique for separating complex mixtures. In its second "wet nutrient" or "chicken soup" mode, a rich nutrient of 19 organic compounds was added as an additional stimulus to metabolic activity and the products detected in the same manner.

The third experiment, headed by Norman Horowitz of California Institute of Technology, was termed the "pyrolytic release" experiment. Based on the philosophy that it was best to test for Martian organisms under conditions known to exist on Mars, only radioactively tagged carbon dioxide and carbon monoxide, gases known to exist in the Martian atmosphere, were added to the Martian soil. Any Martian organisms would assimilate these gases and convert them to organic matter. After 120 hours of incubation, the soil chamber was to be heated to 625°C to pyrolyze the organic matter (break its compounds into their constituent parts) and release the volatile organic products.

All three experiments sought to detect metabolic activities. Oyama's wet nutrient mode was the most Earth-like approach. Horowitz's was the most Mars-like, and Levin's fell in between. Two of the three experiments, Horowitz's and Levin's, gave presumptive positive results. However, a gas-chromatograph mass spectrometer (GCMS) experiment, run by Klaus Biemann, detected no organic molecules down to parts per billion. Without organic molecules, the building blocks of life, it was impossible to have life.

After running numerous experiments, all of the principal investigators except Levin agreed that the Martian soil was chemically, but not biochemically, active. Employing Occam's razor, which says that the simplest explanation must be invoked for any observation, they concluded that the surface of Mars was undergoing a chemical reaction involving superoxides that would release oxygen when exposed to water vapor.

In effect, Mars had an ozone layer at its surface. Levin disagreed, and even 25 years later, he forcefully argued that the GCMS experiment was not as sensitive as thought, and that a biological interpretation of his data was possible.

The scientific consensus, however, was that life had not been detected on Mars, and although subsequent spacecraft, including the *Mars Odyssey* and the *Mars Exploration Rovers Spirit* and *Opportunity*, found evidence of past water on Mars, no evidence for life itself was found. Although the *Viking* results slowed the search for life on Mars, in 2008 it remained a major preoccupation of the discipline of astrobiology.

Steven J. Dick

See also: Astrobiology, Viking

Bibliography

American Geophysical Union, *Scientific Results of the Viking Project* (1977).
Steven J. Dick and James E. Strick, *NASA and the Development of Astrobiology* (2004).
Barry DiGregorio et al., *Mars: The Living Planet* (1997).
Norman Horowitz, *To Utopia and Back: The Search for Life in the Solar System* (1986).

Mercury

Mercury is the closest planet to the Sun. Known to the classical Greeks in antiquity, Mercury's orbit became the focus of attention in the nineteenth century as the precession of its perihelion could not be explained by Newtonian gravitation theory. While some astronomers searched for a planet closer to the Sun, notionally called Vulcan, that perturbed Mercury's orbit, eventually its orbital anomalies were explained by Albert Einstein's general theory of relativity in the early twentieth century.

By the 1950s astronomers still knew little about the planet. The consensus was that the planet's rotation period matched its solar orbit period of 88 days, that it had little or no atmosphere, and that its surface was volcanic. In 1965 radar astronomy studies showed that Mercury's rotation period was 58.65 days, or two thirds of its orbital period.

Mariner 10 flew by Mercury three times in 1974–75, photographing 45 percent of the planet, uncovering a heavily cratered planet with no atmosphere, but with a magnetic field. The magnetic field was a surprise to scientists, because Mercury's small size implied more rapid internal cooling than the other inner planets, and hence a solid core with no magnetic field. In 2007 results of ground-based radar studies indicated the possibility of a molten core. In 1991 radar studies provided evidence that Mercury's permanently shadowed polar craters could contain water ice, which rekindled interest in the planet.

NASA launched the *MESSENGER* (*Mercury Surface, Space Environment, Geochemistry, and Ranging*) spacecraft in August 2004. The first of three scheduled flybys occurred on 14 January 2008, when *MESSENGER* passed 200 km above Mercury. The data revealed a dynamic system with a history of volcanism and an iron-rich liquid outer core. At least 60 percent of the planet's mass is in the core, twice as much as the other terrestrial planets. Cooling of the core has led to contraction of the entire planet, and creates a dipolar magnetic field that interacts with the surface and exosphere. The second flyby on 6 October 2008 revealed 30 percent of the surface previously unseen, at which point spacecraft had imaged 95 percent of the planet. The remaining flyby will shape *MESSENGER*'s orbit to allow it to enter orbit around Mercury in March 2011.

Katie J. Berryhill

See also: *Mariner 10*

Bibliography

Applied Physics Laboratory *MESSENGER*. http://messenger.jhuapl.edu/index.php.
David Leverington, *Babylon to Voyager and Beyond* (2003).
Special Issue *Science*, 4 July 2008.

Micrometeorite Study Missions

Micrometeorite study missions collected information at various distances from Earth about the number, size, and composition of natural, interplanetary particles with a diameter less than 1 mm (0.04 in). Properly called micrometeoroids until they survive entry into the atmosphere, such dust particles travel through space at cosmic velocities and can damage spacecraft. In Douglas Aircraft Company's Project RAND satellite feasibility studies of 1946, mathematician George Grimminger and astronomer Fred Whipple seriously discussed the possibility of destructive meteoroid collisions with Earth-orbiting satellites. Whipple suggested that acoustic analysis of meteors striking the skin of V-2 rockets might reduce uncertainties about the energy and frequency of these poorly understood objects. Plans for human spaceflight compounded the urgency of collecting data from which to model the meteoroid environment, assess the risk of impacts, and design protective structures.

Already in the 1930s scientists had begun collecting information and constructing models of the interplanetary-dust environment based on optical and radar measurements, plus actual micrometeorite deposits from Antarctica. Astronomical studies of the solar F-corona and of zodiacal light established that both of those phenomena resulted from scattering and diffraction of sunlight by interplanetary particles. Photometric studies from this time onward furnished points of comparison with measurements of the micrometeoroid environment obtained from the first Earth-orbiting artificial satellites. Radar observations, especially from high-power, large-aperture systems like the Arecibo Observatory in Puerto Rico and the Jicamarca Radio Observatory in Peru, both built during the early 1960s, augmented Earth-orbiting satellite data on meteoroids.

Numerous suborbital, orbital, and interplanetary flights, especially during the 1960s, carried sensors for direct measurement of micrometeoroid impacts. Even the nitrogen-filled body of *Sputnik*, the world's first artificial satellite, served as a crude micrometeoroid detector, because a puncture would have released the gas and, in turn, would have affected pressure and temperature readings. The most effective early satellites to collect micrometeoroid data in the late 1950s were *Explorer 1*,

Sputnik 3, and *Vanguard 3*. They employed various acoustic, wire-grid, piezoelectric, photosensitive, pressurized-cell, or erosion detectors to accumulate data on micrometeoroid impacts. *Explorer 8*, *Midas 2*, *Ranger 1*, *Samos 2*, *Mariner 2*, the first three Orbiting Geophysical Observatory satellites, *Geminis 9* and *12*, five *Lunar Orbiters*, *Pioneers 8* and *9*, *Orbiting Solar Observatory 3*, *Space Electric Rocket Test 2*, Soviet *Lunas 1* and *2*, plus the United Kingdom's *Prospero* and *Ariel 2* were among the many orbital or interplanetary flights in the 1960s and early 1970s that had other primary missions but, nonetheless carried micrometeoroid detection or collection experiments. Suborbital Aerobee and Nike Apache sounding-rocket launches, plus X-15 rocket-powered aerospace plane flights, contributed additional information. Three Aerobee micrometeoroid missions, each with a recoverable payload, were especially unique: the Venus Flytrap on 6 June 1961; the Inflatable Micrometeoroid Paraglider on 10 June 1964; and the Luster sampling experiment during the Leonid meteor shower in November 1965.

Only 10 orbital flights had micrometeoroid sampling as a primary mission. In conjunction with modeling the near-Earth meteoroid environment for the Apollo program, NASA launched three Explorer S-55 and three Pegasus satellites. *Explorer 13*, launched from Wallops Island, Virginia, in August 1961, hosted a variety of micrometeoroid sensors: pressurized cells, steel-covered grids, copper-wire cards, cadmium-sulfide cells, and impact detectors. In its brief orbital lifetime of 2.5 days, *Explorer 13* measured several hundred impacts but no penetrations of any of its sample materials. Based on a reassessment of the meteoroid environment using the *Explorer 13* results, NASA modified the *Explorer 16* micrometeoroid satellite. Launched on 16 December 1962, *Explorer 16* carried stainless-steel segments of three different thicknesses—0.001, 0.003, and 0.006 in—which experienced six, one, and no micrometeoroid penetrations, respectively, during the satellite's operational lifetime of 221 days. *Explorer 23*, the third S-55 mission, launched on 6 November 1964 for an operational life of 365 days, during which the data on puncture rates conformed to what scientists had collected from other flights.

Scout rockets launched the Explorer S-55 satellites, but Saturn 1 rockets launched the Pegasus series of micrometeoroid satellites. *Pegasus 1*, the first active payload launched by the Saturn system, went into orbit on 16 February 1965, followed by similarly designed *Pegasus 2* and *3* on 25 May and 30 July 1965, respectively. Each Pegasus carried 416 flat-plate capacitor detectors, each measuring 20 × 40 in and mounted on 208 penetration-sensitive panels affixed to its 96 ft long × 14 ft wide wings. Engineers chose three different thicknesses of aluminum—0.0015 in, 0.008 in, and 0.016 in—to compare puncture rates. The on-orbit results conformed essentially to ground-based predictions for the thicker materials and to other satellite measurements for the thinnest material.

Four subsequent micrometeoroid satellite missions collected data engineers could use in designing safer spacecraft for longer-duration human spaceflights. The short-lived *Radiation Meteoroid Satellite* (*RMS*) launched on 9 November 1970 from Wallops Island on the same Scout rocket as the *Orbiting Frog Otolith*, with separation of the two 30 in diameter spacecraft after they achieved orbit. *RMS* tested an improved detector system of thin-film capacitors for better measurement of meteoroid velocity in the near-Earth orbital environment. *Explorer 46* or the *Meteoroid Technology Satellite* (*MTS*), launched on 13 August 1972, carried a double-wall stainless-steel meteoroid bumper experiment that continued into January 1975. Based on data from pressurized cells on the *Explorer 46* bumper experiment, meteoroids punctured 51 of 94 cells during the course of 899 days. *MTS* also carried meteoroid penetration and velocity sensors, but spacecraft difficulties necessitated shutting down those experiments after only two weeks.

The next mission dedicated primarily to studying micrometeoroids was the *Long Duration Exposure Facility* (*LDEF*), which the crew of Space Shuttle *Challenger* delivered to orbit on 7 April 1984 and *Columbia* retrieved 2,076 days later on 12 January 1990. Built by NASA's Langley Research Center and originally called the Meteoroid and Exposure Module, the school bus–size *LDEF* carried more than 40 different experiments focused on micrometeoroids and debris. Government, university, and commercial investigators from the United States and eight other nations conducted the various experiments. The *LDEF* exposed to the micrometeoroid and debris environment in low-Earth orbit most of the materials—metals, polymers, composites, ceramics, and glasses—of interest to spacecraft designers. Before the *LDEF*'s retrieval, investigators realized that quantification of the cumulative impact history to the greatest extent possible required supplementing data from the satellite's dedicated instrumentation with a systematic survey of all exposed surfaces. During February–April 1990, a Micrometeoroid and Debris Special Investigation Group at Kennedy Space Center optically scanned and photo-documented all exposed surfaces on the *LDEF*, finding approximately 4,600 individual impact craters. NASA's Johnson Space Center archived the team's documentation, along with *LDEF* material samples of special interest, for future use by qualified investigators.

On 9 February 1994, the Space Shuttle *Discovery* crew released Germany's small *BremSat* into orbit to measure micrometeoroid volume and dust flow at very low orbital altitudes above Earth. Financed by the German Space Agency and the Federal Ministry of Research and Technology, *BremSat* was developed by the University of Bremen's Center of Applied Space Technology and Microgravity in cooperation with Orbitale Hochtechnologie Bremen-System and Hypersonic Technology Göttingen. The 140 lb satellite carried a Munich Dust Counter, the same kind of sensitive plasma impact counter as Japan's *Hiten* that launched in January 1990 and crashed onto the lunar surface in April 1993. Before its reentry into Earth's atmosphere on 12 February 1995, *BremSat* signaled more than 11,600 high-velocity impacts.

Although no spacecraft after *Bremsat* had micrometeoroid sampling as a primary experiment, a substantial number carried micrometeoroid impact detectors. Scientists and engineers subjected others, like Japan's *Space Flyer Unit*, which the Space Shuttle *Endeavour* retrieved and returned to Earth in January 1996 after a 10 month exposure in space, to extensive postflight analysis in the laboratory. That study yielded evidence of more than 700 hypervelocity impacts. A multinational, multiagency experimental satellite, *Space Technology Research Vehicle 2*, which the United Kingdom and the U.S. Air Force, NASA, and the Ballistic Missile Defense Organization developed and launched in June 2000, carried micrometeoroid and debris detectors among its instruments. Several micrometeoroid detectors built and operated by university students also flew on spacecraft, including one on Stellenbosch University's *SunSat*, a South African microsatellite, in 1999–2001 and another—the University of Colorado's Student Dust Counter—on *New Horizons*, which NASA launched in January 2006 for an encounter with Pluto in 2015.

Rick W. Sturdevant

See also: National Aeronautics and Space Administration

Bibliography

Nicholas L. Johnson and David M. Rodvold, *Europe and Asia in Space 1993–1994* (1995).
Long Duration Exposure Facility Archive System. http://setas-www.larc.nasa.gov/LDEF/.
NASA Technical Report Server. http://ntrs.nasa.gov/search.jsp.
Hajime Yano, "Japanese Contribution to In Situ Meteoroid and Debris Measurement in the Near Earth Space," *Earth, Planets and Space* 51, no. 11 (1999): 1233–46.

Multi-Planet Missions

Multi-planet missions are robotic planetary missions designed to visit more than one planet. Astronomers had known since the late eighteenth century that cometary orbits were affected by close approaches to Jupiter. The idea that a spacecraft's velocity could be changed in a similar way appeared in the 1931 science fiction novel *Brigands of the Moon* by Ray Cummings and a 1954 mathematical paper by Derek Lawden published in the *Journal of the British Interplanetary Society*. In 1961 Michael Minovitch, then a graduate student working at the Jet Propulsion Laboratory (JPL), developed the idea of gravity assist, in which a spacecraft is precisely aimed close to a planet, which exchanges some of its momentum with the spacecraft, changing the spacecraft's velocity and altering its trajectory. This allows spacecraft to be launched from Earth to distant planets using less powerful boosters than would otherwise be required.

NASA used gravity assist for the first time in 1970 to carry the crippled *Apollo 13* capsule around the Moon into a free return trajectory that would bring the astronauts back to Earth. The first use of gravity assist in planetary exploration was during the *Mariner 10* mission, which in 1974 used Venus's gravity to guide it toward Mercury. Later that year, the *Pioneer 11* spacecraft used Jupiter's gravity to propel it toward Saturn.

In 1965 JPL engineer Gary Flandro proposed the idea of using gravity assist to take advantage of a rare arrangement of the outer planets. This arrangement, which occurs once in about 175 years, would occur in 1976–78, allowing a grand tour of Jupiter, Saturn, Uranus, and Neptune. *Voyager 1* flew only to Jupiter and Saturn, but *Voyager 2* visited all four giant planets, arriving at Neptune 12 years after launch. A direct Earth-Neptune flight would have taken 30 years.

Later uses of gravity assist in planetary exploration included the *Ulysses* spacecraft, which used Jupiter's gravity to propel it out of the plane of the solar system to observe the Sun's poles; *Galileo*, which used both Venus and Earth to get to Jupiter;

and *Cassini*, which flew past Venus, Earth, and Jupiter on its way to orbit Saturn. Two spacecraft currently en route to their destinations as of 2007 are also using gravity assist. *New Horizons* will encounter Jupiter on its way to Pluto, cutting its travel time by at least three years. *MESSENGER* (*Mercury Surface, Space Environment, Geochemistry, and Ranging*) flew by Earth in 2005 and Venus in 2006 and will encounter Venus one more time and Mercury three times in order to match the velocity of Mercury to enter orbit.

Katie J. Berryhill

See also: Orbital Mechanics

Bibliography

JPL Basics of Space Flight. http://www2.jpl.nasa.gov/basics/index.php.

Mariner 10. *Mariner 10*, the first spacecraft to visit Mercury, was the first spacecraft to take close-up images of Venus. In addition, it was the first to use the gravitational attraction of one planet, in this case Venus, to direct it to another. *Mariner 10* was managed and operated by the Jet Propulsion Laboratory and built by Boeing.

Mariner 10 *image of the jumbled, chaotic terrain antipodal to the Caloris Basin on Mercury. This region is on the exact opposite side of the planet from the major impact which formed the Caloris Basin. Scientists theorize that the seismic waves generated by the impact were focused to this point as they travelled through the planet, and that this is responsible for the chaotic appearance of this region. (Courtesy NASA/Jet Propulsion Laboratory)*

In December 1969 NASA approved the *Mariner 10* mission for launch during the 1973 launch window. Shortly after approval, it became clear that, if launch occurred around the middle of that window, the spacecraft could visit Mercury not once but three times as, after its first Mercury flyby, its orbital period would be almost exactly twice that of the planet. Unfortunately, Mercury rotates three times for every two of its years, and so the same half of the planet would be illuminated at each intercept.

The main problem facing the *Mariner 10* designers was the extra heat the spacecraft would receive when flying as close to the Sun as Mercury. As a result, they included a large sunshade and placed most of the scientific instruments on the shadow side. The most important of these instruments, a pair of television cameras, was mounted on a scan platform that enabled them to follow their target during each flyby. These cameras were each fitted with a 1,500 mm telephoto and a 62 mm focal length wide-angle lens and a filter wheel, which included ultraviolet, blue, and yellow filters.

The 503 kg *Mariner 10* was launched on 3 November 1973 by an Atlas/Centaur. It suffered a number of problems en route to Venus, including a failure of the heaters for the television cameras, which meant the cameras had to be left permanently switched on. There were also problems with *Mariner 10*'s scan platform, computer, radio transmission system, power supply, and roll gyros. But fortunately none of these problems proved terminal, and *Mariner 10* completed its full mission.

On 5 February 1974 the spacecraft flew within 5,800 km of Venus. The ultraviolet images, which had the most contrast, showed chevron-shaped cloud patterns that rotated around Venus once every four days. While these clouds had been observed fleetingly from Earth, *Mariner 10*'s images produced far more detail, allowing scientists to analyze Venus's atmospheric dynamics for the first time.

Mariner 10 then flew within 703 km of Mercury on 29 March 1974, followed by two further Mercury intercepts at about six-month intervals. The third encounter was the closest, at 327 km. Images showed a surface that looked, at first sight, like that of the Moon with a large number of impact craters, although extensive lunar-like maria were noticeably absent. In addition, the distance of the ejecta blankets from the crater rims on Mercury was less than on the Moon, because of the planet's higher surface gravity.

Mercury's atmosphere was found to consist mainly of atomic helium and hydrogen at a pressure of about 10^{-15} bar. It was not really an atmosphere in the usual sense, but rather a local concentration of the solar wind caused by Mercury's gravitational attraction. In fact, lack of wind erosion on Mercury's craters indicated that the planet had not had an appreciable atmosphere for most of its lifetime.

The largest feature on Mercury was the 1,300-km-diameter Caloris basin, ringed by mountains about 2,000–3,000 m high and up to 100 km across. Antipodal to Caloris is an area called the "weird terrain," consisting of hills and mountains up to 1,800 m high arranged in a strange pattern. They appeared to have been formed by the shock wave generated by the Caloris impact.

Surprisingly, despite its small size, *Mariner 10* found Mercury to have a significant magnetic field and a magnetosphere, although it had no measurable radiation belts.

David Leverington

See also: Jet Propulsion Laboratory, Venus

Bibliography

James A. Dunne and Eric Burgess, *The Voyage of Mariner 10 Mission to Venus and Mercury* (1978).

Pioneer. Pioneer was a series of spacecraft launched by the United States between 1958 and 1978. The name Pioneer was initially applied to several early programs of lunar probes launched by the U.S. Army, U.S. Air Force (USAF), and NASA, only two of which were successful. *Pioneers 0–4* were funded by the Advanced Research Projects Agency and had no mid-course correction capability, which made their political goal of reaching the Moon before the Soviet Union problematic. They nonetheless made some important discoveries about the regions between Earth and the Moon, such as the discovery of different layers of the Van Allen belts. *Pioneers 0*, *1*, *2*, *5*, *P-3*, *P-30*, and *P-31* were built by Space Technology Laboratories (STL) of TRW and launched by the USAF, while *Pioneers 3* and *4* were built by Jet Propulsion Laboratory and launched by the U.S. Army. NASA managed all Pioneers after the first failed launch and found itself scrambling to reschedule experiments after all the early mishaps.

In 1962, NASA approved a new project, also called Pioneer, after Charles F. Hall proposed a series of solar probes. The new Pioneer project, which was Ames Research Center's first foray into major space programs, eventually comprised four spacecraft in solar orbit (*Pioneers 6–9*), which formed the first space weather network, plus *Pioneers 10* and *11*, the first spacecraft to fly by Jupiter and Saturn, respectively, and the *Pioneer Venus Orbiter* and *Pioneer Venus Multiprobe*. TRW's STL built *Pioneers 6–11*, while Hughes Aircraft Corporation built the two Venus craft. All Pioneer spacecraft were spin-stabilized. They proved and inspired technologies for later probes, including the Voyagers and *Galileo*. The project emphasized simple, tested designs, a concept that created some long-lived spacecraft. *Pioneers 6–9* all monitored the Sun and the solar wind well beyond their design life; NASA used the network to provide advance warning to the Apollo astronauts of potentially deadly solar flares. *Pioneer 6*, designed for a six-month mission, returned data successfully when NASA contacted it in 2000 to celebrate its 35th anniversary, making it NASA's oldest operational spacecraft. On *Pioneer 10*'s 30th birthday, NASA contacted the spacecraft and received a return signal, including science data, from far beyond Pluto's orbit. Overall, the Pioneers were among the most economical and successful NASA spacecraft.

Katie J. Berryhill

See also: Advanced Research Projects Agency, Ames Research Center, *Pioneer 10* and *11*, *Pioneer Venus*, TRW Corporation

Bibliography

Mark Wolverton, *The Depths of Space* (2004).

Pioneer Spacecraft and Missions (Part 1 of 2)

Spacecraft	Launch Vehicle	Launch Date	Launch Result	Significance
Pioneer 0	Thor-Able	17 Aug 1958	First stage exploded	Mission: lunar orbit. Built by TRW; launched by USAF; named *Pioneer 0* after launch failure.
Pioneer 1	Thor-Able	11 Oct 1958	Second stage failure; insufficient velocity to reach Moon; reentered atmosphere after two days	Mission: lunar orbit. Built by TRW; launched by USAF, but first NASA mission; returned data on radiation belts.
Pioneer 2	Thor-Able	8 Nov 1958	Launch vehicle failure, reentered atmosphere	Mission: lunar orbit. Built by TRW; launched by USAF; small amount of data returned.
Pioneer 3	Juno II	6 Dec 1958	Launch vehicle failure	Mission: lunar flyby. Built by JPL; launched by U.S. Army; discovered second radiation belt.
Pioneer 4	Juno II	3 Mar 1959	Successful	Mission: lunar flyby. Built by JPL; launched by U.S. Army; first U.S. spacecraft to escape Earth's gravitational field; second spacecraft to fly by Moon (it was supposed to impact); first spacecraft in solar orbit.
Pioneer 5	Thor-Able	11 Mar 1960	Successful	Built by TRW, launched by USAF; made first measurements of interplanetary magnetic field; first space weather report, signaling Earth 4–8 hours before solar storm arrived.
Pioneer P-3	Atlas-Able 4	26 Nov 1959	Payload shroud ejected prematurely. Spacecraft destroyed.	Mission: lunar orbit. Built by TRW, launched by USAF.
Pioneer P-30	Atlas-Able 5	25 Sep 1960	Faulty second stage caused spacecraft reentry.	Mission: lunar orbit. Built by TRW, launched by USAF.
Pioneer P-31	Atlas-Able 5	15 Dec 1960	Launcher exploded during ascent.	Mission: lunar orbit. Built by TRW, launched by USAF.

(Continued)

Pioneer Spacecraft and Missions (Part 2 of 2)

Spacecraft	Launch Vehicle	Launch Date	Launch Result	Significance
Pioneer 6 (aka *Pioneer A*)	Thor-Delta E	16 Dec 1965	Successful	Built by TRW; managed and operated by NASA ARC; oldest NASA spacecraft extant in 2005; created (with *Pioneer 7–9*) the first space weather network.
Pioneer 7 (aka *Pioneer B*)	Thor-Delta E	17 Aug 1966	Successful	Built by TRW; managed and operated by ARC; part of first space weather network.
Pioneer 8 (aka *Pioneer C*)	Thor-Delta E	13 Dec 1967	Successful	Built by TRW; managed and operated by ARC; part of first space weather network.
Pioneer 9 (aka *Pioneer D*)	Thor-Delta E	8 Nov 1968	Successful	Built by TRW; managed and operated by ARC; part of first space weather network.
Pioneer E	Thor-Delta L	27 Aug 1969	Launch vehicle failure; failed to orbit	Built by TRW; managed and operated by ARC.
Pioneer 10 (aka *Pioneer F*)	Atlas-Centaur TE364-4	2 Mar 1972	Successful	Built by TRW; managed and operated by ARC; first spacecraft to fly by asteroid belt and Jupiter; continued operating for at least 30 years.
Pioneer 11 (aka *Pioneer G*)	Atlas-Centaur TE364-4	5 Apr 1973	Successful	Built by TRW; managed and operated by ARC; first spacecraft to fly by Saturn.
Pioneer 12	Atlas-Centaur	20 May 1978	Successful	Built by Hughes Aircraft Company; managed and operated by ARC; *Pioneer Venus Orbiter*; long-term observations of Venus; radar map of 93 percent of surface; global maps of atmosphere.
Pioneer 13	Atlas-Centaur	8 Aug 1978	Successful	Built by Hughes Aircraft Company; managed and operated by ARC; *Pioneer Venus Multiprobe*; carrier and four probes measured temperature, pressure, density, and composition of Venus's atmosphere.

***Pioneer 10* and *11*.** *Pioneers 10* and *11* were the first spacecraft to visit the outer planets. They were originally designed to monitor the solar wind outside Earth's orbit, but in 1969 their missions were changed to fly by Jupiter instead. Managed by NASA's Ames Research Center and built by TRW, the 260 kg *Pioneer 10* was launched toward Jupiter by an Atlas-Centaur rocket on 3 March 1972, followed 13 months later by the nearly identical *Pioneer 11*. If *Pioneer 10* failed in its Jupiter mission, then *Pioneer 11* would repeat that mission. Otherwise, mission planners had had several options for *Pioneer 11*, including intercepting Jupiter along a different trajectory that would allow it to also fly by Saturn. After the Voyager program was approved in 1972, the Pioneers performed important "pathfinder" functions to aid these more sophisticated missions.

To get to Jupiter, both spacecraft had to fly through the asteroid belt with the unknown risk of a collision. In addition, high-energy particles within Jupiter's very active magnetosphere could possibly destroy solid-state devices on board the spacecraft. Neither of these hazards proved terminal, and *Pioneer 10* flew past Jupiter on 4 December 1973, just 21 months after launch. Its success allowed NASA to retarget *Pioneer 11* to fly by Saturn. *Pioneer 11* intercepted Jupiter almost exactly a year after *Pioneer 10*, and then flew by Saturn on 1 September 1979, successfully flying through the plane of its rings.

Little could be done to design the Pioneers to protect them from these hazards, apart from choosing components that were relatively insensitive to charged particles and

Instruments on *Pioneers 10* and *11*

Name	To Measure
Vector helium magnetometer	Magnetic fields from 1×10^{-7} to 1.4 gauss.
Fluxgate magnetometer	Magnetic fields up to 10 gauss.
Plasma analyzer	Density and energy of ions (up to 18,000 eV) and electrons (up to 500 eV). Jupiter's bow shock.
Charged particle composition instrument	Lightweight elements from hydrogen to oxygen in interplanetary space. Protons and electrons close to Jupiter.
Cosmic ray telescope	High-energy particles (including protons up to 800 MeV and electrons up to 1 MeV).
Geiger tube telescope	Intensity, energy spectra and angular distribution of electrons and protons in Jupiter's magnetosphere
Trapped radiation detector	Type, angular distribution and intensity of particles trapped by Jupiter.
Asteroid/meteoroid detector	Solid material from asteroid-sized to microscopic particles.
Meteoroid detector	Mass and energy of solid particles, of mass $> 10^{-6}$ gram.
Ultraviolet photometer	Hydrogen and helium in Jupiter's atmosphere
Imaging photopolarimeter	Imaging in red and blue light. Width of raster lines 0.03°. Best resolution images of Jupiter about four times sharper than from Earth.
Infrared radiometer	Jupiter's heat energy balance.

shielding those that were. In fact, the main difference between the design of *Pioneers 10* and *11* and those of previous interplanetary spacecraft was in the area of power generation. Solar arrays were not practical because of the large distances from the Sun. Instead each spacecraft carried four radioisotope thermoelectric generators.

Pioneer 10 first detected high-energy electrons from Jupiter some distance in front of its bow shock, indicating that they had crossed this shock wave. As the spacecraft continued to approach Jupiter, it crossed both Jupiter's bow shock and magnetopause. But three days later, an increase in the solar wind caused Jupiter's bow shock to overtake the spacecraft.

Jupiter's radiation belts were found to be about 10,000 times as intense as Earth's Van Allen belts, and the dipole moment of Jupiter's magnetic field was about 19,000 that of Earth. The angle between Jupiter's magnetic and spin axes was about 10º, with the dipole offset about 0.10 radii from the center. Jupiter's rapid rotation caused the outer part of its magnetic field to be non-dipolar, with a concentration toward Jupiter's equatorial plane.

Pioneer 10 detected helium in Jupiter's atmosphere for the first time and showed that it consisted of about 99 percent hydrogen and helium. This, like the ratio of helium to hydrogen measured on Jupiter, is similar to that of the Sun, indicating that Jupiter has changed little since forming from the solar nebula. Pioneer confirmed that Jupiter emits about twice as much energy as it receives from the Sun.

At Jupiter, *Pioneer 11* provided similar information to *Pioneer 10*. Before the Pioneer intercepts, no magnetosphere had been detected for Saturn. *Pioneer 11* showed that Saturn's magnetic field was dipolar with a magnetic moment about 540 times that of Earth. Titan was found to be generally within Saturn's magnetosphere, but varying pressure of the solar wind caused this magnetopause to move back and forth across Titan, sometimes leaving the satellite exposed to the solar wind.

Pioneer 11's images of Saturn and its satellites showed virtually no more detail than visible from Earth. One new ring was found, the narrow F ring, just outside the A ring, and a new small satellite, Epimetheus.

David Leverington

See also: Jupiter, Saturn

Bibliography

Robert S. Kraemer, *Beyond the Moon: A Golden Age of Planetary Exploration 1971–1978* (2000).
David Morrison, *Voyages to Saturn* (1982).
David Morrison and Jane Samz, *Voyage to Jupiter* (1980).

Voyager. Voyager was conceived in the mid-1960s as a "Grand Tour" mission in which one spacecraft would fly past Jupiter, Saturn, Uranus, and Neptune, using the gravity of one planet to accelerate and redirect it on to the next. This gravity-assist technique was championed by Michael Minovitch, a University of California at Los Angeles graduate student, who was working at NASA's Jet Propulsion Laboratory (JPL). Previously, travel to the outer solar system had been thought to require very powerful boosters. Then in 1965 Gary Flandro, also at JPL, found that the four planets

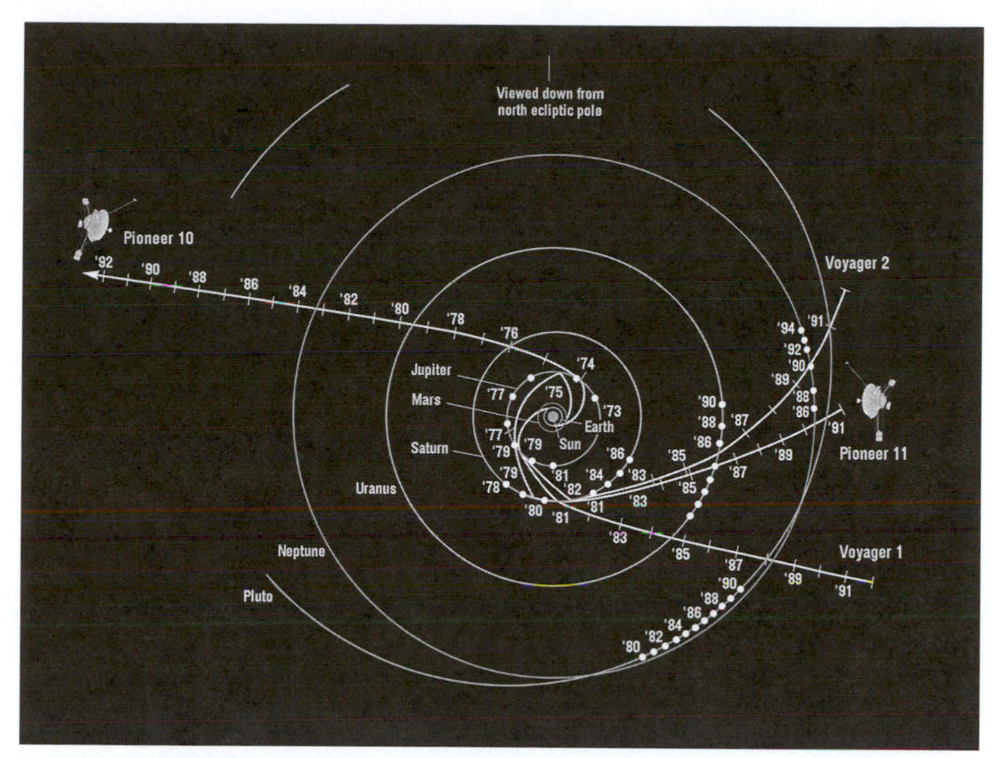

Orbits of the Voyager 1 *and* 2 *spacecraft compared with those of* Pioneer 10 *and* 11. *(Courtesy NASA/Ames Research Center)*

were perfectly aligned for an optimum gravity-assist trajectory if launch took place during the period 1976–80. Such an alignment, which occurred only once every 175 years, would enable the mission time from Earth to Neptune, for example, to be reduced from about 30 to 12 years.

In 1969 a NASA Working Group suggested that the Grand Tour mission could be split in two, with one spacecraft visiting Jupiter, Uranus, and Neptune and the other Jupiter, Saturn, and Pluto, supplemented by a Jupiter entry probe and Jupiter orbiter. NASA was under severe budgetary constraints, and the scientific community was split. Some scientists supported the Grand Tour concept, but others felt that it was too risky, as it required a newly designed, sophisticated spacecraft to operate for more than 10 years. Instead they proposed a mission dedicated to the study of Jupiter and its moons.

NASA finally approved a simplified version of the original Grand Tour spacecraft. Two spacecraft were proposed, based on the Mariner design, to be called *Voyager 1* and *2*. *Voyager 1* was to fly past Jupiter (1979) and Saturn (1980), and *Voyager 2* was to fly past Jupiter (1979), Saturn (1981), and possibly Uranus (1986) and Neptune (1989). *Voyager 2* was designed to visit Uranus and Neptune, even though there was no funding for such visits at that time.

The 815 kg Voyager spacecraft, built and operated by JPL, were much more sophisticated than their predecessors, *Pioneers 10* and *11*. In particular, a more powerful computer system allowed all spacecraft functions, except for trajectory changes, to

Instruments on *Voyagers 1* and *2*

Name	To Measure
Imaging science	Wide-angle (200 mm, field of view 3°) and narrow-angle (1,500 mm) television cameras to image both the planets and their satellites. Best resolution of Jupiter and Saturn, 20 km and 5 km, respectively.
Ultraviolet spectrometer	Atmospheric composition and structure. Distribution of ions and neutral atoms near planets. (Spectral range 50–170 nm.)
Infrared interferometer spectrometer	Atmospheric composition, including molecular hydrogen/helium ratio, thermal structure and dynamics. Surface composition and thermal properties of satellites. Composition, thermal properties, and particle sizes in Saturn's rings.
Photopolarimeter	Atmospheric aerosols, distribution of sodium around Jupiter, composition and particle sizes in Saturn's rings, and satellite surface texture
Low-energy charged particle detectors	Distribution, composition, and flow of energetic ions and electrons from 10 KeV to 30 MeV
Magnetometer	Planetary magnetic fields, magnetospheres, and effects of satellites
Planetary radio astronomy	Detect planetary radio emissions, plasma densities, and solar bursts
Plasma spectrometer	Magnetospheric ion and electron distribution from 10 eV to 6 KeV
Plasma wave experiment	Plasma waves, thermal plasma density profiles, and wave particle interactions
Cosmic-ray detectors	Distribution, composition and flow of nuclei of atomic number up to 30, from 0.5–500 MeV

be controlled automatically on board. This was essential as the round-trip time for signals between Neptune and Earth would be more than eight hours.

A key element of the *Voyager 1* mission was a close intercept of Titan, Saturn's largest satellite, which was thought to have an atmosphere similar to that of early Earth. Scientists considered this intercept so important that, if it failed, *Voyager 2* would be retargeted for a close flyby of Titan, even though this would prevent it from visiting Uranus and Neptune.

Voyager 2 was launched on 20 August 1977 by a Titan IIIE/Centaur TC-7 rocket to fly past Jupiter in July 1979. *Voyager 1* was launched two weeks later on a slightly different trajectory so that it would overtake *Voyager 2* en route to Jupiter.

By December 1978 *Voyager 1* images of Jupiter were already better than those taken from Earth. At the end of January, still 35 million km from Jupiter, it started taking images once every 96 seconds through each of three color filters. This enabled a full-color image to be produced on the ground every 4.8 minutes. These were then assembled into a movie sequence covering 10 Jupiter "days," showing its very active atmosphere in motion. *Voyager 1* crossed Jupiter's bow shock on 28 February 1979 and recrossed that and Jupiter's magnetopause several times throughout the next few days, as their positions responded to the varying intensity of the solar wind.

Pioneer 11 had detected a reduction in the number of high-energy particles near its closest approach to Jupiter, which had been attributed to an undiscovered ring or satellite. So, just before *Voyager 1*'s closest approach, a single exposure was taken looking to the side of Jupiter, where a very faint ring could be seen. Before the *Voyager 1* encounter, scientists generally thought that Io would look like a reddish version of Earth's Moon, but covered with sulphur-coated impact craters. However, as *Voyager 1* closed in, it resolved a fresh-looking surface of volcanic calderas and rivers of lava. Eight of the volcanoes were found to be active during the flyby.

Systematic imaging of Saturn by *Voyager 1* started on 25 August 1980, some 80 days before closest encounter. The images revealed cloud features, which showed that the equatorial jet stream was much broader than on Jupiter, with a maximum wind velocity of about 1,100 mph, or about three times that on Jupiter.

As *Voyager 1* approached Saturn, radial "spokes" were imaged on the B ring, and each of the main rings was found to consist of hundreds of thin concentric rings. Even the Cassini division contained rings. Two new satellites were found, one just inside and one just outside the narrow F ring, which appeared to be shepherding or stabilizing the ring. Another small satellite was found orbiting Saturn just 800 km outside the outer edge of the A ring, apparently restricting its outward expansion. The narrow F ring that had been discovered by *Pioneer 11* was found to be three intertwined or braided rings, apparently contravening the laws of dynamics.

Although *Voyager 1*'s intercept with Titan was successful, its images were virtually featureless because Titan was completely covered in cloud. Nevertheless, *Voyager 1* measured Titan's surface-level atmospheric pressure as about 1,500 mb at a temperature of 92 K. Its main constituents appeared to be nitrogen and methane.

In the meantime, *Voyager 2* had flown past Jupiter and was on its way to fly past Saturn in August 1981. Because of *Voyager 1*'s success at Titan, NASA announced in January 1981 that *Voyager 2* would fly by Uranus and Neptune after Saturn. The resulting gravity assist maneuver would cause the spacecraft to cross Saturn's ring plane just 1,200 km outside the newly discovered G ring.

Although *Voyager 2* performed well during its encounters with Jupiter and Saturn, it was not in the best of health. For example, one axis of the scan platform had jammed during the Saturn encounter and, although the fault had been corrected, NASA decided to restrict the scanning speed to avoid further problems at Uranus. Instead, engineers decided to pan the cameras by rotating the whole spacecraft. Starting and stopping the tape recorder also caused a jitter that affected image quality, but this was solved using very short compensatory bursts from the spacecraft's thrusters.

There was a problem with the transmission of data from *Voyager 2* at Uranus, because of its great distance from Earth, and the gradual reduction in the power available from Voyager's Radioisotope Thermal Generator power source. As a result,

NASA decided to improve the ground receiving facilities. In particular, the 64 m diameter Deep Space Network Canberra antenna in Australia, which was due to cover the most important part of the Uranus intercept, was linked to the 64 m Parkes radio telescope and to two 34 m dishes. In addition, JPL engineers reprogrammed the two onboard Flight Data Subsystem computers to perform image data compression.

Voyager 2 approached the Uranian system in a trajectory almost perpendicular to the orbital plane of its satellites, as these are almost perpendicular to the ecliptic. This meant that the whole close encounter sequence would last for little more than five hours, compared with 34 hours for the Jupiter system and even longer for Saturn. The trajectory of *Voyager 2* would take it through Uranus' ring plane between Miranda, the nearest known satellite to the planet, and the outermost ring.

As Voyager approached Uranus in early December 1985, scientists realized that its satellites were not exactly where they were expected to be, because the planet was about 0.25 percent heavier than previously thought. So spacecraft operators slightly modified Voyager's trajectory with a 14-minute rocket burn; otherwise, Voyager would also have missed Neptune by about 4 million km!

Five days before closest approach to Uranus, Voyager detected a strong burst of polarized radio signals, confirming the presence of a magnetic field. Analysis of timing variations in the radio emissions during several weeks showed that the interior of Uranus rotates with a period of 17 hours, 14 minutes. This was the first unambiguous measurement of Uranus' rotation period. On 20 January 1986, four days before Voyager's closest approach to Uranus, two small shepherding satellites were found on either side of the ε ring. No further shepherding satellites were found, so only one of the nine previously known rings had been found to have a pair of shepherds. The highest resolution images of all the Uranian satellites were those of Miranda. Its complex structure was a big surprise with its chevron-shaped feature, two large ovoids, and enormous cliffs up to 20 km high. In fact, Miranda was very different from the expected ancient, cratered world.

Voyager 2 was now on course for its close encounter with Neptune in August 1989. The first question that had to be settled was exactly where to target the spacecraft to ensure a close flyby of Neptune and its only large satellite, Triton. Evidence had started to accumulate in the early 1980s indicating that Neptune may possess one or more partial rings. Clearly the spacecraft had to avoid such rings from a safety point of view. In addition, in 1985, scientists concluded that Neptune was nearly 1,000 km larger than previously thought and its mass was 1.5 percent smaller. The estimate of Neptune's inclination was also increased and Triton's orbit modified. Voyager's new trajectory, taking all these factors into account, would take it only 4,800 km above Neptune's cloud layer and, just over five hours later, it would fly by Triton at a distance of about 40,000 km.

Voyager observations during the encounter phase showed a Great Dark Spot on Neptune, a little like the Great Red Spot on Jupiter. Other features included a second, smaller dark spot called D2, and a small, fast-moving, light-colored feature called the Scooter. Eight days before closest approach, *Voyager 2* detected radio signals from Neptune that were varying with a period of 16.11 hours, which was assumed to be the spin rate of Neptune's interior. Relative to this, Neptune's clouds showed that there was an equatorial jet of 1,100 mph, which is surprisingly large considering that Neptune receives little solar radiation. Scientists concluded that these winds are driven by Neptune's already-known internal heat source.

In early August, about three weeks before closest approach, *Voyager 2* confirmed that Neptune apparently had ring arcs. But, as the spacecraft drew closer to Neptune, it became clear that these ring arcs were just the brightest segments of two complete rings. After its closest approach to Neptune, the spacecraft also imaged a third, diffuse ring and a broad band of relatively large particles and dust.

Two new satellites were found, one just inside each of the two narrow rings. These satellites could not cause clumping of the rings on their own, however, and neither could stop either ring from expanding outward. To do this, a further satellite was required just outside each of the rings, but no such satellites were found.

Voyager arrived at Triton during early summer in its southern hemisphere and revealed a bright, pinkish-white, southern polar cap of nitrogen ice extending three-quarters of the way from the pole to the equator. Dark streaks of up to 150 km long were seen all over the south polar cap, probably the result of geyser-like eruptions, two of which seemed to be in progress during the Voyager encounter.

As of 2006 *Voyagers 1* and *2* were still working well beyond the orbit of Pluto. One is south and the other is north of the ecliptic, participating in the Voyager Interstellar Mission. *Voyager 1* crossed the termination shock and entered the heliosheath in December 2004, and eventually will reach the heliopause. Both spacecraft are being monitored to find out what happens when they leave the influence of the Sun's magnetic field, and of the solar wind, at the heliopause. After that they will be in the interstellar environment, giving the first direct measurements of that, if they survive that long.

David Leverington

See also: Deep Space Network, Jet Propulsion Laboratory

Bibliography

Henry C. Dethloff and Ronald A. Schorn, *Voyager's Grand Tour* (2003).
Ellis D. Miner, *Uranus: The Planet, Rings and Satellites*, 2nd ed. (1998).
Ellis D. Miner and Randii R. Wessen, *Neptune: The Planet, Rings and Satellites* (2002).
David Morrison, *Voyages to Saturn* (1982).
David Morrison and Jane Samz, *Voyage to Jupiter* (1980).

Neptune

Neptune was discovered by Johann Galle and Heinrich d'Arrest at the Berlin Observatory on 23 September 1846, following a request for a search by French mathematician Urbain Le Verrier. Le Verrier had calculated the position of a hypothesized new planet based on discrepancies in Uranus's orbit. British astronomer George Airy challenged Le Verrier's prediction priority, claiming that John Couch Adams

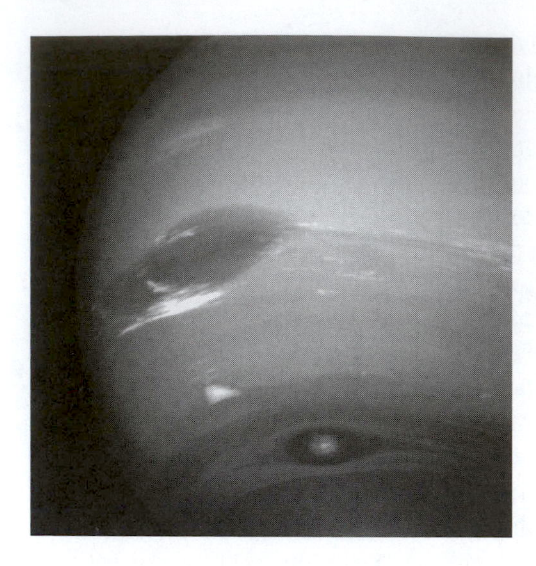

Neptune with the Great Dark Spot left of center, and white methane clouds on its southern edge. Taken by Voyager 2. *(Courtesy NASA/Jet Propulsion Laboratory)*

had predicted the planet's position before Le Verrier, which led to an international controversy between the astronomical communities in France and the United Kingdom. Both claims were given credence, as the British Royal Society awarded Le Verrier the prestigious Copley Medal for 1846 and gave Adams the same award two years later. In the mid-1990s, evidence from some of Airy's rediscovered papers showed that Adams kept modifying his predictions, which complicated the search for the new planet. When he produced the first calculated orbit for Neptune just a few months after its discovery, Adams was surprised to find that its mean distance from the Sun was just 30 AU, rather than the 36 AU assumed by Le Verrier, or the 37 AU assumed by himself.

Less than a month after Neptune's discovery, William Lassell discovered a moon, Triton, close to Neptune. Further work showed its orbit to be circular, inclined at about 30º to the ecliptic, and retrograde. Astronomers believed that as Triton was relatively large and close to Neptune, they both must have been formed at the same time, and so Neptune's spin was probably also retrograde. In 1949 Gerard Peter Kuiper discovered Nereid in a prograde, but elliptical orbit. The discovery of Triton allowed astronomers to estimate Neptune's mass as almost the same as Uranus. Although Neptune was about 50 percent farther from the Sun than Uranus, far infrared measurements in the early 1970s showed that Neptune was warmer than Uranus, indicating that it had an internal heat source.

Angelo Secchi obtained the first spectra of Neptune and Uranus in 1869, showing that both had broad absorption bands, with a particularly broad one in the yellow/orange region, explaining both planets' blue appearance. Rupert Wildt showed in the early 1930s that these bands, and others found since, were caused by methane in the planets' atmospheres. Based on spectra provided by Kuiper, Gerhard Herzberg announced in 1952 the discovery of molecular hydrogen in the atmosphere of Neptune. Before the *Voyager 2* encounter with Neptune in 1989, ammonia and ethane had been detected in trace amounts. Astronomers estimated Neptune's rotation period as about 17.6–17.8 hours prograde, based on observed cloud motions.

The discovery of rings at Uranus (1977) and at Jupiter (1979) meant that Neptune was apparently the only gas giant planet without rings. A flurry of stellar occultation observations in the early 1980s provided no firm evidence for rings, although some of the observations yielded positive detections, but only on one side of the planet. In July 1984 two observatories obtained evidence of a partial ring or ring arc, though its true nature did not become clear until the *Voyager 2* encounter five years later.

In 1978 Dale Cruikshank and Peter Silvaggio detected methane on Triton. In 1983 Cruikshank, Roger Clark, and Hamilton Brown detected nitrogen. In the following year, Lawrence Trafton predicted that Triton would have seasonal caps, probably of methane ice, condensing out of a tenuous methane atmosphere.

Voyager 2 arrived at Neptune in August 1989. It discovered six new moons, all interior to the orbit of Triton. It also found four narrow rings and one or more broad, dusty rings. The outermost (Adams) ring contained five ring arcs, which were strung together over about 10 percent of the ring circumference and were likely the source of most Earth-based positive detections. *Voyager 2* provided vastly improved values for the mass of both Neptune and Triton. The helium mass fraction is 0.32 ± 0.05, identical to that of the Sun, within the given uncertainties. The planet's equatorial radius of 24,764 km is about 3 percent smaller than Uranus, but its higher density (1.638 g cm^{-3}) yields a mass 19 percent higher than Uranus. The ratio of the total radiated heat to that absorbed from the Sun is 2.69, by far the highest of the planets. The magnetic dipole field tilt is large (47°; Uranus has a 59° tilt), and it is offset 55 percent of the way from the planet center to its cloud tops. *Voyager 2* measured the rotation period from radio emissions to be 16.1 hours, although various cloud features rotated around the planet in periods ranging from 15.8–18.3 hours. Detailed images of the atmosphere, the rings, and Triton were obtained. Methane cloud altitudes, relative to the main cloud deck, were found to be about 50 km from shadow measurements. Several storm systems were imaged, including a feature named the Great Dark Spot. Triton's reflectivity is high, and its surface temperature is 30 K, making it the coldest surface in the solar system. Despite its frigidity, *Voyager 2* found active plumes spewing nitrogen gas and trapped surface material about 8 km into a thin nitrogen atmosphere from beneath a nitrogen (not methane) ice cap. There are also strange grooves and a cantaloupe-like terrain on the non-ice-covered surface.

In Earth-based observations since the *Voyager 2* encounter, five additional moons have been discovered, bringing the total to thirteen as of early 2007. Recent images of Neptune from the *Hubble Space Telescope* and from the Hawaii Infrared Telescope Facility show that the Great Dark Spot has disappeared and other storms have appeared, so the atmospheric storms appear relatively short-lived.

Ellis Miner

See also: Voyager

Bibliography
Dale P. Cruikshank, ed., *Neptune and Triton* (1996).
David Leverington, *Babylon to Voyager and Beyond: A History of Planetary Astronomy* (2003).
Ellis D. Miner and Randii R. Wessen, *Neptune: The Planet, Rings, and Satellites* (2002).
Tom Standage, *The Neptune File* (2001).

Pluto and Kuiper Belt Objects

Pluto and Kuiper belt objects are relatively small objects that orbit the Sun, generally outside the orbit of Neptune.

Pluto was discovered in 1930 by Clyde Tombaugh, of the Lowell Observatory in Arizona, while searching for a planet that could explain the apparent discrepancies in the orbits of Uranus and Neptune. As Pluto only appeared as a point source, without a moon, it was impossible to estimate its size and mass. At that time, astronomers thought that Pluto was probably a little smaller than Earth. Over time, however, its probable mass estimates gradually reduced, so it could not have affected the orbit of the other two planets as first thought. Then in 1978 James Christy, of the U.S. Naval Observatory, discovered a moon, later called Charon, very close to Pluto. This enabled astronomers to calculate the mass of Pluto and its moon. Pluto's diameter was subsequently found to be 2,320 km, which is even smaller than Earth's Moon, with Charon's diameter about 1,200 km. In 2005 two very much smaller moons of Pluto were discovered, later called Nix and Hydra, using the *Hubble Space Telescope*.

Pluto and Charon are so similar in size and mass, and so close together, that they present their same face to each other as they orbit their common center of mass in about 6.4 days. Pluto's orbit around the Sun is much more inclined to the ecliptic than that of any other planet, and quite eccentric, so that it sometimes comes closer to the Sun than Neptune. It is also much smaller than any other planet; it is even smaller than seven planetary moons. In fact, some astronomers thought that Pluto might once have been a moon of Neptune.

Because Pluto and Charon are so close to each other, it was impossible to separate their spectra. However, in 1987 there were a series of eclipses of Charon by Pluto, and measurements at that time seemed to indicate that Pluto's surface was covered with methane ice, whereas Charon's surface was covered with water ice. In the following year, astronomers detected a very tenuous atmosphere around Pluto. Five years later, astronomers concluded that Pluto's surface was largely covered with nitrogen ice, not methane ice, while its atmosphere was mainly composed of nitrogen, with small amounts of methane and carbon monoxide.

Pluto reached perihelion in 1989, and astronomers thought that as it receded from the Sun, its surface would get colder and its atmosphere would disappear as it froze onto the ground. In 2002, however, its atmospheric pressure appeared to have increased rather than decreased, and, in 2006 measurements made with the Submillimeter Array on Mauna Kea indicated that Pluto's temperature was 10 K less than expected. It will take some decades for Pluto's atmosphere to condense completely onto its surface.

Kenneth Edgeworth and Gerard Kuiper had suggested in about 1950 that there should be many planetesimals, left over from the formation of the planets, orbiting

GERRIT "GERARD" PETER KUIPER (1905–1973)

Gerard Kuiper was a Dutch-born U.S. astronomer. Known as the father of modern planetary science, in the 1940s he discovered Uranus's Miranda, Neptune's Nereid, identified methane in the atmosphere of Saturn's moon—Titan—and carbon dioxide in the Martian atmosphere. He developed a theory that short-period comets come from a region of minor planets beyond Pluto, postulated earlier by Kenneth Edgeworth. Detections of objects in the region beyond Neptune began in 1991, commonly known as the Kuiper Belt. In 1960 Kuiper founded the Lunar and Planetary Laboratory, where he promoted multidisciplinary approaches to planetary science. He served as a scientist and advisor in various NASA solar system exploration programs, including the Ranger Project Scientist, finding landing sites for Surveyor and Apollo.

Katie Berryhill

the Sun outside the orbit of Neptune. In 1992 David Jewitt and Jane Luu found the first such trans-Neptunian or Kuiper belt object, with the University of Hawaii's 2.2 m telescope. Its average distance from the Sun is about 44 AU, compared with Pluto's 39 AU, and it is about 240 km in diameter. During the next few years, a number of other Kuiper belt objects were found, many of them, like Pluto, locked in a 2:3 resonance orbit with Neptune. These so-called plutinos are thought to define the inner edge of the Kuiper belt. By 2006 astronomers had found about 1,000 Kuiper belt objects.

In 1996 Luu and Jewitt discovered an object with a perihelion distance of 35 AU, and aphelion of 135 AU, giving it an orbital period of about 790 years. Since then many more of these so-called scattered disk objects have been found, with very large and highly eccentric orbits, generally outside the classical Kuiper belt. Astronomers think that these scattered disk objects have probably been scattered out of the plane of the solar system by gravitational interactions with Neptune.

In 2002 a classical Kuiper belt object, now called Quaoar, was discovered by Chad Trujillo and Mike Brown using the Oschin Schmidt telescope on Mount Palomar. Quaoar has an estimated diameter of about 1,200 km. Then, in 2003 Sedna was discovered with a diameter of about 1,500 km. Finally, in 2005 a scattered disk object, later called Eris, was found by Brown, Trujillo, and David Rabinowitz. It has a diameter of 2,400 ± 100 km, or about the same as Pluto. Evidently more similarly sized or larger objects may exist in the outer solar system. As a result, in 2006 the International Astronomical Union devised a new class of object, called a dwarf planet. The initial list of such objects includes both Pluto and Eris.

In January 2006 NASA launched its *New Horizons* spacecraft, which was planned to arrive at Pluto in July 2015, and then fly by one or two Kuiper belt objects.

David Leverington

See also: *Hubble Space Telescope*

Bibliography

John Davies, *Beyond Pluto: Exploring the Outer Limits of the Solar System* (2001).

Alan Stern and Jacqueline Mitton, *Pluto and Charon: Ice Worlds on the Ragged Edge of the Solar System* (2005).

Dwarf Planets. *See* Pluto and Kuiper Belt Objects.

***New Horizons*.** *See* Pluto and Kuiper Belt Objects.

Sedna. *See* Pluto and Kuiper Belt Objects.

Saturn

Saturn is the most distant of the planets known to ancient astronomers, who determined that its repeat cycle (now, a Saturnian year) was equivalent to 29.4 Earth years. When in mid-1610, Galileo Galilei pointed his rudimentary telescope at Saturn, he spotted two large moons of Saturn, situated on opposite sides and not moving around the planet like Jupiter's Galilean moons. He was even more confused when, in the fall of 1612, they had disappeared altogether. Dutch astronomer Christiaan Huygens, who in 1655 discovered Saturn's largest moon, Titan, a few years later, recognized that the large companions of Saturn were the ansae (ears) of an equatorial ring that completely circled the planet. Huygens believed that the ring was solid, even after Italian-born French astronomer Giovanni Domenico Cassini in 1675 discovered a division (soon, the Cassini Division) in the rings, and over time more gaps were reported. In 1705 Cassini suggested that the rings might be composed of numerous small satellites, although in the early nineteenth century most astronomers believed Pierre Simon Laplace's theory that there were numerous, thin solid rings. In 1850 George P. Bond realized that solid rings would be unstable—although he thought they were fluid. In 1857 James Clerk Maxwell showed mathematically that the rings must be composed of small particles in orbit around the planet.

Cassini also discovered four of Saturn's moons: Iapetus in 1671, Rhea in 1672, and Dione and Tethys in 1684. In 1789 William Herschel discovered Saturn's moons Mimas and Enceladus. Several more moons were discovered during the next two centuries: Hyperion by William Cranch Bond, George Phillips Bond, and William Lassell in 1848; Phoebe by William Pickering in 1898; and Janus by Audouin Dollfus in 1966. Epimetheus was discovered in the same year by Robert Walker, but it was thought to be the same object as Janus until Larson and Fountain (who later shared discovery credit with Walker) showed in 1977 that it was a separate moon.

In the 1950s, spectroscopic observations showed the primary gas in Saturn's atmosphere to be hydrogen, with helium the most likely candidate for second in concentration. Ammonia was discovered in 1966, hydrogen sulfide in 1976, and water in 1980, which, together with ammonium hydrosulfide, scientists thought formed Saturn's cloud layers at various depths within the atmosphere. Based on its observed size and on the orbital periods and distances of its moons, Saturn's mean density was estimated to be 0.7 g cm^{-3}, making it the only planet in the solar system with a density less than that of water. With an oblateness of 0.098, its disk is also the most flattened among known planets. As with Jupiter, the only larger planet in the solar system, it is a gas ball that likely has no liquid or solid surface beneath its clouds. From ground-based observations of cloud features, astronomers measured its rotation period to be between 10.24 hours at the equator and 10.67 hours at mid-latitudes.

Pioneer 11 was the first spacecraft to encounter Saturn, in September 1979. It detected a magnetic field and its associated magnetosphere, and discovered the narrow F ring at the outer edge of Saturn's main rings. From a noticeable decrease in magnetospheric particles with distance from the planet, scientists inferred the presence of ring material where Voyager later imaged the G ring.

The closest approaches to Saturn of *Voyagers 1* and *2* occurred in November 1980 and August 1981, respectively. Voyager scientists found five additional moons in 1980, and tiny Pan was found in 1990 using Voyager images. Although a D ring at the inner edge of the ring system had been previously reported, that report was erroneous. Voyager imaged the real D ring and showed that it was too faint to have been seen from Earth. It also revealed radial variations in all the rings. Voyager measured the periodicity of radio bursts that occur as the solar wind interacts with the magnetosphere, finding the rotation period of the magnetosphere to be 10.656 hours, essentially the same as shown by previous mid-latitude cloud features. This implied that Saturn's equatorial winds reached an astounding 900 mph. Radio and infrared data seemed to indicate that the percentage of helium in Saturn's atmosphere is much less than for either Jupiter or the Sun. Later, the Galileo Probe showed Jupiter's helium fraction to be significantly higher than deduced from Voyager measurements. This caused a reevaluation of Voyager's Saturn measurements, which then showed that its helium fraction was equal to that of Jupiter, within measurement uncertainties. Infrared measurements revealed that Saturn emitted 82 percent more energy than it received from the Sun, which may be due to residual heat left over from the time of its formation.

Voyager imaged the surfaces of all the moons known before 1980, except for haze-cloaked Titan, whose surface radius was determined from the cutoff of the *Voyager 1* radio signal as the spacecraft passed behind the moon as seen from radio receivers on Earth. Radio scientists also found Titan's atmospheric pressure near the surface to be 60 percent higher than Earth's sea-level atmospheric pressure. The primary constituents of the moon's atmosphere are nitrogen and methane, and Titan's temperature and pressure, near methane's triple point, implied the presence of liquid methane lakes on the surface. Other major surprises among the moons were the very sharp boundary between dark and light areas on Iapetus, with the dark hemisphere facing forward in its orbit around Saturn, and Enceladus, whose paucity of impact craters and high surface reflectivity (nearly 100 percent) implied a young and perhaps geologically active surface.

The *Cassini-Huygens* spacecraft was inserted into orbit around Saturn on 1 July 2004. The *Huygens Probe*, carried into orbit by the *Cassini Orbiter*, descended to the surface of Titan in January 2005, conducting detailed measurements of its atmosphere during its 2.5-hour descent and continuing to collect data for about

72 minutes after reaching the surface. The descent imagery revealed dry riverbeds, apparently carved by liquid methane rivers, and a surface, whose temperature of 94 K is too cold for liquid water, moistened with, but not flooded by methane liquid. Cassini's radar team found what appear to be large lakes of liquid methane in radar images of the polar regions of the moon. Among the Cassini orbiter results are the discovery of plumes of water and water ices being ejected into the E ring through surface cracks near the south pole of Enceladus, myriads of new and altered ring structures, detailed imaging of Titan's surface, and a longer and somewhat variable magnetospheric rotation period due to mass loading of the magnetosphere by the Enceladus injections and consequent slipping of the magnetosphere relative to Saturn's internal rotation. The *Cassini Orbiter* discovered a number of new moons in and near the rings, some of which are possibly loose rubble piles. Combining those discoveries with many new Earth-based discoveries, Saturn's moon count had reached 56 by early 2007.

Ellis Miner

See also: *Pioneer 10* and *11*, Voyager

Bibliography

Arthur F. O'D. Alexander, *The Planet Saturn: A History of Observation, Theory, and Discovery* (1962).
Henry C. Dethloff and Ronald A. Schorn, *Voyager's Grand Tour* (2003).
Tom Gehrels and Mildred Shapley Matthews, eds., *Saturn* (1984).
Ellis D. Miner et al., *Planetary Ring Systems* (2007).
Linda J. Spilker, ed., *Passage to a Ringed World* (1997).

Cassini-Huygens. *Cassini-Huygens* was a joint mission of the European Space Agency (ESA), NASA, and the Agenzia Spaziale Italiana (ASI—Italian Space Agency) to perform an in-depth study of the Saturnian system, including Titan's atmosphere and surface. The *Cassini Orbiter* was built by the Jet Propulsion Laboratory (JPL), the *Huygens Titan Probe* by Aérospatiale as prime contractor, and ASI provided the high-gain antenna. An innovative scheme was used to permit instrument developers to trade dollars for mass or power (with science management approval), with the result that all three resources remained within allocations. The mission was managed by JPL. JPL also operated the *Cassini Orbiter*; the *Huygens Probe* was operated by ESA's Operations Center in Darmstadt, Germany. The *Cassini-Huygens* mission was one of the most expensive and elaborate planetary missions ever undertaken.

The primary goal of the *Cassini-Huygens* mission was to conduct an in-depth study of the Saturn system, subdivided into five general areas: (1) the atmosphere and surface of Titan, which was the major focus of the mission; (2) the magnetosphere, the charged particle-laden magnetic bubble that surrounds Saturn; (3) the other (icy) moons of Saturn; (4) Saturn's complex ring system; and (5) the atmosphere and interior of Saturn itself. There were 12 scientific instrument packages on the *Cassini Orbiter*, and six on the *Huygens Probe*. In addition to complex imaging systems on both the *Orbiter* and the *Probe*, the *Orbiter* carried a Visible and Infrared Mapping

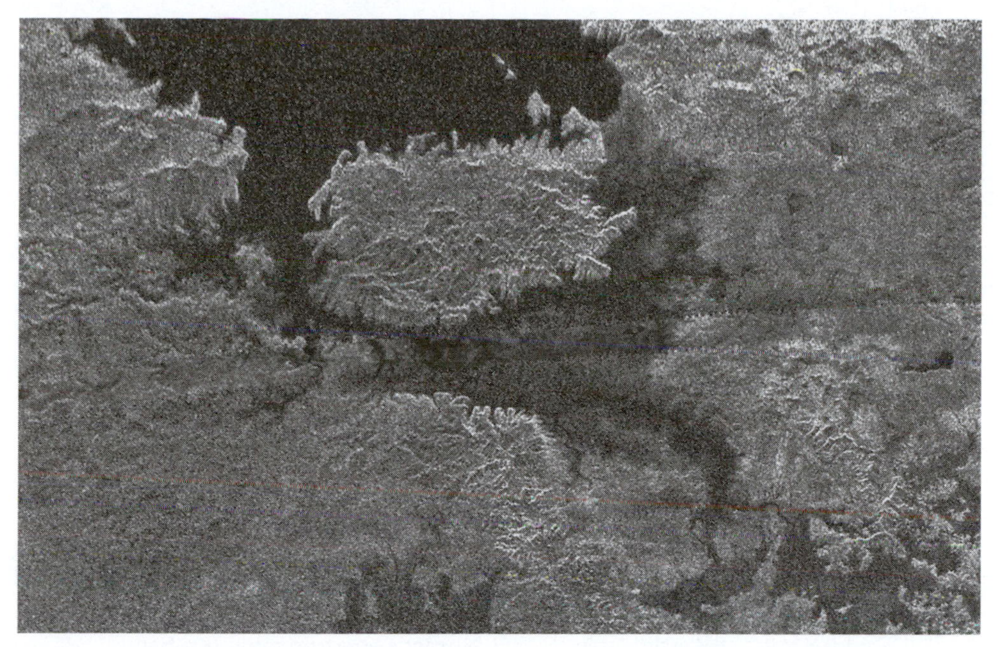

A radar image of an island in the middle of what appears to be a lake of liquid methane, obtained by Cassini*'s radar instrument during a near-polar flyby on 22 February 2007. (Courtesy NASA/Jet Propulsion Laboratory)*

Spectrometer, a Titan radar high-resolution surface mapper, a three-frequency radio science system, and a Magnetospheric Imaging Instrument.

Cassini-Huygens was launched from Cape Canaveral, Florida, atop a Titan IVB-Centaur launch vehicle on 15 October 1997 and entered Saturn orbit 1 July 2004. With a launch mass of 5,722 kg, it was more than twice the weight of the *Galileo* spacecraft and used a similar set of gravity assists to reach Saturn: Venus (twice), Earth, and Jupiter. Nineteen days before arriving at Saturn, the spacecraft flew close to Saturn's distant moon, Phoebe. The spacecraft fired its onboard rocket for 96 minutes to slow it sufficiently to the desired orbit. *Huygens* separated from the orbiter on 24 December 2004 and descended to Titan's surface on 14 January 2005, becoming the first artificial object to land on an outer solar system body. It collected about 150 minutes of data during descent and 72 minutes of data from Titan's surface, exceeding expectations. The *Cassini Orbiter*'s primary mission was planned to continue until 1 July 2008 before entering an extended mission phase.

A short summary of some key findings in each of the focus areas above includes: (1) Titan appears to have large polar hydrocarbon lakes and other surface features sculpted by tectonics, erosion, and winds; (2) a previously unknown radiation belt exists between the rings and Saturn's atmosphere; (3) several new small moons were found, including one within the Keeler Gap in the outer A ring; Enceladus has an atmosphere and active water-ice plumes emanating from fissures in its south polar region; from its low density and the spectral detection of water ice, Phoebe was believed to be an object captured from the icy regions beyond Neptune; Iapetus

***Cassini-Huygens* Investigations (Part 1 of 2)**

Cassini Orbiter	Acronym	Name	Characteristics
	CIRS	Composite Infrared Spectrometer	High-resolution infrared spectroscopy, 7–1,000 μm
	ISS	Imaging Science Subsystem	Photometric charge-coupled device imaging from near ultraviolet (UV) to near infrared (IR), 0.2–1.1 μm. Wide-angle camera (3.5° field of view, FOV), narrow-angle camera (0.35° FOV)
	UVIS	Ultraviolet Imaging Spectrograph	Extreme and far UV imaging spectroscopy, 0.055–0.190 μm. Hydrogen-deuterium absorption cell to measure deuterium. High speed photometer.
	VIMS	Visible and Infrared Mapping Spectrometer	Visible and IR imaging spectroscopy; 0.35–5.1 μm
	RADAR	Cassini Titan Radar	Synthetic aperture radar (Ku-band: 13.8 Ghz). Microwave radiometry (resolution < 5 K). Altimetry (height resolution 100 m)
	RSS	Radio Science Subsystem	Ka-, X- and S-band two-way measurements of frequency, phase, timing, and amplitude.
	MIMI	Magnetosphere Imaging Instrument	Imaged plasma distribution. Measured composition, charge state and energy distribution of energetic ions and electrons. Detected fast neutral particles.
	CAPS	Cassini Plasma Spectrometer	Particle detection and spectroscopy
	CDA	Cosmic Dust Analyzer	Directional flux and mass of dust particles in the mass range of 10^{-16} to 10^{-6} g
	MAG	Dual Technique Magnetometer	Measured magnitude and direction of magnetic fields
	INMS	Ion and Neutral Mass Spectrometer	Mass spectrometry of positive ions and neutrals
	RPWS	Radio and Plasma Wave Spectrometer	Measured electrical and magnetic fields, electron density, and temperature
Huygens Probe	**Acronym**	**Name**	**Characteristics**
	HASI	Huygens Atmospheric Structure Instrument	In-situ atmospheric temperature (50–300K), pressure (0–2 bar), density, winds and turbulence. Atmospheric and surface conductivity and radar reflectivity

Cassini-Huygens Investigations (Part 2 of 2)

Huygens Probe	Acronym	Name	Characteristics
	GCMS	Gas Chromatograph/ Mass Spectrometer	Atmospheric and surface composition
	ACP	Aerosol Collector and Pyrolyser	Aerosol composition
	DISR	Descent Imager and Spectral Radiometer	IR and visible surface imaging (0.35–1.7 μm). Side-looking visible imaging. Argon and methane concentration measurements. Aerosol properties.
	DWE	Doppler Wind Experiment	Wind measurement (velocities 2–200 m/s)
	SSP	Surface Science Package	Surface state and composition at landing site, whether solid or liquid. Gravity 0–100 g and tilt, temperature 65–100 K, thermal conductivity; sound speed; liquid density and refractive index. Atmospheric composition and temperature.

has an enormously high ridge extending for hundreds of kilometers along its dark-side equator; (4) the rings are far more complex than previously thought, and much of the complexity is shaped by embedded moonlets (which clear gaps whose widths are proportional to the moonlet mass) or by large unresolved ring particles (which often create visible wakes called propellers); and (5) mass loading by the ice particles being ejected from Enceladus causes the magnetosphere to slip relative to the internal magnetic field, making Saturn's internal rotation period difficult to measure.

Ellis Miner

See also: European Space Agency, Jet Propulsion Laboratory

Bibliography

ESA *Cassini-Huygens*. http://www.esa.int/SPECIALS/Cassini-Huygens/index.html.

Bram Groen and Charles Hampden-Tuner, *The Titans of Saturn: Leadership and Performance Lessons from the Cassini-Huygens Mission* (2005).

David Harland, *Cassini at Saturn: Huygens Results* (2007).

NASA *Cassini*. http://www.esa.int/SPECIALS/Cassini-Huygens/index.html.

Uranus

Uranus was discovered by William Herschel in 1781; the first planet to be discovered since ancient times. Six years later, he discovered two Uranian moons, Titania and Oberon, with orbits inclined at about 90° to the ecliptic. Anders Lexell estimated that the radius of Uranus's orbit was about 19 AU, or about twice that of Saturn, and

Voyager 2 *image of Uranus's moon Miranda, which surprised planetary scientists with its fragmented and terraced surface. (Courtesy NASA/Jet Propulsion Laboratory)*

Herschel estimated the planet's diameter as 55,000 km, or about four times that of Earth. In 1851 William Lassell discovered two more moons, Ariel and Umbriel, and Gerard Kuiper discovered a fifth moon, Miranda, in 1948.

Herschel noted a slight flattening of the poles of Uranus, caused by a rapid axial rotation. Subsequent attempts to measure the rotation period met with limited success, although before the start of the space age it was thought to be about 10.8 hours. Then in the 1970s, figures of the order of 15–17 hours were proposed.

Angelo Secchi examined the spectrum of Uranus in 1869 and found it crossed by dark broad absorption lines in the blue, green, and yellow/orange parts of the spectrum. The yellow/orange absorption explained the observed blue color of Uranus. The cause of these lines was a mystery until Rupert Wildt showed in the 1930s that they were because of methane. Twenty years later molecular hydrogen absorption lines were identified in Uranus's spectrum. Spectroscopic evidence for trace amounts of ammonia, hydrogen sulfide, and ethane were later found. Infrared and microwave spectra provided effective temperatures that indicated that Uranus, unlike Jupiter, Saturn, and Neptune, had little internally generated heat.

An attempt in 1977 by James Elliot and others to measure the structure of the atmosphere, by photometrically monitoring the light of a star as Uranus passed in front of it, showed that Uranus has a set of narrow rings. About a year later those rings were detected (but not resolved) in infrared image data from Earth. Several additional stellar occultations between 1977 and the 1986 arrival of *Voyager 2* revealed much about the widths, shapes, and precession rates of those rings.

Voyager 2, the only spacecraft to have encountered Uranus, flew by the planet in January 1986. It discovered ten new moons, two additional rings (and a complex ring dust band), a highly tilted magnetic field, and a rotation period of that magnetic field of 17.24 hours. It obtained vastly improved mass values and gravity harmonics for the planet and a helium mass fraction of 0.26 ± 0.06, essentially identical to that of the Sun. The bulk density is essentially identical to that of the larger Jupiter. The amount of heat being radiated from the interior is between 0 and 14 percent that of the reradiated solar energy (compared with 169 percent for Neptune). The magnetic field is not only tilted 59° from the rotation pole, but it is offset from the planet's center by one-third the radius of the planet. As a result, the magnetic field strength at the cloud-tops varies from 0.1 to 0.9 Oe. Detailed images of the atmosphere, the rings, and the five previously known moons were obtained. Of particular interest were the images of Miranda, which showed enormously high cliffs, both young and old terrain, and three concentric ringed features, called coronae.

Subsequent to the *Voyager 2* encounter, 12 additional moons have been discovered (one of which was discovered in 1999 in Voyager images and later confirmed in *Hubble Space Telescope* [*HST*] images), bringing the total to 27 as of early 2007. Two were discovered in images taken at the 200 inch Hale telescope, and seven were found by a team of astronomers at various telescopes around the world. The final two moons and two additional broad rings, much more distant from the planet, were discovered using the *HST*. The relatively bland atmosphere observed by *Voyager 2* in 1986, when the planet's pole was essentially pointed at the Sun, had been replaced by a more active and stormy atmosphere by 2007. The apparent path of the Sun crossed the Uranus equator in 2007.

Ellis Miner

See also: Voyager

Bibliography

Jay T. Bergstralh et al., *Uranus* (1991).
Henry C. Dethloff and Ronald A. Schorn, *Voyager's Grand Tour* (2003).
Ellis D. Miner, *Uranus: The Planet, Rings, and Satellites*, 2nd ed. (1998).

Venus

Venus is the brightest planet in the sky, which, with the naked eye, can be seen for a few hours near sunrise and sunset. Mayan priests (ca. 200–1500) associated its movements with astrological omens, and developed a Venus-based calendar. Motivated by astrology, Babylonian priests (ca. 1600–400 BCE) applied sophisticated mathematical techniques to predict the first appearances and disappearances of Venus. The Greek philosopher Heracleides (fl. 300 BCE) suggested that Mercury and Venus orbited the Sun as the Sun orbited the Earth. Claudius Ptolemy's (fl. 150) Earth-centered theory had the movements of Mercury and Venus centered on a line connecting the Earth and Sun, but not orbiting the Sun.

In 1610 Galileo Galilei's early telescopes revealed that Venus showed phases like the Moon. Its full illumination provided evidence against Ptolemy's theory, and for Nikolaus Copernicus's Sun-centered theory (which had been foreshadowed by Aristarchus) and Tycho Brahe's Earth-centered theory (which had been foreshadowed by Heracleides). In 1761 Mikhail Lomonsov concluded that Venus had an extensive atmosphere, when he observed a thin, luminous ring around the planet when Venus was backlit by the Sun. In the 1920s, photographs taken in ultraviolet light showed variable cloud features.

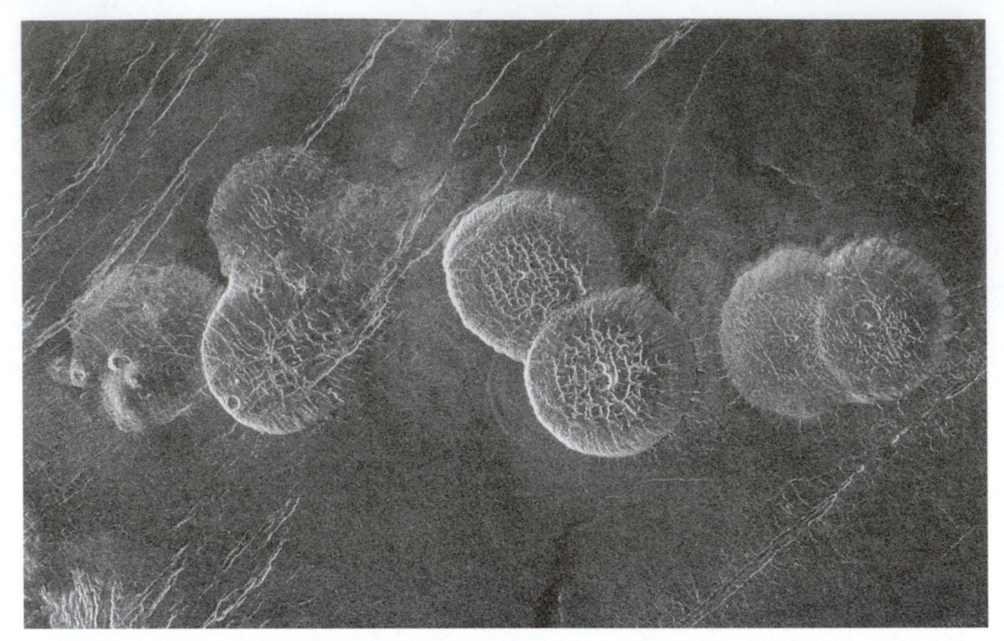

These pancake-like volcanic domes, imaged by Magellan, *are each about 750 meters high and 25 kilometers in diameter. (Courtesy NASA/Jet Propulsion Laboratory)*

Timings of transits of Venus, when Venus passes in front of the Sun, were crucial in determining the distance from the Sun to Earth, the astronomical unit. Astronomers traveled globally to observe these events, starting with the transit of 1761.

Venus's total cloud cover frustrated attempts to determine the rotation period, which yielded estimates ranging from 22 hours to 225 days. In 1958 the Lincoln Laboratory of Massachusetts Institute of Technology made the first attempts to probe through the Venusian atmosphere using ground-based radar. In March 1963 Richard Goldstein and Roland Carpenter of Jet Propulsion Laboratory's Goldstone tracking station announced that Venus had a retrograde rotation period of about 240 days.

Spectroscopic measurements of Venus's atmosphere, which began in the late nineteenth century, were difficult to interpret because of the intervening Earth's atmosphere. In 1932 astronomers at Mount Wilson Observatory, using a near-infrared spectrograph, identified significant quantities of carbon dioxide. In 1939 Princeton University's Rupert Wildt linked this to greenhouse heating and potentially high surface temperatures. By the late 1950s, optical, infrared, and radio telescope observations yielded estimates of surface pressure of 5–100 bars (5–100 times that of Earth), and surface temperature estimates from well under the boiling point of water to more than 300°C. If surface temperatures were lower than water's boiling point, then scientists believed life was possible on Venus, feeding speculation of hot jungles under the clouds.

With the instigation of the space race in the late 1950s and early 1960s, a politically driven robotic spacecraft race targeted Venus as the nearest objective for U.S. and Soviet planetary missions. The U.S. *Mariner 2* flew within 35,000 km of Venus in

1962, determining the cloud tops to be 60–80 km above the surface, whose temperature measured 425°C. It found no magnetic field. After failed attempts in 1962, 1964, and 1965, the Soviet Union's *Venera 4* reached Venus in 1967. It found a weak magnetic field, and hence a bow shock (where Venus's magnetic field deflected the solar wind) only 500 km above the planet's surface. *Venera 4* also carried a lander, which measured atmospheric constituents of 96 percent carbon dioxide, and traces of nitrogen, water vapor, and oxygen. It also measured a temperature of 270°C and pressure of 20 bars when its transmissions ended, which contradicted U.S. results, including that of *Mariner 5*, which arrived at Venus one day after *Venera 4*. Actually, the *Venera 4* lander's last transmissions had stopped 25 km above the surface.

For the next decade, the Soviet Union had Venus to itself while the United States focused on Mars and the outer planets. Its *Venera 7* and *8* spacecraft of 1970 executed soft landings and found what appeared to be volcanic basalt rock. They measured high-altitude wind speeds up to 225 mph and surface winds of 2 mph, and found that the atmosphere shielded Venus's surface from cosmic radiation and from 98.5 percent of the Sun's light. The *Venera 9* and *10* landers of 1975 transmitted the first surface images, showing little erosion and virtually no sand. The *Venera 10* orbiter altimeter measured a relatively flat surface across the planet, as compared to Earth.

In 1978 the U.S. launched the *Pioneer Venus Orbiter* (*PVO*) and the Pioneer Venus Multiprobe, which released four probes into the Venusian atmosphere. The probes confirmed sulphuric acid in Venus's clouds, which had been suggested early by Godfrey Sill, and Andrew and Louise Young, while the orbiter's radar mapped 93 percent of the planet to 20 km surface resolution and 100 m altitude resolution. The *PVO* measured a mainly flat surface, with two major high plateaus and a number of shield volcanoes. The Soviet Union continued with *Veneras 11* and *12* in 1978, *Veneras 13* and *14* in 1981, *Veneras 15* and *16* in 1983, and *Vegas 1* and *2* in 1984. The Vega spacecraft flew by Venus on the way to Halley's Comet, dropping modules with balloons that measured atmospheric circulation. The Veneras discovered chlorine, carbonyl sulphide, and hydrogen sulfide in the atmosphere. The *Venera 13* landing site showed more sand and dust than other locations, but all landing sites exhibited volcanic compositions. *Veneras 15* and *16* radar provided resolution to 2 km and provided data on the north polar region skipped by *PVO*. The U.S. *Magellan* spacecraft, launched in May 1989, used synthetic aperture radar to provide surface resolution mapping to 120 m resolution of 98 percent of the planet.

Magellan's high-resolution images showed the limited occurrence of folded mountains, but no mountain systems as extensive as those on Earth, and no indications of planetwide plate tectonics. The relative lack of craters showed that Venus's surface is young, averaging some 400 million years old. *Magellan* also imaged unique Venusian features, such as "pancake domes" of hardened lava. On Venus, less dense rocks, such as the granites on Earth, have not generally "floated" at higher levels than denser ones, such as terrestrial basalts. On Earth the lighter continents are on average several kilometers higher than the denser ocean floors, resulting in two-level topography. A very large portion of Venus's surface consists of low-lying plains, while extensive elevated regions are fewer and smaller in area compared to Earth. The cause of this difference between the two planets is uncertain.

In April 2006, the European Space Agency's *Venus Express* spacecraft was inserted into Venus orbit, primarily intending to study the mechanisms of Venus's atmosphere. It also provided the first images of the Venusian south pole, which showed a double vortex, as previously seen at its north pole. As of April 2010 its mission was continuing.

Ronald A. Schorn and Stephen B. Johnson

See also: National Aeronautics and Space Administration, Russia (Formerly the Soviet Union)

Bibliography

William W. Kellogg and Carl Sagan, *The Atmospheres of Mars and Venus* (1961).
David Leverington, *Babylon to Voyager and Beyond: A History of Planetary Astronomy* (2003).
Patrick Moore, *The Planet Venus* (1959).
Ronald A. Schorn, *Planetary Astronomy: From Ancient Times to the Third Millennium* (1998).

Early Mariners to Venus. Early Mariners to Venus, comprised *Mariners 1*, *2*, and *5*. The U.S. Venus Program started in 1958 with the decision to build two 169 kg Pioneer spacecraft to be launched during the June 1959 launch window to orbit Venus. Instead they were launched toward the Moon in an attempt to catch up with the Soviet Union, following its success with *Luna 1*.

Having missed the 1959 Venus launch window, NASA planned in 1960 to develop a series of spacecraft called Mariner for launches in the next three Venus launch windows starting in mid-1962. The first two spacecraft, *Mariner A* and *B*, were large, weighing about 1,000 kg each. The third spacecraft, *Mariner R*, was smaller, based on the lunar Ranger design. *Mariners A* and *B* were designed to be launch by an Atlas-Centaur, which had not then been developed, whereas the lighter *Mariner R* could use the already-developed Atlas-Agena. Over the next year, it became clear that the Centaur would not be ready by 1962, so NASA decided, instead, to launch two Jet Propulsion Laboratory (JPL)–designed Mariner R spacecraft, using an Atlas-Agena. These spacecraft, called *Mariner 1* and *2*, were launched to Venus on a flyby mission in 1962.

The launch of *Mariner 1* failed because of problems with the Atlas-Agena. But the causes of the failure were quickly established and *Mariner 2* launched successfully just over a month later on 27 August 1962. At this time, there was considerable debate about the likely conditions on the surface of Venus beneath its blanket of cloud, with surface temperature estimates of between 80°C and 300°C and surface atmospheric pressures up to 100 bar.

Mariner 2, which weighed only 204 kg, included a magnetometer to measure the interplanetary magnetic field, a plasma analyzer and ion chamber to measure the solar wind, an instrument to measure cosmic rays, and a micrometeoroid detector. As the spacecraft flew by Venus, a scanning microwave radiometer would measure Venus's surface temperature and an infrared radiometer would measure its cloud-top temperatures.

Mariner 2 flew by the planet on 14 December at a distance of 35,000 km. Venus's surface temperature appeared to be about 425°C and its surface atmospheric pressure about 20 bar. The cloud tops, with temperatures of about –40°C, were about 70 km above the surface. Venus had no measurable magnetic field and no discernible radiation belts. There were only two micrometeoroid hits measured en route to Venus. The interplanetary magnetic field varied randomly by a factor of five, and the velocity of the solar wind varied from about 400 to 700 km/s.

Mariner 2 was the first spacecraft to successfully intercept a planet. It also gave a timely boost to JPL's credibility, as at that time it was having serious problems with its Ranger spacecraft.

The next U.S. Venus probe, *Mariner 5*, had been a backup for the Mars-bound *Mariners 3* and *4*. But when *Mariner 4* was successful, NASA decided to send a modified *Mariner 5* to Venus with a launch in the mid-1967 window. As it turned out, the Soviet Union was to launch *Venera 4* to Venus just two days before *Mariner 5*.

Mariner 5 was to flyby Venus and go behind the planet, as seen from Earth, while transmitting and receiving radio signals. Scientists hoped the behavior of these signals would enable them to examine the height profile of the planet's atmosphere. In addition, an ultraviolet photometer would detect any atomic hydrogen and oxygen in the upper atmosphere.

Mariner 5 flew by Venus on 19 October 1967, at an altitude of about 4,000 km, just one day after *Venera 4*. Mariner detected a surface temperature of at least 430°C and a surface atmospheric pressure of between 75 and 100 bar. Venus's magnetic dipole moment was at most one-thousandth that of Earth, and its bow shock was just a few hundred kilometers above the surface. Scientists detected an atomic hydrogen corona around the planet, and concluded that its atmosphere consisted of about 80 percent carbon dioxide.

David Leverington

See also: Jet Propulsion Laboratory

Bibliography

Mariner-Venus 1962, Final Project Report (1965).
Mariner-Venus 1967, Final Project Report (1971).
Harold J. Wheelock, *Mariner Mission to Venus* (1963).

Early Soviet Venus Program. The early Soviet Venus program began in 1961 when the Soviet Union made its first attempt to send a spacecraft to Venus. Its first spacecraft, *Sputnik 7*, designed by OKB-1 (Experimental Design Bureau) under Sergei Korolev, got stranded in Earth orbit when the fourth stage of the launcher failed to fire. Eight days later, however, Soviet engineers managed to inject their next spacecraft, *Venera 1*, on its trajectory to Venus. But they lost contact with it just five days later when it was only 1.9 million km from Earth.

The Soviet Union's Venus program was plagued by problems with the SL-6 Molniya launcher, with three Venus-bound spacecraft lost due to launcher failures in both 1962 and 1964. The sequence finished in April 1964, however, when *Zond 1* was successfully launched to Venus. A second course correction failed when the spacecraft was 14 million km from Earth and contact was lost.

In 1965 the Soviet Union tried four more times to send a spacecraft to Venus, but *Venera 2* and *3* failed en route to the planet, while the other two spacecraft never left

Earth orbit. Korolev subsequently handed responsibility for future spacecraft to the Lavochkin design bureau.

On 12 June 1967, the 1,090 kg *Venera 4* was launched to Venus. It had a new thermal control system and a new course correction motor to overcome some of the problems experienced by *Venera 2* and *3*. The *Venera 4* spacecraft also had a 380 kg landing capsule with a parachute system to slow its descent through Venus's atmosphere to just a few m/s on landing.

Four months after launch, *Venera 4* arrived at Venus in good condition and detached its landing capsule. Contact was lost with the capsule about 100 minutes after it hit the top of the atmosphere. The Soviet Union maintained it had reached the surface, but later analysis showed it had failed with about 25 km to go. Nevertheless, the capsule provided the first in situ measurements of Venus's atmosphere. When it stopped transmitting, *Venera 4*'s capsule had measured an atmospheric pressure of 20 bar and a temperature of 270°C. In addition, it detected an atmosphere of about 96 percent carbon dioxide, which was a surprise. *Mariner 5* confirmed a high carbon dioxide concentration the next day during its Venus flyby.

January 1969 saw the launch of the next two Soviet spacecraft to Venus, called *Venera 5* and *6*. Their landing capsules had smaller parachutes than *Venera 4* to enable them to descend more rapidly through Venus's atmosphere, subjecting them to its very high temperature for as short a time as possible. Both capsules, while confirming the scientific results of *Venera 4*, stopped operating at a similar altitude to their predecessor.

The next pair of Soviet spacecraft was launched in August 1970 but only one, *Venera 7*, successfully left its Earth transfer orbit. *Venera 7*'s landing capsule was significantly different from its predecessors, as it was built to withstand a much higher external pressure. In addition, its interior was cooled before the capsule was separated from the main spacecraft, and yet another new parachute system was installed. The latter was only partially successful, and the lander crashed onto the surface at about 20 m/s. Even so, spacecraft controllers detected a feeble signal for 23 minutes.

The *Venera 7* capsule was the first artificial object to land and transmit data from the surface of another planet, measuring the ground-level atmospheric temperature in situ at about 475°C. It did not transmit any atmospheric pressure data from the surface, but extrapolating data transmitted during the capsule's descent gave a ground level pressure of about 90 bar.

The *Venera 7* capsule had landed on the night side of Venus, but the *Venera 8* capsule, launched in March 1972, landed on the daylight side. It measured surface atmospheric temperatures and pressures of 470°C and 90 bar that were, within error, the same as those on the night side. The wind speed varied from 100 m/s at 48 km altitude to just 1 m/s below about 10 km. Only 1.5 percent of the Sun's illumination reached the ground, where a gamma-ray spectrometer indicated that the surface rocks were probably volcanic in origin.

David Leverington

See also: Russia (formerly the Soviet Union)

Bibliography

James Harford, *Korolev: How One Man Masterminded the Soviet Drive to Beat America to the Moon* (1997).

Robert Reeves, *The Superpower Space Race: An Explosive Rivalry through the Solar System* (1994).

Magellan. The *Magellan* spacecraft launched to Venus in 1989 was the result of a program that had started 11 years earlier with NASA's plans to send the VOIR (Venus Orbiting Imaging Radar) spacecraft to Venus in 1985. VOIR was to have included a high-resolution Synthetic Aperture Radar (SAR), similar to that flown on the Earth-orbiting *Seasat*, instead of the simple radar altimeter flown on *Pioneer Venus*. However, budget cuts caused NASA to seriously reduce the mission's scope and insist on maximum reuse of Voyager and Galileo components. As a result, it launched this new smaller spacecraft called *Magellan* on a relatively slow 15-month trajectory to Venus. Launched by the Space Shuttle *Atlantis* and Inertial Upper Stage on 4 May 1989, it was the first planetary spacecraft launched by a Space Shuttle.

Magellan was managed and operated by the Jet Propulsion Laboratory (JPL) and built by Martin Marietta. It had just two experiments on board: a SAR and a radar altimeter designed to have horizontal and vertical resolutions of about 100 and 20 m, respectively. When *Magellan* arrived at Venus, it went into a 290 × 8,460 km near-polar orbit with a periapsis at 9.5° N and a period of 3 h 15 min. Although this highly eccentric orbit complicated image processing, the low altitude, circular orbit originally planned would have required too much on-board fuel. As it was, *Magellan*, including its retrorocket, still weighed 3.5 tons.

JPL lost contact with *Magellan* on 16 August 1990, just six days after orbit insertion around Venus. Although it was reestablished some 14 hours later, contact was lost again five days later before the onboard computer problem was finally resolved.

The first mapping cycle began on 15 September. The SAR antenna operated for 37 minutes per orbit storing data on an onboard tape recorder, before the same antenna was used for data transmission to Earth. This required the antenna to point toward Earth and then back to Venus each orbit. Mapping data was recorded from near the north pole to 54° S on one orbit, and from 72° N to 68° S on the next orbit, with the pattern repeating on alternate orbits.

The first mapping cycle lasted 243 days, as planned, during which time Venus had spun once below the spacecraft's orbit. In all, three complete mapping cycles were completed, allowing mapping of about 98 percent of the surface at a minimum of 100 m resolution. Then engineers reduced the periapsis to 185 km to allow gravitational mapping around periapsis, by measuring the effect of local gravity concentrations on the spacecraft's orbit. Finally, they reduced the apoapsis by an aerobraking maneuver to allow gravitational mapping at higher latitudes.

Most of the 900 impact craters imaged by *Magellan* appeared fresh, indicating that weathering on Venus is a slow process. The lack of any craters larger than about 275 km diameter, and the relative paucity of double-ringed craters, indicated that the surface is relatively young. Crater counts imply that most of the surface appears to be about only 400 million years old, with no features older than about one billion years.

Venus shows clear signs of tectonic activity, although there is no global network of faults like that on Earth. So there has been no obvious plate tectonic activity over the last few hundred million years.

The plains of Venus, which cover most of its surface, showed complex patterns indicating that they are volcanic in nature. They also have thousands of individual shield volcanoes, so it was a surprise when *Magellan*'s gravitational measurements indicated that the rigid lithosphere appears to be at least 30 km thick.

Venus' lava flows indicate that the lava on Venus is sometimes extremely fluid and at others very viscous. In particular, the fluid lava has produced channels over 6,000 km long that look more like rivers on Earth, whereas the viscous lava has produced large pancake-shaped volcanic domes, unlike anything seen on Earth.

David Leverington

See also: Jet Propulsion Laboratory, Martin Marietta Corporation

Bibliography

Andrew J. Butrica, *To See the Unseen: A History of Planetary Radar Astronomy* (1996).
Henry S. F. Cooper, Jr., *The Evening Star: Venus Observed* (1993).
Ladislav E. Roth and Stephen D. Wall, *The Face of Venus: The Magellan Radar Mapping Mission* (1995).

Mariner 2. *See* Early Mariners to Venus.

Mariner 5. *See* Early Mariners to Venus.

Pioneer Venus. *Pioneer Venus* consisted of two spacecraft, the *Pioneer Venus Orbiter* (PVO), and the *Pioneer Venus Multiprobe*. The program was managed by NASA Ames Research Center and the spacecraft built by Hughes Aircraft Corporation.

The 553 kg *PVO* included a 179 kg retrorocket to inject the spacecraft into a 24 hour near-polar orbit around Venus. Its main instrument was a radar altimeter designed to "see" through the all-enveloping clouds and provide the first detailed map of Venus, with a surface resolution of about 20–30 km and an altitude resolution of about 100–200 m. Earth-based radars had already produced maps of Venus, but these were relatively crude and limited to a small part of the planet's surface.

The *Multiprobe* consisted of a carrier spacecraft plus four atmospheric probes, designed to descend to Venus's surface. None were expected to survive impact. The 316 kg main probe and the three identical 90 kg probes were targeted to enter the atmosphere at widely separated locations to get some idea of Venus's atmospheric dynamics.

It was important to keep the mass of the retrorocket on the *PVO* as small as possible, which entailed keeping the approach velocity to Venus relatively low. Such a constraint did not apply to the *Multiprobe*, however, provided the probes could withstand the aerodynamic heating by Venus's atmosphere. As a result, the *PVO* was launched toward Venus on 20 May 1978, some two and a half months before the

Multiprobe, although both spacecraft arrived at the planet within a few days of each other in December 1978.

By the end of its mission, the *PVO* radar had mapped about 93 percent of the surface, showing it to be relatively flat, with about two thirds of the surface within ±500 m of the mean level. There were high- and low-lying areas, however, with the 10,800 m Maxwell Montes, its adjacent 4,000 m high Lakshmi Planum, and the 2,900 m deep Diana Chasma being the most evident. The *PVO* also found, surprisingly, that at 85 km altitude Venus's atmosphere was warmer at the poles than at the equator. The height of the ionopause, at the top of Venus's ionosphere, responded dramatically to variations in the velocity of the solar wind, varying from 250 km when the wind velocity was about 500 km/s to more than 1,500 km when the velocity had reduced to 250 km/s. The *Multiprobe* also added to existing evidence that Venus's clouds were mostly composed of sulfuric acid droplets.

David Leverington

See also: Ames Research Center, Hughes Aircraft Company

Bibliography

J. Kelly Beatty, "Pioneers' Venus: More than Fire and Brimstone," *Sky and Telescope* (July 1979).

R. Fimmel et al., *Pioneer Venus* (1983).

David G. Fischer, "From Heaven to Hell: The Pioneer Venus Story," *Quest* 1, no. 4 (1992): 4–14, 35–36.

Robert S. Kraemer, *Beyond the Moon: A Golden Age of Planetary Exploration 1971–1978* (2000).

Sputnik 7. *See* Early Soviet Venus Program.

Vega. Vega was the name of two Soviet Union spacecraft designed to release an aeroshell into Venus's atmosphere as they flew past the planet en route to Halley's Comet. The aeroshell contained both a lander and a balloon system, the latter based on a French design. The program was managed by the Space Research Institute, the spacecraft designed and manufactured by NPO (Scientific-Production Association) Lavochkin, and the international coordination handled by Interkosmos.

Vega 1, which was launched in December 1984, released its aeroshell two days before closest approach to Venus in June 1985. The balloon canister separated from the lander at an altitude of 61 km, just after the aeroshell had been jettisoned. The radio-transparent Balloon 1 then inflated and settled at an altitude of 54 km, in the middle of the planet's most active cloud layer, where the pressure and temperature were a reasonable 0.54 bar and 32°C. It communicated directly with Earth in a burst mode every half hour. Twenty antennae were used around the world, including those of NASA's Deep Space Network, to receive the signals and track the balloon.

Balloon 1, which was released just north of the equator at about local midnight, continued operating for two days, drifting about a quarter of the way around the planet until its batteries gave out. Its average horizontal velocity was an unexpectedly high

240 km/h. Downward gusts were also stronger than expected at up to 12 km/h, indicating good vertical mixing of the atmosphere.

Balloon 2 was released from the *Vega 2* aeroshell near local midnight, just south of the equator, four days after and at about the same altitude as Balloon 1. It operated for about the same length of time as Balloon 1, and traveled about the same distance. Thirty-three hours after release, Balloon 2 came into a very turbulent region as it passed over a 5 km high mountain. Surprisingly, the turbulence continued for another 2,000 km.

Neither of the two landers included cameras because they landed at night, but both carried a drilling system and an X-ray fluorescence spectrometer to analyze the surface material. As the landers descended, they detected sulfur, chlorine, and possibly phosphorus in the clouds.

The drilling system failed on Lander 1, but that on Lander 2, which landed on the edge of Aphrodite Terra, worked perfectly. It detected a type of rock similar to that found in the Precambrian areas of Earth and in the lunar highlands. It was the oldest yet found on Venus.

David Leverington

See also: Halley's Comet Exploration

Bibliography

Mikhail Ya. Marov and David H. Grinspoon, *The Planet Venus* (1998).
Yuri Surkov, *Exploration of Terrestrial Planets from Spacecraft: Instrumentation, Investigation, Interpretation* (1997).

Venera. *See* Early Soviet Venus Program.

Venera 9–16. *Venera 9–16* were a series of Soviet space probes designed by the Lavochkin Design Bureau and sent to Venus between 1975 and 1983. The Soviet Union had decided to pass the next Venus launch window, after launching *Venera 8*, and concentrate on designing a completely new spacecraft to be launched by their larger Proton rocket. Consequently, the mass of *Venera 9* was more than four times that of its predecessor.

The *Venera 9* lander was mounted inside an aeroshell to protect it during its initial entry into Venus's atmosphere. After releasing the aeroshell, the main spacecraft was to orbit Venus, the first spacecraft to do so, acting as a relay for the lander. This was because the landing site, which was on the daylight side of Venus, was on the far side as seen from Earth. The lander included a camera to take the first photograph of the surface, and the orbiter undertook extensive observations of Venus's cloud system.

Both *Venera 9* and its twin *Venera 10* were launched in June 1975. On arrival, the *Venera 9* lander touched down on the eastern slope of a shield volcano, Rhea Mons, while *Venera 10* touched down at the foot of another shield volcano, Theia Mons. The Soviet Union received data from both landers for about an hour after landing, until their respective relay orbiters disappeared below their local horizon.

The landers detected three layers of cloud centered on heights of 64, 55, and 51 km. Somewhat surprisingly, the clouds were found to be relatively transparent, so their opacity, as seen from Earth, was clearly due to their great depth. The photographs of the surface showed numerous rocks with little or no sand. Detailed analysis indicated that the surface is relatively young and may be still active.

The orbiters showed that the cloud particles were definitely not water droplets, although they were liquid droplets of some sort. Surprisingly the cloud-top temperatures on the night side of the planet were found to be higher than those on the day side. A simple radar system on *Venera 10* showed that the surface of Venus was relatively flat, varying by only a few kilometers in elevation from a perfectly smooth surface.

The orbital geometry was less favorable at the next launch window in September 1978. So engineers needed to reduce the masses of the next two Soviet spacecraft, *Venera 11* and *12*, by about 500 kg. As a result, they replaced the orbiter with a flyby spacecraft, although both Veneras still included a lander.

During their descent, the landers showed that the droplets in the upper-level clouds were composed mostly of chlorine, while those in the other clouds appeared to consist largely of sulphuric acid. The landers also found a relatively high level of argon 36 in Venus's atmosphere, compared with Earth. Scientists thought this might indicate that Venus's atmosphere had generally formed from the original solar nebula. In addition, there was a significant amount of argon 40, possibly due to some planetary outgassing produced by volcanism or tectonic motion. The camera covers did not eject on either lander, and the soil analyzers also failed.

The next pair of Soviet spacecraft, *Venera 13* and *14*, launched in late 1981, consisted of a flyby spacecraft and a landing module. This time the landers each contained a drilling and analysis system, to analyze surface material, and a new panoramic camera system with a color capability.

The surface images transmitted by the *Venera 13* lander showed a sandy landscape littered with sharp rocks. Dust that had blown onto the bottom of the lander moved between successive images. The site looked similar to those of *Venera 9* and *10*, whereas the *Venera 14* site looked different, with much less fine-grained material and with rocks of a more rounded shape, indicating they were older.

Geologists concluded that the material in the drilled samples at both lander locations was basalt. The samples in the rolling plains region of *Venera 13* appeared to be like leucitic basalt, which is often found on the slopes of terrestrial volcanoes. The material in the lowland region of *Venera 14* appeared to be like tholeiitic basalt found on the Earth's ocean floor.

The next two Soviet spacecraft, *Venera 15* and *16*, launched in 1983, had a Synthetic Aperture Radar (SAR) on board. SARs require a large antenna, produce an enormous amount of data, and are relatively power hungry. So, these spacecraft did not carry landers, replacing their normal landing module with the SAR. In addition to the SAR, which had a surface resolution of about 2 km, both spacecraft included a radar altimeter that had a height resolution of about 50 m.

On arrival at Venus, both spacecraft entered highly eccentric, highly inclined orbits around the planet with periods of 24 hours. The imaging was centered on an orbital periapsis of 60° N, limiting imaging to the northern hemisphere down to a latitude of about 18° N. Venus rotated beneath the spacecraft orbital paths completing one revolution in 243 days, thus allowing imaging of all longitudes.

The images showed many ridges and valleys, which appeared to have been formed by the horizontal motion of the crust, while many volcanic domes were found scattered over the surface. Two large volcanic caldera were imaged that may have been the source of the Lakshmi lava. Few impact craters were seen in the lowland regions suggesting they may be young lava plains; crater counts in some of these regions gave an age as young as 600 million years. A 100-km diameter crater was clearly seen on Maxwell Montes, as well as two unusual circular features called Anahit Corona and Pomona Corona, which were thought to have been formed by a mixture of both volcanic and tectonic processes. These two large corona structures and some thirty smaller ones appear to be unique to Venus.

David Leverington

See also: Russia (formerly the Soviet Union)

Bibliography

David Leverington, *New Cosmic Horizons: Space Astronomy from the V2 to the Hubble Space Telescope* (2000).

Mikhail Ya. Marov and David H. Grinspoon, *The Planet Venus* (1998).

Yuri Surkov, *Exploration of Terrestrial Planets from Spacecraft: Instrumentation, Investigation, Interpretation* (1997).

Venus Express. *Venus Express*, the first European Space Agency (ESA) spacecraft to visit Venus, had its origins in ESA's March 2001 "call for ideas" to the European scientific community for a low-cost mission based on the reuse of ESA's *Mars Express* bus. *Venus Express* was chosen from a list of ten possible missions and the program approved in November 2002.

The spacecraft was built by an international industrial consortium led by Astrium, Toulouse, France. The designs of four of its seven scientific instruments were based on *Mars Express* instruments, with the remainder based on those from *Rosetta*, ESA's comet mission. *Venus Express* was launched by a Russian Soyuz-Fregat rocket from Baikonur, Kazakhstan, on 9 November 2005.

Previous Soviet/Russian and U.S. spacecraft had been designed to image the surface of Venus, while most of the atmospheric data was provided by in situ measurements from entry probes. *Venus Express*, on the other hand, was designed to measure the atmosphere, below, within, and above the clouds virtually continuously over two sidereal Venus days of 243 days each. As Venus has no significant magnetic field, the top of its atmosphere interacts directly with the solar wind. *Venus Express* was to observe this interaction, which, scientists thought, would result in the solar wind depleting the top of Venus's atmosphere.

Venus Express arrived at Venus on 11 April 2006, 154 days after launch. A 53 second spacecraft main engine burn put the spacecraft into a provisional orbit around Venus. Smaller burns over the next few days achieved the desired 24 hour, 250 × 66,600 km polar orbit; the highest part of the orbit being over the south pole.

Venus Express has, for the first time, imaged both the horizontal and vertical structure of the double-eyed vortices over both poles. Before *Venus Express*, only the north polar vortices had been detected in two dimensions. The spacecraft has also detected the loss of various ions, including oxygen and helium, from the upper atmosphere. It has provided high-resolution images of the mysterious atmospheric "ultraviolet absorbers," which absorb large amounts of the incident solar ultraviolet; generated images of Venus's sulphuric acid clouds in three dimensions, allowing their dynamics to be studied; and produced temperature maps of the surface and of the atmosphere at different altitudes. *Venus Express* has also imaged the airglow structures on the night side.

In February 2007 ESA approved a mission extension to May 2009.

David Leverington

See also: European Aeronautic Defence and Space Company (EADS), European Space Agency

Bibliography

Andrea Accomazzo et al., *ESA Bulletin* (August 2006): 38–44.
ESA *Venus Express*. http://asimov.esrin.esa.it/SPECIALS/Venus_Express/index.html.
Donald McCoy et al., *ESA Bulletin* (November 2005): 8–41.

Zond 1. *See* Early Soviet Venus Program.

SCIENTIFIC DETECTORS

Scientific detectors operating above the atmosphere have transformed astronomical and Earth sciences. Throughout the years a number of technologies have been employed, and this article mentions only the main ones.

A Geiger counter on board a sounding rocket first detected solar X rays in 1949. Thereafter, for many years, proportional counters were the workhorse detectors for X rays. They were used to detect the first X-ray source outside the solar system, Sco X-1, in 1962. Proportional counters have been flown on most X-ray spacecraft until the mid 1990s, including *BeppoSAX* (launched 1996). With the 1993 launch of the *Advanced Satellite for Cosmology and Astrophysics*, charge-coupled devices (CCDs) began to be used, though this spacecraft also continued to employ proportional counters. The *XMM-Newton* and *Chandra X-ray Observatory* (both launched in 1999), however, depended almost exclusively on CCDs. At shallow angles, X rays reflect from mirrors, and this is used to form X-ray telescopes, producing direct images of the X-ray sky with a position-sensitive proportional counter or a CCD at the focus. Gamma-ray detectors on spacecraft have changed little through the years except for becoming larger. In the scintillation detector, a gamma ray passing through sodium iodide doped with thallium produces a light flash (scintillation). At the shortest wavelengths, scintillations from high-energy electrons produced by the photon provide indirect detection of the gamma rays.

Instruments, similar to those used for X rays, detect and image the shortest wavelength ultraviolet photons, while telescopes related to those employed in the optical region operate at longer wavelengths. Ultraviolet spectra of the Sun were first obtained in 1946 using photographic emulsion combined with V-2, rocket-launched telescopes, the film being retrieved after a parachute return to Earth. Later, image intensifiers equipped with ultraviolet sensitive photoelectron emitters and coupled to photographic emulsion provided greater sensitivity. Spacecraft, such as the *International Ultraviolet Explorer* launched in 1978, had spectroscopes with image intensifiers combined with television cameras. More recently microchannel plates, which detect and intensify the signal from ultraviolet photons by producing electrons that are multiplied many times as they pass through small holes in the plate, are combined with CCDs to provide high sensitivity and resolution imaging—for example, they were used in the Space Telescope Imaging Spectrograph that was installed in the *Hubble Space Telescope* (*HST*) in 1997. Alternatively CCDs, with a fluorescent coating to enhance their ultraviolet sensitivity or by themselves, form the detectors—as with the *HST* Wide Field Planetary Camera 2 (WFPC2) installed in 1993.

With the exception of the *HST*, few spacecraft have been designed for optical (near infrared, visible, and near-ultraviolet light) observations, because these can be made from the ground. The *HST* used a conventional reflecting telescope, and some of its instruments used CCD detectors and avoided the blurring effect of the atmosphere. By 2008 this advantage was being eroded as more terrestrial telescopes were able to correct for atmospheric effects for infrared images in the central field of view. Earth resources and planetary flyby, orbiting, and landing probes mostly obtain optical images by conventional telescopes and detectors (televisions in the early years, CCDs later). Some however, especially the earlier probes and Earth resources missions such as Landsat, used push-broom scanning—a single line of detectors aligned perpendicular to the spacecraft's motion, operating continuously as the spacecraft moves over the surface, building an image along the track covered by the detectors.

Infrared imaging uses conventional reflecting telescopes, although the longer wavelengths permit lower surface accuracies for the mirrors than are needed for optical telescopes. The main development of infrared studies was the improvement of the detectors—increases in sensitivity, extensions of detection limits to longer wavelengths, and the increasing use of arrays. Because no cooling system was needed and the devices were physically robust, early work employed room-temperature bolometers, such as those in the horizon-sensor on board the *Tiros 7* meteorological satellite (launched in 1963). However, cooled detectors were used in later instruments, and in many cases the telescopes were cooled also. Originally semiconductor bolometers made, for example, from germanium doped with gallium covered most of the infrared. Such detectors were still used in 2007 for the longest wavelength work, but at shorter wavelengths have mostly been replaced by photoconductive cells made from various materials, such as indium antimonide and gallium-doped germanium. Arrays giving direct imaging can be constructed from both types of detectors and the size of the arrays has increased markedly. For example, the *Infrared Astronomy Satellite* (launched in 1983) used an array with 62 elements, whereas the arrays on the *Spitzer* spacecraft (launched in 2003) were 1,000 times larger.

There has been little need to place radio telescopes on board spacecraft because radio waves penetrate Earth's atmosphere. However an 8-m radio dish was carried by the Japanese *Highly Advanced Laboratory for Communication and Astronomy* spacecraft (launched 1997) in order to extend the baseline for radio interferometry. The *Cosmic Background Explorer* (launched 1989) and the *Wilkinson Microwave Anisotropy Probe* (launched 2001) performed observations in the microwave region using horn antennae as feeds to conventional radiometers. The radars carried by planetary probes, though, are the main spacecraft application for long wavelength instrumentation. Launched in 1978, the *Pioneer-Venus Orbiter* included conventional radar adapted from military designs that produced images with a surface resolution of about 30 km and a height resolution of 100 m. Almost all subsequent probes, such as *Venera 15* and *16* (Venus, 1983 launch); *Magellan* (Venus, 1989 launch); *Cassini* (Saturn's satellites, 1997 launch); and *Envisat* (Earth, 2002 launch), have used synthetic aperture radars (SAR). Otherwise normal radars, SARs improve their resolution by combining data observed at different points along the spacecraft's orbit. *Magellan* mapped Venus to a surface resolution of 120 m, while *Envisat* achieved a 30 m resolution.

The flux-gate magnetometer has been in use since *Sputnik 3* (launched in 1958) and was, in 2007, the most widely used instrument flown on spacecraft for measuring magnetic fields directly. Usually three such magnetometers aligned perpendicularly to one another are used so that the field strength can be quantified in all three dimensions. Because spacecraft usually carry other instruments that generate their own magnetic fields, magnetometers are normally deployed on the ends of long booms to minimize interference from such sources. The vector helium magnetometer, which detects magnetic fields through its effect on helium atoms, was used for some missions.

The first successful U.S. spacecraft, *Explorer 1* (launched in 1958), carried a Geiger counter in the hope of detecting cosmic rays. Instead it found high-energy electrons, leading to the discovery of the Van Allen Belt around Earth. Instruments of many types have since been used to study the interplanetary medium, the solar wind, and gases emitted by comets. The *International Cometary Explorer* (launched in 1978), for example, carried a simple mass spectrometer to study the composition of the solar wind. *Ulysses* (launched in 1990) used electrostatic analyzers for the same purpose. The *Solar, Anomalous, and Magnetospheric Particle Explorer* (launched in 1992) used a proportional counter, a mass spectrometer, and other instruments to study solar particle emissions, cosmic rays, and magnetospheric electrons, while *Genesis* (launched in 2001) sampled the solar wind by trapping it in aerogels and returning these to Earth. Larger particles have often been detected accidentally when they collide with spacecraft at high speed. Although little knowledge is provided about the particle, which is completely evaporated during the impact, particle densities can be deduced. The *Long Duration Exposure Facility* (launched in 1984), though, was specifically designed to study micrometeorite impacts in addition to other effects of long-term exposure in space.

The material forming the surfaces of the Moon and some planets and satellites has been studied in detail and in a few cases been chemically analyzed by landers and rovers and via sample return missions and crewed missions. Many different analysis techniques and instruments have been used, especially when samples were returned to Earth. Two examples of in situ analysis were the two pairs of Mars landers. The twin Viking landers on Mars (launched in 1975) carried automated laboratories that could subject samples of the surface material to various chemicals, including water, and test for the presence of life (none was found). The Martian rovers *Spirit* and *Opportunity* (launched in 2003) could move around to get to the best positions and carried cameras that showed the surface in sufficient detail for geologists on Earth to use their normal skills to identify and understand the observed geological structures.

Chris R. Kitchin

See also: Astrophysics, Planetary Science

Bibliography

Christopher R. Kitchin, *Astrophysical Techniques*, 5th ed. (2008).
Paul Murdin, ed., *Encyclopedia of Astronomy and Astrophysics* (2001).
James H. Shirley and Rhodes W. Fairbridge, eds., *Encyclopedia of Planetary Sciences* (1997).

SUN

The Sun has been an object of study and speculation from ancient times. For example, Aristotle had taught in the fourth century BCE that the Sun was a perfect body. Since then various astronomers had seen naked eye sunspots, but they were thought to be objects between the Earth and Sun. However, in the early seventeenth century Galileo Galilei, Johann Fabricius, and others observed sunspots with their telescopes, and concluded that they were on the surface of the Sun, and that their movement indicated that the Sun was rotating.

Sunspots were still an enigma in the nineteenth century. Many astronomers thought that they were holes in the photosphere, but were puzzled because they were dark, rather than bright, as they should have been if they revealed the hotter lower layers of the Sun. Then in 1872 Angelo Secchi suggested that matter was ejected from the surface of the Sun at the edges of a sunspot. This matter then cooled and fell back into the center of the spot, so producing its dark central region. Just over 30 years later, Alfred Fowler and George Ellery Hale independently found, using spectroscopy, that sunspots are cooler than the surrounding Sun.

In 1843 Heinrich Schwabe found that the number of sunspots varied with a period of about 10 years. Nine years later, Edward Sabine, Rudolf Wolf, and Alfred Gautier independently concluded that there was a correlation between sunspots and disturbances in Earth's magnetic field.

Richard Carrington discovered in 1858 that the latitude of sunspots changed during the solar cycle. In the following year he found that sunspots near the solar equator moved faster than those at higher latitudes, showing that the Sun did not rotate as a rigid body. Secchi interpreted this so-called differential rotation of the Sun as indicating that the Sun was gaseous. In the same year he also observed two white light solar flares moving over the surface of a large sunspot. About 36 hours later this was followed by a major geomagnetic storm on Earth.

A coronal mass ejection taken from the SOHO *spacecraft on 7 August 2002. (Courtesy Solar and Heliospheric Observatory/European Space Agency/NASA)*

In the nineteenth century the Sun's corona had been found to have a faint continuous spectrum crossed by a number of bright emission lines. Of particular interest was a bright green emission line that had been found by Charles Young and William Harkness during the solar eclipse of 1869. It was originally attributed to iron, but was found later to have a slightly different wavelength. As no known element produced the required line, it was attributed to a new element called coronium.

At that time it was assumed that the temperature of the Sun and its corona gradually reduced from the center of the Sun moving outward. But in 1941 Bengt Edlén showed that coronal emission lines are produced by highly ionized iron, calcium, and nickel at a temperature of at least 2 million K. The "coronium" line, in particular, was due to highly ionized iron. How the temperature of the corona could be so high, when the photosphere temperature is only of the order of 6,000 K, was a mystery.

In 1908 Hale concluded that sunspot spectra at very high dispersion showed that sunspots were the home of very high magnetic fields, of the order of 3,000 G. Hale then started to examine the polarities of sunspots, and found that spots generally occur in pairs, with the polarity of the lead spot, as they crossed the disk, being different in the two hemispheres. This pattern was well established by 1912 when the polarities reversed at the solar minimum. They reversed again at the next solar minimum in 1923, showing that the solar cycle was about 22 years, not 11.

Walter Maunder found in 1913 that large magnetic storms on Earth start about 30 hours after a large sunspot crossed the center of the solar disc. Later work showed that the most intense storms were often associated with solar flares. In 1927 Charles Chree and James Stagg found that smaller storms, which did not seem to be associated with sunspots, tended to recur at the Sun's synodic period of 27 days. Julius Bartels called the invisible source on the Sun of these smaller storms M regions. Both the so-called "flare storms" and the "M storms" were assumed to be caused by particles ejected from the Sun.

In 1942 James Hey and George Southworth independently discovered that the Sun was a radio source. Hey detected intermittent radio emissions in the meter waveband, which appeared to be caused by an active sunspot group. Southworth, on the other hand, detected continuous radio emission in the centimeter waveband, which was attributed to the chromosphere.

In 1948 Robert Burnight, using a V-2 experiment, found the first indication that the Sun may be emitting X rays. Much clearer evidence was found by Herbert Friedman in the following year. Friedman's group continued to study the Sun over the next few years, and found that the majority of soft solar X rays were due to the corona. Then in 1956 they found that solar flares produced high energy X rays, and four years later both radio emissions and X rays were detected from a solar flare by a ground-based receiving station and the *SOLRAD 1* (Solar Radiation) spacecraft. In 1967 the *OSO 3* (Orbiting Solar Observatory) spacecraft measured the spectrum of a solar flare, which indicated that it had heated the local plasma to a temperature of about 30 million K.

Horace Babcock and Harold Babcock studied the magnetic field of the Sun in the 1950s, and found that there were both bipolar and unipolar regions on its surface. In the bipolar regions the magnetic flux leaving the Sun was about equal to that entering it. The Babcocks suggested that ions and electrons leaving the Sun in these regions would follow the field lines and collide over the Sun, generating radio noise and forming prominences and flares. On the other hand, those particles emitted from unipolar regions would stream away from the Sun, never to return. The Babcocks suggested that these unipolar regions were Bartels' M regions that emitted particles during M storms.

Eugene Parker developed his theory of the solar wind, in 1957–58, in which charged particles emitted by the Sun drew magnetic field lines in the corona out into the solar system. Parker envisaged that these charged particles would be emitted by the Sun continuously, and that the field lines would spiral out from the Sun because of the combined effect of radial flow and solar rotation. The particles were first detected outside the influence of Earth's magnetic field by *Luna 2* in 1959, but it could not measure their velocity. Three years later the *Mariner 2* spacecraft found that their velocity ranged from about 400–700 km/s, and occasionally exceeded 1,200 km/s.

In 1964 Norman Ness found, using the *Interplanetary Monitoring Platform 1* spacecraft, that the magnetic field lines in interplanetary space made an angle of about 45° to the Sun-Earth line. But, surprisingly, the magnetic field lines changed direction by about 180° every few days, effectively producing magnetic sectors (regions) of alternate polarity. Ness and John Willcox then found that the interplanetary magnetic field's polarity correlated with that of the Sun's photosphere at the Sun's equator with a lag of about 4.5 days. So the interplanetary magnetic field was an extension of the Sun's via the solar wind. The three-dimensional configuration of the magnetic sectors was unclear, however, but in 1976 *Pioneer 11* found that the sectors disappeared when the spacecraft's heliographic latitude exceeded 15°. This indicated that the dividing lines between sectors were the intersection of the ecliptic and a warped magnetically neutral sheet.

In 1957 Max Waldmeier discovered coronal holes in visible light as gaps in the corona on the solar limb, and in 1973 *Skylab* imaged many coronal holes in X rays.

In the same year, Allen Krieger, Adrienne Timothy, and Edmond Roelof found a good correlation between coronal holes, seen in X rays at the solar equator, and high solar wind velocities, measured by the Vela and *Pioneer 6* spacecraft. This showed that coronal holes were Bartels' M regions, which emitted the fast solar wind. *Skylab* measurements also showed that coronal holes were unipolar regions where the magnetic field lines ran freely into space, vindicating the Babcocks' theory in which energetic particles were emitted from unipolar regions.

Two Helios spacecraft were placed into orbits around the Sun in the mid-1970s. They confirmed earlier measurements showing a slow and fast solar wind with average velocities of about 350 km/s and 750 km/s. The Helios spacecraft confirmed that the fast wind was emitted by coronal holes. Twenty years later the *Ulysses* spacecraft, which orbited the Sun out of the ecliptic, found that around solar minimum the slow solar wind was constrained to equatorial regions. But the fast solar wind, which originated from the Sun's polar and other coronal holes, filled the majority of the heliosphere. In 2007 the *Hinode* spacecraft discovered that the low speed solar wind originated from the edge of an active region next to a coronal hole.

In 1971 the *OSO 7* spacecraft detected clouds of protons leaving the Sun in what were later called coronal mass ejections (CMEs). The clouds were hundreds of thousands of kilometers in diameter, with a temperature of about 1 million K, and were leaving the Sun at velocities of about 1,000 km/s. Numerous CMEs have been observed since then, particularly by *Skylab*, *Solar Maximum Mission* (*Solar Max*), and the *SOHO* (*Solar and Heliospheric Observatory*) spacecraft. Since 2006 the *STEREO* (*Solar Terrestrial Relations Observatory*) spacecraft has imaged CMEs spectroscopically in an attempt to understand their origin and development in detail.

Five-minute solar oscillations had been discovered by Robert Leighton, Robert Noyes, and George Simon in 1960, and in 1973 Robert Dicke found that this was not a local effect but represented vibration of the Sun as a whole. Later work with *SOHO*, GONG (Global Oscillation Network Group), and BiSON (Birmingham Solar Oscillations Network) ground-based solar networks enabled the internal structure of the Sun to be determined by analyzing these vibrations. They provided the first images of structures and flows below the surface, and confirmed that the decrease in angular velocity with latitude, measured at the surface, extended through the convective zone. They found that at the base of the convective zone, about one-third of the way into the Sun, was an adjustment region or tachocline leading to the more orderly radiative interior. The tachocline appeared to be the source of the Sun's magnetic field.

Raymond Davis found in the late 1960s, using the Homestake neutrino detector, that the number of electron neutrinos coming from the Sun was no more than half that expected based on the current theory of solar heat generation. This indicated that there was either something wrong with his neutrino detection system or something seriously wrong with theory. In 1969 Bruno Pontecorvo suggested that some of the electron neutrinos emitted by the Sun may transform to muon neutrinos en route to Earth and so be missed by Davis's detector. Later the existence of tau neutrinos was postulated giving another possible transformation route for electron neutrinos. In 2002 the Sudbury

Neutrino Observatory in Canada, which was sensitive to all types of neutrino, showed that Pontecorvo had been correct, as the total number of all types of neutrinos detected was as predicted by solar nuclear physics. The neutrinos were changing type en route to Earth.

In 1980 Richard Willson showed, using the *Solar Max* spacecraft, that the solar constant *reduced* slightly as large sunspot groups crossed the central solar meridian. It then gradually became clear, over the next few years, that the solar constant *increased* slightly as the Sun approached solar maximum. In 1988 Peter Foukal and Judith Lean explained this apparent inconsistency as due to faculae that overcompensated the blocking effect of sunspots over the course of a solar cycle.

Solar Max also provided valuable information on the development of solar flares, showing that electrons and protons are accelerated simultaneously in flares. Hard X rays brightened simultaneously at both ends of the magnetic loop containing the flare, while soft X rays were emitted by active regions that extended well into the corona. In 1992 the *Yohkoh* spacecraft showed that hard X rays were also emitted from the apex of flaring loops, probably as a result of magnetic reconnection. About 10 years later the *RHESSI* (*Reuven Ramaty High Energy Solar Spectroscopic Imager*) spacecraft confirmed that solar flares were caused by magnetic reconnection.

There has been an ongoing investigation, since the 1940s, into possible mechanisms of heating the solar corona to temperatures of millions of degrees K. In 1997 *SOHO* found evidence of the upward transfer of magnetic energy from the Sun's surface to the corona though a magnetic carpet, in which the observed magnetic flux was found to be highly mobile. This upward transfer of energy was thought to be a major source of coronal heating. Ten years later the *Hinode* spacecraft found that the chromosphere is permeated by Alfvén waves, which are sufficiently strong to accelerate the solar wind and possibly to heat the quiet corona.

David Leverington

See also: European Space Agency, Goddard Space Flight Center, Japan, *Skylab*

Bibliography

Karl Hufbauer, *Exploring the Sun; Solar Science since Galileo* (1991).
Kenneth R. Lang, *The Sun from Space* (2000).
David Leverington, *New Cosmic Horizons; Space Astronomy from the V2 to the Hubble Space Telescope* (2000).

ACE, Advanced Composition Explorer

ACE, Advanced Composition Explorer, was a NASA Explorer mission to study the composition of solar, interstellar, and intergalactic particles. *ACE* provided near real-time information about the solar wind and interplanetary magnetic field and could give 30–60 minutes' advance notice of coronal mass ejections that may cause geomagnetic storms on Earth. The project was managed by Goddard Space Flight Center, and the science payload was managed by the California Institute of Technology's Space Radiation Laboratory.

The Johns Hopkins University Applied Physics Laboratory built the spacecraft. It consisted of two octagonal decks, 1.6 m across and 1 m high, plus four solar arrays to provide power. Magnetometer booms were at the ends of two of the solar arrays. The 785 kg spacecraft was launched 25 August 1997 by a *Delta II* booster. The launch was timed for the solar minimum so that *ACE* was in place to observe the transition to solar maximum.

ACE orbited the Earth–Sun L1 libration point, 1.5 million km from Earth and 148.5 million km from the Sun. This orbit kept the spacecraft between Earth and the Sun, but far enough away so that Earth's magnetic field did not affect it.

ACE carried eight instruments, besides the magnetometer, which allowed it to measure the direction of a particle's motion, in addition to the particle's type, mass, charge, and energy. *ACE* was capable of measuring the properties of particles with masses ranging from hydrogen to zinc and with energies ranging from low-energy solar wind particles to high-energy cosmic rays.

Information collected by *ACE* was important to gain a better understanding of the formation and evolution of the solar system, to learn how solar activity affected Earth, and to improve space weather forecasting. Its data was used, along with that of *SOHO* (*Solar and Heliospheric Observatory*) and the *Geostationary Operational Environmental Satellites* (*GOES*), to produce space weather forecasts. Comparing *ACE* data with that of *Ulysses* gained scientists a better understanding of the heliosphere. The *ACE* mission was originally intended to last two years, with a goal of five years. As of early 2006, it had been in position for eight years, and all but one of the instruments continued to function well. At that time, *ACE* had enough fuel to stay in its orbit until approximately 2019.

Virginia D. Makepeace

See also: Goddard Space Flight Center, Johns Hopkins University Applied Physics Laboratory

Bibliography

California Institute of Technology *ACE*. http://www.srl.caltech.edu/ACE/.
Eric R. Christian and Andrew J. Davis, "*Advanced Composition Explorer* (*ACE*) Mission Overview," in *Advanced Composition Explorer*, 2nd ed., ed. S. B. Jacob et al. (2002).

ACRIMSAT, Active Cavity Radiometer Irradiance Monitor Satellite. *See* Solar Irradiance Missions.

Genesis

Genesis was a NASA sample-return mission to collect solar-wind samples outside Earth's magnetosphere and return them to Earth for analysis. Isotopic and elemental relative abundances of the solar wind provide a cornerstone data set for theories about how, starting ~4.6 billion years ago, the solar nebula transformed into the present solar system. Analysis of *Genesis* samples can use modern solar-wind data to answer questions about the ancient solar nebulae because a preponderance of evidence suggested that the outer layer of the Sun preserves the composition of the early solar nebula.

Built by Lockheed-Martin and managed by the Jet Propulsion Laboratory, *Genesis* was launched aboard a Delta 7326 vehicle from Kennedy Space Center in August 2001.

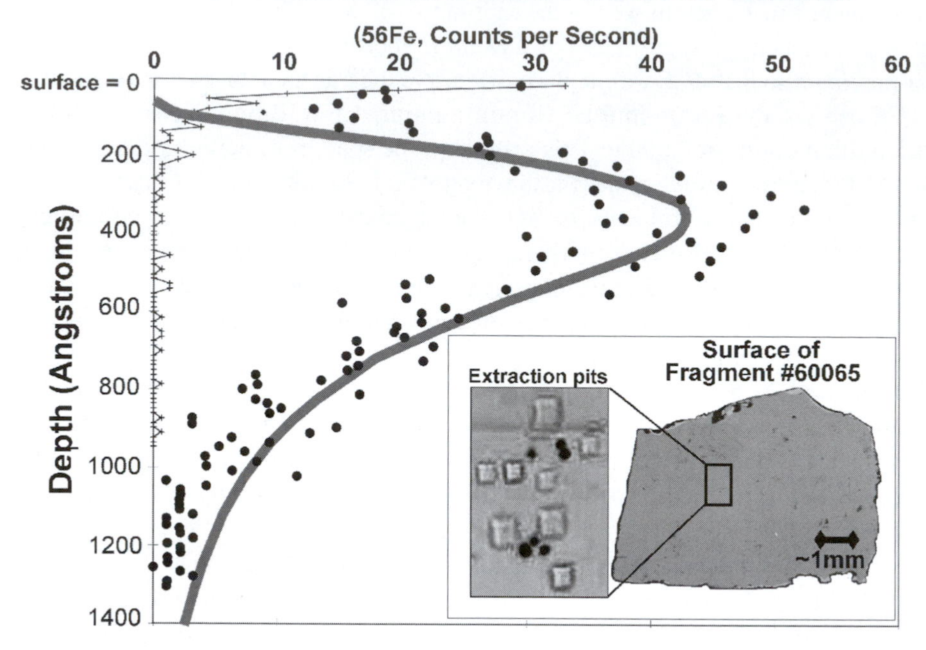

Solar-wind iron captured by Genesis*: the extraterrestrial origin of the iron is demonstrated by comparing exposed with unexposed* Genesis *materials. All measurements were made using Secondary Mass Spectroscopy (SIMS). Filled circles = measurement on flown fragment; thin black line = measurement on identical fragment not exposed to solar wind; thick line = expected solar-wind concentrations estimated using* Genesis *onboard solar-wind monitors. The inset shows a reflected-light photograph of the* Genesis*-flown fragment, and the magnified region shows the "pits" made during SIMS analyses. (Courtesy NASA/Genesis Mission Team)*

Placed in a halo orbit around the L1 Lagrange Point, for 886 days it passively collected solar-wind samples; solar-wind ions impacted *Genesis* collectors at greater than 200 km/sec and buried themselves in specially selected materials. Spaceborne *Genesis* instruments included ion and electron solar-wind monitors, deployable arrays for sampling solar-wind regimes, and an electrostatic mirror that concentrated light elements. The project also funded two ground-based sample analysis laboratories.

After the collection period, the spacecraft closed up and flew a trajectory toward Earth; on close approach, it released its sample-return capsule (SRC), which entered Earth's atmosphere on 8 September 2004. As the result of a design error involving the installation of the gravity switches, the parachute system failed to deploy. On impact, the SRC partially opened and the collectors were shattered, some scattering in the desert. The *Genesis* recovery team salvaged most of the collectors, allowing the *Genesis* preliminary examination team to detect and measure solar wind. In March 2005 Johnson Space Center curatorial staff started allocating solar-wind collectors to the international scientific community for analysis, which was planned to continue for decades.

Analysis of *Genesis* samples progressed slowly because (1) there is little solar wind in each fragment (the entire spacecraft collected the equivalent of a few grains of salt) and (2) terrestrial contamination made these difficult measurements even more challenging. *Genesis* data showed that noble gases in lunar-soil surfaces have a distribution consistent with modern-day solar wind, refuting an earlier theory invoking hypothetical "solar energetic particles" to explain the distribution. Scientists reported that the ratio of isotopes argon 36 to argon 38 is higher than Earth's atmosphere by ~3.42 percent, suggesting that Earth may have lost a portion of its atmosphere early in its history. By early 2009 scientists were measuring the absolute abundances of several elements as well as ratios of the isotopes of important elements, such as oxygen and nitrogen, data that will be used to attain new perspectives on the origin of meteorites, planets, and other solar-system bodies.

Amy Jurewicz, Don Burnett, Eileen Stansbery, Roger Wiens, and Don Sweetnam

See also: Jet Propulsion Laboratory, Johnson Space Center, Lockheed-Martin Corporation, *Stardust*

Bibliography

Genesis Mission NASA–JPL. http://genesismission.jpl.nasa.gov/.

A. Grimsberg et al., "Solar Wind Neon from Genesis: Implications for the Lunar Noble Gas Record," *Science* (2006).

A. Meshik et al., "Constraints on Neon and Argon Isotopic Fractionation in Solar Wind," *Science* (2007).

Space Science Reviews (*Genesis* special issue, 2003).

Helios

Helios was a West German–U.S. project to monitor the Sun and interplanetary medium. The negotiations that led to Helios began in 1965, when the Federal Republic of Germany Chancellor Ludwig Erhard met with U.S. President Lyndon Johnson, and the two leaders agreed to undertake a technically challenging project, which became Helios. Johnson's offer was part of a presidential strategy to improve relations with European allies by offering to collaborate on high-technology space programs to narrow the "technology gap" between the United States and Europe. The West German Federal Ministry for Research and Technology administered the development of two spacecraft, while NASA agreed to contribute two Atlas-Centaur launchers. Messerschmidt-Bölkow-Blohm, which during the course of the project merged with Entwicklungsring Nord (ERNO), designed and built the two spacecraft.

The two spacecraft featured a unique spool shape covered with silvered mirrors to reflect the extreme heat. Each spacecraft carried ten instruments, including a plasma experiment, flux-gate magnetometers, a search coil magnetometer, a plasma wave experiment, cosmic radiation experiments, a low-energy electron and ion spectrometer, a zodiacal light photometer, and a micrometeoroid analyzer.

To test the new Titan 3–Centaur launch vehicle being developed for the Viking Mars missions and to provide greater performance for the overweight, 371-kg Helios spacecraft design, NASA offered to replace the Atlas-Centaur with the new combination. The Germans accepted, and *Helios 1* was launched from Cape Canaveral,

Florida, in December 1974 into a heliocentric orbit with its closest approach near the orbit of Mercury at approximately 0.3 AU, and its farthest point from the Sun at Earth's orbit. *Helios 2* was launched in January 1976, reaching its closest approach in April. The higher heat load, due to its closer approach to the Sun at 0.29 AU, caused the spacecraft to encounter heating problems with its spin thrusters, leading to the failure of its transmitter tube March 1980 after eight perihelion passes. The mission ended in March 1986 due to power system degradation of *Helios 1* after covering a complete solar cycle from solar minimum to solar minimum.

Using data combined from both spacecraft, scientists studied the motions of plasma ejections moving away from the Sun. They found that near the Sun there was significant short period time variability in the solar wind, but at Earth's orbit this short period variability largely disappeared. They observed the effects of coronal mass ejections (CME) on the solar wind. Scientists determined that the shock fronts pushed by CMEs seem to extend far around the Sun, not just in the original direction of the outburst. With the twin spacecraft, scientists were able to detect differences in the propagation of shock fronts in addition to the presence of accelerated particles at different angular distances from the CME. Data from other spacecraft was also combined with *Helios* data. For example, in 1977 a pair of shocks detected by *Helios 2* merged before encountering *Voyager 1* and *2*.

The missions provided important information about the Sun and solar wind, which were important inputs to later solar missions, such as *Ulysses* and *STEREO* (*Solar Terrestrial Relations Observatory*).

Stephen B. Johnson and Katie J. Berryhill

See also: Germany

Bibliography

Robert S. Kraemer, *Beyond the Moon: A Golden Age of Space Exploration 1971–1978* (2000).
Max Planck Institute for Solar System Research. http://www.mps.mpg.de/en/.
Herbert Porsche, ed., *10 Jahre HELIOS 10 Years* (1984).

Hinode

Hinode (Japanese for "sunrise"), known as *Solar B* before launch, was launched on 22 September (UT) 2006 from the Uchinoura Space Center in Japan into a 600 km, Sun-synchronous, polar orbit. It was a Japanese-led mission with U.S., United Kingdom, European Space Agency, and Norway support. *Hinode* was designed to investigate magnetic activity on the Sun, including methods of magnetic field generation, transport, and dissipation. Of particular interest were the mechanisms responsible for heating the corona and for the production of solar flares and coronal mass ejections.

Japan's Institute of Space and Aeronautical Science built the spacecraft. Launched as a follow-up to *Yohkoh*, the 900 kg *Hinode* included three payload instruments: a 50 cm diameter optical telescope, a high resolution grazing incidence X-ray telescope, and an extreme ultraviolet (EUV) imaging spectrometer. The solar optical telescope had an unprecedented resolution of 0.2 arcsec, the X-ray telescope had a resolution three times better than *Yohkoh*, and the EUV imaging spectrometer had a sensitivity 10 times better than *SOHO* (*Solar and Heliospheric Observatory*).

Hinode identified the source region of the low speed solar wind for the first time, by measuring the plasma outflow from the edge of an active region next to a coronal hole. The spacecraft discovered that the chromosphere is permeated by Alfvén waves, which are sufficiently strong to accelerate the solar wind, and possibly to heat the quiet corona. *Hinode* also detected Alfvén waves in the corona and provided new information on how magnetic field lines may cross and reconnect to cause solar flares.

David Leverington

See also: Japan

Bibliography

Publications of the Astronomical Society of Japan Special Issue, 59, SP3 (30 November 2007).
Science Magazine Hinode Special Issue, 318 (7 December 2007).

Interplanetary Monitoring Platform Program

The Interplanetary Monitoring Platform (IMP) program included a network of 10 Explorer-type satellites, sometimes referred to as Interplanetary Monitoring Probes, launched by NASA between 1963 and 1973 primarily for continuous scientific monitoring of interplanetary plasma, cosmic rays, energetic particles, and Earth's distant magnetic field. The program, together with several other NASA satellites, enabled data collection during a complete, 11-year solar cycle.

Manufactured and operated by NASA's Goddard Space Flight Center in Greenbelt, Maryland, with contract support from Westinghouse Aerospace Division for system design, integration, assembly, and launch, IMP satellites had two standard shapes and sizes to accommodate a variety of instruments and experiments. The first model, which included seven satellites, had an octagonal body 2.3 ft wide × 1 ft high, an average weight of 143 lb, and silver-cadmium batteries, plus four solar panels that delivered 38 W of power. Three later IMP satellites featured 16-sided polyhedrons 4.5 ft wide × 5 ft high, weighing an average of 818 lb, with solar cells on their exteriors that delivered 110 W of power. All the satellites were spin stabilized and generally carried magnetometers, Geiger counters, ion and electron probes, plasma probes, and instruments for detection of electric fields, cosmic rays, and micrometeoroids. Engineers fitted two IMP satellites, known as Anchored IMP (AIMP) satellites, with solid-propellant rockets for deceleration into elliptical lunar orbit, but all the others occupied highly elliptical Earth orbits. Delta boosters carried eight IMPs into space from Cape Canaveral and two from Vandenberg Air Force Base, California.

The first mission, designated variously as *Explorer 18*, *IMP A*, or *IMP 1*, launched on 27 November 1963 and carried integrated circuits into orbit for the first time. Other IMP missions included *Explorers 21*, *28*, *34*, *41*, *43*, *47*, and *50*. The two AIMP missions, *Explorers 33* and *35*, which launched on 1 July 1966 and 19 July 1967 respectively, captured interplanetary plasma and geomagnetic measurements in the vicinity of the Moon. *Explorer 35*, otherwise known as *AIMP 2*, *AIMP E*, or *IMP E*, also obtained data on distribution of interplanetary dust, solar and galactic cosmic rays, Earth's magneto-hydrodynamic wake at lunar distances, and the Moon's gravitational field, ionosphere, and radiation environment. Although most of the IMP satellites had an operational life of two to six years, the last one, designated *Explorer 50, IMP J*, or *IMP 8*, went into space on 26 October 1973 and continued as an independent mission until October 2001, after which it operated as an adjunct to the Voyager and Ulysses missions, supplying parametric data on the solar wind.

In addition to providing astronauts with early warning of solar flares during Apollo lunar missions, the IMP satellite network enabled the first accurate measurements of interplanetary magnetism, Earth's magnetosphere, and the shock wave created by interaction of the solar wind with Earth's magnetic field. Data from IMP satellites enabled scientists to map in detail, for the first time, the shape of the magnetotail on Earth's night side.

Rick W. Sturdevant

See also: Goddard Space Flight Center

Bibliography

Goddard Space Flight Center, *Interplanetary Monitoring Platform: Engineering History and Achievements* (1980).

Goddard Space Flight Center. http://www.nasa.gov/centers/goddard/home/index.html.

OSO, Orbiting Solar Observatory

OSO, Orbiting Solar Observatory was a series of eight NASA missions flown during 1962–78 over more than a complete 11 year solar cycle. They observed principally in ultraviolet, X-ray and gamma-ray radiation, which are absorbed by Earth's atmosphere. The OSO program, managed by Goddard Space Flight Center, built on the discoveries of X-ray emissions by the Sun by sounding rockets as early as 1948.

Each OSO spacecraft followed the same basic design, consisting of a wheel section that rotated to stabilize the spacecraft and a despun sail section that pointed constantly at the Sun. Imaging instruments and the solar power cells were on the sail, while attitude control and other instruments were on the wheel section so they could scan the universe and the Sun. Ball Aerospace built the first seven, while Hughes built *OSO 8*.

The instrument suites were not identical across the program, but were increasingly advanced as engineers designed successively more sophisticated telescopes and detectors over the years. *OSO 1* carried 13 instruments, which measured solar radiation from visible through gamma-ray wavelengths, charged and neutral particles, and micrometeorites. Later spacecraft carried fewer, but higher-performing specialized instruments. Over time, experiments focused on higher-resolution studies to determine the location at which certain phenomena were occurring. Later missions included contributions from the United Kingdom, Italy, and France. *OSO 4* carried a solar X-ray telescope that produced low-resolution images of the Sun in soft X rays,

OSO Satellites and Instruments (Part 1 of 2)

Name	Operation Dates	Instruments	Comments
OSO 1	7 Mar 62–6 Aug 63	Thirteen instruments, including Ultraviolet (UV); X-ray and gamma ray monitors; dust particle experiment; solar visible, UV, gamma-ray, and high-energy gamma ray radiation spectrometers; neutron flux; proton-electron flux; Earth horizon sensor; emissivity stability; photoelectric sensor stability	Measured cosmic gamma radiation flux; detected rapid fluctuations in solar X-ray flux and correlation to Earth upper atmosphere temperature and UV radiation intensity.
OSO 2	3 Feb 65–29 Nov 65	UV and astronomical UV spectrometers, white light coronagraph, zodiacal light, high-energy and low-energy gamma-ray monitors, emissivity stability	Successful data return for 4,100 orbits.
OSO C	25 Aug 65		Launch failure.
OSO 3	8 Mar 67–10 Nov 69	X-ray telescope, gamma-ray telescope, particle detector, UV and X-ray spectrometers, solar X-ray detector, technological instrumentation	Detected flare episode from Sco X-1; observed changes in UV spectrum during solar flares and flare temperature of 30 million K.
OSO 4	18 Oct 67–7 Dec 71	Solar X-ray spectroheliograph, spectrometer, and monochromator; X-ray monitor; Geocorona hydrogen Lyman alpha telescope; UV spectrometer; Earth proton-electron telescope	First photographs of corona over entire face of solar disc
OSO 5	22 Jan 69–Jul 75	X-ray monitor and spectroscope, extreme UV spectroheliograph, solar X-ray spectroheliograph (United Kingdom—UK), zodiacal light telescope, solar far UV monitor, low-energy gamma ray scintillation detector	Measurement of X-ray bursts; obtained high-resolution data during a solar rotation.
OSO 6	9 Aug 69–Jan 72	X-ray telescope (Italy), UV scanning spectrometer, zodiacal light polarimeter, solar X-ray emission line spectrometer, solar UV polychromator (UK), high-energy neutron telescope	Study solar bursts and flares; obtained high-resolution solar data during a solar rotation.
OSO 7	29 Sep 71–9 Jul 74	White light and extreme UV coronagraphs, X-ray polarimeter, X-ray and extreme UV	X-ray all-sky survey, detected gamma rays from solar flares; obtained

(Continued)

OSO Satellites and Instruments (Part 2 of 2)

Name	Operation Dates	Instruments	Comments
		spectroheliograph, celestial X-ray telescope, gamma-ray spectrometer, cosmic and X-ray telescopes	high-resolution visible and extreme UV data from corona during a solar rotation.
OSO 8	21 Jun 75–1 Oct 78	Cosmic X-ray spectroscope, high energy and soft X-ray background experiments, cosmic X-ray spectrometer, mapping X-ray heliometer, high-resolution UV spectrometer (France), UV spectrometer, extreme UV detector, high-sensitivity crystal spectrometer and polarimeter	Detected black body spectrum from X-ray bursts; investigated UV and X-ray behavior at chromosphere/corona interface.

and a geocoronal Lyman-alpha telescope, which showed strong activity by hydrogen atoms in Earth's exosphere. The payload advances were also marked by a steady increase in spacecraft mass, from 209–290 kg for *OSO 1* through *6*, then 635 kg for *OSO 7* and 1,066 kg for *OSO 8*. A planned Advanced Orbiting Solar Observatory was canceled, but many of the instruments initially planned were built and flown as part of the Apollo Telescope Mount on the *Skylab* space station.

OSO observations showed that solar flares include hard X-ray events, and that flares can induce nuclear reactions, as indicated by gamma-ray emissions. In December 1971 *OSO 7* observations, coupled with those from ground-based observatories, led to the discovery of coronal mass ejections, the eruption of large bubbles of ionized hydrogen from the solar atmosphere into space. OSO spacecraft also observed numerous extra-solar events, such as a flare from Sco X-1, X-ray bursts and the spectra of these bursts, showing that they showed blackbody characteristics.

Dave Dooling

See also: Ball Aerospace and Technologies Corporation, Goddard Space Flight Center

Bibliography

Karl Hufbauer, *Exploring the Sun: Solar Science since Galileo* (1991).
David Leverington, *New Cosmic Horizons* (2000).

RHESSI, Reuven Ramaty High Energy Solar Spectroscopic Imager

RHESSI, Reuven Ramaty High Energy Solar Spectroscopic Imager, was a NASA Small Explorer Mission to study how solar flares were triggered and how they explosively released energy. Originally named HESSI, the mission was renamed in tribute to Dr. Reuven Ramaty (1937–2001), Goddard Space Flight Center astrophysicist and HESSI coinvestigator, who died a few months before its launch.

The spacecraft, which cost $85 million including the launch, was built by Spectrum Astro, Inc., and a Pegasus XL launched *RHESSI* into a circular low Earth orbit at an altitude of 600 km on 5 February 2002. The mission was managed by Space Sciences Laboratory of the University of California, Berkeley. The only instrument on board *RHESSI* was a high-resolution (2–36 arcseconds) imaging spectrometer capable of detecting X rays and gamma rays from 3 keV to 17 MeV with high temporal (tens of milliseconds) resolution. *RHESSI* operated in conjunction with other ground-based and space-based systems, such as *SOHO* (*Solar and Heliospheric Observatory*) and *TRACE* (*Transition Region and Coronal Explorer*). The primary mission objective was to study the basic physics of particle acceleration and explosive energy released in solar flares, in addition to other cosmic sources of X rays and gamma rays.

A *RHESSI* observation in December 2002 found that gamma-ray bursts were polarized, confirming that powerful magnetic fields were involved in their generation. In 2003 combined observations from *SOHO*, *TRACE*, and *RHESSI* confirmed that coronal mass ejections (CME) and solar flares were related phenomena, as *RHESSI* first detected X rays at the start of a flare, as a magnetic "lid" above the Sun's surface opened. *TRACE* then picked up extreme-ultraviolet signatures of a CME, and both it and *RHESSI* then detected typical flare ultraviolet and X rays. Finally *SOHO* monitored the CME moving away from the Sun. *RHESSI* observations confirmed that bursts of X rays occurred when magnetic field lines reconnected after stretching and breaking in a flare. Along with other spacecraft and ground observatories, *RHESSI* observed the most intense-to-date gamma-ray burst in December 2004, which together were able to pinpoint its genesis at the opposite end of the galaxy in a magnetar (a neutron star with extremely strong magnetic field), with much more powerful flare processes of the kind described above. The *RHESSI* spacecraft was designed to last for two years, but as of 2008, it continued to operate.

Virginia D. Makepeace and Stephen B. Johnson

See also: Goddard Space Flight Center

Bibliography

Goddard Space Flight Center RHESSI. http://hesperia.gsfc.nasa.gov/hessi/.

Robert P. Lin et al., "2005 Sun-Solar System Connections (S^3C) Senior Review Proposal for the Reuven Ramaty High Energy Solar Spectroscopic Imager (RHESSI)" (2005).

Skylab Solar Science

Skylab solar science experiments provided significant scientific results during *Skylab*'s crewed operations in 1973–74. Performing extensive observations of the Sun was one of the primary goals of *Skylab*, the first U.S. space station.

Astronauts operated eight solar science instruments that were mounted on the Apollo Telescope Mount (ATM), which was attached to *Skylab*'s side. The ATM featured its own separate set of solar panels to power the experiments. The instruments included two X-ray telescopes, an X-ray and extreme-ultraviolet camera, an ultraviolet spectroheliometer, an extreme ultraviolet and ultraviolet spectroheliograph, a white light coronagraph, and two hydrogen-alpha telescopes. The ATM was capable of pointing these experiments to within two arc-seconds of their targets on the surface of the Sun. *Skylab* also contained handheld X-ray and ultraviolet spectrograph for use during extravehicular activities.

Using *Skylab*'s solar instruments, astronomers conducted nearly continuous observations of the Sun, without the filtering effects of Earth's atmosphere. *Skylab* crew operations occurred during a period between a solar maximum and the subsequent solar minimum, allowing solar scientists to collect data on both quiescent and active solar behavior. Among the solar observations conducted on *Skylab* were observations of the structure of the solar chromosphere, short-term changes in sunspot structure, the structure and temperature characteristics of solar prominences, and short-scale changes in the structure of the inner corona. *Skylab* also studied coronal holes and confirmed that they were the source of the solar wind. In addition, *Skylab* achieved a number of "firsts" in solar science—the first observation that solar flares occur in small, hot, localized regions and are triggered by small "loops" in an active region, and the first observations of the early stages of a coronal mass ejection.

Michael Engle

See also: *Skylab*

Bibliography

John A. Eddy, *A New Sun: The Solar Results from Skylab* (1979).

SOHO, Solar and Heliospheric Observatory

SOHO, Solar and Heliospheric Observatory, was a cooperative project of the European Space Agency (ESA) and NASA to study the Sun (from its deep core to the outer corona), the solar wind, and the interaction with the interstellar medium. A consortium of European space companies led by prime contractor Matra Marconi Space (later European Aeronautic Defence and Space Astrium) built *SOHO* under ESA management, and 12 instruments were developed by international research groups. NASA launched *SOHO* on 2 December 1995, and it was inserted into a halo orbit around the L1 Lagrangian point in February 1996. NASA was responsible for communications and daily operations, while ESA had overall responsibility for the mission. The focal point for science planning and instrument operations was at NASA's Goddard Space Flight Center.

After ESA canceled two earlier solar mission proposals (GRIST—Grazing Incidence Solar Telescope—in 1981 and DISCO—Dual Spectral Irradiance and Solar Constant Orbiter—in 1983), their experiments were combined into the proposed *SOHO* mission. Recognizing the advantages of coordinating international solar observing efforts, in September 1983, ESA, NASA, and Japan's Institute for Space and Astronautical Science agreed to form the International Solar-Terrestrial Physics Program, to which ESA contributed the *SOHO* and Cluster missions.

SOHO provided the first images of structures and flows below the Sun's surface and of activity on the far side of the Sun. It discovered sunquakes and mapped the evolution of zonal, meridional, and other large-scale flow patterns below the surface. It eliminated uncertainties in the internal structure of the Sun as a possible explanation for the "neutrino problem" (the observed flow of neutrinos from the Sun is roughly half of theoretical predictions). It provided evidence for upward transfer of magnetic energy from the surface to the corona through a "magnetic carpet" (a weave of magnetic loops extending above the Sun's surface) and revealed an extremely dynamic solar atmosphere where plasma flows play an important role. It identified the source regions and acceleration mechanisms of the fast solar wind and discovered the large disparity between electron and ion temperatures in coronal holes. It revolutionized understanding of solar-terrestrial relations and dramatically boosted space weather forecasting capabilities by providing, in a near-continuous stream, a comprehensive suite of images covering the dynamic atmosphere, extended corona, and activity on the far side of the Sun, which can be detected by careful analysis of sound waves traveling from the far side through the solar interior to the front side where they are observed.

The *SOHO* teams set a new standard in providing images and results through the Internet, capturing the imagination of the science community and the public. *SOHO* images and movies became stock footage for news organizations around the world. As a by-product of the efforts to provide real-time data to the public, amateurs came to dominate in *SOHO*'s discovery of hundreds of Sun-grazing comets. *SOHO* observations were available in an open, online archive, a policy that by 2006 resulted in more than 2,500 papers in refereed journals, authored by more than 2,300 scientists worldwide.

Control of the spacecraft was lost in June 1998 due to an unfortunate series of events during a spacecraft maneuver, but restored three months later in a dramatic recovery operation. All 12 instruments were usable, most with no ill effects. Despite subsequent failures of all three gyroscopes (the last in December 1998), new gyroless control software, installed by February 1999, allowed the spacecraft to return to normal scientific operations, providing an even greater margin of safety for the spacecraft.

Bernhard Fleck, Joseph B. Gurman, and Stein V. H. Haugan

See also: Earth Magnetosphere and Ionosphere, European Space Agency, Goddard Space Flight Center, International Solar-Terrestrial Physics Science Initiative, Matra Marconi Space

Bibliography

B. Fleck et al., "The *SOHO* Mission," *Solar Physics* 162 (1995).
NASA *SOHO*. http://sohowww.nascom.nasa.gov/.

SOLAR A. *See Yohkoh.*

SOLAR B. *See Hinode.*

Solar Irradiance Missions

Solar irradiance missions consisted of two NASA missions: *ACRIMSAT* (*Active Cavity Radiometer Irradiance Monitor Satellite*) and *SORCE* (*Solar Radiation and Climate Experiment*). Both spacecraft were built by the Orbital Sciences Corporation.

ACRIMSAT carried just one instrument, the ACRIM III active cavity radiometer, designed to monitor the long-term trend in solar irradiance, the Sun's energy received by Earth. ACRIM I had been launched on the *Solar Maximum Mission* (*SMM*) spacecraft in 1980, producing solar irradiance data until 1989. ACRIM II had been launched on the *UARS* (*Upper Atmosphere Research Satellite*) in 1991, producing similar data until 2001. ACRIM III's design was an improved version of ACRIM II.

Richard Willson, of the Jet Propulsion Laboratory, and colleagues had shown in the 1980s, using ACRIM I, that solar irradiance was reduced by individual sunspots, but that it increased at sunspot maximum. It had been known for some time that there had been virtually no sunspots on the Sun between about 1645 and 1715, and during this period, Europe had experienced very cold weather in the so-called Little Ice Age. So there seemed to be a correlation between sunspots, solar irradiance, and Earth's climate, which it was vital to understand.

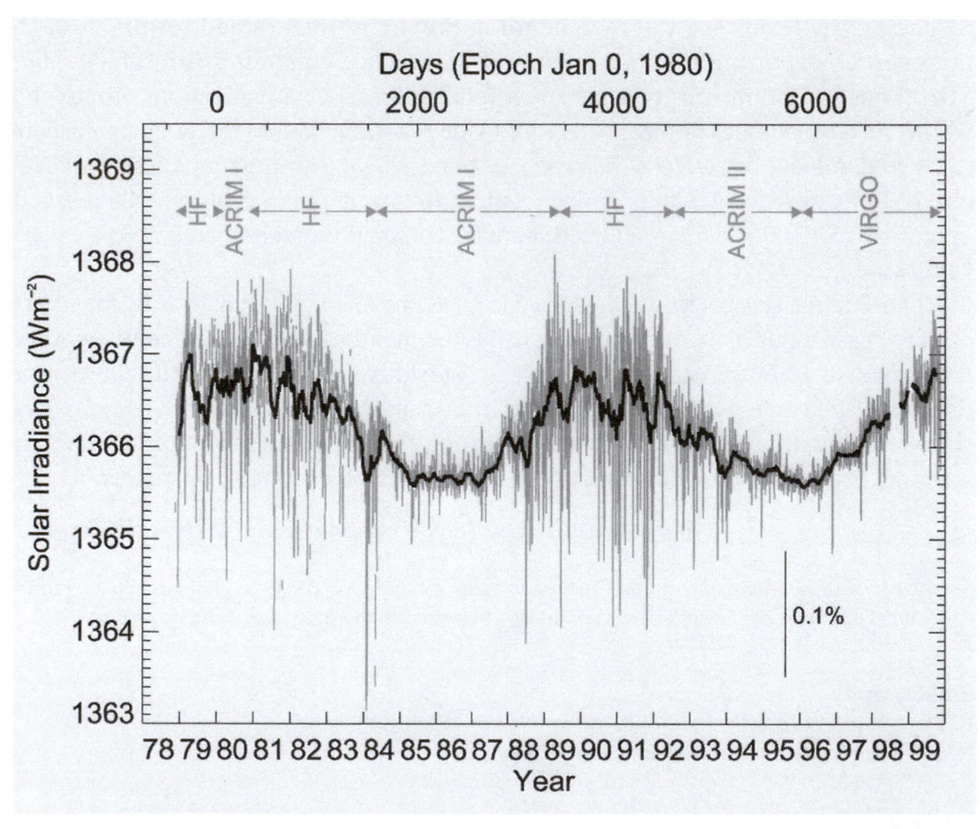

Variations in solar irradiance as measured by various spacecraft. The irradiance varies over an 11-year cycle, where maximum solar irradiance occurs at times of sunspot maximum. (Courtesy National Geophysical Data Center, Boulder, Colorado)

SORCE Instruments

Instrument	Wavebands (nm)	Expected Accuracy	Heritage
Total Irradiation Monitor (TIM)	All visible and infrared (IR)	Absolute irradiance 0.03 percent, relative 0.001 percent per year.	ACRIM III
Spectral Irradiance Monitor (SIM)	Ultraviolet (UV) to IR (200–2,000)	Irradiance accuracy 0.1 percent. Spectral resolution 0.25–33 nm.	New
Solar Stellar Irradiance Comparison Experiment (SOLSTICE)	UV (115–320)	Spectral resolution 0.1 nm. Absolute irradiance 5 percent, relative 0.5 percent.	SOLSTICE on *UARS*
XUV Photometer System (XPS)	X ray and UV (1–34)	Spectral resolution 5–10 nm. Absolute irradiance 20 percent, relative 4 percent.	Solar EUV Experiment on *TIMED, Thermosphere Ionosphere Mesosphere Energetics and Dynamics*

A comparison between the measurements made by ACRIM I and ACRIM II showed that those instruments could not measure solar irradiance accurately enough to determine long-term trends in absolute terms. The solar irradiance had only varied by about 0.1 percent during a solar cycle, and ideally absolute measurements should be made of an order of magnitude better than that, which ACRIM I and II could not do. The expected absolute accuracy of ACRIM III was 0.1 percent, with a stability of about 0.01 percent over the ultraviolet to infrared band from 0.2–2.0 μm.

The 115 kg *ACRIMSAT* spacecraft, with Richard Willson, then of Columbia University, as principle investigator, was launched by a Taurus launcher from the Vandenberg Air Force Base in December 1999. Unfortunately, it was put into the wrong orbit with the wrong spacecraft orientation, which took four months to correct. After that it produced accurate data from its 700 km Sun-synchronous, near-polar orbit.

The 290 kg *SORCE* spacecraft was launched by a Pegasus XL rocket air-dropped over the Atlantic Ocean in January 2003. It was put into a 645 km orbit inclined at 40° to Earth's equator. The spacecraft was operated by the University of Colorado. On board *SORCE* were four instruments (see table above), including a TIM instrument based on ACRIM III, but with considerable improvements. Two other instruments, SOLSTICE and XPS, gave more accurate information in more discrete frequency bands. In the early twenty-first century, scientists were working to produce an accurate record of solar irradiance since about 1975, using ACRIM I, II, and III and TIM instrument data and similar data produced by instruments on other, earlier spacecraft. These included the Hickey-Frieden radiometer on *Nimbus 7*, the Earth Radiation Budget Experiment on *ERBS* (*Earth Radiation Budget Satellite*), and the Variability of Solar Irradiance and Gravity Oscillations instrument on *SOHO* (*Solar and Heliospheric Observatory*).

David Leverington

See also: Jet Propulsion Laboratory, Orbital Sciences Corporation

Bibliography

ACRIM. http://www.acrim.com/.
NASA Headquarters. http://www.nasa.gov/centers/hq/home/index.html.

Solar Maximum Mission

Solar Maximum Mission, or *SMM*, also called *Solar Max*, was designed to observe the Sun, and solar flares in particular, during the 1980–81 solar maximum (a period, roughly 11 years apart, when there are a maximum of sunspots on the Sun). There were seven experiments on the 2,400 kg spacecraft, including the first telescope (HXIS) designed to image the Sun in high-energy X rays. A key feature of the spacecraft was the onboard coordination of instrumental observations in response to solar flares.

Mission Specialist George Nelson (at the upper right corner of the lower solar panel) with the Solar Maximum Mission *satellite prior to returning it to Space Shuttle* Challenger *for repair in April 1984. (Courtesy NASA/Marshall Space Flight Center)*

***Solar Max* Instruments**

Instrument	Acronym	Primary Purpose	Energy Range	Main Result
UVSP	Ultraviolet Spectrometer and Polarimeter	Image flares in far ultraviolet (UV)	115–360 nm	Detected plasma flows along magnetic loops of active regions
HXIS	Hard X-ray Imaging Spectrometer	Image flares at X-ray energies	3.5–30 keV	Showed that hard X rays are generated during impulsive phase of flares, brightening simultaneously at both ends of magnetic loop
C/P	White Light Coronagraph/ Polarimeter	Image corona out to dist. of 6 solar radii	440–660 nm	First high-resolution observations of coronal mass ejections
ACRIM	Active Cavity Radiometer Irradiance Monitor	Measure total irradiance from Sun to high accuracy	UV to infrared	See text
XRP	X-ray Polychromator	Measure soft X-ray spectrum of flares	0.14–2.2 nm	Showed that active regions extend well into the corona
HXRBS	Hard X-ray Burst Spectrometer	Measure hard X-ray spectrum of flares	20–300 keV	Along with HXIS, showed that hard X rays peak before soft X rays in flares
GRS	Gamma-ray Spectrometer	Measure gamma-ray spectrum of flares	0.01–160 MeV	First instrument to detect neutrons emitted by flares

Built and operated by Goddard Space Flight Center, *Solar Max* was the first spacecraft constructed using Fairchild's Multimission Modular Spacecraft bus, which was used on later missions, and could be retrieved by the Space Shuttle. *Solar Max* was launched into a low Earth orbit on 14 February 1980 by a Thor Delta. But less than a year later, the spacecraft lost its fine-pointing capability, reducing it to a non-imaging mission. *Solar Max* was eventually repaired in orbit in April 1984 by astronauts from the Space Shuttle *Challenger*, making it the first spacecraft to be retrieved, repaired, and redeployed in orbit. It collected data until November 1989, covering nearly an entire 11 year solar cycle.

Shortly after its launch, the ACRIM instrument detected a small reduction in the Sun's energy output when two large sunspot groups crossed the central solar meridian. Throughout the next few years the Sun's energy output was found to go through a minimum at solar minimum in data produced by both the *Solar Max* and *Nimbus 7* spacecraft. So although the Sun's energy decreased when individual sunspot groups crossed the solar meridian, it also decreased as the Sun approached sunspot minimum.

Solar Max instruments showed that electrons and protons are accelerated simultaneously in solar flares. Hard X rays are emitted in short bursts, brightening simultaneously at both ends of the magnetic loop containing the flare. Then soft X rays are produced during a longer period of time. The hard X rays are emitted from a region just above the Sun's visible surface, whereas the soft X rays are emitted well into the solar corona.

Solar Max allowed scientists to establish correlations between the behavior of various solar phenomena, including coronal mass ejections, eruptive prominences, and solar flares.

David Leverington

See also: Goddard Space Flight Center, Space Shuttle

Bibliography

Andrew Chaikin et al., *Sky and Telescope* (June 1984): 494–503.

Karl Hufbauer, *Exploring the Sun: Solar Science since Galileo* (1993).

Keith T. Strong et al., eds., *The Many Faces of the Sun: A Summary of the Results from NASA's Solar Maximum Mission* (1999).

SolRad Program

The SolRad program (solar radiation satellite) was one of the earliest and longest-running U.S. scientific satellite programs, monitoring the Sun from outside Earth's atmosphere. While best remembered for having a classified secondary mission as the first electronic intelligence (ELINT) satellite, called *GRAB* (*Galactic Radiation and Background*), SolRad was a successful program that included 10 working satellites orbited (in addition to one inert test payload, three launch failures, and one canceled mission) from 1960–76.

The SolRad/GRAB program was developed by the Naval Research Laboratory (NRL), which had a longtime interest in monitoring solar activity due to its influence on the ionosphere, and hence on radio communications. In 1958, when GRAB was conceived, the NRL's Space Science Division was developing the SolRad concept. Combining SolRad and GRAB saved money and gave the ELINT payload a "cover" mission. SolRad was the first satellite built by the NRL Satellite Techniques branch. Of the 13 SolRads launched, five carried GRAB payloads, and six carried a follow-on intelligence system, code-named POPPY.

The initial SolRad satellite design was highly efficient, incorporating the two missions in one simple, affordable, spin-stabilized spacecraft. The satellite bus was a sphere, 51 cm in diameter, from the Vanguard satellite program. Radiation detectors for the X-ray and Lyman-alpha bands, plus the radar detectors, were fitted in and on the 19 kg satellite. SolRad satellites of the 1970s used a 12-sided cylindrical bus, weighed up to 181 kg, and carried more sophisticated experiments. Their orbits varied from 25.6 to 70.1 degrees of inclination and from low (400–600 km) to high altitudes.

After a SolRad mass simulator was launched as a test on 13 April 1960, the series formally began with the launch of *SolRad 1* along with *Transit 2A* on a Thor Able-Star from Cape Canaveral on 22 June 1960, the first successful launch of two

SolRad Satellites and Instruments (Part 1 of 2)

Satellite	Launch Date	Instruments	Comments
Unnumbered SolRad mass simulator	13 Apr 1960	None	Orbited successfully
SolRad 1 (aka *GRAB 1*)	22 Jun 1960	Two ion chambers, one for the ultraviolet (UV) region and one for the X-ray region, plus GRAB intelligence payload	Successful data return. Launched with *Transit 2A*, the first successful launch of two spacecraft with one launcher by any nation.
SolRad 2	30 Nov 1960	Two ion chambers, one for the UV region and one for the X-ray region, plus GRAB intelligence payload	Launch failure
SolRad 3 (aka *GRAB 2*)	29 Jun 1961	Two X-ray region ion chambers, solar aspect sensor, plus GRAB intelligence payload	Failed to separate from *INJUN 1* satellite and returned limited data.
SolRad 4 (aka *SolRad 4A*)	24 Jan 1962	Eight ion chambers, four for the UV region and four for the X-ray region, plus GRAB intelligence payload	Launch failure
SolRad 4B	26 Apr 1962	Eight ion chambers, four for the UV region and four for the X-ray region, plus GRAB intelligence payload	Launch failure
SolRad 5	Canceled		Canceled
SolRad 6 (aka *SolRad 6A*, *POPPY 2*)	15 Jun 1963	Six ion chambers covering wavebands in the X-ray region, plus POPPY intelligence payload	Because of short-lived orbit during minimal solar activity, reentered 1 Aug 1963 and returned limited data.
SolRad 7A (aka *POPPY 3*)	11 Jan 1964	Five ion chambers covering wavebands in the X-ray region, plus POPPY intelligence payload	Successful data return

(Continued)

SolRad Satellites and Instruments (Part 2 of 2)

Satellite	Launch Date	Instruments	Comments
SolRad 7B (aka *SolRad 6B*, *POPPY 4*)	9 Mar 1965	Six ion chambers covering wavebands in the X-ray region, plus POPPY intelligence payload	Successful data return
SolRad 8 (aka *Explorer 30*)	18 Nov 1965	Two X-ray region Geiger counters, two ion chamber photometers. First SolRad with onboard data memory (failed) and the first with active attitude control to correct precession of the spin-stabilized axis	Successful data return for first nine months, deteriorating for remainder of mission as active stabilization system failed.
SolRad 9 (aka *Explorer 37*)	5 Mar 1968	Five X-ray region ion chambers, one X-ray region Geiger counter, onboard data memory	Successful data return
SolRad 10 (aka Explorer 44)	6 Jul 1971	Seven X-ray region ion chambers, one scintillation counter, two background ion chambers pointed away from Sun, onboard data memory	First SolRad with spin axis pointed to Sun instead of along Sun-satellite line. Successful data return.
SolRad 11A	14 Mar 1976	25 experiments measuring gamma ray bursts, Earth auroral and stellar X-rays, solar charged particle and electromagnetic emissions, extreme UV emission, background (anti-solar direction) X-ray and charged particle emissions	*SolRad 11A* and *11B* placed 180 degrees apart in 120,000-km circular orbit. Successful data return until telemetry system failure in Jun 1977.
SolRad 11B	14 Mar 1976	25 experiments (same as *SolRad 11A*)	*SolRad 11A* and *11B* placed 180 degrees apart in 120,000-km circular orbit. Successful data return, although some experiments failed in Dec 1976.

satellites with one launch vehicle. *SolRad 1* was the world's first orbiting astronomical observatory, and it quickly confirmed the theory that radio communications "fadeouts" were caused by solar X-ray emissions. Subsequent SolRads increased scientists' knowledge of the Sun, establishing the normal energy flux during the absence of solar flares and the effects of all types of solar radiation on Earth's ionosphere. Another example is that *SolRad* data combined with that from other satellites established that large solar eruptions had higher temperatures and shorter wavelengths than small ones.

Most *SolRad* missions used low Earth orbits inclined from 51 to 70.1 degrees. The designations for missions before *SolRad 8* (launched 16 November 1965) were sometimes changed before or after launch, leading to some confusion. Three (*SolRads 8*, *9*, and *10*) were also considered part of NASA's Explorer missions program and had Explorer designations. *SolRad 11A* and *B* were the last two of the series, and the only two ever launched together. These were orbited by a Titan 3C from Cape Canaveral on 15 March 1976. The *SolRad 11A* and *B* mission was also notable for placing its satellites in low-inclination, high-altitude (120,000 km) orbits, spacing them 180 degrees apart to provide for stereoscopic observations. Each carried an identical suite of 25 experiments to monitor the Sun's X-ray, ultraviolet, and energetic particle emissions.

Matt Bille

See also: GRAB, United States Navy

Bibliography

S. W. Kahler with R. W. Kreplin, "The NRL SolRad X-ray Detectors: A Summary of the Observations and a Comparison with the SMS/GOES Detectors," *Solar Physics* 133, no. 71 (1991).

Robert A. McDonald with Sharon K. Moreno, "Grab and Poppy: America's Early ELINT Satellites," National Reconnaissance Office (2005).

Naval Research Laboratory, "SolRad/GRAB," fact sheet (n.d.).

David Van Keuren, "Cold War Science in Black and White: U.S. Intelligence Gathering and Its Scientific Cover at the Naval Research Laboratory, 1948–62," *Social Studies of Science* 31, no. 2 (2001).

SORCE, Solar Radiation and Climate Experiment. *See* Solar Irradiance Missions.

Spartan 201

Spartan 201 (derived from Shuttle Pointed Autonomous Research Tool for Astronomy) was a Space Shuttle–based, free-flying solar observatory used on five Shuttle missions from 1993 to 1998. Manufactured by Swales, Inc., and managed by Goddard Space Flight Center, the *Spartan* satellite was developed specifically to take advantage of the capabilities offered by the Space Shuttle—frequent access to space and the ability to deploy, retrieve, and return spacecraft to Earth for repairs and refurbishment. The ton-and-a-half *Spartan* featured a system for storing scientific data on board (which would then be downloaded once *Spartan* was returned to Earth) and hardware to mount it in the Shuttle's payload bay. *Spartan* could be reconfigured with different scientific instruments and during its operational lifetime was configured not only as a solar observatory but also as a platform for ultraviolet astronomy and as a test platform for an inflatable space antenna. In addition to the five *Spartan 201* missions, the *Spartan* flew on three other Space Shuttle flights in different configurations. Astronauts deployed *Spartan* using the Shuttle's robot arm. During its free flight, *Spartan* would observe the Sun, planets, stars, and other astronomical targets. It was then retrieved with the Shuttle's robot arm and returned to Earth.

The *Spartan 201* series was devoted specifically to solar studies, logging nearly 300 hours of solar observations. The primary objectives of the solar observations conducted by *Spartan 201* were to study solar wind physics (by observing hydrogen, proton, and electron temperature and densities) and to study the structure of the solar corona. *Spartan 201* used two primary instruments to gather data about the Sun—the ultraviolet coronal spectrometer and the white light coronagraph. The data returned by *Spartan 201* was still being analyzed in 2008 by solar scientists, but one important result was that solar coronal hole profiles become less complex approaching solar minimum.

Michael Engle

See also: Space Shuttle

Bibliography

Goddard Space Flight Center Projects Directory, *Spartan 201*: NASA's Mission to Explore the Sun's Corona website.

STEREO, Solar Terrestrial Relations Observatory

STEREO, Solar Terrestrial Relations Observatory, was the third mission in NASA's Solar Terrestrial Probes (STP) program. The main objective of the STP program was to understand the fundamental physical processes operating in the space environment from the Sun, through the solar system, to the interstellar medium. STEREO's mission was to understand the causes and mechanisms of coronal mass ejection (CME) initiation, discover the mechanisms and sites of particle acceleration in the low corona and interplanetary medium, and enable three dimensional models to be produced of the solar wind. To do this, two 630 kg STEREO spacecraft were placed into heliocentric orbits ahead of (*STEREO A*) and behind (*STEREO B*) Earth, thus enabling production of stereoscopic images.

The two STEREO spacecraft, which had been built by Johns Hopkins University's Applied Physics Laboratory, were launched by a Delta II on 26 October 2006 into highly elliptical geocentric orbits. On 15 December mission controllers used the Moon's gravity to inject *STEREO A* into a heliocentric orbit inside that of Earth, which meant that it orbited the Sun faster than did the Earth. As a result *STEREO A* was ahead of Earth in its orbit at a gradually increasing distance. A month later mission controllers used the Moon's gravity to inject *STEREO B* into a heliocentric orbit outside that of Earth, so it was behind the Earth at a gradually increasing distance. The angle *STEREO A*/Sun/*STEREO B* increased at about 44° per year, with Earth approximately halfway between the two spacecraft.

The instruments on both STEREO spacecraft included an extreme ultraviolet imager, two white light coronagraphs to image the solar corona, and two heliospheric imagers to image the space between the Sun and Earth. Other instruments observed the solar wind and radio disturbances traveling away from the Sun.

Simultaneous observations with *STEREO A*, *Wind*, and *Advanced Composition Explorer* in the first half of 2007 showed that *STEREO A* had detected ions streaming off Earth's bow shock "upstream" of Earth up to 4,000 Earth radii away. STEREO provided three-dimensional images of coronal loops, polar coronal jets, and CMEs. The polar coronal jet images of 7 June 2007 clearly showed a helical structure for the first time, helping astronomers to understand the way polar coronal jets were produced. In April 2007 *STEREO A* imaged the collision between a CME and Comet Encke, which caused the comet's tail to be detached. A new tail was visible a few days later.

David Leverington

See also: Johns Hopkins University Applied Physics Laboratory

Bibliography

NASA STEREO. http://www.nasa.gov/centers/hq/home/index.html.
"The Report of the Science Definition Team for the STEREO Mission" (1997).
"STEREO: Beyond 90 Degrees. A Proposal to the Senior Review of Heliophysics Operating Missions" (February 2008).

TRACE, Transition Region and Coronal Explorer

TRACE, Transition Region and Coronal Explorer, was a small spacecraft carrying a high-resolution extreme ultraviolet telescope. Managed and built by Goddard Space Flight Center (GSFC), it was launched on 2 April 1998 on a Pegasus-XL rocket into a Sun-synchronous, low Earth orbit, the first U.S. solar science satellite since the 1980 launch of *Solar Maximum Mission*. A unique aspect of the *TRACE* telescope was that the primary and secondary mirrors had different coatings in each quadrant to optimize performance for different parts of the ultraviolet and extreme-ultraviolet spectrum. The imaging detector was a Charge-Coupled Device, which with the telescope's capabilities provided a spatial resolution of 1 arc-second. The field of view was 8.5 arc-minutes, about 25 percent of the width of the solar disk, so images often were presented as highly detailed composites of the solar disk.

The ultraviolet and extreme ultraviolet wavelengths observed by *TRACE* were selected to correspond with emissions by hot iron and carbon. Images taken in the longest waveband showed the effects of the 6,000 K photosphere, while those in the shortest waveband imaged the multimillion degree corona. *TRACE* mission objectives were to follow the evolution of magnetic field structures from the solar interior to the corona, to investigate the mechanisms of the heating of the chromosphere and corona, and to investigate the triggers and onset of solar flares and mass ejections. *TRACE* observations often were made in coordination with other solar spacecraft, such as the *SOHO* (*Solar and Heliospheric Observatory*), and ground-based observatories to provide complementary data sets. *TRACE*'s ability to take images about every five seconds produced high-resolution movies showing the build-up and release of energy in solar flares and active regions. Because high-energy emissions are linked to gases constrained inside magnetic fields, *TRACE* images outlined the shapes of magnetic structures in active regions as beautiful arcs that expand and brighten during the life of a flare. These helped in building the "magnetic carpet" concept of a fine-scale magnetic structure of the Sun's photosphere.

Dave Dooling

See also: Goddard Space Flight Center

Bibliography

GSFC *TRACE*. http://trace.lmsal.com/.

Ulysses

Ulysses was the first space probe to explore the region of space above the Sun's poles, as a joint project of the European Space Agency (ESA) and NASA. The spacecraft was provided by ESA and the launch by NASA. The idea to send a probe out of the ecliptic (the plane in which the planets orbit the Sun) and over the Sun's poles was first put forward by a group of U.S. scientists in 1959, just two years after the launch of *Sputnik*. Even though the scientific importance of an out-of-ecliptic mission was recognized early in the space age, the technical means to accomplish it became available only in the early 1970s as a result of knowledge gained through the Apollo program and by sending probes to the outer planets.

Planning for *Ulysses* (first called the Out-of-Ecliptic mission and later International Solar Polar Mission—ISPM) began in 1974. The idea was to launch two spacecraft from the Space Shuttle toward Jupiter and to use its gravity to swing them far out of the ecliptic. The two probes, each carrying a suite of scientific instruments, were to orbit the Sun in opposite directions. Launch was foreseen for February 1983, with ESA and NASA each providing one of the probes. The ESA spacecraft was built by a European consortium led by West Germany's Dornier Systems and carried nine instruments: a pair of magnetometers to measure the strength and direction of the heliospheric magnetic field; a suite of sensors to measure the bulk properties of the solar wind, including its elemental and charge-state composition; a suite of energetic particle and cosmic ray detectors; an instrument to measure the natural radio emission from the Sun and planets; a gamma-ray burst detector; instruments to measure interplanetary and interstellar dust grains and neutral interstellar helium atoms.

In 1980 the project received its first setback when, as a result of difficulties with the development of the Space Shuttle, NASA announced a launch delay of two years. Work on the ESA spacecraft was already well advanced and so ESA decided to continue with its development and integration and then, following completion in 1983, to store the spacecraft until the new launch date. In February 1981 funding problems led NASA to cancel its spacecraft and delay the launch of the ESA spacecraft by one more year until May 1986. These decisions, which were taken without consultation with ESA, constituted a unilateral breach of the ISPM Memorandum of

Understanding and soured NASA-ESA relations. An obvious consequence of the change from a dual- to a single-spacecraft mission was the loss of half of the scientific payload, including a number of European experiments. Enough of the original unique scientific objectives remained, however, that ESA decided to continue development of its spacecraft. The launch disruptions caused by the loss of Space Shuttle *Challenger* in January 1986 resulted in a further launch delay.

Finally on 6 October 1990 *Ulysses* began its journey on board Space Shuttle *Discovery.* Following a Jupiter flyby in 1992 *Ulysses* passed over the Sun's polar regions three times, first in 1994–95 and again in 2000–2001 and 2006–8. The first and third polar passes occurred close to solar activity minimum and revealed a relatively simple heliosphere dominated by fast solar wind streams originating in the polar regions. In contrast, the second polar passes occurred near solar maximum, with complex structure found at all latitudes. An important difference between the two solar minimum passes was the reversal of the Sun's magnetic field that occurred in 2001 and its effect on the charged cosmic ray particles.

From its unique orbit inclined at 80 degrees to the ecliptic, *Ulysses* gave scientists important new information about the nature of the Sun's magnetism; the creation and behavior of the solar wind, cosmic rays, and energetic charged particles; the dust and gas that surround Earth's solar system; and some of the most energetic events in the known universe, gamma-ray bursts. In addition to its scientific successes, *Ulysses* also exemplified the benefits and pitfalls of international cooperation in space exploration.

Richard G. Marsden and Edward J. Smith

See also: European Space Agency, Space Science Policy, Space Shuttle

Bibliography

André Balogh et al., *The Heliosphere Near Solar Minimum: The Ulysses Perspective* (2001).

Roger M. Bonnet and Vittorio Manno, *International Cooperation in Space: The Example of the European Space Agency* (1994).

Yohkoh

Yohkoh carried X-ray imaging and spectroscopy instruments to study high-energy activities across the entire solar disk, particularly during solar flares. The Earth-orbit satellite, designed, built, and operated by Japan's Institute of Space and Astronautical Sciences, was called Solar A during development and was renamed *Yohkoh* (Sunbeam) after its successful launch on 30 August 1991. It provided high-quality images until December 2001, when the spacecraft's batteries were drained in the aftermath of a faulty response to a solar eclipse.

Yohkoh's soft X-ray telescope (SXT), contributed by the United States, used grazing incidence optics to focus X rays in the 0.1–4 keV energy range to produce images on a 1,024 × 1,024 pixel charge-coupled detector. A multigrid telescope (acting like a pinhole camera) produced images in the hard X-ray region, 20–80 keV. The two spectrometers, developed in the United Kingdom, produced continuous spectra of the solar disk in the 3 keV–20 MeV range (X rays through gamma rays), and the spectral lines of highly ionized iron, silicon, and calcium. While the latter atoms make up a small component of the Sun's composition, they act as highly visible tracers of solar activity because specific wavelengths correspond to specific temperatures. Also, because the solar disk does not emit at such high energies, these bands allow scientists to observe the entire solar corona continuously in X rays.

Yohkoh provided many remarkable observations. One was the formation of bright S-shaped structures as precursors to coronal mass ejections, which can hurl immense quantities of ionized gas across the solar system. The S shape indicates regions where magnetic energy is concentrated with opposing polarities and may cause an eruption. *Yohkoh* also provided new information about the structure of solar flares, finding high-energy X rays at the bottom of solar flare loops and also at the top, probably at magnetic line reconnection locations. *Yohkoh*'s long life also allowed scientists to track the evolution of the corona across nearly an entire 11 year sunspot cycle.

Dave Dooling

See also: Japan

Bibliography

David Leverington, *New Cosmic Horizons: Space Astronomy from the V2 to the Hubble Space Telescope* (2000).
JAXA *Yohkoh*. http://trace.lmsal.com/.

SPACE SCIENCE ORGANIZATIONS

Space science organizations played a variety of roles in space science research throughout the world. Universities and scientific professional organizations developed and managed the science aspects of space missions and turned scientific data from these missions into researched publications. Government operational groups allocated funding, provided oversight, and managed spacecraft operations for space-based scientific research missions. Government advisory groups guided governments on priorities for space science research funding. Advocacy groups promoted specific missions or technology.

In ancient times priests (as in ancient Babylonia) or natural philosophers (in classical Greece and early modern Europe) were among the first to make long-term observations and predictions of the night sky. As modern society developed, universities, government-sanctioned organizations, and private philanthropists supported space science by building observatories and funding research. Government financial support spawned the need for advisory groups to set priorities and research standards. The dawn of the space age allowed scientists to place instruments outside Earth's atmosphere. Many new organizations developed in response to the need for space-based satellite development, operations and data analysis, and governmental advice.

Government advisory groups set standards for research and record keeping and provided forums for scientists to determine priorities for space science missions. Many of them became national representatives to the International Astronomical

Union (IAU) upon its creation in 1919. The IAU's responsibilities included assigning names to astronomical bodies and surface features; providing guidance on issues such as coordinate systems, timekeeping, and data processing; and sponsoring symposia to harmonize international research efforts. The Royal Society of London was founded in 1660, and from 1919 to the early twenty-first century was the United Kingdom's representative to the IAU, after which time, the Royal Astronomical Society (founded 1820) took over that responsibility. The French Académie des Sciences was established in 1666 and eventually became France's representative to the IAU. Similarly the Russian Academy of Sciences was established by Peter the Great in 1724 and later represented Russia in the IAU. The U.S. Congress established the National Academy of Sciences (NAS) in 1863 to advise the U.S. government on scientific issues. The NAS established its Space Science Board in 1958 to recommend priorities for space-based scientific research. By the early twenty-first century NAS boards were involved in planning most U.S. science missions. The National Academies, which came into being in 1964 with the creation of the National Academy of Engineering, was the U.S. national member of the IAU.

Nations with space programs had government organizations to manage funding and operations of space-based science programs. NASA, the oldest, largest, and most influential government space organization in the world, had its Science Mission Directorate act as the primary funding agent, manager, and sometimes developer of U.S. scientific satellites. This directorate was created as the Office of Space Science and Applications in 1958, until merging in 2004 with the Earth Science enterprise. The Space Science Advisory Committee provided guidance on strategic goals and objectives. The Planetary Protection Advisory Committee, established in 2001, provided guidance on protecting Earth and solar system bodies from biological contamination of one another. The American Association for the Advancement of Science, created in 1848, became an advisory body for NASA and provided a forum for NASA scientists to present their research. The Earth System Science and Applications Advisory Committee advised the NASA administrator and advisory council on strategic goals for Earth science.

Many other nations and groups of nations created organizations similar to NASA. The European Space Research Organisation (ESRO) was the first intergovernmental space science organization for Europe, established in 1964. Its goal was to provide a mechanism for West European nations to collaborate on space science programs. The European Space Agency (ESA), created in 1975, was formed from ESRO, and support of science programs became mandatory for ESA Member States. Its Science Advisory Committee (later its Space Science Advisory Committee) and Science Programme Committee provided direction for ESA's space science program. Long-term programs developed by these groups included Horizon 2000, Horizon 2000+, and Cosmic Vision 2020, and they produced missions such as Hipparcos and XMM-Newton (X-ray Multi Mirror-Newton). Both inside and outside Europe, many nations have their own government organizations that support space science, such as India's Indian Space Research Organisation and Japan's Institute of Space and Astronautical Science.

Government operational groups operate scientific spacecraft. In 2008 two NASA field centers operated most U.S. science satellites: the California Institute of Technology's Jet

Propulsion Laboratory (JPL) and NASA's Goddard Space Flight Center (GSFC). JPL started as a rocket research organization but became the facility in charge of planetary exploration in 1959. By the early twenty-first century it operated all U.S.-controlled planetary exploration satellites and rovers. GSFC was responsible for Earth-orbiting spacecraft, researching the near-Earth environment and developing space-based astronomical observatories, the first of which was the 1968 *Orbiting Astronomical Observatory 1*.

Universities historically supported space science through education and research, with the research primarily funded by government institutions. Universities typically provided scientific instruments and analyzed the scientific data, but occasionally designed entire spacecraft, such as JPL, which became a de facto NASA field center because of its expertise. The need to work together to establish large observatories accessible to their staff and students led seven U.S. universities to form AURA, the Association of Universities for Research in Astronomy in 1957. The goal of AURA was to create ground-based astronomical observing facilities accessible to all members, independent of private sponsorship. AURA built and managed several ground-based observatories and became involved in space-based research in 1981, when it established the Space Telescope Science Institute at Johns Hopkins University to manage the science operations of the *Hubble Space Telescope* (*HST*). It also garnered responsibility to manage operation of the *James Webb Space Telescope*, the *HST* replacement.

Private advocacy groups harness public support for space science and sometimes fund research. The British Interplanetary Society, the oldest such organization, was founded in 1933 to promote astronautics and space exploration. Another important example, the Planetary Society was founded in 1980 to harness public support for planetary exploration and the search for extraterrestrial life in order to influence public policy and sponsor its own research program. This organization promoted the use of robotic rovers as planetary explorers and advocated funding for New Horizons, an explorer mission to Pluto launched in 2006. The Planetary Society published scientific reports, new spacecraft designs, and policy papers in the bimonthly *Planetary Report*.

Finally, many private companies developed space science expertise so as to build scientific instruments for spacecraft, and to provide expertise for developing proposals for scientific missions. In some larger companies these grew to become full-fledged space science or instrument organizations within the company, whereas some small companies were founded as science instrument designers and manufacturers.

The combined efforts of worldwide space science organizations have shaped public understanding of the universe and priorities for space science research.

Kathleen Mandt

See also: Government Space Organizations, Space Advocacy Groups

Bibliography

International Astronomical Union. http://www.iau.org/.
International Dark-Sky Association. http://www.darksky.org.
NASA. http://www.nasa.gov/.

National Academies. http://www.nationalacademies.org/.
Planetary Society. http://www.planetary.org/home/.
Royal Astronomical Society. http://www.ras.org.uk/.
Smithsonian Institution. http://www.si.edu/.
VSOP. http://www.vsop.isas.ac.jp/.

AURA, Association of Universities for Research in Astronomy

AURA, Association of Universities for Research in Astronomy is a consortium of academic universities and nonprofit organizations designed to conduct stellar, solar, and planetary science research. During the early 1950s private institutions, such as the Rockefeller Foundation and the Carnegie Institution, owned the largest optical telescopes in the United States, including the Mount Wilson Observatory and Palomar Observatory in California, so astronomers not affiliated with these private observatories did not have access to these large telescopes. Following the establishment of the National Science Foundation (NSF) in 1950, scientists and astronomers from less favored academic institutions noted that they could utilize funds from the NSF to construct new observatories. In 1957 seven universities entered into a cooperative agreement with the NSF to form AURA in an effort to serve the interests of astronomers. AURA proceeded to build and operate various observatories, including the Kitt Peak National Observatory (KPNO) in Arizona and Cerro Tololo Inter-American Observatory (CTIO) in Chile. By 1976 AURA was also responsible for managing the operations of the Sacramento Peak Observatory (SPO) in New Mexico, which involved monitoring solar activities. In 1981 NASA selected AURA to create the Space Telescope Science Institute at Johns Hopkins University, which managed the operation of and data collected by the *Hubble Space Telescope*. NASA planned for it to perform the same function for the James Webb Space Telescope. In 1982 AURA consolidated the astronomical research programs of the National Solar Observatory, SPO, CTIO, and KPNO to form the National Optical Astronomy Observatory (NOAO).

Following the construction of the Gemini Observatory by an international consortium consisting of the United States, United Kingdom, Canada, Chile, Australia, Argentina, and Brazil, AURA was selected to manage the operations of the twin telescopes located in Hawaii and Chile, as part of NOAO. In January 2001 AURA established the New Initiative Office to monitor the development of a Giant Segmented Mirror Telescope.

Kevin Brady

See also: *Hubble Space Telescope*

Bibliography

Association of Universities for Research in Astronomy. http://www.aura-astronomy.org/.
Frank K. Edmondson, *AURA and Its U.S. National Observatories* (1997).
W. Patrick McCray, *Giant Telescopes: Astronomical Ambition and the Promise of Technology* (2006).

Goddard Space Flight Center

The Goddard Space Flight Center (GSFC) located in Greenbelt, Maryland, serves as the lead NASA center for developing and operating robotic, Earth-orbiting satellites.

Before the establishment of NASA in October 1958, Senator J. Glenn Beall of Maryland announced that the federal government planned to construct a laboratory in Greenbelt for space research. Following the creation of NASA, personnel from Project Vanguard and scientists from the Naval Research Laboratory transferred to the Beltsville Space Center, which initially oversaw the operations of the Space Task Group at the Langley Research Center in Virginia. In May 1959 NASA renamed the facility Goddard Space Flight Center in honor of American rocket pioneer Robert H. Goddard.

During the 1960s, GSFC developed a variety of weather and communications satellites including the Tiros and the Echo satellite. By the mid-1960s, GSFC scientists focused their research on astronomy satellites, which culminated in the launch of the Orbiting Astronomical Observatory in 1968. Aside from conducting satellite research, GSFC also served as the hub for a tracking-and-communications network known as the Spaceflight Tracking and Data Network, consisting of more than 20 different international sites designed to monitor crewed and robotic spacecraft for NASA. GSFC also managed the Thor-Delta booster program.

As the Apollo program concluded in 1972, GSFC experienced a reduction of its workforce and suffered from budgetary constraints. Despite these setbacks, personnel conducted Earth science research and developed new satellites. In 1972, GSFC scientists placed the first Earth Resources Technology Satellite in orbit. Four years later GSFC continued its efforts to explore the universe by launching the International Ultraviolet Explorer into space.

Following the development of the Space Shuttle in the early 1980s, the new reusable orbiter altered the way in which GSFC personnel designed and built satellites. By utilizing the Space Shuttle's payload bay as a way of carrying satellites into orbit, GSFC engineers were no longer constrained by the weight and size limitations associated with using expendable rockets to launch satellites. With this advantage, GSFC personnel developed the Compton Gamma Ray Observatory, which weighed more than 17 tons.

Meanwhile, budgetary constraints caused NASA to consider closing its Wallops Station in Virginia, one of the oldest launch facilities. GSFC had previously utilized the research station for its sounding rocket projects and aircraft programs, so the space agency saved the Wallops Island facility by consolidating it with GSFC in 1982. The Wallops Flight Facility continued to be a vital part of GSFC by supporting suborbital programs.

By 1993, GSFC personnel assisted a team of astronauts and engineers from Johnson Space Center in Houston, Texas, to develop and implement a plan to repair the *Hubble Space Telescope*. The hard work paid off, as the servicing mission was a success. GSFC continued to expand scientists' knowledge of Earth by launching satellites into orbit including *Landsat 7*, *QuikScatterometer*, *Terra*, Polar-Orbiting Operational Environmental Satellite Program spacecraft, and the Geostationary Operational Environmental Satellites.

Kevin M. Brady

See also: National Aeronautics and Space Administration

Bibliography

NASA Goddard Space Flight Center. http://www.nasa.gov/centers/goddard/home/index.html.

Alfred Rosenthal, *Venture into Space: Early Years of Goddard Space Flight Center* (1985).

Lane E. Wallace, *Dreams, Hopes, Realities: NASA's Goddard Space Flight Center: The First Forty Years* (1999).

International Astronomical Union

The International Astronomical Union (IAU) was created in 1919 to foster international cooperation in astronomy, under the auspices of the International Research Council, organized in 1919 as a forum for the exchange of scientific and technical knowledge. The IAU evolved from the International Union for Cooperation in Solar Research, which held its first meeting in 1904. As of 2006, there were 62 member nations and approximately 9,000 individual members from 65 countries.

Astronomers worldwide report their discoveries, such as new comets or asteroids, to the IAU's Central Bureau for Astronomical Telegrams, which then sends the news in IAU circulars. Related to this is the Minor Planet Center, founded in 1947 to collect asteroid position reports to refine orbital data. The IAU is perhaps best known as the authority on nomenclature, responsible for assigning names to celestial bodies and surface features through various working groups, which set policies for each type of object. The IAU's authority for naming astronomical objects relies on tradition and agreement of its member nations and has no legal power to prevent abuses, such as the attempts to market naming rights of stars.

As the world's astronomical community voice, the IAU is a permanent observer on the United Nations Committee on the Peaceful Uses of Outer Space. It speaks out on such matters as space debris, light pollution, and protection of the frequency bands needed for radio astronomy. Considerable IAU efforts are also directed toward improving astronomy education worldwide.

Katie J. Berryhill

See also: United Nations (UN) Committee on the Peaceful Uses of Outer Space

Bibliography

Johannes Andersen, "Discover the International Astronomical Union," *Mercury* 29 (January–February 2000): 32–38.

Adriaan Blaauw, *History of the IAU* (1994).

Jet Propulsion Laboratory

The Jet Propulsion Laboratory of the California Institute of Technology (Caltech), in Pasadena, California, is the premier U.S. organization for planetary exploration. Its origin, however, was in rocket research led by graduate student Frank Malina in the late 1930s. Successful tests of rockets to help aircraft take off in 1939 led eventually to its formal constitution as an Army facility in June 1944, managed under contract by Caltech and led by Malina.

For its first two decades, JPL's specialty was rockets and missiles. During World War II, JPL developed a variety of unguided rockets, expanding to guided missiles during the 1950s. Its engineers developed two guided missile systems for the Army, Corporal and Sergeant. These were complete, mobile weapon systems, and in the process of developing them, JPL created a new engineering discipline known as systems engineering.

The 1957 launch of *Sputnik* led to a revolution at JPL. The lab had cooperated with the Army Ballistic Missile Agency to launch a small satellite on one of the Army's Jupiter-C rockets. This became America's first artificial satellite, *Explorer 1*, on 31 January 1958. The Army transferred JPL's facilities and Caltech contract to NASA in December 1958. This university arrangement, unique among NASA facilities, caused occasional friction among all three organizations, as NASA at times desired more direct control over JPL, and at other times, such as in the 1970s and 1980s, allowed JPL to expand its contracts with other agencies such as the Army and the Department of Energy.

In 1959, NASA assigned JPL the lead role in planetary exploration, along with the *Ranger* lunar impact probes and the *Surveyor* landers. While ultimately successful, a series of failures on these two programs led to strengthening of JPL's project management and of NASA oversight. JPL's first planetary success was *Mariner 2*, which flew by Venus in 1962. Perhaps the two most dramatic missions of this early period were *Mariner 4*, whose images showed a cratered, Moon-like Mars, and *Mariner 9*, which showed dramatic images of volcanic cones and massive canyons, with tantalizing evidence that water might have once flowed on the Martian surface but had since disappeared. JPL's last Mariner mission, 1974's *Mariner 10*, was the only spacecraft, as of 2007, to explore Mercury. Because of its planetary role, JPL also became the builder and manager of the Deep Space Network that communicates with planetary spacecraft.

With the planetary exploration budget shrinking in the 1970s, the lab diversified into Earth science and astronomy. JPL developed specialties in atmospheric chemistry, physical oceanography, geophysics, and infrared astronomy. In addition to its spacecraft-building role, the lab developed expertise in remote-sensing instruments. JPL carried out four highly productive missions during this period. Its twin Voyagers carried out flybys of Jupiter and Saturn, and *Voyager 2* explored Uranus and Neptune. JPL built the orbiters for NASA Langley Research Center's two Viking missions to Mars and operated the spacecraft.

New missions essentially ceased until 1989, when JPL sent *Galileo* and *Magellan* to Jupiter and Venus, respectively. JPL's diversification continued, as it developed an integrated combat management system for the Army, the All-Source Analysis System.

The arrival of *Magellan* at Venus in 1990 began JPL's resurgence in planetary exploration. *Magellan* produced the first complete map of the planet's shrouded surface. JPL followed it with *Galileo* and *Cassini* missions to Jupiter and Saturn and a host of missions to Mars, including successes such as *Mars Pathfinder*, *Mars Global Surveyor*, and Mars Exploration Rovers, and disappointments such as *Mars Observer* and *Mars Polar Lander*, whose failure was attributed to overzealous cutbacks in budgets and testing. JPL sent missions to examine the interior of cometary nuclei and to collect particles of the solar wind and return them to Earth. It built cameras to repair the *Hubble Space Telescope* and orbited infrared and ultraviolet telescopes. JPL developed important missions to Earth, including the first global ocean circulation measurements and new weather prediction instruments.

Erik M. Conway

See also: Deep Space Probes, Planetary Science, Systems Engineering, United States Army

Bibliography

William E. Burrows, *Exploring Space: Voyages in the Solar System and Beyond* (1990).
Clayton Koppes, *JPL and the American Space Program* (1982).
Bruce C. Murray, *Journey into Space: The First Thirty Years of Space Exploration* (1989).
Peter Westwick, *Into the Black: JPL and the American Space Program, 1976–2004* (2007).

Lunar Receiving Laboratory. *See* Apollo Science.

Civilian and Commercial Space Applications

Civilian and commercial space applications are practical uses of space, implemented by either nonmilitary government institutions or privately owned enterprises. These uses include communications, navigation, burial services, commercial space launch, insurance, and Earth observation for a variety of purposes, including weather observation, land use planning and regulation, geological prospecting and surveying, mapmaking, and monitoring of environmental issues.

Before the first satellite launches in the late 1950s, military and civilian planners and researchers predicted that space would be useful as a location for a communications relay and as a vantage point for weather monitoring. Sounding rocket experiments confirmed the ability of satellites to provide useful pictures of cloud formations. Military experiments proved that the Moon could be used to relay communications signals, and that artificial satellites would be much better for this purpose. Engineers, such as Arthur C. Clarke, and researchers, such as John Pierce of American Telephone and Telegraph (AT&T) Company, recognized that three satellites in geosynchronous orbit could provide global communications coverage, except at the poles.

The 1950s–60s featured a three-way collaboration and competition among the U.S. military, government nonmilitary, and commercial organizations to investigate the feasibility of communications satellites (comsats). Upon its founding in October 1958, NASA inherited several military organizations and projects, including the *Echo 1* comsat, which was a large aluminum balloon that would passively reflect electromagnetic waves. The U.S. military initially focused on active repeater satellites, which would receive, amplify, and retransmit signals. AT&T supported Pierce's ideas by providing ground tracking stations for *Echo 1* and by developing *Telstar 1*, which AT&T paid NASA to launch in 1962. *Telstar 1* transmitted the first transatlantic telephone call and television broadcast. The John Kennedy presidential administration was adamantly against AT&T extending its near-monopoly on telephone communications to satellite communications, and it encouraged contracts with other companies to develop geosynchronous satellites, such as Hughes's Syncom, to create competition with AT&T. Along with Congress, the Kennedy administration also pressed for the creation of the Communications Satellite Corporation (COMSAT) to manage the development of a worldwide comsat system. This became the International

Telecommunications Satellite Organization (Intelsat), an international consortium that operated a system of geosynchronous comsats.

Foreign experts observed these U.S. developments and began to formulate their responses. One response was to organize themselves to negotiate as a group with the United States for participation in Intelsat and, later in the 1970s in the maritime equivalent, the International Maritime Satellite Organisation (Inmarsat). Another response was to develop competing systems. The Soviet Union developed the Molniya satellite series and formed the Intersputnik consortium with its communist allies. In the 1960s–70s, western European nations began developing national satellite systems and, through the European Space Research Organisation and later the European Space Agency, integrated European systems. These experimental satellite systems led to the creation of operational comsats by the 1980s, deployed through the European Telecommunications Satellite (Eutelsat) organization and domestic systems operated by various national governments. The 1970s–90s saw the steady expansion of comsat development and deployment by national governments around the world. While larger nations (such as China, Japan, and India) developed comsats, less-developed nations (such as Indonesia, Mexico, and Turkey) and other groups of nations (such as the Arab nations) purchased comsats from U.S. or European comsat manufacturers.

Private companies made profits in satellite communications as manufacturers or operators. In the United States, Hughes, Ford Aerospace, General Electric, and Radio Corporation of America (RCA) became prominent comsat manufacturers, while Alcatel and Matra Marconi in Europe competed with them for comsat sales to national governments, international consortia, and private operators. Private operators began to develop in the 1970s–80s, including existing companies, such as Western Union, and new companies, such as Pan American Satellite, Société Européenne des Satellites (SES), DIRECTV, and Echostar. By the 1990s–2000s private companies were offering comparable services at lower cost than national governments and consortia, which helped to drive a wave of privatization, including Intelsat and Inmarsat. In the meantime, military and intelligence services around the world continued to deploy comsat systems, but they also found it cost effective to contract for services from the commercial operators. This provided a significant source of income for commercial operators.

The success of satellite communications created a demand for launch services. Because they did not wish to depend on the United States, several European nations, led by France, developed an independent launch capability, successfully deployed in 1979 with Ariane's first launch. This broke the U.S. space launch monopoly. In the 1980s a vigorous competition to launch commercial payloads erupted among Arianespace, a commercial space launch company founded to market and operate Ariane, NASA's Space Shuttle, and, after 1984, private U.S. companies McDonnell-Douglas (Delta), General Dynamics (Atlas), and Martin Marietta (Titan). NASA withdrew from the commercial marketplace after the 1986 *Challenger* accident. The U.S. companies were joined in the late 1980s–90s by China's Great Wall Corporation, which marketed Long March launch services, and several Russian and Ukrainian organizations, which marketed their launchers. Often joining with western companies in joint ventures, these new competitors could underbid the western companies due to their much lower wages and capital costs. The United States countered by limiting the number of launches that these new companies were allowed to launch, because the U.S. and European companies manufactured the satellites. All these companies had close relationships with their national governments through direct or indirect subsidies or government partial or complete ownership.

In the 1990s a large spike in projected launches of medium-orbit satellite constellations, such as Iridium, led to the creation of a number of new private corporations to develop new launch vehicles to meet the demand. However, the medium orbit constellations proved unprofitable and these new communications companies, along with the new launcher companies, went bankrupt. By the early 2000s no purely private company had successfully developed and operated a launch vehicle without government funding.

Commercial satellite communications, and the competition to launch them, prompted the need for space insurance. While governments could build another satellite if the original spacecraft or its launch failed, private companies and international consortia, such as Intelsat, did not have the resources to do this. Instead they purchased insurance, initially from large insurers of unusual events or artifacts, such as Lloyd's of London. Since the 1972 United Nations Convention on International Liability for Damage Caused by Space Objects required that governments take responsibilities for damages caused by launches and satellites from their nations, these nations had to set limits to damages for which private companies would be held liable. For major accidents, such as could be caused by a launch vehicle landing on a city, private companies would pay up to the government-set limit, and then the government would pay any damages beyond that. Since the 1960s, European insurers have dominated the space insurance business, with U.S. insurers holding a large, but minority, share of policies.

In the 1960s NASA's Tiros series proved the utility of weather satellites. The U.S. Weather Bureau decided that satellite weather monitoring was a necessary capability and it drew from the military's Defense Meteorological Satellite Program design (which in turn drew from Tiros) instead of NASA's advanced Nimbus satellites for its Tiros Operational Satellite series. As in other applications, the Soviet Union soon followed with its Meteor series that used high inclination orbits, demonstrated by the mid-1960s. Meteor enabled weather observation at the Soviet Union's near-polar latitudes. For much of the world, the most effective weather monitoring satellites were placed in equatorial geosynchronous orbits, because these systems provided continuous observations from a single location. The United States, Europe, Japan, and India developed geosynchronous satellites for this purpose. In the early 1970s the Coordination Group for Meteorological Satellites was formed, which, along with the World Meteorological Organization, provided a mechanism for worldwide (including communist nations) coordination and sharing of weather data. These cooperative arrangements provided temporary services from one nation or group of nations to another when a geosynchronous satellite failed. The United States continued to operate separate civilian and military polar-orbiting weather satellites through the twentieth century, but by the twenty-first century, U.S. leaders were developing the next generation of polar-orbiting satellites, which would be a single system serving both military and civilian purposes. As of the early twenty-first century, weather observation remained a civilian-dominated government-provided service around the world.

Observing the Earth from space was also useful for nonweather purposes. The first land remote sensing satellites were the U.S. Corona and Soviet Zenit reconnaissance spacecraft. These were secret in the 1960s, but their technologies were known to a number of engineers and researchers outside military and intelligence services. In 1972 the U.S. *Landsat 1* spacecraft was launched, pioneering land remote sensing for civilian purposes. Landsat images were sold at a relatively low price and were used for a variety of purposes, including scientific studies, commercial geological prospecting, civilian and commercial mapmaking, and government studies and regulation of urban and rural environments. Because some of its images were used by commercial companies, by the late 1970s the U.S. government decided to privatize the program. While some continued to argue that Landsat imagery should be provided as a public service, like weather imagery, by the mid-1980s the program was privatized, with the Earth Observation Satellite Company (EOSAT) marketing the imagery. To cover operational costs, EOSAT raised image prices significantly, which created a backlash among users, many of whom could no longer afford to buy the imagery. The launch of the first French SPOT (Satellite Pour l'Observation de la Terre) satellite, and its commercial marketing through SPOT Image, created stiff competition for the imagery market. The combination of foreign competition with lower prices and higher resolution, and complaints of domestic users who could no longer afford Landsat imagery, pushed Congress to return Landsat to government control in 1992.

This legislation also contained other provisions that allowed the development of purely commercial high-resolution satellites to compete with SPOT. Several U.S. companies began to develop private high-resolution systems—*Ikonos 1* from Space Imaging (which had acquired EOSAT) was the first to orbit in 1999. Several other U.S. and non-U.S. companies and organizations were selling both high resolution imagery (roughly 1 m), and lower-resolution imagery such as that offered by Landsat and SPOT. These high-resolution systems broke the secret military-intelligence monopoly of high-resolution imagery, since individuals, corporations, and nations could purchase on the open market images of almost anywhere in the world.

Like Earth remote sensing, space-based navigation also evolved from military roots. The U.S. Navy developed and deployed its Transit satellites to assist nuclear submarines in determining their positions, which enabled submarine-launched ballistic missiles to accurately plot a trajectory to their targets in the Soviet Union. By the late 1960s it became clear that civilians could also use these signals to find their positions, and the U.S. Navy made the signals available to civilian and commercial users. These early uses included applications that did not require continuous rapid updates, such as offshore petroleum exploration, shipping, surveying, and space applications, such as the ocean recovery of U.S. manned spacecraft. The Soviet Tsikada navigational satellites, first launched in 1976, were used for both military and civilian purposes.

Commercial applications took off with the next generation of navigational satellites, the U.S. Global Positioning System (GPS), and the Soviet Global Navigation Satellite System (GLONASS). While the first generation of navigational satellites could provide only horizontal accuracies of roughly 100–200 m within a few hours, GPS precision signals could initially provide continuous horizontal accuracies of 22 m and 27.7 m vertically. In addition, GPS signals were accurate in time to within 200 nanoseconds. In 2000 the precision signals were made permanently available to civilian and commercial users. While improved accuracies were helpful, the continuous nature of the signals made GPS useful for many more commercial applications than Transit. Commercial and civilian applications grew rapidly, such that by the 2000s commercial navigation based on GPS signals was a multibillion-dollar industry, far outpacing the military applications that were the reason for GPS's existence. A few examples of new applications included aircraft navigation, trucking, animal tracking, automobile navigation, and recreational applications, such as determining the precise yardage from the golfer's position to the hole. Accuracy of GPS signals could be, and was, enhanced by a variety of Earth-based signals that enabled even more applications. The commercial success of space-based navigation led the Europeans in the 2000s to begin development of a commercial system known as Galileo.

Businesses were involved with all these civilian and commercial applications in addition to almost all other space endeavors, including military and intelligence applications, science, and human spaceflight. In capitalist economies, businesses usually developed and manufactured the spacecraft and launch vehicles. These were delivered to the customer, which could either be the government or another company or organization (such as Intelsat or a university). The customers typically operated the systems. Some customers were other private companies, such as PanAmSat, DIRECTV or Sirius Satellite, which provided services to their customers, including individuals, corporations, nonprofit organizations or governments.

The first companies in what would become the space industry were newly founded rocket companies, such as Aerojet and Reaction Motors, created during World War II. After World War II, as missile development became a larger business, existing large aviation companies and a few other high-technology companies won government contracts to build rockets and their payloads. These included well-known aviation firms, such as North American Aviation, Boeing, Martin, Douglas, Lockheed, and de Havilland. Other high-technology firms, such as AT&T, General Electric, and Rolls Royce, also entered the new field of rocket and missile development through their expertise in electronics and aircraft engines.

After the first experimental satellites were launched in the late 1950s–60s, several other companies entered the space industry, often to develop spacecraft, including Hughes, Grumman, Marconi, Philco, RCA, Selenia, Alcatel, Mitsubishi, Ball Brothers, Matra, and Dornier. Several aircraft companies developed space expertise, including Chrysler (which manufactured the Redstone rockets), Ford (which purchased Philco and manufactured comsats), General Motors (which eventually purchased Hughes and its comsat business), Daimler (which acquired much of the West German space industry by the 1980s), and Saab (which manufactured various components for European space systems in which it was a partner). Many new businesses formed explicitly with space business as their core competence, including TRW, COMSAT, SES, Arianespace, Pan American Satellite, Earth Observing Satellite Corporation, SPOT Image, Antrix, Orbital Sciences, XM, and DigitalGlobe, to name a few. Celestis was a unique space company that placed cremains (cremated remains) into space. Space Adventures became a small but successful space tourism company.

Civilian and commercial space applications, and the many businesses that provided services to these and all other space endeavors, continued to grow in importance throughout the space age. They provided services used every day by people and nations around the world, such as weather forecasts, communications, and navigation and have been a significant part of the global infrastructure.

Stephen B. Johnson

MILESTONES IN THE DEVELOPMENT OF CIVILIAN AND COMMERCIAL SPACE APPLICATIONS

1897 Founding of Marconi Company, Ltd.

1912 Founding of Lockheed Corporation.

1916 Founding of Pacific Aero Products Company, in 1918 renamed The Boeing Airplane Company.

1917 Founding of Glenn L. Martin Company.

Founding of Dornier GmbH.

1921 Founding of Douglas Aircraft Company.

1928 Founding of North American Aviation.

1929 Founding of Grumman Aircraft Engineering Corporation.

1934 Founding of Mitsubishi Heavy Industries, Ltd.

1935 Founding of Hughes Aircraft Company.

Founding of Hawker Siddeley as a merger of Hawker Aircraft, Armstrong Siddeley, and Armstrong Whitworth Aircraft.

1939 Founding of McDonnell Aircraft Corporation.

1945 Founding of Matra (Mécanique Aviation Traction).

15 October: Arthur C. Clarke proposes that three satellites in geosynchronous orbit around Earth could provide communications for the entire globe.

1946 2 May: Project RAND publication of "Experimental World Circling Spaceship," which describes the feasibility of spacecraft and their potential applications.

1953 Founding of Ramo-Wooldridge Corporation.

1954 A U.S. Navy Aerobee sounding rocket photographs a previously undetected hurricane.

General Dynamics Corporation purchases Convair.

1956 December: Ball Brothers acquires Control Cells Company and forms the Ball Brothers Research Corporation to enter the space business.

1958 October: Thompson Products merges with Ramo-Wooldridge Corporation to form TRW.

18 December: *SCORE* (*Signal Communication by Orbiting Relay Equipment*) carries the first communications payload into orbit.

1959 Radio Corporation of America forms its Astro Electronics Division.

7 August: Launch of *Explorer 6*, which sends first television photograph of Earth from space.

1960 1 April: *Tiros 1* launched, the first dedicated weather satellite.

13 April: *Transit 1B* orbited, the first successful navigation satellite.

12 August: *Echo 1*, the first passive communications satellite, placed into orbit.

4 October: Launch of *Courier 1B*, the first active repeater communications satellite.

1961 Founding of Entwicklungsring Nord (ERNO), a joint venture of Focke-Wulf and Weserflug.

12 December: Launch of *OSCAR 1*, the first amateur communications satellite.

1962 U.S. Congress passes the Communications Satellite Act.

10 July: The first transatlantic television signal sent from the United States to France via satellite transmitted through AT&T's privately owned *Telstar 1*.

1963 Incorporation of Communications Satellite Corporation.

First major international meeting on space radio frequency regulation by the International Telecommunication Union (ITU).

26 July: *Syncom 2*, first successful geosynchronous communications satellite, placed into orbit.

1964 The U.S. Weather Bureau made responsible for operational weather satellites.

Foundation of Intelsat.

28 August: Launch of *Nimbus 1*.

28 August: First test flight of the Soviet Meteor weather satellites as *Kosmos 44*.

1965 6 April: Launch of *Early Bird*, Comsat's first communications satellite, including the first space insurance policy.

23 April: First successful launch of the Soviet Molniya 1 communications satellite series.

1966 7 December: Launch of *Applications Technology Satellite 1* (*ATS 1*), which demonstrates frequency division multiple access communications and a black-and-white spin-scan camera, later used for meteorology.

Merger of North American Aviation and Rockwell Standard to form North American Rockwell Corporation.

1967 The World Meteorological Organization (WMO) and the International Council of Scientific Unions agree to sponsor the Global Atmospheric Research Program (GARP).

April: merger of Douglas Aircraft Company and McDonnell Aircraft Company into McDonnell-Douglas Corporation.

5 November: Launch of *ATS 3*, which demonstrates the first color spin-scan camera, creating the first color image of Earth's entire disk.

1968 Foundation of Messerschmidt-Bölkow-Blohm from Junkers, Messerschmidt and portions of Heinkel companies.

1969 Formation of Radio Amateur Satellite Corporation (AMSAT).

14 April: Launch of *Nimbus 3*, which carries sounding instruments for atmospheric profiles of temperature, water vapor, and ozone concentrations.

1970 1 January: Société Nationale Industrielle Aérospatiale created from a merger of Sud Aviation, Nord Aviation, and the Société d'Etudes et de Réalisation d'Engins Balistiques.

1972 The U.S. Federal Communications Commission authorizes U.S. private companies to launch privately owned communications satellites.

First meeting of the Coordination of Geostationary Meteorological Satellites (CGMS) group.

The Liability Convention makes states liable for damages caused by launches from their soil.

23 July: Launch of *Landsat 1*, the first civilian land remote sensing spacecraft.

9 November: Launch of *Anik A1*, the first Canadian communications satellite.

11 December: Launch of *Nimbus 5*, which carries a microwave sensor that can see through clouds.

1973 Intelsat's permanent agreement negotiated.

1974 13 April: Launch of *Westar 1* for Western Union, the first U.S. private domestic satellite since *Telstar*.

30 May: Launch of *ATS 6*, which demonstrates communications technologies and experiments with direct-to-home broadcasting.

9 July: Launch of the first Meteor Priroda, the first civilian Soviet land remote sensing satellite.

19 December: Launch of *Symphonie A*, a Franco-German experimental communications satellite.

1975 16 October: First launch of a U.S. Geostationary Operational Environmental Satellite.

26 November: Launch of *Fanhui Shi Weixing 0-1*, the first Chinese remote sensing satellite.

1976 Foundation of Arabsat, the Arab Satellite Communications Organization.

Foundation of Inmarsat, the International Maritime Satellite organization.

8 July: Launch of Indonesia's *Palapa 1* communications satellite.

1977 Establishment of Eutelsat, the European Telecommunications Satellite organization.

Founding of the nationalized British Aerospace from British Aircraft Corporation, Hawker Siddeley, and Scottish Aviation.

14 July: Launch of Geostationary Meteorological Satellite, Japan's first geostationary weather satellite.

23 November: Launch of *Meteosat 1*, the first European geosynchronous meteorological satellite.

1978 26 April: launch of *Heat Capacity Mapping Mission*, the first of three Application Explorer Missions.

27 June: Launch of *Seasat*.

24 October: Launch of *Nimbus 7*, the first spacecraft to measure atmospheric pollutants.

1979 Signing of Memorandum of Understanding between Canada, France, the United States, and the Soviet Union to develop a space-based search-and-rescue system.

7 June: Launch of *Bhaskara 1*, the first Indian Earth resource satellite.

24 December: First launch of Ariane 1.

1980 Foundation of Arianespace, the world's first commercial space transportation company.

Foundation of NESDIS, the U.S. National Environmental Satellite, Data, and Information Service.

1981 British Aerospace becomes the private corporation British Aerospace Public Limited Company, with the government selling over 51 percent of its shares.

1982 Foundation of Spot Image, the first private company to market remote sensing imagery.

Foundation of Orbital Sciences Corporation.

10 April: Launch of *INSAT 1A*, the first of the Indian National Satellite series.

1983 The U.S. government announces that the Global Positioning System will be available for use by civil airlines.

5 April: The Space Shuttle deploys *Tracking and Data Relay Satellite 1*.

1984 Celestis Incorporated receives a license to place cremated human remains in orbit.

The U.S. government allows private companies to compete with Intelsat for international satellite communications services.

Foundation of Pan American Satellite Corporation.

U.S. Congress passes the Land Remote Sensing Commercialization Act, to remove the U.S. government from operation of Landsat and marketing its data.

Foundation of SPACEHAB, Incorporated.

11 April: First satellite retrieval and repair on orbit during Space Transportation System (STS) 41C.

1985 NASA establishes its program to create Centers for the Commercial Development of Space.

China offers Long March as a commercial launch vehicle.

Earth Observation Satellite Company (EOSAT) wins contract to operate Landsat and market its imagery.

General Motors purchases Hughes Corporation.

1986 Foundation of EUMETSAT.

Martin Marietta acquires the space assets of the former Radio Corporation of America from General Electric.

22 February: Launch of *SPOT 1*, first French land imaging satellite.

1987 19 February: Launch of *Momo 1*, the first Japanese Earth remote sensing satellite.

1988 Foundation of Asiasat, the first privately owned Asian communications satellite operator.

Creation of the Office of Space Commerce in the U.S. Department of Commerce.

17 March: Launch of *IRS 1A*, the first Indian operational remote sensing satellite.

6 September: Launch of *Feng Yun 1A*, the first Chinese meteorological satellite.

11 December: Launch of SES's *ASTRA 1A*, the first privately owned European direct-to-home broadcasting satellite.

1989 Foundation of Deutsche Aerospace AG from the merger of Messerschmidt-Bölkow-Blohm, Dornier, and Daimler-Benz.

Merger of Matra Espace with the United Kingdom's General Electric Company Marconi Space Systems to form Matra Marconi Space.

1990 Loral Space and Communications purchases Ford Aerospace Corporation.

Alenia Spazio formed from merger of Aeritalia Space Systems Group and Selenia Spazio.

1991 The U.S. government offers the Global Positioning System for use to all nations.

17 July: Launch of *European Remote Sensing 1*, the first European Earth remote sensing mission.

1992 India creates Antrix Corporation to sell IRS imagery.

The U.S. Land Remote Sensing Act transfers control of Landsat to U.S. government control and allows development of private imaging satellites.

Foundation of Worldview Corporation to develop a private imaging satellite.

10 August: launch of *TOPEX/Poseidon*, the first ocean surface topography mission.

1993 Martin Marietta purchases General Electric Aerospace.

17 December: launch of Hughes's *Direct Broadcast Satellite 1*, which inaugurates U.S. digital satellite television broadcasting.

1994 Martin Marietta purchases General Dynamics Corporation's Space Systems Division.

Northrop purchases Grumman to form Northrop-Grumman Corporation.

Matra Marconi Space purchases British Aerospace Space Systems and Ferranti Satcomms.

1995 Creation of International Launch Services, a joint venture of Lockheed Martin and Khrunichev, to market and launch Atlas and Proton launch vehicles.

Creation of Sea Launch Corporation, a joint venture of Boeing, Rocket and Space Corporation Energia, State Design Office Yuzhnoe, and Kvaerner, to market and operate the Zenit launcher from a seagoing platform.

March: Lockheed Martin Corporation formed as a merger of Lockheed and Martin Marietta.

April: Global Positioning System declared fully operational.

4 November: Launch of *Radarsat 1*, first Canadian sea and ice monitoring satellite.

1996 Boeing acquires Rockwell International's space business.

Merger of Lockheed and Martin Marietta to form Lockheed Martin Corporation.

Boeing and Lockheed Martin form the United Space Alliance joint venture to operate the Space Shuttle.

1997 Boeing merges with McDonnell-Douglas Corporation.

6 January: Celestis Incorporated places the first cremated human remains in orbit aboard *Lunar Prospector*.

1999 Foundation of BAE Systems from a merger of British Aerospace and Marconi Electronic Systems.

August: Iridium goes bankrupt.

24 September: Launch of Space Imaging's *Ikonos*, the first private imaging satellite.

14 October: Launch of *China-Brazil Earth Resource Satellite* (*CBERS*) *1*.

2000 January: Boeing acquires Hughes Electronics Corporation's space business to become Boeing Satellite Systems.

1 May: The U.S. government orders the end of encryption of high-precision GPS signals.

May: Matra Marconi Space merges with DaimlerChrysler Aerospace to form Astrium Company.

10 July: Foundation of EADS (European Aerospace and Defence Systems) Corporation through the merger of Aérospatiale-Matra, Dornier GmbH and DaimlerChrysler Aerospace AG, and Construcciones Aeronáuticas SA (CASA).

2001 Lockheed Martin acquires a controlling stake in Communications Satellite Corporation.

18 March: Launch of XM Satellite Radio's *Rock* satellite.

18 July: Intelsat becomes a private corporation.

2002 The European Union and European Space Agency agree to develop the Galileo space navigation system.

December: Northrop Grumman acquires TRW Corporation.

2004 20 September: launch of *EDUSAT*, India's first dedicated educational satellite.

2005 Alcatel Alenia Space formed from a merger of Alenia and Alcatel space operations.

United Launch Alliance formed as a joint venture of Lockheed Martin and Boeing.

2006 GeoEye formed from a merger of Orbimage and Space Imaging.

Stephen B. Johnson

See also: Convention on International Liability, Earth Orbiting Robotic Spacecraft, Earth Science, Economics of Space, Expendable Launch Vehicles and Upper Stages, Military Communications, Navigation, Reconnaissance and Surveillance, Space Politics, Tourism

Bibliography
Roger Handberg, *International Space Commerce: Building from Scratch* (2006).
Stephen B. Johnson, "Space Business," in *Space Politics and Policy: An Evolutionary Perspective*, ed. Eligar Sadeh (2002), chap. 13.
Pamela E. Mack and Ray A. Williamson, "Observing Earth from Space," in *Exploring the Unknown, Volume III: Using Space*, ed. John Logsdon et al. (1998), 155–328.
Donald H. Martin et al., *Communication Satellites*, 5th ed. (2007).

CIVILIAN SPACE APPLICATIONS

Civilian space applications are nonmilitary, government-owned, or government-controlled satellites that provide services for practical applications, such as communications, weather forecasting, land-use planning, and geological surveying.

The utility of Earth-orbiting satellites for weather forecasting and communications was predicted by early researchers, such as the authors of the 1946 Project RAND study, "Preliminary Design of an Experimental World-Circling Spaceship." In the 1950s sounding rockets and early satellites, such as *Explorer 6*, provided images of clouds from space, including a hurricane in the Caribbean Sea not previously detected. NASA launched *Tiros 1*, the first weather satellite, in 1960. Its images proved useful to weather forecasters, leading to a series of Tiros satellites and NASA's experimental Nimbus series. Experimental civilian communications satellites (comsats) were orbited in the same period, with the *SCORE* (*Signal Communication by Orbiting Relay Equipment*) satellite in December 1958, the *Echo 1* passive comsat in August 1960, and the Syncom series, first orbited in 1963, that proved the viability of geosynchronous comsats.

In the United States and in the Soviet Union, both military and civilian organizations found utility in weather forecasting and satellite communications. This led to dual-use systems in the Soviet Union, such as Molniya and Meteor, while in the United States the separation of military from civilian functions, mandated by President Dwight Eisenhower and later administrations, led to a division between military and civilian applications for meteorology and communications. Tiros originated as an Army project, though it was transferred to NASA. Military satellite meteorology was served by the Defense Meteorological Satellite Program, with civilian operations handled by the Environmental Science Services Administration and later the National Oceanic and Atmospheric Administration. The U.S. military continued a satellite telecommunications program for its secure, global communications needs, while the U.S. civilian telecommunications program was largely delegated to commercial providers.

Several nations followed with the development of their own civilian communications and weather satellites. European nations, Japan, China, and India led the way with domestically developed systems for both purposes deployed in the 1970s–80s, and sometimes by purchasing foreign-manufactured comsats. Developing nations, such as Indonesia and Mexico, did not have the capability to develop their own satellites and therefore purchased foreign-manufactured comsats in the same period. Several nations

> **VERNER SUOMI (1915–1995)**
>
> Suomi, a University of Wisconsin–Madison professor and the "father of satellite meteorology," developed the spin-scan camera, which revolutionized the study of Earth's weather from space, forming the basis for imagery in television weather reports. He developed an instrument, launched on *Explorer 7* in 1959, to measure the radiation budget of Earth, inaugurating the era of satellite meteorology. Suomi conceived a computer system capable of handling enormous data from weather satellites, which soon came into global use. He also was involved with missions to study the atmospheres of other planets.
>
> *Katie J. Berryhill*

(and groups of nations), including the United States, Europe, India, Russia, and Japan contributed meteorological satellites and their data to form a worldwide meteorological system, coordinated by the World Meteorological Organization and bilateral national ties. While communications and weather monitoring functions were usually separated into different satellite systems, in some cases these were combined into a single satellite platform. Prominent examples include NASA's Applications Technology Satellites of the 1960s–70s and the Indian National Satellite series, orbited in the 1980s–2000s.

Civilian remote sensing came into being with the launch of *Landsat 1* by the United States in 1972, followed by the Soviet launch of the Meteor Priroda and Resurs F systems in 1974–75. The Landsat program found a variety of uses, both practical and scientific, for remote sensing data. This led to debates in the United States regarding whether Landsat should be privatized. In the 1980s the argument for privatization won out, and Earth Orbiting Satellite Corporation took over operations. However, the United States reversed course in 1992, with the U.S. government taking back control of the future *Landsat 7*. Landsat's success spurred competition, with France's government developing SPOT (Satellite Pour l'Observation de la Terre), first launched in 1986. France created Spot Image to market SPOT imagery, successfully grabbing market share from Landsat. The Indian Resource Satellite program, the European Remote Sensing program, and others followed by the 1990s. Their successes spurred commercial remote sensing companies, which began orbiting their satellites in the late 1990s–2000s.

While private, commercial competition was a significant factor for satellite communications starting in the 1970s, and for remote sensing by the 1980s, governments continued to exclusively control the operation of weather satellites into the early twenty-first century.

Stephen B. Johnson

See also: Arabsat, Commercial Space Applications, Earth Science, Economics of Space Communications, Economics of Space Reconnaissance and Remote Sensing, Inmarsat, Intelsat, Military Communications, Space Telecommunications Policy

Bibliography

John Logsdon et al., eds., *Exploring the Unknown, Volume III: Using Space* (1998).

Donald H. Martin et al., *Communication Satellites*, 5th ed. (2007).

Abraham Schnapf, ed., *Monitoring Earth's Ocean, Land, and Atmosphere from Space: Sensors, Systems, and Applications* (1985).

Civilian Meteorological Spacecraft

Civilian meteorological spacecraft have been orbited since the early years of the space age. In 1946 U.S. engineers working for the military Project RAND suggested that Earth-orbiting spacecraft could be used, among other things, to observe cloud patterns from space to facilitate short-range weather forecasting. In the following year a camera mounted on a V-2 rocket gave some idea of what could be done, and in 1954 an image taken by a Navy Aerobee sounding rocket showed a major hurricane in the Gulf of Mexico that had been completely missed by conventional weather detectors. The Tiros meteorological satellite program started shortly afterward, with *Tiros 1* being launched on 1 April 1960 as the world's first successful meteorological satellite.

Ten Tiros experimental spacecraft (see table on page 331) were launched into low Earth orbits between 1960 and 1965. Early Tiros included simple television cameras to produce black-and-white images in the visible waveband, but later spacecraft also included infrared radiometers to measure sea surface and cloud top temperatures.

Even before *Tiros 1* was launched, NASA was defining an operational meteorological satellite system. It consisted of six polar-orbiting satellites called Nimbus to provide high resolution images and four geostationary satellites to provide full-disk Earth images. There then followed a dispute with the Weather Bureau on the design and control of the satellites. This was resolved in 1964, with the Weather Bureau made responsible for any operational system, and NASA given responsibility for developing new instrument and spacecraft designs. As result, NASA developed the experimental Nimbus series of spacecraft, while the Weather Bureau, which became part of the Environmental Science Services Administration (ESSA) in 1965, produced a series of ESSA operational spacecraft, based on the Tiros design. The Tiros, ESSA, and early Nimbus spacecraft just carried imaging instruments. But *Nimbus 3*, launched in 1969, carried sounding instruments that could measure the vertical temperature, water vapor, and ozone concentration profiles of the atmosphere.

Weather systems can change rapidly, which was a problem for meteorologists using Tiros, ESSA, and Nimbus polar orbiting spacecraft, as they were not permanently located over one point on Earth to observe these changes. Geostationary spacecraft, on the other hand, are always observing the same disk of Earth but, being 36,000 km above Earth's surface, their resolution is relatively poor. However, this is not important for observing the movement of large weather systems.

Geostationary spacecraft were potentially useful for communications as well as meteorology. So NASA devised a series of experimental geostationary Applications Technology Satellites (ATS) to test both meteorological and communications systems.

Civilian Meteorological Spacecraft (Part 1 of 2)

Low Earth Orbit			
Spacecraft	**Meaning of Acronym**	**Launched**	**Country/Region**
Tiros 1–10	Derived from Television Infrared Observation Satellite	1960–1965	United States
Nimbus 1–7		1964–1978	United States
Meteor—first generation, 35 launched		1964–1977	Soviet Union
ESSA 1–9	Environmental Science Services Administration	1966–1969	United States
Tiros M and *NOAA 1–5*	National Oceanic and Atmospheric Administration	1970–1976	United States
Meteor—second generation, 25 launched		1975–1993	Soviet Union/ Russia
Tiros N and *NOAA 6, 7,* and *12*		1978–1991	United States
Advanced *Tiros N* or *NOAA 8–14*, excluding *12*		1983–1994	United States
Meteor—third generation, 7 launched		1984–1994	Soviet Union/ Russia
FY 1A–1D	Feng Yun	1988–2002	China
POES or *NOAA 15–17*	Polar-orbiting Operational Environmental Satellite	1998–2002	United States
Meteor 3M		2001	Russia
NOAA N		2005	United States
MetOp A	Meteorological Operational	2006	Europe
Geostationary Orbit			
Spacecraft	**Meaning of Acronym**	**Launched**	**Country/Region**
ATS 1–6	Applications Technology Satellite	1966–1974	United States
SMS 1 and *2*	Synchronous Meteorological Satellite	1974–1975	United States
GOES 1–13	Geostationary Operational Environmental Satellite	1975–2006	United States
GMS or *Himawari 1–5*	Geostationary Meteorological Satellite	1977–1995	Japan
Meteosat 1–7		1977–1997	Europe
INSAT—first generation	Indian National Satellite	1982–1990	India
INSAT—second generation		1992–1999	India

(Continued)

Civilian Meteorological Spacecraft (Part 2 of 2)

Geostationary Orbit			
Spacecraft	**Meaning of Acronym**	**Launched**	**Country/Region**
GOMS 1/Elektro 1	Geostationary Operational Meteorological Satellite	1994	Russia
FY 2A–2D	Feng Yun	1997–2006	China
INSAT—third generation		2000–2003	India
Meteosat 8–9 or *MSG 1–2* (second generation)	Meteosat Second Generation	2002–2005	Europe
Kalpana 1		2002	India
MTSAT 1R and *2* (*Himawari 6* and *7*)	Multi-functional Transport Satellite	2005–2006	Japan

ATS 1, launched in 1966, produced a new black-and-white image every half hour covering Earth from 52° N to 52° S, while *ATS 3*, launched the following year, produced color images covering Earth from pole to pole, also every half hour. Both spacecraft were used to study the development of weather systems, measure horizontal cloud velocities, and improve tropical storm predictions.

In 1967 the World Meteorological Organization (WMO) and the International Council of Scientific Unions agreed to sponsor the Global Atmospheric Research Program (GARP), and its First GARP Global Experiment (FGGE) planned for 1976. Five geostationary satellites were to be approximately equally spaced around Earth's equator during FGGE; two from the United States (Geostationary Operational Environmental Satellites—*GOES East* and *West*), one from Europe (*Meteosat*), one from the Soviet Union (*Geostationary Operational Meteorological Satellite—GOMS*) and one from Japan

Image taken on 25 August 1992, by the NOAA GOES 7 *weather satellite of the Americas and Hurricane Andrew as it makes landfall on the Louisiana coast. (Courtesy NASA/Marshall Space Flight Center)*

(*Geostationary Meteorological Satellite—GMS*). The start of FGGE, which lasted one year, was eventually delayed to 1 December 1978.

In 1972 representatives of the United States, the European Space Research Organisation (ESRO; from 1975 the European Space Agency—ESA), and Japan, and observers from WMO and the Joint Planning Staff for GARP met in Washington to discuss compatibility between geostationary meteorological satellites, in preparation for FGGE and other cooperative programs. It was intended to make the communications interfaces between geostationary meteorological spacecraft and their ground stations identical, so that any ground station could operate any spacecraft from any organization. This 1972 meeting was the first meeting of the Coordination of Geostationary Meteorological Satellites (CGMS). In the following year the Soviet Union (later Russia) and WMO became full members of CGMS, and later India, EUMETSAT, China, France, Korea, and the Intergovernmental Oceanographic Commission became members. In 1992 CGMS's area of interest was extended to cover low orbiting in addition to geostationary meteorological spacecraft, and its name changed to the Coordination Group for Meteorological Satellites (still CGMS).

The success of the meteorological instruments onboard *ATS 1* and *3* encouraged NASA to develop a geostationary spacecraft devoted to meteorology. The operational prototypes, *Synchronous Meteorological Satellites* (*SMS*) *1* and *2* were launched in 1974 and 1975, and the almost identical operational *GOES 1–3* spacecraft were launched during the period 1975–78. The color images on *ATS 3* were not as useful as originally foreseen, and so the imagers on *SMS 1* and *2* and on *GOES 1* to *3* produced black-and-white images in the visible and infrared wavebands. During FGGE *GOES 2*, *SMS 1*, and *SMS 2* were at various times the GOES East spacecraft at about 75° W, while *GOES 3* was the GOES West spacecraft at about 135° W.

In 1975 NASA moved *ATS 6* over India for a one-year experiment to demonstrate the usefulness of geostationary spacecraft for communications, broadcasting, and meteorological purposes. The experiment was a great success, encouraging India to start its own multipurpose geostationary satellite system, with meteorological instruments carried on some of the spacecraft. The first series of spacecraft, called *INSAT 1A–1D* (Indian National Satellite), was built in the United States, but India built the second series, starting with *INSAT 2A*, which was launched in 1992. In 2002 India launched its first geostationary spacecraft, called *Kalpana*, dedicated to meteorology.

In 1969 France had started to develop a geostationary meteorological spacecraft called *Meteosat*. In parallel ESRO was also designing a meteorological system consisting of a polar orbiter and a geostationary spacecraft. Eventually in 1972 the ESA Member States agreed to adopt *Meteosat* as a European program instead of continuing with their own. The design of *Meteosat* was similar to that of SMS, but it had the significant addition of a second infrared channel optimized for detecting water vapor. Eventually nine Meteosats were launched. The early spacecraft were ESA spacecraft, but the later ones were procured by ESA to EUMETSAT's requirements, with EUMETSAT operating them in orbit after 1995.

Japan adopted a different approach to Europe in purchasing its GMS geostationary spacecraft, which was similar to SMS, from the United States. NASA launched *GMS 1* into its geostationary orbit position at 140° E in June 1977 and launched the

ESA *Meteosat 1* spacecraft into its position on the Greenwich meridian (at 0°) in November 1977. Unfortunately the Soviet Union withdrew its funding for *GOMS* to cover the costs of covering the 1980 Moscow Olympics. So to fill the gap in the geostationary orbit coverage for FGGE, the United States agreed to move *GOES 1* to 60° E, where it was operated by ESA. *GOMS* was finally launched in 1994, but numerous problems caused it to be withdrawn from service in 2000. China also launched a series of geostationary spacecraft, called *FY* (*Feng Yun*) *2A–2D*, to a position at 105° E starting in 1997. Their sensor complements and data relay capabilities were similar to those of the other geostationary spacecraft.

In the meantime, the last American ESSA spacecraft had been launched into a Sun-synchronous polar orbit in 1969. Then in 1970 NASA launched the first of a new series of spacecraft called Tiros M into a polar orbit. The second spacecraft in the series was called *NOAA 1* (National Oceanic and Atmospheric Administration), which had taken over responsibility in 1970 for the National Weather Service, the successor to the Weather Bureau. Then in 1972 NASA launched *NOAA 2*, in which the earlier television cameras had been replaced by scanning radiometers providing visible and infrared images. *NOAA 2* also included the first operational instruments for sounding the atmosphere.

In 1973 a national space policy study was carried out in the United States to examine the possibility of combining the civil and military meteorological satellite programs. It concluded that combining the programs would not significantly reduce costs; although NOAA was directed to use the Block 5D Defense Meteorological Satellite bus for its next generation of polar orbiters. The first of these, called *Tiros N*, was launched in 1978. It was the first of the Tiros series to carry a microwave sounder, which could measure atmospheric temperature profiles in the presence of clouds. *Tiros N*, and the following spacecraft in the series, *NOAA 6*, operated with the five geostationary satellites during FGGE.

In 1978 the United Kingdom and France provided instruments to fly on *Tiros N* and subsequent NOAA polar orbiters. Then in the early 1980s, the Ronald Reagan U.S. presidential administration put pressure on NOAA to reduce the number of American polar orbiting meteorological satellites. This prompted NOAA to start talks with ESA and EUMETSAT with the idea of Europe providing not just instruments, but some polar orbiting meteorological satellites. Initially it was suggested that meteorological instruments could fly on the ESA polar platform that was, at that stage, part of the International Space Station program. But when the platform was canceled in 1992, ESA and EUMETSAT agreed to produce a number of polar orbiting meteorological spacecraft called MetOp (Meteorological Operational). *MetOp A* was launched in 2006 with a payload of European and American instruments onboard.

The end of the Cold War and pressures to reduce the U.S. budget deficit reopened the question in the early 1990s as to whether it would be beneficial to combine the U.S. civilian and military meteorological systems into a common system. This time the answer was positive. NOAA would manage the program, the Department of Defense (DoD) would acquire the spacecraft, and NASA would develop new instruments. As a result the existing NOAA polar orbiting spacecraft would be

Operational Civilian Meteorological Spacecraft November 2006

Sun-Synchronous Polar Orbiting Satellites (in order of equatorial crossing time)			
Satellite	**Country/Region**	**Launched**	**Local Equatorial Crossing Time (N is Northward, S is Southward)**
NOAA 18	United States	2005	13.40 N
NOAA 16	United States	2000	15.30 N
NOAA 12	United States	1991	17.10 N
NOAA 15	United States	1998	17.40 N
FY 1D	China	2002	18.50 S
NOAA 14	United States	1994	21.30 N
MetOp A	Europe	2006	21.30 N
NOAA 17	United States	2002	22.10 N

Geostationary Satellites (in order of orbital position)			
Satellite	**Country/Region**	**Launched**	**Orbital Position**
GOES 9	United States	1995	160° E
MTSAT 2	Japan	2006	145° E
MTSAT 1R	Japan	2005	140° E
FY 2C	China	2004	105° E
INSAT 3A	India	2003	94° E
FY 2A	China	1997	87° E
Kalpana 1	India	2002	74° E
Meteosat 5	Europe	1991	63° E
Meteosat 7	Europe	1997	58° E
Meteosat 6	Europe	1993	10° E
Meteosat 9	Europe	2005	0° W
Meteosat 8	Europe	2002	4° W
GOES 10	United States	1997	60° W
GOES 12	United States	2001	75° W
GOES 13	United States	2006	105° W
GOES 11	United States	2000	135° W

Note: In early 2007 there were no operational Russian meteorological satellites in orbit.

phased out, and new spacecraft procured as part of the National Polar-orbiting Operational Environmental Satellite System (NPOESS). NPOESS and Europe's MetOp spacecraft would be designed to operate as an integrated system.

The Soviet Union launched its first meteorological spacecraft into an inclined orbit in 1964, followed by four generations of Meteor spacecraft. As with the American spacecraft, the television cameras of the early spacecraft were replaced by scanning radiometers. Some of the later Meteors also carried ozone sounders, and *Meteor 3M*, launched in 2001, also carried microwave radiometers. But by late 2006, after launching almost 70 meteorological spacecraft, Russia had no operational meteorological spacecraft in orbit.

China, on the other hand, did not launch its first meteorological spacecraft, *FY 1A*, until 1988. It, and its immediate successor, failed relatively quickly. Later China launched two more polar orbiters, *FY 1C* and *1D*, both of which were operational in late 2006.

David Leverington

See also: China, European Space Agency, Indian National Satellite Program, Japan, Meteorological and Climate Studies, National Aeronautics and Space Administration, National Oceanic and Atmospheric Administration, Russia (Formerly the Soviet Union)

Bibliography

Helen Gavaghan, *The History of EUMETSAT, European Organisation for the Exploitation of Meteorological Satellites* (2001).

Janice Hill, *Weather from Above: America's Meteorological Satellites* (1991).

J. Krige et al., *A History of the European Space Agency 1958–1987*, vol. 2, *1973–1987* (2000).

Pamela E. Mack and Ray A. Williamson, "Observing Earth from Space," in *Exploring the Unknown*, vol. 3, *Using Space*, ed. John Logsdon et al. (1998), 155–328.

Abraham Schnapf, ed., *Monitoring Earth's Ocean, Land, and Atmosphere from Space: Sensors, Systems, and Applications* (1985).

Coordination Group for Meteorological Satellites. The Coordination Group for Meteorological Satellites (CGMS) was formed in 1972 as the Coordination of Geostationary Meteorological Satellites organization when representatives of the European Space Research Organisation, Japan, the United States, and observers from the World Meteorological Organization (WMO) met in Washington, DC, to discuss compatibility among geostationary meteorological satellites. In the following year, the Soviet Union and WMO became full members. Then in 1992 the scope of the organization was extended to include geostationary and polar-orbiting meteorological satellites, and its name changed to the Coordination Group for Meteorological Satellites. By 2006 its membership included organizations from China; Europe; France; India; Japan; Korea; Russia; the United States; the Intergovernmental Oceanic Commission of the United Nations Educational, Scientific, and Cultural Organization; and the WMO.

At the beginning of the twenty-first century, CGMS provided a forum for the exchange of technical information before and after the launch of geostationary and polar-orbiting meteorological satellites. These exchanges helped to establish compatible telemetry interfaces, intercalibration of sensors, and common processing algorithms for meteorological images. As a result, CGMS members have, from time to time, used one of their in-orbit spare satellites to cover for the launch delay or failure of other members' satellites and so avoid any gaps in global coverage. For example, in 1985 the U.S. National Oceanic and Atmospheric Administration moved *Geostationary Operational Environmental Satellite 4* (*GOES 4*) to cover the loss of the Data Collection Service throughout Europe from the European Space Agency's *Meteosat 2*. Then in

CGMS Membership (January 2006)

Agency	Acronym	Country/Region	Joined
China Meteorological Administration	CMA	China	1989
European Organisation for the Exploitation of Meteorological Satellites	EUMETSAT	Europe	1987
European Space Agency	ESA	Europe	1972, rejoined 2003
French Space Agency	CNES	France	2004
India Meteorological Department	IMD	India	1979
Intergovernmental Oceanographic Commission of the United Nations Educational, Scientific, and Cultural Organization	IOC/UNESCO	World	2001
Japan Aerospace Exploration Agency	JAXA	Japan	2003
Japan Meteorological Agency	JMA	Japan	1972
Korea Meteorological Agency	KMA	Korea	2005
National Aeronautics and Space Administration	NASA	United States	2003
National Oceanic and Atmospheric Administration	NOAA	United States	1972
Russian Federal Service for Hydrometeorology and Environmental Monitoring	ROSHYDROMET	Russia	1973
Russian Federal Space Agency	ROSCOSMOS	Russia	2003
World Meteorological Organization	WMO	World	1973

1991 ESA moved *Meteosat 3* to the western Atlantic to provide cover for NOAA's aging *GOES 7* satellite.

David Leverington

See also: National Oceanic and Atmospheric Administration

Bibliography

Consolidated Report of CGMS Activities, 10th ed. (2003).
Report of the 33rd Meeting of the Coordination Group for Meteorological Satellites (2005).

Environmental Science Services Administration (ESSA) Satellites. Environmental Science Services Administration (ESSA) satellites, also known as Tiros Operational Satellites (TOS), were operational weather satellites launched by the United States between 1966 and 1969. ESSA was a unit of the U.S. Department of Commerce formed by merging the functions and personnel of the U.S. Weather Bureau, which was responsible for U.S. civil weather satellite programs, with those of the Coast and Geodetic Survey and the Central Radio Propagation Laboratory of the National Bureau of Standards in 1965. ESSA was incorporated into the new National Oceanic and Atmospheric Administration (NOAA) in 1970.

Satellites in the TOS series, built in a cylindrical configuration similar to *Tiros 9*, were built by Radio Corporation of America, launched by NASA and then turned over to ESSA when they became operational. These were designated *ESSA 1* through *ESSA 9* and were launched into Sun-synchronous orbits, operating in pairs. *ESSA 1*, launched 3 February 1966, carried two wide-angle black-and-white television cameras, which had flown on earlier Tiros spacecraft, and a flat plate radiometer. *ESSA 2*, launched 28 February 1966, was the first satellite with operational Automatic Picture Transmission (APT), allowing pictures from the satellite to be received by any ground station with appropriate equipment. *ESSA 3*, launched 2 October 1966, provided two high-resolution television cameras, which had been used on *Nimbus 1* in 1964, and a flat plate radiometer, while *ESSA 4*, launched 26 January 1967, offered two independent APT channels. *ESSA 5* and *6*, launched 20 April 1967 and 10 November 1967 respectively, essentially duplicated the capabilities of the preceding pair.*ESSA 7* and *8*, launched 16 August 1968 and 15 December 1968 respectively, added S-band transmission capability. The series ended with *ESSA 9*, launched on 26 February 1969.

Three ESSA Command and Data Acquisition Stations (located at Gilmore Creek, Alaska; Wallops Station, Virginia; and Suitland, Maryland) received signals from ESSA satellites. These were relayed to the National Environmental Satellite Service at Suitland, where the data was analyzed and relayed to U.S. and foreign weather consumers.

Pictures from ESSA satellites formed the basis for the first international exchange of space-based weather data, when ESSA sent photos to the Soviet Hydrometeorological Service in 1966. ESSA received at first only diagrams based on analysis of Soviet weather satellite observations—photos from Meteor satellites were later provided.

John D. Ruley

See also: National Oceanic and Atmospheric Administration

Bibliography

Janice Hill, *Weather from Above* (1991).
P. Krishna Rao et al., *Weather Satellites: Systems, Data, and Environmental Applications* (1990).
NOAA History. http://www.history.noaa.gov/.

EUMETSAT. EUMETSAT, derived from "European Meteorological Satellite," a European organization founded in 1986 to take over responsibility for the Meteosat system of geosynchronous meteorological satellites from the European Space Agency (ESA). ESA, however, continued to manage the program and operate the satellites for EUMETSAT.

Initially the Meteosat program had been approved in 1972 by ESRO (European Space Research Organisation), later ESA, Member States as a preoperational program. Then, in 1975 the Meteosat Protocol was approved, which required ESA to operate the Meteosat system until three years after the first successful launch, which took place in November 1977. This protocol was later extended, but it soon became clear that a new European organization was required to develop the Meteosat system according to the requirements of the various European meteorological agencies. This new organization was EUMETSAT, which by the early twenty-first century represented 19 European Member States.

The 1986 agreement between ESA and EUMETSAT covered the period up to the end of November 1995. So in 1991 EUMETSAT decided to establish its own ground control system, to take over from the ESA system that had been in place since the launch of *Meteosat 1* in 1977. Formal control of all Meteosat spacecraft was handed over to EUMETSAT on 1 December 1995. At that time *Meteosats 5* and *6* were operational, in September 1997 joined by *Meteosat 7*. Two Meteosat second-generation spacecraft, *Meteosat 8* and *9*, were successfully placed in orbit in 2002 and 2005, respectively.

Since 1998 ESA has developed a polar-orbiting satellite, MetOp (Meterological Operational), in collaboration with EUMETSAT, European industry, CNES (French Space Agency), and the U.S. National Oceanic and Atmospheric Administration. Three MetOp Satellites, the first of which launched in October 2006, and two others scheduled for launch in 2010 and 2015, were to be part of EUMETSAT's Polar System to monitor changes in Earth's weather, climate, and environment.

David Leverington

See also: European Space Agency, National Oceanic and Atmospheric Administration

Bibliography

Helen Gavaghan, *The History of EUMETSAT: European Organisation for the Exploitation of Meteorological Satellites* (2001).
Lars Prahm et al., *The 20th Anniversary of EUMETSAT: A Living History* (2006).

Feng Yun. Feng Yun (FY—Wind and Cloud) is the name for two series of Chinese polar-orbiting and geosynchronous weather satellites designed and built by the Shanghai Academy of Space Technology (SAST). The National Satellite Meteorological Center has developed and operated ground systems and data distribution from Feng Yun since its creation in 1971.

SAST drew on Soviet and Western experience in its weather satellite design. Feng Yun 1 satellites were three-axis stabilized craft to be placed into Sun-synchronous orbits to achieve a repeated ground track daily. To place them into polar orbit, the Chinese developed the more powerful Long March 4 booster and a new launch facility at Taiyuan, from which all were launched. *FY 1A*, launched in September 1988, lasted only 38 days. *FY 1B*, launched in September 1990, carried two balloons to measure the density of the upper atmosphere. It survived for about one year, while its upper stage exploded one month after launch, creating hundreds of pieces of orbital debris. Both *FY 1A* and *1B* carried five-channel scanning radiometers. *FY 1C*, considered

Feng Yun Spacecraft

(A) Sun-Synchronous Polar Orbiters, 860 km altitude					
Spacecraft	**Launch Date**	**Main payload**	**Wavebands (μm)**	**Ground resolution**	**Comments**
FY 1A	Sep 1988	VHRSR (Very High Resolution Scanning Radiometer)	Visible 0.48–0.53, 0.53–0.58, 0.58–0.68 0.73–1.1 (near IR—infrared) 10.5–12.5 (IR)	All channels 1.1 km in the high resolution mode and 4 km in the Automatic Picture Transmission mode	Failed after 38 days
FY 1B	Sep 1990	Same as *FY 1A*			Failed after one year
FY 1C	May 1999	MVISR (Multichannel Visible and IR Scan Radiometer)	Visible 0.43–0.48, 0.48–0.53, 0.53–0.58, 0.58–0.68 IR channels 0.84–0.89, 0.90–0.97, 1.58–1.64, 3.55–3.95, 10.3–11.3, 11.5–12.5		
FY 1D	May 2002	Same as *FY 1C*			

(B) Geostationary Satellites, Operational Spacecraft					
Spacecraft	**Launch Date**	**Main payload**	**Wavebands (μm)**	**Ground resolution (km)**	**Comments**
FY 2A	Jun 1997	Visible Infrared Spin Scan Radiometer (VISSR)	0.50–1.05 (Vis/Near IR) 6.3–7.6 (IR/Water Vapor) 10.5–12.5 (IR)	1.44 5.8 5.8	Placed at 105° E Partial failure April 1998 Retired in April 2000
FY 2B	Jun 2000	Same as *FY 2A*			Placed at 105° E
FY 2C	Oct 2004	5 channel VISSR	0.55–0.9 (Vis/Near IR) 3.5–4.0 (IR) 6.3–7.6 (IR, Water Vapor) 10.3–11.3 (IR) 11.5–12.5 (IR)	1.25 5 5 5 5	Placed at 105° E First operational spacecraft
FY 2D	Dec 2006	Same as *FY 2C*			Placed at 80.6° E

Tables by David Leverington

the first operational satellite, was orbited in May 1999, and *FY 1D* was launched in May 2002, each with 10-channel scanning radiometers. In January 2007, China destroyed *FY 1C* in orbit in its first successful antisatellite weapon test.

China also developed spin-stabilized geosynchronous meteorology spacecraft, the Feng Yun 2 series, all of which were launched on Long March 3 boosters from Xi Chang. Feng Yun 2 spacecraft carried a multichannel scanning radiometer. *FY 2A*, *2B*, and *2C* were placed at 105° E longitude, while *FY 2D* was inserted to 80.6° E longitude to better cover western China. In April 1994, during a propellant loading operation, the first (unnamed) Feng Yun 2 exploded, killing one technician and injuring 31 others. After a major redesign, *FY 2A* was launched in June 1997, operating normally for 10 months before major problems limited its returned images to six per day. *FY 2B* replaced it in June 2000. *FY 2C* in October 2004 and *FY 2D* in December 2006 were orbited to enhance coverage for the 2008 Olympic Games.

Stephen B. Johnson

See also: China, Long March

Bibliography

Federation of American Scientists. http://www.fas.org/.
Brian Harvey, *China's Space Program: From Conception to Manned Spaceflight* (2004).

Geostationary Meteorological Satellite. Geostationary Meteorological Satellite (GMS) system was a series of five Japanese weather satellites, nicknamed Himawari (sunflower). Operated by Japan's National Space Development Agency (NASDA—later Japan Aerospace Exploration Agency) and Japan Meteorological Agency (JMA), the satellites operated in geosynchronous orbit at 140° E, with the previous satellite at 120° E as a backup, and provided data to the World Meteorological Organization's World Weather Watch Programme.

GMS 1 launched onboard a Delta booster from Kennedy Space Center in 1977. NASDA launched the remaining four satellites from Tanegashima Space Center on Japanese boosters. *GMS 2* and *3* used N-II rockets, in 1981 and 1984 respectively. *GMS 4* launched in 1989 onboard an H-1 and *GMS 5* on an H-II in 1995.

Built by Hughes Aircraft Corporation as prime contractor to Nippon Electric Company, the satellites were spin-stabilized. The primary instrument was the U.S.-built Visible Infrared Spin Scan Radiometer (VISSR), of similar design to that used on the early American Geostationary Operational Environmental Satellite (GOES) spacecraft. The instrument, which included four visible and one infrared detector, imaged along a north-south line, while the 100-rpm rotation of the satellite moved the instrument from west to east. The mirror moved on the north-south axis with each rotation, building a full disk image of Earth in 25 minutes. The VISSR instrument observed cloud cover day and night and measured radiant and solar-reflected energy, allowing inference of wind fields and cloud heights. *GMS 5* carried an improved VISSR, with an additional infrared detector and a water vapor detector. The first four satellites carried the Japanese-built Space Environment Monitor to monitor solar particles.

In 1999 *GMS 5*'s successor, *Multi-Functional Transport Satellite 1* (*MTSAT 1*), failed to reach orbit. Subsequently there were problems with the construction of its replacement, *MTSAT 1R*, so JMA negotiated with the National Oceanic and Atmospheric Administration to temporarily move *GOES 9* to 155° E in 2003 to provide coverage. *MTSAT 1R* launched from Tanegashima in February 2005 and took over as the Asia-Pacific region's primary source of geostationary weather data.

Katie J. Berryhill

See also: Japan

Bibliography

Anthony R. Curtis, *Space Satellite Handbook* (1994).
P. Krishna Rao et al., *Weather Satellites: Systems, Data, and Environmental Applications* (1990).

Geostationary Operational Environmental Satellites. Geostationary Operational Environmental Satellites (GOES) are U.S. civilian geosynchronous weather satellites that provide continuous monitoring of Earth. From their fixed spot above Earth (35,800 km or 22,300 miles) along the orbital arc, they provide warnings for immediate weather events such as tornadoes, flash floods, hail storms, and hurricanes plus rain or snow precipitation estimates. The GOES program grew out of the 1960s Tiros and Nimbus satellites. NASA supervised construction of GOES satellites, but once in orbit and declared operational, the National Oceanic and Atmospheric Administration's National Environmental Satellite, Data, and Information Service assumed control. The collected information is provided primarily to the National Weather Service but also to other government agencies and departments, other countries' weather services, and to the private sector. The private sector analyzes the data for its specific customer needs. The program was originally called the Synchronous Meteorological Satellite (SMS) program, renamed GOES with the orbiting of *GOES 1* in 1975. The early satellites through *GOES 7* were spin stabilized but that provided incomplete coverage (about 10 percent), a problem solved by employing three-axis stabilized spacecraft, which allowed 100 percent continual coverage. The GOES program from its inception has been built by two major contractors: Philco–Ford Aerospace/Space Systems Loral completed the *SMS 1* and *2*, *GOES 1–3*, and *GOES 8–12*, while Hughes Space and Communications/Boeing Satellite Systems completed the *GOES 4–7* and *GOES 13* (or *N*), *O*, and *P*. By 2006, 13 GOES satellites had been placed in orbit.

The United States has normally kept two operational satellites in orbit, a GOES East at 75° W longitude for the Western Atlantic and GOES West at 135° W longitude for the Eastern Pacific basin. When possible, a third GOES satellite is kept in orbit as a spare. GOES satellites operate in conjunction with Polar Operational Environmental Satellites to provide complete coverage of weather events.

Two primary instruments on the GOES satellites provide weather information: the imager and the sounder. The former is a multichannel instrument that measures

GOES 8 *image of Hurricane Mitch off the northeast coast of Honduras, 26 October 1998. (Courtesy National Oceanic and Atmospheric Administration/NESDIS Operational Significant Event Imagery)*

radiant energy and reflected solar energy from Earth and its atmosphere. The latter, first flown on *GOES 4*, measures the vertical temperature and moisture profile of the atmosphere, surface and cloud top temperatures, and ozone. Other instruments included are a search-and-rescue transponder, data collection system, and a space environment monitor for coverage of the near-Earth environment. GOES satellites have been slowly moved east or west to provide greater coverage of the Atlantic or Pacific to assist other weather services whose satellites have encountered difficulties. The Japanese and Europeans likewise provided expanded coverage when U.S. GOES satellites have had difficulties, as GOES West did in the late 1990s, while the United States provided coverage for Japan when its *Geostationary Meteorological Satellite 5* satellite failed in 2003.

Roger Handberg

See also: National Environmental Satellite, Data, and Information Services (NESDIS); National Oceanic and Atmospheric Administration; Search and Rescue

Bibliography

GOES Project Science. http://goes.gsfc.nasa.gov/.

John H. McElroy and Ray A. Williamson, "The Evolution of Earth Science Research: NASA's Earth Observing System," in *Exploring the Unknown, Vol. 6: Space and Earth Science*, ed. John M. Logsdon (2004).

ITOS, Improved Tiros Operational System. *See* NOAA Polar Orbiting Satellites.

Meteor. Meteor is a series of weather satellites launched by the Soviet Union and Russia since 1964. The go-ahead for the development of meteorological satellites came in a government decree issued in October 1961. The design of the satellites was entrusted to the All-Union Scientific Research Institute of Electromechanics (VNIIEM) in Moscow, headed by Andronik Iosifyan.

All Meteors consisted of an upper cylinder, containing support systems, and a lower cylinder, carrying the instrument packages. An innovative system of electric flywheels was used to orient the satellites in space, whose mass ranged from 1,280 to 2,350 kg. The satellites were placed into high-inclination near-circular orbits with altitudes between about 600 and 1,200 km. The same bus was also used for the *Meteor Priroda* and Resurs-O remote sensing satellites.

The first-generation satellites carried a television camera, a scanning infrared radiometer, and instruments to study Earth's radiation budget. Test flights using the Vostok booster began in August 1964 (*Kosmos 44*). The first weather pictures were returned by *Kosmos 122* in mid-1966. The first satellite to be officially announced as Meteor was launched in March 1969, and the system became operational that same year. In the early 1970s the television camera was replaced by a scanning visible radiometer sending back images in Automatic Picture Transmission mode. Thirty-five first-generation Meteors were successfully launched between 1964 and 1977.

A second-generation model known as *Meteor 2* made its debut in 1975. It featured an improved spacecraft bus, two scanning visible radiometers, two scanning infrared radiometers, and a radiation measurement system. Beginning in 1982 the Soviet Union gradually switched launches from the Vostok to the Tsiklon 3 rocket. Twenty-five satellites were orbited between 1975 and 1993.

In 1984 the Soviet Union introduced *Meteor 3*, which carried virtually the same instrument complement as *Meteor 2* but had a special truss structure at the base to accommodate additional payloads. Several satellites flew ozone monitoring instruments, including the U.S.-built Total Ozone Mapping Spectrometer launched in 1991. The last of seven satellites went into orbit in 1994.

Because of the Russian economic crisis, it was not until 2001 that another Russian meteorological satellite left Earth. *Meteor 3M*, launched by a Zenit rocket, carried a more versatile payload than its predecessors, including some of the cameras flown on Resurs remote sensing satellites and the NASA-built Stratospheric Aerosol and Gas Experiment III instrument for atmospheric studies. Unfortunately, the meteorological payload was lost after two years due to a transmitter failure.

Bart Hendrickx

See also: Resurs-O

Bibliography

Bart Hendrickx, "A History of Soviet/Russian Meteorological Satellites," *Journal of the British Interplanetary Society* 57, no. 1 (2004): 56–102.

Standard *Meteor* (1) Instrument Suite

Instrument	Number of Spectral Bands	Band Wavelengths (μm)	Ground Swath (km)	Ground Resolution (km)
MR-600 television camera	1	0.5–0.7	1,000	1.25 × 1.25
Lastochka infrared radiometer	1	8–12	1,100	15 × 15
Actinometric equipment	3	0.3–12	2,500	50 × 50

Standard *Meteor 2* Instrument Suite

Instrument	Number of Spectral Bands	Band Wavelengths (μm)	Ground Swath (km)	Ground Resolution (km)
MR-900 scanning telephotometer	1	0.5–0.7	2,100	2
MR-2000 scanning telephotometer	1	0.5–0.7	2,600	1
BCh-100 scanning infrared (IR) radiometer	1	8–12	2,800	8
Scanning IR radiometer	8	11.1–18.7	1,000	37
RMK radiation measurement system		No data available		

Standard *Meteor 3* Instrument Suite

Instrument	Number of Spectral Bands	Band Wavelengths (μm)	Ground Swath (km)	Ground Resolution (km)
MR-900B scanning telephotometer	1	0.5–0.8 μm	2,600	1.0 × 2.0
MR-2000M scanning telephotometer	1	0.5–0.8 μm	3,100	0.7 × 1.4
Klimat scanning IR radiometer	1	10.5–12.5 μm	3,100	3 × 3
174-K scanning IR radiometer	10	9.65–18.7 μm	1,000	42
RMK radiation measurement system		0.15–3.1 MeV (electrons)		

***Meteor 3M* Number 1 Instrument Suite**

Instrument	Number of Spectral Bands	Band Wavelengths (μm)	Ground Swath (km)	Ground Resolution (km)
MR-2000M1 scanning telephotometer	1	0.5–0.8 μm	2,900	1.5 km
Klimat scanning IR radiometer	1	10.2–12.5 μm	3,100	1.7 km
SFM-2 ultraviolet spectrometer	4	0.2–0.51 μm	–	–
MSU-E high-resolution tele-photometer	3	0.5–0.9 μm	76	38 × 38 m
MSU-SM medium-resolution tele-photometer	2	0.5–1.1 μm	2,240	225 m
MIVZA microwave radiometer	5	22–94 GHz	1,700	25–110 km
MTVZA microwave radiometer	21	18.7–183 GHz	2,600	12–75 km
SAGE-3 spectrometer	11	0.29–1.55 μm	–	–
KGI-4S	5	0.17–3.2 MeV (electrons) 5–40 MeV (protons)	–	–

Meteosat. Meteosat was a series of European geosynchronous meteorological satellites, the first of which, *Meteosat 1*, was launched by a Delta rocket from Cape Canaveral in November 1977. Subsequent satellites in the series were launched by European Ariane launchers from French Guiana.

Observing trends in the United States and the Soviet Union, in the late 1960s some European meteorologists desired to use observations from orbiting spacecraft to assist in providing more accurate weather forecasts. At the same time, the World Meteorological Organization (WMO) also suggested to its European members that they emulate the superpowers and launch meteorological satellites. However, not all meteorologists were convinced as to their usefulness, preferring the system they knew, rather than using expensive and unproven new techniques.

Nevertheless, spurred in part to keep pace with NASA, CNES (French Space Agency) decided to start a meteorological spacecraft program called Meteosat. In 1971 the French delegate to the European Space Research Organisation (ESRO) suggested that it be adopted as a European program with CNES technical support,

Images from Meteosat 1 *taken on 15 October 1979 in the visible, infrared, and water vapor bands (from left to right). In the infrared image black is hot and white is cold. So the Sahara desert in North Africa is black and the highest clouds are the whitest. In the water vapor image the whitest areas show the highest atmospheric water vapor content, with the small white areas generally being thunderstorms. (Copyright 2010 EUMETSAT)*

which meant that it would be managed by an ESRO team in France. It would also reduce the cost to France while increasing its user base among the European meteorological services. The ESRO Member States concurred, and ESRO took on the project. In 1973, ESRO selected Aérospatiale as the spacecraft prime contractor, with Matra responsible for the imaging radiometer. The ground processing system for the images was developed in house by ESRO, later the European Space Agency (ESA).

Meteosat was the European contribution to a planned global system of geosynchronous meteorological satellites to be launched in the 1970s. This global system was designed to contribute to the World Weather Watch (WWW) of the WMO. A series of five geosynchronous satellites were to be placed in orbit, equally spaced over the equator. Two orbital slots were to be covered by the United States, and one each by Europe, Japan, and the Soviet Union. Their specifications were coordinated by CGMS (Coordination of Geostationary Meteorological Satellites). However, the Soviet Union (later Russia) did not fill its orbital slot over the Indian Ocean until 1994, and this was variously covered, in the interim, by Meteosat and other satellites. By 2006 the system included satellites from China, Europe (Meteosat), India, Japan, Russia, and the United States.

The Meteosat radiometer produced images in the visible, infrared, and water vapor channels, with resolutions of between 2.5 and 5 km. The images were refreshed every 30 minutes. In addition, the spacecraft was used to transmit processed images from a ground processing facility at the European Space Operations Centre in Germany, in addition to acting as a data relay satellite, retransmitting data received from automatic data collection platforms. The Meteosat processing center also extracted various meteorological parameters from the images, including sea surface temperatures, cloud top heights, and cloud motion winds.

As of 2007 there had been seven first-generation and two second-generation Meteosats launched. The second generation spacecraft produced images in 12 spectral channels instead of three and retransmitted images every 15 minutes instead of every 30.

Meteosat Spacecraft

Spacecraft	Launched	Deorbited	Comments
Meteosat 1	23 Nov 1977		Radiometer failed November 1979
Meteosat 2	19 Jun 1981	Dec 1991	
Meteosat 3	15 Jun 1988	Nov 1995	Refurbished prototype spacecraft. In-orbit spare March 1989. Moved to 75° W February 1993 to replace failed American GOES spacecraft
Meteosat 4 (MOP 1) (Meteosat Operational 1)	6 Mar 1989	Nov 1995	EUMETSAT's first operational spacecraft
Meteosat 5 (MOP 2)	3 Mar 1991		Moved to 63° E in early 1998. Made available in February 2005 for use in global tsunami warning system
Meteosat 6 (MOP 3)	20 Nov 1993		In-orbit spare September 1997. Rapid scanning service started September 2001
Meteosat 7 (MTP) (Meteosat Transition Program) [between MOP and MSG]	2 Sep 1997		Drift to 63° E started July 2006 to take over from *Meteosat 5*
Meteosat 8 (*MSG 1*) (Meteosat Second Generation 1)	28 Aug 2002		Became operational January 2004
Meteosat 9 (*MSG 2*)	21 Dec 2005		

ESRO, then ESA, managed the Meteosat program following its adoption as a European program in 1972. Initially it was a research program, as it was unclear how the images received by Meteosat could be turned unambiguously into useful data, like cloud top heights and wind data. During this period the European meteorological services interacted with ESA via a technical advisory committee, but as the program progressed, the need for a much more coordinated approach from the users became obvious, particularly if the program was to reach an operational status. ESA, as a development organization, preferred to install new technologies on its systems, while meteorologists wanted proven, reliable methods. As a result EUMETSAT was created in 1986 to manage the Meteosat program via ESA and develop the Meteosat products that the meteorological users required. In December 1995, EUMETSAT took control of the Meteosat first-generation satellites in orbit.

David Leverington and John Krige

See also: Aérospatiale, European Space Agency (ESA)

Bibliography

Helen Gavaghan, "The History of EUMETSAT: European Organization for the Exploitation of Meteorological Satellites" (2001).

John Krige and Arturo Russo, *A History of the European Space Agency 1958–1987* (2000).

E. A. Trendelenburg et al., *ESA Bulletin* (December 1977).

National Oceanic and Atmospheric Administration. National Oceanic and Atmospheric Administration (NOAA), within the U.S. Department of Commerce (DOC), was created by executive order in 1970. It is the oldest scientific agency in the U.S. government by virtue of its legacy agencies incorporated into the new organization. The most important of these for space history were the U.S. Coast and Geodetic Survey (C&GS), the U.S. Weather Bureau (USWB), and the Central Radio Propagation Laboratory (CRPL). These three separate agencies had first been coupled in 1965 in the Environmental Science Service Administration (ESSA). In 1970 all were melded into NOAA, encompassing the National Ocean Service, the National Weather Service, and the Office of Oceanic and Atmospheric Research (OAR). The vast data sets they engendered triggered the creation of the National Environmental Sciences, Data, and Information Service (NESDIS). These four are major line offices of NOAA, along with the National Marine Fisheries Service, and NOAA Marine and Aviation Operations.

The CRPL was established in 1946 in Boulder, Colorado, evolving from the Radio Laboratory Section of the National Bureau of Standards, established in 1913. John Howard Dellinger, who investigated solar effects on radio communications, was its chief from 1919 to 1946, and chief of the CRPL from 1946 to 1948. When the CRPL became part of ESSA, it was renamed Institute for Telecommunication Sciences and Aeronomy (ITSA). Two years later, ITSA was divided into the Aeronomy Laboratory and the Institute for Telecommunications Sciences. The latter became part of the National Telecommunications and Information Administration, while the Aeronomy Laboratory became part of ESSA, and from 2005, part of NOAA's Earth System Research Laboratory. The C&GS played a major role in establishing the geodetic reference frame (the Figure of the Earth, the model of Earth's size, shape, and orientation) and the North American Datum (the geopositioning system centered on that continent) on which the major part of the U.S. space program has been based. The Survey's geodetic research was complemented by research in terrestrial magnetism and gravity and other fundamental Earth properties that became paramount when the United States began to launch missiles and rockets. C&GS surveyed the Atlantic Missile Range (Cape Canaveral) and the Pacific Missile Range (Vandenberg Air Force Base) and determined local gravitational deflections from the vertical at the founding of these launch sites. Outside the U.S. mainland, the Survey had throughout many decades created hydrographic and topographic maps of U.S. island possessions, had primary responsibility for determining the U.S.-Canadian border, and developed the primary geodetic network for Alaska, the Aleutian Islands, and their surrounding seas. This proved critical for development of space launching and landing sites and also many space-related aspects of the Distant Early Warning Line and the entire progress of the Cold War, since trans-Arctic Soviet and American ballistic missile flights were at its core.

Starting in the early 1960s, C&GS began a program of satellite triangulation using simultaneous observation of satellites from precisely sited camera observatories; their data complemented the Doppler effect data of The Johns Hopkins University's Applied Physics Laboratory's later Transit satellite program. In 1965 in collaboration with the Department of Defense (DoD) and NASA, the Survey developed the World-Wide Satellite Triangulation Network, a global network of BC-4 camera observatories positioned using the 40 m diameter PAGEOS (Passive Geodetic Earth Orbiting Satellite) balloon satellites. Positioning by visual observation was succeeded by electronic systems. Precise positional distancing by radio phase shift from distant stellar radio waves (Very Long Baseline Interferometry) eventually led to positioning by simultaneous resolving of distances by phase-comparison from active satellite-based systems, as in the Global Positioning System. Later geodetic work involved radar altimetry from satellite platforms, such as *TOPEX-Poseidon*. The geodetic science of the Survey made the occupation of space possible; later satellite altimetry allowed the Office of Coast Survey to characterize the bottom of the ocean from space.

The USWB, which began in 1871 as part of the Army Signals Corps, was civilianized in the Department of Agriculture in 1891, and transferred to the DOC in 1940. The USWB's primary focus was meteorological, hydrological, and climactic prediction and history, but assessment of agricultural production and problems, both national and global, was retained as a major focus. Its major role in space history was based on the development and operation of satellites and their sensors, in three overlapping groups.

The first were Sun-synchronous orbit satellites, including the Tiros series of the early 1960s, followed by the ESSA series from 1966 to 1969, and then Nimbus. Nimbus pioneered instruments to capture vertical distribution of temperature, moisture, and cloud penetration. The second group included the geostationary satellites beginning with the *Applications Technology Satellite 1* (1966). Following the experimental Synchronous Meteorological Satellites *SMS 1* and *SMS 2* (1974–75), NASA and NOAA began the Geostationary Operational Environmental Satellite (GOES) program, using a program model in which NASA performed the development phase of the platform and sensors, and then NOAA assumed operational control and data management. *GOES 1* was launched in 1975, and *GOES 13* was launched in 2006. NOAA maintained two GOES satellites in geosynchronous orbit, called East and West, for weather observation geared to North America. The third class of satellites focused on land cover observations. The quality and utility of the data from Nimbus inspired the creation of the first generation of Earth Application Satellites, including Landsat. The Advanced Very High Resolution Radiometer introduced on *Tiros N* (1978) provided global coverage in the visible and infrared bands at 1 km resolution every 12 hours. The capabilities of that data for global land cover characterization were the foundation for the next generation of sensor systems. By the early 2000s the NASA Earth Observing System (EOS) demonstrated sufficient multi-instrument experimental capabilities that NOAA, NASA, and DoD combined these into a single unified program and platform, the National Polar-orbiting Operational Environmental Satellite System (NPOESS).

The CRPL, the third major component of the agencies that became NOAA, was the major federal facility for basic research on the interactions between radio waves and the nature of the ionosphere and space. Its Tropospheric Telecommunications Division concentrated on the environment up to 10 miles above Earth. The Aeronomy Division researched the upper atmosphere and ionosphere. The Space Environment Forecasting Division concentrated on solar events and how to forecast them and their impacts, especially on the near-space environment of satellites and crewed flight. With the transition from ESSA to NOAA, CRPL and its research melded into NESDIS and OAR.

John Cloud

See also: Space Geodesy

Bibliography

Karl Hufbauer, *Exploring the Sun: Solar Science since Galileo* (1991).
Eileen L. Shea, "A History of NOAA," NOAA. http://www.history.noaa.gov/legacy/noaahistory_1.html.

Nimbus. Nimbus was an early NASA satellite program designed to develop instruments for remote sensing. It had been initially conceived by NASA and the Weather Bureau in 1959 as the first operational weather satellite program to follow the Tiros program. But design delays and disputes between NASA and the Weather Bureau about the system design caused the Weather Bureau to withdraw from Nimbus. Shortly after the launch of *Nimbus 1*, the Bureau became part of the Environmental Science Services Administration (ESSA). There, it developed ESSA satellites for the first U.S. operational weather satellite program. However, it also used Nimbus data to produce weather forecasts.

Nimbus was designed by the Goddard Space Flight Center and integrated by the General Electric Company. It used a multimission bus, with a payload platform, to carry the remote sensing instruments, so payload instruments could be changed from one spacecraft to another without redesigning the bus. Nimbus was three axis stabilized, pointing the instruments continuously at Earth as it orbited in a Sun-synchronous, near-polar orbit. This resulted in complete Earth coverage twice every 24 hours, at about local noon and midnight. Early Nimbus satellites were devoted to meteorology, but the emphasis gradually changed to remote sensing as the program progressed.

The 374 kg *Nimbus 1* spacecraft was launched by a Thor-Agena from Vandenberg Air Force Base on 28 August 1964 with a payload of three meteorological instruments. The Advanced Vidicon Camera System (AVCS) produced images in the visible waveband with a resolution of about 1 km at nadir. The High Resolution Infrared Radiometer (HRIR) produced infrared images, allowing temperature estimates to be

Nimbus Spacecraft

Nimbus Spacecraft Designation	Spacecraft Mass (kg)	Launcher Launch Date	Mission Terminated	Perigee (km)	Apogee (km)	Payload Instruments (number of/ wavebands used)	Comments
1	374	Thor Agena B 28 Aug 1964	23 Sep 1964	429	937	3 Visible (VIS) and Infrared (IR)	Launcher malfunction. Spacecraft failed after less than one month.
2	414	Thor Agena B 15 May 1966	18 Jan 1969	1,093	1,176	4 VIS, IR, far IR	Onboard recorders failed after a few months.
B	570	Thor Agena D 18 May 1968	–	–	–	7	Launch failure.
3	576	Thor Agena D 14 Apr 1969	22 Jan 1972	1,071	1,130	7 Ultraviolet (UV), VIS, IR, far IR	Horizon scanner failed compromising spacecraft operations after 25 Sep 1970.
4	620	Thor Agena D 8 Apr 1970	30 Sep 1980	1,088	1,099	9 (UV), VIS, IR, far IR	Progressive failure starting 14 Apr 1971.
5	770	Delta 11 Dec 1972	29 Mar 1983	1,088	1,101	6 VIS, IR, far IR, microwave	
6	829	Delta 2910 12 Jun 1975	29 Mar 1983	1,101	1,114	9 (UV), VIS, IR, far IR, microwave	
7	832	Delta 2910 24 Oct 1978	14 Feb 1995	943	957	8 (UV), VIS, IR, far IR, microwave	

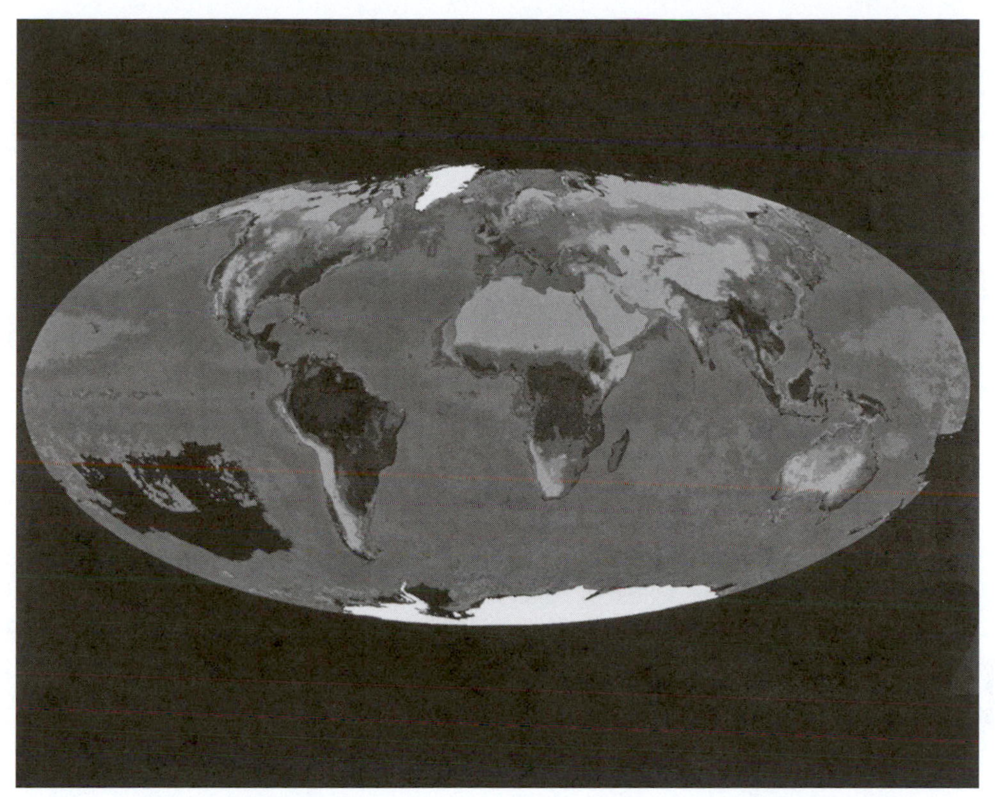

Composite image of the global biosphere constructed from images taken by Nimbus 7 *from November 1978 through June 1980. This is the first global image of the global biosphere constructed from space imagery. (Courtesy NASA)*

made of the ground and cloud tops, and the Automatic Picture Transmission (APT) System produced real-time visible images with a 3 km resolution. AVCS and HRIR images were stored onboard and read out when the spacecraft was within range of the Command and Data Acquisition (CDA) stations. The APT images could be received by anyone with a relatively simple ground receiving system.

NASA had intended to place *Nimbus 1* in a 1,000 km circular orbit, but problems with the launcher resulted in a 429 × 937 km orbit, which reduced the time that the spacecraft was within range of the CDA stations. Less than one month after launch, the solar array drive locked in position, rendering the spacecraft inoperable.

Launched in May 1966, *Nimbus 2* had the same payload as its predecessor but with the addition of another infrared instrument. This enabled the effect of water vapor and ozone on Earth's heat balance to be studied. The onboard tape recorders failed a few months after launch, but real-time data was produced until January 1969 when the mission was terminated.

During the next 12 years, a further five Nimbus spacecraft carried more and more complex experiments, so when *Nimbus 7* was launched in 1978, the total spacecraft mass had more than doubled, while the payload mass had increased by about sixfold. The later payloads operated in all wavebands from ultraviolet to microwave, with the emphasis changing from imagers to sounders. The results were used in meteorology, oceanography, hydrology, geology, geography, cartography, and agriculture.

Among the highlights of the program, *Nimbus 1* provided the first global images of cloud systems, *Nimbus 2* produced ocean temperature measurements. *Nimbus 3* generated atmospheric temperature profiles and was able to locate and interrogate remote sensing platforms. *Nimbus 4* produced the first global maps of ozone distribution. *Nimbus 5* mapped and measured sea ice, and *Nimbus 6* produced data on Earth's radiation budget. *Nimbus 6* was also able to transmit real-time data to ground via a geostationary satellite (*Applications Technology Satellite 6*). *Nimbus 7* was the first satellite to measure natural and human-made atmospheric pollution.

In 1985 scientists of the British Antarctic Survey showed that the ozone layer above Antarctica was being depleted. Scientists quickly realized that the Total Ozone Mapping Spectrometer (TOMS) on *Nimbus 7* should show this effect, but it apparently did not. This was because a filter had been put on the raw TOMS data to eliminate spurious high and low measurements. Fortunately, the raw data was still available, and analyzing it confirmed the British results. In fact, the TOMS instrument showed that there was a hole above the whole of Antarctica that was gradually getting deeper with time.

David Leverington

See also: Goddard Space Flight Center

Bibliography

I. S. Haas and R. Shapiro, "The Nimbus Satellite System: Remote Sensing R&D Platform of the 1970s," in *Monitoring Earth's Ocean, Land, and Atmosphere from Space—Sensors, Systems, and Applications*, ed. Abraham Schnapf (1985).

Janice Hill, *Weather from Above* (1991).

Nimbus Program History: Earth-Resources Research Satellite Program (2004).

NOAA, National Oceanic and Atmospheric Administration Polar Orbiting Satellites. NOAA, National Oceanic and Atmospheric Administration polar orbiting satellites followed the earlier U.S. Tiros and ESSA (Environmental Science Services Administration) civilian weather satellites. The NOAA craft can be classified into four groups: eight Improved Tiros Operational System (ITOS) satellites launched during 1970–76, five Tiros N spacecraft launched during 1978–91, six Advanced Tiros N (Tiros ATN) weather satellites orbited from 1983 to 1994, and four Polar Operational Environmental Satellites (POES) first launched in 1998. NASA's Goddard Space Flight Center managed development of all these satellites, while Radio Corporation of America (RCA) and its successors—General Electric (1986), Martin Marietta (1993), and Lockheed Martin (1995)—built them. NOAA operated the satellites. All were placed into circular Sun-synchronous polar orbits from Vandenberg Air Force Base to ensure consistent, repeated coverage daily. The ITOS series was placed into 1,463-km orbits, while all later NOAA polar orbiters went into lower 840-km orbits, which improved instrument ground resolution.

Improved Tiros Operational Satellites Table 1

Typical orbit 1,463 km circular Sun-synchronous polar. Typical spacecraft mass 308 kg.

Spacecraft Designation	Launcher— Launch Date
Tiros M or *ITOS 1* (NASA Prototype)	Delta N 23 Jan 1970
NOAA 1	Delta N 11 Dec 1970
ITOS B Launch Failure	Delta N 21 Oct 1971

INSTRUMENTS
2 APT Cameras, similar to instrument on *ESSA 2*, *4*, *6*, and *8*
2 Advanced Vidicon Camera Systems (AVCS), similar to on *ESSA 3*, *5*, *7*, and *9*
Low Resolution Flat Plate Radiometer (FPR), similar to on *ESSA 3*, *5*, *7*, and *9*
Solar Proton Monitor (SPM)
2 Scanning Infrared Radiometers

Improved Tiros Operational Satellites Table 2

Typical orbit 1,463 km circular Sun-synchronous polar. Typical spacecraft mass 342 kg.

NOAA 2	Delta 15 Oct 1972
ITOS E Launch Failure	Delta 16 Jul 1973
NOAA 3	Delta 6 Nov 1973
NOAA 4	Delta 15 Nov 1974
NOAA 5	Delta 29 Jul 1976

INSTRUMENTS
2 Very High Resolution Radiometers (VHRR)
2 Vertical Temperature Profile Radiometers (VTPR), a development of Nimbus SIRS sounder
2 Scanning Radiometers (SR)
Solar Proton Monitor (SPM)

Whereas the previous ESSA satellites provided repeatable coverage of any given point on Earth every 24 hours, the ITOS series doubled the coverage by repeating coverage every 12 hours, providing visible and infrared imagery, cloud top and surface temperatures. They were all launched on two-stage Delta rockets. *ITOS 1* carried Advanced Vidicon Camera Systems to gather daytime visible images and Scanning Radiometers for nighttime infrared imagery and temperature measurements. They also carried a Flat Plate Radiometer and a Solar Proton Monitor (SPM) to detect solar storms. *NOAA 2* through *NOAA 5* dispensed with the vidicon cameras and added Very High Resolution and Vertical Temperature Profile Radiometers.

Tiros N Configuration

Typical orbit 840 km circular Sun-synchronous polar. Typical spacecraft launch mass 1,420 kg, on orbit mass 745 kg, of which payload mass 230 kg.

Tiros N (NASA Prototype)	Atlas F 13 Oct 1978
NOAA 6 or *NOAA A*	Atlas F 27 Jun 1979
NOAA B Launcher Problem	Atlas F 29 May 1980
NOAA 7 *NOAA C*	Atlas F 23 Jun 1981
NOAA 12 *NOAA D*	Atlas E 14 May 1991

INSTRUMENTS

Advanced Very High Resolution Radiometer (AVHRR), a development of VHRR on earlier NOAAs

Tiros Operational Vertical Sounder (TOVS), consisted of three instruments:

1. High Resolution Infrared Sounder 2 (HIRS 2), a development of HIRS on *Nimbus 6*
2. Stratospheric Sounding Unit (SSU), provided by the United Kingdom, similar to PMR on *Nimbus 6*
3. Microwave Sounding Unit (MSU), similar to SCAMS on *Nimbus 6*

Space Environmental Monitor (SEM), a development of SPM on earlier NOAAs

Data Collection and Platform Location System, also called ARGOS, provided by France

Advanced Tiros N or Tiros ATN

Typical orbit 840 km circular Sun-synchronous polar. Typical spacecraft launch mass 1,710 kg, on orbit mass 1,025 kg, of which payload mass 370 kg.

NOAA 8 or *NOAA E*	Atlas E 23 Mar 1983
NOAA 9 or *NOAA F*	Atlas E 12 Dec 1984
NOAA 10 or *NOAA G*	Atlas E 17 Sep 1986
NOAA 11 or *NOAA H*	Atlas E 24 Sep 1988
NOAA 13 or *NOAA I*	Atlas E 9 Aug 1993
NOAA 14 or *NOAA J*	Atlas E 30 Dec 1994

INSTRUMENTS

AVHRR, TOVS (that is, HIRS 2, SSU, and MSU), SEM, ARGOS, as above

Search and Rescue Satellite Aided Tracking (SARSAT), provided by France and Canada

Earth Radiation Budget Experiment (ERBE) on *NOAA 9* and *10* only

Solar Backscattered Ultraviolet Radiometer 2 (SBUV 2), a development of SBUV on *Nimbus 7*, on *NOAA 9*,
11, *13*, *14* (in "morning" orbits)

NOAA Polar Operational Environmental Satellite (POES)

Typical orbit 840 km circular Sun-synchronous polar. Typical spacecraft launch mass, for *NOAA 15–17*, 2,230 kg, on orbit mass 1,480 kg. Spacecraft on orbit mass of *NOAA 18*, 1,420 kg.

NOAA 15 or *NOAA K*	Titan II 13 May 1998
NOAA 16 or *NOAA L*	Titan II 21 Sep 2000
NOAA 17 or *NOAA M*	Titan II 24 Jun 2002
NOAA 18 *NOAA N*	Delta 7320 20 May 2005
NOAA 19 *NOAA N Prime*	Delta 7320 6 February 2009

INSTRUMENTS
AVHRR extended
HIRS 3
SEM 2
ARGOS modified
SARSAT modified
SBUV 2
Advanced Microwave Sounding Units (AMSU) A and B, the latter provided by the United Kingdom.

NOAA 18 and *19*: No AMSU B. Microwave Humidity Sounder from EUMETSAT. HIRS 3 replaced by HIRS 4.

Table provided by David Leverington

The launch of *Tiros N* (N = next generation) in October 1978 inaugurated the Advanced Very High Resolution Radiometer (AVHRR) and Tiros Operational Vertical Sounder (TOVS) instruments, which improved the capabilities for infrared imagery and temperature and humidity profiles from the surface to space, respectively. The French-provided ARGOS system allowed the Tiros N series to collect, store, and transmit data from balloons and buoys. Tiros N spacecraft, which included *Tiros N* and *NOAA 6*, *B*, *7*, and *12*, used a spacecraft bus derived from RCA's Defense Meteorological Satellite Program 5D bus design, since the U.S. military had developed a highly capable polar-orbiting spacecraft design. All were launched on Atlas boosters.

The Advanced Tiros N (ATN) series, orbited using Atlas E launchers, tested new capabilities. All Tiros ATN satellites carried Search and Rescue Satellite Aided Tracking system equipment in support of the international COSPAS-SARSAT program (the International Satellite System for Search and Rescue), which detected and relayed distress signals from ships, aircraft, and land travelers. This system, which included Russian, European, and Indian spacecraft, had saved more than 20,500 lives as of 2006. *NOAA 9* and *10* carried the Earth Radiation Budget Experiment, which along with the Earth Radiation Budget Satellite measured Earth's energy balance. *NOAA 9*, *11*, *13*, and *14* included the Solar Backscatter Ultraviolet Radiometer 2 to detect atmospheric ozone.

POES improved the capabilities of the Tiros ATN sensors and also added two Microwave Sounding Units (one provided by the United Kingdom), which measured atmospheric temperature and humidity profiles even in cloudy regions. *NOAA 15*, *16*, and *17* were launched on Titan II boosters and *NOAA 18* and *19* (the final in the series) on Delta 2 launchers. POES spacecraft send their data to NOAA's Command and Data Acquisition (CDA) ground stations. Images and data for any region of Earth, never more than six hours old, are distributed by the National Environmental Satellite, Data, and Information Service (NESDIS).

NOAA 18 was the first in a series of polar-orbiting satellites in the Initial Joint Polar System program with EUMETSAT. *NOAA 18* and *19* carry a EUMETSAT instrument. In return, EUMETSAT is carrying NOAA instruments on its Meteorological Operational Satellites.

To reduce costs, NOAA and Department of Defense weather satellites are being merged into a single National Polar-orbiting Operational Environmental Satellite System (NPOESS) under NOAA. The first is expected to be launched in 2011.

David Leverington, Stephen B. Johnson, and Anthony R. Curtis

See also: Defense Meteorological Satellite Program, Radio Corporation of America, Lockheed Martin Corporation

Bibliography

Air Force Handbook for Congress, Space Section, Key Air Force Space Programs, *National Polar-orbiting Operational Environmental Satellite System (NPOESS)* (1997).

Stanley Kidder and Thomas Vonder Haar, *Satellite Meteorology: An Introduction* (1995).

Ralph Taggart, *Weather Satellite Handbook* (1996).

NPOESS, National Polar-orbiting Operational Environmental Satellite System. NPOESS, National Polar-orbiting Operational Environmental Satellite System, is a program to develop U.S. satellites in low polar orbits, serving civilian and military needs by monitoring global weather, atmosphere, oceans, and land and space environment conditions. From the early 1960s the U.S. government maintained two separate operational weather satellite systems. Civilian weather satellites were operated by the Department of Commerce National Oceanic and Atmospheric Administration (NOAA) weather satellites. Military weather satellites were operated by the Department of Defense (DoD) through the U.S. Air Force Defense Meteorological Satellite Program.

Officials recognized that capabilities of the two systems overlapped and debated merging them into a single system. Changes in world politics in the 1990s and declining budgets prompted the National Space Council in 1992 to recommend combining the separate systems. In 1993 congressional interest and recommendations led NOAA, DoD, and NASA to assess the idea. They submitted to Congress in 1994 an implementation plan, which President William Clinton endorsed and signed. That year the three organizations created an Integrated Program Office (IPO) in NOAA to develop, manage, acquire, and operate NPOESS. NOAA was assigned management of the program, DoD satellite acquisition, and NASA new technology development.

NPOESS Sensors and Payloads

NPOESS Sensors	Acronym	Supplier	Heritage
Cross-track Infrared Sounder	CrIS	ITT	New
Visible/Infrared Imager Radiometer Suite	VIIRS	Raytheon	MODIS
Advanced Technology Microwave Sounder	ATMS	Northrop Grumman	New
Conical Scanning Microwave Imager/Sounder	CMIS	Boeing	DMSP
Ozone Mapping and Profiler Suite	OMPS	Ball	New
Radar Altimeter	ALT	Alcatel (France)	JASON, TOPEX/Poseidon
Earth Radiation Budget Suite	ERBS	Northrop Grumman	NOAA
Total Solar Irradiance Sensor	TSIS	University of Colorado	SORCE
Space Environment Sensor Suite	SESS	Ball—others	DMSP, POES, GOES
Advanced Data Collection System	ADCS	CNES (France), Northrop Grumman	POES
Search and Rescue Satellite Aided Tracking	SARSAT	CNES (France), DND (Canada), Northrop Grumman	POES
Aerosol Polarimetry Sensor	APS	Raytheon	New
Survivability Sensor	SS	Northrop Grumman, Sandia	DMSP

The program planned to establish a Joint Polar Satellite program with EUMETSAT's Meteorological Operational Satellite system.

In addition to monetary savings through reduction in development costs, number of satellites launched, and consolidation of ground systems and satellite operations, NPOESS expanded U.S. capabilities to observe, assess, and predict the total Earth system of atmosphere, ocean, land, and space environment. The NPOESS operational configuration was planned for three satellites in Sun-synchronous polar orbits, each four hours apart in coverage. Each satellite would use the same spacecraft bus, but a different set of instruments, so that combined they provided data on six primary parameters: atmospheric vertical moisture, atmospheric temperature profiles, imagery, sea surface temperatures, sea winds, and soil moisture. The system, integrated by Northrop Grumman Corporation, would also monitor other parameters, including relevant measurements of clouds, aerosol particles, ocean waves, ozone, solar irradiance, and near-Earth plasmas, among others.

The program planned a series of risk-reduction projects, including sensors on the WINDSAT spacecraft, launched in 2003, and on aircraft. By 2007 IPO's first spacecraft, NPOESS Preparatory Project, was under construction for NASA at Ball Aerospace and scheduled to launch in 2009. After program costs ballooned from $6.8 billion to $13.8 billion, the program was restructured in 2006, reducing the number of satellites that Northrop Grumman had to deliver from five to two. The first NPOESS launch was planned for 2011.

Stephen B. Johnson and Anthony R. Curtis

See also: Defense Meteorological Satellite Program, Goddard Space Flight Center, Northrop Grumman Corporation

Bibliography

NPOESS. http://www.history.noaa.gov/legacy/noaahistory_1.html.

Presidential Decision Directive/NSTC-2, "Convergence of U.S.-Polar-Orbiting Operation Environmental Satellite Systems" (5 May 1994).

Tiros. Tiros, whose name derived from Television and Infrared Observation Satellite, was a series of 10 experimental weather satellites launched by NASA between 1960 and 1965. All the spacecraft carried black-and-white television cameras to produce cloud-cover images of Earth in daylight. In addition, some spacecraft carried infrared radiometers to enable meteorologists to estimate cloud top temperatures and investigate Earth's heat balance.

In 1951 a RAND Corporation study investigated the feasibility and potential utility of weather satellites. Three years later, Harry Wexler, director of research for the U.S. Weather Bureau (USWB), followed with a paper titled "Observing the Weather from a Satellite Vehicle." Wexler became an important proponent of the USWB move to satellite technology.

The Tiros program derived from a Radio Corporation of America (RCA) study for the U.S. Air Force (USAF) secret reconnaissance program. RCA failed to win the USAF contract, so it approached the Army Ballistic Missile Agency to develop a satellite with a television camera for meteorology or reconnaissance. When the USAF was assigned the reconnaissance mission exclusively, this program, called Janus 2, focused on meteorology. In April 1959 NASA acquired Janus 2, renamed Tiros. While *Explorer 6* transmitted the earliest experimental television picture of Earth's cloud cover in 1959, *Tiros 1* was the first true weather satellite. In April 1960 the 270 lb *Tiros 1*, built by RCA, was launched to a roughly 700 km high orbit. Within hours, government scientists showed U.S. President Dwight D. Eisenhower a photo of clouds from space.

Tiros 1 and *2* each carried two television cameras, one with a wide angle and one with a narrow angle field of view for more detail. However, meteorologists did not find the narrow-angle camera useful, so it was replaced on later spacecraft by another wide-angle camera. The first four Tiros satellites had orbit inclinations of about 48°, later satellites (*5–8*) improved coverage by increasing inclinations to about 58°, and then (*9–10*) to 96°, a Sun-synchronous polar orbit that provided complete daily coverage of the entire Sun-illuminated portion of Earth. *Tiros 6* performed the first

Tiros Spacecraft (Part 1 of 2)

Tiros Satellite Number	Spacecraft Mass (kg)	Launch Year	Orbit Altitude (km)	Inclination of Orbit to Equator	Payload Instruments	Comments
1	120	1960	720	48°	Television (TV) camera with wide-angle (104°) lens. TV camera with narrow-angle lens (12°).	Six-hour elapsed time from image taking to user. Spacecraft failed after 78 days.
2	130	1960	670	48°	TV cameras as in *Tiros 1*. Five-channel medium resolution infrared (IR) scanning radiometer. Two-channel nonscanning low resolution IR radiometer.	Both radiometers produced generally unreliable results.
3	130	1961	780	48°	Redundant TV cameras with wide-angle (104°) lens. Five- and two-channels IR radiometers as in *Tiros 2*. Low-resolution omnidirectional heat budget radiometer.	Observed all six major hurricanes of the 1961 hurricane season. Five- and two-channel radiometers problematic. Omnidirectional radiometer worked normally.
4	130	1962	770	48°	Same as *Tiros 3*.	Five- and two-channel radiometers problematic. Both TV cameras failed within four months.
5	130	1962	590 × 970	58°	TV camera with wide-angle (104°) lens. TV camera with medium-angle lens (78°).	Elliptical orbit due to launcher problem. Medium-angle camera failed 17 days after launch. The other camera worked well for 11 months.
6	130	1962	700	58°	Same as *Tiros 5*.	Medium-angle camera failed two months after launch. The other camera worked well for 13 months.

(Continued)

Tiros Spacecraft (Part 2 of 2)

Tiros Satellite Number	Spacecraft Mass (kg)	Launch Year	Orbit Altitude (km)	Inclination of Orbit to Equator	Payload Instruments	Comments
7	135	1963	630	58°	Redundant TV cameras with wide-angle (104°) lens. Five-channel medium-resolution IR scanning radiometer. Low-resolution omnidirectional heat budget radiometer. Langmuir probe (to measure electron density and temperature).	Visible images transmitted for 2.5 years. Langmuir probe failed after one month. Omnidirectional radiometer failed after three months.
8	120	1963	730	58°	Single wide-angle (104°) TV camera. Wide-angle (108°) camera and Automatic Picture Transmission (APT) system.	APT system transmitted real-time images to local APT stations.
9	140	1965	700 2,600	96°	Redundant TV cameras with wide-angle (104°) lens.	First "wheel configuration" spacecraft. Spacecraft placed in wrong orbit. One of the cameras failed after 10 weeks.
10	130	1965	790	99°	Same as *Tiros 9*.	First spacecraft funded by U.S. Weather Bureau. Two-hour elapsed time from image taking to user.

Table by David Leverington

experiments to detect snow cover from space. In 1963 *Tiros 8* carried the first automatic picture-transmission system. The new design relayed all spacecraft imagery continually to ground receiving stations.

All the early Tiros spacecraft were spin stabilized, with the cameras, which pointed through the spacecraft's baseplate, aligned parallel to that spin axis. As the spacecraft orbited Earth, the spacecraft's spin axis, and hence its cameras, pointed to the same point in space. As a result, the cameras only imaged Earth for part of every orbit. This problem was solved for *Tiros 9* by orienting its spin axis perpendicular to its orbital plane and placing the cameras 180° apart on the sides of the spacecraft, so the cameras imaged Earth as the spacecraft "rolled" around its orbit. This became the standard configuration for the later Environmental Science Services Administration spacecraft that followed Tiros. The last in the NASA experimental series was *Tiros 10*, launched 1 July 1965.

From the advent of the Tiros meteorological satellite, scientists and governments have monitored weather conditions globally from space, and weather satellite images have become staples of daily news and are an essential public service. The initial Defense Meteorological Satellite Program satellites were based on the Tiros design, as well as a series of civilian weather satellites from the Environmental Science Service Administration Satellites launched in the late 1960s and its successors operated by the National Oceanic and Atmospheric Administration, launched from the 1970s through the early 1990s.

David Leverington, Stephen B. Johnson, and Anthony R. Curtis

See also: Defense Meteorological Satellite Program, Radio Corporation of America (RCA), United States National Weather Service

Bibliography

Janice Hill, *Weather from Above: America's Meteorological Satellites* (1991).

Pamela E. Mack and Ray A. Williamson, "Observing Earth from Space," in *Exploring the Unknown, Volume III: Using Space*, ed. John Logsdon et al. (1998), 155–77.

Abraham Schnapf, "The Tiros Meteorological Satellites—Twenty-Five Years: 1960–1985," in *Monitoring Earth's Ocean, Land, and Atmosphere from Space: Sensors, Systems, and Applications*, ed. Abraham Schnapf (1985), 51–70.

Tiros ATN (Advanced Tiros N). *See* NOAA Polar Orbiting Satellites.

Tiros N. *See* NOAA Polar Orbiting Satellites.

Tiros Operational Satellites. *See* Environmental Science Services Administration Satellties.

United States National Weather Service. The United States National Weather Service (NWS), headquartered in Silver Spring, Maryland, and its predecessors, has provided weather forecasts as a public service since its foundation as the General Weather Service in 1870 when Congress directed the Secretary of War to forecast storms over the Atlantic and Gulf coasts and the Great Lakes. The service was assigned to the Army Signal Corps, which performed the task until it was moved to a new civilian Weather Bureau (WB) in the Department of Agriculture in 1891. When the looming war in Asia and Europe prompted the Army and Navy to establish their own weather services, civilian forecasting was transferred to the WB in the Department of Commerce (DOC) in 1940. In 1965 the WB became part of a new Environmental Science Services Administration (ESSA) along with the U.S. Coast Guard and Geodetic Survey, with climatology separated into a new Environmental Data Service. Two years later the WB was renamed NWS. The National Oceanic and Atmospheric Administration (NOAA) was formed in 1970 in the DOC, with ESSA as one of its major components.

By the early twenty-first century, NWS operated the National Meteorological Center at Camp Springs, Maryland; National Hurricane Center at Coral Gables, Florida; and National Severe Storms Forecast Center at Kansas City, Missouri.

The space age added satellites to NWS forecasting capabilities, starting with *Tiros 1* in 1960. Satellite data, enhanced with improved computing capabilities and ground measurements, provided increasingly accurate short- and long-term forecasts. In particular, forecasting of severe weather, such as hurricanes, was significantly improved. By the 1990s Geostationary Operational Environmental Satellites (GOES) and Polar Operational Environmental Satellites (POES) were operated by NOAA National Environmental Satellite, Data, and Information Service (NESDIS). NESDIS supported a variety of functions, including aviation and maritime safety. Its search-and-rescue instruments helped save 10,000 people. NESDIS also operated satellites in the Department of Defense's Defense Meteorological Satellite program. NWS represented the United States in the World Meteorological Organization.

Anthony R. Curtis

See also: National Environmental Satellite, Data, and Information Service (NESDIS); United States Department of Commerce; National Oceanic and Atmospheric Administration

Bibliography

Janice Hill, *Weather from Above: America's Meteorological Satellites* (1991).
NOAA History. http://www.history.noaa.gov/.

Civilian Remote Sensing

Civilian remote sensing generally refers to remote sensing from space of Earth's land and ocean surface and atmosphere for civilian purposes over the near to long term. In this article it excludes meteorology, solid Earth, and Sun-Earth interactions, which are covered elsewhere.

NASA was the world's first organization to start a civilian remote sensing program from space when it launched experimental sensors on its Nimbus spacecraft. Early Nimbus spacecraft were devoted to developing meteorological sensors, but in 1969 *Nimbus 3* carried sensors to measure solar ultraviolet radiation and to determine the atmosphere's chemical constitution and temperature profile. This enabled meteorologists to construct simple atmospheric models, which had both short-term weather and long-term climatological applications.

Nimbus 4 was the first spacecraft to measure the vertical distribution of ozone globally, and *Nimbus 5* was the first to carry a microwave sensor, which could see through clouds. In 1978 *Nimbus 7* carried the coastal zone color scanner that measured chlorophyll concentrations and was the first spacecraft to measure both natural and human-made atmospheric pollution. Over about 10 years, NASA and the scientific community had developed instruments and techniques using the Nimbus spacecraft that, in one form or another, were still in use in 2007.

In parallel, NASA had started in the late 1960s to develop a land remote sensing spacecraft program. Unfortunately, unlike weather, the potential user base was spread among many different organizations. NASA, being a technological organization, wanted to build a state-of-the-art high-tech spacecraft to test the application of new technologies. The potential users, on the other hand, wanted a more conservatively designed spacecraft using technologies that they understood. Eventually a compromise experimental program, soon called Landsat, was approved, with visible and infrared imagers. By the early twenty-first century, seven Landsats had been launched in a program that had operated for more than 30 years.

Imaging of the Landsat type is important for most land remote sensing purposes. But this caused a major problem with image processing for Landsat in the early days. This was because the images were being downloaded from the spacecraft in an essentially continuous stream, with no opportunity for the processing system to catch up if it fell behind. In addition, most of the images required some form of distortion correction, which caused a large problem with the relatively simple computers of the time.

In the late 1970s there was a push from the Jimmy Carter U.S. presidential administration to transfer operation and funding of the Landsat program to private industry, and in 1985 the Earth Observing Satellite Corporation (EOSAT) took over the operational Landsat program. However competition from the higher-performing French SPOT system caused major financial problems for EOSAT, and the United States was eventually forced to renationalize Landsat.

The sea is a dynamic environment and is far more important than the land and atmosphere in driving the world's weather and climate. Yet scientists knew little about it in a global context until the advent of space remote sensing. At that time, oil companies were deploying large oil rigs on the continental shelf, but climate data was sketchy, with little data on maximum wind speeds, wave heights, and ice conditions, so the rigs had to be built with large safety margins. In 1969 a conference was held in Williamstown, Massachusetts, to define what was required from an ocean remote sensing spacecraft system. However, it was not until 1975 that the program, later called Seasat, was approved, like Landsat, as an experimental system.

Seasat included a number of microwave instruments in its payload, including the first space-borne Synthetic Aperture Radar (SAR), which could image the ocean surface through clouds. Unfortunately, *Seasat* operated for only just over three months. But in this time it demonstrated the enormous potential of ocean remote sensing from space, providing the first global views of the ocean and sea floor topography, plus data on wind speed, wave height, sea surface temperatures, and ice conditions. However, as with Landsat, the length of time it took to analyze the raw data made it less useful to the users.

Some Earth Resource and Climatological Spacecraft and Shuttle Payloads (Part 1 of 7)

Spacecraft	Meaning of Acronym	Launched	Mass (kg)	Country	Purpose/Comments
Nimbus 3 and *4*		1969, 1970	580, 620	United States (USA)	Development of remote sensing instruments. *Nimbus 3* was the first Nimbus spacecraft to carry a non-imaging instrument.
Landsat 1, *2*, and *3*		1972, 1975, 1978	800, 950, 960	USA	First American Earth Resource spacecraft. Dedicated to land observations.
Nimbus 5		1972	770	USA	Development of remote sensing instruments. First Nimbus to carry a microwave sensor.
Meteor 1–18 (Meteor-Priroda)		1974	3,800	Soviet Union	First of a series of five experimental Soviet Earth resource spacecraft. (The last one, *Meteor 1-31*, was launched in 1981.)
Nimbus 6		1975	830	USA	Development of remote sensing instruments.
HCMM/AEM 1	Heat Capacity Mapping Mission/ Applications Explorer Mission	1978	120	USA	Land measurements. Produces land thermal maps to determine rock types, measure plant-canopy temperatures, soil moisture effects, thermal effluents, and temperature of snow fields.
Seasat		1978	2,300	USA	Measured sea state, sea temperature, atmospheric water vapor, and clouds.
Nimbus 7		1978	830	USA	Development of remote sensing instruments.
SAGE/AEM 2	Stratospheric Aerosol and Gas Experiment	1979	150	USA	Aerosol and ozone measurements.

Some Earth Resource and Climatological Spacecraft and Shuttle Payloads (Part 2 of 7)

Spacecraft	Meaning of Acronym	Launched	Mass (kg)	Country	Purpose/Comments
Bhaskara 1 and *2*		1979, 1981	440 each	India	Experimental spacecraft. Research in hydrology, forestry, geology, and oceanography.
Kosmos 1076 and *1151* (Okean E series)		1979, 1980	1,950 each	Soviet Union	Ocean, particularly ice monitoring.
SIR A, *B*, and *C*	Spaceborne Imaging Radar	1981, 1984, 1994	3,000–11,000	USA	Shuttle-borne Synthetic Aperture Radars. Applications to agriculture, including crop identification and soil moisture, terrain mapping in the tropics, and oceanography.
Landsat 4 and *5*		1982, 1984	1,940, 1,950	USA	First Landsats with Thematic Mapper.
FSW 0-4	Fanhui Shi Weixing	1982	1,800	China	Recoverable Test Satellite, mapping, first real-time imagery.
Kosmos 1500 and *1602* (Okean 0E series)		1983, 1984	1,950 each	Soviet Union	Ocean, particularly ice and fisheries monitoring.
ERBS	Earth Radiation Budget Satellite	1984	2,500	USA	Measured Earth's radiation budget over time. (Spacecraft operational for 21 years.)
Geosat		1985	640	USA	Measured sea surface heights.
Resurs 01, 1–4		1985–1998	1,500–2,000	Soviet Union/Russia	Land and ocean.
SPOT 1, *2*, and *3*	Satellite Pour l'Observation de la Terre	1986, 1990, 1993	1,830, 1,870, 1,910	France	Land imaging. Ozone and aerosol measurement (*SPOT 3* only).
Kosmos 1766 (Okean O1)		1986	1,950	Soviet Union	Ocean, particularly ice and fisheries monitoring. (First of eight operational spacecraft similar to *Kosmos 1500*).

(Continued)

Some Earth Resource and Climatological Spacecraft and Shuttle Payloads (Part 3 of 7)

Spacecraft	Meaning of Acronym	Launched	Mass (kg)	Country	Purpose/Comments
Kosmos 1870		1987	18,500	Soviet Union	All-weather observations of the ocean, ice, and land.
Momo 1 and *1B* (MOS 1 and 1B)	Marine Observation Satellite	1987, 1990	750 each	Japan	Monitored ocean currents, sea surface temperatures, ocean chlorophyll, atmospheric water vapor, and land vegetation.
IRS 1A and *1B*	Indian Remote Sensing	1988, 1991	980 each	India	India's first operational Earth Observation satellites.
SSBUV	Shuttle Solar Backscatter UV Spectrometer	1989–1996	410	USA	Flew on eight shuttle flights. Measured vertical profiles of atmospheric ozone concentration.
Almaz		1991	18,500	Soviet Union	As *Kosmos 1870*.
UARS	Upper Atmosphere Research Satellite	1991	7,000	USA	Comprehensive survey of chemicals (including carbon dioxide, ozone, and chlorofluorocarbons) in the upper atmosphere. Measured stratospheric winds and temperatures and variations in solar energy.
ERS 1	European Remote Sensing	1991	2,300	Europe	All-weather observation of the ocean and land, coastal waters, and ice sheets.
Fuyo (JERS)	Japanese Earth Resource Satellite	1992	1,340	Japan	Land surveys, agriculture, forestry, fisheries, environmental preservation, coastal surveillance, and disaster monitoring.
ATLAS 1, *2*, and *3*	Atmospheric Laboratory of Applications and Science	1992, 1993, 1994	3,800–6,900	USA/multinational	Shuttle-borne remote sensing laboratories measured the constitution of Earth's atmosphere and total solar irradiance.

Some Earth Resource and Climatological Spacecraft and Shuttle Payloads (Part 4 of 7)

Spacecraft	Meaning of Acronym	Launched	Mass (kg)	Country	Purpose/Comments
TOPEX/Poseidon	Topography Experiment	1992	2,400	USA/France	Measured height of ocean to high accuracy, to study ocean circulation.
Landsat 6 and *7*		1993, 1999	2,200 each	USA	*Landsat 6* built during privatization era. *Landsat 7* built with government money.
LITE	Lidar In-Space Technology Experiment	1994	2,000	USA	Shuttle-borne experiment. Measured atmospheric structure, including vertical structure of clouds.
IRS P2 and *P3*		1994, 1996	870, 930	India	First Indian launched Earth Resource Spacecraft.
ERS 2		1995	2,400	Europe	As *ERS 1*, plus ozone measurements.
IRS 1C and *1D*		1995, 1997	1,350 each	India	Second generation instruments.
Radarsat 1		1995	2,750	Canada	Sea and land ice monitoring, environmental pollution monitoring, oceanography, and forestry
Midori I (ADEOS I)	Advanced Earth Observing Satellite	1996	3,500	Japan	Atmosphere and ocean observation and measurement.
Mir-Priroda		1996	19,700	Russia/USA	Measured atmospheric impurities, sea surface temperatures, sea–atmosphere energy exchange, structure and moisture of clouds, geological structures.
TOMS-EP	Total Ozone Mapping Spectrometer, Earth Probe	1996	250	USA	Global mapping of total ozone.
TRMM	Tropical Rainfall Measuring Mission	1997	3,500	USA/Japan	Measured rainfall in the tropics and subtropics, and Earth radiation budget.
SPOT 4		1998	2,500	France	Land imaging.
FORMOSAT 1 (was *ROCSAT 1*)	Formosa Satellite (Republic of China Satellite)	1999	400	Taiwan (Formosa)	Ocean remote sensing, plus communications and ionospheric physics.

(Continued)

Some Earth Resource and Climatological Spacecraft and Shuttle Payloads (Part 5 of -365)

Spacecraft	Meaning of Acronym	Launched	Mass (kg)	Country	Purpose/Comments
OCEANSAT 1 (*IRS P4*)		1999	1040	India	Ocean color monitoring.
QuikSCAT	Quick Scatterometer	1999	970	USA	Measured surface wind vectors over the oceans.
Terra (*EOS AM 1*)	Earth Observing System	1999	5,100	USA	Flagship satellite of NASA's EOS. Made land, ocean, and atmospheric measurements.
ACRIMSAT	Active Cavity Radiometer Irradiance Monitor Satellite	1999	115	USA	Measured total solar irradiance.
Okean O		1999	6,400	Russia	Oceanography, ice observation, and monitoring oil pollution.
CBERS 1 (*ZY 1A*) and *CBERS 2* (*ZY 1B*)	China Brazil Earth Resources Satellite	1999, 2003	1,500	China/Brazil	Land imager.
KOMPSAT 1 (*Arirang 1*)	Korea Multipurpose Satellite	1999	470	Korea	Land and ocean monitoring.
NMP/EO 1	New Millennium Program, Earth Observing	2000	570	USA	Technology demonstration for follow-on Landsat. Flew in formation with *Landsat 7*.
Shuttle Radar Topography Mission		2000	13,600	USA/ Germany/ Italy	High resolution topography.
SAC C	Satelite Aplicanciones Cientificas	2000	470	Argentina/ multinational	Study of terrestrial, marine, and atmospheric environment
Jason 1		2001	500	France/USA	*TOPEX/Poseidon* follow-on to measure height of ocean surface to high accuracy.
Meteor 3M		2001	2,500	Russia/USA	Land and ocean imaging, atmospheric aerosol and ozone profiles, space weather, and meteorological observations.

Some Earth Resource and Climatological Spacecraft and Shuttle Payloads (Part 6 of 7)

Spacecraft	Meaning of Acronym	Launched	Mass (kg)	Country	Purpose/Comments
SPOT 5		2002	3,100	France	Land imaging.
ENVISAT	Environmental Satellite	2002	8,200	Europe	Comprehensive Earth observation mission for land, atmospheric, and ocean studies.
Haiyang 1 (*HY 1*)		2002	360	China	Ocean observation, including chlorophyll concentration, sea surface temperature, sedimentation, and ocean pollution.
Aqua (*EOS PM 1*)		2002	3,100	USA	Atmosphere, ocean, land, and ice observations and measurements.
Midori II (*ADEOS 2*)	Advanced Earth Observing Satellite	2002	3,700	Japan	Atmosphere, ocean, and ice observation and measurement.
ICESat (*EOS LAM*)	Ice, Cloud, and Land Elevation Satellite	2003	970	USA	Measured ice sheet mass balance, cloud and aerosol heights, land topography, and vegetation characteristics.
RESOURCESAT 1 (*IRS P6*)		2003	1,360	India	Crop discrimination for agriculture.
SORCE	Solar Radiation and Climate Experiment	2003	290	USA	Measured variations in solar output and in Earth's radiation budget. Follow-on from *ACRIMSAT*.
Aura (*EOS Chem 1*)		2004	3,000	USA	Measured ozone and air quality.
PARASOL	Polarization and Anisotropy of Reflectances for Atmospheric Science coupled with Observations from a Lidar	2004	110	France	Cloud and aerosol measurements.
Shiyan 1 (*SY 1*)/*Tansuo 1* (*TS 1*) and *SY 2* (*TS 2*)		2004, 2004	200 each	China	Land imaging.

(Continued)

Some Earth Resource and Climatological Spacecraft and Shuttle Payloads (Part 7 of 7)

Spacecraft	Meaning of Acronym	Launched	Mass (kg)	Country	Purpose/Comments
Sich 1M		2004	2,200	Russia/ Ukraine	Ice and land monitoring. Atmospheric profiling.
FORMOSAT 2 (was *ROCSAT 2*)		2004	760	Taiwan (Formosa)	Environmental monitoring, agriculture, urban planning, and ocean surveillance. Military applications.
CARTOSAT 1		2005	1,560	India	Stereoscopic mapping mission.
Monitor E		2005	700	Russia	Land resources and ecology, geological mapping. (This is the first of a series of new Russian Earth observation spacecraft.)
Daichi (*ALOS*)	Advanced Land Observation Satellite	2006	4,000	Japan	Cartography, environmental, and disaster monitoring and resource surveying.
ESSP/Cloudsat	Earth System Science Pathfinder	2006	1,000	USA	Measured altitude, structure, and properties of clouds.
ESSP/CALIPSO	Cloud-Aerosol Lidar and Infrared Pathfinder Satellite Observations	2006	560	USA/France	Cloud and aerosol measurements. (Spacecraft flew in formation with *Cloudsat*.)
Yaogan 1 (*JB 5 1*)		2006	2,700	China	Surveying, crop yield estimates, and disaster prevention and relief.
KOMPSAT 2 (*Arirang 2*)		2006	800	Korea	Land imaging for mapping, urban planning, and hazard management. Also used for military purposes.
Resurs-DK1		2006	6,600	Russia	Resource management and environmental pollution of the sea and land. Includes provision of high-resolution surface images on a commercial basis.

In the wake of the *Challenger* accident, the Ride Report recommended in 1987 that America should invest in Mission to Planet Earth (MTPE) to understand Earth and its environment as an interactive system on all timescales. The spacecraft envisaged included four large Sun-synchronous polar platforms (two provided by the United States, one by Europe, and one by Japan) and a series of low-altitude spacecraft, supplemented by retained Shuttle payloads.

Although the United States embraced the MTPE concept, the program soon underwent major changes, with the cancellation of the polar platforms. However, by the 1990s the United States developed such an MTPE program (in 2007 called Earth Science Enterprise) with the Earth Observing System (EOS) satellites, supported by the EOS Data and Information System (EOSDIS), being the largest element.

The Soviet Union's first civilian remote sensing spacecraft were developed in the early 1970s. It was difficult for such a secretive country to start a nonmilitary spacecraft program that would allow the outside world to view its landmass. But the Soviet Union knew that American military spy satellites had been imaging its country for many years, and this activity would soon spread to the civilian area with American plans to launch Landsat. Nonetheless, the Soviet Union could still not bring itself to produce high-resolution images of its country for general consumption. So it developed two types of spacecraft in parallel. One, Meteor Priroda, produced low-resolution images in real time, and the other, Fram, produced higher-resolution images on film that was returned to Earth for evaluation. This latter system, which was partly military, enabled Soviet authorities to control nonmilitary access to the images.

A number of Meteor Priroda spacecraft were launched throughout the next 10 years, but in 1985 the Soviet Union launched the first of its next-generation, civilian remote sensing spacecraft called Resurs O. These provided moderate resolution images, which the Soviet Union offered for sale on the commercial market. In 2006 Russia launched Resurs DK, which was the first Russian spacecraft to provide high-resolution images commercially.

The first Chinese remote sensing spacecraft of the mid 1970s used recoverable film capsules, but in 1982 China launched its *FSW 0-4* spacecraft that provided real-time images. Its remote sensing program concentrated on mapping spacecraft throughout the next two decades, but in 2002 China launched its first marine survey spacecraft, *Haiyang 1*. Four years later, China launched *Yaogan 1*, its first spacecraft to carry a SAR for high-resolution, all-weather imaging.

India in the 1960s, when it began to consider a space program, was a poor country with a large population, a large land area, and a large coastline. With relatively limited resources, India devoted its attention to applications programs like remote sensing and communications, which could potentially make a real difference to the lives of its people. Its early remote sensing spacecraft, *Bhaskara 1* and *2* launched in 1979 and 1981, showed what could be done, and since then India has launched a number of remote sensing spacecraft. All its early spacecraft were devoted to land observation, but in 1999 India launched its first ocean remote sensing spacecraft mainly to support its fishing industry.

Japan is a relatively small country with a large population and a large fishing industry. So it was natural that its first remote sensing spacecraft, *Momo 1*, launched in 1987, should be devoted to oceanography. Japan is also susceptible to earthquakes, so its first land remote sensing spacecraft, *Fuyo*, launched in 1992 was designed, among other things, to measure vertical land deformations in attempt to provide early warning of earthquakes. Later Japanese remote sensing spacecraft studied the ocean, land, sea ice, and atmosphere. Another spacecraft, *Daichi*, was used to produce accurate maps of Japan.

In 2007 France had the largest national space program in Western Europe, built on its desire to be independent of the United States and to develop high-tech industries, like nuclear power and space. In the 1970s France started to develop its land remote sensing spacecraft, SPOT, and tried to persuade European Space Agency (ESA) Member States to Europeanize it. The Member States rejected that in 1977. Nevertheless, France continued with the program, launching the first SPOT spacecraft in 1986. Its success, and that of its later variants, caused the United States to abandon its attempt at privatizing Landsat.

Having rejected the Europeanization of SPOT, ESA Member States decided to develop their own remote sensing spacecraft. This was to become *ERS 1* (European Remote Sensing), launched in 1991, as primarily an ocean monitoring spacecraft. As such it was complementary to SPOT. *ERS 2*, launched four years later, was similar to its predecessor, but with the addition of an ozone monitoring instrument.

ESA had been working on a polar platform as part of the *International Space Station* while working on ERS. But this polar platform, like those of the United States, had been canceled. In ESA the polar platform mission was split between a remote sensing mission, called Envisat, and a meteorological mission. *Envisat* at 8 tons was, at that time, the heaviest Earth resource spacecraft to date. Orbited in 2002 it was used to detect forest fires, deforestation, floods, ozone depletion, and marine pollution.

In the mid 1970s Canada was interested in gas and oil exploration in the ice-infested seas north of its mainland. The area is vast, and the pack ice moves frequently, so air reconnaissance was of limited use. As a result, Canada decided to develop a space-based observation system called Radarsat, using a SAR optimized for ice detection. *Radarsat 1* was launched in 1995, and its successor, *Radarsat 2*, was launched in 2007.

In the 1990s and early 2000s, a number of less-developed countries began remote sensing spacecraft programs, in particular Taiwan with FORMOSAT, Brazil (and China) with CBERS, Korea with KOMPSAT, and Argentina with SAC C.

All countries are naturally interested in their own landmasses and any surrounding sea areas, and many of the space programs mentioned above have been designed to monitor these. But Earth's atmosphere and global ocean are two resources that are of common concern to all nations. The response of the atmosphere and ocean to both human-generated and naturally occurring pollution and to changes in solar energy output is of concern to all. A number of spacecraft have been launched, and will be launched, to understand the total Earth system of atmosphere, ocean, land, and polar ice masses. This whole system has to be monitored and understood if humankind is to have any chance of recognizing and combating the threat of global climate change.

David Leverington

See also: Civilian Meteorological Spacecraft, Commercial Remote Sensing, Earth Science, Reconnaissance and Surveillance

Bibliography

Ghassem Asrar and Jeff Dozier, *EOS: Science Strategy for the Earth Observing System* (1994).

Pamela E. Mack, *Viewing the Earth: The Social Construction of the Landsat System* (1990).

Claire L. Parkinson et al., *Earth Science Reference Handbook: A Guide to NASA's Earth Science Program and Earth Observing Satellite Missions* (2006).

Abraham Schnapf, ed., *Monitoring Earth's Ocean, Land, and Atmosphere from Space—Sensors, Systems, and Applications* (1985).

Application Explorer Missions. Application Explorer Missions (AEM) was a series of three Earth-observing spacecraft including *Heat Capacity Mapping Mission* (*HCMM*), *Stratospheric Aerosol and Gas Experiment* (*SAGE*), and *Magnetic Field Satellite* (*Magsat*). *HCMM*, also known as *AEM1* and *Explorer 58*, was a 134 kg spacecraft launched by a Scout D from Vandenberg Air Force Base, California, on 26 April 1978 into a 600 km, Sun-synchronous orbit. Built by Boeing and managed by Goddard Space Flight Center (GSFC), its mission was to measure Earth's surface temperature at high spatial resolution (500 m^2). Operating until 30 September 1980, it carried a heat capacity mapping radiometer (.55–1.1 and 10.5–12.5 μm). Thermal surface maps were used for a number of purposes, including identifying mineral deposits and monitoring soil temperature cycles and snow fields.

NASA launched *SAGE*, also known as *AEM2* and *Explorer 60*, from Wallops Island, Virginia, on 18 February 1979 on a Scout D. The 147 kg spacecraft, built by Boeing and managed by Langley Research Center, was in a 548 × 660 km, 55° orbit and operated until November 1981. The four-channel (385, 450, 600, and 1,000 nm wavelength) Sun photometer measured the effect of changes in solar radiation on atmospheric aerosols and ozone by observing the Sun through the atmosphere at sunrise and sunset. In conjunction with the Stratospheric Aerosol Measurement instrument on *Nimbus 7*, scientists obtained nearly global coverage. NASA used *SAGE* to track the spread of volcanic particles following the eruptions of Mount St. Helens and La Soufrière. Future versions of the SAGE instrument later flew on the *Earth Radiation Budget Satellite* and *Meteor 3M*.

AEM3, called *Magsat* and *Explorer 61*, was launched 30 October 1979 on a Scout G rocket from Vandenberg. A joint GSFC/U.S. Geological Survey mission built by the Johns Hopkins University Applied Physics Laboratory (using spare parts from *Small Astronomy Satellite 3*), the 181 kg spacecraft was in a 350 × 551 km Sun-synchronous orbit. During its seven-and-one-half month mission, it acquired the first quantitative survey of Earth's magnetic field from low Earth orbit and a global map of magnetic anomalies in Earth's crust, using a vector magnetometer and a scalar magnetometer.

Katie J. Berryhill

See also: Boeing Company; ERBS, *Earth Radiation Budget Satellite;* Explorer; Goddard Space Flight Center; Meteor; Nimbus

Bibliography

Robert A. Langel and William J. Hinze, *The Magnetic Field of the Earth's Lithosphere: The Satellite Perspective* (1998).

M. P. McCormick et al., "Satellite Studies of the Stratospheric Aerosol," *Bulletin of the American Meteorological Society* 60 (September 1979): 1038–46.

Nicholas M. Short and Locke M. Stuart, *The Heat Capacity Mapping Mission (HCMM) Anthology* (1982).

Cartosat. *See* Indian Remote Sensing Satellites.

CBERS, China Brazil Earth Resources Satellite. *See* Chinese Remote Sensing Satellites.

Chinese Remote Sensing Satellites. Chinese remote sensing satellites include the Fanhui Shi Weixing (FSW) series of recoverable satellites, the Jian Bing (JB) 4 and 5 satellites, the Haiyang and Zi Yuan (ZY) Earth resources satellites, the international China–Brazil Earth Resources Satellite (CBERS) program, and the Tansuo and *Beijing 1* microsatellites.

Approved in 1965, FSW spacecraft were developed by the Chinese Academy for Space Technology, and the Long March 2 launcher was developed to place them into orbit. Like U.S. and Soviet reconnaissance satellites of the period, FSW satellites were designed to take film images from space and return them to Earth, though it is not known with certainty if the mission was specifically military. The FSW program entailed three series of spacecraft—FSW 0, 1, and 2, and was the ancestor of the JB program, which had its first launch in 2003. The program proved challenging, as China had to develop technologies significantly beyond its capabilities of the time, such as nonablative shielding to survive reentry.

The first launch attempt in November 1974 engendered a catastrophic launch pad explosion. One year later *FSW 0-1* was launched and successfully recovered after a three-day flight. Eight further FSW 0 satellites were orbited by August 1987.*FSW 0-4* (September 1982) extended the mission length to five days and added charge-coupled device (CCD) cameras for real-time imagery retrieval. *FSW 0-9* was the last FSW 0 mission and the first to fly microgravity materials and biology experiments, including two experiments for the French company Matra. The FSW 1 series constituted five spacecraft launched from 1987 to 1993. These were primarily mapping and microgravity experiment satellites, some of which (*FSW 1-2*—Germany and *FSW 1-4*—Japan) carried foreign payloads. *FSW 1-3* was the first Chinese mission with animals on board. The three FSW 2 spacecraft, launched from 1992 to 1996, extended missions to 18 days, increased payload capability to 350 kg, and could maneuver in orbit. They also continued microgravity materials, biology, and mapping experiments.

The first of the JB satellites (sometimes referred to as the FSW 3 series), *JB 4A1*, was launched in November 2003 from Jiuquan on a Long March 2D booster. These satellites improved pointing accuracy for better on-orbit mapping and performed a variety of scientific experiments. In August and September 2004 and again in August 2005, JB 4 launches occurred in pairs roughly four weeks apart, which Chinese authorities have not explained. The JB 5 series orbited China's first synthetic aperture radar, capable of producing high resolution imagery at night and in any weather. It was first launched in April 2006 and was used for military and civilian purposes.

China's State Oceanic Administration developed marine survey satellites, Haiyang, first launched with the *Feng Yun 1D* meteorological satellite in May 2002. Each

Chinese Remote Sensing Satellites (Part 1 of 2)

Name	Meaning	Launcher	Launch Dates	Orbit Altitude and Inclination	Comments
Fanhui Shi Weixing (FSW) 0	Recoverable Test Satellite	Long March (CZ) 2	1974–1987	~200 × 400 km, 57–70°	Ten launches, one launch failure, remote sensing, last mission included materials processing
FSW 1		CZ 2	1987–1993	~200 × 300 km 63°	Five launches, one recovery failure, remote sensing and microgravity experiments
FSW 2		CZ 2	1992–1996	170 × 350 km 63°	Three successful launches
FSW 3/Jian Bing (JB) 4	JB = Pathfinder	CZ 2	2003–2005	140–300 × 160–550 km 63°	Five successful launches, surveying, mapping, improved pointing control, last four launched in pairs, each one month apart
JB 5/Yaogan	Yaogan = Remote Sensing	CZ 4B	2006–2007	650 km near-circular ~98°	Three successful launches, synthetic aperture radar
CBERS/Zi Yuan (ZY) 1	CBERS = China Brazil Earth Resources ZY = Earth Resources	CZ 4B	1999–2007	775 km circular Sun-synchronous ~98.5°	Three successful launches; cameras: —high resolution (20 m) —wide field (260 m) —multispectral (80 m)
ZY 2		CZ 4B	2000–2004	490 km near-circular Sun-synchronous ~97.5°	Three successful launches, possible same cameras as CBERS, possible military reconnaissance

(Continued)

Chinese Remote Sensing Satellites (Part 2 of 2)

Name	Meaning	Launcher	Launch Dates	Orbit Altitude and Inclination	Comments
Haiyang 1, 2 (HY)	Ocean	CZ 4B	2002, 2007	790 km near-circular 98.5°	Two successful launches, ocean remote sensing, two-band color scanner, four-band scanner
Tansuo 1, 2 (TS)/ *Shiyan 1, 2* (SY)	Tansuo = Exploration Shiyan = Experiment	CZ 2C	2004	600–700 km near circular ~98° inclination	Two successful launches, 10 m stereo resolution camera
Beijing 1/ *Tsinghua*	Beijing = China's capital Tsinghua = Tsinghua University	Kosmos 3 M	2005	683 × 700 km 98.1	Land resources and topography, 4 m panchromatic, 32 m multispectral, part of Disaster Monitoring Constellation

spacecraft carried a 10-band color scanner and a four-band CCD scanner to observe ocean temperatures, pollutants, sea ice, and other ocean phenomena.

The China Academy of Space Technology developed non-recoverable Earth observation satellites, initially in the joint CBERS program with Brazil. *CBERS 1* was launched in October 1999 on a Long March 4B from Taiyuan, followed four years later by *CBERS 2*. These two craft each carried a CCD camera with 20 m resolution, a multispectral infrared scanner, and a wide field imager, being used for a variety of land-use observations and research programs. China and Brazil signed a new agreement soon after the launch of *CBERS 2*, increasing Brazil's contribution by 20 percent to achieve parity with China on the next two spacecraft, the next of which launched in 2007. In the meantime China launched its domestic version of CBERS, the ZY 2 spacecraft, which were placed in lower orbits than CBERS, therefore enabling higher resolution imagery. These craft may have military as well as civilian functions.

China also developed and operated remote sensing microsatellites. Two Tansuo satellites were developed by Harbin Institute of Technology and the Chinese Research Institute of Space Technology. Launched in 2004, the two Tansuo (sometimes referred to as Shiyan) spacecraft each carried stereo cameras. The *Beijing 1* (sometimes called Tsinghua) satellite, built by Surrey Satellite Technology of the United Kingdom for Beijing Landview Mapping Information Technology Company, was launched with several other spacecraft on a Kosmos 3 M launcher from Plesetsk Cosmodrome in October 2005 and was China's contribution to the international Disaster Monitoring Constellation.

Stephen B. Johnson

See also: Brazil, China, Long March

Bibliography

Phillip S. Clark, "Development of China's Recoverable Satellites," *Quest* 6, no. 2 (1998): 36–43.
Brian Harvey, *China's Space Program: From Conception to Manned Spaceflight* (2004).
China Defence Today. Sinodefence.com. http://sinodefence.com.

Civilian Ocean Surface Topography Missions. Civilian ocean surface topography missions used the *TOPEX/Poseidon* (Topography Experiment) and *Jason 1* spacecraft to investigate the circulation of oceans and their interactions with the atmosphere. The *TOPEX/Poseidon* spacecraft was launched by an Ariane 42P from Kourou, French Guiana, in August 1992. *Jason 1* was launched by a Delta II from Vandenberg Air Force Base, California, in December 2001. Both their orbits were circular, 1,340 km altitude, inclined at 66° to the equator, with a repeat cycle of 10 days.

In 1979 NASA's Jet Propulsion Laboratory had begun developing the TOPEX spacecraft to measure local deviations in ocean height in order to detect ocean currents. The technique had been developed using radar altimeters on the *GEOS 3* (*Geodynamics Experimental Ocean Satellite*) and *Seasat* spacecraft, launched in 1975 and 1978 respectively. At about the same time, France's Space Agency, CNES, was planning a similar mission called Poseidon. In 1983 the United States and France decided to combine their missions into the TOPEX/Poseidon mission. The United States provided the spacecraft bus and four instruments and was responsible for spacecraft operations. France provided two instruments and the Ariane launcher. *TOPEX/Poseidon* was to become the second major satellite in NASA's Mission to Planet Earth, with 38 principal investigators from nine countries.

The 2.4-ton *TOPEX/Poseidon* spacecraft used a Fairchild Multi-Mission Spacecraft bus. The primary instrument on board was an American dual-frequency radar altimeter, based on *Seasat* and *Geosat* designs. A second frequency had been added for *TOPEX/Poseidon* to provide ionospheric corrections. A single-frequency French radar altimeter was also included for trial purposes. Although with less performance than the American instrument, it was much lighter and used much less power. The remainder of the satellite payload was mainly to determine the position of the spacecraft as accurately as possible, so that the position of the local ocean surface illuminated by the radar altimeter could be determined in absolute terms, rather than just as a satellite to ocean range measurement.

TOPEX/Poseidon provided ocean heights to an accuracy of 4.2 cm. It helped scientists to determine the speed and direction of ocean currents and to understand how the oceans transport heat from Earth's equatorial regions toward the poles. It provided valuable information on the creation and decline of El Niño and La Niña, and their effects on the world's weather and climate, and helped scientists improve global ocean tide models and understand the role of tides in ocean mixing. The spacecraft's design life was three years, with a goal of five years. It lasted 13 years, until it was decommissioned in January 2006.

TOPEX/Poseidon Instruments

Instrument	Operating Frequency	Measurements and Main Applications	Comments
Dual-Frequency NASA Radar Altimeter (NRA)	13.6 GHz (Ku band) 5.3 GHz (C band)	Distance of sea surface from spacecraft (including ionospheric correction), sea surface wind speed, wave height.	Based on *Seasat* and *Geosat* designs, but used two frequencies instead of one.
Single-Frequency Solid State Radar Altimeter (SSALT, or POSEIDON)	13.6 GHz (Ku band)	Same parametric measurements as NRA, but SSALT required ionospheric corrections from the DORIS instrument.	Provided by France. Experimental instrument to validate improved solid sate technology, resulting in a much lower power consumption and mass than NRA.
Three-Frequency TOPEX Microwave Radiometer (TMR)	18, 21, and 37 GHz	Measured sea surface brightness temperature and tropospheric water vapor. Provided water vapor corrections to satellite range measurements.	
Laser Retroreflector Array (LRA)		Precise tracking data.	
Dual-Frequency Doppler Tracking System Receiver (DORIS)	2,036 and 401 MHz	Precise tracking data. Also measured temperature, humidity, and atmospheric pressure to provide atmospheric corrections. Provided ionospheric corrections to SSALT data.	Provided by France.
Global Positioning System Demonstration Receiver (GPSDR)	12.2 and 1.58 GHz	Precise tracking data.	Experimental receiver.

The 500 kg *Jason 1* spacecraft was designed to continue the *TOPEX/Poseidon* ocean current and tidal measurements. Its instrument payload was similar to that of *TOPEX/Poseidon*, except there was only one radar altimeter: a two-frequency instrument provided by France, called Poseidon 2, which was based on the single-frequency instrument on *TOPEX/Poseidon*. France provided the Alcatel Proteus bus, and the United States provided the launcher.

Jason 1 measured the ocean heights to an accuracy of 3.3 cm. It also flew in tandem with *TOPEX/Poseidon* for nearly three years, which allowed small-scale ocean phenomena to be studied, such as coastal tides and ocean eddies.

David Leverington

See also: France, Jet Propulsion Laboratory

Bibliography

JPL missions. http://www.jpl.nasa.gov/missions/index.cfm.
NASA Headquarters Earth science missions. http://science.nasa.gov/earth-science/missions/.

Earth Resources Observation Satellite. *See* Israel.

Earth Resources Observation System (EROS) Data Center. The Earth Resources Observation System (EROS) Data Center (EDC) is the national archive for images of Earth's land surface acquired by remote sensing satellites and aircraft, administered by the U.S. Geological Survey of the U.S. Department of Interior (DOI).

The center was created as the site for receiving and distributing data from the *Landsat 1*, which was launched in 1972. At the time, experts wanted a site in the central United States, separated from large cities whose electromagnetic interference could interfere with the facility's ability to receive satellite data transmissions. After political maneuvering by the South Dakota congressional delegation, the DOI selected a site near Sioux Falls, South Dakota. Groundbreaking for the new facility took place in 1972.

The center began receiving raw Landsat data in 1975. Prior to that time NASA's Goddard Space Flight Center (GSFC) processed the data, which EDC later stored and distributed. In 1977, the center was authorized to develop the capability to perform basic processing to put Landsat data into a more useful form, taking over some of those functions from GSFC. Since data sales on computer tapes was much lower than anticipated, EDC encouraged sales by training potential customers, mainly from foreign countries, the DOI, and state governments. The extensive use of inexpensive but powerful computers and the birth and growth of the World Wide Web in the 1990s made EDC data accessible to a vast base of potential clients. EDC responded by housing one of the largest computer complexes in the Department of the Interior.

The center stored data from NASA, the U.S. Department of Defense, the U.S. Environmental Protection Agency, the U.S. Agency for International Development, the International Council of Scientific Unions, and the U.S. intelligence community agencies. It had some 600 government and contractor employees, including workers at a field office in Anchorage, Alaska, and at NASA Ames Research Center, Moffett Field, California. As of 2008, the center's archive managed more than three million satellite images of Earth and nine million aerial photographs, including those from Landsat and the Corona spy satellite program. These are used to study a wide range of natural hazards, global environmental change, and economic development and conservation issues.

Anthony R. Curtis

See also: Corona

Bibliography
Pamela Mack, *Viewing the Earth: The Social Construction of the Landsat Satellite System* (1990).
EROS Data Center. http://eros.usgs.gov/.
U.S. Geological Survey Remote Sensing. http://www.usgs.gov/science/science.php?term=981.

Envisat. *See* European Remote Sensing Program.

European Remote Sensing Program. The European Remote Sensing Program includes the European Space Agency's (ESA) *European Remote Sensing 1* (*ERS 1*), *ERS 2*, and *Envisat* satellites, and a series of Earth Explorer missions.

In 1977 the French government suggested that the French SPOT (Satellite Pour l'Observation de la Terre) program, primarily devoted to land observation, should be taken over as a European program. Only the Belgian and Swedish delegates to ESA supported this Europeanization proposal, so it was abandoned, and ESA began developing its own remote sensing program.

An image of Iceberg B-15A during its collision with the Drygalski ice tongue in April 2005, breaking off a piece. This iceberg was at birth the world's largest free-floating object measuring 27 × 122 kilometers. The ice tongue is a projection of an Antarctic glacier. The image was constructed from data taken by ESA's Envisat *SAR. (Courtesy European Space Agency)*

By 1980 ESA's proposed program consisted of two ocean monitoring spacecraft, similar in type to the U.S. *Seasat* spacecraft, and an advanced land applications satellite, complementary to SPOT. Both the ocean spacecraft, called ERS, and the land spacecraft, called Land Application Satellite System (LASS), would carry a synthetic aperture radar (SAR) for all-weather observing. ERS was the priority mission, however, and the LASS mission was later dropped for technical, financial, and political reasons.

ERS 1 was produced by a European industrial consortium led by Germany's Dornier Company as prime contractor. The spacecraft used a modified SPOT bus, carried an all-weather SAR to image both the ocean and land, and a suite of microwave and infrared instruments to observe and detect changes in the ocean, land, coastal waters, and ice sheets. These instruments were designed to measure wave heights and sea surface winds, detect ocean currents and atmospheric water vapor, and measure sea surface temperatures to an unprecedented accuracy of 0.2 K.

The 2.4-ton *ERS 1* spacecraft was 11.5 m long, with a laterally-mounted 10 m long SAR antenna, and a 12 m long solar array that provided 2.4 kW of power. The spacecraft was launched into a 790 km near-circular orbit by an Ariane 4 launcher from Kourou, French Guiana, in July 1991. Its orbit was Sun-synchronous, near-polar, with a 35 day repeat cycle.

The SAR produced a large amount of image data, which could not be stored onboard. So it was downloaded to ground stations in real time at a rate of 105 Mbps. Onboard recorders, with a capacity of 6.5 Gbits, were provided to record the data from all the other experiments. This was relayed to ground by a separate transponder link at 15 Mbps. There were three ESA ground stations in Europe, but more than two dozen foreign stations were also authorized to download data. Data processing and archiving facilities were in France, the United Kingdom, Germany, and Italy.

ERS 2, launched in April 1995, was almost identical to *ERS 1*, with the significant addition of the Global Ozone Monitoring Experiment to measure the concentration of ozone and other trace gases in the atmosphere. This enabled the mapping of ozone holes at high latitudes in both hemispheres, which showed how the atmosphere was responding to the reduction of ozone depleting substances agreed in the Montreal Protocol.

ERS 1 operated successfully for nine years, three times its expected lifetime, and *ERS 2* was still functioning in 2007.

The next operational ESA remote sensing spacecraft, *Envisat*, was to provide synergistic ocean, land, atmospheric, and ice cap measurements. Initially ESA had envisaged a Polar Platform (PP) as an element of the International Space Station program. But in 1991 ESA decided to cut the PP from the space station program and use it as an autonomous spacecraft. In the following year, this was split into Envisat (an Earth resources mission) and MetOp-A (Meteorological Operational, a meteorological mission).

The 8.2-ton *Envisat* spacecraft consisted of three basic elements: a Service Module (SM), a Payload Module split into a Payload Carrier and Payload Equipment Bay (PEB), and the payload of ten instruments (see table on page 384). The SM provided power generation and storage, attitude and orbit control, bus data handing,

ERS Instruments

Instrument	Wavelength/Waveband	Main Measurements/Uses	Comments
SAR[1]	5.3 GHz (C band)	(1) Image Mode. Resolution 26 m cross track, 6 m along track of both ocean and land. (2) Wave Mode. Measured wavelength and direction of ocean waves.	Image data acquired for maximum of 10 minutes per orbit. Recording not available for image mode.
Wind Scatterometer (SCATT)[1]	5.3 GHz (C band)	Wind Mode, measuring wind speed and direction of ocean waves.	
Radar Altimeter (RA)	13.8 GHz (Ku band)	Wave and ocean surface height to within a few centimeters. Surface wind speed. Topographic measurements over ice sheets.	
Along Track Scanning Radiometer (ATSR)	Infrared Radiometer (IRR) 1.6 to 12 μm. Visible channels added for *ERS 2*. Microwave Sounder (MWS) 23.8 and 36.5 GHz	(1) IRR. Sea surface temperature to better than 0.2 K in cloud free areas. Cloud top and land surface temperatures. (2) MWS. Total water vapor content to correct RA's measurements.	*ERS 1*. One of the four IRR channels failed in May 1992. *ERS 2*. Visible channels added to enable vegetation cover to be determined.
Global Ozone Monitoring Experiment (GOME)	240 to 790 nm	Column measurements of ozone and other trace gases.	On *ERS 2* only.
Precise Range and Range Rate Experiment (PRARE)	8.5 GHz (X band) and 2.25 GHz (S band)	All weather satellite to ground distance measurement.	
Laser Retroreflector	Infrared	Satellite to ground distance measurement to calibrate RA and improve spacecraft orbit determination.	

[1]The SAR and Wind Scatterometer formed the Active Microwave Instrument (AMI).

Envisat Instruments

Instrument	Wavelength/Waveband	Main Measurements/Uses	Comments
Advanced Synthetic Aperture Radar (ASAR)	5.3 GHz (C band)	Sea ice, snow, and land ice mapping. Ocean waves. Coastal protection and pollution, soil moisture, agriculture, forest, and ship traffic monitoring.	Twelve minutes of onboard recording. Near real-time data products available. Improved version of the *ERS 1* and *2* SARs.
Radar Altimeter 2 (RA-2)	13.6 GHz (Ku band) and 3.2 GHz (S band)	Ocean currents, global and regional sea level changes, ice sheet elevation	MWR and DORIS used to correct data. Improved RA compared with *ERS 1* and *2*.
Advanced Along Track Scanning Radiometer (AATSR)	Seven channels from 0.55 to 12.0 μm	Sea surface temperatures. Vegetation biomass, moisture, and health.	Provided continuity with ATRS on *ERS 1* and *2*.
Global Ozone Monitoring by Occultation of Stars (GOMOS)	250 to 930 nm	Simultaneous monitoring of ozone and other trace gases, water vapor, aerosols and temperature distribution in the stratosphere	Self-calibrating instrument
Laser Retroreflector (LRR)	Passive laser reflectors	Precise ranging data	Corner cubes. Identical to *ERS 1* and *2* device.
Microwave Radiometer (MWR)	23.8 and 36.5 GHz	Atmospheric water vapor and liquid water to provide distance correction for Radar Altimeter	
Medium Resolution Imaging Spectrometer (MERIS)	Spectral bands from 390 to 1040 nm	Ocean color to assess water quality. Detection of phytoplankton, dissolved organic material, and suspended sediment.	Push-broom design. Mission follows on from U.S. Coastal Zone Color Scanner.
Michelson Interferometric Passive Atmospheric Sounder (MIPAS)	Spectrometer range 4.1 to 14.6 μm, spectral resolution about 0.006 nm	Simultaneous measurement of ozone and 20 atmospheric trace gases, including the oxides of nitrogen, CFCs, and aerosols	Atmospheric limb sounder
Scanning Imaging Absorption Spectrometer for Atmospheric Chartography (SCIAMACHY)	0.24 to 2.4 μm	Measurement of trace gases, aerosols and clouds	Nadir and limb sounding
Doppler Orbitography and Radio positioning Integrated by Satellite (DORIS)	Beacon frequencies 400 MHz and 2 GHz	Very precise orbit determination	First launched on *SPOT 2* and *TOPEX/Poseidon*. The *Envisat* DORIS was second-generation equipment.

and telemetry. It was based on the *SPOT 4* bus but with an enlarged structure and a new solar array. The PEB contained the payload dedicated support systems for power distribution, thermal control, instrument control, data handling, and communications. The payload weighed almost as much as the whole of the *ERS 1* spacecraft. British Aerospace, Bristol (later European Aeronautic Defence and Space Astrium) was the Polar Platform prime contractor.

Envisat was launched on 1 March 2002 by an Ariane 5 into a 790 km, Sun-synchronous, near-polar orbit. It was designed to communicate either direct to ground or via ESA's *Artemis* geosynchronous communications satellite. Data was recorded by two 70 Gbit solid state recorders, and both the real time and recorded data was transmitted to ground at 100 Mbps. Unfortunately, problems with *Artemis*'s 2001 launch delayed the satellite link for 18 months.

ESA's remote sensing satellites have detected forest fires, changing farm use, deforestation, plant disease, floods, ice melt, ozone depletion, air and marine pollution, ocean storm surges, El Niño, and ocean currents. As an example, Johannessen et al. used ERS radar altimeter data during the period 1992–2003 to examine the change in height of the Greenland ice sheet. They found that it had increased by 6.4 cm/yr in the interior but had decreased by about 2.0 cm/yr at the margins. This and similar data is essential for assessment of global warming.

In the late 1990s, ESA reassessed its Earth Observation program, leading to approval in 1999 of its Living Planet program. This had two main components: the Earth Explorer missions, which were science- and research-driven using 600–1,200 kg sized spacecraft, and more operationally oriented missions under the Earth Watch designation. The latter included meteorological missions in collaboration with EUMETSAT, and new missions, called Global Monitoring for Environment and Security (GMES), carried out in collaboration with the European Commission.

David Leverington

See also: European Aeronautic Defence and Space Company; SPOT, Satellite Pour l'Observation de la Terre

Bibliography

G. Duchossois, "ERS-1: Mission Objectives and System Description," in *Monitoring Earth's Ocean, Land, and Atmosphere from Space—Sensors, Systems, and Applications*, ed. Abraham Schnapf (1985).

Envisat Special Issue, *ESA Bulletin* 106 (June 2001).

ESA. http://earth.esa.int/ers/.

Fanhui Shi Weixing. *See* Chinese Remote Sensing Satellites.

Haiyang. *See* Chinese Remote Sensing Satellites.

Indian Remote Sensing Satellites. The Indian Remote Sensing Satellites (IRS) program provided the world's largest constellation of commercial remote sensing satellites in operation as of 2008. It started with the purpose of supporting the Indian national economy in the areas of agricultural water resources, forestry and

Indian Remote Sensing Satellites (Part 1 of 4)

Satellite	Launch Vehicle, Launch Date	Earth Imaging Sensors	Spatial Resolution (m)	Swath Width (km)	Repeat Cycle (days)	Remarks
Bhaskara 1	Kosmos 3M (Soviet Union) 7 June 1979	Two television cameras and microwave radiometers	Television cameras operated in visible (0.6 μ) and near-infrared (0.8 μ) to collect data related to hydrology, forestry, and geology. Satellite microwave radiometer operated in 19.24 GHz, 22.235 GHz, and 31.4 GHz for study of ocean state, water vapor, and liquid water content in the atmosphere.			India's first experimental Earth observation satellite. Mission completed in 1989.
Bhaskara 2	Kosmos 3 M (Soviet Union) 20 November 1981	Two television cameras and microwave radiometers	Satellite and instruments same as for *Bhaskara 1*			India's second experimental Earth observation satellite. Mission completed in 1991.
IRS 1A	Vostok (Soviet Union) 17 March 1988	LISS 1 LISS 2 A/B	72.574 × 2	148	22	India's first operational Earth observation satellite. Mission completed in 1992.
IRS 1B	Vostok (Soviet Union) 29 August 1991	Satellite and instruments same as for *IRS 1A*				India's second operational Earth observation satellite. Mission completed in 1999.
IRS 1E	PSLV D1 20 September 1993	Satellite could not be placed in orbit due to launch vehicle failure.				The first developmental flight of the indigenous Polar Satellite Launch Vehicle (PSLV) failed and plunged into the Bay of Bengal.

(Continued)

Indian Remote Sensing Satellites (Part 2 of 4)

Satellite	Launch Vehicle, Launch Date	Earth Imaging Sensors	Spatial Resolution (m)	Swath Width (km)	Repeat Cycle (days)	Remarks
IRS P2	PSLV D2 15 October 1994	LISS 2 M	32 × 37	66 × 2	24	First indigenously placed Earth orbiting (EO) satellite by PSLV launch vehicle. Mission completed in 1997.
IRS-1C	Molniya (Russia) 28 December 1995	LISS 3	23.5	142	24	First Indian EO satellite with three sensors, namely PAN, LISS 3, and WiFS. Satellite in operation in 2006.
		PAN	5.8	70	24	
		WiFS	188	804	5	
IRS P3	PSLV D3 21 March 1996	WiFS	188	804	5	In addition to the sensors, the satellite carried a Modular Opto-elecronics Scanner (MOS) developed by the German Space Agency, DLR, and an X-ray astronomy instrument. Satellite in operation in 2006.
IRS 1D	PSLV C1 29 September 1997	Satellite and instruments are identical to those of *IRS 1C*				Due to underperformance of the fourth stage of PSLV, onboard fuel was used to place the satellite in a near circular orbit. Satellite in operation in 2006.

Indian Remote Sensing Satellites (Part 3 of 4)

Satellite	Launch Vehicle, Launch Date	Earth Imaging Sensors	Spatial Resolution (m)	Swath Width (km)	Repeat Cycle (days)	Remarks
IRS P4 (*Oceansat 1*)	PSLV C 226 May 1999	OCM	360 × 236	1420	2	Primary mission is oceanography application to carry out ocean color monitoring for conducting fisheries survey and development of a fisheries model for Indian Ocean region. Satellite in operation in 2006.
MSMR	105 × 68, 66 × 43, 40 × 26, 34 × 22 (km)	1360	2			
TES	PSLV C322 October 2001	Technology experimental satellite that demonstrated in orbit control, high torque reaction wheels, solid state recorder, X-band phased array antenna, and improved satellite positioning system for future satellites.				First Indian remote sensing satellite with 1 m resolution. Satellite in operation in 2006.
IRS P6	(*Resourcesat 1*)	PSLV C517 October 2003	LISS 4	5.8	70	24
	Primary mission is agricultural application to allow multiple crop discrimination and species level discrimination. Also carried a solid state recorder to store images and send for later transmission. Satellite in operation in 2006.	LISS 3	23.5	140	24	

(Continued)

Indian Remote Sensing Satellites (Part 4 of 4)

Satellite	Launch Vehicle, Launch Date	Earth Imaging Sensors	Spatial Resolution (m)	Swath Width (km)	Repeat Cycle (days)	Remarks
	AWiFS	70	740	5		
IRS P5 (*Cartosat 1*)	PSLV C6 5 May 2005	PAN A	2.5	30	5	First civilian stereographic mapping satellite. Primary application is three-dimensional digital terrain mapping. Satellite in operation in 2006.
		PAN F	2.5	30		
IRS P7 (*Cartosat 2*)	PSLV C7 10 January 2007	PAN	<1	9.6	4	Mapping, land use.
IRS P8 (*Cartosat 2A*)	PSLV C9 28 April 2008	PAN	<1	9.6	4	Mapping, land use.
IRS P9 (*IMS 1*)	PSLV C9 28 April 2008	Multi-Spectral, Hyper-Spectral	37 m MS, 506 m HS	151 MS, 129 HS		Experimental minisatellite. Multi-spectral camera intended for foreign use, hyper-spectral by Indian users.

LISS—Linear Imaging Self-scanning Sensor
MSMR—Multifrequency Scanning Microwave Radiometer
OCM—Ocean Color Monitor with eight spectral bands
PAN—Panchromatic camera
WiFS—Wide Field Sensors

ecology, geology, watersheds, marine fisheries and coastal management, and advanced research in Earth sciences. The IRS series of satellites was designed and built by the Indian Space Research Organisation (ISRO) Satellite Center based in Bangalore, while the ISRO Space Application Center at Ahmedabad fabricated the remote sensing payloads and cameras. The National Remote Sensing Agency at Hyderabad was responsible for acquisition, processing, and the supply of data to the various programs that used IRS data for optimal use of the country's natural resources and water management. However, IRS data remained underutilized because of inadequacies in regional- and district-level ground facilities and interagency coordination. The National Natural Resources Management System, an autonomous organization under the aegis of the Indian Department of Space, addressed these issues. International customers could obtain IRS data from the U.S. Space Imaging Corporation.

Vikram Sarabhai, the father of the Indian space program, suggested in the early 1960s that India should use space research to benefit the broader population. As a result the program became applications-driven, concentrating on remote sensing, communications, and meteorology.

The IRS program began with the launch of two experimental Earth observation satellites, *Bhaskara 1* and *Bhaskara 2*, in 1979 and 1981 by Soviet launchers. The payloads consisted of television cameras and radiometers. The first operational remote sensing satellites, *IRS 1A* and *IRS 1B*, were launched in 1988 and 1991 on Soviet Vostok launchers. Both satellites carried identical instruments operating in four spectral bands, with a resolution of up to 36.25 m and orbital repeats every 22 days. The satellites provided vital environmental data for more than a decade. *IRS 1C*, launched on a Russian Molniya in 1995, and *IRS 1D*, launched on India's Polar Satellite Launch Vehicle (PSLV) in 1997, had identical second-generation instruments with resolutions of up to 5.8 m. The first developmental flight of the PSLV D1, carrying the *IRS 1E* satellite, failed on launch in 1993 due to an attitude control software malfunction. However, the second and third developmental flights of the PSLV successfully launched *IRS P2* and *IRS P3* in 1994 and 1996.

Oceansat 1, India's first spacecraft to measure the physical and biological aspects of the ocean, was launched by PSLV C2 in 1999. Two years later another PSLV launched the *Technology Experiment Satellite* (*TES*), the first Indian remote sensing satellite with 1 m resolution. To address land and water resources, *Resourcesat 1* was placed in Sun-synchronous orbit in 2003. In addition to three remote sensing cameras, the satellite carried a solid state recorder to store images taken by cameras, which could be received later by the ground stations. *Cartosat 1*, launched in 2005, carried two sensors with 2.5 m resolution, with a repetition of 5 days, enabling three-dimensional digital terrain mapping and cartography applications. The experimental *Indian Minisatellite 1*, launched in April 2008 with *Cartosat 2B*, assessed the utility of small satellites for remote sensing. A Radar Imaging Satellite based on active antenna technology capable of all weather and day or night operation to complement the existing optical payloads of the IRS satellites was planned for launch in late 2008 or 2009.

Venkatesan Sundararajan

See also: India, Indian Launch Vehicles

Bibliography

"Eyes in the Sky," *New Scientist*, 19 February 2005.

Government of India, Department of Space, *Annual Report* (2004–5).

Brian Harvey, The Japanese and Indian Space Programmes: Two Roads into Space (2000).

Frank Morring Jr., "India's Prolific Space Program," *Aviation Week & Space Technology*, 22 November 2004.

Jian Bing. *See* Chinese Remote Sensing Satellites.

Landsat. Landsat, a series of U.S. Earth observation satellites, first launched in 1972. The program became important for global change research because of its continuous and growing archive of data and controversial because of its complex politics.

By the mid-1960s, the United States had imaged Earth from space with weather and reconnaissance satellites, showing the utility of these civilian and military applications. Encouraged by images of Earth taken by Mercury astronauts, in 1964 NASA's Office of Manned Space Science funded the Department of the Interior (DOI), the Army Corps of Engineers, the Department of Agriculture (USDA), and several universities to study the potential of remote sensing of Earth for civilian purposes. Along with Gemini imaging experiments, the results were promising, leading to further studies and an aircraft remote sensing program to begin sensor testing and evaluation.

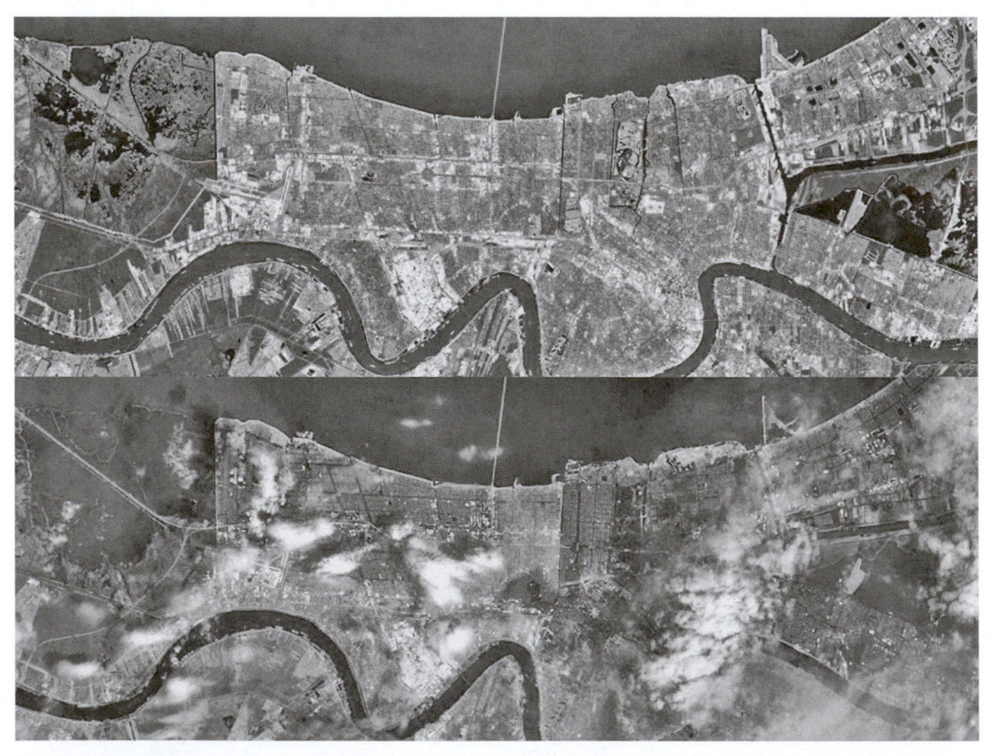

Two Landsat 7 *images of New Orleans, before (top) and after (bottom) flooding from Hurricane Katrina in August 2005. The darkened areas in the city in the bottom image are flooded. (Courtesy U.S. Geological Survey Center for Earth Resources Observation & Science)*

Perceiving that NASA did not give high priority to the program, by 1966 the DOI announced that it would develop its own simple operational satellite system. NASA wanted a more complex, experimental spacecraft to advance new technologies. In the resulting bureaucratic compromise, NASA accelerated development of an experimental satellite based on DOI specifications.

During 1967 NASA focused on assessing sensors, ultimately selecting a Return Beam Vidicon (RBV), which was favored by the U.S. Geological Survey (USGS) in the DOI, and a Multi-Spectral Scanner (MSS), favored by the USDA. Efforts to include a photographic sensor were quashed by the Department of Defense, to protect its classified mapping and reconnaissance activities.

The RBV was a television (TV) camera, with a much finer line structure than a normal TV camera. There was one camera for each color of blue-green, yellow-red, and near infrared (up to 0.83 μm). The MSS used a scanning mirror and needed just one instrument to cover its complete spectral range, producing images in three bands, similar to the RBV, plus a fourth band from 0.8–1.1 μm in the infrared (IR). Both the RBV and MSS had a ground resolution of about 80 m. Higher performance instruments had already been developed for the U.S. military, but their use was vetoed for security reasons.

Getting program approval proved difficult, as the demands of Apollo and the Vietnam War competed with social programs for funding. The Bureau of the Budget (BoB) demanded that NASA prove cost effectiveness, which proved problematical in part because of the difficulty of identifying the diverse community of users, who in turn were reluctant to invest in what might turn out to be a short-term program. The BoB approved one experimental satellite, initially called *Earth Resource Technology Satellite 1* (*ERTS*), and a limited ground processing system. These funding constraints led NASA to examine whether *ERTS* could use an existing spacecraft bus to carry its instruments. As a result, in July 1970 NASA selected General Electric and its Nimbus bus to build the spacecraft, renamed *Landsat 1*, and its backup, *Landsat 2*. NASA's Goddard Space Flight Center (GSFC) managed the development program and operated the spacecraft.

Landsat 1 was launched in July 1972 by a Delta 900 rocket from the Vandenberg Air Force Base and placed into a Sun-synchronous 916 km near-circular, near-polar orbit. The spacecraft completed 14 orbits per day, with a repeat cycle every 18 days. Within a month, however, power problems had caused NASA to shut down the RBV. The MSS continued operating exceptionally well until the spacecraft was turned off in January 1978.

To collect, process, archive, and distribute Landsat data, the DOI selected a location near Sioux Falls, South Dakota, for the Earth Resources Observation Data Center (EDC), far from major sources of radio interference. Limited funding led to a division of labor between GSFC, which did the initial data processing, and EDC, which performed further processing, data storage, and distribution. Producing a color image for the RBV proved difficult as each of the three cameras (one for each color) had different image distortions, which needed correcting before integration into a single image. For *Landsat 1* this was not a major problem, as the RBV only operated for one month, but it was a serious problem for *Landsat 2*, which operated from 1975 to 1982. Annual data sales in the 1970s remained fairly steady under 400,000 frames, about 20 percent of initial expectations.

Landsat Satellites

Landsat	Mass (kg)	Launch Date	End of Service	Orbit Altitude (km)	Payload Instruments	Wavebands	Ground Resolution (m)	Comments
1	816	23 Jul 1972	6 Jan 1979	912	Return Beam Vidicon (RBV) Multi Spectral Scanner (MSS) Data Collection System (DCS)	Visible and Very Near Infrared (IR) Visible and Near IR	80 80	RBV turned off 6 Aug 1972.
2	953	22 Jan 1975	25 Feb 1982	913	RBV MSS DCS	Visible and Very Near IR Visible and Near IR	80 80	
3	960	5 Mar 1978	31 Mar 1983	906	RBV MSS DCS	Visible (Panchromatic) Visible, Near and Thermal IR	30 80	Thermal IR band failed shortly after launch.
4	1,940	16 Jul 1982	Placed on standby Jan 1986	692	MSS Thematic Mapper (TM) GPS	Visible and Near IR Visible and IR	80 30	Both direct to ground TM downlink transmitters failed within first seven months.
5	1,950	1 Mar 1984	Operational Feb 2007	692	MSS TM GPS	Visible and Near IR Visible and IR	80 30	Link to TDRS failed in 1987. MSS turned off 1995.
6	2,200	5 Oct 1993	5 Oct 1993		Enhanced TM	Visible and IR.	15 (PAN) 30 (MS)	Ruptured fuel line caused loss of spacecraft during launch sequence.
7	2,200	15 Apr 1999	Operational Feb 2007	705	Enhanced TM Plus	Visible and IR	15 (PAN) 30 (MS)	Scan Line Corrector failed May 2003.

Table by David Leverington

Landsat 3, launched in 1978, was similar in design to its predecessors, but its RBV was converted into a higher resolution panchromatic instrument. This avoided registration problems in the ground processing of color images. Its MSS also added a thermal IR band (10.4–12.6 μm), but unfortunately, the thermal IR detectors lost sensitivity shortly after launch.

Landsat 4 was different from previous Landsats, with a new higher-powered spacecraft bus, built by Fairchild, and a new primary instrument, the Thematic Mapper (TM). The new communications system, capable of 85 Mbps data transmission (more than five times that of previous satellites), could transmit data through NASA's new Tracking and Data Relay Satellites (TDRS) and direct to ground stations. Planning to use the planned suite of three TDRS spacecraft, *Landsat 4* had no onboard data recorders. The TM was a seven-band scanner with a 30 m ground resolution in the six optical to mid-IR bands, and 120 m in the Thermal IR band. Initially, the Office of Management and Budget (OMB) (successor to BoB) proposed eliminating *Landsat 4*'s MSS. Protests from the users, who pointed out the benefit of continuity of imagery from one instrument design, caused the OMB to back down, and the MSS was reinstated.

Landsat 4 was launched by a Delta 3920 in July 1982, into a lower orbit than its predecessors, providing a 16 day repeat cycle. Within seven months both of its TM's direct downlink transmitters had failed, leaving it dependent on the TDRS link. Unfortunately, that was not yet available because of a launch delay to *TDRS 1*. Even when *TDRS 1* was launched a few months later, the downlink transmitter loss resulted in a significant reduction in the quantity of TM data. Solar array problems also caused a significant reduction in power.

These problems were solved for *Landsat 5*, which was otherwise identical to *Landsat 4*. Launched in March 1984, *Landsat 5*'s link to the TDRS failed in 1987, after which data could only be downloaded when the spacecraft was in line of sight of a ground receiving station.

The launch of *Landsat 6* failed in October 1993 because of a ruptured fuel line feeding its apogee kick motor, but *Landsat 7* successfully reached orbit, following a launch by a Delta II in April 1999.

Landsat 7's payload consisted of an Enhanced Thematic Mapper Plus (ETM^+) which, compared with the TM, had a 15-m resolution panchromatic band added, the Thermal IR band's resolution had been increased from 120 m to 60 m, and the instrument was much better calibrated, improving the ability to detect scene changes over time. In 2003 the ETM^+'s Scan Line Corrector failed, which caused the image to be crossed by black lines. These images were improved by adding data from earlier spacecraft passes.

Throughout the 1970s, NASA, DOI, OMB, and presidential administrations debated whether Landsat was an experimental or operational system. If experimental, it was best operated by NASA, but if operational, it was best operated by an agency such as the USGS or by private industry, funded by paying customers. In 1979 President Jimmy Carter decided that Landsat was to be operational and that the National Oceanic and Atmospheric Administration (NOAA) should have responsibility for operational Earth satellites, including both Landsat and its existing weather satellites. NASA GSFC would still be responsible for the development, launch, and initial checkout of the *Landsat 4* spacecraft, after which time NOAA would operate it. Landsat, however, was to eventually be transferred to private industry. Many people opposed this move, arguing that Landsat operations and data should be provided as an inexpensive (to users) government service, such as weather satellites. Nevertheless, the Ronald Reagan presidential administration pushed ahead with privatization, and in October 1985 the Earth Observation Satellite Corporation (EOSAT), a joint venture of Hughes Aircraft Corporation and Radio Corporation of America (RCA) took over the operational Landsat program from NOAA.

To cover costs and make a profit, EOSAT increased image charges by some 600 percent. High prices discouraged purchases, which led to gaps in data coverage, because the company was reluctant to spend money on data collection without paying customers. This angered scientists and reduced the program's political support. Then, France launched its more capable SPOT (Satellite Pour l'Observation de la Terre) Earth resource spacecraft in 1986. With the French company, SPOT Image, selling its higher quality images for lower prices and other nations launching their own remote sensing satellites, the market for Landsat images slipped badly, and in 1992 Congress mandated government development and operation of *Landsat 7*. NASA developed *Landsat 7*, while NOAA eventually operated the spacecraft, and DOI distributed imagery. In 1998 management of *Landsat 4* and *5* was shifted from NOAA to USGS, and in 2001 EOSAT returned operational control of the two spacecraft to the government, ending the era of privatization.

Difficulties with *Landsat 7*'s ETM^+ starting in 2003 highlighted the need for a replacement spacecraft, which encountered continuing political problems. The initial Landsat Data Continuity Mission received a single commercial bid in 2004, which the government rejected, resulting in a directive that year to put Landsat instruments on the National Polar-orbiting Operational Environmental Satellite System. In December 2005 this directive was overturned, and the government committed to procuring a separate spacecraft under the Landsat Data Continuity Mission program. As of early 2007 the acquisition program was proceeding, with NASA in charge of spacecraft acquisition and checkout, and the USGS charged with development of the ground systems, spacecraft operation, and data processing. The importance of Landsat's data for global change research, with its continuous data sets dating to 1972, continued to generate sufficient political support to keep the program alive.

David Leverington, Stephen B. Johnson, and John D. Ruley

See also: Goddard Space Flight Center, EOSAT, Earth Observation Satellite Company

Bibliography

P. K. Conner and D. W. Mooneyhan, "Practical Applications of Landsat Data," in *Monitoring Earth's Ocean, Land and Atmosphere from Space—Sensors, Systems, and Applications*, ed. Abraham Schnapf (1985).

Pamela E. Mack, *Viewing the Earth: The Social Construction of the Landsat System* (1990).

Edward W. Mowle, " 'Landsat D and D'—The Operational Phase of Land Remote Sensing from Space," in *Monitoring Earth's Ocean, Land and Atmosphere from Space—Sensors, Systems, and Applications*, ed. Abraham Schnapf (1985).
GSFC Landsat. http://landsat.gsfc.nasa.gov/.

National Environmental Satellite, Data, and Information Service. The National Environmental Satellite, Data, and Information Service (NESDIS) was created in 1980 as a line office of the National Oceanic and Atmospheric Administration (NOAA). Its major responsibilities were to manage the major system of U.S. Earth science application satellites; to organize, archive, and provide access to the vast quantities of data these satellite systems produced, in addition to much of the non-satellite data produced and distributed by NOAA; and to pursue research in new technologies and applications for Earth science applications in space and on Earth.

NESDIS was most associated with satellite operations and satellite-derived data, but its origins were in the divisions of NOAA's legacy agencies directed to the organization and management of new classes and volumes of information and data produced by computer systems. In 1951 the Weather Bureau's weather tabulation center moved from New Orleans, Louisiana, to Asheville, North Carolina, and became the National Weather Records Center. In 1965 the three major component agencies within the Department of Commerce of the U.S. Coast and Geodetic Survey, the Weather Bureau, and the Central Radio Propagation Laboratory of the National Bureau of Standards were joined into a new Earth Science Services Administration (ESSA). The National Weather Records Center then became the Environmental Data Center complex with the transfer of the U.S. Coast and Geodetic Survey's Seismology Data Center to Asheville in 1966. That year also saw the launch of a NASA experimental satellite *Applications Technology Satellite 1* with early imaging and weather broadcast systems onboard. In 1970 ESSA was changed to NOAA and the separate legacy agencies within ESSA were redistributed and renamed. The new satellite management and data management enterprises merged into the National Environment and Satellite Service (NESS) within the Office of Oceanic and Atmospheric Services. Joint NASA/ESSA (later NOAA) experimental satellites continued until 1974–75, when NASA's *Synchronous Meteorological Satellites 1* and *2* were launched as operational geostationary weather satellites. These satellites were the prototype for NOAA's Geostationary Operational Environmental Satellites (GOES). The first NOAA-owned and operated geostationary satellite *GOES 1* was launched on 16 October 1975. The first NOAA-funded polar-orbiting environmental satellite was launched in June 1979. The National Environmental Satellite Center was established to manage the new satellite systems, while the Environmental Data Service was designated as the office to manage data from the satellites and other data flows previously designated.

In 1979 as part of major upheavals in U.S. satellite programs and proposed privatization of such programs as the U.S. Geological Survey's Landsat program, NOAA was briefly assigned management responsibilities for all U.S. civilian operational remote sensing programs. In response to this major increase in mission scale and objectives (even though it was subsequently rescaled and diminished), NESS was removed from the Office of Oceanic and Atmospheric Services and elevated to become a NOAA line agency, directed by a NOAA assistant administrator, and renamed NESDIS. It continued in 2008 as the operator of the largest Earth science applications satellite system and the greatest gatherer and distributor of Earth sciences data.

John Cloud

See also: Applications Technology Satellites, Geostationary Operational Environmental Satellites (GOES)

Bibliography

NESDIS. http://www.nesdis.noaa.gov/.

Oceansat. *See* Indian Remote Sensing Satellites.

Radarsat. Radarsat is a series of Canadian-built remote sensing satellites utilizing synthetic aperture radar (SAR) to image Earth's surface in any lighting or weather conditions. Radarsat images are used to monitor sea ice; map floods and forest cover; detect oil spills; search for mineral, oil, and natural gas deposits; and facilitate urban planning.

In 1975 the Canadian government began discussing the potential use of spacecraft radar to monitor ice conditions in the far north to support gas and oil exploration, providing wider and more continuous coverage than aircraft. After careful analysis of available aircraft- and spacecraft-based data, Canada decided to use a C-band (5.3 GHz) SAR, instead of the L-band (1.28 GHz) used by the U.S. Seasat program. After approval of the program in 1980, the 2,750 kg *Radarsat 1* was built by Spar Aerospace. NASA provided a free launch atop a Delta II rocket on 4 November 1995 from Vandenberg Air Force Base, California, into a Sun-synchronous, near-polar 798 km circular orbit in exchange for rights to receive the data at approved ground stations. This orbit provides daily coverage above 70° north latitude, crucial for ice observations in Canada's far north. It provides a 24 day repeat cycle and provides swath widths from 50–500 km, and resolutions from 8–100 m, respectively. The spacecraft, which used C-band (5.3 GHz) radar, remained operational in 2008, well beyond its planned five-year lifetime, and had mapped both the Arctic and Antarctic regions. Radarsat International Corporation (in 2008 MacDonald Dettwiler and Associates Ltd. or MDA Geospatial Services International) was created in 1989 to process, market, and distribute Radarsat data.

Radarsat 2, which was launched on 14 December 2007 on a Soyuz launcher from Baikonur, was a joint venture of the Canadian Space Agency and MDA, with MDA the prime contractor and an Italian spacecraft bus. NASA withdrew from the second Radarsat, apparently because of security concerns about its 3 m maximum resolution. *Radarsat 2* looked on both sides of its orbit, instead of *Radarsat 1*'s single side.

Chris Gainor

See also: Canada, MacDonald Dettwiler and Associates Ltd.

Bibliography

Howard Edel et al., "The Canadian RADARSAT Program," *Quest* 11, no. 4 (2004): 48–53.

Chris Gainor, "The Chapman Report and the Development of Canada's Space Program," *Quest* 10, no. 4 (2003): 3–19.

Edward J. Langham, "Radarsat Enters Phase B," in *Monitoring Earth's Ocean, Land and Atmosphere from Space—Sensors, Systems and Applications*, ed. Abraham Schnapf (1985).

Canadian Space Agency Radarsat-1. http://www.asc-csa.gc.ca/eng/satellites/radarsat1/default.asp.

Resurs-DK. Resurs-DK was a series of Russian high-resolution digital remote sensing satellites. Although first announced in 1996, the first of the satellites was not launched until 15 June 2006, after numerous delays.

Resurs-DK was designed by the Central Specialized Design Bureau (TsSKB) in Samara. DK are the initials of Dmitry Kozlov, who headed TsSKB from 1958 to 2003. Resurs-DK was a 7.93 m long satellite with two solar panels. It had a mass of 6,670 kg, which included 900 kg of propellant for on-orbit maneuvers and a 1,200 kg payload. Resurs-DK was believed to have been derived from the Neman digital reconnaissance satellites introduced in 1986.

The satellite's remote sensing payload was called Geoton-L and operated in four different wavebands in the visible and near-infrared. The Geoton-L complex was capable of imaging an area of 700,000 m^2 in 24 hours. Data could be transmitted to Earth in real time or dumped from an onboard recorder with a capacity of 768 Gbit. Resurs-DK received navigational support from Russia's Glonass navigation satellites.

According to Russian officials, Resurs-DK was a "dual-use" satellite, collecting images for civilian and military purposes. Resurs-DK had considerable reserves in mass and power supply, enabling it to carry other experiments besides the main payload. The first Resurs-DK satellite was launched with two additional payloads, an Italian magnetic spectrometer called RIM-PAMELA, to measure the energy spectrum of antiprotons and positrons in cosmic radiation, and a Russian spectrometer and particle detector called ARINA, to study electromagnetic field fluctuations that precede earthquakes.

Resurs-DK1 was launched by a Soyuz rocket from the Baikonur cosmodrome into a 200 × 370 km orbit inclined 70.0° to the equator. Later the satellite was transferred to an elliptical 360 × 600 km orbit. Other inclinations being considered for future

Geoton-L Remote Sensing Payload

	Number of Spectral Bands	Band Wavelengths (μm)	Ground Swath (km)	Ground Resolution (m)
Panchromatic	1	0.58–0.80	Up to 28 km at nadir, 360 km altitude	Up to 1 m
Multispectral	3	0.50–0.60 0.60–0.70 0.70–0.80	Up to 28 km	Up to 3 m

satellites were 64.8°, 64.9°, and 70.4°. The satellites had a design lifetime of at least three years.

Bart Hendrickx

See also: Russian Imaging Reconnaissance Satellites

Bibliography

Herbert Kramer, "Resurs-DK1 (Resurs—High Resolution 1)," eoPortal website, 2007.
eoPortal. http://www.eoportal.org/.

Resurs-F. Resurs-F was a series of general-purpose remote sensing satellites launched by the Soviet Union and Russia between 1975 and 1999 and based on the Vostok spacecraft and Zenit reconnaissance satellites. Presumably inspired by the impending launch of the first U.S. Landsat satellite, the Soviet government issued a decree in December 1971 calling for the development of two types of civilian remote sensing satellites, one like Landsat that would stay in orbit for a long period of time and send back relatively low-resolution pictures in real time (see Resurs-O), and another that would remain aloft for about two weeks and return high-resolution exposed film to Earth in a recoverable capsule.

Weighing about 6 tons, the satellites were designed and built by the Central Specialized Design Bureau (TsSKB) in Kuibyshev (later Samara). Like the Zenit spy satellites, they consisted of a service module, a forward-mounted engine unit for orbital maneuvers, and a spherical capsule containing the camera equipment that returned to Earth. One difference between these vehicles and the original Vostok was that the service module contained a solid-fuel rather than liquid-fuel deorbit engine. All satellites were launched by the Soyuz rocket from Plesetsk, most of them into low 81°–82° inclination orbits.

A first generation of satellites, with the program name Fram, flew 26 missions between 1975 and 1985. The onboard remote sensing complex consisted of five cameras (black-and-white and color) operating in five different regions of the electromagnetic spectrum.

The second-generation satellites (Resurs-F1), approved by a government decree in May 1977, were virtually identical to Fram, but had a more advanced camera complex, allowing them to make higher-resolution images. The payload consisted of one or more high-resolution cameras (SA-20M/KFA-1000) and topographic cameras (SA-34/KATE-200). A total of 50 satellites successfully reached orbit between 1979 and 1993. A modified version (Resurs-F1M), carrying the same types of cameras but using different film, flew two missions in 1997 and 1999.

In the mid-1980s, TsSKB developed a modernized version called Resurs-F2. The major external difference with its predecessor was the presence of two solar panels on the section connecting the return capsule to the orbital maneuvering unit. This increased the maximum flight duration to 30 days. Resurs-F2 satellites carried a single SA-M camera, combining the high resolution of the SA-20M with the multispectral capability of the SA-34. Ten satellites were launched between 1987 and 1995. Plans for an improved Resurs-F2M never materialized.

Some serial Zenit spy satellites flew for civilian purposes. This was the case for at least five Zenit-2M/Gektor satellites in 1975–79 and at least four Zenit-8/Oblik satellites between 1984 and 1994. The latter, carrying the high-resolution KFA-3000 cameras, were sometimes unofficially referred to as Resurs-F3 and Resurs-T.

Bart Hendrickx

See also: Zenit

Bibliography

Phillip Clark, "Recoverable Satellites Flown at 81.3–82.6° Orbital Inclinations," *Journal of the British Interplanetary Society* 50, no. 1 (1997): 13–24.

Yevgeniy Lukashevich, "The Space System 'Resurs-F' for the Photographic Survey of the Earth," *Russian Space Bulletin* 1, no. 4 (1994): 2–4.

Resurs-O. Resurs-O was a series of civilian remote sensing satellites launched by the Soviet Union and Russia between 1974 and 1998 and based on the Meteor weather satellites. In all, three generations of satellites were flown, only the last of which was announced as Resurs-O. The "O" stands for the Russian word *operativnyi* ("efficient, expedient"), referring to the fact that the satellites transmitted images in real time.

Probably inspired by the impending launch of the first U.S. Landsat satellite, the Soviet government issued a decree in December 1971 calling for the development of two types of remote sensing satellites, one like Landsat that would stay in orbit for a long period of time and send back relatively low-resolution pictures in real time, and another, eventually called Resurs-F, that would remain aloft for about two weeks and return high-resolution exposed film to Earth in a recoverable capsule.

The Landsat-type satellites, initially called Meteor Priroda, were designed and built by the All-Union Scientific Research Institute of Electromechanics (VNIIEM) in Moscow. Like the Meteor satellites, they consisted of an upper cylinder containing support systems and a lower cylinder carrying the instrument packages. It was the first Soviet spacecraft to use a system of flywheels to orient the satellites in space.

The first-generation satellites were based on the first-generation Meteor bus. Five satellites were launched in this series between 1974 and 1981. The first two were placed into standard Meteor orbits from Plesetsk, but the final three were launched

Meteor Priroda and Resurs-O Satellites

Generation	Program Name	Launched Between	Number of Launches	Comments
First generation	Meteor Priroda	9 Jul 1974 and 10 Jul 1981	5	Officially announced as "Meteor" and "Meteor Priroda." Launched by the Vostok rocket.
Second generation	Meteor Priroda	18 Jul 1980 and 24 Jul 1983	2	Officially announced as "Meteor" and "Kosmos." Launched by the Vostok rocket.
Third generation	Resurs-O	3 Oct 1985 and 10 Jul 1998	4	Officially announced as "Kosmos" and "Resurs-O." First two launched by the Vostok rocket, the last two by the Zenit rocket.

Meteor Priroda and Resurs-O Instruments

Instrument	Number of Spectral Bands	Band Wavelengths (μm)	Ground Swath (km)	Ground Resolution (m)
MSU-M	4	0.5–1.1	1,930	1,000
MSU-S	2	0.58–1.0	1,380	240
MSU-SK	5	0.5–1.1 10.4–12.6	600	170 600
MSU-E	3	0.5–0.9	45	45
Fragment	8	0.4–2.4	85	80

into Sun-synchronous orbits from Baikonur, becoming the first Soviet satellites to use that type of orbit. All were equipped with two multichannel scanning radiometers called MSU-S (for low-resolution images) and MSU-M (for medium-resolution images). Some of the satellites also carried additional payloads built in East Germany and Bulgaria.

The second-generation satellites, employing the Meteor 2 bus, were seen as experimental precursors of the Resurs-O satellites, approved by a government decree in May 1977. There were just two launches in 1980 and 1983. Besides the traditional MSU-M and MSU-S cameras, the satellites featured an experimental set of multichannel scanning radiometers (MSU-E, MSU-SK, and Fragment), each of which used a different scanning technique and had different resolutions.

The third-generation series, derived from the Meteor 3 satellites, saw four launches between 1985 and 1998. The key payloads were the MSU-E and MSU-SK radiometers. Images from the two final satellites were offered commercially on the international market. The second satellite also tested a Travers synthetic aperture radar antenna under development for the Okean satellites; the final satellite had a meteorological camera and several international payloads for solar and radiation studies.

Bart Hendrickx

See also: Russia (formerly the Soviet Union)

Bibliography

"The Resurs-O1 Series Spacecraft," *Russian Space Bulletin* 5, no. 4 (1998): 10–13.

Yuriy Trifonov, "The Russian Space Earth Observation System RESURS-O," *Space Bulletin* 1, no. 2 (1993): 11–13.

Seasat. *Seasat* was the first spacecraft dedicated to the remote sensing of Earth's oceans. It included the first spaceborne synthetic aperture radar (SAR), which operated by sending pulsed radar signals to Earth and analyzing their complex return pattern to produce an image. The other instruments were a radar altimeter, a radar scatterometer, a scanning microwave radiometer, and a visible and infrared radiometer (VIRR). All but the VIRR worked at microwave wavelengths and so could see through clouds. The SAR data rate was too high for onboard recording, but the data from all the other sensors could be recorded on board.

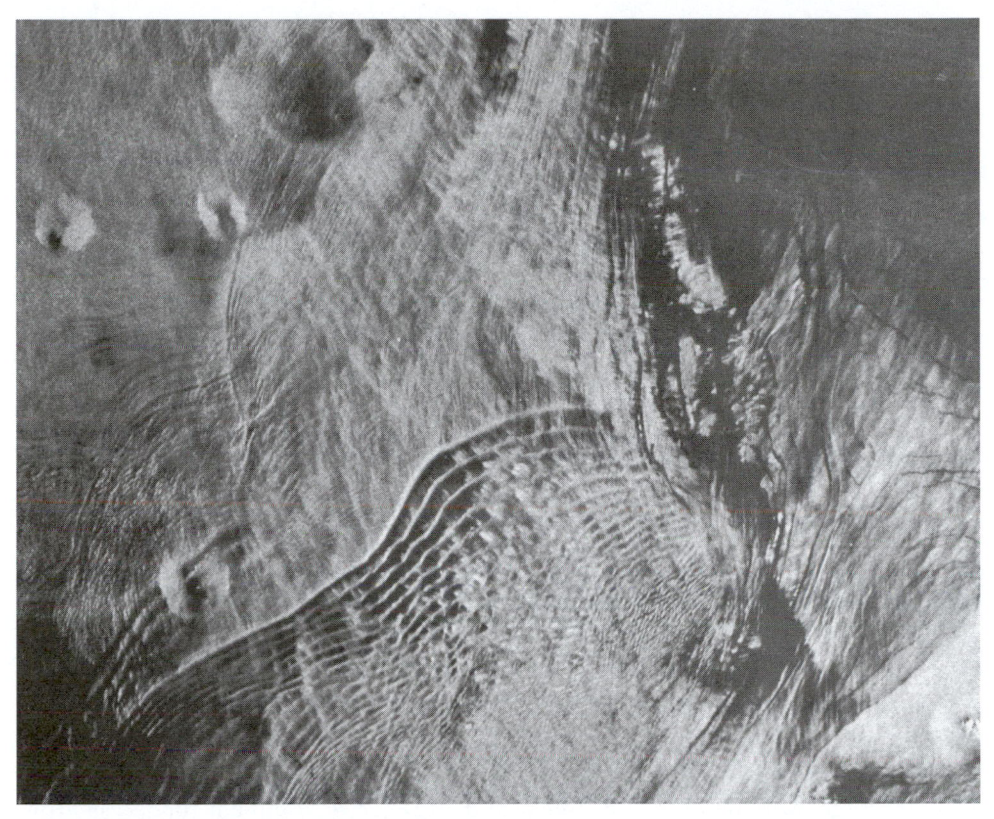

A Seasat *SAR image of the sea surface heights and wave patterns of the Gulf of Mexico, northeast of the Yucatan Peninsula. (Courtesy NASA/Jet Propulsion Laboratory)*

Prospective *Seasat* users defined their preliminary requirements in 1969, but the program was not approved until 1975. *Seasat*, managed by NASA's Jet Propulsion Laboratory, was an experimental or proof-of-concept spacecraft.

The 2,270 kg spacecraft was built by Lockheed, with Ball Aerospace providing the 10.7 m × 2.1 m SAR antenna. It was launched in June 1978 from Vandenberg Air Force Base into an 805 km circular, near-polar orbit by an Atlas Agena. The spacecraft bus, including the solar array, was located at one end of the spent Agena rocket, with the sensor platform, including the SAR antenna, at the other end, making the spacecraft 13 m long.

Seasat observed 95 percent of Earth's oceans every 36 hours. The VIRR failed 62 days after launch, and the whole of the spacecraft failed 44 days later because of a short circuit in its power supply system. Nevertheless, *Seasat* was highly successful during its brief lifetime, producing a wealth of data that took years to analyze. It provided the first global views of ocean and ocean floor topography and showed the enormous potential of microwave remote sensing from space, provided the data could be made available in near real-time. For example, wind speed, wave height, and ice condition information is invaluable for routing ships. In addition, longer-term information on ocean conditions would enable oil rig operators to design their rigs more optimally for the expected sea conditions.

The next U.S. Earth-oriented SAR missions were Shuttle Imaging Radar-A (SIR-A) in 1981 and SIR-B three years later. Both used spare Seasat hardware.

David Leverington

See also: Jet Propulsion Laboratory, Lockheed Corporation

Bibliography

S. W. McCandless Jr., "Seasat—A Retrospective," in *Monitoring Earth's Ocean, Land, and Atmosphere from Space—Sensors, Systems, and Applications*, ed. Abraham Schnapf (1985).

D. R. Montgomery, "Seasat Data Applications in Ocean Industries," in *Monitoring Earth's Ocean, Land, and Atmosphere from Space—Sensors, Systems, and Applications*, ed. Abraham Schnapf (1985).

SPOT, Satellite Pour l'Observation de la Terre. SPOT, Satellite Pour l'Observation de la Terre, a French-led Earth observation program, with Belgian and Swedish participation. CNES (French Space Agency) was the system's prime contractor, with the spacecraft prime contract going to Matra, later Matra Marconi Space, and later yet European Aeronautic Defence and Space (EADS) Astrium. Originally called Système Probatoire Pour l'Observation de la Terre (but changed by the mid-1980s to its current name), SPOT satellites typically offered higher-resolution imagery (10 m) than U.S. Landsat satellites (30 m) of the same period. SPOT imagery was made available through SPOT Image, a company chartered in 1982 to distribute its imagery. Scientific researchers and the French government received free access to SPOT imagery, while civilian users purchased it. Because the French government funded satellite development, pricing did not reflect these costs. The resulting competitive market for commercial remote sensing data broke the U.S. Landsat monopoly.

SPOT 1 was launched into a Sun-synchronous polar orbit on an Ariane booster from the European Space Agency Kourou, French Guiana, launch site in February 1986. The primary sensors were dual high-resolution visible and near-infrared scanners capable of operating in either panchromatic (black and white) or three-band multispectral modes and delivering a resolution of 10 or 20 m, respectively.

SPOT 2 and *SPOT 3*, launched in January 1990 and September 1993 respectively, offered similar optical imaging capability to *SPOT 1*. A radio-based positioning system, Doris, was added beginning with *SPOT 2* that allowed determination of the satellite location to an accuracy of 10 cm; this was continued on later SPOT satellites.

SPOT 4, launched in March 1998, offered significantly enhanced capability with a four-band scanner and a low resolution (1 km) vegetation-monitoring instrument that competed with the Advanced Very High Resolution Radiometer instrument carried by U.S. polar orbiters. It also carried a laser telecommunications package. *SPOT 5*, orbited in May 2002, offered high-resolution geometric imaging up to 2.5 m when operated in panchromatic mode. Multispectral stereoscopic imagery resolution improved to 10 m in the visible and 20 m in the infrared. As of 2007, *SPOT 4* and *5* were fully operational and *SPOT 2* partially functioning. France planned to replace the SPOT satellites with a new constellation of higher-resolution, smaller Pleiades satellites, beginning in 2009.

Despite its nominally civilian character, SPOT imagery had military implications. The availability of high-resolution civilian imagery from non-U.S. sources, such as

SPOT, vitiated the policy of shutter control under which the U.S. government attempted to restrict access to such imagery in politically or militarily sensitive situations. During the 1991 Gulf War, this issue led to diplomatic pressure to prevent the sale of SPOT imagery to Iraq, while at the same time purchasing and successfully using this data for coalition military purposes.

SPOT satellite designs influenced several later programs. The *European Remote Sensing 1* and *2* satellites used a variant of the *SPOT 1* bus. *SPOT 4* bus variants migrated to the French Helios military observation satellites and the service modules for the European *Envisat* remote sensing and MetOp low-Earth orbit meteorological satellites.

John D. Ruley

See also: European Aeronautic Defence and Space Company (EADS), France, Helios

Bibliography

Thomas M. Lillesand and Ralph W. Keifer, *Remote Sensing and Image Interpretation* (2000).
NASA Commercial Space Transportation Study (1994).
Jon Trux, "Desert Storm: A Space-Age War," *New Scientist* (1991).
CNES-SPOT. http://www.cnes.fr/web/CNES-en/1415-spot.php.

Tansuo. *See* Chinese Remote Sensing Satellites.

Technology Experiment Satellite. *See* Indian Remote Sensing Satellites.

TOPEX/Poseidon. *See* Civilian Ocean Surface Topography Missions.

Zi Yuan. *See* Chinese Remote Sensing Satellites.

Civilian Space Communications

Civilian space communications are nonmilitary, government-owned, or government-controlled satellites that provide communication services for nations or groups of nations.

Most early civilian communications satellites (comsats) were placed into orbit to test the feasibility of satellite communications or new technologies to improve their capabilities. Examples include the NASA-funded *Echo 1* and Syncom (synchronous communications) satellites of the early 1960s, which proved the viability of satellite communications in general and of geosynchronous comsats in particular. NASA followed these successes with the Applications Technology Satellite series, which proved out successively more capable communications technologies, along with remote sensing capabilities. In the 1970s, France and West Germany developed and launched the experimental Symphonie satellites, Italy orbited its *Sirio* comsat, and the European Space Agency launched its Orbiting Test Satellites. The Soviet Union, with its huge landmass and isolated polar locations, developed the Molniya dual-use (military-civilian) comsats that used polar orbits.

After the feasibility of comsats and their various new technologies was proven, many nations developed or purchased satellites that would be operated by national

Artist's conception of Syncom. Syncom 2 *was the first experimental communications satellite to reach geosynchronous orbit. (Courtesy NASA/Goddard Space Flight Center)*

governments or international organizations to provide domestic communications services. The U.S. Tracking and Data Relay Satellite System was developed to provide space-based communications for the Space Shuttle at costs significantly less than NASA's worldwide ground tracking network. Intelsat (International Telecommunications Satellite Organization), Inmarsat (International Maritime Satellite Organisation) and Arabsat (Arab Satellite Communications Organization) were international civilian consortia to provide international communications services. Most nations assigned their ongoing national operations to domestic government postal and telecommunications departments. Examples of government-operated domestic comsats include Telecom (operated by France Telecom), *Kopernikus* (operated by Deutsche Bundespost), Turksat (operated by the Turkish Ministry of Post and Telecommunications), Multifunctional Transport Satellites (operated by Japanese Civil Aviation Bureau), Indonesia's Palapa satellites (operated by Perumtel), the Insat series (operated by India's Department of Space), and Mexico's Morelos satellites (operated by Satmex).

While government control was prevalent in the early decades of the space age, by the 1990s the relative efficiency and profit-making prowess of private comsat operators created pressure on governments to privatize their comsat networks. In the case of government organizations or international consortia intending to compete for business in the United States, the 2000 Open-market Reorganization for the Betterment of International Telecommunications (ORBIT) Act required that Intelsat and Inmarsat could operate in the country only if fully privatized. These events were significant factors in the privatization of Intelsat, Inmarsat, and Satmex, among others. Nonetheless many national civilian comsat systems continued to operate in the early twenty-first century.

Stephen B. Johnson

See also: Arabsat, Commercial Space Communications, Economics of Space Communications, Inmarsat, Intelsat, Military Communications, Space Telecommunications Policy

Bibliography

Joan Lisa Bromberg, *NASA and the Space Industry* (1999).
Roger Handberg, *International Space Commerce: Building from Scratch* (2006).
Donald H. Martin et al., *Communication Satellites*, 5th ed. (2007).

Anik. *See* Canadian Communications Satellites.

Applications Technology Satellites. Applications Technology Satellites (ATS) were a series of satellites designed and launched by NASA during the mid-1960s for testing new technologies in space. ATS spacecraft pioneered new technologies for communications, weather monitoring, and other space research. The NASA ATS experimental satellites tested various crucial satellite technologies used by satellite builders and operators into the twenty-first century. The NASA Goddard Space Flight Center managed the ATS program until the mid-1980s, when Lewis Research Center took over management of the program. Hughes Aircraft Company built *ATS 1* through *5*. All ATS satellites were launched from Cape Canaveral, Florida, the first three on Atlas-Agenas, and next two on Atlas-Centaurs, and the last on a Titan IIIC.

ATS 1, the first civilian spacecraft to provide meteorological images from geosynchronous orbit, was launched in December 1966 and operated until 1985. It first demonstrated "frequency division multiple access" communications technologies in space, which successfully combined multiple ground-uplink communications channels into a single-downlink channel. It also demonstrated a black-and-white, spin-scan camera, which progressively built an image of Earth by gathering one 3.2 km line of data from each spacecraft revolution, changing its camera angle slightly for the next revolution until it constructed an image of Earth's entire disk. Variations of this design were used on many later meteorological spacecraft.

ATS 2 carried communications and space-weather experiments, in addition to a test of gravity-gradient stabilization in a medium-altitude (6,000 nautical miles) orbit for the U.S. Department of Defense (DoD). Launched in April 1967, it failed to achieve a stable orbit because of an Agena upper-stage failure. Some data was returned before atmospheric reentry occurred in 1968.

Launched in November 1967, *ATS 3* carried a color version of the spin-scan camera, returning the first full disk color images of Earth from space. In addition to providing useful images of hurricanes, *ATS 3* carried communications, imaging, and propulsion experiments, pioneering the use of hydrazine thrusters. *ATS 3* far outlived its three-year design life, providing communications support for polar research stations and emergency response until being shut down in May 2004.

The first three ATS satellites were spin stabilized. By contrast, *ATS 4* was to demonstrate three-axis, gravity-gradient stabilization at geosynchronous altitudes for the DoD. After its failed August 1968 launch, *ATS 5* also failed prematurely one year later because of a thruster problem in the parking orbit. NASA ground controllers were able to command *ATS 5* into a geosynchronous orbit ahead of schedule, but failed to cease the satellite's rotation. Nonetheless testing occurred with some of *ATS 5*'s experiments, such as an L-band communications payload, ion propulsion testing, and various physics and space weather experiments.

The U.S. Congress canceled the ATS program in 1973 to prevent federal government competition with privately funded communications satellite industry research and development activities. Only the second-generation *ATS 6* satellite, built by Fairchild Space and Electronics, survived the cancellation and was launched in 1974. Much larger than its predecessors, *ATS 6* hosted a large 9 m parabolic antenna and high-powered solid-state communications electronics in addition to new spacecraft subsystem technologies. The *ATS 6* design demonstrated technologies later used in NASA's Tracking and Data Relay Satellite System and private-sector communications satellites. In 1975–76, NASA and the Indian Space Research Organisation operated *ATS 6* for the Satellite Instructional Television Experiment, which supplied educational and entertainment television broadcasting to 2,400 remote villages.

David Hartzell

See also: Commercial Satellite Communications, Goddard Space Flight Center, India

Bibliography

Daniel R. Glover, "NASA Experimental Communications Satellites, 1958–1995," in *Beyond the Ionosphere*, ed. Andrew Butrica (1997).

NASA Experimental Communications Satellites. http://history.nasa.gov/commsat.html.

ARTEMIS, Advanced Relay Technology Mission. *See* European Communications Satellites.

Broadcast Satellite Experimental. *See* Direct Broadcast Satellites.

Canadian Technology Satellite. *See* Direct Broadcast Satellites.

Echo 1. *Echo 1*, the world's first telecommunications satellite, was launched by NASA on 12 August 1960. The satellite had its origin in 1954 when John Pierce of Bell Laboratories suggested that an Earth-orbiting satellite could reflect transmitted radio signals from one point on the globe to another. Pierce determined that the satellite should be 100 ft in diameter and should orbit Earth at a distance of 1,000 miles. Two years later Pierce saw a drawing for an inflatable balloon satellite that National Advisory Committee on Aeronautics (NACA) engineer William J. O'Sullivan Jr. had envisioned for testing atmospheric drag. Pierce suggested combining the projects, and in 1957 NACA gave its approval.

The newly created NASA adopted the program in 1958, supplying the satellite, while Bell and the Army-funded Jet Propulsion Laboratory (JPL) would have responsibility for the transmission and reception of radio signals using the satellite. General Mills received the contract to produce an inflatable satellite made of bonded strips of Mylar coated with vaporized aluminum, and the G. T. Schjeldahl Company developed a process for securely sealing the strips together. A series of suborbital tests in 1959–60 proved the satellite's viability. The first launch attempt in May 1960 failed, but the second attempt achieved orbit three months later. Officially christened *Echo 1*, technicians at JPL's Goldstone, California, facility bounced a message recorded by President Dwight Eisenhower off the satellite. Personnel at Bell Laboratories's Holmdel, New Jersey, site received the message, thus proving the viability of communications via satellite. During its lifetime, *Echo 1* was used for numerous experiments, including the transmission of television signals. Because it could easily be seen by the naked eye, *Echo 1* gave U.S. prestige a significant boost globally.

Donald C. Elder

See also: American Telephone and Telegraph (AT&T) Company, Jet Propulsion Laboratory

Bibliography

Donald C. Elder, "Something of Value: Echo and the Beginnings of Satellite Communications," in *Beyond the Ionosphere: Fifty Years of Satellite Communication*, ed. Andrew J. Butrica (1997).

Craig B. Waff, "Project Echo, Goldstone, and Holmdel: Satellite Communications as Viewed from the Ground Station," in *Beyond the Ionosphere: Fifty Years of Satellite Communication*, ed. Andrew J. Butrica (1997).

EDUSAT. *EDUSAT*, the Indian Space Research Organisation's (ISRO) first thematic satellite, dedicated entirely to provide distance educational services for rural India. Launched into the geostationary orbit by Geostationary Satellite Launch Vehicle on its first operational flight in September 2004, *EDUSAT* was specifically configured for the audiovisual medium and carried five Ku-band transponders providing spot beams, one Ku-band transponder providing a national beam, and six Extended C-band transponders with national coverage beam.

Built and operated by the ISRO Satellite Center, *EDUSAT*'s utilization was a collaborative project with the Ministry of Human Resource Development and Indira Gandhi National Open University. The concept for *EDUSAT* was in part derived from the successful demonstration of beaming educational and health-related programs through the Satellite Instructional Television Experiment conducted using the American *Application Technology Satellite 6* in 1975–76. Starting in 1983 India utilized transponders from its own Indian National Satellite System satellites for telecast of educational programs. *EDUSAT*'s primary purpose was to establish dedicated two-way connectivity among India's renowned urban educational institutions and a large number of under-equipped rural institutions.

Given India's regional diversity in culture and languages, educators had to prepare content in 18 official languages and install satellite receivers in thousands of

The Indian Space Research Organisation's Geosynchronous Satellite Launch Vehicle carrying the EDUSAT *lifts off from the Satish Dhawan Space Centre near Sriharikota, India, in 2004. (Courtesy AP/ Wide World Photos)*

rural institutions to help reduce the nation's estimated 35 percent illiteracy rate in 2005.

Venkatesan Sundararajan

See also: Education, India, Indian National Satellite Program

Bibliography

ISRO *EDUSAT.* http://www.isro.org/scripts/sat_EDUSAT.aspx.

Federal Communications Commission. The Federal Communications Commission (FCC) is an independent U.S. government agency responsible to Congress. Established by the Communications Act of 1934, the FCC regulates nongovernmental interstate and international communications by radio, wire, television, satellite, and cable. FCC jurisdiction covers the 50 states, the District of Columbia, U.S. possessions, and U.S.-owned satellites in outer space.

The FCC is directed by five commissioners, only three of whom may be members of the same political party, appointed by the president and confirmed by the Senate for five-year terms. The commissioners supervise FCC activities, delegating responsibilities to staff units and bureaus. There are six operating bureaus and 10 staff offices. Bureau responsibilities include processing applications for licenses to construct and operate radio transmitting equipment and other filings, conducting investigations, developing and implementing regulatory programs, and taking part in hearings. Several bureaus and offices are involved in regulating space electromagnetic uses.

The FCC first engaged in radio frequency meetings for space use during the International Telecommunication Union's (ITU) Plenipotentiary Conference of 1959, when a standard frequency for satellite tracking was established. The first major international meeting on space radio regulation was the ITU's Extraordinary Radio Administrative Conference in Geneva, Switzerland, in 1963. Numerous subsequent conferences involving space communications matters were held at increasingly frequent intervals during the last third of the twentieth century. The meetings concentrated mainly on allocating radio spectrum to accommodate the expanding use of outer space for 17 separate space radio service categories. From 1985 to 1988, the ITU held a two-part World Administrative Radio Conference on the Geostationary Satellite Orbit and the Space Services Using It. In addition to assigning space radio frequency bands for use in countries and regions, this conference assigned specific "parking slots" by country for use of the geostationary orbit.

The International Bureau represents the FCC in satellite and international matters, including international system licensing. With the U.S. Department of State, the National Telecommunication Information Administration (for governmental radio matters), the FCC participates on behalf of the United States in the international coordination, planning, and regulatory functions of the ITU in Geneva.

The Media Bureau regulates amplitude modulation and frequency modulation radio, television broadcast stations, multipoint distribution (cable and satellite, including direct broadcast satellites), and instructional television fixed services.

Stephen Doyle

See also: Commercial Space Communications, Satellite Radio

Bibliography

FCC. http://www.fcc.gov/.

Indian National Satellite Program. The Indian National Satellite program (INSAT) consists of a series of four generations of multipurpose geosynchronous satellites. Originally conceptualized for educational purposes, INSAT went through many changing requirements and evolved to serve three main purposes: telecommunications, meteorology, and a wider broadcasting network. Driven by Vikram Sarabhai's vision of harnessing science and technology for national development, the INSAT program has been a success.

Setting the stage for INSAT was the U.S.-India Satellite Instructional Television Experiment (SITE), using NASA's *Applications Technology Satellite 6*, carried out in India in 1975–76, which established a basis for satellite telecommunications in India, servicing about 2,400 remote villages. This brought national and international recognition to the Indian Space Research Organisation (ISRO), although SITE faced problems of its own. Due to the Indian bureaucracy, the national television broadcasting provided limited selective channels, and SITE, being purely educational, did not catch the attention of the rural audience. ISRO benefited from international collaboration with France, Germany, and the Soviet Union. When INSAT became operational, Indian television

INSAT Satellites

Satellite	Date of Launch	Results
INSAT 1A	April 1982	Abandoned in September 1982, attitude control propellant exhausted.
INSAT 1B	August 1983	Service completed as planned.
INSAT 1C	July 1988	Abandoned in November 1989, transponders lost due to power system failure.
INSAT 1D	June 1990	Seriously damaged 10 days before launch. Ford Aerospace repaired the fully insured satellite at a reported cost of $10 million.
INSAT 2A	July 1992	Service completed as planned.
INSAT 2B	July 1993	Service completed as planned.
INSAT 2C	December 1995	Service completed as planned.
INSAT 2D	June 1997	Prematurely terminated in October 1997 due to power supply failure.
INSAT 2DT	February 1992	Acquired from Arabsat on 26 November 1997.
INSAT 2E	April 1999	Operational.
INSAT 3B	March 2000	Operational.
INSAT 3C	January 2002	Operational.
INSAT 3A	April 2003	Operational.
INSAT 3E	September 2003	Operational.
INSAT 4A	December 2005	In service. Enhanced direct to home broadcasting.
INSAT 4C	July 2006	Launch failed. Attempted indigenous launch.
INSAT 4B	March 2007	In service.
INSAT 4CR	September 2007	In service.

programming expanded to include entertainment and political and intellectual debates, bringing competitiveness to Indian broadcasting.

First launched in April 1982, Ford Aerospace of the United States built the first generation of INSATs—*1A*, *1B*, *1C*, and *1D*—that were either launched by a Delta rocket, the Space Shuttle, or an Ariane vehicle, out of which *1A* and *1C* were abandoned due to technical problems. The second generation of INSATs—*2A*, *2B*, *2C*, *2D*, and *2E*—first orbited in July 1992 and were indigenously built by ISRO, integrated by the ISRO Satellite Center in Bangalore and launched by an Ariane vehicle from Kourou, French Guiana, because India did not have launch capabilities powerful enough to put a 2,500 kg satellite into orbit. The second generation had improved communications capacity in newer frequencies and an improved radiometer for weather observation, with *INSAT 2C* and *INSAT 2D* being communications-only spacecraft, though *INSAT 2D* was a failure. *INSAT 2E* was one of the most advanced satellites in terms of technological and electronic software. The INSAT 3 series, launched in March 2000, aimed to achieve continuity of services, increase capacity, and improve capabilities. *INSAT 3D*, projected to launch in 2007–8, carries a six-channel imager and 19-channel sounder.

The INSAT 4 series, the first of which, *INSAT 4A*, was launched in December 2005, provided enhanced capacity for direct-to-home television and communications services. It carried 12 high-power Ku-band transponders and an additional payload of 12 C-band transponders. *INSAT 4C* was destroyed during launch in July 2006, during the first attempt to use an Indian launcher to place an INSAT spacecraft into orbit.

INSAT satellites are controlled from the INSAT Master Control Facility at Hassan in Karnataka. The S-band is aimed at mobile communications, the C-band is used for television broadcasting, and the Ku-band supports business communications among India's four major cities: Delhi, Mumbai, Calcutta, and Madras. Despite constantly changing design needs and increasing technological complexity, INSAT emerged in the early twenty-first century to be the largest domestic communications system in the Asia-Pacific region. By 2006 INSAT satellites were used for many purposes, such as domestic long-distance telecommunications, meteorological observation and data collection services, direct satellite television and radio broadcasting to community television receivers in rural and remote areas, and satellite-aided search-and-rescue services.

Shubhada Savant

See also: India

Bibliography

GlobalSecurity. www.globalsecurity.org.

Brian Harvey, *The Japanese and Indian Space Programmes: Two Roads into Space* (2000).

ISRO. www.isro.org.

Raman Srinivasan, "No Free Launch: Designing the Indian National Satellite," in *Beyond the Ionosphere: Fifty Years of Satellite Communication*, ed. Andrew J. Butrica (1997).

International Telecommunication Union. The International Telecommunication Union (ITU) coordinates international and cross-national usage of the electromagnetic spectrum and establishes rules and standards for international telephony and telegraphy. In May 1865 continuing rapid expansion of national telegraph networks prompted 20 European nations to adopt the International Telegraph Convention, establishing the International Telegraph Union (ITU). Following the patent of the telephone in 1876 and the subsequent expansion of telephony, the ITU began in 1885 to create international regulations governing telegraphy and telephony.

The invention of wireless telegraphy (radio communications) in the 1890s prompted similar discussions to standardize and regulate radio communications. In 1906 the first International Radiotelegraph Conference established the first International Radiotelegraph Convention, which contained the first regulations governing wireless telegraphy.

The 1927 International Radiotelegraph Conference internationally allocated frequency bands to various radio services existing at the time (fixed, maritime and aeronautical mobile, amateur, broadcasting, and experimental). At a 1932 Madrid conference, the Union's members decided to combine the International Telegraph Convention of 1865 and the International Radiotelegraph Convention of 1906 to form the International Telecommunication Convention. The institutional names were combined into the International Telecommunication Union in January 1934. Under an agreement with the newly created United Nations (UN), the ITU became a UN specialized agency in October 1947.

In 1959 the ITU set up a study group responsible for studying space radio communications. In 1963 an Extraordinary Administrative Radio Conference for space communications allocated the first frequencies to the emerging space services. Subsequent conferences made further allocations and created regulations governing satellite use of the radio-frequency spectrum and associated orbital slots for what eventually became 17 separate space radio services.

In 1989 with the gradual globalization and liberalization of world telecommunications markets, the Nice Plenipotentiary established a community of experts to evaluate ITU structures, operation, working methods, and resources. In 1992 the Additional Plenipotentiary Conference in Geneva restructured the ITU into three sectors corresponding to its three main areas of activity: Telecommunication Standardization (ITU-T), Radio Communication (ITU-R), and Telecommunication Development (ITU-D). By 2005 the ITU had organized development of fully interconnected and interoperable terrestrial and space networks and services. The fastest growing area of global telecommunications services since 1975 has been the use of radio for space services, especially voice and television relay services.

Stephen Doyle

See also: Economics of Space Communications, Space Telecommunications Policy

Bibliography

George A. Codding, *The International Communication Union: An Experiment in International Cooperation* (1972).

George A. Codding with Anthony M. Rutkowski, *The International Telecommunication Union in a Changing World* (1982). *Global Communication in the Space Age: Toward a New ITU* (1972).

Rita Lauria White and Harold M. White Jr., *The Law and Regulation of International Space Communication* (1988).

Intersputnik. Intersputnik was founded in the early 1970s as the socialist rival to the capitalists' Intelsat organization. Intersputnik comprised the Soviet Union, the socialist countries of eastern and central Europe, and other countries friendly to the Soviet Union, such as Afghanistan, Algeria, Cuba, Laos, Mongolia, Syria, Vietnam, and Yemen. Intersputnik mirrored in the communications field the Interkosmos program of cooperation in space science for the socialist countries.

The Intersputnik network relied mainly on the Molniya 3 series of communications satellites and its Orbita stations of 12 m diameter dishes (Orbita 1), later transmitting through smaller dishes 7 m across (Orbita 2). Later the 24-hour satellites of the Gorizont and Raduga series were also used.

Intersputnik relayed television, telephone, facsimile, and newspaper links. With the collapse of the Soviet Union in 1991, Intersputnik withered as a separate organization and the Russian Federation no longer had the resources to run a separate international communications organization. Russia joined the European Telecommunications Satellite organization, Eutelsat, in 1994.

Brian Harvey

See also: Interkosmos

Bibliography

Novsoti Press Agency, *Economic Uses of Space Technology* (1985).

MARECS, Maritime European Communications Satellite. *See* European Communications Satellites.

Marots, Maritime Orbital Test Satellite. *See* European Communications Satellites.

Molniya. Molniya (lightning), a series of Russian dual-purpose communications satellites, which utilizes a highly elliptical orbit with an altitude of 500 × 40,000 km and an inclination of 63–65°. A single satellite supports continuous communications for eight hours per revolution, while uninterrupted global coverage is ensured by a constellation of eight satellites divided in two groups positioned in orbital planes offset by 45°. Two different constellations of Molniya 1T and Molniya 3 satellites were in service as of 2008.

While all Molniya satellites share the same general design: the sealed cylindrical body, 4.4 m long and 1.4 m in diameter, an average mass of 1,650 kg, crowned by six

The Molniya 1 *telecommunications satellite. (Courtesy ITAR-TASS/Sovfoto)*

solar panels spanned for 8.2 m, the later models employed state-of-the-art electronics. The satellite was capable to correct its orbit via a rocket engine, compressed gas thrusters, and a variable speed gyroscope. The latter unique system, first employed by Molniya 1, later became standard for the majority of the Soviet spacecraft. Throughout four decades, Molniya satellites improved their orbital life span from one to five years. The satellites were launched by a Molniya M variant of an R-7 rocket, initially from Baikonur and later from Plesetsk, but starting in 1989 only from Plesetsk.

The Molniya program began in 1960 with the intent to provide television, telephone, and telegraph communications to the eastern part of the Soviet Union, where only rudimentary ground communications infrastructure existed at the time. OKB-1 (Experimental Design Bureau), the leading Soviet rocket and space organization near Moscow, designed *Molniya 1*, but in 1965 all further system development was transferred to OKB-10 in Zheleznogorsk, known later as the Applied Mechanics Corporation. The highly elliptical orbit was chosen instead of a more efficient geosynchronous orbit, because it required less energy for launch; in the 1960s the Soviet Union did not have a rocket powerful enough to place a 2,000 kg satellite into geosynchronous orbit.

After two failed attempts in 1964, the first successful Molniya 1 launch came on 23 April 1965. By late 1967 the system achieved operational status with Orbita (orbit) receiving stations quickly built around the country. It was an instant success; more than 20 million people in remote areas gained access to modern communications. In 1970–75 several space-based military secure communications systems were developed, including the Korund (corundum) system for command and control of the Strategic Forces. After that time Molniya 1 was used primarily by the military and in 1983–87 it was replaced by a new Molniya 1T satellite. One experimental satellite, *Molniya 1S*, was placed in a geosynchronous orbit by a Proton rocket in 1974. Meanwhile, civilian communications were supported by a separate constellation of Molniya 2 satellites, first launched in 1971 and by 1977 replaced by Molniya 3. Molniya satellites became a foundation for the Unified System of Satellite Communications (ESSS), which by 1975 combined various military and civilian space communications. With the advent of Soviet geosynchronous satellites in the late 1970s, most civilian ESSS channels were transferred there, leaving the Molniya 3 constellation for military use. Since 2001 the Molniya 3K was being phased in as its replacement. The Russian military consider Molniya satellites an important

Molniya Satellites, 1964–2005

Satellite	Years of Launches	Number of Launches
Molniya-1	1964–1987	71
Molniya-1T	1983–2004 (in service)	29
Molniya-1S	1974 (experimental)	1
Molniya-2	1971–1977	19
Molniya-3	1974–2003 (in service)	54
Molniya-3K	2001–2005 (new, in service)	2

supplement to geosynchronous satellites and a survivability factor for its global communications in case of a major conflict, as these satellites are easier to launch and faster to replace than their geosynchronous counterparts.

A total (including failures) of 176 Molniya satellites were launched from 1964 to 2005, most under their real names, with only a few announced as Kosmos.

Peter A. Gorin

See also: Kosmos, Russian Space Forces, Russian Space Launch Centers, Russian Launch Vehicles

Bibliography

Bart Hendrickx, "The Early Years of the Molniya Program," *Quest* 6, no. 4 (1998): 28–36.
Nicholas L. Johnson and David M. Rodvold, *Europe and Asia in Space, 1993–1994* (1995).
Russia's Arms Catalog, vol. 6, *Missiles and Space Technology* (1998).

Nimiq. *See* Canadian Communications Satellites.

Olympus. *See* European Communications Satellites.

Orbital Test Satellite. *See* European Communications Satellites.

OSCAR, Orbital Satellite Carrying Amateur Radio. *See* Amateur Communications Satellites.

Palapa. *See* Regional Satellite Communications Operators.

Search and Rescue. Search and rescue is an important function spun off from communications satellite technology and incorporated onto several weather, navigation, research, and communications satellites. The Cospas-Sarsat system used radio equipment onboard various satellites to receive and relay distress messages from emergency transmitters carried on ships, aircraft, and with people on Earth. Sarsat is short for Search and Rescue Satellite-Aided Tracking. Cospas is a Russian acronym for Space System for the Search of Vessels in Distress.

Amateur radio operators (also called "hams") pioneered the satellite search-and-rescue technique in the 1970s, on board their privately built *Orbiting Satellite Carrying Amateur Radio* (*OSCAR*) *7* satellite. The Radio Amateur Satellite Corporation, a global network of hams that built and operated communications and science satellites, proved the Cospas-Sarsat concept could work when *OSCAR 7* received low-power radio signals from the ground and repeated them to NASA's Goddard Space Flight Center in Maryland in December 1975. *OSCAR 7* showed that a weak signal from the ground could be tracked by satellite to within two to four miles of an emergency site.

Based on a memorandum of understanding signed in 1979, Canada, France, the Soviet Union, and the United States developed the Cospas-Sarsat network using American and Soviet government spacecraft. It was declared operational in 1985. By 2005 Algeria, Argentina, Australia, Brazil, Chile, India, Indonesia, Italy, Japan, New Zealand, Nigeria, Norway, Pakistan, People's Republic of China, Peru, Republic of Korea, Saudi Arabia,

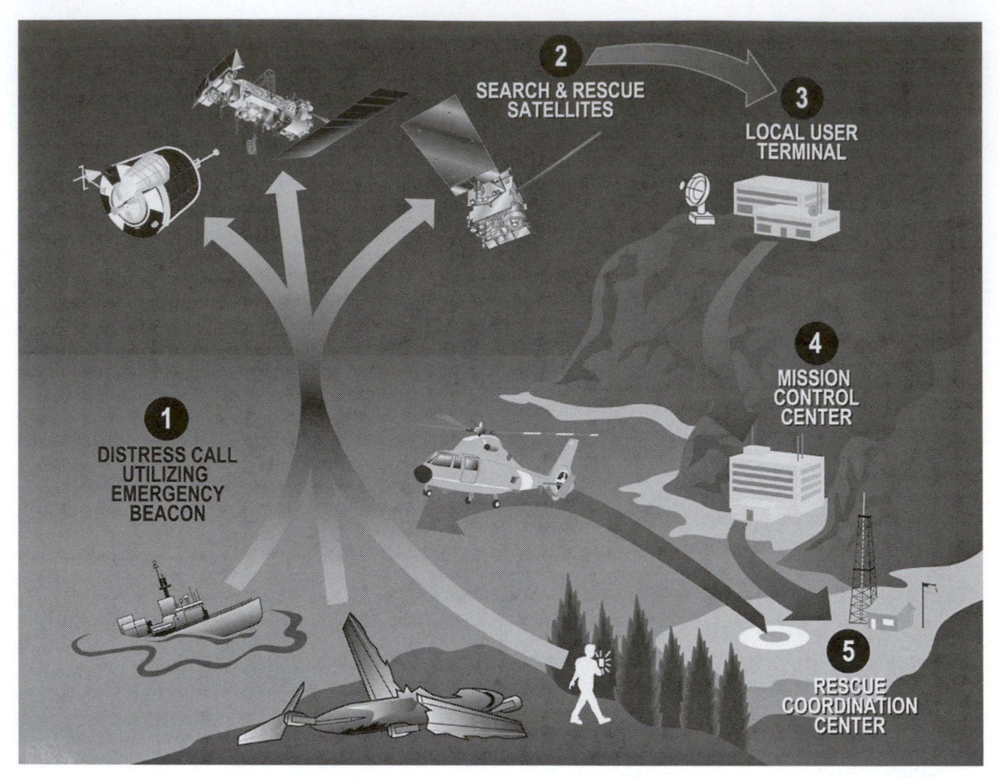

Satellite-aided search and rescue system concept. (Courtesy NASA/Goddard Space Flight Center)

Singapore, South Africa, Spain, Thailand, the United Kingdom, and Vietnam also were providing receiving stations on the ground to monitor the clutch of Russian and American Cospas-Sarsats to trace ships in trouble at sea, downed aircraft, and lost individuals. In addition Denmark, Germany, Greece, Madagascar, the Netherlands, Sweden, Switzerland, and Tunisia used the system. Russia and the United States provided satellites, which also carried equipment supplied by Canada and France.

While carrying out their primary tasks, the Cospas-Sarsat satellites continuously monitored international radio distress frequencies. The portable emergency alarm equipment carried by ships, boats, planes, cars, and individuals is known as Emergency Locator Transmitters and Emergency Position-Indicating Radio Beacons (EPIRB). The early emergency transmitters suffered a high false-alarm rate because sometimes users turned them on accidentally. Later EPIRBs transmitted an identification code so ground stations could phone to verify an emergency. When an emergency signal is transmitted from the ground or ocean, a Cospas-Sarsat passing overhead picks up the signal and relays it to a search-and-rescue team on the ground. The team uses information in the relayed distress signal to determine the location of the person or vehicle in trouble. Rescuers are dispatched to save the victims. Authorities credit the rescue system with having saved nearly 20,000 people around the world, and more than 5,000 in the United States, from shipwrecks, plane crashes, and even a dog sled race that went awry.

Low-altitude Earth Orbit Search and Rescue Satellites

LEOSAR satellites past, present, and planned		
Satellite Name	**Launch**	**Status**
Cospas-1 Cosmos 1383	June 1982	Decommissioned
Cospas-2 Cosmos 1447	March 1983	Decommissioned
Cospas-3 Cosmos 1574	June 1984	Decommissioned
Cospas-4 Nadezhda-1	July 1989	Not in continuous operation
Cospas-5 Nadezhda-2	February 1990	Decommissioned
Cospas-6 Nadezhda-3	March 1991	Decommissioned
Cospas-7 Nadezhda-4	July 1994	Decommissioned
Cospas-8 Nadezhda-5	December 1998	Decommissioned
Cospas-9 Nadezhda-6	June 2000	In operation
Cospas-10 Nadezhda-7	September 2002	Decommissioned
Cospas-11	Planned launch 2006	
Cospas-12	Planned launch 2007	
Sarsat-1 NOAA-8	March 1983	Decommissioned
Sarsat-2 NOAA-9	December 1984	Decommissioned
Sarsat-3 NOAA-10	September 1986	Decommissioned
Sarsat-4 NOAA-11	September 1988	Decommissioned
Sarsat-5 NOAA-13	August 1993	Failed after launch
Sarsat-6 NOAA-14	December 1994	In operation
Sarsat-7 NOAA-15	May 1998	In operation
Sarsat-8 NOAA-16	September 2000	In operation
Sarsat-9 NOAA-17	June 2002	In operation
Sarsat-10 NOAA-N	May 2005	In operation
Sarsat-11 METOP-1	Planned launch 2010	

The Cospas-Sarsat satellites in low polar orbits are known as Cospas-Sarsat Low-altitude Earth Orbit System for Search and Rescue. There usually are two Russian Cospas and two U.S. Sarsat spacecraft on duty. Cospas satellites from Russia flew in Sun-synchronous, near-polar orbits at 1,000 km altitude and were equipped with Russian search-and-rescue receivers. Polar Operational Environmental Satellites, supplied by the United States and flying similar orbits at about 850 km, were equipped with search-and-rescue receivers supplied by Canada and France. Each satellite made a complete orbit of Earth around the poles every 100 minutes. Each viewed a 6,000 km swath of Earth. To a viewer on Earth, each satellite crossed the sky in about 15 minutes. The U.S. National Oceanic and Atmospheric Administration operated Sarsat.

Anthony R. Curtis

Geostationary Search and Rescue Satellites

GEOSAR satellites past, present, and planned			
Satellite Name	**Source**	**Launch**	**Status**
GOES-9	USA	May 1995	In operation
GOES-10	USA	April 1997	In operation
GOES-11	USA	May 2000	Spare in orbit
GOES-12	USA	July 2001	In operation
INSAT-3A	India	April 2003	In operation
INSAT-3D	India	Planned launch 2007–8	
MSG-1	Europe	August 2002	In operation
MSG-2	Europe	Planned launch 2005–6	
Electro-L GOMS N2	Russia	Planned launch 2006	

The constellation of higher, stationary-orbit satellites called Geostationary Search and Rescue was composed of Geostationary Operational Environmental Satellites provided by India National Satellite and the United States, Meteosat Second Generation satellites from EUMETSAT, and Electro-L satellites from Russia.

See also: Amateur Communications Satellites; Geostationary Operational Environmental Satellites (GOES); Meteosat; NPOESS, National Polar-orbiting Operational Environmental Satellite System; National Oceanic and Atmospheric Administration

Bibliography

Cospas-Sarsat. www.cospas-sarsat.org.
National Oceanic and Atmospheric Administration. www.noaa.gov.
United States Coast Guard Navigation Center EPIRB Information. http://www.navcen.uscg.gov/marcomms/gmdss/epirb.htm.

SIRIO, Satellite Italiano Ricerca Industriale Operativa. *See* Italy.

Symphonie. Symphonie was a bilateral satellite communications program of France and West Germany from the late 1960s to the mid-1980s. Difficulties in obtaining launches for Symphonie from the United States gave added emphasis to the desire of the Europeans to develop their own launcher, under the European Space Vehicle Launcher Development Organisation (ELDO), and later contributed to the creation of the European Space Agency (ESA) and enhanced support for France's proposed Ariane rocket.

During the 1960s, ELDO was developing a heavy launcher, but it was unreliable, so Symphonie developers approached NASA in 1968. NASA eventually agreed to launch Symphonie satellites, provided that they were not used commercially. This policy stemmed officially from the U.S. position to prevent competition to the Intelsat space telecommunications consortium by not launching any competing communications satellites. The policy also promoted U.S. economic dominance of space communications.

Without launch capability of its own and needing telecommunications satellites to meet the exploding demand for overseas communications, Europe was disadvantaged by this arrangement. U.S. intransigence regarding the launch of Symphonie strengthened French arguments that Europe needed independent space capability, including launchers. The Europeans subsequently modified the European Space Research Organisation to form ESA in 1975. As part of the deal to create ESA, France led the development of the Ariane launcher to ensure Europe would not be held to U.S. policy.

Symphonie was developed by a six-company Franco-German consortium led by Aérospatiale in France and Messerschmitt-Bölkow-Blohm in Germany. NASA launched *Symphonie A* and *B* in 1974 and 1975 into geosynchronous orbits on Delta rockets from Cape Canaveral. The two satellites operated until the mid-1980s, performing a variety of communications experiments for France, Germany, and, from 1976, for several African and Asian countries. The three-axis stabilized spacecraft provided 600 one-way voice circuits or one color television channel.

Llyn Kaimowitz and Stephen B. Johnson

See also: Ariane, European Space Agency, Intelsat, Space Telecommunications Policy

Bibliography

John Krige and Arturo Russo, *A History of the European Space Agency, 1957–1987, Volume I* (2000).

Lorenza Sebesta, "U.S.-European Relations and the Decision to Build Ariane, the European Launch Vehicle," in *Beyond the Ionosphere: Fifty Years of Satellite Communication*, ed. Andrew J. Butrica (1997).

William Triplett, "The French Succession," *Air and Space Magazine*, April–May 1996.

Syncom. Syncom was the first geosynchronous communications satellite. It originated at Hughes Aircraft Company in the late 1950s after the U.S. Air Force canceled the company's F-108 interceptor program. This left people at Hughes looking for work. Frank Carver, manager of the radar lab, asked Harold A. Rosen, a PhD engineering graduate of the California Institute of Technology, to look for something the lab could do in space. Rosen and young Harvard-trained physicist/engineer Donald D. Williams looked for a new challenge. Williams suggested that they could design a navigation system using geosynchronous orbit (GEO) satellites. Rosen felt that GEO was more suitable for communications satellites. They were spurred on by an article they read in the March 1959 issue of the *Proceedings of the Institute of Radio Engineers* by John R. Pierce and Rudolf Kompfner, of American Telephone and Telegraph (AT&T) Company, suggesting that GEO satellites were too complicated at that time.

Rosen and Williams thought otherwise. They quickly designed a 25 lb geosynchronous satellite suitable for launch on a Scout rocket from Christmas Island in the Pacific Ocean. Hughes began designing the satellite and building components.

However, the company could not find a communications company to team with. NASA was building a medium-Earth-orbit (MEO) system (*Relay*) as was AT&T (*Telstar*). The Department of Defense (DoD) was building a 1,500 lb geosynchronous satellite (*Advent*). Prodded by the DoD, NASA decided to fund a heavier Delta-class satellite: Syncom (75 lb).

On 14 February 1963—a few months after the launch of *Relay 1*—*Syncom 1* was launched. The apogee kick motor (AKM) fired and the satellite fell silent. Using photographs taken from telescopes, engineers concluded that *Syncom* was in geosynchronous orbit, but a structural failure had destroyed its ability to communicate. On 26 July 1963, NASA launched *Syncom 2*, which performed well. A more powerful version of the Delta rocket launched *Syncom 3* on 19 August 1964. The extra power allowed the AKM to place *Syncom 3* in a geostationary orbit, which appeared to stay in exactly the same spot as seen from Earth. Syncom's success provided the justification for the Communications Satellite Corporation to launch an experimental/operational GEO satellite in April 1965. All successful communications satellite systems since then have been geosynchronous—although MEO and low-Earth-orbit systems have been attempted.

David J. Whalen

See also: Hughes Aircraft Company

Bibliography

Boeing Satellite Systems. http://www.boeing.com/defense-space/space/bss/.
David J. Whalen, *The Origins of Satellite Communications 1945–1965* (2002).

TDRSS, Tracking and Data Relay Satellite System. TDRSS, Tracking and Data Relay Satellite System was a space communications network operated by NASA Goddard Space Flight Center. To provide constant and high-performance communications for space missions, TDRSS employed a fleet of Tracking and Data Relay Satellites (TDRS) in geostationary orbit around the globe.

Starting in the mid-1960s, NASA began studies of a space-based relay network of at least three geostationary satellites to provide nearly constant orbital coverage. Such a system would allow the agency to shut down a number of expensive ground stations, while boosting performance. Due to budget constraints, NASA eventually proposed to lease rather than buy TDRSS; Congress approved in May 1974. In December 1976, NASA awarded the $800 million prime contract, which would not be paid until the system was operational, to Western Union Space Communications, with TRW building the satellites along with ground computers and software. NASA and commercial users were to share the system, but Western Union was unable to find sufficient commercial customers. Through a series of transactions in the 1980s, the TDRSS contract changed hands several times until 1990, when NASA acquired full ownership.

TRW continued as the satellite manufacturer for *TDRS 1–7*. Hughes Aircraft Company built the second generation *TDRS 8–10*. The first-generation satellites provided communications services in the S- and Ku-bands, using five antennas: two single access, one 30-element multiple access, one omni, and one space-to-ground link. The second generation satellites again provided five antennas, with the single-access antennas enhanced to include faster Ka-band communications, and the multiple access antenna expanded to include 32 receive and 15 transmit elements.

The Space Shuttle deployed, in 1983, *TDRS 1*, the first satellite of the system. In 1986 the Space Shuttle *Challenger* carried *TDRS 2*, but the Shuttle exploded shortly after launch. Space Transportation System (STS)-26 deployed *TDRS 3* in 1988, and the following year STS-29 installed *TDRS 4* into orbit, completing the minimum TDRSS constellation needed for full coverage. Once completed, TDRSS delivered communications to orbiting missions at least 85 percent of the time for polar-orbiting missions and 100 percent of the time for equatorial-orbiting missions like the Space Shuttle. As of 2008 nine TDRS satellites were in orbit, providing communications to missions using advanced technologies, such as multiple-access, and Ku-, Ka-, and S-band communications.

TDRSS consisted of space, ground, and user segments. The TDRS satellites comprised the space segment, and the ground segment consisted of two ground stations located at White Sands, New Mexico, and a backup station in Guam. The Goddard Space Flight Center located in Greenbelt, Maryland, maintained the network operations center, also part of the ground segment.

Various NASA missions utilized TDRSS throughout their lifetimes and comprised the user segment. In addition to the Space Shuttle, the *Hubble Space Telescope*, Landsat, *Seasat*, and Terra sent back large amounts of data to the ground via TDRSS. By agreement with NASA, other U.S. government agencies utilized TDRSS, such as Department of Defense intelligence-gathering satellites.

David Hartzell and Stephen B. Johnson

See also: Hughes Aircraft Company, Military Communications, Space Shuttle, TRW Corporation

Bibliography

NASA. http://www.nasa.gov/.

Sunny Tsiao, *"Read You Loud and Clear!" The Story of NASA's Spaceflight Tracking and Data Network* (2007).

COMMERCIAL SPACE APPLICATIONS

Commercial space applications include launch, telecommunications, insurance, navigation, remote sensing, tourism, and burial services. There have also been studies of potential future commercial applications, including solar power, mining, and materials processing.

From 1945 it was clear that space provided an excellent arena for global communications relays with three satellites in geosynchronous orbit. American Telephone and Telegraph (AT&T) Company recognized the potential. It participated with NASA in the Echo communications satellite (comsat) project and launched its own *Telstar* comsat in 1962. The U.S. government funded several comsat demonstration projects from 1958 to 1963, which created competitors to AT&T, and then the Communications Satellite Act of 1962 created Communications Satellite (Comsat) Corporation to lead the development of a worldwide system of Earth-circling comsats. Comsat represented the United States in the negotiations to create the International Telecommunications Satellite Organization (Intelsat), the international consortium to operate the system, and became the manager of Intelsat operations. By the 1970s Intelsat had fielded a profitable global system, and several nations, businesses, and international organizations were developing their own systems. During the next two decades satellite communications grew steadily, and with new applications, such as satellite television and radio, it became by far the largest and most profitable commercial space application.

The demand to put comsats in space spurred international competition to launch them. This led to the creation of the commercial space launch sector, though the companies and organizations involved were closely related to their national governments (or in the case of Arianespace, the international European Space Agency) either through partial or complete government ownership, as government organizations themselves, or as the beneficiaries of government-developed launch vehicles. By the 1990s international competition to launch commercial payloads was the norm, with launch services offered from the United States, Europe, Russia, Ukraine, China, India, and Japan.

International treaties determined that nations were responsible for potential damages caused by spacecraft launched and operated by their governments or corporations. This and the ever-present possibility of launch or satellite failure led to the development of the space insurance sector, which grew along with the commercial comsat industry.

The commercial satellite navigation and remote sensing sectors evolved directly from military roots, the Transit and Global Positioning Systems (GPS) in the case of the former, and reconnaissance satellites for the latter. Commercial space-based navigation grew to a multibillion-dollar industry in the 1990s–2000s, as a result of many applications that use GPS signals. Commercial remote sensing developed in the same period, as private customers purchased civilian remote sensing satellite images. Influenced by international competitors, such as France's Spot Image, the U.S. government allowed the development of commercial remote sensing companies, which placed high-resolution satellites into orbit in the late 1990s–2000s. Image sales grew, but by the early 2000s remained insufficient to cover the high cost of satellite development.

Space tourism began in the early 2000s with private citizens paying to fly on the Russian Soyuz to visit the *International Space Station* and the potential of suborbital flights and orbiting hotels in the future. Celestis provided the service of placing cremains (cremated remains) into space for a fee. Research into materials processing showed some promise, but by the 2000s commercial success remained elusive. The potential to transmit solar power from space to Earth, or to gather resources, such as rare metals or helium-3 from space, remained as tantalizing studies.

Stephen B. Johnson

See also: Civilian Remote Sensing, Civilian Space Communications, Convention on International Liability, Economics of Space, Microgravity Science, Space Business, Space Law, Space Politics, Tourism

Bibliography

Joan Lisa Bromberg, *NASA and the Space Industry* (1999).
Roger Handberg, *International Space Commerce: Building from Scratch* (2006).
Stephen B. Johnson, "The Political Economy of Space," in *Societal Impact of Spaceflight*, ed. Stephen J. Dick and Roger D. Launius (2007), chap. 9.

Burial Services

Burial services became a new use of outer space in 1997 with the launch of the cremated human remains (cremains) of 24 individuals into space on a Pegasus launcher by Celestis, Inc.

In the early twenty-first century, only Celestis, Inc., offered space burial services. It acquired the name and basic concept from the Celestis Group of Florida, which had received a license in 1984 from the Office of Commercial Space Transportation to place cremains in long-term orbit through launches made on the Conestoga rocket developed by Space Services Corporation. The attorney general of Florida posed a legal challenge, wanting Celestis Group to adhere to the same rules as cemeteries, including acreage, roads, and other constraints. While the Group did eventually win the case, it went out of business. Other entrepreneurs then approached the Group's remaining members to obtain rights to the name, forming Celestis, Inc., in 1994. The new company's 1997 "Founders Flight" included the cremains of television's *Star Trek* creator Gene Roddenberry and physicist Gerard O'Neill. Several other launches followed. Celestis merged with Space Services Inc. in 2004 and was based in Houston, Texas.

The individual tubes with cremains are packaged together to form a small satellite or payload, which then hitchhikes on another launch. What happens next depends on the type of service chosen: from a low-Earth-orbit flight, where the cremains burn on reentry, to sending cremains to the Moon or farther, such as the ashes of Eugene Shoemaker, whose cremains went to the Moon aboard the *Lunar Prospector* satellite. As of 2005 more than 100 people had used this service.

Stephen B. Johnson

See also: Orbital Sciences Corporation

Bibliography

Space Services Inc. http://www.spaceservicesinc.com/.

Centers for the Commercial Development of Space

Centers for the Commercial Development of Space were established by NASA in 1985 under the Office of Commercial Programs to help encourage a broad range of industry to become involved with space-based research and development efforts.

Since inception, efforts to provide for the commercial development of space underwent several changes.

NASA split the program into two separate but complementary groups. The first group promoted industry research, and NASA placed these efforts under the Space Product Development Program. The centers that remained under the program became known as Commercial Space Centers. Industry selected research under this program, receiving no direct NASA funding. The second group of centers, which retained the title of Centers for the Commercial Development of Space (CCDS) focused on research of mutual interest to NASA and to industry. With the CCDSs, NASA often funded the research in conjunction with industry.

In the early 2000s the two groups of centers, with one exception, were recombined under the Space Product Development Program as Commercial Space Centers. In 2004 NASA changed the name of these centers to Research Partnership Centers, and it placed renewed emphasis on research that would benefit NASA and industry as opposed to purely commercial research chosen and funded by industry. As of 2008 the centers were managed under NASA's Innovative Partnerships Program.

The centers, which were usually research groups within U.S. academic institutions, were subject to periodic outside audits. If there was insufficient industry interest in a center, NASA would close it. NASA also created new centers in response to government and/or industry interest. The number of centers fluctuated from roughly 10 to 20 at any given time, with funding typically in the range of $20–30 million per year. The centers included addressed a variety of topics such as materials processing, space power, communications technologies, remote sensing, robotics, biophysical applications, and space food.

Stephen B. Johnson

See also: Economics, Universities

Bibliography

Mark Nall et al., "Space Product Development: NASA Partnering with Industry for Out of This World Results," *23rd International Symposium on Space Technology and Science* (2002).

NASA Space Partnership Development. http://www.nasa.gov/offices/ipp/partnership_devel/index.html.

Commercial Navigation

Commercial navigation includes the commercial aspects of space-based global navigation systems. Commercial navigation companies provide end-user products, usually receivers with specialized software that determine location, velocity, and timing information. The sector supplies consumers in other markets including aviation, marine, surveying, transportation, recreation, and safety.

The navigation sector grew rapidly at the end of the twentieth century. Based initially on World War II developments, such as the Long Range Navigation (LORAN) terrestrial-based radio system for aircraft and ship navigation, navigation systems moved to space with the U.S. Navy's Transit satellites, first deployed in 1960. Initial end-user systems, such as those developed by Trimble Navigation Corporation, founded in 1978, used LORAN signals for continuous real-time navigation. A number of companies developed techniques and systems to utilize the Transit system, for uses including offshore oil exploration, marine shipping and surveying. The U.S. Global Positioning System (GPS) soon provided space-based navigation signals to terrestrial users free of charge and Trimble quickly introduced end-user navigation products utilizing GPS signals. Advances in electronics miniaturization allowed for accurate, inexpensive portable receivers that typically resolved a user's global location down to a few meters and came integrated with digital mapping technology. This portability and map integration enhanced the quality of navigation devices, which in turn motivated consumer demand, resulting in widespread adoption and a multi-billion dollar market by the twenty-first century.

The U.S. military developed space-based navigation services like Transit and GPS to aid with operations, logistics, and weapons deployment. Most military vehicles, marine vessels, and airplanes were equipped with GPS receivers and applications, such as precision-guided munitions, and this increased the number of GPS receivers sold to the military. During the First Gulf War in 1990–91, however, the U.S. military found itself with insufficient numbers of portable receivers and purchased thousands of commercial units from Trimble. The resulting successes and publicity jump-started the commercial market.

By the mid-1990s the commercial demand exceeded military demand for receivers, with the automobile navigation segment generating the largest portion of navigation revenues by 2003. During the late 1990s navigation companies and automobile manufacturers made arrangements to install receivers into new vehicles. Mobile GPS receivers with embedded digital maps could provide spoken directions and location services to drivers. Sales of handheld devices to the consumer market became the largest segment by 2006 and typically included GPS devices sold for recreational purposes, such as hiking and fishing. Portable high-precision GPS receivers became readily available through retail stores at commodity prices.

GPS navigation also augmented traditional fields like civil surveying and mapping by increasing the accuracy of measurements and reducing the time and costs associated with surveying activities. The civil segments demanded subcentimeter accuracy, so manufacturers of GPS surveying devices responded by increasing the accuracy of the technology, benefiting the entire sector.

Tracking services became a widely used application, including the tracking of cargo, packages, and vehicles. Location information was monitored from a centralized location, which could subsequently manage the deployment of resources. The inclusion of GPS technology into mobile phones permitted emergency response officials to locate individuals in distress, but as of 2005, privacy concerns restricted deployment and adoption of this capability. Public safety services, such as police and rescue services, utilized GPS for tracking mobile units deployed in the field, and wearable GPS tracking devices were utilized to track criminals on parole or house arrest. Various new applications led to a rise in the Original Equipment Manufacturing sector that specialized in developing GPS chipsets for resale to other manufacturers that would develop complete end-user products.

The aviation and marine sectors primarily used GPS as a navigational aid, permitting efficient routing and providing real-time velocity information. As an alternative to the aging air traffic control system, the concept of "free-flight" became possible, by which airplanes could use GPS and satellite communications for navigational services, collision avoidance, and enhanced visibility.

Many manufacturers of navigation equipment existed by 2003, contributing to a highly competitive $16 billion market. Trimble Navigation was a market leader throughout the late twentieth century. In 1984 it marketed the first commercial GPS products to support scientific research and geodetic survey applications. By 2004 the vast majority of its revenue came from the sale of survey and construction-related GPS products. Trimble also developed a fleet management service to assist with the tracking and shipping.

Thales Navigation marketed navigation products through its various divisions. Thales' Magellan brand supplied consumer GPS products, which included handheld and automobile devices. Other divisions provided surveying products and GPS components. Garmin International, established in 1989, provided products for the commercial sector, including portable aviation and marine GPS receivers, which it integrated with communications transceivers enabling users to communicate location information.

Japanese manufacturers supplied approximately 44 percent of commercial receivers worldwide. European manufacturers supplied approximately 23 percent of commercial receivers by the start of the twenty-first century, and the lure of further receiver profits was a significant spur for the Europeans to develop the Galileo system, whose deployment was to start by 2009. Commercial space navigation was projected in 2008 to continue to grow at a rapid pace for the near future.

David Hartzell

See also: Global Positioning System, Navigation, Transit

Bibliography

Trimble. http://www.trimble.com/.
U.S. Department of Commerce, *Trends in Space Commerce* (2001).

Galileo (Navigation System). *See* Commercial Navigation.

Commercial Remote Sensing

Commercial remote sensing began in 1972 with the launch of the first civilian remote sensing satellite, *Landsat 1*, developed by NASA. Quickly after its launch, Landsat's data became popular among its user community, leading to the development of a commercial remote sensing market with the U.S. Geological Survey as the single supplier of data with customers globally. With its follow-on satellites through 1970s, the Landsat program's success attracted the attention of both domestic and foreign policy makers. U.S. politicians called for transferring Landsat from experimental to operational and then to commercial status. Outside the United States other emerging spacefaring nations recognized the value of satellite remote sensing and decided to develop their own indigenous imaging satellites for commercial and civilian purposes.

In 1984 the U.S. Congress approved the Land Remote Sensing Commercialization Act to commercialize the Landsat program. One year later the U.S. government contracted with Earth Observation Satellite Company (EOSAT) to operate the existing Landsat satellites and to construct and operate its successors. However as EOSAT had to raise prices to meet its legal and financial obligations, Landsat data sales declined in the late 1980s.

This decline was accelerated when, in February 1986, CNES (French Space Agency) launched its first remote sensing satellite, *Satellite Pour l'Observation de la Terre 1* (*SPOT*), which was technically more capable than Landsat. To market *SPOT 1* data, France created a public-private partnership, Spot Image, in 1982. Unlike EOSAT, Spot Image did not have to make profit from data sales to meet *SPOT 1*'s operational expenses or develop follow-on satellites, as the French government was responsible for these. This enabled Spot Image to market SPOT data at lower prices than Landsat, leading to SPOT's domination of the remote sensing market. France launched *SPOT 2* in January 1990, which further increased SPOT data sales.

Amid these developments, in the mid-1980s, the Soviet Union offered remote sensing data to international users from Resurs and Okean environmental satellites, collecting medium-resolution digital imagery. However because of the lack of marketing experience and strict government control concerning the sales of satellite imagery, initial sales efforts floundered. Only in the early 1990s did sales begin to improve. In 1987 the Soviet Union offered 5 m resolution photographic imagery collected by the Kosmos series military satellites. Although sales were slow, its impact on the market was nonetheless huge, because its high-resolution data had significant dual-use (military and civilian) implications. In 1991 Sovinformsputnik, established by the Russian Aviation and Space Agency, offered 2 m resolution imagery from Kosmos satellites, which fired a resolution race in the commercial market and created significant security and economic concerns among U.S. politicians and aerospace industry leaders. These contributed to the U.S. government's decision to remove legal and political restrictions on commercialization of high-resolution imaging systems.

In the 1990s several others, including Canada, the European Space Agency (ESA), and India joined the commercial remote sensing market by offering data from their indigenous civilian imaging satellites. India commercialized Indian Remote Sensing (IRS) data in 1992 through Antrix Corporation, established to market Indian space products and services. To increase worldwide IRS data sales, Antrix contracted with EOSAT in 1995 to market IRS data collected outside India. Taking over EOSAT in 1996, Space Imaging renewed the contract in 2004 and received the right to market data from Indian new generation high-resolution satellites *Resourcesat 1* and *Cartosat 1*. Canada joined the commercial remote sensing market in 1995 with the Canadian Space Agency's (CSA) *Radarsat 1*. Radarsat International was established in 1989 as a wholly owned subsidiary of MacDonald Dettwiler and Associates Ltd. (MDA) and other government and private partners from Canada and the United States to market *Radarsat 1* data. *Radarsat 1*'s success led to the development of *Radarsat 2*, funded by CSA and MDA. After the 2007 launch MDA owned and operated the satellite and marketed the data and in return provided data to CSA at production cost. ESA launched three satellites for scientific purposes but also having commercial uses: *European Remote Sensing 1* (*ERS 1*) launched in July 1991, *ERS 2* in April 1995, and *ENVISAT* in March 2002. Scientists bought data at production cost or free of charge, while others paid for the data. Eurimage and SARCOM Consortium distributed ESA satellite data for commercial uses.

Besides government-sponsored civilian satellites, in the early 1990s private industry began to fund high-resolution imaging satellites. After the failure of Landsat commercialization, the U.S. government passed the Land Remote Sensing Policy Act of 1992, transferring Landsat back to government control. The Act also allowed development of private imaging satellites. With French success in the market and the Russian release of high-resolution imagery, in 1994 the William Clinton presidential administration allowed U.S. companies to develop high-resolution imaging systems. Several U.S. companies (Space Imaging, Orbimage, and DigitalGlobe) initiated programs to build high-resolution remote sensing satellites and offer imagery commercially. The first truly commercial high-resolution satellite, *Ikonos*, was launched in September 1999 by Space Imaging, and several other private satellites launched in the following years by the other companies.

Due to the high cost of satellite development and strong national security ramifications of high-resolution systems, purely private entries into the market remained difficult in the early twenty-first century, and existing operators required major government data purchases and satellite development subsidies.

Incigul Polat-Erdogan

See also: Reconnaissance

Bibliography

John C. Baker et al., eds., *Commercial Observation Satellites: At the Leading Age of Global Transparency* (2001).

Michael Krepon et al., eds., *Commercial Observation Satellites and International Security* (1990).

Mary Wagner, "Russia: Making Remote Sensing History with a Little Help from Friends," *Earth Observation Magazine*, November 1995.

Private Remote Sensing Corporations. Private remote sensing corporations supply commercial and civilian remote sensing data to the global market. The former markets data from its own satellites, whereas the latter markets data from satellites owned by government entities. In 2008 the corporations marketing their own data included GeoEye, DigitalGlobe, and ImageSat International. These companies, which must cover the development costs of their satellites from data sales, had significant financial difficulties in the early twenty-first century despite their technical successes.

GeoEye was created in 2006 from the merger of Space Imaging and Orbimage. Space Imaging was established in 1994 by its major shareholders Lockheed Martin and Raytheon/E-Systems. Acquiring Earth Observation Satellite Company in 1996, the company launched two satellites: *Ikonos 1*, destroyed in a launch failure in April 1999, and the 1 m resolution *Ikonos* launched in September 1999. After losing the competition for U.S. National Geospatial-Intelligence Agency (NGA) funding to build a next-generation satellite, it was purchased by Orbimage in September 2005 for $58.5 million. Orbimage was established as a wholly-owned subsidiary of Orbital Sciences Corporation (OSC) in 1993. It launched three satellites: *OrbView 2*, a 1 km resolution ocean monitoring satellite launched in August 1997; *OrbView 4* destroyed during the launch in September 2001; and *OrbView 3* launched in June 2003. The failure of *OrbView 4* and the numerous delays of *OrbView 3* led Orbimage to file for bankruptcy in April 2002. In 2003 it emerged from bankruptcy, separating from OSC. Winning NGA funding in 2004, Orbimage began building a 0.25 m resolution satellite, which after the formation of GeoEye was called *GeoEye 1*, launched in September 2008.

WorldView was established in 1992. In January 1995 Ball Aerospace & Technologies Corporation merged its commercial remote sensing efforts with WorldView, and from the merger was created EarthWatch, which was renamed DigitalGlobe in 2001. It developed three satellites: *EarlyBird 1* failed on orbit in December 1997, *QuickBird 1* failed to achieve the orbit in November 2000, and 0.6 m resolution *QuickBird* launched in October 2001. Winning a competition for funding from the NGA in 2003 to build a next-generation remote sensing satellite, DigitalGlobe launched its 0.5 m resolution *WorldView 1* satellite in September 2007.

Israel-based West Indian Space was formed in January 1997 by a partnership between Israeli Aircraft Industries and California-based Core Software Technology. Changing its name to ImageSat International a year later, the company launched 0.82 m resolution *Earth Resource Observation Satellite 1A* (*EROS 1A*) in December 2000. In 2005 it began developing a new series of satellites: *EROS B* was launched in April 2006, and *EROS C* was planned for a 2010 launch.

Private companies marketing government satellite data included Antrix (India); MacDonald Dettwiler and Associates Ltd. (MDA) Geospatial Services (Canada); Sovinformsputnik (Russia); and Spot Image (France). The first of these was Spot Image, established in 1982 to market imagery products derived from CNES's (French Space Agency) SPOT (Satellite Pour l'Observation de la Terre) series satellites. Its major shareholders in 2005 were CNES and SPOT satellite manufacturer European Aeronautic Defense and Space Company. Besides SPOT, the company owned marketing rights to data from *FORMOSAT 2*, a Taiwanese imaging satellite. Together with seven partners, Spot Image created the SARCOM consortium in 2000 to market data from the European Space Agency's Envisat and European remote sensing satellites.

MDA Geospatial Services became the exclusive worldwide distributor of Canadian Space Agency's *Radarsat 1*, launched in November 1995, and *Radarsat 2*, launched in December 2007. In exchange for providing one quarter of development funds for *Radarsat 2*, MDA Geospatial Services will own the satellite and provide data to the Canadian government at cost.

Sovinformsputnik was established in 1992 by the Russian Aviation and Space Agency to market archived satellite imagery taken by the former Soviet Union's reconnaissance satellites. Antrix was established in 1992 as a commercial arm of the Indian Space Research Organisation. It is the exclusive distributor of the Indian remote sensing series satellites.

Incigul Polat-Erdogan

See also: EOSAT (Earth Observation Satellite Company), Landsat, MacDonald Dettwiler and Associates Ltd., National Geospatial-Intelligence Agency

Bibliography

John C. Baker et al., eds., *Commercial Observation Satellites: At the Leading Age of Global Transparency* (2001).

DigitalGlobe. http://www.digitalglobe.com/.

GeoEye. http://www.geoeye.com/CorpSite/.

ImageSat International. http://www.imagesatintl.com/.

Remote Sensing Value-added Sector. The remote sensing value-added sector consists of organizations that produce and sell products derived from space or airborne remotely sensed data. The sector adds value to raw images by processing data into information for human interpretation, often by creating maps, augmenting features, indicating changes, or illustrating data. Other products included custom data processing and analysis and geographic information system (GIS) software allowing end users to integrate data from various sources for analysis and mapping purposes.

Many companies engaged in the sector had roots predating the space age. Before space-based remote sensing, aerial platforms collected a majority of data used by the sector for creating maps and analyzing Earth's surface and for applications such as agriculture and urban planning. In the 1960s, space-based remote sensing platforms, such as Corona, allowed civilian U.S. government geographers to create unclassified (but detailed) maps based on classified intelligence photographs. Value-added companies then created new products based on the unclassified maps, which were sold to the public in the form of maps. Weather forecasters also generated some of the first remote sensing value-added products based on data gathered from weather satellites owned and operated by the U.S. government.

The U.S. Geological Survey (USGS) Landsat program became a critical source of data for the value-added sector in the 1970s. The USGS sold raw Landsat data directly to users in addition to hard-copy prints of the images. Value-added companies would often purchase data from the USGS and perform additional processing for end consumers. The U.S. government's attempt to commercialize the Landsat program in 1984 allowed the Earth Observation Satellite Company (EOSAT) to establish itself as a critical member of the remote sensing value-added sector. EOSAT became the sole provider of the new *Landsat 4* and *5* data and generated additional revenue by creating value-added products. The Commercialization Act of 1984 required EOSAT to be cost competitive with other companies in the value-added sector, as it had immediate access to Landsat data for producing products. Space Imaging, the first provider of 1 m spatial resolution data from space, purchased EOSAT in 1996 and became the United States's largest supplier of remotely sensed data and value-added products.

The rapid increase in computing technology during the late twentieth century bolstered the demand for GIS software, as users could quickly process raw data into new products using small, inexpensive personal computers. GIS companies emerged and developed software for image processing and geographic analysis, and by 2000 the GIS sector generated approximately $1.2 billion of revenue. Established in 1969, the Environmental Systems Research Institute (ESRI) offered services for collecting and managing geographic information and by the 1980s, ESRI focused on developing computer tools to allow users to collect, manage, and analyze their own data with a software suite known as ArcGIS. Leica Geosystems produced another popular image processing and GIS software suite known as ERDAS IMAGINE during the early twenty-first century.

By 2005 many companies and government agencies worldwide were providing value-added products to the market. These products became critical for business in sectors such as agriculture, defense and public safety, insurance, construction, mapping, and various other sectors. In general the value-added sector consisted of many small companies providing products and services whose revenues collectively greatly exceeded the revenues raised by selling the raw data.

David Hartzell

See also: Corona, Landsat

Bibliography

ESRI. http://www.esri.com/.
Leica Geosystems. http://www.leica-geosystems.us/en/index.htm.

Commercial Satellite Communications

Commercial satellite communications became the most economically important space application of the twentieth century. While there are some mentions of satellite communications before World War II—including geosynchronous communications satellites—these became more numerous immediately after the war. The best known was an article by Arthur C. Clarke in the October 1945 issue of *Wireless World*. Clarke was then an officer in the Royal Air Force and a member of the British Interplanetary Society and soon became a renowned science fiction author and science popularizer. Clarke envisioned three large crewed space stations in 24-hour geosynchronous orbit above the Americas, Asia, and Europe, broadcasting radio and television programs globally. The station was staffed because someone had to change the vacuum tubes. The science fiction community contributed several other articles throughout the next few years: articles by Eric Burgess in 1949 and John R. Pierce (writing as J. J. Coupling) in 1952.

During this time the U.S. Army Air Force asked the Douglas Aircraft RAND organization to study the possibility of Earth satellites. Its May 1946 report, "Preliminary Design of an Experimental World-Circling Spaceship," among other conclusions discussed the use of an Earth-circling satellite as a communications relay station. Other RAND reports followed in the late 1940s and early 1950s. By 1954 the U.S. Air Force was considering the development of reconnaissance satellites, in addition to communications and weather satellites. This increased interest indirectly led to another article by Pierce that examined almost all the possible configurations of a communications satellite: active or passive (a mirror), low Earth orbit (what was later called medium Earth orbit), and high 24-hour (geosynchronous) orbit.

The launch of *Sputnik* in October 1957 changed the earlier visions from mere possibilities to potential realities. By the end of the decade, the Department of Defense (DoD), NASA, and the American Telephone and Telegraph (AT&T) Company were

An inflation test of the Echo 1 *passive communications satellite, whose successful launch and operation in 1960 helped prove the viability of artificial Earth satellites as communication relays. (Courtesy NASA/Langley Research Center)*

planning communications satellites—as were Radio Corporation of America (RCA), General Electric, International Telephone and Telegraph (ITT), General Telephone and Electronics, Lockheed, and Hughes Aircraft Company. The DoD was first to launch in 1958 with *SCORE*, (*Signal Communication by Orbiting Relay Satellite*) a communications payload attached to an Atlas rocket. NASA and AT&T were second to launch in 1960 with *Echo 1*, a passive reflective balloon satellite. A few days after the *Echo 1* launch, DoD launched the first Courier satellite.

By 1960 it was clear—at least to some—that commercial satellite communications were feasible. AT&T was prepared to fund an active medium Earth orbit satellite system; Hughes Aircraft Company was looking for a telecommunications company partner; and ITT was beginning to spend money. RCA, General Electric, and Lockheed were unwilling to fund a satellite communications system, but they were ready to support a government-funded effort.

The initial experimental satellites—*Telstar*, *Relay*, and *Syncom*—provide a glimpse into the arguments made at the time. *Telstar*, funded and built by AT&T, was intended for a medium-altitude (10,000 km) polar orbit. A constellation of 40–50 of these satellites would provide global coverage. The distance (10,000 km) was not too great, so the radio signal delay would only be about 1/10 of a second for the roundtrip from the ground, to the satellite, to the ground, back to the satellite, and back to the ground.

Relay was similar but was built by RCA under a NASA contract. *Syncom* was different. Hughes had originally designed and built a prototype using its own funds. NASA then funded a larger Delta-class geosynchronous *Syncom*, which was only half the mass of *Telstar* and *Relay*. Few believed that the attitude and orbit control required for geosynchronous orbit could be implemented in such a tiny satellite. Many doubted that the longer, half-second signal delay would be acceptable to telephone users.

Many people in government felt that allowing a private organization to operate this system—especially a monopolistic organization like AT&T—would be inappropriate. By contrast the Federal Communications Commission (FCC) argued that the appropriate model was the transoceanic submarine telephone cable organizations—a series of bilateral and multilateral agreements in which AT&T was dominant, but not a monopoly. After much argument, Congress passed the Communications Satellite Act of 1962, which formed the Communications Satellite Corporation (Comsat). The company was incorporated in 1963 and had a successful initial public stock offering in 1964. As a result of the 1962 Act, Intelsat was formed in 1964, with Comsat acting as the "interim" manager for the consortium and the United States indirectly controlling the organization. Other countries fought this control, and the 1971 Definitive Intelsat Agreements established rules to prevent U.S. dominance. By the end of the 1970s, Comsat no longer "managed" the organization but was still the most influential member.

Comsat launched *Early Bird* (*Intelsat I*) in 1965 and the Intelsat II series in 1967. These geosynchronous satellites had no international content and were—to some extent—experimental. The Intelsat III series was the first to have significant Intelsat input into the specifications and a modest amount of international content. The Intelsat III series was specifically designed to be launched into either geosynchronous or medium altitude orbit. By the time the Intelsat IV series was being designed, it was clear that geosynchronous orbit was the proper choice. It was also clear that the "commercial" C-band (6 GHz up and 4 GHz down) was the proper frequency choice—despite concerns about interference from and to terrestrial microwave systems. The Soviet Union did not join Intelsat, preferring to remain outside the capitalist organization and to launch the high-inclination Molniya satellites that provided coverage of the high latitudes of the Soviet Union. To provide a service similar to Intelsat to the communist bloc, Soviet leaders created Intersputnik.

About the time that the Intelsat III series was being launched, Canada launched its communications satellite. This was allowed, but discouraged, under the Intelsat agreements. After the Canadian *Anik* was launched in 1972, several U.S. private communications satellite systems were launched in the mid 1970s. Unlike the Intelsat satellites, which carried telephone signals, the U.S. satellites soon carried mostly television signals.

Space telecommunications policy continued to change the industry as much as the technology. The Intelsat agreements were multilateral treaties among the members, not commercial agreements. Although these government-directed monopolistic policies were originally "imposed" by the United States, it later tended to favor free markets. One of the policy success stories in terms of providing services was the maritime satellite agreements that resulted in the creation of London-based Inmarsat in 1982.

These satellites provided L-band (1.6 GHz) communications services to ships at sea. Unlike Intelsat, the Soviet Union and communist nations joined Inmarsat.

A policy disaster was the Aerosat system, which was canceled amid confusion and anger in 1977. NASA and the European Space Research Organisation had been discussing an aeronautical satellite for some time, as had Comsat and the International Air Transportation Association. Discussions went on for a decade until a deal was agreed on in the mid-1970s and a satellite contract awarded to General Electric. At that point the U.S. Federal Aviation Administration withdrew its financial support on the grounds that the airlines were not interested.

Starting in the 1970s and accelerating in the 1980s, other countries began to launch domestic communications satellites. Some seemed to be justified only by national pride. Others fulfilled clear national needs, such as Indonesia's Palapa series, which was crucial for the nation's thousands of islands. In the 1990s satellite filings at the International Telecommunication Union (ITU) exceeded 1,000 per year—far in excess of the 10–20 satellites to be launched in a normal year. The launch rate increased in the 1990s to more than 30, but this was not sustainable. Given that geosynchronous communications satellites can be placed no closer than two degrees of longitude, there are only 180 slots available. Since the late 1960s satellite lifetimes have increased from 18 months to 15 years. A "normal" launch rate would be about 12 per year.

In the 1990s several nongeosynchronous satellite communications systems were begun. Several of these were in low Earth orbit (LEO) from about 750–1,500 km above Earth. Iridium, Globalstar, and Orbcomm were examples of LEO constellations launched to provide satellite telephone service. Prices were high—higher than Inmarsat—and orders of magnitude higher than typical cellphones. In addition they required international regulations that allowed cross-border movement of what were in effect two-way radios. All the companies went bankrupt and eventually exited bankruptcy; however, in 2008 it remained unclear if they would survive.

The geostationary and geosynchronous systems suffered after the 1999 Internet business collapse and the LEO bankruptcies, but they still earned significant amounts of money. The success of digital television compression techniques decreased the demand for transponders, but the lowered cost for each channel brought in additional broadcasters. The telephony market that had been the mainstay of Intelsat virtually disappeared. Data transmission—especially Very Small Aperture Terminal (VSAT) data systems—grew steadily throughout the years but did not compensate for the transponder glut and the overall softening of the market. In general fewer transponders were leased at lower prices.

The wave of telecommunications privatization initiatives (as the result of U.S. congressional action and the desire of the postal, telephone, and telegraph organizations to acquire needed cash) and the lowered stock prices of satellite operating companies made these companies likely targets for buyouts by private investors. This privatization placed Intelsat, Inmarsat, Panamsat, New Skies, and other satellite companies into the hands of private equity investors.

The potential success stories of the early twenty-first century were the satellite radio companies (XM and Sirius) and the direct broadcast television companies (DIRECTV and Echostar). While these companies had limited profitability at that time, their revenues may eventually exceed the $100 billion mark.

Regulation of satellite communications has usually taken a back seat to the politics of satellite communications. The ITU is the United Nations body that registers satellite filings and records the results of administration-to-administration (country-to-country) coordination efforts. Individual countries must use their own regulatory bodies—such as the FCC in the United States—to enforce ITU regulations and their own national rules and regulations, as the ITU works by consensus. National regulatory bodies are ultimately dominated by national governments and national interests. While the United States has often been opinionated in its approach to regulating satellite communications, other countries have seen the large (often American) satellite operators as cash cows. Some countries charge more for an Earth station license than the cost of the Earth station and transponder lease.

In the early twenty-first century, satellite communications appeared to have been the biggest benefit of the space race—and certainly the most ubiquitous.

David J. Whalen

See also: Global Positioning System, Navigation, Transit

Bibliography

Andrew J. Butrica, ed., *Beyond the Ionosphere: Fifty Years of Satellite Communication* (1997).
David J. Whalen, *The Origins of Satellite Communications 1945–1965* (2002).

Amateur Communications Satellites. Amateur communications satellites have been built privately and launched by groups of amateur radio operators around the world. By 2008 more than 70 had been launched.

These spacecraft are used for communications on amateur radio frequencies and science research. They relay voice, Morse code, and digital computer signals. Many have message boards. Some are science craft for propagation tests, ionosphere research, meteor sounding, radio astronomy, radiolocation, and space photography.

Project OSCAR (Orbital Satellite Carrying Amateur Radio), a California amateur radio group, built the first amateur radio satellite, *OSCAR 1*, which was launched 12 December 1961 as the first parasitic (secondary) payload on *Discoverer 36*, an Thor-Agena launch vehicle carrying a primary payload of a *Corona KH-3* reconnaissance satellite. Project OSCAR built and flew four satellites. Since that time the majority of amateur satellites have been called OSCAR. Amateur radio OSCARs are not the same as U.S. Navy Oscar navigation satellites. Russians call their amateur satellites Radiosputniks. The ground segment of the OSCAR satellites are ham operators around the world that use the satellites when they are in view to communicate over the horizon.

A group of amateurs formed the Radio Amateur Satellite Corporation (AMSAT) in 1969. Headquartered in Washington, DC, AMSAT remains the major international group in ham space activity in the early twenty-first century. AMSAT's first flight was *OSCAR 5* built by Australian students and launched from California in 1970.

OSCARs and Radiosputniks have been financed through donations of time, hardware, and funds from ham radio operators in Argentina, Australia, Belgium, Brazil, Canada, Finland, France, Germany, Israel, Italy, Japan, Mexico, Russia, Saudi Arabia,

Amateur Satellite Launches through February 2006 (Part 1 of 5)

Date	Name	Origin	Innovation
1961 Dec 12	OSCAR 1	Project OSCAR	Transmitted Hi HI and temperature.
1962 Jun 2	OSCAR 2	Project OSCAR	Similar to OSCAR 1.
1965 Mar 9	OSCAR 3	Project OSCAR	First solar powered, first to relay signals.
1965 Dec 21	OSCAR 4	Project OSCAR	First U.S.-Soviet amateur link.
1970 Jan 23	Australis-OSCAR 5	Melbourne Univ.	Australia, launched from USA. First remotely controlled amateur sat.
1972 Oct 15	AMSAT-OSCAR 6	AMSAT	Sat-to-sat comm via AO7, medical relays.
1974 Nov 15	AMSAT-OSCAR 7	AMSAT	Stored & forwarded messages.
1978 Mar 5	AMSAT-OSCAR 8	AMSAT	Message store & forward.
1978 Oct 26	Radiosputnik 1	USSR	First Russian amateur satellite, launched with RS 2.
1978 Oct 26	Radiosputnik 2	USSR	Russian amateur satellite, launched with RS 1.
1980 May 23	AMSAT Phase 3A	AMSAT	Ariane rocket launch failed.
1981 Oct 6	UoSAT-OSCAR 9	Univ. of Surrey, U.K.	First on-board housekeeping computer.
1981 Dec 17	Radiosputnik 3	USSR	Launched with RS 4-8.
1981 Dec 17	Radiosputnik 4	USSR	Launched as part of the RS 3-8 flotilla.
1981 Dec 17	Radiosputnik 5	USSR	Robot communicator for Morse code contacts with amateurs on ground.
1981 Dec 17	Radiosputnik 6	USSR	Launched as part of the RS 3-8 flotilla.
1981 Dec 17	Radiosputnik 7	USSR	Robot communicator for Morse code contacts with amateurs on ground.
1981 Dec 17	Radiosputnik 8	USSR	Launched as part of the RS 3-8 flotilla.
1982 May 17	ISKRA 2	USSR	Launched from Salyut 7 space station.
1982 Nov 18	ISKRA 3	USSR	Hand launched by cosmonauts tossing out of Salyut 7 space station.
1983 Jun 16	AMSAT-OSCAR 10	AMSAT	Lower orbit than intended.
1984 Mar 1	UoSAT-OSCAR 11	Univ. of Surrey, U.K.	Camera, speech synthesizer, dust detector, Geiger counter.
1986 Aug 12	Fuji-OSCAR 12	JARL	First Japanese amateur satellite.
1987 Jun 23	Radiosputnik 10	USSR	Piggyback, along with RS 11, on Soviet navigation satellite.

Amateur Satellite Launches through February 2006 (Part 2 of 5)

Date	Name	Origin	Innovation
1987 Jun 23	Radiosputnik 11	USSR	Piggyback, along with RS 10, on Soviet navigation satellite.
1988 Jun 15	AMSAT-OSCAR 13	AMSAT	Near-Molniya orbit.
1990 Jan 22	UoSAT-OSCAR 14	Univ. of Surrey, U.K.	One of 6 microsatellites launched together. Cosmic particle detector.
1990 Jan 22	UoSAT-OSCAR 15	Univ. of Surrey, U.K.	One of the 6 microsatellites. Failed in orbit.
1990 Jan 22	AMSAT-OSCAR 16	AMSAT	One of the 6 microsats. Digital comms and message store & forward.
1990 Jan 22	DOVE-OSCAR 17	AMSAT-Brazil	One of the 6 microsats. Partial success.
1990 Jan 22	WEBERSAT-OSCAR 18	Weber Univ. students	One of the 6 microsatellites. CCD camera.
1990 Jan 22	LUSAT-OSCAR 19	AMSAT-Argentina	One of the 6 microsatellites. Packet radio store-and-forward.
1990 Feb 7	Fuji-OSCAR 20	JARL	Second Japanese amateur satellite.
1991 Jan 29	Radiosputnik 14	AMSAT-U & AMSAT-DL	a.k.a. AMSAT-OSCAR 21. Russian-German project. Message reader.
1991 Feb 5	Radiosputnik 12	Russia	Piggyback, along with RS 13, on Russian navigation satellite.
1991 Feb 5	Radiosputnik 13	Russia	Piggyback, along with RS 12, on Russian navigation satellite.
1991 Jul 17	UoSAT-OSCAR 22	Univ. of Surrey, U.K.	Camera and digital messaging system.
1992 Aug 10	KITSAT-OSCAR 23	South Korea	Camera and digital messaging system, similar to UO 22.
1993 May 13	Arsene-OSCAR 24	France	First French amateur satellite.
1993 Sep 26	KITSAT-OSCAR 25	South Korea	Camera, science instruments. Launched with IO 26, AO 27, PO 28.
1993 Sep 26	Italy-OSCAR 26	Italy-AMSAT	Launched with KO 25, AO 27, PO 28. Similar to AO 16, LO 19.
1993 Sep 26	AMRAD-OSCAR 27	AMRAD	Launched with KO 25, IO 26, PO 28.
1993 Sep 26	PoSAT-OSCAR 28	Portugal	Launched with 3 others. Camera, star sensor, GPS, cosmic ray test.
1994 Dec 16	Radiosputnik 15	Russia	Similar to RS 3-8.
1995 Mar 28	UNAMSAT 1	Natl. Univ. of Mexico	Russian launcher failed.
1995 Mar 28	TechSat 1a	Israel	Russian launcher failed.

(Continued)

Amateur Satellite Launches through February 2006 (Part 3 of 5)

Date	Name	Origin	Innovation
1996 Aug 17	Fuji-OSCAR 29	JARL	Third Japanese amateur satellite. Digitalker message reader.
1996 Sep 5	Mexico-OSCAR 30	Natl. Univ. of Mexico	Twin of UNAMSAT 1. Meteor sounder.
1997 Nov 4	Radiosputnik 17a	Russia	Hand launched from Mir station on 40th anniversary of Sputnik 1.
1998 July 10	Thai-Microsat-OSCAR 31	Thailand	TMSAT 1 launched by Russia. Camera, GPS, digital comms.
1998 Jul 10	Gurwin-OSCAR 32	Israel	TechSat 1b launched by Russia.
1998 Oct 24	SEDsat-OSCAR 33	U of Alabama-Huntsville	Camera, remote sensing.
1998 Oct 30	PANSAT-OSCAR 34	Naval Postgrad School	Digital message system and communications.
1998 Nov 10	Radiosputnik 18	Russia	Hand launch from Mir station. Sputnik 41. Prior RS 17 was Sputnik 40.
1999 Feb 23	SUNSAT-OSCAR 35	South Africa	Univ. of Stellenbosch. Communications and experiments.
1999 Apr 21	UoSAT-OSCAR 36	Univ. of Surrey, U.K.	2 cameras, GPS receiver, resistojet propulsion, Internet protocols.
2000 Jan 27	ArizonaState-OSCAR 37	Arizona State Univ.	One of five picosats launched together. ASUsat1. Digital comms.
2000 Jan 27	OPAL-OSCAR 38	Stanford Univ.	One of five picosats. Orbiting Picosat Automatic Launcher.
2000 Jan 27	StenSAT	Stensat Group USA	One of five picosats. CubeSat, a 12 cu. In.c, 8.s oz. picosatellite.
2000 Jan 27	Thelma and Louise	Santa Clara Univ.	One of five picosats. a.k.a. Thunder and Lightning, ParaSat, a picosat.
2000 Jan 27	Weber-OSCAR 39	Weber Univ.	One of five picosats. JAWSAT-Joint Air Force Weber Sat.; carried OPAL.
2000 Sep 26	SaudiSat-OSCAR 41	King Abdulaziz SRI	Launched by Russia on converted ICBM along with SO 42, MO 46.
2000 Sep 26	SaudiSat-OSCAR 42	King Abdulaziz SRI	Launched with SO 41, MO 46.
2000 Sep 26	Malaysian-OSCAR 46	Malaysia	a.k.a. TiungSAT 1.
2000 Nov 16	AMSAT-OSCAR 40	AMSAT	Largest, most complex, most powerful to date. Ten years to build.

Amateur Satellite Launches through February 2006 (Part 4 of 5)

Date	Name	Origin	Innovation
2001 Sep 30	Starshine-OSCAR 43	Project Starshine USA	37 in. ball covered with 1,500 mirrors polished by 40,000 students.
2001 Sep 30	Nav-OSCAR 44	U.S. Naval Academy	a.k.a. PCsat. Launched with SO 43 NO 45 & PICOsat from Alaska.
2001 Sep 30	Nav-OSCAR 45	Washington U. & Stanford	a.k.a. Sapphire. Infrared, camera, voice.
2002 Mar 20	Radiosputnik 21	Russia-Australia	Russian-Australian educational project launched from space station.
2002 May 3	BreizhSAT-OSCAR 47	AMSAT-France	a.k.a. IDEFIX CU1. Picosatellite, read messages.
2002 May 3	BreizhSAT-OSCAR 48	AMSAT-France	a.k.a. IDEFIX CU2. Picosatellite, read messages.
2002 Dec 20	AATiS-OSCAR 49	AMSAT-Germany	a.k.a. SAFIR-M. Cooperative educational project with schools.
2002 Dec 20	SaudiSat-OSCAR 50	King Abdulaziz SRI	Launched by Russia on converted ICBM along with AO 49.
2003 Jun 30	DTUsat	Tech. Univ. of Denmark	One of six CubeSAT nanosatellites launched by Russia.
2003 Jun 30	AAU CubeSat	Univ. Aalborg, Denmark	One of the six CubeSAT nanosatellites. Camera.
2003 Jun 30	CanX 1	Univ. of Toronto, Canada	One of the six CubeSATs. Horizon tracker, star sensor, GPS.
2003 Jun 30	Quakesat	Stanford Univ.	One of the six CubeSATs. Listened for earthquake radio signals.
2003 Jun 30	CUTE 1	Tokyo Inst. Tech., Japan	One of the six CubeSAT nanosatellites.
2003 Jun 30	CubeSat XI-IV	Univ. of Tokyo, Japan	One of the six CubeSAT nanosatellites. Camera photographed Earth.
2003 Sep 27	Radiosputnik 22	Mozhaisky Military Academy	a.k.a. Mozhayets 4. Training satellite, beacon.
2004 Jun 29	AMSAT-OSCAR 51	AMSAT	a.k.a. Echo. Launched from Russia. Voice and data communications.
2005 May 5	VuSat-OSCAR 52	AMSAT-India	a.k.a. HamSat. First amateur radio satellite from India.
2005 Aug 3	PCSat 2	U.S. Naval Academy	Deployed from International Space Station (ISS)

(Continued)

Amateur Satellite Launches through February 2006 (Part 5 of 5)

Date	Name	Origin	Innovation
2005 Oct 27	XO-53/SSETI Express	Europe	Carried three picosats; Student Space Exploration & Technology Initiative
2005 Oct 27	CubeSat XI-V	University of Tokyo	One of three CubeSAT nanosatellites.
2005 Oct 27	UWE 1	University of Wurzburg	One of three CubeSAT nanosatellites.
2005 Oct 27	Ncube 2	University of Oslo	One of three CubeSAT nanosatellites.
2006 Feb 3	AMSAT-OSCAR 54	ARISS/AMSAT	a.k.a. Suitsat. Space suit with radio gear, deployed from ISS

Other Amateur Radio Space Communications Systems

1983–1999	SAREX	USA	Shuttle Amateur Radio Experiment
2000–present	ARISS	USA	Amateur Radio aboard the International Space Station

Many additional launches 2006–2008 http://www.amsat.org/amsat-new/satellites/history.php

South Africa, South Korea, the United Kingdom, and the United States. They often received free rides to space as ballast on government rockets carrying other satellites to orbit. Amateurs developed a few communications satellites for nonprofit public service corporations. These amateur-related satellites were often launched alongside amateur radio satellites.

The number of amateur satellites has grown dramatically, from four launched in the 1960s to nearly two dozen launched in the first five years of the twenty-first century. The growth in the number of amateur satellites reflects the spread of space technologies around the world, and their significantly decreased cost, due to the miniaturization of computers, communications equipment, imaging and navigational instruments. The lowered costs made it feasible for more universities to build satellites as student educational projects. Most remained in orbit and many were in use in 2008.

Anthony R. Curtis

See also: Corona

Bibliography

American Radio Relay League, *The ARRL Satellite Anthology*, 5th ed. (1999).

Anthony R. Curtis, ed., *Space Satellite Handbook*, 3rd ed. (1994).

Martin Davidoff, *The Radio Amateur's Satellite Handbook*, rev. 1st ed. (1999).

"The Extraordinary History of Amateur Radio Satellites," Space Today Online. http://www.spacetoday.org/Satellites/Hamsats/HamsatsBasics.html.

American Private Communications Satellites. American private communications satellites emerged in the 1970s after U.S. policy analysts reviewed Canadian efforts to build its own domestic communications satellite system. Rethinking U.S. dedication to a "single global system," telecommunications policy analysts reasoned that domestic communications satellites might provide benefits unavailable from Intelsat. Intelsat provided service across the oceans, whereas a domestic satellite would provide service over land masses. This meant that the services would also be different. Intelsat mostly provided telephone service, whereas domestic satellites could provide private telephone service and perhaps television distribution service. This should have been obvious, as a 1945 article by Arthur C. Clarke discussed television and radio broadcasting, and in 1965 American Broadcast Corporation (ABC) and Hughes Aircraft Corporation proposed a domestic television distribution satellite that would save ABC millions of dollars it otherwise paid American Telephone and Telegraph (AT&T) to distribute network programming.

The Federal Communications Commission (FCC) pondered the problem of domestic satellites—after all, the Communications Satellite Corporation (Comsat) and Intelsat had a monopoly embodied in the Communications Satellite Act of 1962. The Richard Nixon presidential administration changed that in 1970 by insisting on an "open skies" approach: all technically and financially competent players could launch their own satellites. In 1972 the FCC authorized all companies that had filed in 1970 to launch their systems. First to launch was Western Union with *Westar 1* and *Westar 2* in 1974. Radio Corporation of America (RCA) launched a more sophisticated satellite—*RCA Satcom 1*—in 1975 and RCA *Satcom 2* in 1976. AT&T/Comsat launched two Comstars in 1976. While all three systems were nominally launched for the purpose of servicing corporate "private line" customers, Western Union and RCA soon became distributors of cable television programs—something AT&T was precluded from doing because of anti-monopoly agreements with the FCC. The cable revolution was made possible by satellites that provided cheap distribution.

A burst of new U.S. satellite operators arrived in the 1980s. The first of these was Satellite Business Systems (SBS), a new incarnation of the Microwave Communications, Inc./Lockheed filing of 1970, with new partners: International Business Machines (IBM), Comsat, and Aetna. Its first satellite, *SBS 1*, was the first Hughes HS 376 and the first all Ku-band (14/12 GHz frequency) satellite. It was designed to provide high-speed data links for large companies. Others followed: *Galaxy 1* (Hughes) in 1983, *Spacenet 1* (Southern Pacific) in 1984, *Gstar 1* (General Telephone and Electronics—GTE) in 1985, and *PanAmSat 1* in 1988. PanAmSat was the first of the "separate systems" authorized to compete with Intelsat.

By the start of the twenty-first century, only RCA American Communications (Americom) survived in anything like its original organization. It absorbed GTE Spacenet in 1994, which itself had already absorbed Southern Pacific. In 2001 Société Européenne des Satellites (SES) bought Americom. Loral purchased AT&T in 1997, while Hughes bought Western Union's satellite organization and merged with PanAmSat in 1997. Intelsat bought Loral's domestic fleet in 2004. In 2006 Intelsat and PanAmSat merged. Many of these mergers were the result of unsuccessful companies being absorbed by successful companies. In other cases the potential market power of combined companies seemed to make sense.

In the 1990s three new businesses arose: low Earth orbit (LEO) mobile satellites, direct-to-home (DTH) television services, and DTH satellite radio. The LEO operators all went through bankruptcy. However, the "new" Iridium, Globalstar, and Orbcomm, all rebounded by the early twenty-first century. DTH television became a multibillion-dollar business. DTH radio became a surprise success, with challenges similar to DTH television.

In 2005 the American Fixed Satellite Service operators were: PanAmSat (bought by News Corp and then sold to private equity investors), SES Americom (owned by the Grand Duchy of Luxembourg and others), Loral Skynet, and Intelsat (also owned by private equity investors). The Intelsat/PanAmSat merger made it the largest satellite communications company, owning about one-fourth of all communications satellites. The merger trend will probably continue in the twenty-first century.

David J. Whalen

See also: American Telephone and Telegraph (AT&T) Company, Federal Communications Commission, Hughes Aircraft Company, PanAmSat Corporation

Bibliography

George A. Codding Jr., *The Future of Satellite Communications* (1990).
Loral Skynet. http://www.loralskynet.com/index2.asp.
Robert S. Magnant, *Domestic Satellite: An FCC Giant Step* (1977).
Intelsat. http://www.intelsat.com.
SES Americom. http://www.ses.com/ses/welcome/.

Arabsat

Arabsat, Arab Satellite Communications Organization, was established in 1976 to design, execute, and operate the first satellite communications system for the Arab world.

Arab Ministers of Information and Culture agreed in 1967 that it should establish a satellite communications network to help integrate nations in the League of Arab States. This became legally possible in 1973 when the permanent agreement for Intelsat (International Telecommunications Satellite Organization) provided for the establishment of regional satellite systems, leading to collaboration of 21 nations from Morocco to the Arabian peninsula to create Arabsat. After establishing the financial basis for the organization and the satellite requirements, Aérospatiale won the contract to build the first three Arabsat satellites in 1981. The first two were placed in orbit in 1985, by Ariane and Space Shuttle launches, and the third in 1992. By 1988 personnel from Arab nations took over full operational control of the system. Aérospatiale also won the contract for the second-generation satellites, launched in 1996, while Alcatel built the third-generation satellite launched in 1999. Because of a partial failure of *Arabsat 3A*, the organization leased the *Panamsat 5* and Eutelsat *Hotbird 5* satellites. Arabsat planned to place two Arabsat 4 satellites, with 36 transponders, built by European Aeronautic Defence and Space Astrium, into orbit in 2006.

By 2005 Arabsat provided a variety of services, including television, telephone, and data transmission. Revenues by 2002 had reached $147 million.

Stephen B. Johnson

See also: Aérospatiale, Alcatel, European Aeronautic Defence and Space Company (EADS)

Bibliography

Arabsat. www.arabsat.com.

Atlantic Bird. *See* Eutelsat.

Canadian Communications Satellites. Canadian communications satellites have been a priority for the Canadian government since the late 1960s, when it identified communications satellites as an urgent national priority to serve far-flung settlements in Canada's north and to provide television and radio services in both English and French, Canada's two official languages. Canadian satellites made it easier for regulators to control the content of Canadian television to protect Canadian producers in the face of easily available U.S. television shows.

Canada's interest in communications dates back to its first satellites, *Alouette* and *Integrated Satellite Interface System* (ISIS), which were built to collect data on the ionosphere's role in reflecting radio waves. In 1969 the government created Telesat Canada to operate Canada's communications satellites. Canada became the first country to have a domestic geosynchronous communications satellite when *Anik A1* (Inuit for "little brother") was launched from Cape Canaveral by a Delta rocket on 9 November 1972. Telesat has launched 15 Anik communications satellites and 2 Nimiq (Inuit for "to bring together") direct-to-home (DTH) broadcast satellites into geosynchronous orbit. Canada also backed the North American mobile satellite system (MSAT). The Canadian government cooperated with NASA and the European Space Agency to build the Communications Technology Satellite or *Hermes*, which was launched in 1976. *Hermes* pioneered three-axis stabilization and Ku-band broadcasting and made the first DTH television broadcast, a hockey game, in 1978.

Although the Canadian government planned to build the three Anik A satellites in Canada, Telesat decided to give the contract to Hughes Aircraft in California, with subcontracting work for Canadian firms. After some controversy, the Canadian government supported Telesat's decision. Since then U.S. contractors have built most Canadian communications satellites. The Canadian government encouraged the development of its own space industry in the 1970s–80s by supporting the aspirations of Spar Aerospace of Toronto to become a prime contractor for communications satellites. Spar became prime contractor for two Anik D and two Anik E satellites, the MSAT satellite, and two Brasilsat communications satellites. Spar later sold its space operations, and in 2005 the Space and Technology Division of EMS (Electromagnetic Sciences) Technologies Inc. in Montreal along with Canadian firms such as Com Dev International and MacDonald Dettwiler and Associates Ltd. were major subcontractors for communications satellites built by American and European contractors. The Canadian government supported Canadian satellite subcontractors through a variety

Canadian Communications Satellites

Name	Launch Date	Launch Vehicle	Transponders	Comments
Anik A1	9 Nov 1972	Delta	12 C-band	Canada's first communications satellite (comsat)
Anik A2	20 Apr 1973	Delta	12 C-band	
Anik A3	7 May 1975	Delta	12 C-band	
Hermes	17 Jan 1976	Delta	2 Ku-band	First Ku-band comsat, first three-axis stabilized comsat
Anik B	15 Dec 1978	Delta	12 C-band 6 Ku-band	Built by Radio Corporation of America (RCA)
Anik D1	26 Aug 1982	Delta	24 C-band	First comsat with Canadian prime contractor Spar
Anik C3	12 Nov 1982	Shuttle *Columbia*	16 Ku-band	
Anik C2	18 Jun 1983	Shuttle *Challenger*	16 Ku-band	
Anik D2	8 Nov 1984	Shuttle *Discovery*	24 C-band	
Anik C1	12 Apr 1985	Shuttle *Discovery*	16 Ku-band	
Anik E2	4 Apr 1991	Ariane	24 C-band 16 Ku-band	
Anik E1	26 Sep 1991	Ariane	24 C-band 16 Ku-band	
MSAT 1	20 Apr 1996	Ariane	6 L-band	Satellite dedicated to mobile communications
Nimiq 1	20 May 1999	Proton	32 Ku-band	Canada's first direct-to-home broadcast satellite
Anik F1	21 Nov 2000	Ariane	36 C-band 48 Ku-band	
Nimiq 2	30 Dec 2002	Proton	32 Ku-band 2 Ka-band	
Anik F2	17 Jul 2004	Ariane	24 C-band 32 Ku-band	
Anik F1R	8 Sep 2005	Proton	24 C-band 32 Ku-band	First Anik built in Europe
Anik F3	9 Apr 2007	Proton	24 C-band 32 Ku-band 2 Ka-band	

of research programs that helped these firms maintain leadership in technologies for communications satellites and ground terminals.

Chris Gainor

See also: Canada, Hughes Aircraft Company, Telesat Canada

Bibliography

Chris Gainor, "The Chapman Report and the Development of Canada's Space Program," *Quest* 10, no. 4 (2003).
Doris H. Jelly, *Canada: 25 Years in Space* (1988).

Direct Broadcast Satellites. Direct broadcast satellites (DBS) were probably first conceived by Arthur C. Clarke in his 1945 article in *Wireless World*. In this article he proposed large geosynchronous communications satellites broadcasting radio and television programs direct to homes. He assumed that a radio broadcast—similar to frequency modulation (FM) radio—would require a 1.2 KW transmitter. This was enormous power for a communications satellite channel. He envisaged nuclear power onboard the satellite to support multiple FM radio and television broadcasts.

The earliest communications satellites of the 1960s had power levels ranging from less than 100 W to as much as 1 KW in the late 1970s. Not surprisingly direct-to-home broadcasts (DTH, DBS, and broadcast satellite service—BSS) were seen as unlikely because of the high power levels required. There were some 8- to 10-foot backyard antennas, but these were not classified as direct-to-home. As early as 1965 the American Broadcasting Company recognized the benefit of distributing network television via satellite. By the time U.S. domestic satellites were available in the mid-1970s, their biggest use was to distribute television programs—not just network programming to network affiliates, but also new programming for cable television headends.

The first DTH satellites were experiments: NASA's *Applications Technology Satellite 6* (*ATS 6*) (S-band) in 1974, Canada's *Communications Technology Satellite* (*CTS*) (Ku-band) in 1976, and Japan's *Broadcast Satellite Experimental* (*BSE*) (Ku-band) in 1978. Unlike *ATS 6* and *CTS*, *BSE* was sponsored by a broadcaster: Japan Broadcasting Company (NKH)—programming began immediately. A few years later India orbited *Indian National Satellite 1A* (*INSAT*) that provided S-band broadcasts for Doordarshan, the national broadcaster. During this time period, the International Telecommunication Union (ITU) had established BSS characteristics for Europe and Asia at WARC-77 (World Administrative Radio Conference) and for the Americas at RARC-83 (Regional Administrative Radio Conference). Orbital slots had 32 channels 24 MHz wide. Orbital slots had to be 9–10 degrees apart—minimizing interference from adjacent satellites. Large countries often had several orbital slots allocated, while small countries often had only a few channels allocated at a shared orbital slot.

In the late 1980s and early 1990s, several companies put up DBS consisting of a few 100 W transponders (channels), each carrying a single analog television program. The first American attempt to do this was a financial and management failure, Communications Satellite Corporation's (Comsat) Satellite Television Corporation (STC), but others succeeded—especially Luxembourg's Société Européenne des Satellites (SES) Astra system. Part of the reason for STC's failure was the ubiquity of cable systems providing dozens of channels when STC could only provide three. An equally compelling reason was the promising development of compressed digital video and very high-powered satellites. Instead of a single 3-transponder satellite carrying three analog programs, a pair of 16-transponder satellites occupying a single orbital slot could provide 128 (32×4) compressed digital television programs. *DBS 1* was launched 17 December 1993 for Hughes' DIRECTV. *Echostar 1* was launched two years later. Throughout the ensuing decade, these two companies absorbed many of their competitors and concentrated in a few orbital slots, providing coverage for the United States. These satellites serviced millions of television households generating billions of dollars in revenues. In 2005 DIRECTV had revenues of more than $13 billion, Echostar had revenues of over $8 billion, and BSkyB (on SES Astra) had revenues of more than $6 billion.

The politics of satellite television (especially BSS) have been challenging. Some countries are sensitive to the saturation of their media with foreign culture—especially U.S. culture. Many Fixed Satellite Service satellites have been jammed because of offensive programming. This was particularly true of Iran. When the ITU established the rules for BSS, the allocations were for beams that covered national territory only—not to include adjacent countries.

David J. Whalen

See also: Hughes Aircraft Company, SES

Bibliography

Arthur C. Clarke, "Extra-Terrestrial Relays," *Wireless World* (October 1945).
Leland L. Johnson and Deborah R. Castleman, *Direct Broadcast Satellites* (1991).
Donald H. Martin, *Communications Satellites*, 4th ed. (2000).

Eurobird. *See* Eutelsat.

European Communications Satellites. European communications satellites came into operational commercial service in the mid-1980s, delayed by a perceived lack of need and a lack of a consensus about the direction that a cooperative communications satellite program should take.

Europe's earliest experiments with satellite telecommunications began in the late 1950s when scientists at the United Kingdom's Jodrell Bank Observatory transmitted spoken messages to the United States by reflecting them off the Moon as a passive relay. France also participated in early artificial satellite experiments by receiving

When ESA's Olympus *satellite was launched on 12 July 1989 it was the world's largest telecommunications satellite. (Courtesy European Space Agency)*

signals bounced by American Telephone and Telegraph (AT&T) Company in the United States off the *Echo* satellite in 1960. Despite similar experimental programs in several countries, though, no European country was interested initially in serious commercial satellite development. Communications services in Europe were provided through land-based cables by the government of each nation, none of which saw any financial advantage in investing huge amounts of money in the development of satellites, launchers, and ground stations.

This view changed by the mid-1960s, as European countries realized their economic and technological disadvantages vis-à-vis the United States and the Soviet Union and gradually worked toward a regional integration of industry and commerce. Ten European countries joined in 1964 to form the European Space Research Organisation (ESRO) to promote European collaboration in scientific space research. Seven European countries joined that same year to form the European Space Vehicle Launcher Development Organisation (ELDO). In 1963 the United States initiated discussions about the development of an international telecommunications satellite system. Banding together to negotiate as a group instead of singly, western European nations created European Conference on Satellite Communications (French CETS) in response, leading in 1964 to the formation of Intelsat, the International Telecommunications Satellite Organization, an international consortium to manage the international communications satellite (comsat) system. CETS began discussions with ESRO that year to begin comsat design studies, but some ESRO members objected, fearing this would lead the organization away from scientific research, which was

ESRO's charter. Others believed that land-based networks were adequate for Europe's needs.

Intelsat's launch of *Early Bird* (*Intelsat 1*) in 1965 proved the commercial viability of satellite telecommunications. Along with discussions in CETS, France and Germany (*Symphonie*), Italy (*SIRIO*—Italian Satellite for Operational Industrial Research), and the United Kingdom were each pursuing their own national communications satellite programs, leading to complex negotiations among these nations and other smaller European countries, whose only hope of developing comsats was through an integrated European program. In 1971 ESRO members agreed to fund the Orbiting Test Satellite (OTS) program, which awarded the satellite development contract to the MESH consortium headed by Hawker Siddeley Dynamics. *OTS 1* was destroyed in a failed 1977 launch, but *OTS 2*, which provided 7,200 telephone circuits, succeeded the next year. European governments also agreed to develop an operational satellite system based on OTS, the European Communications Satellite (ECS) system.

Providing a launch capability to put European comsats in orbit was another perplexing problem. *Early Bird*'s launch proved that comsats would be placed primarily in geosynchronous orbit, but ELDO's launcher was not powerful enough to accomplish this. This led to a new series of Europa II launchers to provide a minimal geosynchronous capability. Unfortunately none of ELDO's launchers succeeded. The launcher problem took on urgency as the Franco-German Symphonie comsat project, formed in 1967, tried to determine how it was going to get its first satellite into orbit. Without a European launcher, the French and West German governments had to negotiate with the United States. Wanting to prevent competition with Intelsat and preserve its dominance in space communications, the United States refused to launch *Symphonie* or other European communications satellites unless they remained experimental only. This fueled France's determination to build a European launcher, which ultimately became Ariane.

In the so-called "second package deal" of 1973 that created the European Space Agency (ESA) from the wreckage of ELDO and the foundation of ESRO, France gained its Ariane launcher to place European comsats in orbit, and the United Kingdom won support for the Marots (Maritime OTS) program, which eventually became the Maritime European Communications Satellite (MARECS) series based on the more capable ECS design. ESA awarded the Marots prime contract to Hawker Siddeley Dynamics. The first satellite, *Marots A*, was launched by an Ariane rocket in 1981. *MARECS B* was lost in a launch failure in 1982, and a replacement, *MARECS B2*, was successfully orbited in 1984. ESA leased both operational MARECS satellites to Inmarsat, the International Maritime Communications Satellite Organisation. MESH consortium members British Aerospace (which acquired Hawker Siddeley) and Matra used their experiences to win contracts for *Inmarsat 2*, *Telecom* (France), *Skynet 4* (United Kingdom), *NATO 4* (North Atlantic Treaty Organization), *Hispasat* (Spain), and the private transatlantic system, Orion.

Anticipating the need for an operator for the ECS satellites, a consortium of national communications providers in Europe formed Eutelsat, the intergovernmental European

Major European Civilian and Commercial Satellite Programs (Part 1 of 3)

Country and Operator	Satellite Program Name	Initial Launch Date	Manufacturer	Areas Served	Purpose
France and Germany (CNES—French Space Agency—and Gesellschaft fuer Weltraum-forschung)	Symphonie	19 December 1974	CIFAS consortium, including Aérospatiale, Thomson CSF, Siemens, AEG Telefunken, MBB	Europe, Africa, South America, and the east coast of North America	Telephone, color television (TV) transmission
Italy	Sirio	25 August 1977	Consiglio Nazionale de Richerche (CNR)	Europe	Signal propagation and communications experiments
Europe (ESA)	OTS	11 May 1978	British Aerospace	Europe	Telephone
Europe (ESA)	MARECS	20 December 1981	British Aerospace	Various oceans	Mobile maritime communications
Europe (ESA and Eutelsat)	ECS, renamed Eutelsat	16 June 1983	Various contractors	Europe, Middle East, Africa, Asia	Telephone, fax, data, land mobile services; TV and radio programming
France (France Telecom)	Telecom	4 August 1984	Designed by CNES; Matra Marconi	France and its various territories	Telecommunications to French overseas territories, military communications, TV programming, and data transmissions
Germany (Deutsche Bundespost Telekom)	TV-Sat	21 November 1987	Eurosatellite and MBB	Germany and western Europe	Direct broadcast TV

(Continued)

Major European Civilian and Commercial Satellite Programs (Part 2 of 3)

Country and Operator	Satellite Program Name	Initial Launch Date	Manufacturer	Areas Served	Purpose
France (Télédiffusion de France)	TDF	28 October 1988	Eurosatellite and Aérospatiale	France, nearby European countries, North Africa	Direct broadcast TV and radio
Luxembourg (SES GLOBAL)	Astra	11 December 1988	GE Astro-Space	Western Europe	Direct broadcast TV
Sweden (Nordiska Satellitak-tiebolaget)	Tele-X	2 April 1989	Eurosatellite and Aérospatiale	Northern Europe	Direct broadcast TV, data transmission, teleconferencing services
Germany (Deutsche Bundespost)	DFS (Deutscher Fernmelde Satellit) Kopernikus 1	5 June 1989	R-DFS consortium with ANT and MBB-ERNO	Germany and nearby countries	Telephone and data transmissions, TV and radio broadcasts
Europe (ESA)	Olympus (also called L Sat)	12 July 1989	British Aerospace	Europe	Direct broadcast TV, teleconferencing services, data transmission
United Kingdom (UK) (British Satellite Broadcasting Ltd.)	BSB (also called Marco Polo)	27 August 1989	Hughes Space and Communications Group	United Kingdom	Direct broadcast TV
Italy (Agenzia Spaziale Italiana)	Italsat	15 January 1991	Selenia Spazio	Italy	Telephone
Spain (Hispasat S.A.)	Hispasat	10 September 1992	Matra Marconi	Europe, North America, and South America	Civil, military, and government voice and video communications

Major European Civilian and Commercial Satellite Programs (Part 3 of 3)

Country and Operator	Satellite Program Name	Initial Launch Date	Manufacturer	Areas Served	Purpose
Norway (Telenor Satellite Services)	Thor	First Thor purchased in-orbit from UK in September 1992	Hughes Space and Communications Group	Europe	Direct broadcast TV and telephone services
Sweden (Nordiska Satellitak-tiebolaget)	Sirius	First Sirius purchased in-orbit from UK in 1994	Hughes Space and Communications Group	Scandinavia and Greenland	Direct broadcast TV, data distribution
France (Alcatel)	Europe*Star	29 October 2000	Alcatel	Europe. South Africa, India	Direct-to-home video, Internet, and high-speed data transmission
United Kingdom	ICO	19 June 2001	Hughes Space and Communications Group / Boeing Satellite Systems	worldwide	Mobile telephone and Internet services
Netherland (New Skies Satellite)	NSS	16 April 2002	Lockheed Martin	Europe, North America, Asia, Australia	Video and Internet services
Cyprus/Greece (Astra TechCom)	Hellas Sat	13 May 2003	Astrium	Europe, North Africa, Middle East	Direct-to-home voice and video

Telecommunications Satellite Organization, in 1977. When the ESA's ECS series was successfully launched, beginning in 1983, the satellites were turned over to Eutelsat and renamed for that organization. Eutelsat was granted "permanent status" as an intergovernmental organization in 1985 and was incorporated as a private company in 2001.

A third-generation comsat program, ultimately known as Olympus, emerged from complex negotiations in the late 1970s. European nations wanted to take advantage of the new capabilities of the Ariane launcher and to explore the possibilities of television broadcasting via satellite. Olympus was approved in 1981, to provide two direct broadcast channels, communications experiments with small Earth terminals, and two other experimental payloads. British Aerospace became the prime contractor, leading to its successful launch in July 1989. One of its solar arrays failed in January 1991, and in August 1993 the spacecraft was lost.

Germany and France declined to participate in Olympus, favoring a bilateral program for four television broadcast satellites, called TV-Sat in Germany, and Télédiffusion de France (TDF) in France. Messerschmidt-Bölkow-Blohm (MBB) was the German prime contractor, and SNIAS (Société Nationale Industrielle Aérospatiale, later Aérospatiale) was the French prime. TV-Sat's first satellite failed in November 1987, and its backup satellite succeeded in August 1989. The TDF spacecraft were both successful, launched in October 1988 and July 1990. Sweden, Norway, and Finland purchased a TV-Sat/TDF spacecraft, which they called *Tele-X*, orbited successfully in 1989. SNIAS and MBB formed the Eurosatellite group, which went on to win contracts for *Eutelsat 2* and *Arabsat*. MBB and Entwicklungsring Nord went on to develop *Kopernikus*, a comsat for the German postal service.

All these high-powered European systems were made essentially obsolete by the private company Societé Européene des Satellites (SES), based in Luxembourg. The SES Astra satellites, purchased from the U.S. Radio Corporation of America, used much more advanced receivers and lower-power satellites to deliver television broadcasts at lower costs than the government-sponsored systems. Government telecommunications organizations used the European systems, but these could not compete commercially with SES for consumer business.

ESA's next major telecommunications satellite was *ARTEMIS*, the *Advanced Relay Technology Mission*, built by Alenia Spazio and launched in July 2001. This spacecraft provided capabilities to test radio and laser inter-satellite links, and provided 662 L-band voice channels. By 2006 it was used as a testbed for networking and navigation technologies in preparation for the European Galileo navigational system.

Stephen B. Johnson and Llyn Kaimowitz

See also: Ariane, European Space Agency, European Space Research Organisation, SES

Bibliography

J. Krige and A. Russo, *A History of the European Space Agency, 1958–1987* (2000).

National Space Science Data Center Spacecraft Database. http://nssdc.gsfc.nasa.gov/nmc/SpacecraftQuery.jsp.

Lorenza Sebesta, "U.S.–European Relations and the Decision to Build Ariane, the European Launch Vehicle," in *Beyond the Ionosphere: 50 Years of Satellite Communication*, ed. Andrew J. Butrica (1997).

Hot Bird. *See* Eutelsat.

Inmarsat. Inmarsat (International Maritime Satellite Organisation) was created in 1976 to establish a new maritime communications satellite system. This new system would improve maritime communications, assist in improving communications for distress and safety of life at sea, the efficiency and management of ships, maritime public correspondence services, and radio-determination (navigational) capabilities.

Inmarsat Satellites

Name	Launch Site	Date	Manufacturer	Service Area
Marisat 1	Cape Canaveral, Florida	19 Feb 1976	Hughes	Atlantic Ocean
Marisat 2	Cape Canaveral, Florida	10 June 1976	Hughes	Pacific Ocean
Marisat 3	Cape Canaveral, Florida	14 Oct 1976	Hughes	Indian Ocean
Marecs A	Kourou, French Guiana	19 Dec 1981	British Aerospace	Atlantic Ocean
Marecs B	Kourou, French Guiana	9 Nov 1984	British Aerospace	Pacific Ocean
Inmarsat 2 F1	Cape Canaveral, Florida	30 Oct 1990	Matra Marconi	Atlantic Ocean
Inmarsat 2 F2	Cape Canaveral, Florida	8 Mar 1991	Matra Marconi	Indian Ocean
Inmarsat 2 F3	Kourou, French Guiana	16 Dec 1991	Matra Marconi	Pacific Ocean
Inmarsat 2 F4	Kourou, French Guiana	15 Apr 1992	Matra Marconi	Atlantic Ocean
Inmarsat 3 F1	Cape Canaveral, Florida	3 Apr 1996	Astro Space	Indian Ocean
Inmarsat 3 F2	Baikonur, Russia	6 Sep 1996	Astro Space	Atlantic Ocean
Inmarsat 3 F3	Cape Canaveral, Florida	18 Dec 1996	Astro Space	Pacific Ocean
Inmarsat 3 F4	Kourou, French Guiana	3 Jun 1997	Astro Space	Atlantic Ocean
Inmarsat 3 F5	Kourou, French Guiana	4 Feb 1998	Astro Space	Mediterranean
Inmarsat 4 F1	Cape Canaveral, Florida	11 Mar 2005	Astrium	Indian Ocean
Inmarsat 4 F2	Odyssey Launch Platform	8 Nov 2005	Astrium	Atlantic Ocean
Inmarsat 4 F3	Cape Canaveral, Florida	Planned 2008	Astrium	Pacific Ocean

Headquartered in London, United Kingdom, Inmarsat began operations in 1982, initially leasing capacity on the U.S. Marisat satellites and on the European Space Agency Marecs satellites. From 1990 Inmarsat operated its own spacecraft to provide maritime communications. These satellites went through several generations, with significant improvements in capabilities. By 2006 Inmarsat 4 series satellites provided 19 wide and 228 narrow spot beams, with data speeds up to 492 kilobits per second.

Inmarsat's obligations to provide maritime distress and safety services via satellite were enshrined in 1988 amendments to the Safety of Life at Sea Convention (SOLAS), which introduced the Global Maritime Distress and Safety System (GMDSS). Ships sailing in specified sea areas were required to carry Inmarsat communications equipment for distress and safety calls and to receive navigational warnings. In 2005 the Inmarsat system was the only mobile satellite system recognized by SOLAS Contracting Governments for use in the GMDSS.

In 1998 considering the expanding commercial and competitive capabilities satellite communications offered, and to remove direct governmental presence from the operating organization, Inmarsat's Assembly of Member Governments agreed to terminate government ownership and privatize Inmarsat from April 1999 on. This resulted in establishment of two entities: (1) Inmarsat Global Ltd., a public limited company that forms the commercial arm of Inmarsat and (2) International Mobile Satellite Organization (IMSO), an intergovernmental body established to ensure that Inmarsat continued to meet its public service obligations, including obligations relating to the GMDSS. IMSO replaced Inmarsat as observer at International Maritime Organization meetings.

During the last two decades of the twentieth century, Inmarsat gradually broadened its services to include land mobile service, high-speed data systems, and aeronautical services in addition to its maritime mobile communications satellite services. Inmarsat revenues in 2006 were $501 million, and the total company workforce was slightly more than 500 employees.

Stephen Doyle

See also: Mobile Communications Satellites Services, Space Telecommunications Policy

Bibliography

Inmarsat. http://www.inmarsat.com/.

Edward J. Martin, "The Evolution of Mobile Satellite Communications," in *Beyond the Ionosphere: Fifty Years of Satellite Communication*, ed. Andrew J. Butrica (1997).

Intelsat. Intelsat (International Telecommunications Satellite Organization) was created in 1964 under interim arrangements to provide international satellite telecommunications. The initial 1964 interim agreements provided that each member organization would provide investment and have votes in proportion to the traffic provided on the system by each country. The United States, at that time, generated or received close to half of all the communications satellite services in the world. While many

Intelsat Space Segment Historical Overview (Part 1 of 2)

Space Segment Generation	Transmission Capacity	First Launch	Last Launch	Coverage Available	Launch Site(s)	Manufacturer
Early Bird Intelsat 1 One satellite	240 voice or one television (TV) channel	6 April 1965	6 April 1965	North Atlantic	Florida	Hughes
Intelsat 2 Four satellites	240 voice or one TV channel	26 October 1966	27 September 1967	Atlantic and Pacific basins	Florida	Hughes
Intelsat 3 Eight satellites	1,200 voice or voice/TV mix	19 September 1968	23 July 1970	Global	Florida	TRW Corporation
Intelsat 4 Eight satellites	2,400 voice or voice/TV mix	26 January 1971 One failed	22 May 1975	Global	Florida	Hughes and others in nine nations
Intelsat 4A Nine satellites	Added multiple spot beams	25 September 1975 One failed	31 March 1978	Global	Florida	Hughes and others in nine nations
Intelsat 5 Nine satellites	12,000 voice and two TV channels	6 December 1980 One failed	9 June 1984	Global	Seven Florida Two Kourou	Ford Aerospace
Intelsat 5A Six satellites	15,000 voice and two TV channels	22 March 1985 One failed	27 January 1989	Global	Three Florida Two Kourou	Ford Aerospace
Intelsat 6 Five satellites	120,000 voice and three TV channels	27 October 1989 One rescued	21 October 1991 7 May 1992	Global	Two Florida Three Kourou	Hughes and others in six nations
Intelsat K One satellite	Up to 32 TV channels	10 June 1992	10 June 1992	Atlantic basin	Florida	GE AstroSpace
Intelsat 7 and 7A Eight satellites	22,000 voice and three TV channels	22 October 1993 One failed	15 June 1996	Global	Two Florida Five Kourou One Xi Chang	Space Systems/ Loral

(Continued)

Intelsat Space Segment Historical Overview (Part 2 of 2)

Space Segment Generation	Transmission Capacity	First Launch	Last Launch	Coverage Available	Launch Site(s)	Manufacturer
Intelsat 8 and 8A Six satellites	22,000 voice and three TV channels	1 March 1997	18 June 1998	Global	Florida	Aérospatiale
Intelsat Americas Five satellites	20,000+ voice and up to 40 TV channels	24 May 1997	23 June 2005	Americas	Three Baikonur One Kourou One Sea Launch	Space Systems/ Loral
Intelsat 9 Seven satellites	20,000+ voice or up to 40 TV channels	9 June 2001	25 February 2003	Global	Six Kourou One Baikonur	Space Systems/ Loral
Intelsat 10-2 One satellite	High capacity, specialized	16 June 2004	16 June 2004	Europe	Sea Launch	Astrium
Intelsat IA-8 One satellite	High capacity, specialized	23 June 2005	23 June 2005	Americas	Sea Launch	Space Systems/ Loral

Additional Intelsat satellites have been purchased and/or sold on Earth or in orbit, as needed.

member nations participated in Intelsat through their national government telecommunications monopolies, Canada, Japan, the United States, and others participated through their private international carriers. The Communication Satellite Corporation (Comsat) represented the United States and was selected to be the interim manager of the Intelsat consortium. This established a crucial precedent for private corporations to be able to operate in space. The interim agreements were renegotiated, starting in 1969, which led to a permanent agreement in 1973. The gradually maturing Intelsat system, made fully operational in 1969, enabled people, businesses, and governments to communicate instantly, reliably, and simultaneously, from and to every part of the globe. As the organization expanded and world telecommunications increased, the U.S. share in the organization decreased steadily.

With its space segment (satellites) located in the geostationary orbit above the Atlantic Ocean, the Pacific Ocean, and the Indian Ocean, Intelsat enabled real-time global services for telephone, telegraph, television, data systems, and message traffic. In 2006 Intelsat was a significant enabler of the global Internet. Earth stations, which interconnect through Intelsat satellites, were owned and operated by organizations within the countries in which they operated. Intelsat enabled global television transmission of some memorable events in modern history, while providing global communications infrastructure that anyone might use any day without even knowing it. By broadcasting worldwide the video signals of Neil Armstrong's first steps on the Moon, providing the direct communications ("Hot Line") between the White House and the Kremlin at the height of the Cold War, and distributing to global television events of every Olympics since 1968, Intelsat dramatically and continually demonstrated the value of satellite communications.

By the time Intelsat celebrated its 10th anniversary in 1974, its membership had grown from the original 11 countries to 86 countries. By 1975 Intelsat could carry 5,000 international telephone circuits and had the capacity of 20,000 voice circuits and five television channels of available in-orbit capacity. In June 1978, via Intelsat, about one billion people in 42 countries watched television coverage of the World Cup men's football matches in Argentina. Total viewers set a new high for television coverage, surpassing the record set during the 1976 Olympics. In July 1978 Intelsat participated in the earliest demonstration of the global Internet, using the *Intelsat 4A* satellite among Etam in West Virginia, Goonhilly Downs in the United Kingdom, and Tanum in Sweden, with land-line connections between Norsas in Norway and University College in London. In August 1984 the United Nations signed an agreement with Intelsat to provide global satellite capacity for peacekeeping and emergency relief activities. On 1 January 2000 Intelsat carried a record volume of programming—more than 30 television channels representing more than 1,650 hours—for the millennium transition, with cascading delivery of the first television images of the year 2000 to millions of viewers across every time zone on the globe as it rotated the new century into being, time zone by time zone.

By the late 1990s competition from private companies was eroding Intelsat's market position, leading to pressures to change its status as a quasi-government, quasi-private organization to a purely private company. In 1998, the company spun off five in-orbit satellites along with various assets to the new company, New Skies Satellites, as part of its privatization efforts. On 18 July 2001 Intelsat ended its problematic quasi-governmental, quasi-private composition and became a private company—Intelsat, Ltd.—after 37 years as an intergovernmental/private mixed organization. As a private company, Intelsat focused on maximizing value to its customers and shareholders, in addition to providing global communications capability. In July 2006 Intelsat merged with PanAmSat, with the new Intelsat controlling a combined fleet of over 50 satellites.

In 2008 Intelsat, Ltd., was the world's leading provider of global satellite communications services, supplying video, data, and voice communications in more than 200 countries and territories. Intelsat services were used by leading multinational corporations, national governments, and major broadcast providers globally, including major broadcasters in the United States.

Stephen Doyle

See also: Comsat Corporation, PanAmSat Corporation, Syncom

Bibliography

Andrew J. Butrica, ed., *Beyond the Ionosphere: Fifty Years of Satellite Communication* (1997).
Jonathan Galloway, *The Politics and Technology of Satellite Communications* (1972).
Intelsat. www.intelsat.com.
Joseph Pelton, *Globaltalk* (1984).
David Whalen, *The Origins of Satellite Communications* 1945–1965 (2002).

Mobile Communications Satellite Services. Mobile communications satellite services offered voice or data capabilities to terrestrial users via space-based satellite networks. Terrestrial subscribers accessed the communications network through small handheld transceivers or mobile communications terminals. Gateways provided connectivity to terrestrial networks, so subscribers could literally call around the world.

Predecessors to the twenty-first-century commercial mobile communications satellite service providers dated to the 1960s when several organizations began parallel investigation of various mobile satellite communications methods. In 1963 the U.S. Federal Aviation Administration, NASA, Communications Satellite Corporation (Comsat) and various airline companies envisioned a geosynchronous satellite communications system Aerosat to provide very high frequency (VHF) communications to aircraft, but the system never materialized. During the early 1970s the U.S. Air Force's Tactical Communications Satellite (TACSAT) demonstrated the feasibility of small, mobile communications terminals. Later the U.S. Navy developed a follow-on system known as Fleet Satellite Communications, which expanded on the TACSAT concept by providing a satellite network for two-way communications between naval vessels and fleet command.

International Maritime Satellite (Inmarsat) provided the first commercial mobile communications satellite service. Conceived by the United Nations International Maritime Consultative Organization in the early 1960s, Inmarsat established services in the early 1980s and profitably operated into the twenty-first century. It provided voice, fax, and data services through geostationary satellites to seagoing vessels and to briefcase-sized portable terminals.

Envisioned as an alternative to localized terrestrial cellular networks, large low Earth orbit (LEO) mobile communications satellite services offered the advantage global coverage provided by large satellite constellations. Users could access the satellite network with small, handheld phones from places where terrestrial networks and geostationary satellites could not reach, such as at sea or at the polar regions. Perceived demand for mobile satellite communications service appeared high in the late 1980s, despite the initial lack of adoption for terrestrial cellular networks. Service providers needed to orbit large constellations of satellites to provide continuous worldwide coverage, which created a huge demand for satellite launch services in the late 1990s. Euphoria surrounding the information age of the 1990s and the Internet spawned additional interest in mobile satellite communications. However, by the early 2000s demand for service was so minimal that the two major service providers filed for bankruptcy. The services proved to be too expensive and provided little bandwidth to subscribers, resulting in minimal services such as voice-calls, paging, and low-speed data transfers; high-speed broadband services were not achievable.

Established by Motorola in 1987, Iridium Limited Liability Corporation (LLC) provided the first mobile communications satellite services in 1998. Iridium launched an LEO constellation consisting of 66 interlinked satellites and 12 terrestrial gateway stations throughout the world. Iridium LLC went bankrupt shortly after offering services, and Iridium Satellite LLC acquired the assets and continued to provide service. Iridium Satellite's primary customer was the U.S. Department of Defense (DoD), which often operated in areas not covered by terrestrial communications networks. The DoD signed agreements with Iridium Satellite and was believed to be the largest customer during the early 2000s.

Globalstar Telecommunications, a joint venture between Qualcomm and Loral, offered portable voice and data services in 1999. The Globalstar network consisted of a 48-satellite LEO constellation that employed a technology known as "bent-pipe" communications, where satellites simply relayed signals from subscribers to the nearest terrestrial ground station. Inter-satellite communications was not possible with Globalstar's satellite network, thus coverage was unavailable in the polar and ocean regions, which eliminated Globalstar as a solution for some naval and maritime uses. Globalstar filed for bankruptcy in 2001 and emerged from bankruptcy in 2004, at which time Thermo Capital Partners LLC acquired a majority of Globalstar's assets.

Another company, OrbComm, operated a network of 30 LEO satellites specifically designed to offer near-real time "store and forward" data messaging services to mobile and stationary users via small, low-bit rate mobile terminals. OrbComm's service focused on communicating with remote, fixed sites and mobile tracking services. Real-time telephony and voice was not possible.

As of 2008, Inmarsat, Iridium Satellite, and Globalstar offered voice, paging, and data services to consumers, while OrbComm offered data forwarding services. The proliferation of localized terrestrial cellular networks into major population areas stifled the adoption of mobile satellite services, and the high cost of the mobile satellite services limited their use to governments, military, and maritime commercial sectors.

David Hartzell

See also: Inmarsat, Iridium

Bibliography

Andrew Butrica, ed., *Beyond the Ionosphere* (1997).
Globalstar. www.globalstar.com.
Iridium Satellite. www.iridium.com.
OrbComm. www.orbcomm.com.

Regional Satellite Communications Operators. Regional satellite communications operators provided communications services for regional geographic areas. Governments and nations established services during the 1970s with the intent to integrate sparsely populated peoples throughout a large geographic region where terrestrial communications did not exist or were too expensive to deploy. Neighboring nations often joined to provide communications services and to develop indigenous space technology. By the 1980s some private corporations were able to sustain profitable businesses performing these same services in areas not otherwise covered by government systems.

In the 1960s regional communications satellites were forbidden by the initial Intelsat agreement so as to provide the organization with a worldwide monopoly, thus ensuring its financial success. However, the final Intelsat agreement of 1971 allowed regional communications satellites if they did not conflict with Intelsat, and by the 1980s several privately held regional satellite operators emerged.

In the 1970s European countries developed a regional communications satellite network to deliver television and telephone services and to gain greater technological skills in the space sector. The European Space Agency built and launched Europe's first five communications satellites managed by the European Telecommunication Satellite (Eutelsat) organization that provided the first regional satellite communications service to Europe in 1983. The second generation of satellites (EUTELSAT 2) focused on direct-to-home television services and extended coverage to northern Africa and the Middle East. In 1995 Eutelsat began providing digital television, data services, and mobile services with its HOTBIRD satellites. Eutelsat incorporated in 2001 and continued to expand its reach to African and Asian customers.

The Palapa satellite fleet provided domestic communications services throughout the island country of Indonesia, and later throughout Asia. Established by the government of Indonesia, Palapa satellites provided critical communications services to a country of more than 6,000 inhabited islands, which had proven difficult to interconnect via terrestrial telecommunications technology. Palapa ground stations were controlled and operated by the government-owned communications company Permutel, later known as PT Telekomunikasi Indonesia (Telkom). In 1976 the first Hughes-built Palapa A satellite was launched and by the late 1980s, the Palapa system provided regional coverage through parts of Asia and Australia. The Palapa series satellites B, C, and D augmented the system in the late twentieth century. As the Palapa system carried higher volumes of commercial and private communications, Satelindo (Satelit Palapa Indonesia), a private-sector entity, was permitted to procure, launch, and operate the two Palapa C series satellites, both launched in 1996. In 2007 Indosat (the parent organization of Satelindo) ordered *Palapa D*, built by Thales Aerospace, with a launch date scheduled in 2009. Palapa gained international attention when NASA's Space Shuttle (STS-41-B) delivered *Palapa B2* and Western Union's *WESTAR* to low Earth orbit. A shared booster module failed to insert both satellites into a higher orbit and Shuttle mission STS-51-A later retrieved both satellites.

The Arab Satellite Communications Organization (Arabsat) launched its first two regional communications satellites in 1985. Twenty-one Arab nations formed Arabsat in 1976 to help socially and culturally integrate Arab nations by delivering television, telephone and data services to northern Africa and the Middle East. By 2008 Arabsat launched and operated five satellites and leased additional transponder time from commercial telecommunications satellite providers, providing service to over 164 million television viewers.

Several private regional satellite operators merged and became part of global satellite systems. PanAmSat, a private regional satellite operator established by Rene Anselmo, began carrying regional television services to North America in 1989. PanAmSat effectively broke Intelsat's government monopoly and opened the international market for competition in the United States. Ironically, it merged with Intelsat in 2006 and became part of Intelsat's global system. In 1984 privately owned Société Européenne des Satellites (SES) established regional satellite services for Europe and offered direct-to-home television and radio services. SES later merged with GE Americom, another privately held regional satellite operator in the Americas, combining capabilities to provide near-global service.

Malaysia East Asia Satellite (MEASAT), initially provided broadcasting and telecommunications services in 1996 to India, the Philippines, and Australia via two satellites, *MEASAT 1* and *2*. As of 2008, MEASAT offers service to 80 percent of the world's population, broadcasting to over 145 countries including India, the Philippines, Australia, Europe and Africa via a satellite fleet consisting of *MEASAT 3* and *3a*, and *AFRICASAT 1* and *1a*.

Several Asian regional systems kept their focus on regional service. AsiaSat became the first privately owned regional satellite communications operator to provide services throughout Asia and the Pacific rim. Formed in 1988, AsiaSat launched its first satellite in 1990, and as of 2008, it owned three satellites, *AsiaSat 2*, *AsiaSat 3S*, and *AsiaSat 4*, in geostationary orbit providing television, telephone, and data services throughout the Asia–Pacific rim region. AsiaSat's primary shareholders were the China International Trust and Investment Corporation Group and General Electric

Company. Asia Cellular Satellite (ACeS) system, the first regional provider of dedicated mobile communications, offered portable telephony services. ACeS began offering service in 2000 via its geostationary satellite *Garuda 1*.

David Hartzell

See also: Arabsat, Eutelsat, Intelsat

Bibliography

Arabsat. www.arabsat.com.
AsiaSat. www.asiasat.com.
Andrew Butrica, ed., *Beyond the Ionosphere: Fifty Years of Satellite Communication* (1997).
Donald H. Martin, *Communications Satellites 1958–1995* (1996).
Eutelsat. www.eutelsat.com.
MEASAT. www.measat.com.

Satellite Radio. Satellite radio is a method of information delivery (typically audio programming) that is broadcast directly to stationary, mobile, or portable terrestrial receivers from space-borne relay satellites. Satellite radio service providers offered hundreds of high-fidelity digital audio channels, including news, talk radio, and music programming. Services were often subscriber-based and required the purchase of a subscription and receiving electronics.

Habiba Maalim, center, and other girls at Khadija Umul Mumminin girls' primary school in Mandera in northeastern Kenya, listen to an English lesson broadcast on a WorldSpace satellite radio, 6 November 2002. WorldSpace was set up by Noah Samara, an Ethiopian American, in 1990 in a bid to help spread information in some of the world's poorest countries. (Courtesy AP/Wide World Photos)

Two U.S. providers of satellite radio service were created in the 1990s: SIRIUS Satellite Radio Inc. and XM Satellite Radio Inc., both of which offered satellite radio services throughout the lower 48 states. A third company, WorldSpace Satellite Radio, was based in the United States and offered service to Africa, Asia, Europe, and the Middle East.

In 1997 the U.S. Federal Communications Commission (FCC) made two slots of the S-band portion of the radio spectrum available for nationwide satellite digital audio radio services (SDARS), and four U.S. companies were bidding for space in the SDARS spectrum. In a closed FCC auction, the American Mobile Radio Corporation and CD Radio were the top bidders, each paying approximately $90 million and $83 million respectively.

Most satellite radio programming was commercial free. Providers charged a fee to cover infrastructure and programming costs. Before FCC licensing, there was concern that satellite radio providers would harmfully compete with freely available terrestrial radio stations, but the technology for broadcasting local programming through satellite radio did not exist at the time. During the first years of the twenty-first century, satellite providers developed methods of providing local traffic and weather reports to major metropolitan areas, causing further contention and complaints to the FCC. Local terrestrial radio stations pursued new digital broadcasting techniques in the early twenty-first century to compete with satellite radio.

Founded in 1992 as the American Mobile Radio Corporation, XM (No Modulation) launched its first geostationary satellite, *Rock*, in March 2001. A second satellite, *Roll*, was launched two months later. Both Boeing Model 702 spacecraft were launched by Sea Launch, LLC, and services started nationwide in November 2001. Space Systems/Loral won a follow-on satellite contract. By July 2007 XM had more than 7.5 million U.S. subscribers. It was granted a broadcasting license by the Canadian Radio-television and Telecommunications Commission (CRTC).

Originally founded as CD Radio, SIRIUS was a recipient of the FCC's S-band digital audio radio service license. SIRIUS launched three geosynchronous Loral 1300 satellites in 2000 on Proton rockets to provide service, which began in July 2002, to U.S. customers. It had more than five million U.S. subscribers by July 2007. CRTC granted SIRIUS a license to provide Canadian service.

Founded in 1990, WorldSpace began offering services in 1999 via its satellite *AfriStar*, which was launched in October 1998. *AfriStar* provided service to Africa and southern Europe. WorldSpace launched its second satellite, *AsiaStar*, in March 2000, which provided satellite radio services to a majority of Asia, India, and the Middle East. Matra Marconi Space manufactured both satellites, which were launched by Arianespace. WorldSpace technology used the L-band portion of the radio spectrum and provided programming content to empower and enhance information access to developing nations. Through a partnership with XM, WorldSpace content was provided to XM subscribers in the United States. WorldSpace acquired funding from developing country governments and had more than 170,000 subscribers worldwide by March 2008.

XM and SIRIUS made agreements with automotive manufacturers to install mobile satellite radio receivers in new vehicles, which aided in the deployment and adoption of satellite radio technology. All three providers offered portable, battery-powered receivers, which was critical to providing coverage for remote, underserved areas. In February 2007 the two companies proposed a merger. In March 2008 the U.S. Department of Justice ruled in favor of the merger, and by June 2008 the Federal Communications Commission was assessing the merger.

David Hartzell

See also: Federal Communications Commission

Bibliography

FCC. www.fcc.gov.
SIRIUS. www.sirius.com.
Space.com. www.space.com.
WorldSpace. worldspace.com.
XM. www.xmradio.com.

Commercial Space Launch

Commercial space launch has become an increasingly important and ever-present aspect of the space industry. Since the emergence of commercial launch providers in the 1980s, the industry has been affected by increases in private investment by entrepreneurs, such as the development of new launch vehicles and fluctuations in the demand for commercial launches, primarily due to the use of geosynchronous orbit satellites instead of low-Earth-orbiting satellites. This was evident in the sharp rise in launches in the latter half of the 1990s, followed by an even deeper decline in the first half of the next decade.

Because launch technology has ramifications for national security, technology development, communications, exploration, and national prestige, eight nations had developed commercial space launch capabilities by 2006. Public-private partnership models provided the main mechanisms to transfer technology from government to private or quasi-private entities using an established licensing and regulatory organization. By doing this, governments sustained areas of space commercialization that would not be able to survive on a purely commercial basis, lowered the risk to investors, and sometimes reduced government expenses by providing profit-making mechanisms and organizations to defray costs.

The first generation of U.S. and Russian commercial launchers—Delta, Atlas, Titan, and Soyuz—was initially intercontinental ballistic missiles (ICBM) developed for military use in the late 1950s and early 1960s. These missiles were soon converted to commercial launchers. While the U.S. government funded private contractors to develop and manufacture the launchers, both it and the Soviet Union controlled all launches and launch facilities. When companies like Intelsat or Anik wanted to launch a satellite, they had to pay the U.S. government to perform the service.

This continued in the early 1980s until the U.S. Space Shuttle began deploying commercial payloads. Because of NASA's commitment to launch all possible payloads on the Shuttle, it set prices far lower than actual Shuttle operations costs.

Consequently U.S. expendable launcher manufacturers (Martin Marietta, General Dynamics, McDonnell-Douglas, and Vought) began shutting down their launcher production lines. By 1985 the U.S. Department of Defense challenged this policy and began development of the Titan IV to provide an alternative to the Shuttle. Also, during the first term of the Ronald Reagan presidential administration, Congress passed the Commercial Space Launch Act, which established the Office of Commercial Space Transportation in the Department of Transportation to aid in the licensing of commercial space launches. With the 1986 *Challenger* disaster and the grounding of the Space Shuttle, NASA was soon banned from commercial launches, and U.S. expendable vehicle manufacturers began marketing their launchers to payload operators. In February 1988 the Reagan administration further required that the government purchase launch services.

By this time Arianespace, a European company established in March 1980 under an intergovernmental agreement, led the commercial market. Development of Arianespace's launch vehicle, the Ariane, was funded by the European Space Agency (ESA), with the first Ariane launch in 1979. Participating in the creation of Arianespace were 36 of Europe's key aerospace and avionics manufacturers, 13 major European banks, CNES (French Space Agency), and ESA. Arianespace, the first commercial space launch company, set its prices largely by comparison to expected Space Shuttle prices. ESA continued launch vehicle development and then delivered the designs to Arianespace to market, build, and operate. When the Shuttle fleet was grounded in 1986, Arianespace was in an ideal position to gain a commanding position in the commercial

An Ariane 5G *launch vehicle ready to lift off from Kourou, French Guiana, in 2004. Arianespace, which operated Ariane, was the first successful commercial space launch company. (Courtesy European Space Agency)*

space launch market, and by the mid-1990s European and U.S. vehicles had reached rough equivalence.

In the mid-1990s, Russia and China had entered the commercial launch market. The collapse of the Soviet Union and the subsequent economic collapse of Russia and other former Soviet states spurred the Russians and the Ukrainians to sell launch services. They soon began marketing the Soyuz launcher, through the privatized RKK Energia, and the Proton launcher, marketed by the government organization Khrunichev. These efforts evolved into several international joint ventures to market Russian and Ukrainian launchers. International Launch Services was established in 1995 as a joint venture of Lockheed Martin and Khrunichev to market Proton launch vehicles and U.S. Atlas launchers. The Sea Launch consortium united in 1995 four companies from Norway, Russia, Ukraine, and the United States to market the Zenit 3SL. Eurockot, formed in 1995, was operated by European Aeronautic Defence and Space Company (EADS) Space Transportation and Khrunichev State Research and Production Space Center and marketed the Rockot, a launcher based on the Russian UR-100N ICBM. Starsem was founded in 1996 by the Russian Federal Space Agency, Samara Space Center, EADS Space Transportation, and Arianespace to commercialize the Soyuz launcher.

China began to offer commercial launch services with its Long March series of launchers in 1985. China's vehicles captured a few commercial launches per year, largely because its prices were significantly lower than comparable U.S. and European launchers. China successfully launched 27 foreign-made satellites from 1985 to 2000. However by 2000 a U.S. embargo about improper technology transfer, Chinese launch failures, and collapse of the medium Earth orbit satellite market led to a reduction in Chinese commercial launches. China's prices were competitive compared to other countries, but it was limited by international controls and a higher-than-normal failure rate.

Japan and India also developed commercial space launch capabilities. Japan launched its first satellite in 1970 and offered its H2 launcher commercially in the early twenty-first century. India operated a Polar Satellite Launch Vehicle, which made several commercial flights, and by 2006 was also developing a Geostationary Launch Vehicle.

In the twentieth century there were a number of attempts to build private launch capabilities. One of the first was the Otrag project conducted in the mid 1970s by a German company and funded by Libya. The project was eventually terminated because of pressure from the U.S., German, and Soviet governments to prevent development of a Libyan ballistic missile. Space Services Incorporated (SSI) was the next to try a privately developed, privately operated launch system. Its first attempt was called the Percheron, which failed in a flight test in August 1981. SSI was more successful with its Conestoga design, made from Minuteman rocket parts. This rocket was successfully tested in September 1982, but this did not prevent SSI's bankruptcy due to lack of investors and customers.

Throughout the 1980s–90s, several entrepreneurs tried to develop private launch systems, believing that the projected demand from satellite cellular systems, which

PETER HENRY DIAMANDIS
(1961–)

(Courtesy AP/Wide World Photos)

Trained at Harvard as a medical doctor, Diamandis holds undergraduate and graduate degrees from the Massachusetts Institute of Technology in aerospace engineering. In 1987 he helped found the International Space University, dedicated to providing interdisciplinary space education. Strongly interested in space tourism, he cofounded the nonprofit X Prize Foundation in 1995, motivating private parties to explore reusable space launch vehicles. Similar motives led him to help create several for-profit and nonprofit space-related organizations, such as Space Adventures and Zero Gravity Corporation.

David Hartzell

would have required hundreds of small satellites, would sustain the private space launch industry. When that demand collapsed, so did the private launch companies like Beal Aerospace, Kistler Aerospace, Rotary Rocket, and Pioneer Rocketplane.

A more successful development of commercial space ventures was the Ansari X Prize, a $10 million prize to be given to the first company to design, build, and develop a craft capable of launching three people 100 km into space, returning them safely, and then repeating the feat with the same vehicle within two weeks. On 4 October 2004, *SpaceShipOne*'s success substantiated the X Prize founders' hopes of fostering private spaceflight.

By July 2006 several companies, including Space Adventures, Virgin Galactic, Armadillo Aerospace, Blue Origin, XCOR Aerospace, Rocketplane Limited, and others were planning to offer services in the space tourism industry. The first steps were to be short suborbital flights, with plans for orbital systems and orbiting hotel accommodations. NASA's November 2005 announcement supporting future commercial launch of propellant and cargo into low Earth orbit provided further encouragement to private launch companies.

Andrew Englen

See also: Ariane, Atlas, China, Delta, Europe, India, Russia (Formerly the Soviet Union), *SpaceShipOne*, United States

Bibliography

Joan Lisa Bromberg, *NASA and the Space Industry* (1999).

Federal Aviation Administration, Associate Administrator for Commercial Space Transportation. http://www.faa.gov/about/key_officials/nield/.
Roger D. Launius and Dennis R. Jenkins, eds., *To Reach the High Frontier: A History of U.S. Launch Vehicles* (2002).

United States Department of Transportation. The United States Department of Transportation (DOT) became involved with spaceflight in 1984 when President Ronald Reagan designated DOT, as opposed to the Department of Commerce, as the lead agency for dealing with commercialization of launch vehicles in the United States. The Commercial Space Launch Act (CSLA) in October 1984 codified the status of the DOT.

Regulation of commercial launch vehicles rose from the realization in the early 1980s that NASA's Space Shuttle would not be able to provide a timely launch schedule. The explosion of the Space Shuttle *Challenger* in 1986 solidified the need for a commercial space launch capability in the United States. The DOT established the Office of Commercial Space Transportation (OCST) within the office of the transportation secretary, but then transferred it to the Federal Aviation Administration (FAA) as the associate administrator for Commercial Space Transportation (AST) in 1995.

Congress charged the OCST and the AST to oversee the licensing of commercial launch vehicles in the United States. Launch vehicles operated by and for the U.S. government were exempt from CSLA. Before the establishment of OCST, a commercial launch vehicle required permits and clearances from more than one dozen federal agencies to conduct a launch; after OCST, permission was needed from only one agency. The AST, through the FAA, licensed launches of commercial expendable and reusable launch vehicles and suborbital rockets. U.S. law required licensees to carry $500 million in liability insurance, and they must allow for review of the vehicle, launch, and reentry sites, in addition to production, manufacturing, and testing facilities used by the licensees or their contractors to ensure flight safety. DOT oversight of commercial launch vehicles allowed the United States to vie in a competitive international launch vehicle market. As launch vehicle technology becomes increasingly affordable, and commercial human spaceflight becomes a reality, the role of the AST and the FAA in commercial spaceflight will remain important.

Howard Trace

See also: Commercial Space Launch

Bibliography

Federal Aviation Administration, Commercial Space Transportation. http://www.faa.gov/about/office_org/headquarters_offices/ast/.
T. L. Johnson, "Commercialization of Space Transportation—Exploring the Impact of the National Space Transportation Policy" (1998).
U.S. Congress, Office of Technology Assessment, *The National Space Transportation Policy: Issues for Congress* (1995).

Space Insurance

Space insurance has been essential for all commercial and many government satellites because of the high risks of launching and operating satellites and, in some cases, government regulations. High insurance rates, which have varied from 5 to 30 percent, but historically averaging 15–20 percent of the satellite cost, show that space operations remain risky. Nations and companies can self-insure, where the organization funds another satellite and launch to replace one that fails, or purchase insurance. Government regulations require that private entities purchase liability insurance to cover the potential damage caused to others, such as debris from a failed launch raining down on a city.

The space insurance market began with the 1965 launch of *Intelsat 1*, better known as *Early Bird*. The coverage provided third-party liability and prelaunch insurance (covering construction, shipment, storage, testing, and time on a launch vehicle before ignition), not insurance for the satellite itself. Aviation underwriters provided this policy and, reluctantly, expanded their coverage to include the actual launch for *Intelsat 2* in 1968. Since then, two more types of property damage coverage have evolved: launch and in-orbit. Launch, historically the riskiest phase, lasts from ignition to testing a satellite after it has reached its desired orbit. In-orbit (also called on-orbit) coverage ranges from a few months to several years and was pioneered in 1975 by the Radio Corporation of America Astro Electronics for its Satcom satellites.

Because the 1967 Outer Space Treaty and 1972 Liability Convention made states liable for damages caused by all their launches from their countries, national governments have set minimum liability insurance requirements for everyone launching a vehicle. For example, the U.S. government requires private liability insurance to $500 million and covers liability from $500 million to $2 billion. Liability premiums are low (approximately 0.1 percent of value), reflecting the low probability of use.

Three unique features of space insurance are its quickness, totality of returns, and the small size of the market. For underwriters, launch insurance offers the advantage of immediate profit, especially compared to long-term coverage like buildings. Most losses, however, are total. Either the satellite reaches orbit and operates, or it does not. Because fewer than 200 orbiting satellites and approximately 20 launches annually were insured as of 2006, one or two losses can drastically harm the industry's finances. When an Ariane 5 explosion destroyed two satellites in December 2002, insurers lost $600 million.

Major Satellite Insurance Firms in 1998 and Single Launch Maximum

Assicurazioni Generali S.p.A. (Italy)	$120 million
Assurances Générales de France (France)	$95 million
Le Reunion Spatiale (France)	$95 million
Marham Space Consortium/Lloyd's of London (United Kingdom)	$80 million
Brockbank Insurance Services (United States)	$75 million
Intec/AXA (United States/France)	$65 million

Space Insurance Premiums and Claims (in Millions of Dollars)

Year	Premiums	Claims	Net
1968	0.1	0	0.1
1969	0.4	0	0.4
1970	0.7	0	0.7
1971	0.6	0	0.6
1972	1.1	0	1.1
1973	0.7	0	0.7
1974	5.9	0	5.9
1975	3.4	0	3.4
1976	7.2	0	7.2
1977	9.8	29.1	–19.3
1978	9.2	0	9.2
1979	12.9	91.2	–78.3
1980	9.4	0.2	9.2
1981	32.5	0.7	31.8
1982	60.8	91.6	–30.8
1983	82.4	19.1	63.3
1984	139.6	300.7	–161.1
1985	169.6	333	–163.4
1986	108.4	85.6	22.8
1987	94.9	32.5	62.4
1988	157.6	134.2	23.4
1989	215.8	46.7	169.1
1990	336.6	407.5	–70.9
1991	323.8	147.3	176.5
1992	417.3	281.4	135.9
1993	351.8	207	144.8
1994	612.7	779	–166.3
1995	849	343	506
1996	812	639.9	172.1
1997	1028	609	419
1998	833	1548	–715
1999	821	592	229
2000	1023	1606	–583
2001	571	770.8	–199.8
2002	810	317.5	492.5
2003	600	352.9	247.1
2004	896	765	131

Reflecting the slow commercial exploitation of space, not until 1974 did another organization, the International Maritime Satellite Organization, buy launch insurance. The next year, instead of self-insuring, Indonesia and Canada purchased insurance for their Palapa and Anik satellites.

In the early 1980s marine and nonmarine underwriters entered this market, only to encounter major losses in 1982–84. The role of the Payload Assist Module in the *Palapa B-2* and *Westar 6* failures helped spur the insurance industry's efforts to develop its own data collection and expertise to understand and rate risks and to insist on solutions before writing more insurance. In the early 1990s the insurance industry expanded, enticed by predictions of a boom in telecommunications satellites. The boom did not materialize, but beginning in the late 1990s unexpected quality control problems in some components, such as the Boeing 702 solar power array, shortened the life of several satellites. Insurers, suddenly faced with unexpected payouts, forced manufacturers to increase testing and monitoring to improve satellite quality.

The flow of technical information demanded by insurers to determine risk has often faced a contrary desire for secrecy by governments and companies. The American International Traffic in Arms Regulation was designed to reduce the flow of military technologies outside the United States. The strict State Department reading of this law has restricted the flow of information about U.S. satellites to foreign underwriters since 1998. As European firms have historically underwritten more than half the space insurance market, this restriction has slowed the process of procuring insurance.

Following 1992 and 1995 losses of Long March rockets, insurance industry pressure led Hughes Space and Communications Company to illegally provide information that dramatically increased the Chinese booster's reliability. The U.S. government fined Space Systems/Loral $20 million in 2002 and Hughes $32 million in 2003. When the U.S. Air Force investigated the 1997 failure of a Delta/GPS launch, the insurance industry complained that information restrictions would decrease underwriters' ability to make informed judgments.

Jonathan Coopersmith

See also: China, Commercial Satellite Communications, Commercial Space Launch, Convention on International Liability for Damage

Bibliography

Associate Administrator for Commercial Space Transportation, Federal Aviation Administration, *Special Report: Update of the Space and Launch Insurance Industry* (Washington, DC: Department of Transportation, 1998).

Papers from the annual International Conference on Space Insurance.

Alden Richards, "The Early History of the Satellite Insurance Market," *Quest* 9, no. 3 (2002): 54–59.

Space Resources

Space resources constitute any non-terrestrial materials utilized or potentially utilizable for space activities. The most expansive definition includes items such as sunlight, near-perfect vacuum, and low temperatures in shadows; but in most cases the term applies to physical material. These include raw, unprocessed materials such as rock, ice, atmospheric gas, and meteoritic metal. Space resources also include the processed materials (fuel, water, oxygen, metals) extracted from raw materials. The importance of space resources lies in their potential to substantially reduce the cost of space operations by avoiding the high costs ($1,000s to $10,000s per kg) of lifting needed materials from Earth's surface. Three near-term repositories of space resources were studied in the late twentieth century: near-Earth asteroids/objects (NEA/NEO), the lunar surface, and the Martian surface and moons.

NEO resources have been considered primarily to support large-scale activities in near Earth space. Tens of thousands of NEOs can be easily accessed from low Earth orbit. In 1964 Dandridge Cole (an analyst for General Electric Missile and Space Division) and Donald Cox (a science writer) presented an early science- and engineering-based discussion of asteroid resources; however, poor knowledge of asteroid compositions at that time limited their analysis. In 1977 Massachusetts Institute of Technology researchers Michael Gaffey and Thomas McCord explored the economic value of asteroid mining based on their compositional studies of the minor planets. In the 1979 NASA Summer Study at Stanford University (Space Resources and Space Settlements), Brian O'Leary and colleagues focused on obtaining fuels and water from NEOs but also noted the presence of platinum group metals (PGM) as a potential NEO resource. In 1996 University of Arizona professor John Lewis explored the potential of asteroid PGMs as an economic base for space activities.

Lunar resources are of interest primarily to support activities on the lunar surface or in near-Earth space. In 1977 Princeton physicist Gerard O'Neill proposed using lunar materials to construct large habitats at the Trojan points in the Earth-Moon system. This concept generated a large following and led to the formation of an advocacy organization, the L5 Society. The proposed economic base for such a space-based community was solar power satellites (constructed from lunar materials) selling inexpensive and abundant electric power to customers on Earth. At the same time, pent-up interest among space scientists and engineers led to a series of conferences on space and lunar resources and a series of NASA-sponsored design studies throughout the next 20 years. Support for such activities within NASA in the 1980s–90s came primarily from a small cadre of scientists and engineers (including Wendell Mendell, Mike Duke, David McKay, David Criswell, and others) and did not reflect a high priority for the agency. A major potential lunar product is fuel for transportation to and from the lunar surface and possibly for delivery to low Earth orbit. The probable presence of lunar polar ice, detected by *Clementine* and *Lunar Prospector* spacecraft in the 1990s, increased the viability of lunar fuel scenarios. The Fusion Technology Institute at the University of Wisconsin explored the potential of helium-3 (^{3}He) as an economic basis for lunar settlements. However, the economic potential of ^{3}He depends on establishment of a fusion power industry on Earth, a development that was anticipated—but not realized—by the early twenty-first century.

Martian resources are considered primarily to significantly decrease the cost of future Mars missions by decreasing the total mass transported to Mars (the series of *Case for Mars* volumes produced by the American Astronautical Society since 1984). Fuels (methane and oxygen) can be formed using carbon dioxide from the Martian atmosphere and water from soil or ice deposits. The Martian moons Phobos and Deimos have been proposed as fueling stations, but the uncertainty concerning their actual compositions raises questions concerning the viability of this option.

Space resource utilization studies received a significant boost within NASA after President George W. Bush announced the new Vision for Space Exploration in January 2004, which shifted NASA's plans to send humans to the Moon and Mars.

Michael Gaffey

See also: Lunar Colonization, Space Advocacy Groups

Bibliography

J. Lewis et al., *Resources of Near-Earth Space* (1993).
M. F. McKay et al., *Space Resources* (1992).
W. W. Mendell, ed., *Lunar Bases and Space Activities of the 21st Century* (1985).
G. K. O'Neill, *The High Frontier—Human Colonies in Space* (1977).

Space Solar Power

Space solar power, collecting solar power in space and beaming it to Earth or elsewhere to generate power, has been a much-discussed potential application of space since the 1960s.

In 1968 Peter E. Glaser of Arthur D. Little, Inc., envisioned solar power satellites (SPS) as gigantic arrays of solar cells in orbit that could collect solar energy, convert it to electricity, and beam it through microwaves to antennas on Earth. The antennas would convert the microwaves back to electricity and feed the electricity to the power networks of electrical utilities. The link between orbit and Earth, wireless power transmission, had been pioneered by William C. Brown of the Raytheon Company.

Although Glaser was motivated by projections of resource depletion and potential energy shortages in the twenty-first century, an energy crisis developed sooner. Because of U.S. support of Israel in the 1973 Yom Kippur War, the Organization of Petroleum Exporting Countries embargoed oil shipments to the United States in October 1973. With continental U.S. oil production declining from its peak in 1970, the implications for the economy and national security were ominous. The energy crisis spurred a search for alternate energy sources. Increasing congressional interest in space technology for energy generation prompted the Department of Energy (DOE) and NASA to conduct a comprehensive, three-year Satellite Power System Concept Development and Evaluation Program (CDEP), starting in 1977. The objective of CDEP was to better understand the technical, economic, environmental, and social aspects of the SPS concept.

DOE/NASA set specifications for a Reference System on which CDEP would be based. The Reference System was a national satellite power system, with 60 satellites in geosynchronous earth orbit (GEO) and 60 terrestrial receiving stations. From GEO, the satellites would not be affected by the day/night cycle and would be illuminated by intense sunlight 99 percent of the time. The satellites would be enormous

5 × 10 × 0.5 km structures covered with photovoltaic arrays. Each satellite would have a 1 km diameter microwave antenna capable of transmitting 5 GW of power. A large contingent of astronauts would have built satellite components at low-Earth-orbit construction bases and transferred them to GEO on solar-powered orbital transfer vehicles for final assembly. The system would have required gigantic heavy lift launch vehicles much larger than the Space Shuttle or Saturn V to make hundreds of launches each year. Each satellite would transmit a 2.45 GHz microwave beam to a 10 × 13 km elliptical antenna on Earth. Because of their electrical current–rectifying properties, the receiving antennas were called rectennas. A rectenna resembled a vast field of panels tilted toward one of the satellites in GEO, supplying power to nearby electrical utility grids. CDEP specified technology and infrastructure development phases from 1980 to 2000, with construction to begin near 2000 and completed by 2030.

Because of the potential enormous financial commitment, Congress asked the Office of Technology Assessment (OTA) and the National Research Council (NRC) to independently assess SPS. While OTA's report compared various SPS systems, the NRC found that the Reference System's technologies were too immature, the costs too high, and the health and environmental effects of the microwave beams too uncertain to proceed. With Middle East oil flowing to the United States in greater quantities, government funding for SPS ended at the conclusion of CDEP in 1980.

An alternative to Earth-orbiting SPS, the lunar power satellite (LPS), was proposed in 1984. The LPS concept, as developed by David R. Criswell, consists of photovoltaic arrays on the Moon that send microwave beams directly to Earth or to reflecting satellites in Earth orbit for redirection to selected locations.

The 1990s saw resurgent interest in SPS due to escalating energy use, heightened concerns about global climate change from fossil fuel combustion, and the concentration of most major oil reserves in potentially unstable countries. NASA began a Fresh Look Study of SPS in 1995 to see if technological advances made since 1980 could make SPS economically competitive. The study assessed new versions of SPS (the Sun Tower and Solar Disc) designed to serve regional and international markets. The Fresh Look Study concluded that modular power stations based on new dual-use technologies and autonomous assembly and deployment would cost much less than the monolithic Reference System satellites.

The first economically competitive applications of SPS may be to beam power to developing countries that have little or no existing infrastructure. In addition to providing electricity for terrestrial power grids, SPS could also provide energy for weather and climate control or for other satellites in orbit. SPS may eventually provide power for manufacturing facilities or colonies in space or produce laser beams to propel spacecraft to the outer solar system and interstellar space.

Roger Truesdale

See also: Future Studies

Bibliography

Peter E. Glaser et al., *Solar Power Satellites: A Space Energy System for Earth* (1998).
U.S. Office of Technology Assessment, *Solar Power Satellites* (1981).

United States Department of Commerce

The United States Department of Commerce (DOC) collects information about trends in various commercial space sectors, such as communications, transportation, navigation, and remote sensing, and supports their commercial activities in the United States and abroad.

Commerce Secretary William Verity established the Office of Space Commerce in 1988 to support the former National Space Council. A decade later President William Clinton signed the Technology Administration Act of 1998, establishing the Office of Space Commercialization (OSC) within the DOC Office of the Undersecretary of Technology. In 2004 the OSC was relocated to the National Oceanic and Atmospheric Administration National Environmental Satellite, Data, and Information Service.

The office engaged in policy development, market analysis, and education and outreach. The OSC also served as an advocate of U.S. commercial space interests within the government, providing advice during congressional deliberations on the Commercial Space Act of 1998. Representatives also coordinated commercial space issues with other government agencies, such as establishment of the Remote Sensing Inter-Agency Group to assess commercial remote sensing systems. It tracked trends and developments in the commercial space industry, publishing economic trends about aerospace industry sales, space launch and satellite insurance, and foreign investment in the space industry. On the education and outreach front, the OSC helped commercial space companies validate their business plans and attract additional investment and sales. The OSC also promoted policies that strengthened the international competitiveness of the U.S. commercial space sector in conjunction with the Department of State.

Molly Macauley

Bibliography

Office of Space Commercialization. www.space.commerce.gov.

SPACE BUSINESS

Space business has grown steadily from the beginnings of rocketry through the early twenty-first century. Businesses have been major developers of space technology, and later in the operation of rockets and satellites.

The German Army's A-4/V-2 project to develop the first successful ballistic missile included contracts with private and government-owned businesses. In August 1944 the German Army reconstituted the rocket development team as the government-owned Electromechanical Industries Corporation so as to keep it out of the hands of Heinrich Himmler's SS (Schutzstaffel).

During the 1940s the U.S. military began funding rocket development and several companies took advantage of this new business opportunity. Reaction Motors, Aerojet, and Atlantic Research were founded to develop rocket engines and missiles, while existing companies including Convair, General Electric, North American, Martin, Firestone, and Thiokol received rocket and missile funding. In 1947 Congress passed the Procurement Act, which established the cost-plus contract that paid companies for their expenses plus a profit as the primary legal mechanism for research and development. Because success in the development of new technology was not guaranteed, and the government was usually the only customer, the cost-plus contract was a crucial legal tool to entice companies to work under government contract. In the 1950s more U.S. companies began to develop rocket and space technologies, including Lockheed, Douglas, McDonnell, Boeing, Sperry Gyroscope, Chrysler, Hercules Powder, and Ball Brothers. Grand Central Rocket and TRW corporations were created to develop rockets and missiles.

The U.S. military was not uniform in its industrial relations. The U.S. Air Force (USAF) depended on contractors for technology development and founded nonprofit corporations to perform research and to help manage profit-making corporations. These included RAND (1947), MITRE (1958), and The Aerospace Corporation (1960). The U.S. Army and Navy both had significant in-house technology development organizations, known as arsenals in the Army and design bureaus and research laboratories in the Navy. The Army, and to a lesser degree the Navy, usually performed initial technology research and development in-house, and then contracted to industry for large-scale production. On its founding in 1958, NASA used similar procurement regulations to the U.S. military, and it inherited the differing attitudes to industry. Marshall Space Flight Center, Kennedy Space Center, and the Jet Propulsion Laboratory originated from the Army, had significant in-house capabilities, and preferred to let industry contracts after development was completed. Goddard Space Flight Center, which originated largely from Navy personnel, performed significant in-house development. Johnson Space Center drew many of its leaders from Langley and Lewis Research Centers and used industry development contracts more frequently.

The 1960s saw rapid industry expansion into many new space applications, including navigation, communications, human spaceflight, weapons, observatories and robotic explorers, early warning systems, and reconnaissance. In the United States, most contracts went to previously funded contractors, though in the space communications field several other existing companies won government contracts, including Philco, Radio Corporation of America, and Hughes. By far the largest communications company was American Telephone and Telegraph (AT&T) Company, and it used its own resources to develop the *Telstar* satellite to test the feasibility of space communications. Fearing the expansion of AT&T's near monopoly on telecommunications, President John F. Kennedy's administration and U.S. Congress banned AT&T from developing communications satellites by creating the Communications Satellite (Comsat) Corporation. Comsat became the official U.S. representative to the International Telecommunications Satellite Organization (Intelsat), the new international communications consortium. The formation of Comsat and Intelsat, along

European Consortia, Mid-1960s

MESH	STAR	COSMOS
Matra (France)	SNIAS (France)	Thomson-CSF (France)
ERNO (Entwicklungsring Nord—West Germany)	Messerschmidt-Bölkow-Blohm (MBB—West Germany)	Dornier (West Germany)
Saab (Sweden)	Marconi (United Kingdom)	British Aircraft (United Kingdom)
Hawker-Siddeley (United Kingdom)	Selenia (Italy)	FIAR (Italy)
	ETCA (Belgium)	Fokker (Netherlands)
		Contraves (Switzerland)
		Ericsson (Sweden)

with the provisions of the Outer Space Treaty (1967), ensured that private companies and not merely governments could operate in space, as the Soviet Union had demanded.

In the 1960s European companies won contracts with their national governments and from European organizations, such as the European Space Vehicle Launcher Development Organisation (ELDO), the European Space Research Organisation (ESRO), and later the European Space Agency (ESA). To support the international organizations, European companies formed the consortia MESH, STAR, and COSMOS to ensure a division of industrial funding and responsibilities to match the negotiated national funding allocations. Many of these companies and consortia hired U.S. companies as consultants or formed joint space ventures.

Starting from the mid-1960s, communications satellites funded by national governments and by private companies grew in importance and profitability. By the 1980s–90s the space communications sector was a resounding commercial success with new companies, such as Pan American Satellite (PanAmSat), competing with the international consortia, such as Intelsat, and new applications, such as satellite television and satellite radio, providing the basis for new companies such as DIRECTV, Société Européenne des Satellites, Sirius, and XM. The success of satellite communications created a competition in space launchers to put them into space. Demand for space launch peaked in the late 1990s with the creation of communications satellite constellations in medium Earth orbit (MEO). Several private corporations formed to develop launch vehicles to meet the demand, but the collapse of the MEO satellite model by 2000 led to their bankruptcy. The old comsat consortia, such as Intelsat, were also fully privatized at this time.

The development and initial operations of the U.S. Space Shuttle in the 1970s–80s created both opportunities and problems for U.S. and European nations. Europe had decided to develop the Ariane launcher to gain an independent launch capability, but if successful in reducing launch costs, the Shuttle would make Ariane obsolete. The 1980s proved that far from lowering launch costs, the Shuttle in fact cost much more to operate than conventional launch vehicles, and Ariane, marketed by the first space launch corporation, Arianespace, succeeded in acquiring many lucrative contracts to launch commercial communications satellite payloads. Since the Shuttle's commercial success depended on frequent launches, U.S. policy was to place all payloads on the Shuttle and eliminate the U.S. expendable launcher fleet. The Shuttle's failure to lower costs and the 1986 *Challenger* accident ensured a future for U.S. expendable launch vehicles, and President Ronald Reagan's administration established laws governing commercial space launch vehicles. Soon McDonnell-Douglas, Martin Marietta, and General Dynamics were marketing Delta, Titan, and Atlas commercial launch capabilities. By the early 1990s they were joined by China, which marketed Long March, and Russian and Ukrainian organizations (some private, some government), which marketed Soyuz, Proton, and Zenit as commercial launch vehicles. United Space Alliance was formed as a joint venture of Boeing and Lockheed Martin to provide Shuttle operations and processing services.

Over the course of the space age, one common thread has been the merger of space businesses into ever-larger combinations in the United States and Europe. Early examples include the U.S. merger of Douglas and McDonnell into McDonnell-Douglas in 1967, and in France the creation of Société Nationale Industrielle Aérospatiale (SNIAS) from the 1970 combination of Sud Aviation, Nord Aviation, and Société d'Études et de Réalisation d'Engins Balistiques. Through a series of complicated maneuvers, by the early 2000s the space industry was dominated by several large companies: Lockheed Martin, Boeing, Northrop Grumman, European Aerospace and Defence Systems, and Alcatel Alenia Space.

Despite the seeming dominance of the aerospace giants, new space business opportunities and business continued to emerge. New applications, such as satellite navigation and commercial remote sensing, provided profitable opportunities for companies, such as Magellan Navigation, Qualcomm, Trimble, Spot Image, DigitalGlobe, and GeoEye. Space tourism and space burial were modest profit-making ventures, as implemented by Space Adventures and Celestis corporations.

Stephen B. Johnson

See also: Economics of Space, Nations, Space Politics

Bibliography

Roger Bilstein, *The American Aerospace Industry* (1996).
Joan Lisa Bromberg, *NASA and the Space Industry* (1999).
Roger Handberg, *International Space Commerce: Building from Scratch* (2006).
Stephen B. Johnson, "Space Business," in *Space Politics and Policy: An Evolutionary Perspective*, ed. Eligar Sadeh (2002), chap. 13.

Aérospatiale

Aérospatiale, a French aerospace hardware manufacturer, designed and built aircraft, helicopters, missiles, satellites, launchers, and spacecraft. The Société Nationale Industrielle Aérospatiale was created on 1 January 1970 from a French-government approved merger of Sud Aviation, Nord Aviation, and the Société d'Études et de Réalisation d'Engins Balistiques.

The European Space Agency (ESA) contracted with Aérospatiale on several major projects. Aérospatiale was the integrator for the Ariane launcher in the late 1970s and was the prime contractor of a series of European weather satellites, Meteosats, starting in the late 1970s. The Cassini mission, for which Aérospatiale was prime contractor for the *Huygens* probe, launched from Cape Canaveral in 1997 on its journey to Saturn. The Hermes spaceplane contract was given to Aérospatiale in 1985, but ESA canceled the project in 1992. Aérospatiale was the prime contractor for Anthrorack, a laboratory to evaluate low gravity on the human body that flew on *Spacelab* in 1993, and for the Atmospheric Reentry Demonstrator (launched in 1996) to return humans or payloads to Earth. The company also led development and manufacture of the *Infrared Space Observatory* launched in 1995.

Aérospatiale received contracts from the Arab Satellite Communications Organization and others for Spacebus satellites first launched in 1985, and the meteorological experimental tracking satellite *Eole* (active 1971–74) for CNES (French Space Agency). The Cannes business that manufactured these systems was sold to Alcatel in 1998, becoming Alcatel Space.

In 1999 Aérospatiale merged with Matra Haute Technologie to form Aérospatiale-Matra Missiles, manufacturer of launchers and satellites. In 2000 it partnered with other European companies to form the European Aeronautic Defence and Space Company (EADS), which developed and marketed civil and military aircraft, missiles, space rockets, and related systems. In 2002 the MBDA company was formed as a joint venture of EADS, BAE Systems, and Finmeccanica for the manufacture of missiles from EADS partners Aérospatiale-Matra Missiles, Alenia Marconi Systems, and Matra BAe Dynamics.

Joni Wilson

See also: Cassini, European Space Agency, Hermes, Meteosat

Bibliography

Claude Carlier, *Chronologie Aérospatiale, 1940–1990* (1992).
History of EADS. http://classic.eads.net/800/en/eads/history/history.html.

Alcatel Alenia Space

Alcatel Alenia Space, a division of Alcatel, is one of the world's largest space system contractors and an international supplier of telecommunications, navigation, remote-sensing, meteorological, and scientific satellite platforms. The division developed from three major sources: Alcatel, Aérospatiale's Cannes business, and the Italian company Alenia.

Alcatel's origins date to 1898 in electronics manufacturing. The name Alcatel was adopted in 1985 as the company focused primarily on telecommunications. In the early 1970s, telecommunications was expanding to space-based platforms, and Alcatel followed this trend by developing communications satellites (comsats). Alcatel contributed to Europe's first artificial satellite (*Asterix*, 1962), in addition to the Franco-German Symphonie communications satellites launched in 1974–75. By the 1990s, Alcatel's space division Alcatel Espace had developed its Proteus spacecraft bus, used for a variety of scientific and telecommunications spacecraft.

Aérospatiale's spacecraft capabilities developed primarily at its Cannes facility. These included telecommunications satellites, such as *EUTELSAT 2* and two generations of Arabsat systems. It was the prime contractor for the Meteosat series and the *Infrared Space Observatory.* Alcatel purchased the Cannes business from Aérospatiale in 1998.

Alenia developed as Italy's major space contractor from Aeritalia Space Systems Group, Turin, and Selenia Spazio, Rome. The two companies merged into Alenia Spazio in 1990. Aeritalia developed structural and thermal systems for European launchers and satellites, while Selenia's experience included building Italy's first telecommunications satellite *Sirio 1* and *Italsat*, in addition to communications gear for Intelsat, Meteosat, and other systems. Alenia and Alcatel merged their space operations in 2005, becoming Alcatel Alenia Space.

The European Space Agency worked with Alcatel on construction, integration, and testing of the *Mars Express* and *Venus Express* spacecrafts and the *Huygens* probe. Half of the pressurized living space and numerous components on the *International Space Station* were supplied by Alcatel. The European Geostationary Navigation Overlay Service (EGNOS) and proposed Galileo navigation systems, in addition to numerous climate- and global-observation systems, were based on Alcatel's platforms. Successful development of commercial comsats led Alcatel to develop military comsats for several European nations. In 2003, Alcatel completed the Air Navigation and Telecommunication Space System for Japan, and in 2005, Alcatel became the largest telecom satellite supplier to emerging China. Under agreement with the Regional African Satellite Communication Organization, Alcatel's system planned to cover the whole continent of Africa in 2006, utilizing just one satellite. In 2004, Alcatel Space employed 7,200 people and generated turnover of €1.8 billion.

George S. Sarkisov

See also: *Cassini-Huygens*, Commercial Satellite Communications, Navigation

Bibliography

Alcatel. www.alcatel-lucent.com.

American Telephone and Telegraph (AT&T) Company

The American Telephone and Telegraph (AT&T) Company was created by the American Bell Company in 1885 through a merger of the American Bell Telephone Company (founded 1877) and Western Electric, once a division of Western Union. AT&T provided long-distance telephone services within the United States and soon established divisions outside the country. In conjunction with its regional operating companies, AT&T monopolized local and long-distance telephone markets until the U.S. government forced it to divest its regional Bell operating companies in 1984. After World War II, AT&T participated in the civil, commercial, and defense space sectors through divisions such as Western Electric and Bell Telephone Laboratories (Bell Labs).

Artist's rendering of the Telstar *satellite. (Courtesy NASA/Glenn Research Center)*

During World War II, Bell Labs developed an electronic "predictor," an analog computer that coupled radar-tracking information with antiaircraft guns for automated targeting and firing. Based on this experience, after the war Bell Labs teamed with Douglas Aircraft to develop the automated Nike-Ajax antiaircraft missile system for the U.S. Army, deployed in 1954 by Western Electric, the manufacturing arm of AT&T. The same trio later assisted with the development of a Nike antiballistic missile system and a Nike antisatellite system, known as Program 505 Mudflap.

As early as 1955 AT&T identified orbiting satellites as a method to achieve profitable international communications as an alternative to expensive transoceanic cables. It began developing communications satellite technologies and ground stations to interconnect with the terrestrial telephone network. AT&T hoped that it would be able to create and operate a satellite communications system.

In the early 1960s the John F. Kennedy presidential administration and NASA were concerned that AT&T would monopolize the satellite communications industry as it did with the U.S. telephone industry and acted to prevent an AT&T monopoly. Believing that a satellite communications program should be a public-private relationship, the administration promoted, and Congress passed, the Satellite Communications Act of 1962 that created the for-profit, publicly owned Communications Satellite Corporation (Comsat). Comsat would be responsible for building and launching an operational communications satellite system. Nonetheless, in July 1962 NASA launched AT&T's experimental communications satellite *Telstar*, which was the first successful test of an active communications satellite, relaying telephone and television signals from the United States to Europe. Despite *Telstar*'s success, the Comsat bill banned AT&T from operating its own satellite communications system. AT&T and other telecommunications companies instead purchased shares of Comsat and connected the new Intelsat system into terrestrial communications networks. In the 1960s AT&T also provided systems analysis and engineering support to NASA for the Apollo program through its BellComm division.

David Hartzell

See also: Antiballistic Missile Systems, Antisatellite Systems, Intelsat, Program 505 Mudflap

Bibliography

AT&T. www.att.com.
Joan Lisa Bromberg, *NASA and the Space Industry* (1999).
Robert Buderi, *The Invention That Changed the World* (1996).
Clayton K. S. Chun, "'Nike-Zeus' Thunder and Lightning," *Quest* 10, no. 4 (2003): 40–47.
David J. Whalen, "Billion Dollar Technology: A Short Historical Overview of the Origins of Communications Satellite Technology, 1945–1956," in *Beyond the Ionosphere: Fifty Years of Satellites*, ed. Andrew Butrica (1997).

Americom. *See* SES.

Arianespace

Arianespace was created in 1980 as the world's first commercial space transportation company. Its primary role was to produce, launch, and market the Ariane family of launchers. In 1996 it took a share in a French/Russian joint venture launch company, Starsem, which commercialized the extremely reliable Russian Soyuz vehicle. In 2005 it added the Italian Vega, dedicated to placing small- to medium-sized satellites in orbit. By that year, Arianespace held more than 50 percent of the lucrative international market for satellites launched to geostationary transfer orbit. Its clients were worldwide and included many leading U.S. companies, which accounted for 43 of the first 200 satellites launched by Arianespace. By 1998 it had sold launch services worth about $6 billion. At the end of 2003, the firm, whose shareholders were the companies that built the rocket, had signed contracts for more than 30 launches with Ariane 5, together worth $2.7 billion.

A number of factors accounted for the company's success. The reliability of the Ariane rocket, with a successful first flight in December 1979, reassured customers and reduced the insurance costs of satellites that often cost more than $100 million. The equatorial launch base in Kourou, French Guiana, offered better performance than other launch sites, as it used Earth's maximum rotation velocity. It allowed for a rapid turnaround between flights and was particularly suitable for orbiting telecommunications satellites. The launcher proved cost effective; the research and development costs of the rocket family were borne by European Space Agency member states. This arrangement led to a charge by an independent U.S. launch provider in the 1980s that the firm benefited unfairly from government subsidies. The charge was rejected by the Ronald Reagan presidential administration.

The birth of Ariane was favored, at a time when France had difficulty convincing the other states that Europe needed free access to space with a new launcher succeeding the ill-fated Europa rocket, by the threat of an American embargo on the commercial use of the Franco-German Symphonie telecommunications satellite. An early significant decision was to offer dual launches, thus considerably reducing the launch cost per satellite. This policy was continued. The Ariane 4 family offered great flexibility, with six different versions in various combinations using two basic lateral booster types, one solid and one liquid. U.S. space policy and the *Challenger* accident inadvertently boosted the intrinsic merits of the company and its product. Early in 1980 the U.S. government decided that the reusable Shuttle should be the main American launch vehicle. Production lines for the highly successful Atlas and Delta rockets were shut down. The Shuttle, however, being human-rated, was ill-adapted to the demands of commercial clients and flew fewer missions than originally hoped. When the explosion of the *Challenger* orbiter in January 1986 grounded the fleet for more than two years, Arianespace gained a foothold in a market in which it was temporarily without a competitor and secured and sustained that foothold thereafter. Arianespace shareholder percentages as of 2006 were France (60.12), Germany (18.62), Italy (9.36), Belgium (3.15), Spain (2.01), Netherlands (1.82), Switzerland (2.51), Sweden (2.30), Norway (.10), Denmark (.01).

John Krige

See also: European Space Agency, France

Bibliography

Emmanuel Chadeau, ed., *L'Ambition Technologique: Naissance d'Ariane* (1995).
John Krige et al., *A History of the European Space Agency 1958–1987, Volume II* (2000).

Asia Cellular Satellite. *See* Regional Satellite Communications Operators.

AsiaSat. *See* Regional Satellite Communications Operators.

Astra. *See* SES.

BAE Systems

BAE Systems is an international company engaged in the development, delivery, and support of advanced defense and aerospace systems in the air, on land, at sea, and in space. As of 2006 the company designed, manufactured, and supported military aircraft, combat vehicles, surface ships, submarines, radar, avionics, communications, electronics, and guided weapon systems. It is a technology pioneer with a heritage stretching back to the early 1900s in defense systems through such heritage companies as Marconi, Hawker Siddeley, British Aircraft Corporation, and even centuries back with Royal Ordnance. As of 2005 BAE Systems had major operations across five continents and customers in some 130 countries. The company employed nearly 100,000 people and generated annual sales of approximately $25 billion through its wholly owned and joint-venture operations.

In 2005 BAE Systems was the world's fourth-largest defense contractor. It was formed in 1999 from the merger of British Aerospace (BAe) and Marconi Electronic Systems, the defense and space arm of the General Electric Company (GEC). This merger set the stage for BAE Systems to move into the U.S. defense market. BAE Systems grew to be a leading contractor to the U.S. Department of Defense, with prime positions on various U.S. programs, including the F-35 Joint Strike Fighter, the Bradley Fighting Vehicle, and the 155 mm Lightweight Howitzer.

BAE Systems heritage companies of BAe and Marconi were involved in space programs globally. The Marconi business included communications payloads, active microwave systems, onboard electronics equipment, imaging electronics, and prime contractor roles for *Skynet 2* and *Ariel 5* and *6*. By the twenty-first century, BAe participated in approximately 100 satellite projects. It was prime contractor for many of Europe's science satellites (including *Giotto*), developed Ariane's Spelda carrier, *Spacelab*'s pallets, the Skylark sounding rocket, and pursued its Horizontal Take-off and Landing design until funding from the European Space Agency (ESA) and the United Kingdom government became unavailable in 1989. In addition to work with the European community, BAe was involved in international pursuits, such as solar arrays for the *Hubble Space Telescope* through ESA. BAE Australia and its predecessors were involved in supporting and operating the Woomera Test Range.

In 1994 British Aerospace sold its space interests to Marconi Electronics. Matra Marconi formed a joint venture with European Aerospace and Defence Systems (EADS) in 2000 to form Astrium. After the BAe/Marconi merger formed BAE Systems, BAE was a 25 percent stakeholder in Astrium. In 2003 BAE Systems sold its stake in Astrium and effectively exited the European space arena. Within its defense division, though, it retained a rocket motors division that made satellite apogee engines and orbital maneuvering systems.

At the time of the BAe/Marconi merger, the new entity, BAE Systems began to pursue a strategy of expansion to the U.S. market. A number of U.S. acquisitions (including Tracor, Lockheed Martin's Aerospace Controls and Sanders businesses, DigitalNet, Alphatech, Mevatech, and United Defense) resulted in BAE Systems involvement in military and NASA programs within the United States. BAE Systems also operated one of only two foundries in the United States providing radiation-hardened components for spacecraft, such as the radiation-hardened computers built for the Mars Exploration Rovers.

Stephanie Lyons

See also: European Space Agency, Halley's Comet Exploration, United Kingdom

Bibliography

BAE Systems. www.baesystems.com.

Ball Aerospace and Technologies Corporation

Ball Aerospace and Technologies Corporation descended from a decision in the early 1950s by Ball Brothers Glass Manufacturing Company, a home-canning fruit jar firm based in Muncie, Indiana, to diversify its operations into research and technically oriented ventures. Assessing acquisition of other companies, R. Arthur Gaiser, Ball's research and development director, contacted the University of Colorado's Upper Air Laboratory (UAL) for an independent evaluation of equipment built by Control Cells, Inc., of Boulder, Colorado. UAL members David Stacey and James Jackson found the device deficient, but in the meantime interested Gaiser in the commercial potential of pointing control systems. Under U.S. Air Force contract since 1948, the UAL had constructed servomechanisms to keep spectrographs pointed at the Sun by compensating for rocket oscillation. Ball Brothers acquired Control Cells in December 1956, renamed it Ball Brothers Research Corporation (BBRC), and hired Stacey, Jackson, and several other UAL scientists and engineers to put the company on the "frontiers of space technology." Soon Ball found itself successfully producing and marketing its own pointing controls, spectrographs, and infrared scanners.

In the wake of the nation's new commitment to aerospace in the post-Sputnik era, BBRC acquired several major NASA contracts. In 1959 NASA funded Ball to build the first of seven Orbiting Solar Observatory satellites in the 1960s to determine solar radiation's potential threat to human spaceflight. In the early 1970s BBRC also built a white light coronagraph, along with three other major solar instruments to fly onboard *Skylab*.

From the 1970s Ball continued to grow as a NASA contractor, developing such key systems and components as the Corrective Optics Space Telescope Axial Replacement for the *Hubble Space Telescope*, the flight hardware and ground systems for *Deep Impact*'s collision with deep-space comet Tempel 1, and the advanced optical technology and lightweight mirror system for Hubble's replacement, the *James Webb Space Telescope*. Renamed Ball Aerospace and Technologies Corporation in August 1995, the company, with 3,000 employees and $653 million in 2004 sales, continued in the twenty-first century as an international leader in spacecraft design and construction.

David N. Spires

See also: *Hubble Space Telescope*; OSO, Orbiting Solar Observatory Satellites

Bibliography

Ball Aerospace and Technologies Corporation. www.ballaerospace.com.

David N. Spires, "Walter Orr Roberts and the Development of Boulder's Aerospace Community," *Quest* 6, no. 4 (1998): 5–14.

Boeing Company

The Boeing Company, a global leader in commercial airplane manufacturing and a major defense and space contractor in the twenty-first century was incorporated by William Boeing on 15 July 1916 as Pacific Aero Products Company. By 1918 the

The Boeing Company won the contract to build the manned Dyna-Soar (Dynamic Soaring) orbital spaceplane, which in this 1961 artist's conception clears the launch tower atop an Air Force Titan II launch vehicle. (Courtesy NASA/Marshall Space Flight Center)

company had been renamed the Boeing Airplane Company and was working on its first production order for the U.S. Navy. Boeing developed a variety of military and commercial passenger aircraft.

During World War II Boeing began developing guided missiles. The Ground-to-Air Pilotless Aircraft, the company's first missile, led to the postwar production of the Bomarc missile. During the Cold War, Boeing won the integration contract for the Minuteman missile program, which was the only intercontinental ballistic missile in service in the United States as of 2008. Experience and technology gained from missile development helped Boeing win the contract for the U.S. Air Force's crewed, reusable spaceplane, X-20 Dyna-Soar. The program was canceled in 1963 soon after construction had begun. In the 1960s the company provided overall systems integration for Apollo and loaned NASA 2,000 executives to coordinate activities. Boeing's Lunar Orbiters photographed the Moon so that a safe landing site could be selected, and the company's Lunar Roving Vehicle was used in the last three Apollo missions. Boeing built the first stage of the Saturn V, the S-1C rocket booster. The second and third stages were built by North American Aviation and Douglas Aircraft Company, later acquired by Boeing.

In 1971, due to Apollo's end and an aviation industry recession, Boeing reduced its workforce by more than 50,000. Boeing diversified into new markets: commercial computer products, irrigation, and housing projects management for the federal government, to name a few. The company's *Mariner 10* probe, designed to fly by Mercury and Venus, was launched in 1973. Boeing's missile experience continued with the production of air-launched cruise missiles and short-range attack missiles. In 1977 the country's first Airborne Warning and Control System flew on a 707 airframe.

In the late 1980s Boeing's military and space businesses were combined to form Boeing Defense and Space Group. In the 1990s through 2000, Boeing undertook a number of mergers and acquisitions. The 1996 merger with Rockwell International Corporation's aerospace and defense units, including Rocketdyne, led to the creation of the subsidiary, Boeing North American, Inc., and added the Space Shuttle orbiters and Space Shuttle Main Engine to the company assets. That same year Boeing and Lockheed Martin formed United Space Alliance as a joint venture to conduct all Space Shuttle operations. Boeing merged with McDonnell-Douglas Corporation in 1997, thus acquiring the Delta launcher and Phantom Works, the company's research and development arm. It purchased Hughes Electronics Corporation's space and communications business in January 2000, which became Boeing Satellite Systems. With this purchase came the 376, 601, and 702 communications satellites. Before the mergers and acquisitions, the company assets included the B-2 stealth bomber, Joint Strike Fighter, Inertial Upper Stage booster, the Sea Launch joint venture for launching commercial satellites, and the integration contract for the International Space Station.

In 2002 Boeing merged its space, defense, government, intelligence, and communications business units in a single business called Integrated Defense Systems (IDS). That year IDS, along with Science Applications International Corporation, was awarded the contract for lead system integration of the Army's Future Combat Systems. In 2004 the company's business unit, Connexion by Boeing, made its commercial debut, allowing airplane passengers to use high-speed Internet in flight using the global communications network the unit developed for mobile platforms.

Michael Barboza

See also: Apollo, Ballistic Missiles, Delta, *International Space Station*

Bibliography

Eugene E. Bauer, *Boeing in Peace and War* (1991).
Boeing Company. www.boeing.com.
Joan Lisa Bromberg, *NASA and the Space Industry* (1999).

Com Dev International

Com Dev International Ltd., based in Cambridge, Ontario, was one of Canada's largest space contractors. Founded in 1974 to build satellite components, Com Dev designed and built a variety of equipment from switches to batteries to satellite subsystems and hardware for satellite transmission of Internet signals. Major satellite contractors in various parts of the world used its products and subsystems in remote sensing, space science, military satellites, and communications satellites. Com Dev, which had revenues of about US$100 million in 2004, operated manufacturing facilities in Canada and the United Kingdom, and by 2005 its components had flown in more than 550 satellites.

Chris Gainor

See also: Canada

Bibliography
Lou Dobbs, *Space: The Next Business Frontier* (2001).

Comsat Corporation

Comsat Corporation, Communications Satellite Corporation, was established by the U.S. government in 1962 to develop the communications satellite sector. Comsat launched and operated the first synchronous-orbit commercial communications satellite, *Early Bird*, in 1965.

In the early 1960s concern arose from the John F. Kennedy presidential administration and NASA that aside from American Telephone and Telegraph (AT&T) Company, then the world's largest corporation, the private sector would lack sufficient resources needed to develop and operate communications satellites. The administration feared that AT&T, then developing *Telstar*, the first private telecommunications satellite, would monopolize the satellite communications sector. As an alternative, the administration favored a public-private relationship, and as a result Congress passed the Satellite Communications Act of 1962 that created Comsat, a for-profit, publicly owned corporation, making it responsible for building and managing a communications satellite system and to represent the United States in negotiations with other nations and to create an international consortium to operate the system.

In 1963 Comsat incorporated in the District of Columbia and issued $500 million in shares to the public. Communications carriers could acquire 50 percent of Comsat's shares and private individuals the other 50 percent. By legal intent, AT&T managed to acquire only 29 percent of total shares, ensuring that it could not control Comsat. Other telecommunications companies, including General Telephone and Electrics, International Telephone and Telegraph, and Radio Corporation of America (RCA), acquired the remaining shares.

Comsat contracted with Hughes Aircraft Company to build its first communications satellite, *Early Bird*, which NASA launched into geosynchronous orbit in April 1965. AT&T leased 100 of *Early Bird*'s 240 voice circuits, while NASA leased circuits to support the Apollo program. Comsat's first quarterly report issued in December 1965 listed revenues of $966,000.

Comsat acted as the U.S. signatory in the International Telecommunications Satellite Organization (Intelsat), established in late 1964 to own and operate the global satellite communications system. Comsat initially held 61 percent of Intelsat's voting membership, performed its technical studies, and managed its satellite design, launch, and operations. In 1973 Comsat's relationship was formalized into the Intelsat Management Systems Contractor, managing the introduction of the first four generations of Intelsat satellites. When the United States allowed private companies to launch and operate domestic communications satellites, Comsat teamed with AT&T to launch and operate the Comstar series, first launched in 1976. The PanAmSat Corporation successfully lobbied against Intelsat's government monopoly, and in 1984 the United States allowed other private companies to compete with Comsat-Intelsat for international satellite telephone service.

In 1999 Lockheed Martin acquired 49 percent of Comsat and completed full acquisition in 2001 for $2.7 billion, with approval from the U.S. Congress. Lockheed Martin owned approximately 24 percent of the privatized Intelsat and later sold the Comsat division to Intelsat in 2004 for $90 million.

David Hartzell

See also: Commercial Satellite Communications, Intelsat, Lockheed Martin Corporation

Bibliography

Intelsat. www.intelsat.com.
David J. Whalen, *The Origins of Satellite Communications 1945–1965* (2002).

DigitalGlobe. *See* Private Remote Sensing Corporations.

Douglas Aircraft Company

Douglas Aircraft Company was one of the largest U.S. manufacturers of aircraft and space launch vehicles before it merged with the McDonnell Company. Donald W. Douglas formed the Douglas Aircraft Company in 1921 in Santa Monica, California, and established it as the country's leading commercial and military airplane builder during the 1930s–40s. Douglas developed Thor, the first intermediate range ballistic missile, in the mid-1950s. It was used to launch the first U.S. weather, navigational, and surveillance satellites.

NASA selected Douglas in April 1959 as the prime contractor to build the Delta rocket as a successor to Thor. The first successful Delta flight occurred on 12 August 1960, and its high launch success rate made it one of the country's most reliable space boosters. NASA chose Douglas in April 1960 to build the third stage of the Saturn V rocket (the S-IVB) to take astronauts to the Moon. The S-IVB also served as the second stage of

Douglas-Built Thor/Delta Rocket Milestones

Date	Launch Vehicle	Significance
1 April 1960	Thor-Able	Launched first U.S. weather satellite, *Tiros 1*
13 April 1960	Thor-Able Star	Launched first U.S. navigational satellite, *Transit 1B*
18 August 1960	Thor-Agena A	Launched first successful U.S. surveillance satellite mission, *Discoverer 14*
7 March 1962	Delta	Launched first Orbiting Solar Observatory, *OSO 1*
10 July 1962	Delta	Launched first active communications satellite, *Telstar 1*
6 April 1965	Delta	Launched Communications Satellite Corporation's first satellite (*Early Bird*)

the Saturn IB rocket used during uncrewed Apollo launch tests and the first piloted Apollo mission, *Apollo 7*.

Douglas won the contract to build the U.S. Air Force's Manned Orbiting Laboratory (MOL) in August 1965, but Congress never fully funded the project. Secretary of Defense Melvin R. Laird canceled MOL in June 1969. The failure to secure new, lucrative aircraft contracts forced Douglas to merge with McDonnell in April 1967 to become the McDonnell-Douglas Corporation.

Tim Chamberlin

See also: Apollo, Ballistic Missiles, Delta, Manned Orbiting Laboratory, Saturn Launch Vehicles

Bibliography

Rene J. Francillon, *McDonnell Douglas Aircraft: Since 1920* (1990).
Wilbur H. Morrison, *Donald W. Douglas: A Heart with Wings* (1991).

EOSAT, Earth Observation Satellite Company

EOSAT, Earth Observation Satellite Company was established by the Hughes Aircraft Company and the Radio Corporation of America (RCA) and won the bid to operate the U.S. government Landsat program in 1985. As an attempt to reduce the size of the federal government and commercialize the Landsat program, the Ronald Reagan presidential administration promoted, and Congress passed, the Land Remote Sensing Commercialization Act of 1984 to remove the government from operation of Landsat and sales of its data. The Department of Commerce solicited proposals for the operation of the existing Landsat program and future missions. EOSAT won the contract, which included legal rights to the Earth science data. The government hoped that revenue generated from data sales would allow EOSAT to eventually manage and operate the Landsat program without subsidies.

After winning the contract, EOSAT was required to distribute non-enhanced Landsat data to the public on a nondiscriminatory basis. It could offer value-added Landsat products, but they had to be competitive to other value-added resellers. After the launch of *Landsat 5*, EOSAT significantly increased data prices, which made data inaccessible for many users. These price increases and competition from other remote sensing platforms, such as SPOT (Satellite Pour l'Observation de la Terre), caused Landsat sales, primarily to large corporations and a few government agencies, to decline.

Concern grew within the U.S. government and the United Nations regarding the lack of Landsat data availability to researchers, academic institutions, and developing nations. Many felt that the attempted commercialization of Landsat had failed. In 1992 the Land Remote Sensing Policy Act of 1992 reversed the commercialization decision and again made the government responsible for future Landsat missions. EOSAT maintained rights to the data from the previous Landsat missions. The act also required the government to negotiate with EOSAT to make the data it controlled more available to the public.

Space Imaging, a joint venture among Lockheed Martin, Raytheon, Mitsubishi, and Hyundai, acquired EOSAT in 1996. Between its establishment and acquisition, EOSAT became the largest U.S. supplier of remotely sensed data, selling data and value-added products from various Earth observation missions. Space Imaging launched the *Ikonos 1* spacecraft in 1999, the first commercial imaging satellite to provide 1 m resolution. Space Imaging returned control of *Landsat 4* and *5*, including all data rights, to the government by 2001.

David Hartzell

See also: Landsat; SPOT, Satellite Pour l'Observation de la Terre

Bibliography

EROS Data Center. http://eros.usgs.gov/.

Donald T. Lauer, "The Landsat Program: Its Origins, Evolution, and Impacts," *American Society for Photogrammetry and Remote Sensing* 63, no. 7 (July 1997): 831–38.

European Aeronautic Defence and Space Company (EADS)

European Aeronautic Defence and Space Company (EADS), an aerospace corporation that develops and markets aircraft (civil and military), missiles, space rockets, and related systems. It was created on 10 July 2000 by the merger of Aérospatiale-Matra of France, Dornier GmbH and DaimlerChrysler Aerospace AG of Germany, and Construcciones Aeronáuticas SA (CASA) of Spain. Incorporated in Amsterdam, the Netherlands, EADS has headquarters in Paris, France, and in Ottobrunn, Germany.

Consolidation of the European aerospace industry had been an ongoing concern since prior to World War II, starting out strictly within national boundaries, but by the 1980s occasionally crossing them. However, the mergers of large aerospace companies in the United States, such as Northrop and Grumman in 1993 and Lockheed and Martin Marietta in 1996 caused renewed concern among European leaders that a fragmented European aerospace industry would not survive the strengthened competition of the United States. This provided impetus to discussions among both government bodies and industry management regarding further cross-national consolidation.

Matra Espace, the space and telecommunications division of the Lagardère Group, merged with the United Kingdom's General Electric Company (GEC) Marconi Space Systems in 1989 to become Matra Marconi Space. In 1999 Aérospatiale of France merged with Matra Hautes Technologies, defense division of the French Lagardère Group of companies, to form Aérospatiale-Matra. Matra Marconi's space experience included a variety of communications, science, and remote sensing satellites, while Aérospatiale was prime contractor for the Ariane launcher. Matra Marconi Space merged with DaimlerChrysler Aerospace AG in May 2000 to create the Astrium Company, which after the EADS merger was owned 75 percent by EADS and 25 percent by BAE Systems.

Built by EADS Astrium for the European Space Agency, the Rosetta *probe was scheduled to fly by two asteroids en route to its rendezvous with comet 67P/Churyumov-Gerasimenko in May 2014. (Courtesy European Space Agency)*

Another element of Astrium was the space assets of the former British Aerospace Corporation, whose space assets derived from a prior merger of British Aircraft Corporation and Hawker Siddeley. GEC Marconi purchased British Aerospace Space Systems in 1994 and sold it back to BAE Systems (the successor to British Aerospace) in its purchase of Marconi Electronic Systems in 1999. When BAE Systems sold its European space assets to EADS in 2003, Astrium became wholly owned by EADS.

German industry followed a similar path, starting from many well-known aviation firms, such as Dornier, Focke-Wulf, Heinkel, Junkers, and Messerschmitt, eventually merging into a single state-approved company, Deutsche Aerospace AG (DASA). Dornier GmbH aerospace company had its origins in 1914 as an aircraft manufacturer. It worked on the first German research satellite *Azur* and other craft, such as *Ulysses* and *Cluster*. Junkers, Messerschmitt, and parts of Heinkel by 1968 had consolidated into Messerschmidt-Bölkow-Blohm (MBB), which built the *Highly Eccentric Orbiting Satellite* and *Exosat*, among others. Entwicklungsring Nord (ERNO), a joint venture of Focke-Wulf and Weserflug, formed in 1961, leading to a merger into Vereinigte Flugtechnische Werke (VFW) in 1964. MBB's experience eventually included the Helios solar probes, while ERNO-VFW, for a time combined with Fokker, was prime contractor for the Spacelab program. In 1980 rivals MBB and ERNO joined under the MBB name.

Daimler-Benz began its moves into aerospace in the early 1980s, acquiring a controlling interest in Dornier in 1985. Urged by the German government, in 1989 Daimler-Benz engineered the merger of Dornier, MBB, and other firms, creating DASA. In 1998 Daimler-Benz merged with the U.S. Chrysler Corporation and changed the name of its aerospace operations to DaimlerChrysler Aerospace AG.

The Spanish aerospace company, CASA, was founded in 1923 for the production of aircraft. Later it manufactured space products, including subsystems, such as thermal control systems, cable harnesses, antennas, and deployment mechanisms used on satellites, the Ariane launcher, and the *International Space Station* (*ISS*).

EADS Space in 2006 had several aerospace divisions. (1) EADS Astrium was responsible for satellites such as ESA's *Mars Express Orbiter*, launched in 2003, and *Venus Express*, launched in 2005. Galileo Industries subcontracted with Astrium for the development and construction of the Galileo navigational system, which had its first test satellite launch in December 2005. Astrium had manufacturing plants in France, Germany, Spain, and the United Kingdom. (2) EADS Space Transportation was the prime contractor for Ariane launchers and orbital systems, such as *Columbus* for the *ISS*. It had six sites located throughout France and Germany. (3) EADS Space Services provided satellite services, concentrating on military telecommunications and navigation, such as the Skynet 5 system.

In 2006 EADS employed 116,000 employees working at more than 70 production sites in Australia, Germany, France, Spain, the United Kingdom, and the United States with more than 30 customer service offices. Annual revenues were listed at more than €39.4 billion.

Joni Wilson

See also: Ariane, European Communications Satellites, European Space Agency, France, *Galileo*, Germany, *Hubble Space Telescope*, Italy, Skynet, United Kingdom

Bibliography

European Aeronautic Defence and Space Company, *On the Wings of Time: A Chronology of EADS* (2003).

Eutelsat

Eutelsat, European Telecommunications Satellite organization, the leading European satellite operator working with telecommunications, video broadcasting, networks, mobile communications, and broadband Internet, was created as a provisional intergovernmental company in 1977 with the initial purpose to complete the phone network in Europe. Between 1983 and 1988 the first generation of EUTELSAT 1 satellites was supplied by the European Space Agency, providing the way for satellite television in Europe. In the 1990s Eutelsat strengthened its European presence and pursued expansion in Africa, the American continents, and Asia with its EUTELSAT 2 satellites. After Eutelsat became a private company in July 2001, many of the founding telecommunications companies sold their Eutelsat interests in 2004. The privatized Eutelsat intended to operate worldwide as a leading provider of satellite communications.

EUTELSAT Satellites (Part 1 of 4)

Satellite Name	Launch Date	Manufacturer	Nation/Area Served	Purpose
EUTELSAT 1F1	16 June 1983	ESA	Europe	Retired 1996
EUTELSAT 1F2	4 August 1984	ESA	Europe	Retired 1993
EUTELSAT 1F3	12 September 1985	ESA	Europe	Lost during launch
EUTELSAT 1F4	16 September 1987	ESA	Europe	Retired 2002
EUTELSAT 1F5	21 July 1988	ESA	Europe	Retired 2000
EUTELSAT 2F1	30 August 1990	Aérospatiale (Alcatel Space Industries)	Europe	Retired 2003
EUTELSAT 2F2	15 January 1991	Aérospatiale (Alcatel Space Industries)	Europe	Retired 2005, telecommunications, business services, newsgathering, Internet, multimedia services
EUTELSAT 2F3	7 December 1991	Aérospatiale (Alcatel Space Industries)	Europe	Retired 2004
EUTELSAT 2F4	9 July 1992	Aérospatiale (Alcatel Space Industries)	Europe	Retired 2003
EUTELSAT 2F5	24 January 1994	Aérospatiale (Alcatel Space Industries)	Europe	Lost during launch
HOT BIRD 1	28 March 1995	Alcatel Space Industries	Western Europe	Inclined 2006, cable and satellite television broadcasting, radio
TELECOM 2C	6 December 1995	EADS Astrium	Europe	Acquired from France in 1999, replaced by *ATLANTIC BIRD 3*, television multimedia
TELECOM 2D	8 August 1996	Matra Marconi Space (EADS Astrium)	Northern France	Acquired from France in 1999, leased—television, radio, multimedia services

EUTELSAT Satellites (Part 2 of 4)

Satellite Name	Launch Date	Manufacturer	Nation/Area Served	Purpose
HOT BIRD 2 renamed *EUROBIRD 9*	21 November 1996	Matra Marconi Space (EADS Astrium)	Europe, north Africa, Asia, Middle East	Television, radio
HOT BIRD 3; renamed *EUROBIRD 4*	2 September 1997	Matra Marconi Space (EADS Astrium)	Europe, north Africa, Asia, Middle East	Removed from service October 2006 due to loss of power, television, radio
HOT BIRD 4; renamed *ATLANTIC BIRD 4*	27 February 1998	Matra Marconi Space (EADS Astrium)	Europe, north Africa, Middle East	Television, radio, Internet backbone, multimedia, telecommunications, data
W2	5 October 1998	Alcatel Space Industries	Europe, north Africa, southeast Africa, Middle East	Business services, television, radio
HOT BIRD 5; renamed *EUROBIRD 2*	9 October 1998	Matra Marconi Space (EADS Astrium)	Europe, north Africa, Middle East	Television, radio, broadband
W6	12 April 1999	Alcatel Space Industries	Europe, north Africa, the Sahara, Middle East, Gulf region	Business services, television, radio, telecommunications, multimedia services, Internet
TELSTAR 12	19 October 1999	Space Systems/Loral	Europe, American continents, Caribbean, and south Africa	Leased from Loral Skynet
SESAT 1	18 April 2000	NPO-PM, Alcatel Space	Western Europe, Siberia, sub-Saharan Africa, Asia, and others	Telecommunications, television, radio, Internet
W4	24 May 2000	Alcatel Space Industries	Africa, western Russia, and others	Television, radio

(Continued)

EUTELSAT Satellites (Part 3 of 4)

Satellite Name	Launch Date	Manufacturer	Nation/Area Served	Purpose
Express-A3	24 June 2000	NPO-PM	Europe, Africa, Middle East	Ku-band leased from Russian Satellite Communications Company
W1	6 September 2000	Matra Marconi Space (EADS Astrium)	Europe, north Africa, Middle East, central Asia, Atlantic islands	Business services, television, radio
EUROBIRD 1	8 March 2001	Alcatel Space Industries	Western and central Europe to the Canary Islands	Broadcast digital television, radio
ATLANTIC BIRD 2	25 September 2001	Alcatel Space Industries	American continents, Europe, Africa, western Asia	Television, radio, , broadband and multimedia services
ATLANTIC BIRD 3	5 July 2002	Alcatel Space Industries	American continents, Europe, Africa, western Asia, Middle East, Caribbean	Replaced *TELECOM 2C*, television, radio, Internet backbone, multimedia services, corporate networks
HOT BIRD 6	21 August 2002	Alcatel Space Industries	Europe, north Africa, Middle East	Television , radio,
ATLANTIC BIRD 1	28 August 2002	Alenia Aerospazio	American continents, Europe, north Africa, near Middle East	Replaced *TELECOM 2A*, television, radio, Internet
W5	20 November 2002	Alcatel Space	Western Europe, Middle East, Asia	Business services, television, radio, Internet, multimedia services
HOT BIRD 7	December 2002	EADS Astrium	Europe, north Africa, Middle East	Lost during launch, replaced by *HOT BIRD 7A*
EUROBIRD 3	27 September 2003	Boeing Satellite Systems	Europe, Turkey	Broadcast digital television, Internet
SESAT 2 (12 transponders) (*EXPRESS AM 22* has 12 transponders for Russia)	29 December 2003	NPO-PM, Alcatel Space	Europe, Africa, Middle East, central Asia	Twelve transponders leased from the Russian Satellite Communications Company; telecommunications, television, radio, Internet

EUTELSAT Satellites (Part 4 of 4)

Satellite Name	Launch Date	Manufacturer	Nation/Area Served	Purpose
W3A	16 March 2004	EADS Astrium	Europe, Middle East, sub-Saharan Africa	Telecommunications, business services, television, radio, Internet, multimedia services
HOT BIRD 7A	11 March 2006	Alcatel Alenia Space	Europe, north Africa, Middle East	Television broadcasting, replace *HOT BIRD 1*, and backup *HOT BIRD 2, 3, 4*, radio, interactive services
HOT BIRD 8	5 August 2006	EADS Astrium	Europe, north Africa, Middle East	Television and radio broadcasting, interactive services works with *HOT BIRD 7A*
HOT BIRD 9	2008		Europe, north Africa, Middle East	Digital and new high definition television channels
W2M	2008		Europe, Middle East, Africa, central Asia.	Television broadcasting to data networks and broadband
W2A	2009		Europe, Middle East, Africa, Indian Ocean islands	Video, broadband, telecommunications, and others
HOT BIRD 10	2009		Europe, Africa, Middle East	Cable and satellite broadcasting
W7	2009		Europe, Africa, Middle East, Russia	Television feeds, public/corporate telephony, and data services
KA-SAT	2010		Europe, Mediterranean Basin	Broadband services
W3B	2010		Europe, Africa, Middle East	Broadcast services

In 2007 Eutelsat operated 24 telecommunications geosynchronous orbit satellites, 19 fully owned. The satellites included ATLANTIC BIRD, EUROBIRD, EUTELSAT, Express, HOT BIRD, SESAT, Telecom, Telstar, and the W satellites. Coverage included more than 150 countries in Africa, the American continents, Asia, Europe, India, and the Middle East. Eutelsat offered 2,600 television channels, reaching 164 million households for cable or satellite television, and 1,100 radio stations. With headquarters in Paris, France, 529 people from 27 countries were employed. Revenues in 2007 were €829 million.

Joni Wilson

See also: Direct Broadcast Television Satellites, European Communications Satellites, European Space Agency

Bibliography

Amendments to the Convention and the Operating Agreement of the European Telecommunications Satellite Organization (1995).

Eutelsat. www.eutelsat.com.

General Dynamics Corporation

General Dynamics Corporation (GD) began in February 1899 as Electric Boat Corporation. Its fortunes as a defense contractor followed defense-spending cycles throughout its history. The name was officially changed to General Dynamics in 1952. Following a diversification strategy, GD entered the space arena by purchasing Convair from the Atlas Group, becoming a division of GD in 1954. Convair operated as an independent company under the GD corporate umbrella.

By 1954 Convair received a U.S. Air Force contract to develop and manufacture the Atlas intercontinental ballistic missile. This early project led to 40 years of involvement in the space program as Atlas rockets became invaluable launch vehicles, used for NASA's Mercury program and a variety of commercial and military missions. General Dynamics began commercial sales of Atlas launches in the late 1980s after the *Challenger* accident. The Centaur upper stage, developed in the early 1960s, remained popular in combination with either Atlas or Martin Marietta's Titan boosters. Efforts to modify Centaur for Space Shuttle use in the 1980s came to naught due to safety concerns regarding Centaur's liquid hydrogen fuel.

By 1985 GD formed the Space System Division from Convair. Facing shrinking defense sales at the end of the Cold War in the early 1990s, GD diversified out of the space and defense sectors by selling and restructuring many of its business units. Most notably, Lockheed purchased its Fort Worth Division for $3 billion in March 1993 and Martin Marietta bought its Space Systems Division in 1994 for $209 million. By 1995 the company was again in solid financial shape and began to diversify. It acquired shipbuilding, defense, and computer companies. In 2001 GD bought its way back into the space business by purchasing Motorola's Information Systems Group.

Durand Johnson

See also: Atlas, Centaur

Bibliography
Roger Franklin, *The Defender: The Story of General Dynamics* (1986).
Jacob Goodwin, *Brotherhood of Arms* (1985).
John Wegg, *General Dynamics Aircraft and Their Predecessors* (1990).
Bill Yenne, *Into the Sunset: The Convair Story* (1995).

GeoEye. *See* Private Remote Sensing Corporations.

Grumman Corporation

Grumman Corporation, a major U.S. supplier of civilian and military aircraft, built the Lunar Module (LM) that carried astronauts to the surface of the Moon during Project Apollo.

Leroy R. Grumman, along with four investors, formed the Grumman Aircraft Engineering Corporation in December 1929 in Long Island, New York. By 1937 the company expanded and moved its operations to Bethpage, New York. During World War II the U.S. Navy awarded Grumman contracts worth hundreds of millions of dollars to build several aircraft. The company built military jet aircraft and developed the Gulfstream series of corporate jets in the 1950s before it pursued space projects.

NASA selected Grumman as the prime contractor for the Orbiting Astronomical Observatory (OAO) program in October 1960, leading to four Grumman-built OAO spacecraft that operated in space from 1966 to 1974. Grumman was contracted to build the High Energy Astronomy Observatory, but Congress terminated the project in favor of Skylab. In November 1962 NASA selected Grumman as the prime contractor to build the LM. Ten LMs flew in space from 1968 to 1972. Grumman served as a subcontractor for NASA's Space Shuttle program and constructed the orbiters' main wing assemblies and the vertical stabilizer for the orbiter *Endeavour*. The company changed its name to Grumman Corporation in July 1969. In 1994 Northrop Corporation acquired Grumman for approximately $2.17 billion.

Tim Chamberlin

See also: Apollo

Bibliography
Thomas J. Kelly, *Moon Lander: How We Developed the Apollo Lunar Module* (2001).
Northrop Grumman. www.northropgrumman.com.

Hughes Aircraft Company

Hughes Aircraft Company became one of the largest manufacturers of communications satellites and other space systems. Howard Hughes started an aircraft division of the Hughes Tool Company, which in 1935 became the Hughes Aircraft Company, based in Culver City, California. It developed many aircraft innovations, including flush rivets and retractable landing gear. At the end of World War II, Hughes Aircraft was the largest manufacturer in southern California with 80,000 employees.

In 1948 Hughes formed the Hughes Aerospace Group. Hughes engineers Simon Ramo and Dean Wooldridge developed the MA-1, the first electronic fire control system to combine radar signals directly with analog computer systems. It was used on interceptor aircraft and the first air-to-air missiles—the AIM-4 Falcon. Its success made Hughes Aerospace Group profitable, and Ramo and Wooldridge resigned in 1953 to found Ramo-Wooldridge Corporation, which later became TRW.

In 1961 the Hughes Space and Communications Company was formed, bringing together Hughes Space and Communications Group, the Hughes Space Systems Division, and Hughes Research Laboratories. In 1963 Hughes Space and Communications Company built the NASA-funded *Syncom*, the first geosynchronous communications satellite. It followed this in 1966 with *Applications Technology Satellite 1*, the first geosynchronous weather satellite. The company built the Surveyor landers in support of Project Apollo. By 1995 it was the world's largest supplier of communications satellites, with 40 percent of the market.

General Motors (GM) acquired Hughes in 1985 and in the next 15 years sold off divisions of the company. In 1997 the defense business of Hughes electronics, which operated the remaining aircraft and missile business, merged with Raytheon. Hughes Space and Communications was broken up in 2000 when Boeing bought the El Segundo, California, plant that built telecommunications satellites, becoming Boeing Satellite Systems. The other, more profitable, part of Hughes Space and Communications became DIRECTV.

In 2005 three companies used the Hughes name: the Howard Hughes Corporation, a Nevada real estate development firm; Hughes Research Laboratories; and Hughes Network Systems. Hughes Research Laboratories was jointly owned by GM, Boeing, and Raytheon, working on advanced developments in electronics, information and systems sciences, materials, sensors, and photonics. Hughes Network Systems traced its heritage to the telecommunications satellite products developed by Hughes Space and Communications. It was a leading international provider of broadband satellite network solutions. Hughes Network Systems pioneered the development of high-speed satellite networking and Internet access products and services, which marketed globally under the DIRECWAY® brands.

Paul Woodmansee

See also: Applications Technology Satellites, Commercial Satellite Communications, Direct Broadcast Satellites, Surveyor, Syncom

Bibliography

Boeing Company. www.boeing.com.
Hughes Network Systems. www.hughes.com.
Hughes Research Laboratory. www.hrl.com.

Iridium

Iridium was a satellite-based wireless telecommunications network conceived by Motorola in 1987, early in the cell phone communication era. Its idea was to place a constellation of satellites in low Earth orbit so that anyone on Earth could talk anywhere on a wireless phone.

The original constellation was to have 77 satellites, hence the name Iridium after the element with 77 protons. Later when redesigned to a constellation of 66 satellites, Motorola did not change the name to Dysprosium, the element with 66 protons, as the root of that element means "bad approach" in Latin.

Investing about $5 billion to develop, manufacture, and launch the constellation, Motorola's 66-satellite constellation was in orbit by 1998. In August 1999 the company that was spun off to run the operation, Iridium Limited Liability Corporation (LLC), went bankrupt. Motorola executives had drastically underestimated how inexpensive and easy it was to build many cell phone towers, which unlike satellites, could be upgraded with better equipment at any time. The high costs of the Iridium phone (initially about $3,000 each) and the connection time (about $7 per minute) meant that only wealthy clients in extremely remote locations where no cell phone would operate would need Iridium.

In December 2000 a group of investors formed Iridium Satellite LLC and bought the network for a mere $25 million. The new Iridium could operate at a profit, as it did not have the debt of the older company, while it inherited a number of worldwide users. These included the U.S. Department of Defense, which leased the network for $36 million per year. Connection costs in 2007 were $1.25 per minute.

The Iridium satellite network consisted of 66 operational satellites and 13 orbiting spares, orbiting at 485 miles above Earth in six orbital planes, with 11 operational satellites and at least one reserve in each plane. Each satellite was constantly linked to four others, and messages were sent from satellite to satellite until they reach a subscriber phone. There were two ground link stations: a primarily civilian gateway in Tempe, Arizona, and a military one in Oahu, Hawaii. Boeing conducted satellite operations from Leesburg, Virginia, under contract from Iridium.

Paul Woodmansee

See also: Military Communications

Bibliography

Iridium articles. www.space.com.
Iridium. www.iridium.com.

Lockheed Corporation

Lockheed Corporation began as an aircraft developer in 1912 and entered the space program when the U.S. Air Force (USAF) long-range ballistic missile development proceeded in 1953–54. Lockheed received a USAF contract for the design and construction of the X-17 reentry test vehicles in January 1955 and began developing the Polaris fleet ballistic missile for the Navy one year later. The Poseidon and Trident ballistic missiles followed in later decades. In 1958 the company built seven more X-17 missiles for Project Argus, which tested the effects of nuclear explosions outside the atmosphere.

The development of the Agena upper-stage rocket began in 1956, with the first launch in 1959. This rocket became a versatile USAF space vehicle. One version served as a target for docking experiments during the Gemini program. Last flown in 1987, Agena's record 362 launches were achieved with a success rate exceeding 90 percent and a peak launch rate of 41 launches in one year.

Another version of Agena was used for the Weapon System 117L program, the first U.S. military satellite program. In March 1956 the USAF made Lockheed the prime contractor for the program, which in 1959 split into separate programs for the Corona film-return reconnaissance satellite, the Missile Defense Alarm System satellites, and the Samos radio data-return reconnaissance satellites. Orbited in 1960, Corona flew more than 140 missions through 1972. In January 1957 Lockheed established its Space Systems Division in Palo Alto, California; the division later moved to Sunnyvale.

Early in NASA's Space Shuttle program, Lockheed began manufacturing tiles for the Shuttle's thermal protection system. The company then won the contract to manage ground processing of the Space Shuttle fleet at Kennedy Space Center. It also participated in the activation of Vandenberg Air Force Base for Shuttle operations.

Much of Lockheed's success was driven by production of ballistic missiles, satellites such as Corona and Milstar, and aircraft like the SR-71 Blackbird and F-117A Stealth Fighter. In 1990 the Lockheed-built *Hubble Space Telescope* was delivered to orbit. The "merger of equals" between Lockheed and Martin Marietta formed Lockheed Martin Corporation in March 1995.

Durand Johnson

See also: Corona, High-Altitude Nuclear Tests, Missile Defense Alarm System

Bibliography

Roy A. Anderson, *A Look at Lockheed* (1983).
Walter J. Boyne, *Beyond the Horizon* (1998).
Lockheed Martin. www.lockheedmartin.com.

Lockheed Martin Corporation

Lockheed Martin Corporation (LM) was formed in 1995 from the merger of Lockheed and Martin Marietta and became one of the world's largest defense firms. The end of the Cold War marked a decline in U.S. government defense spending. At a Pentagon briefing in 1993, contractors were told about the downsizing of the defense industry and encouraged consolidation. Both Lockheed and Martin Marietta pursued strategies of acquiring other defense firms to improve their defense market shares and gain

An artist's rendition of a Lockheed Martin–built Space-Based Infrared System Geostationary Earth Orbit satellite. (Courtesy Lockheed Martin)

technical and financial strength. Because their businesses were largely complementary, a merger made sense.

LM continued to acquire other companies and to pursue other cooperative agreements. In 1999 the company bought 49 percent of Communications Satellite Corporation (Comsat) for $2.7 billion and acquired a 30 percent stake in Asia Cellular Service. It created a joint venture with Boeing to create United Space Alliance to service the Space Shuttle in 1995, and in 2005 the two companies formed another joint venture, United Launch Alliance, to provide launch services to the Department of Defense. LM also created a joint venture with Russia's Khrunichev, forming International Launch Services to market Proton and Atlas launchers. LM contributed substantial capital to the X-33 project as part of an agreement between the company and NASA and was awarded the Joint Strike Fighter contract worth an estimated $200 billion before exports. Before the merger, the two companies had major projects, which included Titan, Pershing, and Trident ballistic missiles and Centaur upper stages. LM developed and manufactured *Genesis*, *Gravity Probe B*, portions of the *International Space Station*, *Lunar Prospector*, *Mars Global Surveyor*, *Mars Odyssey*, *Mars Reconnaissance Orbiter*, *Spitzer Space Telescope*, *Stardust*, *Terrestrial Planet Finder*, a number of successful ballistic missiles, and many satellites that provided location data and communications.

By 2005 LM was one of the largest technology companies in the world with more than 126,000 employees. It relied heavily on defense contracting and space enterprises, but diversified to encompass a wide variety of technology services. As of

2005 LM was comprised of 17 major "heritage" companies including General Electric Aerospace, Ford Aerospace, Goodyear Aerospace, Sanders, Radio Corporation of America, International Business Machines Federal Systems, and Unisys Defense. Counting smaller firms, more than 60 were incorporated into the company.

Duke Johnson

See also: Lockheed Corporation, Martin Marietta Corporation

Bibliography

Walter J. Boyne, *Beyond the Horizons: The Lockheed Story* (1998).
Lockheed Martin. www.lockheedmartin.com.

Malaysia East Asia Satellite. *See* Regional Satellite Communications Operators.

Martin Marietta Corporation

Martin Marietta Corporation (MMC) descended from the Glenn L. Martin Company, originally founded in 1917. The company specialized in building bombers through World War II, after which its focus shifted to missiles and space.

Martin's entrée into space began with the Viking sounding rocket built under contract to the Naval Research Laboratory (NRL) starting in 1945. Viking evolved into the Vanguard launcher, which Martin built for NRL, first successfully launched in 1958. However, Martin's Titan series of intercontinental ballistic missiles (ICBM) and space launchers became a primary source of revenue for decades, following the U.S. Air Force's (USAF) selection of Martin in 1955 to build a parallel ICBM to General Dynamics' Atlas. Titan evolved from the Titan I and Titan II ICBMs, to the heavier-lift Titan III launcher developed in the 1960s, to the even larger Titan 34, and finally to the massive Titan IV. NASA used modified Titan II launchers for its Gemini project and the USAF funded the Titan IV to provide an alternate expendable heavy-lift launcher to the Space Shuttle in the late 1980s.

Diversifying to stabilize its revenues, Martin merged with building and construction materials company American-Marietta in 1961 to form MMC. Notable accomplishments include contributions to *Skylab*, the Viking Mars lander, Pershing II ballistic missiles in the late 1970s, the Space Shuttle external tank starting in the late 1970s, NASA's manned maneuvering unit, and winning a Federal Aviation Administration contract to modernize the nation's air traffic control system in 1984. From 1988 to 2000, MMC built the Lacrosse radar imaging satellites for the National Reconnaissance Office. In 1990 the MMC-built *Magellan* spacecraft began mapping Venus.

After the end of the Cold War, MMC President Norm Augustine strengthened its position in the space sector by buying communications satellite manufacturer General Electric Aerospace in 1993 and General Dynamics Space Systems Division, including its Atlas launchers and Centaur upper stages throughout the next year. In 1995 MMC merged with Lockheed to become Lockheed Martin Corporation.

Durand Johnson

See also: *Magellan*, Titan, Vanguard, Viking

Bibliography

Peter F. Hartz, *Merger* (1985).

William B. Harwood, *Raise Heaven and Earth: The Story of Martin Marietta People and Their Pioneering Achievements* (1993).

Lockheed Martin.www.lockheedmartin.com.

Matra Marconi Space

Matra Marconi Space (MMS), a satellite manufacturer, was established in 1990 as a joint venture through a merger of Matra Espace (France), the space and telecommunications division of the Lagardère Group of companies, and General Electric Company (GEC) Marconi Space Systems (United Kingdom or UK). This combining of forces made MMS Europe's leading space contractor.

Matra (Mécanique Aviation Traction) was founded in 1945 to build aircraft. By the early 1960s Matra worked with CNES (French Space Agency) as a satellite prime contractor. In 1986 Matra's space activities were grouped into a subsidiary Matra Espace. Matra Hautes Technologies (MHT) was a separate division, but both became subsidiaries of the Lagardère Group. By 1995 MHT controlled the Lagardère Group's 50 percent interest in MMS. MHT merged with Aérospatiale in 1999 to form Aérospatiale-Matra, which in turn became one of the founding members of European Aeronautics and Defence and Space Company (EADS) in July 2000.

The Marconi Company Ltd. was founded in 1897 as the Wireless Telegraph and Signal Company. It was acquired by the UK's English Electric in 1946. When English Electric and GEC merged in 1968, Marconi became the defense division of GEC, and in 1987 was renamed GEC Marconi, with its space interests concentrated in Marconi Space Systems.

In 1994 MMS purchased British Aerospace Space Systems and Ferranti Satcomms. MMS had after this time two sites in France (Toulouse and Vélizy) and three sites in the United Kingdom (Portsmouth, Stevenage, and Bristol) with about 4,250 employees. Sales in 1999 were $1.4 billion. In that year GEC sold Marconi Electronic Systems, including its space interests in MMS, to British Aerospace, the latter changing its name to BAE Systems as a result. In November 1999 GEC was renamed Marconi Public Limited Company, but had no relationship with MMS.

MMS worked with many nations and companies in the production of various satellites and related space needs. Matra was involved primarily with scientific, Earth observation, and telecommunications satellites for CNES, the European Space Agency, and commercial operators, including *Hipparcos*, SPOT (Satellite

Matra Marconi Space Partial List of Projects (Part 1 of 2)

Satellite or Project	Purpose	Prime Contractor	Nation/ Contractor	Launch Date
Asterix A1	Payload for Diamant launcher	Matra	CNES	1965
Ariel 3	Scientific	BAC(BAe)/ Marconi	United Kingdom	1967
ESRO 2 (2 satellites)	Scientific	Hawker Siddeley (BAe)	ESRO	1967–1968
Skynet 2 (2)	Military communications	Marconi	United Kingdom	1974
TD 1A, Thor-Delta	Scientific	Matra	ESRO	1972
GEOS (2)	Scientific	BAC (BAe)	ESA	1977, 1978
OTS, Orbital Test Satellite (2)	Communications	Hawker Siddeley (BAe)	ESA	1977, 1978
Marots/Marecs (3) Maritime Orbital Test Satellite/ Maritime European Communications Satellite	Maritime communications	BAe	ESA	1981, 1982, 1984
ECS, European Communications Satellite (4)	Communications	BAe	ESA	1983–1988
Giotto	Scientific	BAe	ESA	1985
SPOT 1–5, Satellite Pour l'Observation de la Terre	Civil observation, remote sensing	Matra, MMS, Astrium	CNES	1986, 1990, 1993, 1998, 2002
Telecom 1 (3)	Communications	Matra, MMS	France	1984–1988
Skynet 4A, B, C	Military communications	BAe	United Kingdom	1988–1990
Olympus	Communications	BAe	ESA	1989
Hipparcos	Astrometry	Matra, MMS	ESA	1989
Eurostar 2000 series (20)	Commercial communication	MMS, Astrium	China, Spain, France, Egypt, Argentina	1991–
SOHO, Solar Heliospheric Observatory	Scientific	MMS	ESA	1995
Helios 1A, 1B	Military reconnaissance	MMS	CNES	1995, 1999
Skynet 4D, E	Military communications	MMS	United Kingdom	1998, 1999

Matra Marconi Space Partial List of Projects (Part 2 of 2)

Satellite or Project	Purpose	Prime Contractor	Nation/ Contractor	Launch Date
Skynet 4F	Military communications	Astrium	United Kingdom	2001
ENVISAT, Environmental Satellite	Environmental	MMS, Astrium	ESA	2002
ROCSAT 2, Republic of China Satellite	Earth observation	MMS, Astrium	Taiwan	2004
MetOp A, Meteorological Operational	Meteorological	MMS, Astrium	ESA/ EUMETSAT	2006

Pour l'Observation de la Terre), and *Telecom 1* and *2*. Marconi built spacecraft subsystems and the *Skynet 2* military telecommunications spacecraft before the MMS merger. The former British Aerospace had specialized in scientific and telecommunications satellites, such as *Giotto* and *Orbital Test Satellite*. In May 2000 MMS merged with Germany's DaimlerChrysler Aerospace to become the Astrium Company. When DaimlerChrysler Aerospace participated in the merger with Spain's Construcciones Aeronáuticas Sociedad Anónima and Aérospatiale-Matra that created EADS, Astrium became owned 75 percent by EADS and 25 percent by BAE Systems, responsible for satellite systems, space equipment, and associated ground infrastructure.

Joni Wilson

See also: European Communications Satellites; Helios; Skynet; *SOHO, Solar and Heliospheric Observatory*; SPOT, Satellite Pour l'Observation de la Terre

Bibliography

EADS. www.eads.com.
Lagardère. www.lagardere.com.

MacDonald Dettwiler and Associates Ltd.

MacDonald Dettwiler and Associates Ltd. (MDA) is an information services company based in Vancouver, British Columbia, which became one of Canada's foremost space contractors, with 2006 revenues of US$900 million. The company was founded in 1969 by John S. MacDonald and Werner Dettwiler and won a contract from the Canadian government in 1974 to build the world's first fully transportable Earth observation satellite ground system. In the 1970s–80s, MDA established itself in remote sensing by providing data management software and equipment for remote sensing by satellite and aircraft. MDA became the world's top provider of civilian Earth observation data acquisition centers. MDA widened its work to encompass information management for remote sensing data, land registries, and other information management applications.

In 1999 MDA purchased Radarsat International, later MDA Geospatial Services International, which markets data captured by Radarsat and other remote sensing satellites. In 1998 the Canadian Space Agency selected MDA to build and operate *Radarsat 2*, which was launched in 2007. MDA acquired the space robotics division of Spar Aerospace in 1998 and renamed it MD Robotics. This division manufactured and maintained the Space Shuttle Remote Manipulator System, or Canadarm, and the mobile servicing system for the *International Space Station* and in 2005 was developing other robotic systems. In 1995 MDA was acquired by Orbital Sciences Corporation, but Orbital divested itself of MDA in 2001. In 2008 MDA announced that it was selling its space assets to Alliant Techsystems Inc. (ATK), but after a political controversy the Canadian government blocked the sale.

Chris Gainor

See also: Canada, *International Space Station*, Radarsat

Bibliography

Chris Gainor, "The Chapman Report and the Development of Canada's Space Program," *Quest* 10, no. 4 (2003): 3–19.

McDonnell Company

McDonnell Company built the Mercury and Gemini capsules for NASA's human spaceflight program. James S. McDonnell started the McDonnell Aircraft Corporation in July 1939, which primarily served as an aircraft parts manufacturer during World War II. Based in St. Louis, Missouri, McDonnell became a pioneer, building jet aircraft in the 1940s–50s, including the U.S. Navy's F-4 Phantom fighter, before expanding its business to include spacecraft.

In January 1959 NASA selected McDonnell as the prime contractor to develop the Mercury capsule. McDonnell built 20 single-seat capsules at a cost of $49.3 million. Alan B. Shepard became the first American to fly in space aboard the *Freedom 7* capsule on 5 May 1961. Five more piloted Mercury capsules flew into space from July 1961 to May 1963. McDonnell agreed to build the Gemini spacecraft, which were modifications of the Mercury capsule, in December 1961. It built a total of 12 Gemini capsules at a cost of $696.1 million. The much larger, more complex Gemini capsule accommodated two astronauts for long-duration and rendezvous missions and was adapted for docking and extravehicular activity. NASA flew 10 piloted Gemini capsules from March 1965 to November 1966. McDonnell was also to supply modified Gemini capsules for the U.S. Air Force's Manned Orbiting Laboratory, but the program was canceled in June 1969.

McDonnell's name was shortened to McDonnell Company in 1966, to reflect its various business interests, and changed again after its merger with the Douglas Aircraft Company in April 1967 to become the McDonnell-Douglas Corporation.

Tim Chamberlin

See also: Gemini, Manned Orbiting Laboratory, Mercury

Bibliography

Sanford N. McDonnell, *This Is Old Mac Calling All the Team: The Story of James S. McDonnell and McDonnell Douglas* (1999).

Raymond F. Pisney, *James S. McDonnell and His Company: A Vision of Flight and Space* (1981).

McDonnell-Douglas Corporation

McDonnell-Douglas Corporation, major aerospace manufacturer during the late twentieth century. In April 1967 Douglas Aircraft Company merged with McDonnell Aircraft Corporation to form McDonnell-Douglas Corporation (MDC). Douglas brought experience with Thor-Delta rockets and Apollo third stage, while McDonnell's space expertise revolved around the Mercury and Gemini crewed space capsules. MDC headquarters was in St. Louis, Missouri, and corporate aerospace operations was in Huntington Beach, California.

NASA awarded MDC a contract in August 1969 to convert the third stage of a Saturn V rocket into *Skylab*, launched in May 1973. In the 1970s NASA also awarded

Technicians working in the McDonnell White Room on the Mercury spacecraft, 1960. (Courtesy NASA)

MDC a contract to develop the Orbital Maneuvering System, to maneuver the Space Shuttle in orbit, and the Payload Assist Module, which launched satellites from the orbiter's payload bay. Additionally, MDC attempted to demonstrate the commercial advantages of space exploration by teaming with Ortho Pharmaceuticals to develop the Continuous Flow Electrophoresis device to process and manufacture drugs in a microgravity environment. Astronauts successfully tested it on the Space Shuttle in 1982–83. In 1987 NASA awarded MDC Work Package 2 to develop the integrated truss structure; thermal control systems; Space Shuttle docking systems; airlocks; and guidance, navigation, and control systems for Space Station Freedom. In the later *International Space Station*, NASA subcontracted with MDC to design a pressurized mating adapter connecting the station's U.S. and Russian modules.

MDC continued to develop and manufacture Delta expendable launch vehicles, which had been in existence since 1960. In the mid-1970s MDC added solid rocket boosters to boost heavier payloads into orbit. However, NASA designed the Space Shuttle to supplant all expendable U.S. launchers. NASA stopped funding to Delta, and MDC shut down the Delta assembly line in the mid 1980s. A small company, Transpace Carriers, Inc., attempted to purchase rights to commercialize Delta, but its attempt was foiled by government pro-Shuttle policies, and then, after the 1986 *Challenger* accident, by MDC as it restarted Delta production and received Department of Defense contracts to improve the launcher.

In 1989 the Strategic Defense Initiative Organization awarded MDC a contract to build the reusable DC-X experimental single-stage launch demonstrator, which successfully flew in August 1993. The company was purchased in 1997 by the Boeing Company. Before the merger MDC had approximately 63,873 employees and reported revenue of nearly $2.2 billion from its aerospace operations.

Kevin M. Brady

See also: Delta, *International Space Station*, Space Shuttle, Space Station Freedom

Bibliography

Boeing Company. www.boeing.com.
Joan Lisa Bromberg, *NASA and the Space Industry* (1999).
G. Harry Stine, *Halfway to Anywhere: Achieving America's Destiny in Space* (1996).
Bill Yenne, *McDonnell Douglas: A Tale of Two Giants* (1985).

Mitsubishi Heavy Industries

Mitsubishi Heavy Industries, Ltd. (MHI), has played the major role in the development of Japanese launchers since its early history. In 1934 MHI was formed and was the largest in heavy industry, but because of decentralization of economic power, the company was divided into three companies in 1950. MHI was reestablished in 1964 as the merger of those three companies. By the early twenty-first century, MHI's business field extended to various fields, such as aerospace, shipbuilding, nuclear energy systems, machinery, and power systems.

In the space industry in Japan, MHI has participated in developing and manufacturing various rocket engines. In 1954 MHI started research on liquid rocket engines and by the mid-1970s, both solid- and liquid-fuel launchers—SA, SB, SC, LS-A, and LS-C—were developed. MHI's next major work was the development of engines for N-1 and N-2 rockets in the 1970s–80s, and these were under contract to the National Space Development Agency of Japan. Both N-1 and N-2 were adaptations of U.S. Thor-Delta launcher technology, and these two rockets led to the development of the first Japanese liquid oxygen/hydrogen rocket engine, LE-5, utilized on H-1 rockets for which MHI developed engines.

As the prime contractor and system integrator, MHI played a crucial role in the development and launch of H-2 rockets in the late 1980s–90s. The rockets were capable of delivering 4,000 kg of payload weight into geosynchronous transfer orbit, and H-2 rockets opened the new era in the history of Japanese space development. Since 1995 MHI has developed rocket engines, LE-5B and LE-7A, for H-2A rockets, Japan's main intermediate-size launcher. Since 1995 MHI also has participated in developing the H-2 Transfer Vehicle, a robotic transfer vehicle for the *International Space Station* (*ISS*), to be launched by H-2A. As partners, MHI and Boeing Company have worked on research and development of a next-generation rocket propulsion system, MB-XX. Because of the privatization of launch services, Japan Aerospace Exploration Agency designated MHI as the launch service provider of standard-type H-2A launcher in 2002. In 2003 H-2A Launch Services was offered and managed by MHI. MHI also was one of the major contractors for the Japanese Experiment Module, *Kibo* (hope), for the *ISS*.

Makoto Inomata

See also: Japan

Bibliography

H-2A Launch Services. http://h2a.mhi.co.jp/en/index.html.
Japan Aerospace Exploration Agency. www.jaxa.jp/index_e.html.
MHI. www.mhi.co.jp/en/.

North American Rockwell Corporation

North American Rockwell Corporation (NAR) was formed in 1966 with the merger of North American Aviation (NAA) and Rockwell Standard Corporation. Rockwell Standard began in 1919 and built its reputation in the automotive industry. NAA was incorporated in December 1928 as a holding company for a number of fledgling aviation-related companies, becoming an aircraft manufacturer by the mid 1930s. NAA established its reputation as a small, single-engine airplane provider for the military, building more military aircraft than any other manufacturer. With the end of World War II, NAA began to incorporate new technology, developing the first U.S. swept-wing fighter, F-86 Sabre, in 1947 with the aid of German data.

In 1954 NAA started work on the F-107, a Mach 2 U.S. Air Force (USAF) fighter, and also commenced work on the X-15 rocket plane to conduct research in the hypersonic and spaceflight regimes. The X-15 led to the development of new aircraft systems and structures technologies, which could withstand the high frictional heat, and to improved analysis of human systems. NAA's Space Division was formed in 1960. It evolved from

NAA's Missiles Division, which used captured German V-2 rockets to develop the Navaho cruise missile. It also led to the creation of NAA's Rocketdyne division, among others, which in the 1950s developed engines for the Thor and Atlas ballistic missiles for the USAF. Rocketdyne also built the Redstone rocket's propulsion system.

On 11 September 1961 NASA awarded NAA the contract to build the S-2 rocket engines, the second stage of the Saturn V Moon rocket and soon thereafter the Apollo spacecraft. Throughout Apollo, the company coordinated the services of 20,000 firms and hundreds of universities and scientific laboratories. In 1966 the first uncrewed Apollo spacecraft was launched. Rocketdyne completed qualification tests for the F-1 and J-2 rocket engines to be used in the Saturn V and produced Gemini's control rocket systems. NAR also developed the docking system and docking module used in the Apollo-Soyuz program. NAA's Atomics International Division developed small nuclear power generators for use in NASA's deep space probes. The experience from Apollo provided the basis for NAR's successful bid to build the Space Shuttle orbiters and main engines.

The company was forced to diversify once Apollo came to an end and the demand for rockets decreased. Among its businesses were calculators, waste water management research, inertial navigation equipment for Minuteman and Peacekeeper intercontinental ballistic missiles, and a number of electromechanical devices for different weapons. The company changed its name to Rockwell International in 1973 to reflect its widening range of businesses. Reorganization in 1984 created North American Space Division and the Satellite Division. In 1996 a number of the company's divisions, including Rocketdyne, space systems, and missile systems, were purchased by Boeing. In August 2005 Boeing sold Rocketdyne to Pratt & Whitney.

Michael Barboza

See also: Apollo, Space Shuttle, X-15

Bibliography

Boeing Company. www.boeing.com.
Joan Lisa Bromberg, *NASA and the Space Industry* (1999).
Mike Gray, *Angle of Attack: Harrison Storms and the Race to the Moon* (1992).

Orbital Sciences Corporation

Orbital Sciences Corporation (OSC) opened in 1982 in David Thompson's house with a phone line, a ream of letterhead, and three credit card accounts, two of which belonged to friends Scott Webster and Bruce Ferguson. The idea to create an entrepreneurial space company had begun two years earlier when the three men met as graduate students at Harvard Business School preparing a NASA-sponsored study to assess the risks of commercial space ventures.

With Thompson as president, by 1983 the company had won its first major contract to build an upper-stage booster (Transfer Orbit Stage or TOS), contracting with Martin Marietta to build it, later used to boost the *Mars Observer* and Advanced Communications Technology Satellites. In 1987 a professor of aeronautics at Massachusetts Institute of Technology, Antonio Elias, joined the company, bringing an idea for a new commercial launch vehicle: Pegasus, a small rocket with wings that could launch from

DAVID THOMPSON (1955–)

(Courtesy NASA/Dryden Flight Research Center)

Based on a research project at Harvard Business School, David Thompson cofounded Orbital Sciences Corporation (OSC), established in Dulles, Virginia, in 1982. He served the company as president until 1999 and in 2008 was the chair and chief executive officer. Under his leadership, OSC grew significantly, from six employees in 1983 to more than 3,300 in 2007, becoming a world leader in developing and manufacturing small space and launch systems for government, commercial, and international customers. Perhaps the most significant of OSC's many technical and commercial achievements was the invention of the Pegasus air-launched rocket for small and medium satellites.

Incigul Polat-Erdogan

the belly of an airplane. Funded by the Advanced Research Projects Agency, the first Pegasus launch in 1990 was the first small, relatively inexpensive launch vehicle in more than 20 years. With its TOS booster and a joint venture with Hercules Aerospace Company to develop the Pegasus rockets, OSC raised the private capital needed to develop its rocket and also acquired the Space Data Corporation in Phoenix, Arizona, a leading manufacturer of suborbital rockets, launch facilities, space payloads, and related data systems. OSC became one of the most prominent launchers of suborbital rockets in the world and rose quickly from the ranks of small independent corporations to join the worldwide competition for the launch business.

The company expanded rapidly from 50 employees in 1987 to more than 1,200 by 1991. By 2005 there were 2,700 employees, with headquarters in Virginia and facilities in Arizona, California, and two in Maryland. During the 1990s the company enjoyed continued success with its launch business but failed to realize its goal to create successful businesses in global communications and Earth imaging with its ORBCOMM and ORIMAGE satellites. As opportunities dimmed in the telecommunications industry and in aerospace in general during the decade, OSC was hindered with debt. Both ORBCOMM and ORBIMAGE companies were forced to file for bankruptcy, and OSC sold its interests in several subsidiary businesses to restore financial viability. By 2002 the company had begun to right itself. Even though the commercial space market continued to flounder, OSC won its largest contract ever to build, test, and support ballistic missile interceptor booster vehicles. Through 2004 the company had developed, built, and delivered 476 systems.

Gary Dorsey

See also: Advanced Research Projects Agency, Pegasus

Bibliography
Gary Dorsey, *Silicon Sky* (1999).
Orbital Sciences Corporation. www.orbital.com.

PanAmSat Corporation

PanAmSat Corporation (Pan American Satellite or PAS), was a satellite communications operator that provided service throughout North America and then became the first private operator to provide international communications services. Founded by Rene Anselmo in 1984, PanAmSat intended to provide satellite communications services for television and telecommunications applications. Throughout the 1980s, PanAmSat fought against the government-established Intelsat monopoly for access to the international satellite communications market. While accepting private entities for so-called "regional" satellite markets, U.S. government restrictions and international treaties prohibited private entities from competing in the international market. PanAmSat eventually won approval to provide international services through an aggressive newspaper advertising campaign targeted directly at politicians, including then-President Ronald Reagan, subsequently opening the market for other carriers.

The company launched *PAS 1* over the Atlantic Ocean in 1988, which provided service to the Cable News Network (CNN) for Latin America, among others. During the next few years, PanAmSat launched *PAS 2*, *PAS 3*, and *PAS 4*, in 1995, becoming the first private company to provide communications service around the globe.

Throughout the next few years, PanAmSat's satellite fleet grew through mergers and acquisitions. In 1996 PanAmSat merged with Hughes Electronics Corporation. The newly formed public company named PanAmSat offered service through a fleet of satellites, consisting of PanAmSat's PAS series and Hughes's Galaxy series. Before the merger with PanAmSat, Hughes Electronics acquired International Business Machine's (IBM) Satellite Business Systems fleet and Western Union's Westar fleet, both of which became part of the new PanAmSat fleet.

By 2005 PanAmSat operated a global

REYNOLD ("RENE") ANSELMO (1926–1995)

Rene Anselmo founded PanAmSat with personal funds in 1984. Frustrated with the service and prices associated with Intelsat, the government-owned and operated satellite provider, he attacked Intelsat and its political supporters in Washington, DC, and eventually won approval to operate a private, international satellite communications service. Known for his bold behavior, Anselmo helped cofound Univision and the Spanish Television Network prior to founding PanAmSat. He was famous for his saying, "Truth and technology will triumph over bullshit and bureaucracy!" *Space News* called Anselmo "the maverick who created an industry," awarding him the top spot on the "Top 100 Who Made a Difference." Anselmo died just two days short of PanAmSat's initial public offering.

David Hartzell

fleet of 23 satellites, capable of reaching 98 percent of the world's population, with yearly revenues of $861 million. An economic oversupply of satellite communications services pressured Intelsat and PanAmSat to seek a merger. In July 2006 the merger was completed, with Intelsat acquiring PanAmSat's assets, creating a combined fleet of more than 50 communications satellites.

David Hartzell

See also: Intelsat, Regional Satellite Communications Operators

Bibliography

Intelsat. www.intelsat.com.
Space.com. www.space.com.

Radio Corporation of America (RCA)

Radio Corporation of America (RCA) was founded in 1919 to maintain and control the long-distance communications market to and from the United States. The company operated radio stations, manufactured and sold radio equipment, and conducted research-and-development studies in radio technologies. By the end of World War II, RCA had many inventions in radio communications and television systems, creating the fundamental technology essential for developing space systems.

As a response to the increase in government support for industrial research in the postwar era, specifically in electronics, RCA restructured itself through the 1950s, creating RCA Astro Electronic Division in 1959 in East Windsor, New Jersey. With this division, RCA quickly became one of the leading manufacturers of satellites and related systems, mainly meteorological and communications satellites. RCA developed the first meteorological satellite, Tiros, launched in April 1960, paving the way for future RCA-made meteorological satellites, including Environmental Science Services Administration, National Oceanic and Atmospheric Administration, and Defense Meteorological Satellite Program series satellites. RCA also developed various military satellites in the 1960s, including SAINT (Satellite Interceptor) antisatellites, and Transit and Oscar navigational satellites. In addition, RCA developed communications satellites, including Relay, the Satcom series, and subsystems for many space systems, including Mars Observer, Viking, Dyna-Soar, and the Apollo Lunar Module. In 1975 RCA created RCA-Americom, later SES Americom, to operate RCA-made Satcom series communications satellites, providing broadcasting services for television channels. In 1986 General Electric acquired RCA and sold its satellite manufacturing division to Martin Marietta. The division, becoming an asset of Lockheed Martin after the merger of Lockheed with Martin Marietta in 1995, was permanently closed in 1998.

Incigul Polat-Erdogan

See also: Antisatellite Systems; Environmental Science Services Administration Satellites; Tiros, Television Infrared Observation Satellite Program; Transit

Bibliography

Andrew J. Butrica, ed., *Beyond the Ionosphere: Fifty Years of Satellite Communication* (1997). The David Sarnoff Library.

David Sarnoff Library. http://www.davidsarnoff.org/.

The Museum of Broadcast Communications, Radio Corporation of America. http://www.museum.tv/eotvsection.php?entrycode=radiocorpora.

SES

SES, the first private satellite operator in Europe, began in 1985 in Luxembourg as Société Européenne des Satellites (SES). SES launched its first geostationary satellite,

SES New Skies Satellite

Satellite Name	Launch Date	Manufacturer	Nation/Area Served	Purpose
NSS 513	18 May 1988	Ford Aerospace	Atlantic region	Decommissioned, replaced by *NSS 5*
IS 603	14 Mar 1990	Hughes Space	Atlantic region	Voice data, Internet
NSS K	9 Jun 1992	Lockheed Martin	Atlantic region	Decommissioned, replaced by *NSS 7*
NSS 703	6 Oct 1994	Space Systems Loral	Africa, Asia, Europe	Television and radio programming
NSS 5 (formerly *NSS 803*)	23 Sep 1997	Lockheed Martin	Pacific Ocean region	Private networks, Internet, broadband services
NSS 806	27 Feb 1998	Lockheed Martin	Latin America	Television broadcast
NSS 11 (formerly *APP 1*)	1 Oct 2000	Lockheed Martin	Asia	Television and media communications
NSS 7	16 Apr 2002	Lockheed Martin	Africa, Americas, Europe, Middle East	Broadcast and Internet, replaced *NSS K* and *NSS 803*
NSS 6	17 Dec 2002	Lockheed Martin	Asia	Broadband and direct-to-home broadcast
NSS 10 (formerly *Worldsat 2, AMC 12, Astra 4A*)	3 Feb 2005	Alcatel	Africa, Americas, Europe	Broadcast, Internet, telecom
NSS 8	30 Jan 2007	Boeing	Africa, Asia, Middle East	Rocket exploded on pad/planned broadcast and Internet
NSS 9	Planned late 2008	Orbital	Pacific Ocean region	Replace *NSS 5*
NSS 12	Planned 2009	Space Systems Loral	Africa, Asia, Europe, Middle East	Broadcasts, telecommunications

SES Astra Satellite

Satellite Name	Launch Date	Manufacturer	Nation/Area Served	Purpose
Astra 1A	11 Dec 1988; retired 2 Dec 2004	GE Astro Space	Western Europe	Direct-to-home television
Astra 1B	2 Mar 1991; out of service	GE Astro Space	Western Europe	Direct-to-home television
Astra 1C	12 May 1993	Hughes	Europe	Direct-to-home television and radio
Astra 1D	1 Nov 1994	Hughes	Western Europe, Canary Islands	Digital broadcast and radio
Astra 1E	19 Oct 1995	Hughes	Western Europe	Direct-to-home digital television and radio
Astra 1F	8 Apr 1996	Hughes	Germany and Central Europe	Direct-to-home television and radio
Astra 1G	2 Dec 1997	Hughes	Germany and Central Europe	Direct-to-home television and radio; also used for satellite Internet (with direct video broadcasting modem) and free-to-air television and radio channels
Astra 2A	30 Aug 1998	Hughes	United Kingdom and Ireland	Direct-to-home television
Astra 1H	18 Jun 1999	Hughes	Germany and Central Europe	Direct-to-home television and radio
Astra 2B	14 Sep 2000	Hughes	United Kingdom and Ireland; West Africa	Direct-to-home television; telecommunications and Internet
Astra 2D	19 Dec 2000	Hughes	United Kingdom and Ireland	Direct-to-home television
Astra 2C	16 Jun 2001	Boeing Satellite Systems	Germany and Central Europe	Direct-to-home television and radio
Astra 3A	29 Mar 2002	Boeing Satellite Systems	Europe	Direct-to-cable for corporate networks
Astra 1K	25 Nov2002; deorbited in Dec 2002	Alcatel Space	Stranded in parking orbit due to launch failure	Communications
Astra 4A	3 Feb 2005	Alcatel Space	Africa, Europe, North and South America; Atlantic Ocean region	Data and broadcast applications; small-dish reception; known as *AMC 12* in the Americas
Astra 1KR	20 Apr 2006	Lockheed Martin	Europe	Ku-band provides direct-to-home broadcast services
Astra L	4 May 2007	Lockheed Martin	Europe	Direct-to-home broadcast services
Astra 3B	Planned fourth quarter 2009	EADS Astrium	Europe	To replace *Sirius 2*

SES Americom Satellites (Part 1 of 2)

Saztellite Name	Launch Date	Manufacturer	Nation/Area Served	Purpose
Spacenet 4	12 Apr 1991	GE Astro	North Pacific Rim, East Asia, Western United States	C/Ku band for communications to broadcasters, cable programmers, Internet service providers, government agencies, educational institutions, and corporate networks
Americom 1 (*AMC 1*)	8 Sep 1996	Lockheed Martin	North America, 50 United States	C/Ku band serves a variety of broadcast, mobile, educational, private business and government applications
AMC 2	30 Jan 1997	Lockheed Martin	North America, 50 United States	C/Ku band for television broadcasting
AMC 3	4 Sep 1997	Lockheed Martin	North America; 50 United States	C/Ku-band serves cable, radio, and educational programming distribution for education, broadcast, business television, and broadband Internet
AMC 5	28 Oct1998	Alcatel	48 United States	Ku-band for satellite news gathering (SNG), business television, and broadband Internet
AMC 4	13 Nov 1999	Lockheed Martin	North and South America; 50 United States	C/Ku band broadcasting, business television, broadband Internet market
AMC 7, launched as *GE 7*	14 Sep 2000	Lockheed Martin	North America, 50 United States, Caribbean	C-band for cable television
AAP 1, formerly *WORLDSAT 1*	1 Oct 2000	Lockheed Martin	China, Northeast Asia, South Asia	Ku-band
AMC 6	22 Oct 2000	Lockheed Martin	North America	C/Ku-band provides video and SNG, government network
AMC 8, launched as *GE 8*, also known as *Aurora 3*	19 Dec 2000	Lockheed Martin	North America, 50 United States	C-band radio and telecommunications

SES Americom Satellites (Part 2 of 2)

Saztellite Name	Launch Date	Manufacturer	Nation/Area Served	Purpose
AMC 9	7 Jun 2003	Alcatel	North America	C/Ku-band serves television programmers, government agencies, and enterprise networks
AMC 10 and *11*	5 Feb 2004	Lockheed Martin	North America, 50 United States	C-band for cable
AMC 15	15 Oct 2004	Lockheed Martin	North America, 50 United States	Ku/Ka band provides AMERICOM2Home® service
AMC 16	17 Dec 2004	Lockheed Martin	North America; 50 United States	Ku/Ka-band
AMC 12	3 Feb 2005	Alcatel	North and South America, Europe, Africa	C-band provides links to regional satellite systems; also known as Astra 4A
AMC 23, also known as Worldsat 3	29 Dec 2005	Alcatel	Pacific Ocean Region	C/Ku band
AMC 18, support for *AMC 10* and *11*	8 Dec 2006	Lockheed Martin	North America, 50 United States	C-band for digital transmission services for cable programming and broadcasting
AMC 14	15 Mar 2008 failed to reach planned orbit	Lockheed Martin	North America, 50 United States	Active phased array (APA) payload consisting of a receive mode APA antenna, highest levels of redundancy on core components such as amplifiers, receivers, commanding beam, and computer control systems. DISH Network services
AMC 21	Planned third quarter 2008	Alcatel	50 United States, Caribbean, Southern Canada, Mexico, Gulf of Mexico, Central America	Ku-band 125° West

Automatic Satellite Tracking Research Antenna (*ASTRA*) *1A*, in 1988, which provided direct-to-home satellite broadcasting of analog television and radio channels in Europe. In 1996 commercial digital television began with the launch of *ASTRA 1E*.

In 1999 SES acquired a stake in Asia Satellite Telecommunications Holding Ltd., and the next year it acquired stakes in Nordic Satellite (operator of Sirius Satellite Radio) and in Embratel Satellite Division (operator of Star One). In 2001 SES Global was created after SES's acquisition of Americom, a leading fixed satellite services operator in the United States, incorporated in 1973 with headquarters in New Jersey. Americom's government services division provided secure satellite communications to federal agencies and government contractors. Americom collaborated with Home Box Office in 1975 to develop satellite-to-cable systems. By 2003–4, Americom provided direct broadcasting services and high-definition television to U.S. customers. In 2006 SES Global purchased the Netherlands-based company New Skies Satellites (which was created in 1998 from a restructuring of Intelsat and inherited some of Intelsat's satellites) and changed the name to SES New Skies, offering a network of satellites primarily in Africa, Asia, the Middle East, and South America. SES Americom, SES Astra, and SES New Skies were wholly owned operating companies of SES Global, capable of reaching 99 percent of the world's population. This included audiovisual broadcasting, cable network feeds, Internet trunking, and Internet protocol multicast, in addition to corporate networks, network facilities, and telecommunications services. In 2007 SES Global was renamed SES.

SES Americom reached virtually every household in the United States, while SES Astra offered approximately 2,000 television and radio channels to 109 million European households in 35 countries in 2008. The SES New Skies satellite network broadcast hundreds of channels to government and individual consumers throughout the world. SES reported revenues of €1,258 million in 2005 with an employee base of 1,102.

Joni Wilson

See also: Europe

Bibliography

Donald H. Martin et al., *Communication Satellites*, 5th ed. (2007).
SES Americom. http://ses-americom.com/americom_2008/index.php.
SES ASTRA. www.ses-astra.com.
SES NEW SKIES. www.ses-worldskies.com/worldskies/.
SES. http://www.ses.com/ses/welcome/.

Sea Launch Company

Sea Launch Company, a marine-based heavy-lift commercial launch provider, was formed in April 1995 as a privately financed commercial consortium consisting of four international partners: the Boeing Commercial Space Company of the United States, Kværner of Norway, Rocket and Space Corporation Energia of Russia, and State Design Office Yuzhnoye of Ukraine. The company was created with the intent

The 21,500-ton Sea Launch platform, a former oil platform. (Courtesy AP/Wide World Photos)

to provide a reliable, cost-effective, heavy-lift commercial launch capability. Utilizing a converted oil drilling platform called Odyssey, Sea Launch could transport a modified Zenit 3SL rocket and its payload to a designated location near the equator to maximize launcher effectiveness. By launching on or near the equator, the company minimized the propellant needed to deliver a payload to orbit, which ultimately increased the overall lifespan of the spacecraft. The modified Zenit 3SL rocket could launch a payload weighing 13,227 lbs to a geosynchronous orbit. Because the launch platform was marine-based, Sea Launch was able to launch its booster to any inclination without the risk of launching over populated areas and independently of land-based launch range scheduling constraints. The company processed payloads from Long Beach, California.

The first Sea Launch payload was successfully launched on 27 March 1999, carrying a demonstration satellite. In the first 25 launches through January 2008, Sea Launch experienced only two failures that kept its payload from achieving the proper orbit. Although the company had shown notable achievements, fluctuations in the demand to launch large geosynchronous commercial communications satellites should remain consistent enough to support Sea Launch. As of 2008 the company was planning to provide a land-based launch capability for the Zenit rocket to launch from Baikonur. Land Launch was a venture between Sea Launch and Space International Services, located in Moscow, Russia, to provide a viable medium-weight launch capability to commercial customers.

Robert Kilgo

See also: Commercial Space Launch, Rocket and Space Corporation Energia

Bibliography

Sea Launch. http://www.boeing.com/special/sea-launch/.

SIRIUS Satellite Radio. *See* Satellite Radio.

Space Systems/Loral

Space Systems/Loral (SS/L), a wholly owned subsidiary of Loral Space and Communications Limited, became a major manufacturer of communications and environmental satellites, subsystems, and payloads. Located in Palo Alto, California, and employing more than 2,000 people, the company provided orbital testing, procured insurance and launch services, and managed mission operations.

SS/L evolved from Palo Alto–based Western Development Laboratories (WDL), created in 1957 by Philco Corporation to work on the Discoverer program, the cover name of the secret Corona reconnaissance program. In 1961 Ford Motor Company bought Philco Corporation and combined WDL with its existing Aeronutronic division in 1963 under the name of Philco-Ford to strengthen its participation in defense and space projects. Philco-Ford developed a wide variety of communications satellites for customers in the United States and abroad, including the Department of Defense with its Initial Defense Satellite Communications System, the North Atlantic Treaty Organization, and the International Telecommunications Satellite Organization (Intelsat). It also developed advanced meteorological satellites, including the Geostationary Operational Environmental Satellites (GOES) 1–3 series. In 1976 Philco-Ford became Ford Aerospace Corporation and Communications, then Ford Aerospace Corporation in 1988.

In 1990 Loral Space and Communications Ltd. bought Ford Aerospace for $715 million and renamed its satellite portion Space Systems/Loral. SS/L continued building communications and meteorological satellites. It delivered satellites and payloads for worldwide customers, including the Intelsat 7 and 9 series, Insat for India, *MTSAT* (*Multi-function Transport Satellite*) for Japan, *Agila* for Philippines, *Apstar* for China, and many others for commercial, civilian, and military customers in the United States and for cross-national alliances. SS/L also built *GOES 8–12*, next-generation meteorological satellites for the National Oceanic and Atmospheric Administration. In addition, SS/L developed other space products, including a nickel-hydrogen battery system on board the *International Space Station*, which is the largest power storage system on orbit.

In July 2003 SS/L's parent company, Loral, filed for Chapter 11 bankruptcy protection, causing significant drops in SS/L's revenues. However as Loral emerged from the bankruptcy, SS/L began to return to its original situation. In 2005 with an annual capacity to build 12–15 geostationary satellites, SS/L was still a leader in its field, producing more than 220 satellites since its predecessor WDL's establishment in 1957.

Incigul Polat-Erdogan

See also: Geostationary Operational Environmental Satellites (GOES), Indian National Satellite Program, Initial Defense Communications Satellite Program, Intelsat

Bibliography

Space Systems/Loral. www.ssloral.com.

SPACEHAB, Incorporated

SPACEHAB, Incorporated, founded in 1984, was the first company to develop, own, and operate habitable modules for use in space. These consist of human-rated pressurized and unpressurized modules designed for use in the Space Shuttle's cargo bay and the *International Space Station* (*ISS*).

In 1990, SPACEHAB signed its first contract, the Commercial Middeck Augmentation Module, with NASA in response to a U.S. government invitation to private companies to test the potential of the Shuttle as a research laboratory. Four missions were completed between June 1993 and May 1996. In 1995, the year it went public, SPACEHAB won a second contract for seven logistics missions to the Russian space station *Mir*. A third multi-mission contract was awarded in 1997 to support *ISS* research and logistics missions.

In 1997 SPACEHAB purchased the operating assets of Astrotech Space Operations, Incorporated, from Northrop Grumman Corporation. Astrotech offered customers an alternative to government-owned facilities, integrating payloads, for customers, that launch on expendable launch vehicles, including Atlas, Delta, Pegasus, and Taurus, and secondary payloads on the Space Shuttle. Astrotech maintained payload-processing facilities at Florida's NASA Kennedy Space Center and Cape Canaveral Air Force Station and at the Vandenberg Air Force Base in California. Astrotech also provided facilities maintenance support to Sea Launch Corporation in Long Beach, California. Facilities at Titusville, Florida, process 5 m satellites and payload fairings.

In 1998 SPACEHAB acquired Johnson Engineering Corporation (JE) in Houston, Texas. JE managed training operations and facility engineering at the Johnson Space Center (JSC) Neutral Buoyancy Laboratory and Space Vehicle Mockup Facility, where astronauts trained for space walks and flights aboard the Space Shuttle and *ISS*. The JE name changed in April 2003 to SPACEHAB Government Services when SPACEHAB lost the contract for flight crew training but received a subcontract from ARES Corporation to provide configuration management services.

Space Media, Incorporated (SMI), was a majority-owned subsidiary of SPACEHAB that created space-themed content for education and commerce. Services ranged from building comprehensive exhibits for museums and providing astronauts for product endorsements, to selling caps, T-shirts, and mission patches. In 2000 SMI acquired The Space Store, an online retail operation. The Store offered retail products through its website, thespacestore.com, and a store located near JSC.

In 2003 Space Shuttle *Columbia*'s last mission was also the debut of SPACEHAB's Research Double Module, which considerably expanded Space Shuttle laboratory/research facilities and was a precursor for *ISS* experiments. The loss was both financially and materially substantial for SPACEHAB.

James Paul Ward

See also: *International Space Station*, Kennedy Space Center, Space Shuttle, Vandenberg Air Force Base

Bibliography

Lou Dobbs, *Space: The Next Business Frontier* (2002).
John McLucas, *Space Commerce* (1991).
SPACEHAB Annual Reports.

Telesat Canada

Telesat Canada was incorporated in 1969 to build and operate Anik communications satellites, which the Canadian government saw as necessary for communicating with all parts of the country in both official languages, English and French. Canada had many remote settlements with poor communications at the time, and the Canadian government was anxious to establish bilingual services to stave off nationalism in the French-speaking province of Quebec.

Anik A1, launched in 1972, was the world's first domestic communications satellite in geosynchronous orbit. Since that time, Telesat has launched 14 other Anik satellites and two Nimiq direct-to-home broadcast satellites. In the early years of the company, Telesat was owned jointly by the Canadian government and by private owners under government control. The private shares were held by telecommunications companies that differed with the Canadian government about the use of satellite versus ground transmission equipment that affected Telesat's profitability. Telesat moved to private control when the Canadian government sold its shares in 1992. In late 2006 Telesat's longtime owner Bell Canada Enterprises Inc. sold the firm to a major Canadian pension fund and Loral Space and Communications Inc. Known as Telesat, the firm was based in Ottawa, Ontario, offered television and telecommunications systems, and was a major consultant for satellite telecommunications providers. It also managed assets for other satellite telecommunications providers and produced systems engineering software for satellite operators.

Chris Gainor

See also: Canada, Canadian Communications Satellites

Bibliography

Chris Gainor, "The Chapman Report and the Development of Canada's Space Program," *Quest* 10, no. 4 (2003): 3–19.

TRW Corporation

TRW Corporation (derived from Thompson-Ramo-Wooldridge), one of the major U.S. space contractors in the latter twentieth century, was formed as a merger of Ramo-Wooldridge Corporation and Thompson Products in October 1958. With encouragement from the Department of Defense (DoD), Simon Ramo and Dean Wooldridge left Hughes Aircraft Corporation in 1953 to found Ramo-Wooldridge. The company acquired the systems engineering contract for the Atlas intercontinental ballistic missile (ICBM) program, which expanded to include the U.S. Air Force's (USAF) early ballistic missile and space programs. Rapid expansion required capital to pay for new facilities, leading to negotiations with an automotive firm, Thompson Products, which became an investor. The new company, Thompson Ramo Wooldridge, Incorporated, was besieged with complaints from the aviation industry regarding its insider role for the USAF. This led to congressional investigations between 1958 and 1960, formation of the nonprofit Aerospace Corporation to take over the systems engineering and technical direction role for the USAF on future programs, and TRW's conversion on the ballistic missile programs to a systems engineering and technical assistance contractor. TRW diversified into a variety of businesses outside the space industry, including credit reporting, consulting applications, and continuing its automotive business. In 1965 the company officially became TRW Inc.

By the early 1960s, TRW's Space Technology Laboratory (STL) had already built some of the first American spacecraft, including *Explorer 6* and *Pioneer 5*. STL won a number of NASA spacecraft contracts, including three Orbiting Geophysical Observatories, High Energy Astronomy Observatories, and *Pioneer 10* and *11* probes to Jupiter and Saturn. For the DoD, TRW built the Vela Hotel nuclear detection spacecraft, the second-generation communications constellation, Defense Satellite Communications System II, and several blocks of the Defense Support Program early-warning

SIMON RAMO (1913–)

Simon Ramo, an aerospace engineer, physicist, and business leader, was recognized as the chief scientist of the U.S. Intercontinental Ballistic Missile (ICBM) program. After working at Hughes Aircraft Corporation with Dean Wooldridge on the innovative Falcon missile, he cofounded Ramo-Wooldridge Corporation (R-W), which became the systems engineering and technical direction contractor for the Air Force ICBM program. The methods that he developed and implemented in the ICBM program helped establish systems engineering as a discipline. He cofounded TRW, a merger of R-W and Thompson Products, and helped the company become one of the world's largest corporations in the space arena.

Incigul Polat-Erdogan

system spacecraft. TRW's communications satellite work grew with the development of the U.S. Navy's Fleet Satellite Communications System, NASA's Tracking and Data Relay Satellite System in the 1970s, and the Milstar electronics payload in the 1980s–90s. Smaller but important subsystem contracts, such as the Viking biology package and the Apollo Lunar Module descent engine, came TRW's way during this time. In the late 1980s and 1990s, TRW built NASA's *Compton Gamma Ray Observatory* and *Chandra X-ray Observatory*. TRW has reportedly been active in classified electronics intelligence satellite programs. In the late 1990s and early 2000s, TRW acquired several satellite contracts, including NASA's *James Webb Space Telescope* and *Aqua Earth Observing System* satellite, the DoD's Space-Based Infrared System-Low and the communications payload for the Advanced Extremely High Frequency system, and the National Polar-orbiting Operational Environmental Satellite System.

In December 2002 Northrop Grumman acquired TRW's space business for approximately $7.8 billion.

Stephen B. Johnson

See also: Space Business, Systems Engineering, Systems Management

Bibliography

Davis Dyer, *TRW: Pioneering Innovation and Technology since 1900* (1998).
Northrop Grumman. www.northropgrumman.com.
Simon Ramo, *The Business of Winning and Losing in the High-Tech Age* (1988).

United Space Alliance, LLC

United Space Alliance, LLC, (USA) was formed in 1995 by Rockwell International, later part of Boeing, and the Lockheed Martin Space Operations Company. This was in response to NASA's announcement to consolidate the large number of individual Space Shuttle contracts under one prime contractor. Known as the Space Flight Operations Contract (SFOC), USA was awarded the contract in November 1995 and became NASA's principal contractor for everyday management of the Space Shuttle. USA also became responsible for training and operations planning for the *International Space Station* (*ISS*). SFOC became effective in October 1996.

NASA awarded USA the contract for six years with the option for two two-year extensions, and then consolidated 31 contracts under SFOC in several phases. SFOC's total contract awards in its first 10 years were valued at $16.8 billion, while placing more than $20 billion in government assets under USA control. Under the SFOC, USA's responsibilities included mission design and planning, flight operations, software development and integration, payload integration, integrated logistics, astronaut and flight controller training, and vehicle processing, launch, and recovery. The first USA-managed Space Shuttle mission was Space Transportation System-80, launched in November 1996.

As the second of the two year extension for SFOC was coming to an end, NASA awarded the four-year Space Program Operations Contract (SPOC) to USA in October 2006. This new contract assured USA's role in the Space Shuttle Program through the planned retirement of the Space Shuttle by 2010. Under the same contract, USA also supported various elements of the Constellation Program.

By 2008 USA employed more than 10,000 personnel in Florida and Alabama with its corporate headquarters located in Houston, Texas. These locations included facilities in and near the Kennedy, Marshall, and Johnson space centers.

John Venditti

See also: *International Space Station*, Space Shuttle

Bibliography

United Space Alliance. www.unitedspacealliance.com.

WorldSpace. *See* Satellite Radio.

XM Satellite Radio. *See* Satellite Radio.

Human Spaceflight and Microgravity Science

Human spaceflight and microgravity science evolved together, because human presence in space requires an understanding of microgravity effects on biological organisms, and because the ability of humans to perform microgravity experiments allowed them to become a major element of human spaceflight missions. Human spaceflight was the dream of space visionaries well before spaceflight began, and it quickly became a major goal of the Cold War space race between the Soviet Union and the United States. With the exception of China, other nations have put humans in space aboard Russian and American craft. China placed taikonauts into orbit in the 2000s with its Shenzhou spacecraft. By the early twenty-first century, wealthy individuals could also purchase rides into space.

Fiction writers have imagined humans traveling beyond Earth since the ancient Greeks, using a variety of fanciful or plausibly scientific means. French author Jules Verne was perhaps the most famous. In his 1865 novel *From the Earth to the Moon*, Verne postulated the launch of a space capsule using a giant cannon placed in Florida. His work captured the imagination of millions around the world, including several visionary rocketry pioneers.

In 1903 Russian mathematics teacher Konstantin Tsiolkovsky published *Exploration of Cosmic Space with Reactive Devices*, which discussed how to propel humans to the Moon using liquid-fuel rockets. Human flight was a major inspiration to Robert Goddard in the United States and Hermann Oberth from Romania, both of whom published theoretical treatises on rocketry and began development of rockets. Oberth wrote about space stations, as did Slovenian engineer Herman Potočnik (pen name Hermann Noordung), who envisioned a wheel-like space station in his book *The Problem of Space Travel—The Rocket Motor.*

Between the two world wars, scientists, engineers, and amateur enthusiasts formed rocket societies in the United States, the United Kingdom, Germany, and the Soviet Union. Two engineers from these groups eventually became key figures in the world's first human spaceflight programs: the Soviet Union's Sergei Korolev and Germany's Wernher von Braun. Von Braun had been the technical director of the German A-4 (later called the V-2) ballistic missile program, moving with many members of his rocket team to the United States after World War II under Operation Paperclip. Both the United States and the Soviet Union assimilated and learned from the V-2, in the

Soviet case under Korolev's leadership. With its creation in 1950, Korolev assumed control of Special Design Bureau No. 1 (OKB-1), which was initially devoted to building ballistic missiles for the Soviet military. The U.S. Army placed von Braun in charge of developing the Jupiter ballistic missile at the Redstone Arsenal in Huntsville, Alabama. Both men used their posts to promote human spaceflight.

While he lived, Korolev remained virtually unknown due to the secrecy of the communist regime. By contrast, von Braun became a celebrity, speaking at spaceflight conferences and writing about spaceflight in books and magazines. Von Braun's series of articles for *Collier's* reached millions of readers during the 1950s. He wrote that space stations would be constructed in Earth orbit to serve as observation posts and as way stations for future missions to Mars. Von Braun and his German colleague and author Willy Ley promoted the wheel-concept for space stations because rotation created artificial gravity. American artist Chesley Bonestell captured their ideas in a famous 1952 painting of a circular space station with a central hub.

British engineer and author Arthur C. Clarke also wrote extensively about space stations. In his 1951 book *The Exploration of Space*, Clarke discussed how space stations "would be useful for so many science purposes that their employment as 'filling stations' for rockets might well become of secondary importance." Clarke recognized that space stations had the potential to contribute to astronomical research, remote sensing, biological studies, and as a spacecraft docking port.

Despite dreams of space exploration, military concerns drove rocket development. Nazi Germany's V-2 rocket, and the Cold War ballistic-missile programs of the United States and the Soviet Union, were developed to deliver conventional explosives and later nuclear warheads by traveling through space. Piloted spaceflight was another area with potential military utility. Austrian engineer Eugen Sänger had proposed a rocket-boosted spaceplane bomber during World War II, while in 1945 Soviet engineer Mikhail Tikhonravov proposed a piloted rocket for launching humans into space.

By 1950 the U.S. and Soviet military had started research efforts to understand the effects of spaceflight on humans. Suborbital rocket flights with primates and dogs helped determine the ramifications of weightlessness and cosmic radiation. After World War II, the United States enlisted the help of several German experts brought to the United States. Aeromedical expert Hubertus Strughold was brought in to study the effects of high-altitude flights on pilots, and he became known as the father of space medicine while director of the U.S. Air Force (USAF) School of Aviation Medicine during the 1950s. Bell Aircraft hired Walter Dornberger to develop a piloted space bomber. These ideas led to the USAF Dyna-Soar project, started in 1957 as a single-pilot, reusable spaceplane. By this time the USAF made two high-altitude balloon flights to the edge of space as part of Project Manhigh to investigate the environment of Earth's upper atmosphere and the effects of cosmic radiation on pilots. The X-15 program, which flew from 1959 to 1968, proved the ability of pilots to operate a suborbital spaceplane and further developed key technologies for human spaceflight, such as pressure suits.

In the Soviet Union, OKB-23, which specialized in development of long-range bombers, started its own reusable spaceplane project in 1957, with the support of the Soviet Air Force. There were other crewed, space-related vehicles under development at OKB-23, including a capsule with steering jets for reentry. A few years later Vladimir Chelomey's OKB-52 started development of space launch vehicles and spacecraft, including a winged space vehicle, called the Kosmoplan, and a suborbital, reusable vehicle, called the Raketoplan. This competed with Korolev's OKB-1 Department No. 9, created in 1957 to build a satellite to carry a dog in orbit and a human-piloted orbital capsule.

When the Soviet Union launched *Sputnik* on 4 October 1957, fear and panic resonated throughout the United States, as some political leaders and the media perceived that the Soviet Union had caught up or possibly overtaken the United States as the world's technological leader. Among the U.S. responses to *Sputnik* was the establishment of the civilian National Aeronautics and Space Administration (NASA) in 1958, which quickly began Project Mercury to put a man in space. Korolev reacted by convincing Soviet officials to let OKB-1 build a dual-use capsule that could carry cameras for reconnaissance (Zenit) or a man into orbit (Vostok). After *Sputnik*, Soviet Premier Nikita Khrushchev pressed Korolev to keep the Soviet Union ahead of the United States in space achievements for propaganda purposes. In the early 1960s Korolev succeeded, achieving the first human in space (Yuri Gagarin, 12 April 1961); the first woman in space (Valentina Tereshkova, June 1963); the first three-man mission (*Voskhod 1*, October 1964); and the first space walk or extravehicular activity (EVA) (Alexei Leonov, April 1965). These missions succeeded due to the heavy-lift capability of the R-7 launcher and a willingness to take risks to one-up NASA.

NASA attempted to catch up, but in the Mercury program it was unable to do so. Mercury nonetheless provided prestige value because of NASA's willingness to broadcast its missions on live television, in contrast to the secretive Soviet program. Surpassing the Soviet Union required aiming for a goal that the communist nation could not achieve with its existing technology. Specifically, Wernher von Braun's Saturn launcher program offered a heavy-lift capacity capable of out-muscling Soviet launchers. Responding to the outcry after Gagarin's flight and after the first successful Mercury suborbital flight (Alan Shepard, 5 May 1961), U.S. President John F. Kennedy announced the Apollo program to land a man on the Moon before the end of the decade. Realizing the large gap in time and in technical capabilities between Mercury and Apollo, NASA created the Gemini program to test orbital rendezvous and EVA capabilities. By 1965–66, Gemini proved these capabilities and gained the lead for NASA over the Soviet program. Concerned with the space race's huge costs, in 1963 Kennedy secretly proposed to Soviet leaders a joint program to land men on the Moon. The Soviet Union declined the offer and the Moon race continued.

Korolev's reply to Gemini and Apollo was the Soyuz spacecraft and the N1 launcher, but their development was years behind the United States and was plagued by rancorous divisions among Soviet leaders and design bureaus. Korolev's OKB-1 and Chelomey's OKB-52 fought for control of Soviet lunar programs, leading to competitive efforts that learned little from each other, while Valentin Glushko, the best rocket engine designer, refused to work with Korolev on the N1. Korolev's death in 1966 fatally crippled the Soviet lunar effort.

Both the United States and the Soviet Union decided that their initial seven astronauts and 20 cosmonauts should be selected from the ranks of rigorously trained military test pilots, since nobody knew how humans would be able to perform in space. The early spaceflights proved that humans could function in space but also discovered unexpected problems. Some astronauts and cosmonauts became nauseous with space sickness, which usually subsided after a few hours or days. Another issue was the difficulty of EVAs. Edwin "Buzz" Aldrin on *Gemini 12* was the first to demonstrate effective ways to operate in zero gravity. Both space programs developed life-support systems and space suits that functioned well in space.

In the race to the Moon, both human spaceflight programs assumed, and in 1967 paid for, excessive risk. A fire caused by a spark in the pure-oxygen atmosphere of the Apollo capsule at Kennedy Space Center killed the first Apollo crew during a launch rehearsal. The Soviet Union's best opportunity to catch up with the United States perished with cosmonaut Vladimir Komarov, who died in *Soyuz 1* when a parachute failure led to a crash landing. The accident, along with a series of N1 rocket failures, ended Soviet chances for victory in the manned Moon race.

Neil Armstrong's first steps on the Moon on 20 July 1969 as part of *Apollo 11* signaled U.S. victory in the space race, and later Apollo flights provided new scientific insights about lunar geology. However, popular support for the U.S. space program was fading. The Vietnam War and issues of civil rights, crime, poverty, and pollution became higher priorities. Despite NASA's efforts to continue a robust human spaceflight program, both Congress and President Richard Nixon reduced NASA's budget, which dropped from 4.4 percent to roughly 1 percent of the federal budget by the mid-1970s. Nonetheless, Nixon did not want to preside over the end of NASA's human spaceflight program, and in January 1972 he approved the Space Shuttle as a means to lower the cost of access to space. NASA also gained approval to develop a space station called *Skylab* and the Apollo-Soyuz Test Project, based on leftover Apollo hardware. *Skylab* provided the capability to assess the effect of long-duration space missions on astronauts and to test, improve, and create materials that could not otherwise be done on Earth. The station hosted three crews in 1973–74, which performed a variety of experiments to monitor astronaut metabolic, cardiovascular, and muscular responses in space, along with microgravity biology and materials. *Skylab* missions confirmed the loss of muscle and bone mass on orbit and began the first attempts to counter these effects. In 1975 the Apollo-Soyuz Test Project brought the United States and the Soviet Union briefly together in low Earth orbit. This symbolic gesture of détente was the only significant cooperative human spaceflight effort of the two space powers during the Cold War.

When it became clear that the United States had won the Moon race, Soviet leaders refocused their space program to space stations, falsely claiming that they had never been racing the Americans to the Moon. The Soviet space station effort began as a response to the USAF Manned Orbiting Laboratory (MOL) project, begun in 1963 after the cancellation of Dyna-Soar, to experiment with manned reconnaissance. MOL was terminated in 1969 in favor of cheaper, robotic, reconnaissance satellites, but in the meantime and as a response to MOL, the Soviet Union had authorized Chelomey's military Almaz station, also designed for manned reconnaissance. To compete with *Skylab*, in 1970 the Soviet Union decided to build a civilian Long-Duration Orbital Station (DOS). Together five DOS and three Almaz stations were launched under the name Salyut, with two of the DOS stations failing to reach orbit and never given the Salyut label.

The first crewed space station, *Salyut 1*, was launched in April 1971 and functioned as a civilian research and observation platform. Tragically the first Salyut crew was killed after a valve opened and depressurized its Soyuz capsule during reentry. Spaceflights did not resume for more than two years. Almaz proved less efficient at reconnaissance than robotic satellites and was canceled in the late 1970s after *Salyut 5* (secretly *Almaz 3*). *Salyut 6* introduced two docking ports that enabled automated cargo vehicles to bring supplies and a furnace to perform microgravity materials processing experiments on semiconductors and pharmaceuticals. This capability was expanded to five furnaces installed in the Kristall module of the *Mir* space station, whose first component was launched in 1986. Soviet and Russian crews set a series of long-duration records from the 1970s through the 1990s. By 2009 Sergei Krikalev had spent more time in space than any other person, with two missions to *Mir*, two on the Space Shuttle, and two on the *International Space Station* (*ISS*).

From 1978 to 1981, the Soviet Union opened a seat on its Soyuz spacecraft to other countries within the communist bloc as part of the Interkosmos program. Touted as a way to strengthen international relations, it was a response to NASA's plans to fly Canadian and European astronauts on Space Shuttle missions. A total of nine Interkosmos flights took place. The program set a precedent for six additional international flights to Soviet space stations during the 1980s through bilateral agreements, with France, India, Syria, and Afghanistan.

With the new emphasis on reducing costs, President Nixon had directed NASA to seek more extensive cooperation with U.S. allies on the Space Shuttle program. The United States offered Western European nations and Canada the opportunity to participate. Canada agreed to build a robotic arm. After long negotiations, which also resulted in the creation of the European Space Agency (ESA), the Europeans agreed to participate in the human spaceflight program by building *Spacelab*. West Germany led the Spacelab program to build a microgravity experiment laboratory module that fit in the Shuttle's payload bay. For their participation in the Space Shuttle program, the United States agreed to fly Canadian and European astronauts on the Shuttle, leading to the creation of astronaut programs in Canada and Europe.

The Space Shuttle, which first launched in 1981, could carry seven crewmembers and a large payload and had large wings that enabled it to change its trajectory laterally off its orbital track on reentry. This design, which was a compromise among NASA, USAF, and the Nixon administration, attempted to serve many needs with a

Space Shuttle Columbia *sits on Launch Pad 39A at Kennedy Space Center during pre-launch activities in preparation for the first crewed shuttle flight, STS-1. (Courtesy NASA)*

low development and operations cost. Meeting all these goals proved impossible; the Shuttle's operational costs were far higher than anticipated. This compromised another expected Shuttle benefit, the potential for commercial microgravity research and development. While the Shuttle could support a variety of experiments, the vast majority of them were government funded. Corporations found that space-based experimentation was far more expensive and far too slow compared to ground-based equivalents. The Shuttle's benefits were elsewhere. It provided new and useful capabilities to retrieve and repair spacecraft in low Earth orbit, such as the astronaut repair of *Syncom 3* in 1985 and the servicing missions of the *Hubble Space Telescope* in the 1990s and 2000s. The large crew complement also provided opportunities for many more astronauts, including new mission specialists. NASA sought professionals with advanced engineering and science degrees and greater diversity in its astronaut corps, adding women in 1978.

Spurred by the Shuttle, during the mid-1980s the Soviet Union, France, and Japan pursued reusable spaceplanes. The Soviet Union successfully developed its Buran shuttle, but it flew only once in 1988 before budget pressures resulting from the collapse of the Soviet Union grounded it permanently. France proposed building a reusable spacecraft called Hermes atop an Ariane 5 rocket. Prompted by the *Challenger* explosion in 1986, changes to Hermes' design to improve safety led to delays, which increased its cost and weight. ESA scaled the project back in the early 1990s, and it withered away. About the same time, Japan's space agency pursued development of a robotic prototype spaceplane dubbed "HOPE" (H-2 Orbiting Plane Experimental). It too foundered on budget concerns.

As the Space Shuttle program ramped up in the early 1980s, NASA began considering the development of a space station, which had been the logical complement of a space shuttle since von Braun's *Collier's* articles in the 1950s. NASA convinced President Ronald Reagan, and in 1984 he directed NASA to build a space station. Extending the precedent of the Space Shuttle's international agreements, the Reagan administration directed NASA to work with international partners to lower development costs, improve international relationships, and improve the odds of congressional support. It took until 1988 to develop an intergovernmental agreement among the United States and its space station partners concerning development, construction, and operation. All partners in the newly named Space Station Freedom (SSF) program believed that microgravity research and materials science conducted on the station could lead to breakthroughs in new medicines, better alloys, and other commercial products.

During a celebration recognizing the 20th anniversary of the first Moon landing, President George H. W. Bush endorsed sending astronauts to Mars. With the advice of a newly created National Space Council, Bush in 1989 unveiled the Space Exploration Initiative (SEI). A NASA study placed the cost of the SEI at more then $400 billion, and by mid-1990, Congress and the American public had rejected it as too expensive. Bush tried to revive the project, but given NASA's existing commitments to the Space Shuttle and SSF programs, he did not succeed.

In the late 1980s and early 1990s, the ongoing debate about the SSF's purpose and cost led to several redesigns, with consequent schedule delays and increasing costs without any hardware being built. These issues steadily eroded political support, until by June 1993 the SSF program survived by a single vote a motion in the House of Representatives to kill the program. The program was revived by negotiations between the William Clinton administration and Russia's Boris Yeltsin administration. To support Russia's new democratic government and economy, the Clinton administration proposed to make Russia a full partner in the space station program. This argument resonated with Congress, and the two countries signed an agreement in September 1993 to collaborate on a newly designed space station, which became the *ISS*. The United States and Russia initiated the Shuttle-Mir program so each country's space agency could adjust to working together in preparation for the *ISS*. Cosmonauts flew on the Space Shuttle and astronauts were sent to *Mir* on Soyuz spacecraft. The program was troubled from the start, as NASA was criticized for allowing astronauts to stay on *Mir* without direct oversight for safety, during a time in which several dangerous events occurred, including a fire and a collision with a Progress transport that led to module depressurization. Nonetheless NASA and the Russian Space Agency benefited from working together, overcoming language barriers and differences in assumed risk and decision-making authority.

As *Mir*'s operational lifetime neared its end, the private company MirCorp made an attempt to turn *Mir* into a tourist destination by selling weeklong stays for $20 million per person, but the Russian government decided to deorbit the dilapidated station in March 2001 in favor of the *ISS*. *Mir* had orbited Earth for more than 15 years and had crews totaling 104 men and women.

ISS construction began in November 1998 with the launch of the first module, *Zarya*. Two more modules were added by July 2000, and the first crew, called Expedition 1, took up residence three months later. American millionaire Dennis Tito paid Russia $20 million to become the first tourist on the *ISS* in April 2001. NASA and ESA objected, ostensibly for safety reasons, but eventually backed down, as under terms of the *ISS* agreements Russia had the legal authority to grant Tito access to the Russian segments of the station. By 2003, 16 of 23 assembly flights had put more than 200 tons of hardware into orbit. Construction stopped in February 2003 with the destruction of the Space Shuttle *Columbia* during reentry. The grounding of the Space Shuttle fleet forced NASA to rely on Russian Soyuz and Progress flights to and from the station until the Shuttle flew again in July 2005. ESA and Japan augmented *ISS* resupply capabilities with the development of the Jules Verne Automated Transfer Vehicle (ATV) and the H-2 Transfer Vehicle (HTV). The Jules Verne successfully docked with *ISS* in April 2008, and HTV docked in September 2009.

The loss of *Columbia* made clear the need to replace the aging Shuttle fleet and conduct a fundamental reassessment of the human spaceflight program. In January 2004 President George W. Bush announced a plan to build a human-occupied base on the Moon and then attempt piloted missions to Mars. The new policy directed NASA to complete the *ISS* and then retire the Shuttle, which would release funding for the new program, soon called Constellation. By 2009, NASA was developing the Ares I crew launcher and the Orion crew capsule, which would enable crew flights to the *ISS*. To return to the Moon, NASA planned to develop a heavy-lift Ares V vehicle and a Lunar Surface Access Module (LSAM). Lunar missions would use two flights—an Ares I/Orion and an Ares V/LSAM—to rendezvous in Earth orbit before injection into a translunar trajectory. In lunar orbit, the crew would transfer from Orion to the LSAM to descend to the surface. However, by March 2010 the program was in jeopardy as the Barack Obama administration proposed its cancellation in favor of commercial launch to low Earth orbit.

In the meantime, private entrepreneurs had made significant progress developing commercial spaceflight tourism. Scaled Composites of Mojave, California, launched the first privately funded piloted spacecraft, *SpaceShipOne*, three times in 2004. By the end of 2004 Congress passed modest regulatory standards for the new industry, and the Federal Aviation Administration issued rules for commercial space pilots and space tourists. Congress warned that tougher laws would be instituted in the event that an in-flight accident resulted in severe injury or death. During the next two years, commercial spaceports in four states and around the world were announced in support of planned suborbital tourist flights, and several companies had announced plans to develop tourist vehicles.

China became the third nation to put a human in space in October 2003 with the launch of a single-seat capsule, *Shenzhou 5*. Only after the first piloted mission did the government release information about its activities. In contrast, the second crewed Shenzhou flight in 2005 had two passengers and was broadcast to millions of television viewers. China had upgraded its Long March rockets and modified the Soyuz capsule as a basis for Shenzhou. China's announced goals for human spaceflight included construction of a crewed science laboratory.

In 2006 the Indian Space Research Organisation (ISRO) endorsed a plan for India to move forward with its own human spaceflight program. Indian officials claimed ISRO was capable of putting a human in space within a decade and would even consider a crewed mission to the Moon, but admitted that the country would have to benefit in some way before any official government approval.

Through nearly five decades of human spaceflight, putting humans into space was a highly visible symbol of national prowess dominated by the United States, the Soviet Union, and its successor, Russia. Continuing press coverage confirmed the appeal of human spaceflight, along with its expansion to include astronauts from several nations, most prominently China, which became the third nation with an independent human spaceflight capability. By the 2000s wealthy space tourists could pay to travel to the *ISS* aboard Russian vehicles, with the advent of somewhat less expensive private suborbital flights on the horizon. Space remained a difficult environment in which to operate, but its unique qualities also provided scientific opportunities. While microgravity experimentation on humans, plants, animals, solids, and fluids provided insights useful on Earth, its primary benefits accrued to space exploration, science, and applications.

Tim Chamberlin and Stephen B. Johnson

MILESTONES IN THE DEVELOPMENT OF HUMAN SPACEFLIGHT AND MICROGRAVITY SCIENCE

1903 Russian schoolteacher Konstantin Tsiolkovsky publishes article on rocketry and space exploration.

1921 Tsiolkovsky retires from teaching, begins writing series of articles on theories of spaceflight.

1923 Romanian physicist Hermann Oberth publishes *By Rocket to Space*.

1928 Slovenian engineer Herman Potočnik (Noordung) discusses wheel-like space station in *The Problem of Space Travel—The Rocket Engine*.

1944 1 August: German engineer Eugen Sänger releases final report on rocket-boosted winged spaceplane.

1945 Soviet scientist Mikhail Tikhonravov proposes VR-190 for launching humans into space.

1948 11 June: U.S. Air Force (USAF) begins suborbital biological flights with monkeys as passengers.

1949 9 February: USAF forms Department of Space Medicine; German scientist Hubertus Strughold becomes director.

1951 24 March: Josef Stalin orders creation of OKB-23 (Special Design Bureau) in Moscow.

15 August: Soviet Union begins suborbital biological flights with dogs as passengers.

12 October: Von Braun lectures at First Symposium on Spaceflight in New York.

1952 11 February: *Collier's* magazine publishes concepts for piloted spaceflight discussed at First Symposium on Spaceflight.

17 April: Bell proposes to build piloted BOMI (bomber missile) space bomber for USAF after hiring German scientist Walter Dornberger.

1954 9 July: X-15 Project begins as joint venture of the National Advisory Committee for Aeronautics, USAF, and U.S. Navy.

1955 December: USAF approves high-altitude human-occupied balloon flights under Project Manhigh.

1957 8 March: OKB-1 establishes Department No. 9 to develop piloted spacecraft.

2 June: First flight of Project Manhigh.

3 November: *Sputnik 2* carries first animal, Laika the dog, into Earth orbit.

21 December, USAF issues directive to implement Dyna-Soar (X-20) program.

1958 2 July: Soviet government authorizes development of Raketoplan.

1 October: NASA begins operation; Project Mercury starts.

1959 9 April: NASA announces its first group of astronauts.

28 May: Army Ballistic Missile Agency launches monkeys, Able and Baker, on suborbital spaceflight.

1960 11 January: Soviet Air Force establishes Cosmonaut Training Center.

25 February: Soviet Air Force selects its first group of cosmonauts.

1 March: NASA establishes Office of Life Sciences.

23 June: Soviet leaders approve blueprint for piloted space program.

25 July: NASA designates its advanced manned spaceflight program "Apollo."

1961 12 April: Cosmonaut Yuri Gagarin first human to fly in space.

5 May: Astronaut Alan Shepard Jr. first American to fly in space.

25 May: President John F. Kennedy commits United States to piloted Moon landing.

19 September: NASA establishes Manned Spacecraft Center (later Johnson Space Center).

7 December: NASA reveals plans to begin Project Gemini.

1962 20 February: Astronaut John Glenn Jr. first American to orbit Earth.

1 July: NASA establishes Launch Operations Center (now Kennedy Space Center).

11 July: NASA selects Lunar Orbit Rendezvous as mode for Apollo lunar landings.

1963 16 June: Cosmonaut Valentina Tereshkova first woman to fly in space.

19 July: X-15 Flight 90 is first rocket-powered aircraft to reach space.

3 December: Soviet leaders approve development of Soyuz spacecraft.

10 December: U.S. Department of Defense (DoD) cancels Dyna-Soar (X-20) program; approves start of Manned Orbital Laboratory (MOL).

1964 13 April: Soviet leaders approve Voskhod missions.

3 August: Soviet leaders approve piloted missions to the Moon using Lunar Orbit Rendezvous.

12 October: First multicrewed spaceflight (*Voskhod 1*).

12 October: Soviet Vladimir Chelomey proposes creation of Almaz.

1965 1 January: Soviet government decree cancels Raketoplan, program continues at low level.

18 March: Cosmonaut Alexei Leonov conducts first spacewalk.

1966 14 January: Sergei Korolev, leader of the Soviet space program, dies.

March: OKB-1 becomes TsKBEM; OKB-52 becomes TsKBM.

16 March: First docking in space during *Gemini 8*.

11 May: Vasily Mishin becomes Chief Designer of TsKBEM.

1967 27 January: *Apollo 1* crew dies in launch pad fire.

23 April: Cosmonaut Vladimir Komarov dies during *Soyuz 1* reentry.

1968 27 March: Gagarin dies in training exercise.

24 December: *Apollo 8* crew is first to orbit Moon.

1969 16 January: First successful docking of two piloted Soyuz spacecraft.

21 February: First all-up test of the Soviet N1 rocket fails.

10 June: DoD cancels MOL.

18 July: NASA approves "dry" orbital workshop (Skylab).

20 July: Astronaut Neil Armstrong first human to walk on Moon.

15 September: Space Task Group issues report on post-Apollo space program.

22 October: Leonid Brezhnev announces Soviet goal to build space stations.

1971 19 April: Launch of world's first space station (*Salyut 1*).

7 June: *Soyuz 11* crew first to occupy *Salyut 1* space station.

29 June: *Soyuz 11* crew dies during reentry.

1972 5 January: President Richard Nixon approves Space Shuttle development.

23 November: Last test of Soviet N1 rocket ends in failure.

14 December: Last Apollo crew departs Moon's surface.

1973 3 April: First Almaz space station reaches orbit.

14 May: Launch of *Skylab*.

14 August: European Space Research Organisation signs memorandum of understanding with NASA to build *Spacelab*.

1974 22 May: Valentin Glushko replaces Mishin as TsKBEM Chief Designer.

22 May: TsKBEM merges with KB EnergoMash to form NPO Energiya.

24 June: Glushko cancels Soviet crewed lunar program (N1-L3).

13 August: Glushko cancels N1 program.

1975 15 July: Launch of Apollo-Soyuz Test Project.

1976 7 February: Soviet leaders approve Buran-Energiya Space Shuttle.

1977 12 August: First free-flight test of a Space Shuttle.

1978 16 January: NASA names first women to astronaut corps.

20 January: Use of Soviet automated Progress cargo freighters begins.

2 March: First flight with guest cosmonaut, Vladimir Remek of Czechoslovakia on *Soyuz 28*, through the Interkosmos program.

18 May: First European Space Agency group of astronauts enter service.

1979 11 July: *Skylab* disintegrates after reentering the atmosphere.

1981 12 April: First piloted Space Shuttle mission.

19 December: Soviet government cancels Almaz; bans Chelomey from working on space program.

1983 18 June: Astronaut Sally Ride first American woman to fly in space.

30 August: Astronaut Guion Bluford first African American to fly in space.

28 November: Inaugural mission of *Spacelab-1* during Space Transportation System (STS)-9; Ulf Merbold of Germany first non-U.S. citizen to fly aboard a Space Shuttle.

1984 25 January: President Ronald Reagan approves space station project.

4 February: First untethered extravehicular activity using Manned Maneuvering Unit.

11 April: First satellite retrieval and repair on orbit during STS-41C.

27 August: NASA's Teacher in Space Project begins.

1985 12 April: U.S. Senator Jake Garn first elected official to fly aboard a Space Shuttle.

21 November: Last Salyut crew returns to Earth.

1986 28 January: Shuttle *Challenger* explodes 73 seconds into flight.

20 February: *Mir* reaches orbit.

1988 18 July: President Reagan names Space Station project "Freedom."

28 September: 11 nations join Space Station Freedom project.

29 September: First Space Shuttle flight after *Challenger* disaster.

15 November: Only flight of Soviet space shuttle *Buran*, without crew.

1989 20 July: President George H. W. Bush announces Space Exploration Initiative.

1990 25 April: STS-31 crew deploys *Hubble Space Telescope*.

2 December: Japanese journalist Toyohiro Akiyama is first person to fly to space paid by private corporation.

1991 31 July: United States and Soviet Union sign agreement to conduct joint space activities.

8 December: Soviet Union dissolves.

1992 17 June: United States and Russia renew 1991 agreement to conduct joint space activities.

5 October: NASA and Russian Space Agency approve Shuttle-Mir program.

1993 21 June: First flight of commercially developed *Spacehab* module.

2 September: United States and Russia agree to merge space station programs.

7 November: NASA renames Space Station Freedom the *International Space Station*.

2 December: STS-61 crew repairs *Hubble Space Telescope*.

1994 3 February: Sergei Krikalev first cosmonaut to fly aboard a Space Shuttle.

1995 14 March: Norman Thagard first astronaut to fly aboard a Soyuz.

22 March: Cosmonaut Valeri Polyakov returns to Earth from *Mir* after 437 days in space.

1996 18 May: Creation of the X-Prize.

29 June: First Space Shuttle docking with *Mir*.

1998 29 January: Agreement expands ISS program to 15 nations.

20 November: First *ISS* node (*Zarya* Control Module) reaches orbit.

4 December: First *ISS* assembly flight links *Unity* node and *Zarya*.

1999 23 July: Eileen Collins becomes first female Shuttle commander (STS-93).

2000 2 November: First resident crew commissions *ISS*.

2001 23 March: *Mir* disintegrates after reentering the atmosphere.

28 April: American Dennis Tito is first tourist to fly in space.

2003 1 February: Space Shuttle *Columbia* breaks apart during reentry.

15 October: China's first piloted spacecraft (*Shenzhou 5*) reaches orbit.

2004 January 14: President George W. Bush announces Vision for Space Exploration.

21 June: *SpaceShipOne* is first privately funded spaceflight.

4 October: Mojave Aerospace Ventures wins X-Prize.

2005 26 July: First Space Shuttle flight after *Columbia* disaster.

10 October: Sergei Krikalev logs a total of 803 days, 9 hours, and 39 minutes in space—a world record.

2006 7 November: Indian Space Research Organisation endorses plan for Indian human spaceflight.

2007 8 August: Barbara Morgan becomes first educator astronaut to fly in space (STS-118).

2008 27 September: China conducts its first spacewalk during *Shenzhou 7*.

2009 11 May: Fifth and last *Hubble Space Telescope* servicing mission (STS-125).

2010 1 February: President Barack Obama proposes cancellation of Constellation program.

15 April: Obama unveils new vision for NASA and resurrects plans for Orion spacecraft.

Tim Chamberlin and Stephen B. Johnson

See also: Agreement on the Rescue of Astronauts, Almaz, China, Hypersonic and Reusable Vehicles, Manned Orbiting Laboratory, News Media Coverage, Politics of Prestige, Russia (Formerly the Soviet Union), Space Stations, United States

Bibliography

William E. Burrows, *This New Ocean: The Story of the First Space Age* (1999).

Tim Furniss et al., *Praxis Manned Spaceflight Log 1961–2006* (2007).

John M. Logsdon, ed., *Exploring the Unknown: Selected Documents in the History of the U.S. Civil Space Program, Volume VII, Human Spaceflight: Projects Mercury, Gemini, and Apollo* (2008).

Asif A. Siddiqi, *Challenge to Apollo: The Soviet Union and the Space Race, 1945–1974* (2000).

BIOMEDICAL SCIENCE AND TECHNOLOGY

Biomedical science and technology involves the application of natural sciences and engineering to the structure and behavior of living organisms. It has played a key role in human spaceflight because of the emphasis on human health and the stresses associated with living and working in microgravity.

Long before the first piloted spaceflight, the world's major powers conducted tests to determine how military pilots responded to flight conditions, leading to the creation of aviation and later aerospace medicine. The U.S. military established a department of space medicine in 1949 within the Air Force School of Aviation Medicine, which monitored manned stratospheric flights. These showed that pilots required life support systems and newly developed pressurized suits. The Soviet Union began a small spaceflight biomedical program in 1949 within the Soviet Air Force's Institute of Aviation Medicine. Animals, such as primates and dogs, were used during the 1950s to help determine the physical effects that resulted from reduced pressure, confined spaces, isolation, and other factors. Animal testing created a baseline from which doctors could evaluate human pilots.

With the creation of human spaceflight programs in the United States and the Soviet Union in the late 1950s, both nations formed committees to select astronauts and cosmonauts. Along with physicians, high-ranking officials within the U.S. and Soviet space programs helped determine the criteria for selection and training of astronauts and cosmonauts because of the high-pressure politics of the space race. Candidate selection involved a rigorous screening process and onerous medical tests. Both countries established extensive training programs to prepare for piloted spaceflight and staffed offices to monitor astronauts' and cosmonauts' health and behavior.

Despite the intense preparation for each spaceflight, U.S. and Soviet crews experienced difficulties adapting to spaceflight and suffered from motion sickness, disorientation, and sinus infections. These problems, while bothersome, were never serious threats to overall mission success. NASA formed the Office of Life Sciences in 1960, and the Soviet Union created the Institute for Biomedical Problems in 1963 as a foundation for each country's biomedical space-research programs. With the advent of space stations and increasing mission lengths, engineers and doctors developed and assessed the effectiveness of various countermeasures against the negative effects of zero gravity, such as muscle and bone density loss. These countermeasures, including exercise equipment and regimens, proved increasingly effective.

Through the 1960s, keeping astronauts and cosmonauts alive in space was no simple task. In the 1960s, U.S. spacecraft used pure-oxygen atmospheres, while Soviet spacecraft provided nitrogen-oxygen atmospheres. Engineers designed chemical processes to remove carbon dioxide from cabin atmospheres. Waste management began simply with plastic bags for liquid and solid waste, but by the 1970s evolved to include a commode. Space travelers cleaned themselves with wipes, with the exception of *Skylab* astronauts, who utilized a shower whose water was vacuumed up. Food evolved with special packaging, including freeze drying and squeeze tubes. NASA pioneered food safety programs in the 1960s for astronauts; these soon migrated to the U.S. food industry. *Skylab* and Salyut space stations incorporated refrigerators and food warmers. For early missions, water was not reused, but later space stations including *Mir* and the *International Space Station* (*ISS*) included water-recycling equipment.

Space suits were vital to the safety of astronauts and cosmonauts and were adapted for extravehicular activity (EVA). Both the United States and the Soviet Union developed full-pressure suits used in their initial EVAs, which utilized tethers that supplied air. For Apollo, NASA developed suits suited for lunar surface operations, in which astronauts performed their tasks in a dusty, one-sixth-gravity environment. The Space Shuttle program required the development of reusable and resizable space suits that could be used by astronauts of various sizes. The Soviet Union developed the Sokol suit, for use on Soyuz, and the Orlan suit, for EVAs on *Mir* and *ISS*. Both nations learned to perform longer and more complex EVAs to build and repair space stations and retrieve and repair satellites on Shuttle missions. The *ISS* construction and *Hubble Space Telescope* servicing missions in particular required many complex EVAs.

From the earliest speculations about spaceflight, visionaries and designers recognized that the development of a closed loop ecological system would be necessary for long-term habitation in space or on other planets and moons. Soviet research to investigate the viability of closed loop systems began in the 1960s, with animals and then humans, at the Institute for Biomedical Problems and the Institute of Biophysics. NASA performed its own experiments in the 1970s. In the 1990s Biosphere 2 was established and operated near Tucson, Arizona, using private funds. While these experiments proved the viability of some of the necessary processes, by the early 2000s, fully closed loop ecological systems remained far from operational deployment in space, though some recycling of air and water had been deployed on the *ISS*.

Tim Chamberlin and Stephen B. Johnson

See also: Human Spaceflight Programs, Space Stations

Bibliography

Charles Bourland et al., “Food Systems for Space and Planetary Flights,” *Nutrition in Space Flight and Weightlessness Models* (1999).

Gary L. Harris, ed., *The Origins and Technology of the Advanced Extra-Vehicular Space Suit* (2001).

Maura Phillips Mackowski, *Testing the Limits: Aviation Medicine and the Origins of Manned Space Flight* (2006).
Paul O. Wieland, *Designing for Human Presence in Space: An Introduction to Environmental Control and Life Support Systems (ECLSS), Appendix I, Update—Historical ECLSS for U.S. and U.S.S.R./Russian Space Habitats*, NASA/TM-2005-214007 (2005).

Aerospace Medicine

Aerospace medicine is the study of adaptive human response to the environments of aviation and spaceflight. Paul Bert conducted the first medical studies of human response to flight during the mid-to-late nineteenth century. He invented the first altitude chamber and is considered to be the first flight surgeon based on his medical assistance to balloonists. Bert determined that even at low altitudes, humans responded to changes in temperature, barometric pressure, and oxygen availability. He described the causes of altitude sickness, oxygen poisoning, and the bends. In 1903, powered flight added motion sickness to defined medical problems of flight.

Germany, the United Kingdom, and the United States revised medical standards for pilots during World War I by introducing a physical screening process. These standards dramatically reduced pilot-error fatalities among army pilots, as did aviation medicine training provided to medical personnel overseeing the pilots by Theodore Lyster, the father of American military aviation medicine.

Faster and higher-altitude flights introduced the problems of *g*-forces and the need for a breathing apparatus and pressurized cabin. Advances in flight, especially during World War II, confirmed that aviation medicine practitioners must understand aircrews, their environment, and the machines in which the aircrews worked. Germany's *Luftwaffe* conducted research of flight conditions associated with jet- and rocket-powered aircraft, and also test-flew crewed missiles. The U.S. military monitored the effects of high-altitude, long-duration flights and the problems associated with high-altitude bailouts and frostbite. After World War II, several of Germany's leading aeromedical experts, including Hubertus Strughold, known as the father of space medicine, were brought to the United States under Operation Paperclip.

In 1949, the U.S. Air Force (USAF) School of Aviation Medicine created a department of space medicine and appointed Strughold its director. By the late 1950s, military high-altitude balloon flights were providing tests of closed life-support systems and pressurized flight suits. At about the same time, aeronautics and astronautics in the United States became linked by the term aerospace in technology and medicine.

In 1957 the National Advisory Committee for Aeronautics (NACA) set out a blueprint for a space life-sciences program, placing the selection, skill, and survival of the crew as its top priority. When NACA became the National Aeronautics and Space Administration (NASA) in 1958, its Project Mercury had one basic goal: prove that humans could live and work in space and return safely.

HUBERTUS STRUGHOLD (1898–1986)

Hubertus Strughold, a German physician, is considered the father of space medicine. Strughold was director of the German Luftwaffe's aeromedical institute during World War II. He immigrated to the United States after the war under Project Paperclip—a U.S. military operation to recruit Germany's top scientists. Strughold joined the staff of the U.S. Air Force School of Aviation Medicine in 1947 and became director of the school's department of space medicine in 1949. In 1951 he postulated the region where space begins to affect human functions. Strughold wrote more than 180 papers on space medicine.

Tim Chamberlin

NASA's Space Task Group at Langley Field, Virginia, established criteria for astronaut selection. It defined astronaut duties as first to survive in space, then to perform prescribed tasks during spaceflight, serve as backup for automatic controls, make scientific observations, and monitor environment and flight systems. During its early formation, NASA used USAF personnel, facilities, and expertise to screen test pilots as astronaut candidates. Not until 1 March 1960 was NASA's Office of Life Sciences established to oversee all biological and medical programs within the agency.

Soviet cosmonauts were also being vetted in 1959. With requirements similar to those of the United States, the Central Committee of the Soviet Union Communist Party and the Ministry Council of the Soviet Union sought pilots for the 1961 launch of the *Vostok 3KA* spacecraft. Nikolai P. Kamanin, assistant to the commander-in-chief of the Air Forces for Outer Space and later head of the cosmonaut office, established the methods used to select and train cosmonauts. He was best known for his successful efforts with Vladimir Yazdovskiy, head of flight biomedical operations, to secure a place for the cosmonaut in piloting Soviet spacecraft.

The Soviet satellite *Sputnik 2* was the first spacecraft to carry a living creature, the dog Laika, along with a host of biosensors to monitor her progress, signaling the true beginning of space biology and medicine. Using aircraft flights to simulate weightlessness, cosmonauts trained in the basics of space medicine were taught to eat and drink in microgravity as well as to write, work, and maintain visual orientation. Scientists monitored Yuri Gagarin's heart rate and breathing during his historic *Vostok 1* flight in April 1961 and also used a television camera to watch his activities. Gherman Titov's flight on *Vostok 2* was the first reported instance of space motion sickness (SMS) and the first time a human slept in space. Valentina Tereshkova, the first woman in space, flew on *Vostok 6* in 1962. She demonstrated that both genders can adapt to the microgravity environment. The Vostok flights showed that survival in space was possible and that SMS did not impair mission success. Scientists noted postflight orthostatic intolerance, a drop in blood pressure accompanied by a dizzy feeling that results from fluid shifts during return to normal gravity, but did not consider it critical.

NIKOLAI PETROVICH KAMANIN (1908–1982)

Nikolai Kamanin, a general in the Soviet Union Air Force, supervised the planning of all piloted space activities during the pinnacle of the Soviet human spaceflight program. From the late 1950s until his retirement in 1971, he held high-ranking military positions and oversaw the selection and training of cosmonauts. His roles included Deputy Chief of the Air Force's General Staff for Combat Preparations, First Deputy Chief for Space, and the Air Force Commander-in-Chief's Aide for Space. Entries from Kamanin's personal diaries, published after his death, gave the public unique insight into the Soviet piloted space program, which had gone mostly undocumented for decades.

Tim Chamberlin

NASA flew chimpanzees Ham and Enos in 1961 during *Mercury 2* and *Mercury 5*, respectively. The U.S. Navy had successfully flown spider monkeys even earlier, from 1958 to 1961. Two of those were Sam and Miss Sam, named for the USAF School of Aviation Medicine. While these flights demonstrated that there were no significant health effects from launch, weightlessness, or landing, they also demonstrated the need for adequate mission familiarization in accurate simulators for astronaut and ground support. Between the flights of the chimpanzees, astronauts Alan Shepard Jr. and Virgil "Gus" Grissom flew suborbital missions.

The flight of humans during Project Mercury provided the U.S. space program experience with life-support systems and simplified medical procedures during recovery. NASA researchers developed the astronaut survival kit and training regimen, used postflight medical exams to judge the effects of weightlessness, and used flight simulators to develop medical operations procedures and timing. NASA flight surgeons found that fatigue was a serious problem preflight and postflight (though sleep in microgravity was found to be mostly normal). The flight surgeons developed diets for consumption in space and methods and technology for continuous monitoring of body functions by telemetry. A blood pressure monitor was introduced on *Mercury 6* and *Mercury 7*, and personal dosimeters were added on *Mercury 8* and *Mercury 9* to measure radiation. *Mercury 9* also documented postflight orthostatic intolerance.

The Soviet *Voskhod* missions achieved the first flight of a physician in space (Boris Yegorov on *Voskhod 1* in October 1964) and the first extravehicular activity (Alexei Leonov on *Voskhod 2* in March 1965).

The multiday missions of the Gemini program in 1965 proved that humans could live and work in space long enough to get to the Moon and back. During *Gemini 4*, Edward White successfully tested the use of a life-support unit outside the spacecraft. During *Gemini 5*, Gordon Cooper and Charles "Pete" Conrad successfully tested, for the first time, the use of a pneumatic cuff to reduce orthostatic intolerance. They also tested in-flight exercise procedures. Frank Borman and Jim Lovell spent nearly two

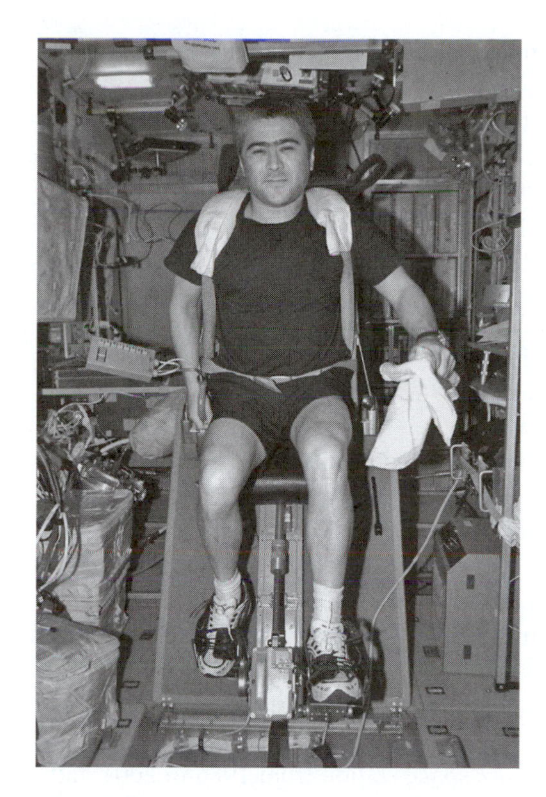

Cosmonaut Salizhan S. Sharipov, a flight engineer during Expedition 10 *on the* International Space Station, *exercises on the Russian VELO Ergometer in the Zvezda Service Module. (Courtesy NASA)*

weeks aboard *Gemini 7* and conducted the first in-flight sleep study, alternating work and sleep shifts to accommodate experiment schedules.

The crew of *Apollo 7*, launched in October 1968, suffered from an in-flight illness that led to establishment of preflight pathogen testing for assigned crew and backups. After Borman suffered from stomach flu during *Apollo 8* in December 1968, NASA created a program to minimize crew clinical health risks before and during a flight. The possibility of contamination from lunar samples in later missions also led to quarantine procedures for both crews and samples. The Apollo Light Flash Investigation solved a medical puzzle during *Apollo 15*, which revealed that flashes of light visible to astronauts in microgravity were due to galactic cosmic radiation impacting the optic nerve. The overall medical findings from the Apollo program showed that less-than-optimal food consumption caused postflight dehydration and weight loss requiring up to one week of recovery. Other medical findings included decreased orthostatic and exercise tolerance postflight, mild cardiac arrhythmias, and decreased red blood cell mass and blood plasma volume.

In 1967–68, Moscow's Institute for Biomedical Problems held the world's first yearlong test of life-support systems. Soviet researchers worked out a number of microgravity countermeasures for use on long-duration flights, including physical exercise, electrostimulation, and prophylactic salts for orthostatic intolerance, among others.

The Soviet Union launched the first of a series of *Salyut* space stations in April 1971. These served as orbiting labs for extensive biomedical studies that emphasized crew health monitoring and scientific experiments. *Salyut* flights helped develop microgravity countermeasures, such as four-day exercise cycles, use of passive stress suits, and the addition of a lower-body negative pressure device. The *Soyuz 14* crew that visited *Salyut 3* in July 1974 followed an expanded exercise program that reduced postflight recovery time. A docking port added to *Salyut 6*, which was staffed from 1977 to 1981, extended missions up to 185 days by allowing the resupply of extra consumables. On *Salyut 7* (1982–86), cosmonauts conducted biomedical experiments that focused on improving health monitoring and microgravity countermeasures. Some of these missions lasted as long as 237 days. During this period, a rehabilitation program was developed for cosmonauts on long-duration flights. On *Salyut 7*, the *Soyuz T-6* crew performed the first electrocardiogram in 1982 during a joint mission between France and the Soviet Union.

Physiology investigations during all three *Skylab* flights from 1973 to 1974 confirmed that physical fitness was not a determining factor in adaptability to microgravity, and that while effects such as fluid shifts occurred immediately, muscle and bone loss accrued over time. Experiments showed that cardiovascular deconditioning is adaptive and related to blood flow volume, in-flight exercise does reduce postflight recovery time, and caloric requirements in space are similar to those on Earth. Many of these experiments resumed during the Apollo-Soyuz Test Project in 1975, where the first joint medical experiments were conducted between the United States and the Soviet Union.

International studies conducted on the *Mir* space station from 1986 to 2001 gathered detailed medical information, including the radiation environment and dosimetry data, for cosmonauts who stayed for periods up to 438 days. From 1995 to 1998, *Mir* hosted seven astronauts who worked closely with cosmonauts in orbit and scientists on Earth on medical experiments that extended work done previously on *Skylab*, the Space Shuttle, and *Salyut*. This cooperation enabled collaborative efforts among American and Russian space medicine researchers that thrived during development of the *International Space Station* (*ISS*) program.

The U.S. Space Shuttle program provided Western space life scientists with a shirt-sleeve laboratory in low Earth orbit and the capacity to fly larger crews. This led to the use of crew teams and round-the-clock schedules for more complex multidisciplinary missions. Arnauld Nicogossian, NASA's chief of medical operations in the late 1970s, established the Longitudinal Study of Astronaut Health in 1992. This long-term, ongoing investigation into human health in the space environment was designed to protect the health of future space travelers by establishing a database for the health risks of spaceflight in comparison to a control group of non-astronauts. Veteran astronaut and U.S. senator from Ohio, John Glenn Jr., a crewmember on Space Transportation System (STS)-95 in 1998 and age 77 at the time, took part in a series of experiments that examined bone and muscle loss, balance disorders, and sleep disturbances by comparing the data to information used during his first flight in 1962.

Beginning with *Expedition 2* in 2001, the *ISS* continued international bioastronautics experiments, many using second- or third-generation hardware initially developed for use on *Skylab* or the Space Shuttle. Looking forward to possible missions to Mars and long-term habitation of the Moon, much of this research was refocused in 2004 to further President George W. Bush's Vision for Space Exploration. The advancement of space medicine has helped to refine selection criteria for astronauts and cosmonauts and to expand flight crew medical training and the availability of in-flight medical care. It has focused research efforts on areas such as radiation protection, postflight rehabilitation, spacecraft habitability, and psychology and group dynamics.

Karen Murphy

See also: Apollo, Gemini, Mercury, *Mir*, Salyut, *Skylab*

Bibliography

Susanne E. Churchill, ed., *Fundamentals of Space Life Sciences* (1997).

Maura Phillips Mackowski, *Testing the Limits: Aviation Medicine and the Origins of Manned Space Flight* (2006).

Arnauld Nicogossian et al., *Space Biology and Medicine* (2004).

Asif A. Siddiqi, *Challenge to Apollo: The Soviet Union and the Space Race, 1945–1974* (2000).

Astronaut Training

Astronaut training broadens proficiencies, sharpens skills, hones decision making, and molds individual competence into effective teamwork among people who already are highly trained pilots, engineers, scientists, and educators. Although not as inherently dangerous as spaceflight, astronaut training is hazardous enough to have claimed several lives.

Most NASA astronauts in the 1960s to mid-1970s were elite test pilots from the military services. Their prior training and flight experience were considered advantageous for the unfamiliar realm of spaceflight. After a thorough medical screening, psychological evaluation, and physical endurance testing, the selected pilots (all male) were designated astronauts.

This first generation of astronauts flew the Mercury, Gemini, and Apollo spacecraft and also assisted in designing and testing them, playing an active test-pilot role in the development of flight hardware and procedures. They trained intensely in flight simulators at the Johnson Space Center (JSC) in Houston, Texas, to master the operations of the new spacecraft. They also trained for such activities as spacewalks and lunar roving vehicle excursions.

To familiarize the astronauts with the sensations of microgravity and build their skills for working in space, NASA developed various simulated microgravity environments. A modified KC-135 tanker aircraft (known as the Vomit Comet), flying roller coaster–like parabolas, served as an airborne lab where trainees experienced a few seconds of weightlessness. Training for spacewalks (extravehicular activity or EVA) occurred underwater in neutral buoyancy tanks (the 40 ft deep Neutral Buoyancy Simulator at Marshall Space Flight Center in Huntsville, Alabama; and the 12 ft deep Weightless Environment Training Facility and, since 1995, the 40 ft

Astronauts, left to right, Richard O. Covey, Steven R. Nagal, and George D. Nelson are on a training flight aboard NASA's KC-135, also known as the "Vomit Comet." The tanker aircraft used a special parabolic maneuver to create brief periods of microgravity, affording a preview of spaceflight. (Courtesy NASA/Johnson Space Center)

deep Neutral Buoyancy Lab, both NASA facilities in Houston). Here an astronaut whose spacesuit was weighted to float without sinking or rising experienced the most realistic simulation of weightlessness during hours of practicing EVA tasks. Air-bearing mobility labs and lunar terrain mockups also proved useful for certain training activities.

Early astronauts received training in science as it related to their mission activities. The Apollo astronauts had detailed training in geology and field research methods to prepare for their missions on the Moon. In the mid-1960s NASA recruited scientists for lunar and Earth orbital missions. Those who were not yet pilots underwent basic flight training. However, the primary focus of the scientists' training was research to be carried out on the Moon or in the *Skylab* space station.

NASA selected the first class of astronaut candidates for the Space Shuttle era in 1978, with further recruitments every two to four years. These groups included pilots and scientists (called mission specialists), reflecting the dual roles of crew members on Shuttle missions. More than half the males and all but three females were selected to be mission specialists, all with advanced degrees in science or engineering. Unlike the earlier astronauts, who were full-fledged astronauts on selection, Shuttle astronauts were candidates until they completed a multidisciplinary curriculum of classes and self-study, in addition to participatory technical exercises. Each class of new astronaut candidates trained together to become eligible for spaceflight; thereafter, pilots focused on piloting and mission specialists on science and engineering competencies until they were assigned to a mission crew. Each mission crew then spent at least one year training together before flight.

In the first year, astronaut candidates mastered an intense orientation to Shuttle systems and operations and to the space sciences and principles of spaceflight. They also completed survival training on water and land, became SCUBA certified, learned about space suits, and gained microgravity experience aboard the KC-135 aircraft.

On completing this basic training, candidates were designated astronauts and started one year of advanced training in a series of simulators. These progressed from single systems to complex multifunction simulators that enabled the astronauts to train for flight operations from prelaunch through landing in addition to orbital operations, such as payload deployment and retrieval, and rendezvous and docking.

On completion of advanced training, astronauts were declared eligible for flight and received technical assignments within NASA until they were assigned to a mission crew. These included, for example, working on upgrades to software or orbiter systems, serving as astronaut support personnel for prelaunch activity, serving as capsule communicators in Mission Control, or specializing in EVA or satellite deployment. During this time, astronauts also had to maintain proficiency in prior training.

On assignment to a mission, each crew jointly embarked on mission-specific training tailored to the planned maneuvers, objectives, and payloads for its flight. This training involved more self-study, laboratory, and simulator work, in addition to sessions at payload developers' and researchers' sites globally. If the mission required EVA or use of the Remote Manipulator System (RMS) arm, the designated crew members trained in the neutral buoyancy facility and RMS simulator. Astronauts headed to the *International Space Station* (*ISS*) received training in *ISS* simulators. The commander and pilot spent hundreds of hours in cockpit simulators and the Shuttle Training Aircraft (a Gulfstream modified to handle like an orbiter) to perfect their skills.

This culmination of training stressed realistic simulations and confident problem solving as the keys to mission success. The crews learned to master every experiment, EVA task, and routine activity, and also to deal with any possible problems. In the final weeks before the mission, crew training became increasingly integrated with the Mission Control Center teams. Every critical phase of the mission was simulated in a "dress rehearsal" of the mission timeline. The simulation director peppered the scenarios with unexpected situations to exercise the mission team's problem-solving and decision-making abilities.

NASA managed astronaut training from JSC, where it maintained a cadre of instructors in all the spaceflight disciplines. Most classroom and simulator training occurred there. Supplementing that core staff were contractors in Houston and other NASA centers, in addition to university-based scientists and payload personnel around the world. Astronaut training involved extensive travel and in-depth study with many instructors.

The value of simulation in preparation for flight, validated by the early astronauts, has continued to be the crux of crew training. In the Shuttle-ISS era, some new training techniques became available as technologies matured and as spaceflight became a more international venture. Astronaut training benefited from advances in computer capabilities and animation for higher-fidelity simulations. Virtual reality visualization became a standard training technique, especially for EVA and space station assembly tasks.

Cosmonaut training generally involved the same kinds of methods and facilities as astronaut training, but geared to the Soyuz spacecraft, which has been Russia's primary piloted spacecraft since 1967. The Gagarin Cosmonaut Training Center (TsPK) in Star City, Russia, has served as the country's main center for spaceflight instruction since the early 1960s. Before Soyuz, early cosmonaut training was centered on survival and emergency skills, adapting to microgravity, scientific experiments, and, later, EVA techniques. Even though the Soviet-built Vostok and Voskhod spacecraft were highly automated, pilots learned how to maneuver the vehicle to change its orbit inclination and reentry inclination. It was not until the advent of Soyuz that cosmonauts also focused training on rendezvous-and-docking techniques. Cosmonaut training for lunar landings never reached the extent of the astronauts' training because no Soviet-piloted lunar mission received immediate priority. Instead training shifted to the Almaz and Salyut space station programs. To prepare for long periods of isolation on space stations, cosmonauts underwent psychological training to maximize crew compatibility, and medical and physical training to prepare their bodies for the effects of long-term exposure to microgravity.

The Apollo-Soyuz Test Project in 1973–75 brought cosmonauts and astronauts together for the first time to train for and conduct a joint rendezvous and docking in space. Since the early 1990s the U.S. and Russian space agencies cooperated on joint Shuttle missions and joint occupancy of *Mir* and *ISS*. This prompted the addition of Russian language and culture training for astronauts and others with a number of NASA personnel in residence at TsPK.

Astronauts from Canada, Europe, Japan, and elsewhere received training through their space agencies in addition to NASA and the Russian Space Agency before flying on the Space Shuttle, Soyuz, or a space station. Chinese astronauts, also called taikonauts, initially trained in Russia as Chinese facilities were being constructed in the early 2000s to launch the country's Shenzhou space vehicle. As of 2008 Thomas Reiter of Germany was the only non-Russian to have been pilot-certified for the Soyuz.

Valerie Neal

See also: Astronauts, Cosmonauts

Bibliography

Henry S. F. Cooper Jr., *Before Lift-Off: The Making of a Space Shuttle Crew* (1987).
Rex D. Hall et al., *Russia's Cosmonauts: Inside the Yuri Gagarin Training Center* (2005).
Tom Jones, *Skywalking: An Astronaut's Memoir* (2006).

Closed Loop Ecological System

Closed loop ecological system (CLES) is an environment in which elements are continuously regenerated from waste products to their initial resource state as a result of biophysical and chemical conversion cycles, without addition of external resources into the environment. Scientists have studied its application for use in space habitats since the beginning of modern rocketry.

Soviet scientist Konstantin Tsiolkovsky first suggested the idea of an autonomously functioning CLES as an approach to life support for long-term space missions in *Investigation of Space by Jet Propulsion Apparatuses* and "Life in the Interstellar Environment," one of several scientific articles he published during his retirement in the 1920s–30s. Soviet geochemist Vladimir Vernadsky defined and proved the existence of closed material cycles for air, water, and chemical elements after he developed biospherics as a new scientific discipline in works he published in 1926 under the title *Biosphere*.

As the Soviet Union and the United States began human spaceflight programs in the late 1950s, a variety of theoretical and experimental studies advanced CLES research. Evgeni Shepelev conducted some of the first experiments that involved CLES at the Institute for Biomedical Problems (IMBP) in Moscow. From 1960 to 1961 he led a study that monitored the effects of micro-algae on rats and dogs inside a closed test facility for periods of up to seven days. Scientists at the U.S. Air Force School of Aviation Medicine conducted similar work in 1961 with monkeys linked to an algae tank via gas exchange for up to 50 hours, without any adverse effects to the animals. The results showed that photosynthetic organisms and animals could support one another in a closed air-loop environment. Shepelev spent a day inside an enclosure with Chlorella (an algae species) in 1961—the first CLES test with a human.

I. A. Terskov shepherded the BIOS-1 and BIOS-2 studies at the Institute of Biophysics (IBP) at Krasnoyarsk, Soviet Union, from 1965 to 1972. These were the first experiments to recycle water and oxygen in a closed system with a human subject. In addition to algae, plants such as lettuce, radishes, and wheat were grown within artificial, closed environments. In 1969 Shepelev directed a study that monitored three test subjects, sometimes referred to as the first bionauts, who lived inside a CLES about the size of a trailer for one year. While food and water were stored ahead of time, plants were grown for use as additional nourishment, and almost all the oxygen and water consumed was recycled. NASA conducted a four-person, 90-day, ground-based chamber test at Langley Research Center in 1970 that resulted in complete regeneration of water and air with no resupply, which set the standard to be used in follow-up research.

BIOS-3, steered by Terskov and Josef Gitelson at IBP from 1972 to 1991, included three separate, closed loop experiments that lasted between four and six months. The results from this work helped determine requirements for life support (amount of water, air, and food) necessary for a person to survive during a long-duration mission in space. The BIOS-3 studies indicated that CLES does not cause adverse effects to a person's health (physiological and psychological) and does not harm adaptation back to Earth's normal environment, and that human control of the system was necessary for it to successfully function. BIOS-3M and BIOS-3M Eco, two additional studies conducted at IBP from 1991 to 1996, helped determine CLES stability limits depending on life-support system configuration and the amount of human control.

U.S. engineer John Allen was the driving force behind Biosphere 2, the largest artificial CLES ever created. (Scientists have referred to Earth's biosphere, more than 3.5 billion years old, as Biosphere 1.) Eight crew members tried to live inside a sealed three-acre glass-domed complex in Arizona from 1991 to 1993 with more than 3,000 species of plants and animals. The small crew could not maintain the delicate balance of the massive facility's closed environment, which suffered when oxygen and carbon dioxide levels inside the complex became unstable. Despite being criticized by the scientific community for not being set up like a typical research lab, Biosphere 2 identified the weaknesses and arduous nature of controlling a large CLES and proved that its immense size and complexity could not be sustained. After the two-year study ended, the primary role of the facility was changed to support research of Earth's ecosystems, not space habitats.

Japan became actively involved with CLES research by constructing a closed ecological experimental facility in Aomori in 1992. The facility was created to develop an advanced system for life support that simulated basic Earth biomes and physiochemical regeneration technologies.

NASA constructed BioHome, a facility at Stennis Space Center in Mississippi used since 1990, to test the integration of bioregenerative technologies, including plants as indoor air purifiers and aquatic plants for waste processing. BioHome was not designed as a completely closed system, but it contributed to research of more efficient CLES technologies that could minimize overall system sizes and weight.

A team of NASA scientists at Johnson Space Center led by Don Henninger conducted four CLES tests in three phases as part of the Lunar–Mars Life Support Test Project from 1995 to 1997. Test subjects lived in sealed chambers for specified periods of time to demonstrate various CLES capabilities, such as regeneration of air and water for up to four persons. The project also tested physiochemical life support equipment for use on the *International Space Station* (*ISS*). Plans for additional CLES research under NASA's advanced life support project BioPlex were terminated as a result of *ISS* cost overruns.

During the early 2000s as part of President George W. Bush's Vision for Space Exploration, the Bioengineering Branch in the Life Sciences Division at NASA Ames Research Center began development of advanced life support technologies for use in regenerative life support systems necessary for long duration human spaceflight missions.

Vadim Rygalov

See also: *International Space Station*, *Spacelab*

Bibliography

Susanne Churchill, *Fundamentals of Space Life Sciences* (1997).
Peter Eckart, *Spaceflight Life Support and Biospherics* (1996).
I. I. Gitelson et al., *Man-Made Closed Ecological Systems* (2003).

Cosmonaut Training. *See* Astronaut Training.

Environmental Control and Life Support System

The Environmental Control and Life Support System (ECLSS) has enabled and maintained habitable environments for crews living and working in space since the first piloted spaceflight in 1961. Included in the ECLSS are subsystems for atmosphere revitalization (breathable air); pressure, temperature, and humidity control; heat rejection (including equipment cooling); fire detection and suppression; food and water supply and management; and waste management. Improvements to the ECLSS developed by the United States and Russia (including the former Soviet Union) occurred as missions in space grew longer and spacecraft became more complex.

Spaceflights conducted by NASA during the Mercury, Gemini, and Apollo programs from 1961 to the early 1970s lasted from 15 minutes to upward of two weeks. The breathable air inside U.S. spacecraft of this time period consisted of 100 percent oxygen. After Apollo, astronauts breathed a mixture of nitrogen and oxygen. Carbon dioxide (CO_2) was removed using lithium hydroxide canisters, while crew cabin fans provided ventilation. Cabin condensing heat exchangers were used for cooling and humidity control, and external radiators were utilized for heat rejection. Crews had to rely on one another's senses for fire or smoke detection. In the event of a fire, astronauts were instructed to use water from the food rehydration gun and had the capability to depressurize the cabin by manually opening the cabin outflow valve. Apollo (and later Skylab) spacecraft were equipped with a portable aqueous gel extinguisher. Potable water was stored and not processed for reuse, except on the Apollo Command Module, where water was a byproduct of the spacecraft's fuel cells. Food was introduced in squeezable tubes during *Mercury 6* in 1962 and in the form of bite-sized cubes on successive Mercury missions. Astronauts later used freeze-dried, thermostabilized, or irradiated foods packaged in special pouches. For human waste management, an in-suit collection bag during Mercury flights stored urine until the end of a mission. The bags were put in use after astronaut Alan Shepard relieved his bladder inside his space suit during *Mercury 3* following a prolonged delay on the launch pad. During Gemini and Apollo flights, feces were collected in bags and stored for disposal on Earth. Bags had to be taped to buttocks to use. Urine was either collected or vented overboard.

The ECLSS introduced by the Soviet Union for its fleet of spacecraft (Vostok, Voskhod, Soyuz, and the Salyut space stations) rivaled U.S. technology and incorporated slightly different processes. Stays aboard Vostok, Voskhod, and Soyuz flights lasted from hours to as much as a week. Crews lived on Salyuts for as long as five months. The air inside Soviet spacecraft was roughly a three-to-one mixture of nitrogen and oxygen, and carbon dioxide was removed through a potassium hydroxide reaction in the vehicle's oxygen regenerator. Each spacecraft was equipped with a liquid-air-condensing heat exchanger. Cosmonauts used a dehumidifier to control humidity, which contained a silica gel drying agent impregnated with lithium chloride and activated carbon. Cosmonauts depended on their senses for fire or smoke detection. Carbon dioxide detectors, which doubled as smoke detectors, were introduced for Salyut missions. Potable water was stored and not processed for reuse. Crews used heaters and refrigerators to prepare and store specially packaged food. During Vostok, Voskhod, and Soyuz flights, urine and feces were entrained in an air stream and collected for disposal at the end of a mission. On Salyut missions, feces were collected in hermetically sealed metal or plastic containers, which were ejected to space about once per week. The urine collector, separate from Salyut's main commode, was a cup-and-tube device with a disposable plastic insert and filter.

Flights to the first U.S. space station (*Skylab*) in 1973–74, where extended stays in low Earth orbit lasted upward of 84 days, required modifications to the ECLSS. This included a multi-bed molecular sieve for carbon dioxide removal, which was later put in use by the Soviet Union on the *Mir* space station and the *International Space Station* (*ISS*). *Skylab* was the first U.S. spacecraft to have freezers, refrigerators, and food warmers and use ultraviolet sensors for fire detection. Feces were collected in gas-permeable bags attached under a form-fitting seat, then vacuum dried and stored. Urine was collected using individual receivers, tubing, and disposable collection bags. Skylab was unique in that it provided a hand shower for the crew; the water was vacuumed up.

The Space Shuttle required several changes to the ECLSS for more than 100 flights, beginning in 1981, that typically lasted one to two weeks. The Shuttle was equipped with lithium hydroxide canisters to filter out CO_2, and an amine-based regenerable system was used for long-duration missions. For pressure, temperature, and humidity control, centralized cabin liquid/air condensing heat exchangers were introduced (and also used on the *ISS*). A Flash Evaporator System provided total heat rejection for the Shuttle during ascent (above 120,000 ft) and reentry (down to 100,000 ft) and supplementary heat rejection during orbital operations. Ionization smoke sensors were used for fire detection, and Halon 1301 served as a fire suppressant. The Shuttle received a commode/urinal system. Feces were collected in a commode storage container, where they were vacuum-dried and held until return to Earth. Urine was sent to a wastewater tank and vented to space when full.

The ECLSS on *Mir*, the Soviet Union's third-generation modular space station launched in 1986, was derived from Salyut, its predecessor. Stays on *Mir* sometimes lasted more than one year. Changes to *Mir*'s ECLSS included optical sensors for fire detection, the addition of three water purification systems, and a commode for urine and feces collection.

The ECLSS for the U.S. segment of the *ISS*, which was occupied by its first resident crew in 2000, was outfitted with multi-filtration and ion exchange sorbent beds and catalytic oxidation for potable water. Crews stayed aboard for six to seven months. The ECLSS also could process urine for water reclamation using vapor compression distillation. Extensive changes where incorporated into the *ISS* fire detection system. The U.S. segment received photoelectric smoke detectors mounted in ventilation ducting, payload racks, and throughout the *Destiny* module. The Russian segment also received improved optical sensors. The U.S. segment was equipped with a commode/urinal system, where feces were collected in a bag and compacted in a cylindrical canister for storage and disposal in a Progress module, along with other trash. The Russian segment had a separate commode/urinal system. On *ISS*, as well as earlier space missions, crew generally cleaned themselves with wet wipes and towels. The towels as well as soiled clothing were disposed as trash.

Details about China's piloted spaceflights in 2003, 2005, and 2008 and its crewed spacecraft, Shenzhou, have mostly been kept from the public, but the design of Shenzhou closely resembles the Soyuz. During Shenzhou's development, China received equipment and expertise from Russia's space agency.

Kathy Daues and Tim Chamberlin

See also: Human Spaceflight Programs

Bibliography

Helen W. Lane and Dale A. Schoeller, *Nutrition in Spaceflight and Weightlessness Models* (2000).

Paul O. Wieland, *Designing for Human Presence in Space: An Introduction to Environmental Control and Life Support Systems (ECLSS), Appendix I, Update—Historical ECLSS for U.S. and U.S.S.R./Russian Space Habitats*, NASA/TM-2005-214007 (2005).

Extravehicular Activity

Extravehicular activity, or EVA, is more commonly known as space walking, a term coined by reporters following the first EVA in 1965. During that EVA, cosmonaut Alexei Leonov became the first human to venture into the hostile environment of space wearing a space suit. Using an inflatable airlock attached to the side of his Voskhod spacecraft, Leonov left crewmate Pavel Belyayev in the crew cabin, and spent 10 minutes floating in space. While Leonov's historic spacewalk proved that humans could function outside a spacecraft, it also highlighted some problems with EVA. Leonov experienced difficulty positioning and maneuvering around the spacecraft, and he also found that the stiff pressurized space suit made it difficult to move—a problem that nearly caused his death. When Leonov attempted to reenter the airlock, he had to loosen a glove attachment and allow a little air to bleed out of his suit so that he could bend his body enough to get back into the airlock. Soviet Union cosmonauts would not attempt another EVA until the joint *Soyuz 4* and *5* missions in January 1969.

During the U.S. Gemini project, NASA's EVA experience began with Ed White's 20-minute spacewalk during the *Gemini 4* mission in June 1965. Subsequent Gemini missions saw astronauts perform a variety of increasingly difficult tasks during EVA, culminating with Edwin "Buzz" Aldrin's spacewalk during the *Gemini 12* mission in November 1966. But U.S. EVA efforts were not trouble-free. Astronauts found that it was virtually impossible to accomplish an EVA without footholds and handholds to restrain their bodies in the weightless free-fall environment. During the *Gemini 9* mission, astronaut Eugene Cernan became so fatigued during his EVA that he was almost unable to make it back into the spacecraft. From that point on, multiple restraint devices, Velcro pads, and handholds were added to all U.S. spacecraft to assist astronauts while they performed EVAs. Cernan's experiences taught NASA that astronauts should spend more time training in simulated weightless environments before flight. By the end of the Gemini project, astronauts had learned how to operate special tools, maneuver efficiently, and even perform simple construction tasks while outside the friendly confines of their spacecraft. Aldrin's successful spacewalk resolved the difficulties that astronauts had previously encountered.

During Apollo, the majority of EVAs performed by U.S. astronauts were on the Moon's surface, where astronauts experienced one-sixth the gravity on Earth. The first Moonwalk during *Apollo 11* lasted about two and a half hours. By the end of the Apollo program, however, astronauts would perform multiple Moonwalks during a single Moon-landing mission, with some lasting more than seven hours. The Apollo astronauts discovered that EVAs on the surface of the Moon presented unique problems. Since they were not weightless, astronauts did not have to worry about body restraints or handholds like their Gemini counterparts. But they did have to contend with dust and dirt, with learning to walk in the lightweight gravity of the Moon, and with the fatigue of fighting bulky pressurized suits while trying to pick up rocks and deploy experiments. Along the way, astronauts also performed four weightless EVAs—one in Earth orbit and three in deep space while returning to Earth from the Moon.

Since the end of the Apollo project, every U.S. and Soviet EVA has occurred in Earth orbit. During the Skylab program in 1973–74, astronauts conducted EVAs to repair broken hardware, install new hardware, and retrieve scientific experiments and materials. Skylab demonstrated conclusively that, given the proper tools and training, astronauts could perform useful and vital work during EVAs. *Skylab 1* astronauts essentially saved the space station following a near disastrous loss of its thermal shield and one of its two solar panels during launch. During several EVAs, astronauts Pete Conrad, Joseph Kerwin, and Paul Weitz installed a replacement heat shield and, using cutters and saws, deployed the remaining solar wing that had become stuck when the heat shield pulled away during launch.

Following the *Soyuz 4* and *5* EVAs in 1969, it would be eight years before cosmonauts performed another EVA. Beginning with *Salyut 6* in 1977, and continuing until the end of the Mir program, the Soviet Union conducted more than 40 EVAs during which they installed hardware, repaired failed components, and tested space-welding techniques, a task performed in 1984 by Svetlana Savitskaya, the first woman to conduct a space walk.

Meanwhile Space Shuttle astronauts conducted a variety of EVAs—demonstrating space construction techniques, satellite repair and maintenance, including the *Hubble Space Telescope* servicing missions, and retrieving errant satellites using a free-flying rocket backpack known as the Manned Maneuvering Unit. With the advent of the *International Space Station*, EVAs had become an integral part of the construction and maintenance of orbiting spacecraft.

Michael Engle

See also: Gemini, *International Space Station*, Salyut, *Skylab*, Space Shuttle, Voskhod

Bibliography

David S. F. Portree and Robert C. Treviño, *Walking to Olympus: An EVA Chronology* (1997).

Space Food

Space food has been freeze-dried, thermostabilized, irradiated, and left in its natural form to supply astronauts and cosmonauts with the necessary nutrients to maintain their health in microgravity. Specialized packaging of food has been required to withstand the rigors of spaceflight and to permit its consumption in microgravity. Adequate nutrition has countered the adverse physiological affects of space travel that include bone and muscle loss and radiation damage, while familiar and appetizing foods have proven essential for crew morale and psychological well-being.

When cosmonaut Yuri Gagarin (*Vostok 1*) and astronaut Alan Shepard (*Mercury 3*) ventured into space, it was unknown whether humans could swallow and digest food in microgravity. Space food development began in the United States with highly engineered foods that met rigid requirements imposed by the spacecraft design and short mission durations of the Mercury and Gemini programs. The lack of adequate bathroom facilities and limited food storage capacity promoted the development of low-fiber diets to reduce fecal output. Food for the Soviet Union's Vostok and Voskhod programs were engineered to meet spacecraft requirements and consisted of pureed foodstuffs in aluminum tubes and bite-size dried meats and fish. As missions lengthened, space food systems evolved.

Gagarin, the first human in space in 1961, was also the first cosmonaut to eat in space. John Glenn became the first astronaut to eat in space when he consumed applesauce from a tube on the third Mercury mission in 1962. Other foods introduced on later Mercury missions included bite-size cubes of a high-calorie mixture of protein, high-melting-point fat, sugar, and fruit or nuts. Food research and development emphasis during Mercury was on calorie-dense, nutritious, and palatable food. Since these missions were brief, no provisions were made for specific food storage on board the spacecraft.

Specialized food and packaging were designed for Gemini flights, which led to the beginning of the Hazard Analysis Critical Control Point (HACCP) system adopted worldwide by the food industry. HACCP, developed by NASA, the U.S. Army Natick Center, and the Pillsbury Company, focused on preventing hazards that could cause food-borne illnesses by testing raw materials throughout the food chain to the finished product. Crews ate bite-size cubes or foods squeezed from tubes. Cubed foods used during the Gemini and Apollo programs contained a gelatin coating to reduce crumbs. Crackers and cookies that were consumed were small so they could be eaten by placing the entire portion in the mouth. Freeze dried powders and artificially flavored beverage powders were used for drinks. Some of the artificially flavored beverages, such as Tang, were used for all missions from Gemini through the International Space Station (ISS) program.

During Apollo, dehydrated food was used to conserve space and weight, especially because water was available as a by-product from the Apollo Command Service Module's fuel cells. Apollo astronauts were the first to use utensils, retort pouches (also called wet packs), and irradiated food products. Foods eaten during Apollo included bacon squares, turkey with gravy, orange drink, and shrimp cocktail. The spoonbowl package was developed during Apollo as a solution to the problem of direct package-to-mouth consumption and was used on *Skylab*, during the Apollo-Soyuz Test Project, and the first four Space Shuttle missions.

Skylab was the first U.S. spacecraft to have freezers, refrigerators, and food warmers, which provided astronauts with a palatable and varied diet chosen from 72 foods. This included ice cream, filet mignon, lobster, and chilled beverages. A dining table had built-in food heaters with timers for advanced preparation of food. Because the food eaten on board *Skylab* was launched with the station, some of it was more than two years old when consumed. Most of the food was packaged in aluminum cans to maintain a two-year shelf life. The most extensive metabolic study ever undertaken by NASA in space took place on *Skylab* and provided the agency with its baseline nutritional information.

The Soviet space station programs, Salyut and Mir, utilized food heaters, refrigerators, and hot water. The refrigerator on *Mir* had limited use due to electrical power constraints. The *Mir* dining table concept, which had built-in food heaters, was used for the *ISS*. To heat food, individual food packages were inserted into heater wells on the dining table.

The number of food choices for Space Shuttle flights was greater than that for all previous missions. Astronauts tasted and selected from 150 food items, which included beef steak, tortillas, scrambled eggs, sweet-and-sour chicken, and banana pudding. NASA dietitians analyzed each astronaut's menu choices for content and recommended substitutions to ensure a balanced supply of nutrients during flight. Crewmembers had the opportunity to exchange meal items for snacks or other foods from the Shuttle's pantry. As a result, actual in-flight dietary intake did not always match the nutritionally balanced menu planned before flight. The food system designed for the Space Shuttle did not have freezers or refrigerators because of the short duration of planned missions and the lack of storage room and electrical power. Water as a by-product from the orbiter's fuel cells helped rehydrate food. Astronauts ate meals from open containers on a meal tray. A single package design was used for food and beverages. The Shuttle's galley was equipped with a convection oven to heat water and food to serving temperatures. Due to the hazards involved in a closed environment, cooking food has never been attempted in space. Foods only have been heated to serving temperatures.

During Shuttle-Mir spaceflights, a system that combined U.S. and Russian menus and foods was used on *Mir* during shared operations. This combined system was incorporated for use on the *ISS*. Although similar preservation techniques were used, both countries selected and packaged foods differently. NASA used flexible pouches for most of its thermostabilized foods, while Russian foods were stored in steel cans that required a can opener. The Russian beverage package had a drink tube that had to be handheld, while NASA's beverage package had a straw with a clamp to allow hands-free consumption. Russian space food included buckwheat gruel, cottage cheese with nuts, prunes stuffed with nuts, and bream in tomato sauce. U.S. space food included breakfast sausage links, seafood gumbo, beef stroganoff, and shrimp cocktail. Russian menu planning recommended four meals per day, while NASA recommended three with a pantry for snacks. The *ISS* food menu started with four meals per day and was later converted to three. A frozen food system with a combination microwave convection oven was planned for the *ISS*, but was canceled when the Habitation Module was removed from the station's construction plans.

Charles Bourland

See also: Apollo, Apollo-Soyuz Test Project, Gemini, *International Space Station*, Mercury, *Mir*, Salyut, Shuttle-Mir, *Skylab*, Space Shuttle, Voskhod, Vostok

Bibliography

Charles Bourland et al., "Food Systems for Space and Planetary Flights," in *Nutrition in Space Flight and Weightlessness Models* (1999).

Malcolm C. Smith et al., "Apollo Food Technology," in *Biomedical Results of Apollo* (1975).

Space Suit

A space suit consists of a pressurized garment and life-support system that protects a person from the hostile environment of space. The earliest reference to space suits was by the Russian teacher Konstantin E. Tsiolkovsky (1857–1935) in his writings on space travel. Today's space suits evolved from early aviation high-altitude pressure suits. The U.S. aviation pioneer Wiley Post worked with the B. F. Goodrich Company to develop the first pressure suit during flights in 1934. As aircraft continued to reach higher altitudes, both the U.S. Air Force and the U.S. Navy developed improved high-altitude partial and full-pressure suits.

The first human in space, Soviet cosmonaut Yuri Gagarin, wore the SK-1 pressure suit. During Project Mercury, astronaut Alan Shepard became the first American in space and wore the silver painted space suit known as the Mark IV pressure suit developed for the Navy by Goodrich. NASA selected the David Clark Company of Worcester, Massachusetts, to develop two space suits for Project Gemini. The first version was a full-pressure suit, and the second had an outer thermal and micrometeoroid garment for extravehicular activities (EVA). Both versions had to operate in the vacuum of space, because the entire Gemini spacecraft depressurized when astronauts opened the hatch to perform a space walk. NASA contracted with the AiResearch Company to develop the first EVA Life Support System (ELSS) for the Gemini space suit. The ELSS provided pressure regulation controls and connections to a 25 ft umbilical attached to the spacecraft. Soviet cosmonaut Alexei Leonov had performed the world's first EVA on 18 March 1965 using the Berkut space suit. During *Gemini 4* on 3 June 1965, astronaut Edward White performed the first U.S. EVA using the G1C space suit. Important technological advances to space suit flexibility and life-support systems, such as carbon dioxide removal, helmet fogging prevention, humidity control, and human thermal control systems, were made during Project Gemini.

In 1961 NASA selected United Technologies Hamilton Standard as the prime contractor for lunar space suit development. Hamilton Standard was responsible for the development of the life-support system and the integration of the pressure garment provided by the International Latex Corporation (ILC). Lunar space suits included the Apollo Block 1 suit, designated the A1C, which utilized a David Clark pressure garment and was used during test flights. The Apollo Block 2 suit, called the A1L, used the ILC pressure garment and was used to perform EVAs in Earth orbit, in deep space during return flights from the Moon, and on the lunar surface. These suits led to the development of the Extravehicular Mobility Unit (EMU), an integrated lunar space suit with the Hamilton Standard–provided Portable Life Support System (PLSS) and the ILC-provided pressure garment. On 20 July 1969 Apollo astronauts Neil Armstrong and Edwin "Buzz" Aldrin walked on the Moon using the EMU.

As follow-on to the Apollo program, *Skylab* was the first U.S. spacecraft to incorporate an airlock so astronauts could conduct EVAs without having to depressurize the entire spacecraft. Engineers modified Apollo A7LB pressure garments for use on *Skylab*. This modified Apollo suit used for *Skylab* EVAs was known as the *Skylab* EMU. Modified Apollo A7LB suits were again used during the Apollo-Soyuz Test Project.

The Space Shuttle program presented new challenges in space suit development requiring the use of reusable space suits that could be resized between flights to fit a variety of astronauts. During test flights in 1981–82, Shuttle crewmembers wore David Clark full-pressure suits in case of cabin depressurization or seat ejection. On subsequent flights the ejection seats were removed, and the crew did not wear full-pressure suits because confidence grew in the performance of the Shuttle. After the *Challenger* accident in 1986, the use of the David Clark full-pressure suit for crew escape during launch and landing resumed on all Shuttle flights. The shuttle EVA suit was a new, ILC-designed modular pressure garment with a waist entry and a Hamilton Standard–modified Apollo PLSS.

Soviet/Russian space suits evolved in parallel to U.S. space suits to support its human space programs. The Research Development and Production Enterprise Zvezda, founded under the name Plant No. 918, designed and built the Soviet/Russian space suits. The SK-1 was used only during Vostok flights. The Berkut space suit was a soft EVA suit that used an umbilical system and a portable life-support system. The Soviet Union also had designed a lunar space suit known as the Krechet. This space suit had a semi-rigid upper torso and a rear-entry hatch for donning the suit. Work on this suit ceased when Soviet plans to land on the Moon ended.

Russian space suits improved during the Salyut and Mir space station programs. The Russians used the Sokol suit on all Soyuz spacecraft, including flights to the *International Space Station* (*ISS*). The Orlan space suit was designed specifically for conducting EVAs from the Salyut and *Mir* space stations and was used on the *ISS*. From 1993 to 1994, Russia and the European Space Agency (ESA) cooperated in the development of a prototype EVA suit to be used on the proposed European Hermes space plane. This suit, known as the Hermes EVA Suit 2000, was similar to the Orlan. ESA canceled this space suit program when it canceled Hermes.

During the launch of China's first human spaceflight on 15 October 2003, the Chinese used a space suit derived from the Soyuz's Sokol suit. On 27 September 2008,

Taikonaut Zhai Zhigang wore a newly developed Chinese Feitian space suit while conducting the country's first EVA during *Shenzhou 7*.

Robert C. Trevino

See also: Gemini, Salyut, Space Shuttle

Bibliography

Isaak P. Abramov and A. Ingemar Skoog, *Russian Spacesuits* (2003).

Gary L. Harris, ed., *The Origins and Technology of the Advanced Extra-Vehicular Space Suit* (2001).

Lillian D. Kozloski, *U.S. Space Gear: Outfitting the Astronaut* (1993).

Space Walking. *See* Extravehicular Activity.

HUMAN SPACEFLIGHT CENTERS

Human spaceflight centers have been built in the former Soviet Union and Russia, the United States, Europe, and China as bases for cosmonaut, astronaut, and taikonaut training, space vehicle research and development, and launch operations and control. Soviet and U.S. centers were built in the 1960s in support of each country's ambitions to put the first human in space and send humans to the Moon as part of the Cold War space race. The centers were expanded during the next four decades in support of space station development and other human spaceflight programs, primarily Soyuz and Space Shuttle. China, which emerged as a leader in the satellite launch industry, constructed its facilities at the start of the twenty-first century.

The Soviet Union was the first country to lay down infrastructure in preparation for human spaceflight. The Cosmonaut Training Center (TsPK) in Zvezdny (Star City), northeast of Moscow, Russia, was approved for construction in January 1960 to train the first group of military pilots selected to fly in space. By 1963 the center's rapid expansion culminated with the appointment of Yuri Gagarin, the first human to fly in space, as its deputy director. The center was renamed in Gagarin's honor after his death in 1968. Cosmonauts and astronauts trained at TsPK during the next 40 years as the Soviet Union became actively involved with international missions to Salyut space stations as part of its Interkosmos program, the Shuttle-Mir program, and construction of the *International Space Station* (*ISS*).

Linked with TsPK is the Baikonur Cosmodrome in Kazakhstan, where (as of 2010) every Soviet and Russian piloted spaceflight has been launched. Baikonur became the central launch hub for Soviet human spaceflight missions in the early 1960s primarily because it already hosted the R-7 launch pad used to orbit the world's first artificial satellite, *Sputnik*. Through 1975 a series of military-run centers supported piloted spaceflight operations until the civilian-run Flight Control Center (TsUP) in Podlipki, near Moscow, was handed control.

Following the creation of NASA and the start of the Mercury program in October 1958, the United States sought a permanent location for a center to accelerate human spaceflight. NASA announced in September 1961 that Houston, Texas, would be the site for the Manned Spacecraft Center, renamed Johnson Space Center in 1973. It hosted the Mission Control Center (MCC) for human spaceflight operations, in addition to being the home center for astronauts. The MCC became one of the most recognized areas of any NASA center as televised footage of mission controllers was shown to millions of viewers around the world. It supported the Space Shuttle, Space Station Freedom, and ISS programs.

Kennedy Space Center (KSC) has served as the launch operations center for NASA since the early 1960s. Located on Merritt Island, Florida, KSC had the advantage of launching over the Atlantic Ocean, away from populated areas. Its tropical climate and unique facilities, such as the gargantuan Vehicle Assembly Building used for the Saturn V and Space Shuttle, has attracted millions of visitors over the years.

Marshall Space Flight Center, located near Huntsville, Alabama, has served since the 1960s as the central location for development of rocket and propulsion systems for the Mercury, Apollo, and Space Shuttle programs.

The European Astronaut Centre was founded in 1990 in Cologne, Germany. As the home of the European Astronaut Corps, it defines and implements astronaut selection, training, and performs medical monitoring of European astronauts in flight for missions to the *ISS*.

China has used the Jiuquan Satellite Launch Center (JSLC) located in the Gobi desert in Inner Mongolia as its primary human spaceflight center. Mission control and other logistical support for China's three taikonaut-piloted Shenzhou missions launched in 2003, 2005, and 2008 took place at JSLC. Construction of an additional center for testing China's piloted spacecraft began in 2005 at the Shenzhuang Industrial Zone in Shanghai. China's State Council and Central Military Commission also approved in September 2007 construction of the Wenchang Satellite Launch Center, in Hainan Province near Wenchang, to support future rocket launches capable of carrying space stations and enabling crewed lunar missions. Chinese officials cited Wenchang's close proximity to the equator, which improves the efficiency of orbital launches, as the reason for its selection.

Tim Chamberlin and Stephen B. Johnson

See also: China, European Space Agency, National Aeronautics and Space Administration, Russia (Formerly the Soviet Union)

Bibliography

Henry C. Dethloff, *Suddenly, Tomorrow Came—: A History of Johnson Space Center* (1993).

Andrew J. Dunar and Stephen P. Waring, *Power to Explore: A History of Marshall Space Flight Center, 1960–1990* (1999).

Brian Harvey, *China's Space Program: From Conception to Manned Space Flight* (2004).

———, *Russia in Space: The Failed Frontier?* (2001).

Kenneth Lipartito and Orville R. Butler, *A History of the Kennedy Space Center* (2007).

Cosmonaut Training Center

Cosmonaut Training Center (Star City) (TsPK), named after Yuri A. Gagarin, the first human to fly in space, has prepared flight crews for the rigors of spaceflight for more than four decades, using airborne and underwater weightlessness simulators, a large centrifuge for research in acceleration overloads, and working mockups of each Soviet spacecraft and the Russian segments of the *International Space Station*.

The Soviet Air Force established TsPK on 11 January 1960 in Zvezdny (Star City), Moscow, to prepare the first group of 20 cosmonauts for the Vostok program. Yevgeny Karpov, an Air Force physician, was the center's first director. By the mid-1960s TsPK became a major research institute with a large staff of scientists and the ability to prepare multiple crews for different missions simultaneously.

Training included learning all theoretical aspects of spaceflight, and because two or three crews were prepared for each mission, qualification tests determined primary and backup crews. Cosmonauts were trained to survive emergency situations that could occur in space and after abnormal landings. Since Soviet spacecraft were designed to return from space by parachuting into remote areas, survival training included isolation in the wilderness and simulated water landings. Apart from the crew's performance, utmost attention was paid to its psychological stability and compatibility. Depending on the mission, training lasted from a few months up to two years.

TsPK trained cosmonauts for all Soviet- and Russian-piloted space missions since 1961, and international crews from 24 countries including France, Germany, Japan, the United Kingdom, and the United States. Chinese astronauts trained at TsPK during the mid-1990s for the Shenzhou program. TsPK has also held responsibility for training mission controllers for Russia's Flight Control Centers located near Moscow. In 2009, the Russian Defense Ministry handed over operations and management of TsPK to the Russian Space Agency (Roskosmos).

Peter A. Gorin

See also: Cosmonauts, Russian Federal Space Agency

Bibliography

Asif A. Siddiqi, *Challenge to Apollo: The Soviet Union and the Space Race, 1945–1974* (2000).

Johnson Space Center

Johnson Space Center (JSC) was established by NASA as the Manned Spacecraft Center in September 1961 and is located about 20 miles southeast of Houston, Texas. The center's Mission Control Center served as NASA's primary command outpost for human spaceflight, and JSC also served as the forward base for astronaut training, spacecraft development, and testing facilities. The center is located on 1,000 acres of land donated by Rice University and an additional 620 acres purchased by the

The Johnson Space Center in Houston, Texas, is home to several NASA facilities, including Mission Control Center. (Courtesy NASA/Johnson Space Center)

government. Construction of the center's facilities was completed at a cost of $60 million in June 1964.

Robert R. Gilruth, chair of the original Space Task Group, served as the center's first director where he oversaw 25 piloted spaceflights through *Apollo 15*. The center was named for President Lyndon B. Johnson one month after his death in January 1973. With the completion of Project Apollo, JSC continued its role as NASA's operations hub during the Skylab project in 1973 and the Apollo-Soyuz Test Project in 1975.

The Lunar Sample Laboratory Facility was built at JSC in 1979 for $2.5 million to permanently house the 842 lbs of lunar rocks, core samples, pebbles, and regolith brought back from the six Apollo missions that landed on the Moon. The 14,000 sq ft facility was constructed to secure the materials in a clean and climate-controlled environment.

In support of the Space Shuttle program, the JSC complex expanded during the 1980s–90s to more than 100 buildings to house new simulators, such as the Shuttle Orbiter Trainer and Manipulator Development Facility. The Space Station Control Center was added in November 1991, in addition to the Neutral Buoyancy Laboratory in 1995, in preparation for construction of the *International Space Station (ISS)*. JSC has served as the primary operations center for the *ISS* since the first resident crew arrived at the station in November 2000.

In 2007 NASA built a multimillion-dollar climate-controlled building near the JSC entrance to enclose and put on display one of the three remaining Saturn V boosters leftover from Project Apollo.

Melinda Marsh

See also: Apollo, Astronauts, *International Space Station*, Space Shuttle

Bibliography

Henry C. Dethloff, *Suddenly, Tomorrow Came—: A History of the Johnson Space Center* (1993).
NASA JSC. http://www.nasa.gov/centers/johnson/home/index.html.

Kennedy Space Center

Kennedy Space Center (KSC) was established by NASA as the Launch Operations Center in July 1962 and is located on Merritt Island, Florida, just north of the rocket launch complex at Cape Canaveral Air Force Station. The center acquired its present name in November 1963 after the assassination of President John F. Kennedy. Before and during its development, launch operations for the Mercury and Gemini programs were conducted from KSC, while the actual launches took place at Cape Canaveral.

The sprawling 139,699-acre facility was expanded to launch Saturn V rockets to the Moon during Project Apollo. Construction of twin Saturn V launch pads, a launch control center and the massive 525 ft Vehicle Assembly Building (VAB) took three years to complete at a cost of $194 million. Kurt H. Debus, a German rocket scientist brought to America by the U.S. Army after World War II, was the center's first director and supervised its development. KSC hosted 13 Saturn V launches from 1967 to 1973 at Launch Complex 39A, six of which sent humans to the surface of the Moon. The complex was also the launch site for the Skylab program in 1973 and the Apollo-Soyuz Test Project in 1975.

During the 1970s KSC built the Orbiter Processing Facility and Shuttle Landing Facility to support the Space Shuttle program, and it modified the VAB and launch complex to accommodate Space Shuttle launch operations. From April 1981 to April 2010, 131 Shuttle missions were flown from KSC. In 1995 construction of the center's Space Station Processing Facility was completed to prepare Shuttle payloads for the *International Space Station*.

Since its formation, KSC has also been the lead center for managing the launch of planetary spacecraft and satellites from Cape Canaveral and maintained the processing facilities for government and commercial payloads placed in orbit. KSC is also the most toured of NASA's 10 centers and includes a visitor's center that has been open to guests since 1966.

Joel Powell

See also: Apollo, *International Space Station*, Space Shuttle

Bibliography

Charles D. Benson and William Barnaby Faherty, *Moonport: A History of Apollo Launch Facilities and Operations* (1978).
Kenneth Lipartito and Orville R. Butler, *A History of the Kennedy Space Center* (2007).

Launch Operations Center. *See* Kennedy Space Center.

Manned Spacecraft Center. *See* Johnson Space Center.

Marshall Space Flight Center

Marshall Space Flight Center (MSFC), located southwest of Huntsville, Alabama, got its start in February 1956 when the Huntsville Arsenal and the Redstone Arsenal combined to form the Army Ballistic Missile Agency (ABMA), commanded by John B. Medaris. In July 1960, with President Dwight D. Eisenhower's approval, management of ABMA facilities was transferred to NASA to establish MSFC, named for George C. Marshall, U.S. Army chief of staff during World War II. German rocket pioneer Wernher von Braun became the center's first director.

Work on Project Mercury began at MSFC in 1958, when it was still the ABMA. Von Braun and a team of MSFC engineers designed and developed the Mercury-Redstone rocket used to launch the first American into space, astronaut Alan Shepard, on 5 May 1961. MSFC engineers also developed the Saturn V booster and the lunar rover used during Project Apollo.

During the 1970s the Research Projects Laboratory, one of eight laboratories at MSFC, studied extensively the feasibility of various NASA projects, including the High Energy Astronomy Observatories, the Large Space Telescope, and the Apollo Telescope Mount. The Skylab workshop was produced at MSFC, and the center's Mission Operations Office played an active role troubleshooting the space station's tumultuous launch in May 1973.

During the next three decades, MSFC managed several of NASA's major projects, including development of the Space Shuttle's solid rocket boosters, external tank and main engines, and the *Hubble Space Telescope*. MSFC also provided technical assistance for Spacelab, a scientific research module that fit inside the orbiter's cargo bay, and support for its integration into Shuttle missions. The Spacelab Mission Operations Control Center was established in 1990 at MSFC to supervise Spacelab experiments, continuing in that role for the *International Space Station*.

The U.S. Space and Rocket Center (USSRC), located adjacent to MSFC, was established as a tourist attraction in 1970 to showcase MSFC's achievements, while the U.S. Space Camp next to the USSRC provided technical experiences for children starting in 1982. In response to the 2004 Vision for Space Exploration, MSFC was assigned management of new crew and cargo launch vehicles to replace the Space Shuttle.

Melinda Marsh

See also: Apollo, *Hubble Space Telescope*, Mercury, Redstone, Saturn Launch Vehicles, *Skylab*, Space Camps, Space Shuttle, Von Braun, Wernher

Bibliography

Andrew J. Dunar and Stephen P. Waring, *Power to Explore: A History of Marshall Space Flight Center, 1960–1990* (1999).

MSFC. http://www.nasa.gov/centers/marshall/home/index.html.

Mission Control Center

The Mission Control Center (MCC) has served as NASA's command headquarters for human spaceflight operations. The MCC has provided continuous support for piloted space vehicles during ascent, on-orbit activities, reentry, and landing.

Originally known as the Mercury Control Center at Kennedy Space Center during the early 1960s, the MCC evolved into "Mission Control" and was permanently established as part of NASA's Manned Spacecraft Center (renamed Johnson Space Center) in Houston, Texas, in 1964 in preparation for the Gemini and Apollo programs. Under the direction of Chris Kraft, NASA's first flight director, the MCC was housed in a three-story, windowless structure called Building 30. In addition to staff offices, the facility included simulation rooms and a Mission Operations Control Room (MOCR) on the second and third floors.

The MOCR resembled a tiered auditorium where flight controllers and other personnel monitored various onboard systems. This included a flight dynamics officer, guidance control officer, and flight surgeon. An astronaut served as capsule communicator (CAPCOM), the official liaison between ground and flight crews. A flight director supervised mission operations and assumed overall responsibility for its success.

The second-floor MOCR served as the primary command and control area for *Gemini 4–12*, and the Apollo lunar landings where data from the launch vehicle, spacecraft, and ground systems was processed and displayed on large screens and computer consoles. The room was also the operations hub for the Skylab program (1973–74) and the Apollo-Soyuz Test Project (1975). The third-floor MOCR was dedicated for missions with military payloads during the Space Shuttle era.

The MCC was modified during the 1980s to support commercial Space Shuttle flights to low Earth orbit. Technology evolved in the MCC, moving in the 1990s from centralized mainframe computers to UNIX-based, decentralized workstations. The MOCR remained in service until the mid-1990s, when it was replaced with two new Flight Control Rooms (FCRs) to handle the Space Shuttle and *International Space Station* (*ISS*) programs. A "White" FCR was created for Space Shuttle operations and a "Blue" FCR for station operations. In 2000 the Training Flight Control Room, "Red" FCR, was added to the MCC to support Shuttle and *ISS* missions through simulations and testing. As construction of the *ISS* progressed, a larger Blue FCR with more computer consoles was needed to accommodate various flight control disciplines. In 2006 the Blue FCR was relocated to the original area that housed the MOCR and retrofitted with new computers and high-definition screens.

Tim Chamberlin

See also: Apollo, Apollo-Soyuz Test Project, Astronauts, Gemini, *International Space Station*, Mercury, *Skylab*, Space Shuttle

Bibliography

Chris Kraft, *Flight: My Life in Mission Control* (2001).

Gene Kranz, *Failure Is Not an Option: Mission Control from Mercury to* Apollo 13 *and Beyond* (2000).

NASA JSC. http://www.nasa.gov/centers/johnson/home/index.html.

Ken Peek, "History of Human Spaceflight Operations," *Quest* 10, no. 4 (2003): 20–27.

Star City. *See* Cosmonaut Training Center.

Zvezdny Gorodok. *See* Cosmonaut Training Center.

HUMAN SPACEFLIGHT PROGRAMS

Human spaceflight programs were established in the Soviet Union and the United States in the late 1950s. The programs developed rapidly during the 1960s as the countries attempted to gain technological superiority and strategic military advantage over each other and in the process reach the Moon. During the next four decades, programs continued to be costly and complex, and, as a result, as of 2010 China was the only other nation to successfully launch humans into orbit. However, a less expensive alternative to space travel emerged in 2004, when privately funded ventures resulted in the first nongovernmental human spaceflight, opening the door to space tourism.

Human spaceflight was a core element of the space race, which pitted the United States against the Soviet Union in a race to various space "firsts." It began with the Soviet launch of *Sputnik* in October 1957, and for the next three years the United States and the Soviet Union competed with robotic spacecraft. The Soviet Union upped the ante by putting the first man in space (Yuri Gagarin, April 1961) and first woman in orbit (Valentina Tereshkova, June 1963) under the Vostok program, which was spawned as the competitor to the U.S. Mercury program. After Gagarin's flight, U.S. President John F. Kennedy searched for a program that stood a good chance of besting the Soviet Union. This became the Apollo program, announced shortly after the first successful U.S. human spaceflight (Alan Shepard, May 1961). To test needed capabilities for Apollo, such as orbital rendezvous, the United States created Gemini. The Soviet Union responded to Gemini and Apollo by creating the Voskhod and Soyuz programs. Voskhod launched the first multi-crewed mission (October 1964) and first extravehicular activity (March 1965). Gemini milestones included the first successful rendezvous flights (December 1965) and the first piloted docking mission (March 1966). The Soviet response to Apollo and its powerful Saturn V booster included Soyuz, Zond, and the N-1 rocket programs, but ultimately these did not put Soviet cosmonauts on the lunar surface. The United States placed 12 astronauts on the lunar surface from July 1969 to December 1972. The cost of trying to reach the Moon was expensive, both in terms of funding and lives, as both cosmonauts and astronauts died in various mishaps. The Soviet crewed lunar program continued into the mid-1970s, but failures of the N-1 rocket and the limited political benefits of a second-place crewed Soviet landing finally led to its demise.

After the race to the Moon, political and public opinion relegated spaceflight to low Earth orbit. The Soviet Union and the United States had initiated military programs during the 1960s to place cosmonauts and astronauts in space stations for reconnaissance purposes. The U.S. Air Force (USAF) Manned Orbiting Laboratory (MOL) program was canceled in 1969 in favor of less-expensive robotic satellites. The Soviet Almaz military space station program continued, but was kept secret by combining it with the civilian

Astronaut Thomas P. Stafford (left) sits in a Soyuz spacecraft simulator with cosmonaut Andriyan G. Nikolayev during a visit to the Soviet Union as part of a U.S. group working on the Apollo-Soyuz Test Project (ASTP). (Courtesy NASA/Johnson Space Center)

Long-Duration Orbital Station program under the civilian Salyut label in the 1970s–80s. The Soyuz spacecraft ferried cosmonauts to seven Salyut space stations in dozens of flights from 1971 to 1986. Just as Almaz had been created in response MOL, the civilian aspects of the Salyut program had been created to counter NASA's Skylab. Three U.S. missions flew to one *Skylab* station during 1973–74. The Apollo-Soyuz Test Project (July 1975), a joint U.S.-Soviet orbital docking program, briefly brought the nations closer, but did not lead to anything meaningful for another two decades.

In the 1970s NASA focused on development of the Space Transportation System (STS), approved by President Richard Nixon in December 1972. Given limited funding, the Nixon administration required that NASA forge international ties to fund and build portions of the STS. The Europeans eventually agreed to build an experiment module, called Spacelab, while Canada contributed a robot arm called Canadarm. To gain the support of the USAF, NASA agreed to design its payload bay to accommodate large U.S. military spy satellites. When the Space Shuttle started flying in 1981, the Soviet Union responded with its Buran shuttle, which only flew once (November 1988) before being decommissioned. France, with the help of the European Space Agency (ESA), tried to build the Hermes space plane during the 1980s, but cost overruns killed the project. NASA's cooperation with Europe and Canada led to the creation of European and Canadian astronaut corps to fly on the Shuttle. The Soviet Union responded by creating the Interkosmos program to fly researcher-cosmonauts, from its global allies, to Salyut.

With the Space Shuttle operational, President Ronald Reagan approved the U.S. Space Station Freedom (SSF) program in 1984. The station was to rival the Soviet Union's *Mir* space station, a next-generation station with multiple docking and expansion ports, launched in 1986. ESA and Canada, along with Japan, contributed to SSF. After the collapse of the Soviet Union in the early 1990s, it evolved to become a much broader international effort, inviting the newly democratic Russia into the International Space Station (ISS) program.

An integral part of the new U.S.-Russian relationship was the ability of both to allow their human spaceflight programs to adapt to each other's needs. NASA gained valuable knowledge from Russia in building and operating space stations. The *ISS*'s core modules (*Zarya* and *Zvezda*) were derived from *Mir*. Likewise, Russia desperately needed money to help its starving economy, and the funds it received as NASA's chief *ISS* partner were critical to keeping its fledgling space program alive. Following the *Columbia* disaster of February 2003, Russia was responsible for sustaining the *ISS* program until Shuttle flights resumed two years later. Without Russia's assistance, the station's orbit would have deteriorated and its crew would not have received critical supplies via Progress, a spacecraft originally developed for the Salyut program as a cargo ferry.

China successfully launched its first taikonaut into space in October 2003 aboard a Shenzhou space vehicle. Building on the success of its Long March rockets, the Chinese piloted program started in the mid-1990s. China sought help from Russian space officials to build a space capsule; Shenzhou derived from the Soyuz design. Chinese astronauts initially trained in Russia as Chinese facilities were being constructed.

California-based Scaled Composites Corporation launched the first privately funded suborbital mission in June 2004, breaking the monopoly of government-run institutions on human spaceflight. Several new start-up space companies emerged and started the space tourism market. Spaceports in four U.S. states and in countries, such as Singapore and the United Arab Emirates, were planned as a result.

In 2004 NASA initiated Constellation, a long-term, phased program designed to take astronauts to the *ISS*, the Moon, and possibly Mars. The program was predicated on the early retirement of the Space Shuttle in 2010. Parts of the STS were to be evolved for use with Constellation, such as the Shuttle's solid rocket boosters, which are the basis for the first stage of the *Ares 1* crew launch vehicle, and the Shuttle's external tank, planned to be used for the proposed *Ares 5* cargo launch vehicle. In March 2010 the Barack Obama administration proposed its cancellation in favor of commercial launch to low Earth orbit. Two months later Obama directed that NASA design and build a new heavy lift rocket no later than 2015, and to use the Orion crew capsule as a lifeboat to the *ISS*.

The robust human spaceflight programs that characterized the space race of the 1960s and early 1970s gave way to internationally driven programs during the last two decades of the twentieth century. During the first 45 years of human spaceflight, more than 200 piloted missions were flown and 10 space stations were built.

Tim Chamberlin and Stephen B. Johnson

See Human Spaceflight Missions Table at the end of the Human Spaceflight and Microgravity Science section, on pages 686–736.

See also: China, European Space Agency, National Aeronautics and Space Administration, Russia (Formerly the Soviet Union), United States

Bibliography

William E. Burrows, *This New Ocean: The Story of the First Space Age* (1999).
Tim Furniss and David J. Shayler, *Praxis Manned Spaceflight Log 1961–2006* (2007).
David Harland and John E. Catchpole, *Creating the International Space Station* (2002).
Asif Siddiqi, *Challenge to Apollo: The Soviet Union and the Space Race, 1945–1974* (2000).
Robert Zimmerman, *Leaving Earth: Space Stations, Rival Superpowers, and the Quest for Interplanetary Travel* (2003).

3KD. *See* Voskhod.

3KV. *See* Voskhod.

Almaz

Almaz (or Diamond), a series of military space stations built by the Soviet Union, was derived from tensions between the world's two superpowers during the height of the Cold War and was built as a vehicle for cosmonauts to spy on Earth-based targets, primarily U.S. military assets.

The Soviet Union began work on a military space station as early as 1962. Sergei Korolev, head of OKB-1 (Experimental Design Bureau), sketched plans for two modified Soyuz spacecraft to dock in Earth orbit to form a small station, the Soyuz-R, which could be used for military purposes. OKB-1's Branch No. 3 was assigned to develop the idea from 1963 to 1965. During that time, the U.S. Air Force scrapped plans for its X-20 Dyna-Soar spaceplane and replaced it with the Manned Orbiting Laboratory (MOL) project—a modified Gemini spacecraft that would allow astronauts to perform visual reconnaissance of the Soviet Union, among other military objectives. In response to MOL, Vladimir Chelomey, head of OKB-52 (later NPO Mashinostroyeniya or Machine Building Scientific-Production Association) in Moscow, proposed in October 1964 the development of a 20-ton military space station called Almaz. This was formally approved by the Soviet Union's Ministry of Defense in early 1966. OKB-52 served as the prime contractor for the entire Almaz project.

During 1966–67, Almaz underwent several redesigns and its development was divided into two stages. The first included building a station with its own three-person crew return vehicle called the Return Apparatus, which resembled MOL's Gemini-B capsule. The second involved construction of a station that would be launched separately from a Transport Supply Ship (TKS). The TKS would ferry crews and cargo to the station and back to Earth. Eventually, Chelomey chose to drop the Return Apparatus from the Almaz station and incorporate it into the design of the TKS.

The *Apollo 11* Moon landing in July 1969 prompted the Soviet government to shift the focus of the country's space program from landing on the Moon to orbiting space stations. Soviet officials wanted to regain some prestige by placing the world's first

Almaz Orbital Piloted Stations (OPS)

Name	Launch Date	Reentry	Days in Orbit	Highlights
Salyut 2 (*Almaz 1* or *OPS 1*)	3 Apr. 1973	28 May 1973	26	Hull rupture 13 days into mission
Salyut 3 (*Almaz 2* or *OPS 2*)	25 Jun. 1974	24 Jan. 1975	213	Resident crew: *Soyuz 14*
Salyut 5 (*Almaz 3* or *OPS 3*)	22 Jun. 1976	8 Aug. 1977	412	Resident crews: *Soyuz 21, Soyuz 24*

space station in low Earth orbit. Because Chelomey could not guarantee that Almaz would be ready for flight in a couple of years, the Soviet Central Committee and Council of Ministers ordered Chelomey in February 1970 to transfer many of his Almaz hulls to OKB-1 for a quick overhaul with Soyuz hardware. This led to the creation of the Long-Duration Orbital Station program, which became known as Salyut (Salute). While work continued on Almaz, publicly the military stations thereafter were called Salyuts, most likely to hide their military role.

Each Almaz was launched from the Baikonur Cosmodrome in Kazakhstan atop a UR-500 (Proton) rocket. Crews rode a Soyuz 7K-T spacecraft to reach the military space stations since development delays prevented use of the TKS for several years.

The first Almaz, *OPS 1* (*Salyut 2*), reached orbit on 3 April 1973, but its hull ruptured 13 days into its mission, rendering it useless. The second Almaz,*OPS 2* (*Salyut 3*), reached orbit on 25 June 1974 awaiting the crew of *Soyuz 14*, cosmonauts Pavel Popovich and Yuri Artyukhin. The men successfully docked with the Almaz on 5 July, conducted scientific experiments, tested the station's surveillance equipment, and returned to Earth two weeks later. *Soyuz 15*, launched on 26 August, was unable to dock with the second Almaz as a result of a rendezvous system failure, so its mission was aborted after only two days. The second Almaz performed much like an uncrewed reconnaissance satellite the remainder of its lifetime, taking photos automatically. It was deorbited on 24 January 1975 before another crew could visit.

The third Almaz, *OPS 3* (*Salyut 5*), reached orbit on 22 June 1976. Three crews attempted to visit the station, but only two were successful. *Soyuz 21* docked on 7 July, but the crew had to cut short its photoreconnaissance mission after a temporary failure of *Salyut 5*'s onboard systems and a buildup of unknown fumes in the station. Despite their early departure, cosmonauts Boris Volynov and Vitali Zholobov worked on the station for 48 days. *Soyuz 23* flew within 40 m of *Salyut 5* but could not dock because of a guidance system malfunction. *Soyuz 24* cosmonauts Viktor Gorbatko and Yuri Glazkov tested the air inside *Salyut 5* before reactivating the station on 8 February 1977. Their tests revealed no contamination, so the crew carried out a list of reconnaissance and scientific objectives before departing on 25 February. The reconnaissance observations made from *Salyut 5* were conducted in conjunction with Soviet military exercises. Cosmonauts used a large camera called the Agat and were able to take pictures, develop them, and return the film to Earth in a small capsule in about 30 minutes. However, the lag time in recovering the film made it practically useless for tactical purposes. The station was deorbited on 8 August after 412 days in space.

By summer 1978 the Ministry of Defense had slashed funding for Almaz stations in favor of the civilian Salyut program, and work on a fourth military space station was canceled. The Soviet government concluded that reconnaissance satellites were more effective than crewed space stations doing the same work. While the piloted component of Almaz was finished, work on an automated Almaz radar platform (Almaz-T) continued until late 1981, when a government decree ended the program. The Almaz-T was resurrected in 1986, and three military satellites of its class were eventually launched from 1987 to 1991.

Despite having never been fully developed for Almaz, the TKS was later used on *Mir* and became the first component of the *International Space Station* (*ISS*) called the Functional Cargo Block (FGB), or *Zarya*, launched in November 1998. Mir's base block, launched in February 1986, and the *ISS Zvezda* module, launched in July 2000, were also derived from Almaz hulls.

Tim Chamberlin

See also: Manned Orbiting Laboratory

Bibliography

Asif A. Siddiqi, "The Almaz Space Station Complex: A History, 1964–1992," pts. 1 and 2, *Journal of the British Interplanetary Society* 54, no. 11/12 (November–December 2001): 389–416; and 55, no. 1/2 (January–February 2002): 35–67.

Asif A. Siddiqi, *Challenge to Apollo: The Soviet Union and the Space Race, 1945–1974* (2000).

Robert Zimmerman, *Leaving Earth: Space Stations, Rival Superpowers, and the Quest for Interplanetary Travel* (2003).

Apollo

Apollo (1960–75), the third U.S. human spaceflight program, put 12 astronauts on the Moon between 1969 and 1972. When NASA formally announced Project Apollo in July 1960, its stated goals were to accomplish circumlunar flight and advanced Earth-orbit operations before 1970, leading eventually to a crewed lunar landing and a permanent crewed space station. Following President John F. Kennedy's announcement in May 1961 that the United States would attempt to land a man on the Moon and return him safely to the Earth by the end of the decade, that goal became Apollo's primary purpose.

Early concepts for accomplishing a crewed lunar landing focused on three different approaches: a direct flight from Earth to the lunar surface (direct ascent); assembling a large lunar spacecraft in Earth orbit (Earth-orbit rendezvous, or EOR) and then launching it toward a landing on the Moon; and lunar surface rendezvous, which would have astronauts land on the Moon and then walk to a return vehicle that had been previously landed automatically. These concepts required technological capabilities considerably beyond the state of U.S. space technology and involved a high degree of risk. Direct ascent depended on a huge rocket capable of generating nearly 10 million pounds of thrust, while EOR would have required complex operations such as rendezvous and docking and extravehicular activity (EVA), neither of which had yet been demonstrated. Lunar surface rendezvous called for precise lunar landings to ensure that the crew could find and walk to its waiting return vehicle. EOR became NASA's initial preferred option.

In June 1961 John C. Houbolt, an engineer at NASA's Langley Research Center in Hampton, Virginia, presented a novel lunar landing concept to a committee chaired by Bruce Lundin, an associate director at Lewis Research Center. Houbolt proposed lunar orbit rendezvous (LOR) as a method of accomplishing a lunar landing. LOR required two spacecraft: a command module (CM), in which the crew would ride during the majority of the flight, and a lunar module (LM), which would be detached from the CM in lunar orbit and landed on the Moon. The Lundin committee ranked Houbolt's LOR proposal near the bottom of several recommendations, stating that the concept was too risky. It instead chose EOR as the best option because the idea was thought to hold greater potential for rescue operations, if necessary, and the concept had already been embraced by some within the agency as a way to develop potential technologies for an orbiting space station.

NASA engineers did not readily accept the recommendation of the Lundin committee. The controversy about the best method for landing a human on the Moon continued until June 1962, when Marshall Space Flight Center Director Wernher von Braun recommended to NASA's Office of Manned Space Flight that LOR be adopted. The next month, NASA selected Houbolt's idea as its most feasible approach to accomplish a crewed Moon landing by the end of the decade. The agency would use the two-stage Saturn 1B booster for Apollo test flights and the larger, three-stage, Saturn V rocket, outfitted with five massive F-1 engines built by North American Rocketdyne, for missions to the Moon.

The Apollo spacecraft's main components were comprised of a CM, an LM, a service module (SM), and a launch escape system. NASA had already chosen North American Aviation as the prime contractor to build the Apollo CM and SM. In November 1962 the agency selected the Grumman Aircraft Engineering Corporation from among nine competing companies as the LM prime contractor.

The CM was cone-shaped; 11 ft, 5 in tall; had a base diameter of 12 ft, 10 in; and weighed 12,250 pounds. It consisted of three main sections. A forward compartment held two reaction control engines and components of the landing system. A crew compartment housed the module's controls and systems and seats for three astronauts. The aft section contained reaction control engines and propellant, helium tanks, and water tanks. A heat shield composed of brazed stainless steel honeycomb covered by a phenolic epoxy resin protected all three compartments from the extreme temperatures that resulted during reentry through Earth's atmosphere.

The SM was a 24 ft, 7 in tall cylinder attached beneath the CM and weighed more than 50,000 pounds. It consisted of a service propulsion system (SPS) with a hypergolic engine, helium and fuel-oxidizer tanks, fuel cells, and stored cryogenic oxygen and hydrogen.

The LM stood 22 ft, 11 in tall and weighed more than 33,000 pounds at launch. It contained a two-person crew compartment, midsection, and aft equipment bay. Because the LM was made for two astronauts, one member of each Apollo crew would pilot the CM while the others rode the LM to the lunar surface. The LM was equipped with an ascent stage that functioned as a separate spacecraft that rendezvoused and docked with the CM after leaving the surface of the Moon.

From 1964 to 66, Project Gemini developed the necessary techniques required for Apollo missions—rendezvous and docking, EVAs, and long-duration spaceflight—and gave astronauts and flight directors vital operational experience.

By 1967 NASA hoped to launch the first crewed Apollo spacecraft, but *Apollo 1* astronauts Virgil "Gus" Grissom, Edward White and Roger Chaffee were killed after a fire broke out in their CM during a launch pad test rehearsal at Kennedy Space Center (KSC) in Florida on 27 January. NASA established a review board the next day to investigate the accident. Its findings, released on 5 April, indicated that the likely cause of the fire was a faulty wire that sparked and ignited the pure-oxygen environment of the CM, starting a fire that quickly filled the spacecraft with deadly fumes. All three astronauts died from asphyxia due to inhalation of toxic gases.

While the tragedy was a major setback to the U.S. human spaceflight program, the Soviet Union had its share of misfortune trying to eclipse America's achievements in space. Already suffering from its own technical delays, the Soviet Union's first crewed Soyuz flight—its new maneuverable crew capsule—ended in failure and the death of cosmonaut Vladimir Komarov on 24 April 1967. Less than a year later, the communist superpower sustained another devastating loss when its most famous cosmonaut, Yuri Gagarin, died in a training accident. By the late 1960s, repeated test failures of the Soviet N-1 Moon rocket made it improbable that the Soviet Union would send cosmonauts to the lunar surface before the United States.

It took NASA nearly 20 months to recover from the Apollo test fire and launch *Apollo 7*, the first crewed Apollo mission. During this period, North American Aviation redesigned the Apollo CM to make it safer and more fire resistant. *Apollo 7*, launched on 11 October 1968, was the only piloted Apollo flight to reach space atop a Saturn 1B booster. Each mission to the Moon thereafter left KSC with the Apollo spacecraft mated to a Saturn V. *Apollo 7* was a full-up test flight of the Apollo command and service modules (CSM) in Earth orbit, but without the LM. Astronauts Walter Schirra Jr., Donn Eisele, and Walter Cunningham successfully completed the 10-day mission, the first time three Americans flew in space together. During the flight, the crew fired the Apollo spacecraft's main engine, the SPS, eight times. The mission also featured the first live television broadcast from a U.S.-crewed spacecraft. *Apollo 7* splashed down in the Atlantic Ocean southeast of Bermuda.

Originally *Apollo 8* was to be a test flight of the LM in Earth orbit. However, in mid-1968 it became apparent to Apollo Program Director George Low that LM development problems would likely postpone the flight from November 1968 to

February or March 1969. Also, it was unclear at the time if the Soviet Union was close to attempting a piloted circumlunar flight with a modified Soyuz spacecraft. As a result, in August 1968 Low proposed that the *Apollo 8* crew instead fly a lunar orbital mission without the LM. NASA's top management viewed the proposal with skepticism, but eventually concluded that it would reduce the possibility of the Soviet Union getting there first and was the only way to ensure that Kennedy's goal would be accomplished by the end of the decade.

Apollo 8 was launched on 21 December. Astronauts Frank Borman, James Lovell, and William Anders entered lunar orbit at an altitude of approximately 69 miles on 24 December—the first humans to leave the gravitational pull of Earth and see the far side of the Moon. During the mission, the crew transmitted televised images of the lunar surface and read biblical verses from Genesis. Anders probably took the first Earthrise photograph during the crew's fourth orbit. The image is often viewed as a catalyst that started the environmentalist movement in the United States. On 25 December, after 10 orbits around the Moon, *Apollo 8* left lunar orbit, landing in the Pacific Ocean two days later.

Apollo 9 was launched on 3 March 1969 into Earth orbit with astronauts James McDivitt, David Scott, and Russell Schweickart on board to conduct the first flight test of the LM. During its 10-day mission, the *Apollo 9* crew performed exhaustive tests of the LM (nicknamed *Spider*) during free flight and while docked to the Apollo CSM (*Gumdrop*). The crew also conducted the program's first EVAs.

Apollo 10 was launched on 18 May to test the performance of the LM in lunar orbit in preparation for the first Moon landing. On 22 May, a day after reaching the Moon, astronauts Eugene Cernan and Tom Stafford entered the LM (*Snoopy*) and separated from the CSM (*Charlie Brown*). Astronaut John Young remained to pilot the CSM. Cernan and Stafford steered the LM to within about 10 miles of the lunar surface. From this lower orbit, they observed the *Apollo 11* landing site, conducted tests of the LM's landing radar, took stereo photographs of the Moon, and tested the LM's descent and ascent stage engines. Following their low altitude tests, Cernan and Stafford maneuvered *Snoopy* into a higher orbit and rendezvoused and docked with *Charlie Brown*. The *Apollo 10* crew returned to Earth on 26 May, splashing down on target in the Pacific Ocean. All of the techniques required for a lunar landing mission had been demonstrated.

Apollo 11 signified the culmination of thousands of civil service employees and subcontractors working together to achieve a common goal. More than 3,000 journalists and photographers from around the world swarmed KSC in preparation for the first lunar landing mission. The Cape was packed with hundreds of thousands of onlookers in what became the most anticipated spaceflight ever.

Apollo 11 was launched on 16 July 1969 and entered lunar orbit after three days in space. On 20 July, astronauts Neil Armstrong and Edwin "Buzz" Aldrin Jr. entered the LM (*Eagle*) and undocked from the CSM (*Columbia*). Astronaut Michael Collins watched the *Eagle*'s descent to the Moon from inside *Columbia*.

Following a harrowing descent to the lunar surface, during which a series of computer overloads and communications difficulties threatened the mission, Armstrong and Aldrin landed on the lunar surface in the Sea of Tranquility. Near the final stage of the descent, the crew realized the planned landing site was strewn with boulders, prompting Armstrong to take manual control and steer *Eagle* to a successful landing

NEIL ALDEN ARMSTRONG
(1930–)

(Courtesy NASA/Marshall Space Flight Center)

Neil Armstrong, an astronaut, was the first human to walk on the Moon on 20 July 1969. He flew 78 combat missions in the Navy during the Korean War and became a test pilot for the National Advisory Committee on Aeronautics in 1955, where he piloted seven rocket-powered X-15 flights. Armstrong was among the first civilians selected as an astronaut in 1962. He commanded *Gemini 8* to the first-ever docking of two vehicles in space in 1966 and successfully responded to a major thruster problem that spun the spacecraft so fast it nearly caused the astronauts to black out. This strong performance earned the respect of his colleagues and superiors and was a major factor in his selection to be the commander of *Apollo 11*. He retired from NASA in 1971.

Tim Chamberlin

with less than 30 seconds of fuel remaining. Armstrong relayed the first words from the Moon, "Houston, Tranquility Base here, the *Eagle* has landed."

On the evening of 20 July, Armstrong became the first human to set foot on the lunar surface, where he uttered the famous words, "That's one small step for a man, one giant leap for mankind." In more than two hours on the Moon, Aldrin and Armstrong collected rock and soil samples, set up a variety of scientific experiments, raised a U.S. flag, and spoke with President Richard Nixon. They departed the Moon the next day, rendezvousing and docking with *Columbia*. The *Apollo 11* crew guided its CM to a successful splashdown in the Pacific Ocean on 24 July and spent the next three weeks in quarantine at the Lunar Receiving Laboratory (LRL) at the Manned Spacecraft Center (renamed Johnson Space Center [JSC] in 1973) in Houston, Texas, to ensure that the astronauts did not bring harmful lunar microbes back to Earth. Lunar samples were stored at the LRL until the Lunar Sample Laboratory Facility was built at JSC in 1979 to permanently house the materials.

Apollo 12 lifted off on 14 November destined for the Moon's Ocean of Storms. Although only the second crewed lunar landing mission, *Apollo 12* had the ambitious goal of retrieving parts from the U.S. *Surveyor 3* spacecraft, which landed on the Moon in 1967. *Apollo 12* nearly ended prematurely when lightning struck the spacecraft's Saturn V booster twice during launch, resulting in a loss of electrical power. With instructions from ground controllers, the *Apollo 12* crew restored power and successfully reached lunar orbit. On 19 November, astronauts Charles "Pete" Conrad Jr. and Alan Bean boarded the LM (*Intrepid*) and guided it to a pinpoint landing about

500 feet from *Surveyor 3*. Richard Gordon Jr. watched the descent from the CSM (*Yankee Clipper*). During two EVAs, the landing crew worked on *Surveyor*, collected lunar samples, and set up an experiment package powered by a small nuclear generator, as opposed to the solar-powered instruments on *Apollo 11*. After 31.5 hours on the Moon, Conrad and Bean rendezvoused and docked with the CSM. By 24 November, the *Apollo 12* crew landed in the Pacific Ocean and was transported to the LRL to spend nearly three weeks in quarantine.

Apollo 13 was to be the third crewed lunar landing mission, but the flight almost became the first U.S. disaster in space. Lifted into space on 11 April 1970, *Apollo 13* began its translunar injection about 2.5 hours after launch. On the evening of 13 April, an oxygen tank exploded inside the SM, so NASA aborted the mission. Astronauts James Lovell Jr., John Swigert Jr., and Fred Haise Jr. used the LM (*Aquarius*) as a lifeboat because the CSM (*Odyssey)* had to be powered down to preserve its batteries for Earth atmospheric entry. For four harrowing days the flight controllers and crew struggled to keep life support systems working so the astronauts could return home alive. On 17 April *Odyssey* splashed down in the Pacific with its crew safe, bringing an end to a near disastrous mission that NASA subsequently deemed a "successful failure."

North American Aviation made modifications to the Apollo SM to ensure that such an event was unlikely to reoccur. This delayed the launch of *Apollo 14* by nearly four months. Commanded by astronaut Alan Shepard, who became the only original Mercury astronaut to make it to the Moon, *Apollo 14* was launched on 31 January 1971 and successfully reached *Apollo 13*'s intended landing site in the Fra Mauro highlands on 5 February. Shepard and Edgar Mitchell exited their LM (*Antares*) and explored the lunar surface during two separate EVAs, culminating with Shepard becoming the first "Moon golfer" when he hit a golf ball more than 400 yards. After 33 hours on the lunar surface, both astronauts used the LM ascent stage to rendezvous with CSM (*Kitty Hawk*) pilot Stuart Roosa. On 9 February the crew of *Apollo 14* landed in the Pacific Ocean, and was the last to be quarantined. By then, NASA's Interagency Committee on Back Contamination was convinced that the Moon harbored no dangerous organisms, so future Apollo crews were not quarantined.

Apollo 15 was launched on 26 July, the first of a series of three intensive scientific missions that would fully exploit the capabilities of the Apollo spacecraft and crews. On 30 July, as the CSM (*Endeavour*) orbited overhead, the LM (*Falcon*) landed in the Hadley-Appenine region of the Moon with astronauts David Scott and James Irwin on board. Scott depressurized the LM, opened its top hatch, and stood up to survey the landing site. The intent of the "stand-up survey" was to provide scientists on the ground with a preliminary scientific assessment of the landing site. The next day, both astronauts began the first of three EVAs that featured the first use of the Lunar Rover, a dune buggy–like vehicle that was stowed on board *Falcon*'s descent stage. Using the rover, Scott and Irwin successfully completed their demanding scientific tasks, including exploration of the majestic Hadley Rill, a large canyon of uncertain origin, and collection of samples from the nearby Appenine Mountains. On 2 August, after nearly 67 total hours at Hadley-Appenine, *Falcon* lifted off to rejoin *Endeavour* and astronaut Alfred Worden, who had conducted orbital science activities while Scott and Irwin explored the lunar surface. Just prior to firing their SPS engine to return home, the *Apollo 15* crew

Astronaut Charles M. Duke Jr. explores the rim of Plum crater at the Apollo 16 *Descartes landing site in April 1972. (Courtesy NASA/Johnson Space Center)*

deployed a small subsatellite, which remained in lunar orbit conducting studies of charged particles, electromagnetic fields, and lunar gravity anomalies. During the return voyage, Worden performed the first deep-space EVA to retrieve data cassettes from the SM's scientific instrument module bay. On 7 August *Apollo 15* parachuted to Earth.

Apollo 16 was launched on 16 April 1972 toward the Descartes region of the Moon to explore the lunar highlands for the first time. The mission was nearly canceled due to problems with the SM's SPS engine, but fortunately the problems were resolved, and on 20 April the LM (*Orion*) landed on target. Astronauts John Young and Charles Duke conducted three exhaustive EVAs with the Lunar Rover totaling more than 20 hours, and on 23 April they departed from Descartes and docked with the orbiting CSM (*Casper*) and pilot Thomas Mattingly II. The *Apollo 16* mission concluded with a landing in the Pacific on 27 April.

The final Apollo mission, *Apollo 17*, lifted off just past midnight on 7 December, the first U.S. crewed space mission to be launched in darkness. On board *Apollo 17* was astronaut Harrison Schmitt, the first American scientist to fly in space. Schmitt was a lunar geologist and before his selection for the mission had been involved in lunar geology training for the previous lunar landing crews. On 11 December Schmitt and mission commander Eugene Cernan landed their LM (*Challenger*) in the Taurus-Littrow region and conducted three EVAs during a 75-hour stay on the Moon. While Cernan and Schmitt explored the lunar surface, astronaut Ronald Evans conducted his own scientific investigations from the CM (*America*). Cernan became the last man of the Apollo program to walk on the Moon, and *Apollo 17* returned to Earth on 19 December, marking the end of the Apollo era.

The space agency had planned three additional lunar landing missions, but Congress canceled them as public enthusiasm for the program waned. With much of the nation's attention focused on the Vietnam War and other domestic concerns, NASA's budget had been drastically reduced by the completion of the last Apollo flight. Leftover hardware from the canceled missions was used to carry out NASA's first space station program, Skylab, in 1973–74, and the Apollo–Soyuz Test Project in 1975.

Michael Engle

See also: Apollo Science, Astronaut Training, Astronauts, Boeing Company, Douglas Aircraft Company, Grumman Corporation, North American Rockwell Corporation, Politics of Prestige, Saturn Launch Vehicles

Bibliography

David Baker, *The History of Manned Spaceflight* (1981).
Andrew Chaikin, *A Man on the Moon* (1994).
Michael Collins, *Carrying the Fire: An Astronaut's Journeys* (1974).
David West Reynolds, *Apollo: The Epic Journey to the Moon* (2002).

Apollo Applications Program. *See Skylab.*

Apollo-Soyuz Test Project

Apollo-Soyuz Test Project (ASTP) was the first joint space mission between the United States and the Soviet Union. NASA and the Soviet Academy of Science agreed to discuss the project in 1970 and reached an agreement two years later to launch a joint mission in 1975. Glynn Lunney, NASA flight director, and Konstantin D. Bushuev, deputy chief designer of the former Korolev design bureau (TsKBEM), managed the project.

ASTP required resolution of numerous technical, linguistic, and cultural differences. This included the incompatibility of Apollo and Soyuz spacecraft. NASA developed a docking module with an airlock to allow the crew exchange between the Apollo low-pressure oxygen environment and the normal-pressure atmosphere of the Soyuz.

NASA and TsKBEM selected the crews in 1973, and crew members trained together in both countries. Astronauts Thomas Stafford (commander), Donald "Deke" Slayton, and Vance Brand represented the United States. Cosmonauts Alexei A. Leonov (commander) and Valeri N. Kubasov were their Soviet colleagues.

The mission began on 15 July 1975 with the launch of *Soyuz 19* and *Apollo* spacecraft seven hours later. Both docked in orbit on 17 July. For two days the crews visited each other's spacecraft, worked on joint experiments, and conducted live television reports. The spacecraft were disengaged for two hours to test maneuvering capabilities and redocked before a final separation. *Soyuz 19* landed on 21 July, and the *Apollo* command module splashed into the Pacific on 25 July.

ASTP demonstrated that Cold War rivals were able to overcome years of mutual mistrust and cooperate in a complex technical undertaking.

Peter A. Gorin

See also: Apollo, Soyuz

Bibliography
Walter Froehlich, *Apollo Soyuz* (1976).

Ares. *See* Constellation.

Atlantis. *See* Space Shuttle.

Challenger. *See* Space Shuttle.

Columbia. *See* Space Shuttle.

Constellation

Constellation, the U.S. human spaceflight program initiated in response to U.S. President George W. Bush's call for the Vision for Space Exploration (VSE) in January 2004. By 2008, Constellation comprised a number of systems, which together were to enable several missions: transportation of astronauts to the *International Space Station* (*ISS*), lunar sorties, lunar habitation, and eventually human missions to Mars.

The loss of the Space Shuttle *Columbia* in February 2003 was a catalyst for a major shift in U.S. human spaceflight goals. The *Columbia* tragedy pointed out the need to replace the aging Space Shuttle fleet. In the meantime, the *ISS* had been bedeviled by high costs and schedule slips, while its seemingly routine missions failed to capture the public's attention. By contrast, robotic missions to Mars, such as Mars Pathfinder in 1997 and the Mars Exploration Rovers in 2004, were followed intensely. The VSE aimed to return NASA to exploration.

Shortly after Bush's speech, the administration formed the President's Commission on the Implementation of United States Space Exploration Policy, headed by E. C. "Pete" Aldrich Jr. Issued in June 2004, the commission's report, called "A Journey to Inspire, Innovate, and Discover," supported the new initiative and called for NASA reforms to implement a program of exploration of the Moon, Mars, and beyond. NASA was to implement the new program, but without a significant increase in its budget. In late 2005, NASA Headquarters delegated Constellation program management to Johnson Space Center.

The program initially focused on development of a crew launch vehicle, Ares I, and a crew exploration vehicle, Orion, which would initially be used to ferry astronauts to the *ISS*. Ares I, managed, designed, and integrated by Marshall Space Flight Center, featured a first stage, evolved from the Shuttle Solid Rocket Booster built by ATK (Alliant Techsystems Inc.), and a new Boeing-manufactured upper stage, which housed a J-2X engine built by Pratt and Whitney Rocketdyne and derived from the Apollo program's J-2 engine. Orion, contracted to Lockheed Martin, used the Apollo command module aeroform, but was much larger with the capability of transporting six astronauts. Lunar missions would require two launches: first, an Ares I–Orion launch for the crew, and second, a Saturn V–class Ares V with an Earth Departure Stage (EDS) to boost a lunar surface access module, Altair, into orbit. Orion and Altair/EDS would rendezvous in Earth orbit, and the EDS would propel the stack toward the Moon. In lunar orbit, Altair would separate from Orion, bringing the crew to the lunar surface and eventually back to Orion. The Orion Service Module would return the crew to Earth to enter Earth's atmosphere much like Apollo, using the Crew Module with parachutes.

In February 2010 the Barack Obama administration proposed cancellation of the program in the Fiscal Year 2011 budget, but faced opposition in Congress. Two months later, Obama unveiled a new plan for NASA that included a scaled back version of Orion to serve as an unmanned lifeboat for the *ISS*. Obama favored commercial launch of crew to low Earth orbit, with a variety of potential deep space destinations for the future, including the Moon, Mars, and asteroids.

Stephen B. Johnson

See also: Politics of Prestige, Saturn Launch Vehicles

Bibliography

NASA Constellation. http://www.nasa.gov/mission_pages/constellation/main/index.html.
Frank Sietzen Jr. and Keith Cowing, *New Moon Rising: The Making of the Bush Space Vision* (2004).

Discovery. *See* Space Shuttle.

DOS, Long-Duration Orbital Station. *See* Salyut.

Endeavour. *See* Space Shuttle.

Enterprise. *See* Space Shuttle.

Gemini

Gemini (1962–66), the second U.S. human spaceflight program, bridged the technological gap between the pioneering Project Mercury and the ambitious lunar goals of Project Apollo. NASA officially announced Project Gemini on 3 January 1962 after it became apparent that an unacceptably long flight gap loomed between the completion of Mercury and the initial Apollo missions. Gemini filled that breach, providing essential operational experience for astronauts and flight directors during 10 crewed missions. Program goals included long-duration flights, rendezvous and docking missions, and extravehicular activity (EVA).

Gemini introduced a two-person spacecraft that retained a similar exterior shape to the Mercury capsule but provided a 50 percent larger volume. Owing to its derivation from the Mercury design, the McDonnell Corporation also built Gemini. An adapter module mounted behind the crew section housed the spacecraft's life-support consumables, Orbit Attitude and Maneuvering System (OAMS) thrusters, electrical supplies, and retrorockets for reentry. Launched atop a modified Titan II intercontinental ballistic missile (ICBM), the Gemini spacecraft weighed 8,000 pounds. The Gemini Agena Target Vehicle (GATV), a modified Agena upper stage, served as the program's rendezvous-and-docking target and was launched atop an Atlas ICBM.

Edward H. White II, pilot of Gemini 4, *was the first American to conduct a space walk. (Courtesy NASA/Johnson Space Center)*

Under the direction of the Manned Spacecraft Center (later named Johnson Space Center) in Houston, Texas, 12 Gemini spacecraft were launched. *Gemini 1* and *Gemini 2* were nonhuman flights designed to verify spacecraft and launch vehicle compatibility, validate spacecraft separation from the launch vehicle, and test the integrity of the spacecraft's heat shield.

Human operations began on 23 March 1965, when Virgil "Gus" Grissom and John Young piloted *Gemini 3* on a shakedown flight lasting three orbits. Despite its short duration, the flight marked the first time an astronaut (Grissom) returned to space and, more important, the first time a human-operated spacecraft changed its orbit using the OAMS thrusters, a capability necessary for rendezvous-and-docking missions.

Slated as the first long-duration mission, *Gemini 4* launched on 3 June for a four-day flight. The first EVA in space by Soviet cosmonaut Alexei Leonov two months earlier pressured NASA to add a space walk on *Gemini 4*. On the third orbit of the mission, Ed White exited the spacecraft and floated for 20 minutes on a tether. His euphoric and apparently effortless experience masked the challenges that EVA presented. *Gemini 4* splashed down in the Atlantic, setting a U.S. spaceflight endurance record.

Aiming to double *Gemini 4*'s mission duration, *Gemini 5* lifted off for an eight-day sojourn on 21 August. For the first time, fuel cells powered a human spacecraft, a technology required for Apollo. Despite a problem with the fuel cell that threatened an early termination of the mission, the crew executed a "phantom" rendezvous at a point in space and completed 16 of 17 planned experiments. These experiments included a variety of medical experiments to determine otolith function in weightlessness, cardiovascular conditioning, several visual acuity experiments for the Department of Defense, and terrain and weather photography. In the process, the crew proved humans could withstand a spaceflight long enough for a lunar mission.

Gemini 6 was the first of the rendezvous-and-docking missions. The mission got off to a shaky start when the GATV, which was launched first, exploded shortly after liftoff on 25 October. The mission was scrubbed. NASA decided *Gemini 7* would serve as the replacement target vehicle for *Gemini 6*, which was given a new launch date.

Gemini 7, the last of the scheduled long-duration missions, lifted off on 4 December with Frank Borman and Jim Lovell aboard for a two-week stay in space. Immediately after launch, ground crews reconfigured the pad to erect *Gemini 6* and its Titan launch vehicle. On 13 December, *Gemini 6* was ready, but the Titan's engines ignited for only one second before shutting down when an umbilical connector separated prematurely. Despite mission rules that indicated the crew should eject, Commander Wally Schirra stayed, saving the spacecraft and the mission.

Finally, on 15 December, *Gemini 6* was safely launched in pursuit of Borman and Lovell, who still had three days remaining in their mission. With Tom Stafford working the onboard computer, Schirra successfully maneuvered the spacecraft to the first rendezvous in space. Approaching as close as 1 ft, the two spacecraft flew in formation for several orbits, marking a major accomplishment for the program.

Gemini 8 lifted off on 16 March 1966 with mission goals that included the first docking attempt with the GATV and an ambitious EVA. The GATV preceded the Gemini launch, this time achieving orbit. After pursuing the GATV, Neil Armstrong achieved the first docking in space. The triumph was cut short when a short circuit caused a thruster to fire continuously, sending the spacecraft into an uncontrolled and life-threatening spin. Armstrong disabled the OAMS and used the Reentry Control System (RCS) to stabilize the spacecraft. NASA immediately aborted the mission and its planned EVA because the RCS had been activated. The spacecraft splashed down in the contingency recovery zone in the Pacific Ocean after only 10 hours in space.

Gemini 9 launched on 3 June on a similar mission. Its GATV failed to reach orbit the previous month, so NASA substituted the Augmented Target Docking Adapter (ATDA), basically an Agena docking collar bolted to a Gemini RCS module, as the target for rendezvous and docking. The crew, Tom Stafford and Gene Cernan, performed a flawless rendezvous only to discover the ATDA's launch shroud failed to jettison, obviating any docking attempt. The crew did manage three rendezvous maneuvers and salvaged most of the mission's objectives, but a planned two-hour EVA by Cernan proved less successful. Cernan had difficulty getting into a jet backpack called an Astronaut Maneuvering Unit (AMU), housed in the spacecraft's adapter section. His exertions in donning the AMU overtaxed his spacesuit's environmental control capability. The rest of the EVA was canceled, and Cernan returned to the spacecraft exhausted, providing the first indications that the challenges of EVA were not well understood.

John Young and Michael Collins lifted off aboard *Gemini 10* on 18 July for a dual rendezvous mission. The crew docked with its GATV on the fourth orbit and then used its propulsion system to send the combined spacecraft to a record altitude of 474 miles. Later the crew undocked and executed a rendezvous with *Gemini 8*'s GATV, which was previously left in orbit. Collins exited the spacecraft and, with the assistance of a handheld maneuvering unit, approached the GATV to retrieve a micrometeorite collector. In the process, Collins lost his camera. The EVA was cut short when OAMS propellant ran low.

With two Gemini missions remaining, the program had accomplished its major goals, but problems with space walks remained unresolved.*Gemini 11*, which launched on 12 September, suffered similar EVA problems. The mission's EVA was cut short when Richard Gordon became exhausted working outside the spacecraft.

NASA's last chance to master space walks rested with *Gemini 12*, which lifted off on 11 November with Lovell and Edwin "Buzz" Aldrin aboard. Following routine rendezvous-and-docking maneuvers, Aldrin performed a successful EVA lasting more than two hours. Employing harnesses, Velcro pads, foot restraints, and special tools, Aldrin performed a variety of tasks without overheating his spacesuit, solving the EVA problems. The crew's splashdown after four days in orbit ended the program.

Thomas J. Frieling

See also: Blue Gemini, Extravehicular Activity

Bibliography

David Baker, *The History of Manned Spaceflight* (1981).

Barton C. Hacker and James M. Grimwood, *On the Shoulders of Titans: A History of Project Gemini* (1978).

International Space Station

The International Space Station (ISS) program is a multinational partnership among the United States (represented by NASA) and 15 other nations and represents the largest international scientific endeavor to build and operate a permanently crewed orbiting laboratory.

In January 1984 U.S. President Ronald Reagan instructed NASA to build a space station within a decade. By the time President William Clinton took office in January 1993, the $8 billion budget for what Reagan in 1988 named Space Station Freedom had been consumed—indeed, exceeded—and, despite numerous design studies, NASA had built no hardware. Clinton, having promised to address the budget deficit, ordered a redesign with the objective of cutting the cost of fabricating the station and assembling it in orbit. On 4 June NASA presented three options, each of which, in its own way, offered savings. On 17 June Clinton chose the option of modular buildup that sought to preserve the existing design while reducing the cost of orbital assembly. On 17 August

In December 1998, the crew of STS-88 began construction of the International Space Station, joining the U.S.-built Unity node to the Russian-built Zarya module. (Courtesy NASA)

Boeing was named as prime contractor to supply the station elements. On 7 September NASA delivered its plan for implementing the chosen option, now referred to as Alpha. If the annual budget was assured, NASA officials said it would be possible to initiate orbital assembly in 1998 and complete the station by 2003 at a cost of $19.4 billion. However, geopolitical events overtook NASA's budgetary review.

U.S. President George H. W. Bush and Soviet Union President Mikhail Gorbachev signed an agreement on 31 July 1991 to symbolize the rapprochement between the two superpowers by flying a Soviet cosmonaut aboard the Space Shuttle and launching an American astronaut on a Soviet rocket to the *Mir* space station. With the collapse of the Soviet Union later that year, Boris Yeltsin became the president of newly independent Russia. On 17 June 1992, at their first summit, Bush and Yeltsin agreed to expand the 1991 agreement. The Human Spaceflight Cooperation protocol signed on 5 October called for Shuttles to visit *Mir*. NASA and the newly established Russian Space Agency set about resolving the technical issues. The Clinton administration sought to further expand international space cooperation. On 2 September 1993 Vice President Al Gore and Russian Prime Minister Viktor Chernomyrdin agreed to merge their space station plans to build a single structure. The next month the U.S. House of Representatives expressed its support for including Russia in the project. On 1 November NASA submitted a unified implementation plan that integrated Russian hardware to Alpha. Six days later NASA's Canadian, European, and Japanese partners ratified this revision. The name Space Station Freedom was dropped in favor of the *International Space Station* (*ISS*). On 7 December Russia was formally invited to participate, and the agreement was signed one week later. This increased to 15 the number of nations involved in the project: Canada, Japan, Russia, the United States, and participating members of the European Space Agency (ESA)—Belgium, Denmark, France, Germany, Italy, the Netherlands, Norway, Spain, Sweden, Switzerland, and the United Kingdom. Brazil joined the project in October 1997.

The incorporation of Russian hardware would significantly increase the station's habitable volume and power supply, provide welcome logistical support, and increase its crew from four to six (the original plan had been for eight, but this had been halved during cost-cutting revisions). Most important, whereas financial constraints had obliged NASA to push permanent habitability to the end of the assembly process, the Russians preferred to build their stations by launching the crew quarters first, so their involvement meant that the *ISS* would be habitable almost from the start. NASA estimated that Russia's involvement would shave some $2 billion off the projected cost of Alpha and advance the completion date by two years. In authorizing the project, Congress established a five-year budget totaling $17.4 billion for fabrication of the necessary hardware.

By exploiting the design work for the *ISS*, NASA was able to make rapid progress fabricating its hardware. With the United States funding the project, Boeing hired the Khrunichev State Research and Production Center to manage the contract to build the *ISS* control module (named *Zarya*—Dawn) that was to provide power and attitude control to the nascent facility. Khrunichev was also to build Russia's primary contribution to the project, the service module (*Zvezda*—Star), derived from the *Mir* base block that would serve as the living quarters. But with the Russian economy in crisis, the Russian Space Agency was starved for funds, and the start of the assembly process was postponed.

The volume and mass constraints of the Russian Proton rocket and the Space Shuttle dictated *ISS* design modularity. The Proton placed into low Earth orbit a 20-ton module up to 13 m in length not exceeding 4.2 m in diameter, and because this made its own way to the *ISS*, it had its own power, navigation, attitude control, and orbital maneuvering systems. Although the Shuttle's payload was similar, its bay was able to accommodate payloads 18 m in length × 4.5 m in diameter, its elements were not required to be capable of operating independently, and their mass could be wholly devoted to specific roles on the station.

The first *ISS* element to be launched in November 1998 was *Zarya*. A fortnight later, NASA added *Unity*, a 5.5 m long cylindrical node with six berthing mechanisms, one on each end and four around its periphery. Command and coordination of *ISS* operations took place from the *ISS* Flight Control Room at Johnson Space Center in Houston, Texas, and the Russian Flight Control Center (TsUP) outside Moscow. TsUP was retrofitted with new equipment in 1999 to better integrate and share the monitoring and control of *ISS* operations with NASA.

After many delays, *Zvezda* was added in July 2000, and the first full-time crew (*ISS* commander William Shepherd, Soyuz commander Yuri Gidsenko, and flight engineer Sergei Krikalev) flew in a Soyuz spacecraft three months later to commission it, arriving on 2 November. NASA's *Destiny* laboratory was added to *Unity* in February 2001, followed in July by the primary airlock, named *Quest*, compatible with U.S. and Russian space suits. In September, a second airlock called *Pirs*, equipped to handle only Russian suits, was added to *Zvezda*.

When Congress deleted NASA's planned 8.5 m cargo module for budgetary reasons, Italy agreed to donate a trio of 6.4 m logistics modules for the transfer of six tons of cargo to the *ISS* in exchange for NASA providing Italy research time out of the U.S. allotment. In 2002 the assembly of a truss structure was begun to hold the solar panels that power the station and the radiators that cool its systems, clear of the cluster of pressurized modules in order to provide maneuvering room for Shuttles. A Canadian-built Mobile Service System was added to the truss structure by the end of 2002.

NASA originally envisaged an orbital facility that would serve a multiplicity of roles, but financial constraints restricted the *ISS* to a microgravity research station involving the behavior of materials and biological systems. Early critics of the project argued that the Space Shuttle would not be able to sustain the rate of flights required to assemble the *ISS* within the projected timescale, and indeed the schedule did slip for a variety of reasons, most notably the loss of the Space Shuttle *Columbia* on an unrelated mission on 1 February 2003. While the Shuttle fleet was grounded, the assembly process was placed on hold, and two-person crews sustained by Russian Soyuz and Progress spacecraft were used to maintain and replenish the station. If not for Russian involvement in the project, the station would have had to have been evacuated, which would have left it vulnerable to being lost to a failure that could readily have been corrected by its crew.

Before the *Columbia* accident, seven Shuttle assembly flights remained to reach the "core complete" phase, a point at which other nations' modules could be attached. NASA maintained it would honor its commitment to add ESA's *Columbus* module and Japan's *Kibo* module once Shuttle flights resumed. In January 2004 President George W. Bush instructed NASA to cease Shuttle operations as soon as the *ISS* was complete.

Despite an attempt to resume Shuttle flights in July 2005 after a 28-month standdown, NASA again grounded its Shuttle fleet after the agency discovered additional foam debris problems with the external tank and orbiter during launch. With its assembly at a standstill, the *ISS* continued to rely on Soyuz and Progress vehicles to maintain essential functions. Assembly of the truss resumed in 2006. In November 2007 NASA permanently attached *Harmony*, the station's second cylindrical node, to *Destiny*. The *Columbus* module was attached to *Harmony* in February 2008, and *Kibo* to *Harmony* in June 2008. NASA added a third node, *Tranquility*, in February 2010 to *Unity*. The *Tranquility* node contained an advanced life support system capable of recycling waste water and generating oxygen for the crew, and also housed the observatory module (*Cupola*—Dome). NASA scheduled its final Shuttle flight to dock with the *ISS* in February 2011.

In NASA's budget, the costs attributable to the *ISS* were separated from Shuttle operations (not all of which involved flying to the *ISS*). During *ISS* assembly, NASA's annual budget was approximately $15 billion, of which typically $1.7 billion was used for the *ISS* and $3.3 billion was used for the Space Shuttle. When the *ISS* was authorized in 1993, the direct costs were budgeted at a total of $17.4 billion, but cost overruns inflated this figure so that in October 2000 a $25 billion cap was imposed.

The *ISS*'s primary technological legacy is likely to be proven systems for environmental and other critical functions that will be able to be incorporated into spacecraft designed for human missions into deep space—such as a mission to Mars. Other legacies will probably be mixed, ranging from its massive cost overruns to lessons about the benefits and pitfalls of international cooperation.

David M. Harland

See also: Boeing Company, Brazil, Canada, European Space Agency, Japan, Khrunichev Center, National Aeronautics and Space Administration, Russian Federal Space Agency.

Bibliography

John E. Catchpole, *The International Space Station: Building for the Future* (2008).
David M. Harland, *The Story of the Space Shuttle* (2004).
David M. Harland and John E. Catchpole, *Creating the International Space Station* (2002).

Long-Duration Orbital Station. *See* Salyut.

Manned Maneuvering Unit

The Manned Maneuvering Unit (MMU) was a personal propulsion backpack that attached to the back of the portable life-support system of the Extravehicular Mobility Unit space suit. MMU-equipped astronauts could safely fly untethered up to 400 ft from the Space Shuttle orbiter during a space walk.

NASA tested early versions of personal propulsion backpacks during the Gemini (1965–66) and Skylab (1973–74) programs. Martin Marietta Corporation developed the MMU to expand the Space Shuttle's capabilities, such as retrieving and servicing satellites and conducting inspections of spacecraft. Two MMUs were mounted externally on individual flight support stations in the orbiter's payload bay.

The MMU had 24 small thruster jets that used high pressure, non-contaminating gaseous nitrogen as propellant. The astronaut used rotational and translational hand controllers mounted on the MMU arms for control during extravehicular activity (EVA). Each MMU weighed 338 lb, excluding the flight support station.

Astronaut Bruce McCandless conducted the first demonstration of the MMU on 3 February 1984 during Space Transportation System (STS)-41B. The MMU was used on two other Space Shuttle missions: STS-41C in April 1984 and a successful satellite retrieval mission in November 1984 during STS-51A. After the *Challenger* accident in January 1986, commercial payloads were no longer deployed from the Space Shuttle, eliminating the need for the MMU. NASA decided to terminate the MMU by the mid-1990s and developed a smaller version known as the Simplified Aid for EVA Rescue for use on the *International Space Station* as a self-rescue device in case an astronaut accidentally became untethered.

Robert C. Trevino

See also: Extravehicular Activity, Space Suit

Bibliography
Gary L. Harris, ed., *The Origins and Technology of the Advanced Extra-Vehicular Space Suit* (2001).
Lillian D. Kozloski, *U.S. Space Gear: Outfitting the Astronaut* (1993).

Mercury

Mercury (1958–63), the first U.S. human spaceflight program, placed astronauts into space using bell-shaped capsules mated to modified military Redstone and Atlas rockets. The Dwight Eisenhower presidential administration provided the impetus for Mercury with the formation, in 1958, of NASA. The new space agency started Mercury in order to build the organizational foundation and gather the technical resources necessary for its human spaceflight efforts.

Briefly called Project Astronaut in 1958, NASA selected Mercury as the project's official moniker for the Roman messenger god who wore winged sandals and provided a direct link between the heavens and Earth.

Under the direction of the Space Task Group (which was redesignated the Manned Spacecraft Center after a move to Houston in 1961), Mercury blended the early spacecraft designs of the U.S. Army, Air Force (USAF), and former National Advisory Committee for Aeronautics—alongside human spaceflight ideas proposed by the Defense Department's Advanced Research Projects Agency—under a civilian umbrella.

U.S. officials learned of the Soviet Union's program to launch a human into Earth orbit only after Mercury began. However, the United States surmised that the communist power wanted to put a human into space for propaganda purposes, and some hoped the Air Force's X-15 hypersonic research rocket plane project—conceived before NASA was created—would do it first. But when the national emphasis was placed on Mercury, an X-15 space first faded.

Astronaut Virgil I. Grissom climbs into his Liberty Bell 7 *spacecraft with help from astronaut John H. Glenn Jr. prior to a successful launch on 21 July 1961. The flight was the second U.S. crewed suborbital mission. (Courtesy NASA/Marshall Space Flight Center)*

MAXIME ALLAN FAGET (1921–2004)

Maxime Faget, an aerospace engineer, was one of the primary designers of the Mercury spacecraft. He joined the staff of the Langley Aeronautical Laboratory in Hampton, Virginia, in 1946 and eventually became head of the Performance Aerodynamics Branch. Faget was part of the original Space Task Group at Langley, which formed NASA's Manned Spacecraft Center in 1961. Faget contributed to the designs of every U.S. human spacecraft through the Space Shuttle until his retirement from NASA in 1981.

Tim Chamberlin

Mercury included suborbital and orbital missions to expose astronauts to the microgravity environment of space. As a research-and-engineering enterprise to build national prestige and technological superiority, its goals included placing humans into space before the Soviet Union and returning them safely and gathering new data on space medicine, spacecraft environmental controls, zero-g and multiple-g effects, and space radiation exposure.

The first Mercury missions involved several uncrewed tests of rocket and spacecraft systems in addition to suborbital and orbital missions with animals as passengers, done at the Cape Canaveral Missile Range complex in Florida. Several of these tests were plagued by accidents that delayed the program. In late 1958 the Army Ballistic Missile Agency's Redstone rocket had been earmarked for Mercury suborbital flights and the USAF Atlas booster for orbital missions. Smaller rockets, such as NASA's Little Joe and Scout, were launched from Wallops Island Station, Virginia, and Cape Canaveral, Florida, to test boilerplate Mercury capsules and orbital communications hardware.

The Mercury spacecraft was designed by NASA engineer Maxime Faget and built by McDonnell Aircraft Corporation. A blunt reentry face, a conical afterbody, a cylindrical recovery compartment, an antenna canister, and an escape tower characterized the 26 ft long, single-seat Mercury capsule. It also included a retrorocket pack to slow the vehicle to reenter the atmosphere and a rocket-propelled escape tower. For suborbital missions, the Mercury capsule's heatshield used beryllium to absorb heat. For orbital flights, the heatshield contained an ablative fiberglass and resin composite to withstand higher reentry temperatures. Mercury's conical afterbody used overlapping shingles composed of a refractory metal for dissipating reentry heat. Many of Mercury's basic engineering concepts, excluding the escape tower and retropack, were modified for the Mercury Mark II program, later named Gemini, begun in early 1960. Primary Mercury spacecraft systems for suborbital and orbital missions included communications, attitude control, environmental control, electrical power, explosive devices (for emergency hatch egress), cabin hardware, and sea-based landing and recovery equipment.

After weeks of technical and weather delays, astronaut Alan Shepard Jr., a U.S. Navy test pilot, became America's first human in space on 5 May 1961 with the launch of *Mercury-Redstone 3*. Unlike Yuri Gagarin's secretive, day-long orbital flight of

ALAN BARTLETT SHEPARD JR. (1923–1998)

Alan Shepard, one of Project Mercury's original seven astronauts and a Navy rear admiral, became the first American to travel in space on 5 May 1961 aboard *Freedom 7*. Shepard's suborbital flight lasted 15 minutes and kept the United States in a close space race with the Soviet Union. Shepard was chief of the Astronaut Office from 1963 to 1969 and from 1971 to 1974. He overcame an inner ear disorder to command *Apollo 14* and was the fifth man to walk on the Moon. Shepard retired from NASA and the Navy in 1974.

Tim Chamberlin

Vostok 1 the previous month for the Soviet Union, millions of people around the world watched Shepard's launch and recovery on live television. His spacecraft, dubbed *Freedom 7* after the seven original NASA astronauts selected for Mercury, was launched from Cape Canaveral on a 116.5 mile high suborbital trajectory. The flight lasted 15 minutes and 22 seconds, with Shepard experiencing weightlessness for one-third of the mission. Despite the short flight duration, the nation viewed the mission as a major accomplishment for NASA—a space agency less than three years old.

All remaining Mercury flights were launched at the Cape. Astronaut Virgil "Gus" Grissom repeated Shepard's suborbital flight during *Mercury-Redstone 4* aboard *Liberty Bell 7* on 21 July 1961. Grissom was lofted to an altitude of 118.26 miles. The mission lasted only 15 seconds longer than Shepard's sortie. The otherwise flawless flight of *Liberty Bell 7* was marred after the capsule splashed down in the Atlantic Ocean. The spacecraft's hatch cover blew off prematurely and allowed seawater to flood the capsule's main compartment. Grissom bailed out and swam away from the sinking capsule. A U.S. Marine recovery helicopter pilot attempted to retrieve the sinking craft with a grappling hook while ignoring Grissom bobbing in the choppy water. With water entering his space suit, Grissom almost drowned. A second helicopter rescued the foundering astronaut, but the capsule sank and rested on the sea bottom until it was retrieved in 1999.

The program's first piloted orbital mission, *Mercury-Atlas 6*, was launched on 20 February 1962 with astronaut John Glenn Jr., a U.S. Marine pilot, on board *Friendship 7*. The three-orbit jaunt lasted 4 hours, 55 minutes, and came less than one year after Gagarin's spaceflight. When it was over, there was a tangible boost to American national prestige. Glenn toured the United States as a national hero. Parades and public tributes in his honor exceeded the accolades heaped on U.S. aviator Charles Lindbergh after his historic 1927 solo flight across the Atlantic Ocean.

Mercury-Atlas 6 was also notable for two near setbacks. During the flight, ground controllers noticed that the heatshield of Glenn's capsule had become unlocked and was held in place only by thin retropack straps. Controllers instructed Glenn to reenter the atmosphere with the capsule's retropack attached even though mission procedures called for the pilot to jettison it. Glenn survived reentry, but immediately after passing through peak g-forces, his capsule began oscillating wildly. Glenn reported that the capsule's drogue

JOHN HERSCHEL GLENN JR.
(1921–)

(Courtesy NASA/Kennedy Space Center)

John Glenn, one of Project Mercury's original seven astronauts and a Marine Corps colonel, became the first American to orbit Earth on 20 February 1962 aboard *Friendship 7*. Glenn's flight made him a national hero and placed the U.S. space program on the level of its Cold War rival, the Soviet Union. Glenn resigned from NASA in 1964 and retired from the Marines in 1965. He was elected to the U.S. Senate in 1974, where he served for 24 years. Glenn became the oldest man to fly in space on 29 October 1998 at age 77 aboard the Space Shuttle *Discovery* on Space Transportation System (STS)-95.

Tim Chamberlin

parachute automatically deployed and stabilized the spacecraft. Both incidents alarmed NASA mission controllers because they jeopardized Glenn's safe return.

Astronaut Donald "Deke" Slayton was grounded for a minor heart problem less than a month after Glenn's flight, so astronaut Scott Carpenter stepped in to pilot *Mercury-Atlas 7* and the *Aurora 7* capsule on 24 May 1962. Carpenter duplicated the previous mission's flight plan. *Mercury-Atlas 7* was a success with the exception of a navigation error that placed *Aurora 7* 250 miles off target on reentry and delayed Carpenter's recovery.

Astronaut Wally Schirra, a U.S. Navy test pilot, was selected for the first of two longer-duration Mercury flights. *Mercury-Atlas 8*, which carried the *Sigma 7* capsule named by Schirra for the Greek letter representing "summary" in engineering parlance, was launched on 3 October 1962. Being the only technically flawless Mercury mission, some NASA officials wanted to end the program with Schirra's flight. *Sigma 7* reached the highest altitude of any of the Mercury capsules—176 miles during six orbits. The mission lasted just over nine hours.

Mercury-Atlas 9, the last Mercury mission, was launched on 15 May 1963 and lasted nearly 35 hours during 22 orbits. Astronaut Gordon "Gordo" Cooper conducted several space science experiments aboard the *Faith 7* capsule, including cosmic radiation monitoring and observations of the zodiacal light and nighttime airglow layer. He was also the first human to see the Southern Lights (or aurora australis) from space.

Even before Cooper's flight, NASA officials were considering four additional Mercury missions up to three days each in length. New extended-mission capsules had been ordered from McDonnell Aircraft, but NASA administrators, eager to focus on the Gemini program, abruptly ended Mercury.

While the Soviet Union's first human spaceflight program, Vostok, put the first human in space, Mercury established the foundation that enabled the United States's next human spaceflight programs, Gemini and Apollo.

Lou Varricchio

See also: Astronauts, Atlas, McDonnell Company, Redstone

Bibliography

David Baker, *The History of Manned Spaceflight* (1981).
M. Scott Carpenter et al., *We Seven* (1962).
Robert Godwin, *Freedom 7: The NASA Mission Reports* (2001).
Loyd S. Swenson Jr. et al., *This New Ocean: A History of Project Mercury* (1966).

Mercury Mark II. *See* Gemini.

Mir

The *Mir* (which means peace, commune, or world) space station served as one of the Soviet Union's symbols of scientific and technological accomplishment in addition to the beginning of a long-term cooperative venture between the U.S. and Soviet space agencies.

In February 1970, shortly after cancellation of the U.S. military's Manned Orbiting Laboratory, Soviet leader Leonid Brezhnev issued a resolution calling for the development of a civilian space station (Salyut). Valentin Glushko, Chief Designer of Scientific-Production Association (NPO) Energia (formerly Sergei Korolev's Experimental Design Bureau-1—OKB-1), first proposed development of *Mir* in 1974. The Buran shuttle program, initiated in February 1976 as the Soviet response to the U.S. Space Shuttle program, included plans for development of two Salyut-derived Long-Duration Orbital Station (DOS) modules that could be carried into orbit. The first of these later became the core module (also called base block) of *Mir*, while the second, originally slated as a backup or possible core module for a second *Mir* station, became the *Zvezda* module of the *International Space Station* (*ISS*).

While work on *Mir* began within Energia, the Salyut Design Bureau (KBS) contributed significantly to the development of the DOS modules, especially when Energia was forced to concentrate on the *Salyut 7*, Soyuz-T, Progress, and Buran projects. Major changes made by KBS to the DOS modules included new solar panel technologies, a new automated approach-and-rendezvous system, and the introduction of new life-support systems, precursors to instruments used on the *ISS*. Blueprints of *Mir* were completed by December 1983, but problems limiting its mass and the diversion of funds in 1984 to the Buran program led to construction delays. Despite the lack of progress on a core module, the Soviet Politburo endorsed a plan requiring the launch of *Mir* in time for the 27th Congress of the Soviet Communist Party in February 1986, with complete assembly due by December 1989.

Multiple approaches were used to lower the mass requirement and meet the launch date. Development of the core module received priority over other spacecraft. Equipment to be used in the base block would be delivered later, thereby diminishing the

The Mir *space station: Upper center, Progress re-supply vehicle, Kvant-1 module, and core module; center left, Priroda module; center right, Spektr module; bottom left, Kvant-2 module; bottom center, Soyuz spacecraft; and bottom right, Kristall module and docking module. (Courtesy NASA/Marshall Space Flight Center)*

initial launch mass. *Mir*'s planned orbital inclination was decreased from 65° to 51.6°, reducing the energy required to reach orbit. The thrust of the Proton rocket, used to deliver *Mir* to low Earth orbit, was enhanced by 7 percent.

The politically motivated timing of *Mir*'s launch on 20 February 1986 complicated the first expedition to the station, *Soyuz T-15*, in March. Cosmonauts Leonid Kizim and Vladimir Solovyov embarked on a four-month crash course to plan for the mission, which involved an unprecedented flight to and from *Salyut 7* once on board *Mir* to transfer equipment. With the arrival of the *Soyuz TM-2* crew in February 1987, work on *Mir* began in earnest.

As cosmonauts added additional modules to the station, they also reconfigured their arrangement using a small docking arm—an example of *Mir*'s flexibility compared to earlier Salyut space stations. *Kvant*, the first module added, was originally built to dock with *Salyut 7*. It was converted for use on *Mir* after numerous design delays. *Kvant* housed a science lab, and special gyroscopes mounted outside the module controlled *Mir*'s orientation without using propellant. Launched in March 1987 on an upgraded Proton rocket, the module docked with *Mir*'s base block using a space tug only after cosmonauts Yuri Romanenko and Aleksander Laveykin removed a wayward trash bag from the docking assembly during a spacewalk. Laveykin's post-spacewalk medical checkup revealed possible heart rhythm irregularities. Physicians on the ground, despite no definitive diagnosis, recommended his return on the arrival of the *Soyuz TM-3* crew. With only one month's warning about his longer-than-expected stay aboard *Mir*, veteran cosmonaut Aleksandr Aleksandrov replaced Laveykin, who returned to Earth only to be pronounced fit for duty after examination by heart specialists.

By 1984 KBS chose not to use a space tug for future dockings, in favor of an integrated self-propelled module. This new design formed the basis for the *Kvant-2*, *Kristall*, *Spektr*, and *Priroda* modules to be added to *Mir*. *Kvant-2*, launched in November 1989, contained *Mir*'s primary airlock for extravehicular activity. *Kristall*, added to the station in June 1990, housed materials processing experiments.

After the collapse of the Soviet Union and the resulting economic disorders, Russia sought external funding to keep its space program operating. In July 1993 the United States and Russia signed an agreement to place up to 700 kg of U.S. scientific equipment on *Spektr* and *Priroda*, which restored plans to launch the modules. Signed by U.S. Vice President Albert Gore and Russian Prime Minister Viktor Chernomyrdin, the agreement formed the basis for sustained cooperation in space through the Shuttle-Mir program. The agreement also laid the groundwork for a collaborative effort to build the *ISS*.

Spektr was to have included a military reconnaissance instrument complex called Oktava. This complex may have been the first small part of a Soviet Strategic Defense Initiative envisioned to include a network of Salyut-derived space stations with laser and kinetic weapons. *Priroda* was also intended as a dual-use, remote-sensing, and military-reconnaissance module, but its role changed to also include instruments from France and Eastern Bloc countries. However, with the Soviet Union dissolved, *Spektr* and *Priroda* were eventually rebranded as remote sensing modules for studies of Earth's surface and atmosphere. Six days after its launch in May 1995, *Spektr* became the first module to successfully dock with *Mir* on the first attempt. *Priroda* followed in April 1996. Although generally not considered a module, the Russian-built Shuttle docking compartment carried by the Space Shuttle *Atlantis* during Space Transportation System (STS)-74 in November 1995 was attached to the *Kristall* module, allowing orbiters to safely dock with *Mir*.

Originally initiated to highlight solidarity with socialist partners of the Soviet Union, a guest cosmonaut program that grew from earlier Interkosmos missions provided a key source of income that helped keep the *Mir* program alive during the 1990s. The cash-for-cosmonaut opportunity led to the first privately financed trip to space: Tokyo Broadcasting System reporter Toyohiro Akiyama visited *Mir* in December 1990 in exchange for at least $12 million. While guest cosmonauts did perform experiments, they were treated as less-than-full members of the *Mir* crew, although this practice subsided somewhat toward the later years of *Mir*, particularly when crews including guests trained together for longer periods of time.

COLIN MICHAEL FOALE (1957–)

Michael Foale, born in England, spent more time in space than any other U.S. astronaut. He logged 374 days in orbit. Foale was selected as an astronaut in 1987 and served on six Space Shuttle missions, including a trip to the *Mir* space station, where he spent 145 days in 1997. He participated in four space walks, including the Space Transportation System (STS)-103 servicing mission of the *Hubble Space Telescope* in 1999. Foale was *Expedition-8* commander of the *International Space Station* from 20 October 2003 to 29 April 2004. In November 2004 he was appointed deputy associate administrator for exploration operations at NASA.

Tim Chamberlin

During the Shuttle-Mir program (1994–98), *Mir* crews carried out extensive experiments despite deteriorating conditions on the space station. Although not the first uncontrolled fire in space, the oxygen candle ignition on board *Mir* on 23 February 1997 represented the largest and most dangerous on-orbit fire to date and filled the whole station with smoke, obstructing access to one of two docked Soyuz escape vehicles. In addition to the fire, the crewmembers dealt with life-support system failures, the extreme darkness of power failures, and the near miss of a Progress resupply vehicle during an attempted manual docking. A similar docking attempt on 25 June 1997 resulted in a collision with *Mir*; only by sealing the *Spektr* module did Russian crewmembers Vasili Tsibliyev and Aleksander Lazutkin and U.S. astronaut Michael Foale manage to prevent the loss of *Mir*.

Mir served as a laboratory where cosmonauts and astronauts learned to live and work together in challenging physical and psychological conditions despite language and cultural barriers. Crews on board *Mir* demonstrated how humans adapted to weightlessness after prolonged stays in space. Feats of endurance, such as Valeri Polyakov's continuous duration record of 438 days or Sergei Avdeyev's total duration of 748 days on *Mir* during three missions, demonstrated that humans can survive in weightlessness long enough to reach Mars.

Construction and operation of *Mir* led to significant advances in docking technology, large-scale space construction (the complex weighed 137 tons by 1997), and international cooperation. The difficulties of operating *Mir* reinforced the challenging nature of long-duration spaceflight endeavors and the need for flexibility and adaptation to new circumstances. For the first time in history, humans continually occupied low Earth orbit on board *Mir* for nearly a decade from September 1989 to August 1999.

Russia was reluctant to deorbit *Mir*. With inadequate funding and little political support to mount another effort of similar scale, abandonment of the space station signaled an end to a string of hard-won space successes that began with the flight of Yuri Gagarin in 1961. Keeping the station aloft became an endeavor of increasing complexity due to its aging and damaged infrastructure, and required funds no longer available after the end of the Shuttle-Mir program. The challenging Russian economic climate and the beginning of the ISS program added to the difficulty of sustaining

Mir. Efforts to save the station using funds from wealthy donors and the promise of space tourism failed, and on 23 March 2001 *Mir* reentered Earth's atmosphere, showering its remains over the Pacific Ocean. A month later, American Dennis Tito, originally slated to fly to *Mir*, visited the *ISS* and became the world's first space tourist.

A technical triumph, *Mir* flew 3.5 billion km while housing 104 cosmonauts and astronauts representing more than 12 countries.

Christopher E. Carr

See also: Cosmonauts, Interkosmos

Bibliography

Rex Hall, *The History of Mir, 1986–2000* (2000).
David M. Harland, *The Story of Space Station Mir* (2005).
Robert Zimmerman, *Leaving Earth: Space Stations, Rival Superpowers, and the Quest for Interplanetary Travel* (2003).

Object OD-2 Satellite. *See* Vostok.

Orbital Piloted Station. *See* Almaz.

Orbiter. *See* Space Shuttle.

Orion. *See* Constellation.

Progress

Progress is a Soviet, later Russian, expendable uncrewed spacecraft developed specifically as a cargo ferry for use with Earth-orbiting crewed space stations. The three-module craft was also designed as a simple means for disposing of space station trash. Soyuz rockets have launched all Progress spacecraft.

Progress was based on the Soyuz spacecraft and was first developed in the 1970s for use with the *Salyut 6* space station. The basic Progress design consisted of three modules: a pressurized spherical forward module for crew supplies, a cylindrical fuel compartment, and a flared propulsion module. The first spacecraft was launched on 20 January 1978. Forty-three Progress vessels were used to successfully service cosmonauts on board *Salyut 6* and *Salyut 7*. Progress development after Salyut can be divided into multiple phases: Progress M, M1, M1-3, and M2.

Progress M, an upgraded version of the original spacecraft able to carry a larger payload, was initiated after the launch of *Mir* in 1986. The M version, with a total cargo volume of 6.6 m^3, carried a total of 2,350 kg of various supplies, including water, air, and propellants. After its successful service assisting long-duration Salyut missions, Progress's automated flight-control system was modernized. The first M-version Progress was launched to *Mir* in August 1989.

By the late 1990s, when plans were solidified to build the *International Space Station* (*ISS*), Progress M1 modifications were undertaken because the craft was selected as an *ISS* supply vessel. The M1's middle section—used to carry propellant tanks for use on the *ISS*—was expanded to accommodate more tanks. A dozen oxygen-nitrogen air tanks for the *ISS* crew were added to the Progress exterior, and a new digital flight-control system was included. Progress M1 boasted a sophisticated new navigation system, known as Kurs-MM, for rendezvous and docking. The first M1 version was launched in February 2000. A slightly modified M1, known as M1-3 (1P), was launched in August 2000 in preparation for the first resident crew of the *ISS*.

Progress M2 was a planned heavier version of the Progress M1 with an elongated cargo module capable of ferrying up to 13 tons of station supplies. M2 was to be launched by a Zenit rocket, but financial woes prevented its construction. The Soviet organization NPO Energia, later the Russian aerospace manufacturing firm Rocket and Space Corporation Energia, managed all aspects of the Progress program.

Following the 30-month stand-down of the Space Shuttle program after the *Columbia* disaster in 2003, Progress played a vital role in conjunction with Soyuz flights supplying *ISS* crews with extra supplies and reboosting the station's orbit.

Lou Varricchio

See also: Rocket and Space Corporation Energia

Bibliography

Rex Hall and David Shayler, *Soyuz: A Universal Spacecraft* (2003).
David Harland, *Creating the International Space Station* (2002).
Jean-Marie Luton, "An Improved Soyuz Will Appear in 2005: Soyuz Is a Versatile Launcher for Low, Medium, and High Elliptical Orbits," *Interavia*, 31 October 2002.

Salyut

Salyut, the first space station program established by the Soviet Union, developed into a series of civilian and military orbiting platforms.

In 1962, one year after cosmonaut Yuri Gagarin became the first human in space, Soviet space engineers—under the guidance of Chief Designer Sergei Korolev—sketched plans for a military space station consisting of pressurized modules launched separately and linked in Earth orbit. Korolev's blueprints prefigured *Mir* and the *International Space Station* (*ISS*). A military version, called Almaz (Diamond), was the first of the Salyut stations to be developed but not the first to be successfully orbited. Championed by Premier Nikita Khrushchev, Almaz was conceived in 1964 by Vladimir Chelomey's state-owned company (later NPO Mashinostroyeniya, or Machine Building Scientific-Production Association) in Moscow as a response to the U.S. Manned Orbiting Laboratory program. Between 1967 and 1970, Chelomey manufactured at least three Almaz hulls and related components. The space station program, as originally conceived, was composed of three key elements: a Proton rocket launcher, a space station, and a Transport Logistics Spacecraft for ferrying crews and cargo.

Following the first U.S. lunar landing, *Apollo 11*, Soviet space planners placed a special emphasis on a robust space station program. After July 1969 the Soviet Defense Ministry turned the space station project—called the Long Duration Orbital Station (DOS), but later publicly designated Salyut (Salute)—over to the Korolev Design Bureau (later Energia NPO or Scientific-Production Association). The name Salyut was publicly attached to both the military and civilian space stations, perhaps to blur distinctions between missions. However, the older name Almaz was still applied internally to military Salyuts. After it appeared unlikely that the Soviet N-1 rocket would reach the Moon, the Korolev Design Bureau replaced Chelomey's Transport Logistics Spacecraft design with its own Soyuz, originally built as the lunar spacecraft to fly atop the N-1.

Nine cosmonauts were selected to train for the first space-station missions aboard *Salyut 1*: Georgi Dobrovolsky, Pyotr Kolodin, Valeri Kubasov, Alexei Leonov, Viktor Patsayev, Nikolai Rukavishnikov, Vladimir Shatalov, Vladislav Volkov, and Alexei Yeliseyev. The solar-powered *Salyut 1* used systems borrowed from the Soyuz spacecraft (including a manual deorbit control stick). Inside, the station boasted a large telescope, spectrometer, electrophotometer, exercise treadmill, worktable, and television.

To celebrate the 10th anniversary of Gagarin's flight, a Proton rocket launched *Salyut 1*, consisting of four connected cylinders and weighing 20 tons, into Earth orbit on 19 April 1971. Soviet space authorities hailed the spacecraft as "the world's first space station." The first Salyut and Almaz space stations were designed to operate for three to six months. Three days after the launch of *Salyut 1*, the station's first planned crewmembers—space veterans Shatalov and Yeliseyev, with rookie Rukavishnikov aboard *Soyuz 10*—rendezvoused and docked with the 14.5-m *Salyut 1* for what was supposed to be a month-long visit, but remained connected for only six hours. When the crew attempted to enter the station, the Soyuz's docking hatch failed to open, so the mission was aborted.

Following the failed *Soyuz 10* mission, the Soviet Union was handed its greatest space tragedy since the death of veteran cosmonaut Vladimir Komarov aboard *Soyuz 1* in 1967. After a successful 24 days in space in June 1971, highlighted by popular cosmonaut telecasts beamed to millions of Soviet people, Dobrovolsky, Patsayev, and Volkov prepared to leave the station during *Soyuz 11* and return to Earth. After what seemed to be a normal reentry, a recovery team on the ground opened the hatch of *Soyuz 11* and found the crew dead. A loss of cabin atmosphere, caused by a premature opening of a pressure equalization valve, killed the three-man crew because none of the cosmonauts wore pressure suits.

Following *Salyut 1*, the next three uncrewed Salyut stations were plagued with mishaps. An unnamed civilian station failed to reach orbit in July 1972. *Salyut 2* (the first Almaz station) was destroyed after it unexpectedly depressurized while in orbit in May 1973. That same month another Salyut, *Kosmos 557*, was lost after it failed to reach its proper orbit. Despite these setbacks, the Salyut program moved forward. In the United States, the Apollo Moon missions had ended and the short-lived *Skylab* space station project wrapped up in early 1974.

From the early to mid-1970s *Salyut 3–5* supported five crews spending a combined near-record time in space. In 1975 *Salyut 4* took a back seat to the first joint crewed U.S.-Soviet space mission, the Apollo-Soyuz Test Project. Meanwhile military surveillance and various experiments occupied Salyut space station crews.

A new generation of space stations was introduced with *Salyut 6* (1977–82) and *Salyut 7* (1982–91), the last in the series. Both stations were equipped with two docking ports, instead of only one as in earlier Salyuts, and were designed for long-duration missions. Leonid Popov and Valeri Ryumin's 185-day mission (*Soyuz 35*) was the longest on *Salyut 6*, while Leonid Kizim, Vladimir Solovyov, and Oleg

Atkov's 237-day mission (*Soyuz T-10B*) was the longest on *Salyut 7*. Between both Salyuts, 26 crews lived aboard the stations.

The first Almaz Transport Logistics Spacecraft was launched in 1981 to *Salyut 6*. Three other spacecraft of this type were also flown to *Salyut 7*. These automated spacecraft proved that large modules could be used in the construction of future multi-modular space stations. Automated Progress cargo freighters—later used on *Mir* and the *ISS*—also made their first visits to *Salyut 6* and *Salyut 7*. Twenty-five Progress vessels delivered 25 tons of food, fuel, and other supplies to both stations between 1978 and 1991.

Both *Salyut 6* and *Salyut 7* proved to be safe harbors for the first space station crew-members from countries other than the Soviet Union or United States. Vladimir Remek of Czechoslovakia visited *Salyut 6* in March 1978 aboard *Soyuz 28* as the first international cosmonaut under the Interkosmos program. Other cosmonauts, hailing from Cuba, East Germany, France, Hungary, India, Mongolia, North Vietnam, Poland, and Romania, also visited the stations from June 1978 to April 1984.

Lou Varricchio

See also: Interkosmos, Manned Orbiting Laboratory

Bibliography

Phillip Clark, *The Soviet Manned Space Program* (1988).
Grujica S. Ivanovich, *Salyut: The First Space Station* (2008).
Dennis Newkirk, *Almanac of Soviet Manned Space Flight* (1990).
Asif A. Siddiqi, *Challenge to Apollo: The Soviet Union and the Space Race, 1945–1974* (2000).

Shenzhou

Shenzhou is the name of the human-rated Chinese launch vehicle. The first Shenzhou (divine or sacred vessel) flight, *Shenzhou 1*, launched in November 1999, completing 14 orbits and returning to Earth after 21 hours. This was a hardware-proving flight without a crew on board. Statements first made in 1996 gave 1999 as the year planned for the first launch with a crew on board, to commemorate the 50th anniversary of the founding of the communist state. Depressed finances (perhaps in part because of lost income from anticipated commercial satellite launches) and technical issues made it impossible to keep the original timetable—there was not enough time for hardware-proving flights to assure there would be no disasters for flights that included a crew.

For Shenzhou, China began with a workhorse Soviet Soyuz design and reengineered and upgraded it to a Chinese design. Both spacecraft have a service module housing the propulsion system, a command module, and an orbital module with a docking ring. Both Shenzhou and the Soyuz TM are capable of carrying three taikonauts/cosmonauts. The Shenzhou orbital module, however, had a second set of solar panels, enabling it to remain in orbit independently for prolonged periods. While China considered buying some Soviet equipment and system upgrades, the price was often prohibitive. With no prior human spaceflight experience though, China bought selected Russian systems, including life support. In other cases, China chose to build its own technology to better understand the fundamentals involved.

Shenzhou 2 launched in January 2001, conducting numerous maneuvers before the descent module landed seven days and 108 orbits later. The Chinese ability to maneuver the *Shenzhou 2* independent orbital module surprised Western observers. Life-support systems were tested on the flight, as were guidance and reentry technologies. No pictures of the returned capsule were released; there was a virtual press blackout, leading to Western speculation that there had been landing problems, likely either with parachutes or retro-rockets. The Chinese denied such allegations.

Shenzhou 3 launched on 24 March 2002 and landed in Inner Mongolia (perhaps to better control launch information flow) on 1 April. In the three seats, dummies were wired to medical monitors to test life-support systems, most of which were purchased from Russia. The *Shenzhou 3* forward orbital module, used to hold experiments, could also act as a docking crew transfer module for future Chinese space missions, including docking with another Shenzhou vehicle to form an interim space laboratory. *Shenzhou 3* left the forward module in orbit, likely for future docking tests. It also apparently carried a relatively sophisticated remote sensing payload (medium-resolution imaging spectroradiometer), transmitting high-quality data to Chinese ground stations. The infrared technologies being validated by the instrument potentially had civil and military applications (for military satellites), illustrating the inherently gray nature of most space technologies and the complexity of analyzing intent behind any space program.

Shenzhou 4 launched on 30 December 2002 and returned on 6 January 2003 after 108 orbits. State newspapers and media heralded stories of the spacecraft after its successful landing. Testing maneuverability and life-support systems was the mission priority. With a successful fourth mission, all systems were considered tested, and the Chinese considered themselves ready for launch with a taikonaut on board.

Shenzhou 5, the first with a human pilot, launched from China's Jiuquan launch site on 15 October 2003 and returned 21 hours later after 16 orbits around Earth. This flight carried taikonaut Yang Liwei, making China the third country with human spaceflight capability, joining the United States and Russia.

Shenzhou 6, China's second piloted spaceflight, launched on 12 October 2005 from Jiuquan and returned on 17 October after 76 orbits, landing one-half mile from its target in Inner Mongolia. The mission lasted 115 hours and 32 minutes and had two crewmembers, taikonauts Fei Junlong and Nie Haisheng. Unlike Yang Liwei's flight in 2003 that was shrouded in secrecy, much of *Shenzhou 6* was televised to the country in an unprecedented display of openness. During their stay in orbit, the taikonauts were able to change into lighter, more comfortable space suits for ease of movement while conducting experiments in space, and, for the first time, were able to enter the orbital module.

Shenzhou 7, China's third piloted spaceflight, launched on 25 September 2008 from Jiuquan and returned on 28 September. It was the first Chinese mission to carry three taikonauts, Zhai Zhigang, Liu Boming, and Jing Haipeng. During the mission, Zhai wore a Feitian space suit to conduct a 22-minute space walk, the first ever by a taikonaut.

Joan Johnson-Freese

See also: China

Bibliography

Phillip S. Clark, "The First Flights of China's Shen Zhou Spacecraft," pts. 1 and 2, *Journal of the British Interplanetary Society* 56, no. 5/6 (2003): 160–74; and 57, no. 5/6 (2004): 196–208.

Joan Johnson-Freese, "Space *Wei Qi*: The Launch of *Shenzhou V*," *Naval War College Review* 57 (Spring 2004): 121–45.

Shuttle-Mir

The Shuttle-Mir program was a U.S.–Soviet Union collaboration to fly cosmonauts on the Space Shuttle and allow astronauts long-duration visits on board *Mir*. The program originated from a U.S. desire to engage the Soviet Union in the post–Cold War era through space cooperation.

Known as Phase 1 of a then-proposed three-stage path to construction of Space Station Freedom, Shuttle-Mir enabled NASA to gain practical space station operations experience. Like the Apollo-Soyuz Test Project (ASTP) in 1975, Shuttle-Mir was linked to U.S. foreign policy and national security objectives. Follow-on discussions to the ASTP had led to suggestions of a Salyut-Shuttle program and cooperative robotic planetary missions, but the deteriorating U.S.-Soviet relationship during the administrations of U.S. Presidents Jimmy Carter and Ronald Reagan, in part caused by the Soviet invasion of Afghanistan in 1979, made collaboration politically undesirable. Interest in U.S.-Soviet space cooperation returned after Mikhail Gorbachev rose to power and

The Space Shuttle Atlantis *is connected to the* Mir *space station in July 1995 during STS-71. This was the first Shuttle docking with* Mir. *(Courtesy NASA/Russia Space Agency)*

SERGEI KONSTANTINOVICH KRIKALEV (1958–)

Sergei Krikalev, a Russian cosmonaut, became the first cosmonaut to fly on the Space Shuttle during Space Transportation System (STS)-60 in 1994. His participation in the mission symbolized the beginning of a joint venture between the United States and Russia as partners in space exploration. Krikalev served as an engineer aboard the *Mir* space station during two missions launched in 1988 and 1991. He flew on STS-88 in 1998, the first *International Space Station* (*ISS*) assembly mission, was a member of the first *ISS* expedition crew in 2000, and served as commander of ISS Expedition 11 in 2005. He logged 803 days, 9 hours, and 39 minutes in orbit during six spaceflights—a spaceflight endurance record.

Tim Chamberlin

removed Soviet troops from Afghanistan in 1988 and accelerated after Russia's inheritance of Soviet space assets after the collapse of the Soviet Union in 1991.

At a mid-1992 summit meeting, President George H. W. Bush and Russian President Boris Yeltsin issued a joint statement on cooperation in space, which led NASA Administrator Daniel Goldin and Russian Space Agency (RSA) Chief Yuri Koptev to create the Shuttle-Mir program. The William Clinton presidential administration further expanded the Shuttle-Mir program in 1993 and, motivated by the need to stem the proliferation of missile technology and provide employment opportunities within the Russian aerospace sector, allocated $400 million to Russia. The money covered, in part, expenses to carry U.S. equipment on Russian-built space station modules and to pay for U.S. crew time aboard *Mir*. The cash payment was a change in policy for the United States, which in the past operated solely on a no-exchange-of-funds basis with Russia.

Shuttle-Mir space operations began on 3 February 1994, when Sergei Krikalev became the first cosmonaut to fly on the Space Shuttle during Space Transportation System (STS)-60. On the same date, one year later, cosmonaut Vladimir Titov rode aboard *Discovery* (STS-63) for the first Shuttle approach (to within 10 m) of *Mir*.

Norman Thagard, the first astronaut to enter space on a Russian Soyuz vehicle (*Soyuz TM-21*), became the first U.S. crewmember of *Mir* in March 1995. Thagard was under the command of Vladimir Dezhurov, a lieutenant colonel in the Russian armed forces. With limited communications allocated to U.S. ground controllers and the late arrival of the *Spektr* module that housed much of his scientific equipment, Thagard experienced extreme isolation and a lack of meaningful work. In June 1995, after some delays, Space Shuttle *Atlantis* (STS-71) carried cosmonauts Anatoliy Solovyov and Nikolai Budarin to *Mir*, while returning Thagard and other crewmembers. Thagard had spent 115 days in space, 25 days longer than originally planned due to the late arrival of *Spektr*.

In November 1995 as Space Shuttle *Atlantis* (STS-74) visited *Mir*, astronauts from the European Space Agency, Canada, and the United States worked with cosmonauts in the same space complex for the first time. *Atlantis* delivered a specialized docking module to provide better clearance between the Shuttle and the station during docking.

SHANNON MATILDA WELLS LUCID (1943–)

Shannon Lucid, a U.S. astronaut, logged a total of 223 days in orbit. She was selected as an astronaut in 1978 and flew on six Space Shuttle missions. She served as an engineer on the *Mir* space station for 188 days in 1996, speaking exclusively in Russian, where she performed numerous microgravity experiments. She was the first woman and scientist to receive the Congressional Space Medal of Honor in 1996 for her work on *Mir*, and later served as NASA's chief scientist from February 2002 to September 2003.

Tim Chamberlin

Soon after the arrival of a new cosmonaut crew aboard *Soyuz TM-23*, astronaut Shannon Lucid arrived on *Mir* aboard *Atlantis* (STS-76) in March 1996. With the arrival of the *Priroda* module on 26 April, Lucid began her scientific research program in earnest, working on a flexible schedule—a key difference in comparison to minute-by-minute scheduling of activities on board the Shuttle. While the next cosmonaut exchange had been delayed by more than 40 days because of Russia's difficulties financing Soyuz missions, booster problems delayed the launch of *Atlantis* (STS-79), extending Lucid's stay to 188 days in space. Lucid spent 179 days on *Mir*.

The arrival of astronaut John Blaha with STS-79 in September 1996 marked the first handover from one astronaut to another on *Mir*. Cosmonaut Gennadi Manakov's heart problem, identified just before the planned launch of *Soyuz TM-24* one month earlier, caused RSA to replace the prime crew with backup crewmembers Valeri Korzun and Aleksandr Kaleri. As a result, Blaha, Korzun, and Kaleri began their mission together practically as strangers but lived and worked with little conflict. By the time astronaut Jerry Linenger arrived aboard *Atlantis* (STS-81) in January 1997, Blaha, in addition to carrying out his scientific research program, had developed a new, more efficient approach to future crew transfers.

Linenger, a physician, had already started his biomedical science program when the crew of *Soyuz TM-25* arrived in February 1997. With a crew of six on board, *Mir*'s life-support system had to be supplemented with Vika "candles" that produced oxygen as a by-product of a chemical reaction. Shortly after a routine candle ignition, dense smoke filled *Mir*; Korzun ordered the only accessible Soyuz vehicle readied for evacuation while he and Linenger extinguished an on-orbit fire. Disaster was averted, but other challenges remained: life-support system failures plagued the remainder of Linenger's stay, and an attempt to manually dock the *Progress M-33* resupply vehicle in March 1997 resulted in a near impact. The manual docking, a test designed to save money by reducing future purchases of the expensive autonomous Kurs docking system, required a cosmonaut on board *Mir* to pilot the resupply vehicle using only a video display transmitted from a camera on board the resupply vehicle. Price increases by the Ukrainian supplier of the Kurs system increased financial pressure on Russia to develop an inexpensive alternative.

About one month after astronaut Michael Foale arrived on *Mir* aboard *Atlantis* (STS-84) in May 1997, cosmonaut Vasili Tsibliyev executed a second manual docking attempt; the absence of position and speed information probably contributed to his loss of control, and the *Progress M-34* module smashed into the *Spektr* module, resulting in a slow but persistent leak of air from the station. To plug the leak, the crew removed interior cabling and sealed the *Spektr* module, which resulted in a power crisis; most of *Mir*'s systems were turned off. Foale and his crewmembers worked for days by flashlight in high temperatures to restore power. Later, with the arrival of the *Soyuz TM-26* crew and return of Tsibliyev and Aleksander Lazutkin to Earth, cosmonauts Anatoliy Solovyov and Pavel Vinogradov performed an internal spacewalk into *Spektr*. They installed a new hatch with electrical connections to *Spektr*'s still-functioning solar arrays and recovered some of Foale's equipment.

Despite concern expressed by the U.S. Congress about *Mir*'s safety, NASA Administrator Dan Goldin announced the day before the launch of *Atlantis* (STS-86) in September 1997 that astronaut David Wolf would visit and stay aboard *Mir* as planned. In addition to scientific experiments, Wolf performed maintenance tasks on the increasingly cluttered and aged *Mir*, and for the first time, cast an American election ballot from space. Late in his mission, Wolf and veteran cosmonaut Solovyov conducted a spacewalk to characterize the external condition of the station.

The last astronaut to stay on *Mir*, Andy Thomas, arrived in January 1998 aboard *Endeavour* (STS-89). While he carried out scientific experiments, his Russian crewmembers were mostly occupied upgrading life-support systems, repairing a leak in the air conditioning system, and carrying out other maintenance tasks.

Despite an imminent lack of funds with the end of the Shuttle-Mir program, members of the Russian Duma, among others, argued that *Mir* should be maintained until construction of the *ISS* was well underway. The visit of RSA flight director Valeri Ryumin to *Mir* aboard *Discovery* (STS-91) in June 1998 confirmed that maintenance of *Mir* required the full-time efforts of two to three crewmembers, limiting the viability of the station. After three more Soyuz visits to the station and failed attempts for additional financing, timed thruster firings caused *Mir* to reenter Earth's atmosphere above the Pacific Ocean on 23 March 2001.

During Shuttle-Mir, NASA gained valuable experience in long-duration spaceflight and expertise in how to build, operate, and maintain large space structures. Physiological experiments on board *Mir* clearly demonstrated the need for countermeasures to prevent the debilitating effects of long-term habitation in microgravity, including the psychological challenges of living and working in cramped and isolated conditions. The challenges encountered during the Shuttle-Mir program highlighted the need for flexibility in carrying out programs of long-duration spaceflight. Collaboration between NASA and RSA made routine joint operations between astronauts and cosmonauts practical, laying the groundwork for Russian participation in the *ISS*.

Chris Carr

See also: Astronauts, Cosmonauts

Bibliography
Rex Hall, *The History of* Mir, *1986–2000* (2000).
David Harland, *The Story of Space Station* Mir (2005).
Clay Morgan, *Shuttle-Mir: The United States and Russia Share History's Highest Stage* (2001).
Robert Zimmerman, *Leaving Earth: Space Stations, Rival Superpowers, and the Quest for Interplanetary Travel* (2003).

Skylab

Skylab, the first U.S. space station, evolved from the Apollo Applications Program (AAP), a follow-on program to the Apollo lunar missions that intended to employ Apollo/Saturn technology for a variety of uses. As originally envisioned, AAP would modify the liquid hydrogen tank of a Saturn S-IVB rocket stage after launch into space to serve as an experiments laboratory called the wet workshop. This configuration became known as the cluster concept once more elements were added to the design. Problems with the liquid hydrogen environment of the S-IVB led to the switch to a dry workshop configuration that would be outfitted on the ground and launched atop a Saturn V rocket.

Renamed *Skylab*, the cluster consisted of the Orbital Workshop, manufactured by McDonnell-Douglas, providing living quarters, stowage, and the experiment laboratory; the Instrument Unit built by International Business Machines (IBM) for launch vehicle control; the Airlock Module and Fixed Shroud built by McDonnell-Douglas, providing utilities for environmental control, data distribution, and a hatch for extravehicular activity; the Apollo Telescope Mount (ATM) for solar observation; the Martin Marietta–built Multiple Docking Adapter, providing two docking ports for the Apollo Command and Service Module built by Rockwell International; and control panels for the Earth Resources Experiment Package.

Skylab's objectives included biomedical experiments aimed at evaluating the crew's physiological responses to long-term spaceflight, solar studies using the ATM, remote sensing studies using the EREP package, and materials sciences experiments to grow crystals in zero gravity.

On 14 May 1973 the last Saturn V launch placed *Skylab* into orbit. On ascent the station's micrometeoroid shield ripped off, jamming one solar panel and tearing the other one off the station. The accident deprived the station of vital electricity. The station quickly overheated because the shield provided protection from the Sun. The launch of the first crew was delayed until Mission Control at Johnson Space Center discovered a way to solve the heating problem and to deploy the stuck solar array.

NASA launched the first crew on 25 May, carrying a solar shield and tools to free the jammed solar panel. After docking, the crew of Charles "Pete" Conrad, Joe Kerwin, and Paul Weitz successfully repaired damage to the station. They returned after 28 days, setting a space endurance record.

Two more crews performed a variety of experiments at the station. Alan Bean, Jack Lousma, and Owen Garriott spent 59 days in orbit beginning 28 July. The final crew—Gerald Carr, Ed Gibson, and William Pogue—launched on 16 November and spent 84 days aloft.

Despite its rocky start, *Skylab* proved a success, hosting three crews for a total of 171 days and providing a wealth of data on living and working in space. Despite belated efforts to keep it aloft, *Skylab* reentered the atmosphere on 11 July 1979.

Thomas J. Frieling

See also: Aerospace Medicine, Skylab Solar Science

Bibliography

Leland F. Belew, ed., *Skylab, Our First Space Station* (1977).

W. David Compton and Charles D. Benson, *Living and Working in Space: A History of Skylab* (1983).

David Shayler, *Around the World in 84 Days: The Authorized Biography of Skylab Astronaut Jerry Carr* (2008).

Soviet Manned Lunar Program

The Soviet manned lunar program consisted of two piloted programs—one for flying around the Moon and another for landing cosmonauts on the lunar surface. In summer 1958 space pioneers Sergei P. Korolev and Mikhail K. Tikhonravov first proposed sending crewed spacecraft to the Moon as part of a list of goals for the Soviet space program. Early lunar spacecraft concepts were based on the use of the Soviet R-7 ICBM, which had the capability to put just more than six tons of payload into low Earth orbit. This limited launch capability would require that any lunar spacecraft be assembled in Earth orbit before departing for the Moon, because its weight would greatly exceed the launch capacity of a single R-7.

Soviet leaders did not give official approval to proceed with lunar missions until August 1964. At the time, the Soviet space program was characterized by competing design bureaus, each with its own concept for a crewed lunar mission. Early concepts ranged from aerodynamic, winged spacecraft to capsule designs. The Soviet lunar program was split between two design bureaus: Korolev's OKB-1 (Experimental Design Bureau), which was chosen to build the lunar landing spacecraft (L3), and the design bureau headed by Vladimir Chelomei, OKB-52, which received approval to build the circumlunar spacecraft (LK-1) to be launched aboard the UR-500 (Proton) rocket. For technical and political reasons, the LK-1 program was eventually canceled in 1965, and Korolev's bureau was placed in charge of both the lunar landing and the circumlunar programs.

The weight and complexity of the aerodynamic concepts required to reach the Moon led to the creation of a multimodule spacecraft with a small capsule for atmospheric entry. This circumlunar spacecraft was designated the 7K-L1, or L1 for short, and was designed by OKB-1. The L1 was similar to the Earth-orbital Soyuz and was eventually referred publicly as Zond, possibly to conceal its true purpose. OKB-1's design concept for a lunar landing mission featured an L1 spacecraft, which would be capable of flying two cosmonauts to the Moon and returning them to Earth, and a small Lunei Kabina (Lunar Cabin) also known as the LK, which would carry one cosmonaut to the lunar surface and return the cosmonaut to the orbiting L1 spacecraft.

When docked together the L1 and the LK formed the Lunnei Orbitalnei Korabl (Lunar Orbiting Spacecraft), referred to as the LOK.

Korolev's plan for reaching the Moon used the giant N-1 rocket to launch the LOK. The N-1 was to lift 40–50 tons into low Earth orbit and its follow-on, the N-2, was expected to launch 60–80 tons. The N-1 and N-2 were behemoths, with up to 30 rocket engines in their first stages. This would be their downfall, as the reliability issues associated with so many rocket engines operating simultaneously proved insurmountable.

Although the Soviet Union was the first to send humans into space, it gradually lost ground to the United States in its attempt to reach the Moon. By 1966 NASA's piloted Gemini project surpassed the communist superpower's achievements by demonstrating rendezvous techniques and perfecting space walks. However, each country's space programs experienced setbacks in 1967. Three *Apollo 1* astronauts were killed in January during a launch rehearsal, and *Soyuz 1* cosmonaut Vladimir Komarov died less than three months later after his capsule crash-landed shortly after reentry from an Earth-orbital mission. NASA recovered quickly, and the *Apollo 8* spacecraft flew around the Moon in December 1968. The Soviet lunar program was not as fortunate as it experienced several more technical hurdles.

A series of Soviet robotic circumlunar flights with the Zond spacecraft was planned during 1967–68. The flights were launched from the Baikonur Cosmodrome in Kazakhstan with a Proton rocket and carried plants, tortoises, flies, and worms to analyze the effects of cosmic radiation on living organisms. Two Zond test flights, referred to as *Kosmos 146* and *154*, occurred in March and April 1967. The first successfully tested lunar return velocities, while the second sustained major system failures. A third undesignated test mission in September failed to reach orbit, but the launch escape system worked properly and the Zond entry module was recovered successfully. In November another circumlunar test mission failed during launch with the reentry capsule successfully recovered. The first officially designated test flight, *Zond 4*, was launched into orbit on 2 March 1968 with an apogee of 205,000 miles, more than 80 percent of the distance to the Moon, but the spacecraft was never recovered after its guidance system failed during reentry. A month later a Zond test flight resulted in a launch failure, and the escape system had to be used. Disaster struck in July when a Zond test flight exploded on the launch pad, killing three workers.

On 15 September the Soviet lunar program achieved partial success with the test flight of *Zond 5*, which flew to the Moon and back. *Zond 5*'s guidance system failed during reentry, but the spacecraft's return capsule was successfully recovered. The Soviet Union launched the uncrewed *Zond 6* on 10 November that passed within 1,500 miles of the lunar surface. A system failure on the way back to Earth led to a depressurization of the spacecraft's cabin—a mishap that would have been fatal if a crew had been on board. During reentry, *Zond 6* performed the first successful double-skip maneuver. This technique, where the spacecraft dips into the atmosphere to slow down to suborbital velocity, skips back out, and then reenters the atmosphere a second time later in the trajectory, was planned for piloted lunar vehicles. During *Zond 6*'s landing sequence, the entry capsule's parachute released prematurely as a result of the depressurization problem, and the spacecraft was destroyed on ground impact. A rescue team was able to extract film from the wreckage of *Zond 6*, which was used to publish fantastic images of the Moon and Earth. Despite this small triumph, the long list of technical failures associated with its lunar program eliminated any chance of the Soviet Union being ahead of the United States with a piloted mission to the Moon.

On 20 January 1969 another robotic circumlunar flight was attempted, but the booster failed and the launch escape system had to be activated. Not until *Zond 7* launched on 8 August, after NASA achieved its first manned Moon landing during *Apollo 11*, did a Soviet circumlunar flight succeed without any significant problems. *Zond 8*, launched on 10 October 1970, successfully reached lunar orbit, but by the time the mission was over, the lunar program was no longer a priority. The focus of the country's human spaceflight program shifted toward building and operating space stations in low Earth orbit. Despite this change, four test flights involving the Soviet lunar lander took place in 1970–71, and development of the N-1 rocket continued until 1976. Because the N-series of rockets never performed as planned, the Soviet Union's hopes of landing a human on the Moon were doomed. With the cancellation of the N-1 program, the Soviet Union never attempted a piloted lunar mission.

Michael Engle

See also: Cosmonauts, Russian Launch Vehicles

Bibliography

Rex Hall and David Shayler, *The Rocket Men:* Vostok *and* Voskhod, *The First Soviet Manned Spaceflights* (2001).
Dennis Newkirk, *Almanac of Soviet Manned Space Flight* (1990).
Asif A. Siddiqi, *Challenge to Apollo: The Soviet Union and the Space Race, 1945–1974* (2000).

Soyuz

The Soyuz (Union) program emerged during 1962–64 when Soviet rocket Chief Designer Sergei Korolev sketched plans for the successor to Vostok, the world's first piloted spacecraft built by OKB-1 (Experimental Design Bureau). Soyuz was primarily conceived as a workhorse for circumlunar, lunar-orbiting, and lunar-landing missions.

Early blueprint versions of the Soyuz complex consisted of three components: Soyuz A, a crewed vehicle with orbital, descent, and instrument modules; Soyuz B, a rocket block to be attached to Soyuz A in orbit; and Soyuz V, a tanker to transport liquid fuel and oxidizer to Soyuz B. Only the Soyuz A component survived as the basis for the modern Soyuz spacecraft.

Early crewed Soyuz spacecraft and subsequent versions consisted of three modules—a reentry capsule (or descent module), an orbital module, and a service module (also called the propulsion module). The descent module housed the Soyuz primary instrument panel, crew couches, air and cooling systems, parachute compartment, and recovery beacon. The orbital module included spacecraft instrumentation, radar and navigation gear, a crew hatch (used for extravehicular activities—EVAs), and storage compartments for a variety of mission supplies. On later missions this module sported

Astronaut C. Michael Foale, left, cosmonaut Alexander Y. Kaleri, center, and European Space Agency astronaut Pedro Duque of Spain sit inside a Soyuz TMA-3 vehicle in a processing facility at the Baikonur Cosmodrome in Kazakhstan during a pre-launch inspection in October 2003. (Courtesy Bill Ingalls/NASA)

a docking probe and access hatch to transfer to other spacecraft and space stations. The service module housed propellant tanks, a radio communications antenna, various sensors, and the solar-power array.

Following four Soyuz-derived uncrewed Zond-Kosmos tests from November 1966 to April 1967, the first piloted Soyuz spacecraft was prepared for launch. NASA had already demonstrated EVAs, rendezvous-and-docking procedures, and long-duration flights during Project Gemini, but the deaths of three astronauts during an Apollo ground test in January 1967 brought the U.S. human spaceflight program to a halt. In the wake of this tragedy, Soviet officials hoped to regain the international spotlight with the new Soyuz.

To begin the crewed Soyuz program in spring 1967, the Soviet space agency planned a spectacular joint rendezvous-and-docking mission consisting of two spacecraft, *Soyuz 1* and *Soyuz 2*. *Soyuz 1*, with Vladimir Komarov aboard, was launched on 23 April 1967. Problems with the vehicle forced ground controllers to abort the mission early. During reentry, the descent module's parachutes became entangled and *Soyuz 1* crashed into a field at high velocity in central Russia, killing Komarov. Soviet public reaction to the event was similar to how Americans mourned the death of the three *Apollo 1* astronauts three months earlier.

Following the *Soyuz 1* accident, Soviet space authorities concentrated on uncrewed Soyuz tests for the next 16 months, using the surrogate Zond-Kosmos spacecraft. Less than a week after the safe return of the first piloted Apollo mission, the uncrewed *Soyuz 2(A)* and *Soyuz 3* (with sole passenger Georgi Beregovoi) were launched in October 1968 to repeat the failed rendezvous-and-docking mission of *Soyuz 1–2*. The mission was a minor success. Beregovoi expended nearly all the maneuvering fuel trying to dock with the *Soyuz 2(A)* drone. Despite its shortcomings, *Soyuz 3* helped put the Soviet human spaceflight program back on track. In January 1969, *Soyuz 4* and *5* successfully docked for several days, forming what the Soviet Union dubbed the world's "first experimental space station."

Other Soyuz feats, military and civilian, followed these early flights with the possibility of a lunar mission always on the minds of Western (and some dissident Soviet) observers. However, starting in 1969, catastrophic failures of the secretive N-1—the giant rocket that might have carried a crewed Soyuz to the Moon—effectively ended any chance of a Soviet attempt to send cosmonauts to the Moon. By the time the 17-day-long, two-person *Soyuz 9* flight returned to Earth on 19 June 1970, the expendable Soyuz spacecraft was already relegated to the role of ferry for the Salyut space stations.

The Soviet human spaceflight program suffered a major blow when the three-man crew of *Soyuz 11* died on 30 June 1971 following a reentry accident. The crew's cabin depressurized too soon after problems associated with separation of the Soyuz orbital module. The deaths prompted Soviet space officials to require that cosmonauts wear pressure suits on future flights during launch and reentry. Piloted flights did not resume until June 1973.

The 19th Soyuz flight involved a joint mission with the United States called the Apollo-Soyuz Test Project, also known as the Soyuz-Apollo Experimental Flight in Russia. The mission was announced in advance and broadcast live in the Soviet Union—both firsts for the country. A Soyuz 7K-TM was modified to allow it to dock to an Apollo spacecraft in July 1975. The mission took place without incident.

With the next generation of Salyut space stations in the works, a robotic version of the Soyuz was created to ferry supplies to cosmonauts for longer missions in space. Called Progress, this uncrewed Soyuz variant extended the operational lifetimes of *Salyut 6* and *7*. Progress was introduced in 1978 and flown 25 times to both stations through 1985. The spacecraft was upgraded in later years to support *Mir* and the *International Space Station* (*ISS*).

The improved Soyuz T (Transport) replaced the original Soyuz in the late 1970s. The Soyuz T orbital module was capable of being used for extra storage space while docked to a Salyut station. Several changes were made to the spacecraft's descent module. Window covers could be removed after reentry for improved viewing, and the flight control system included integrated circuitry chips and a digital computer. Soyuz T crews also had the benefit of a new escape tower that had better safety margins. The attitude control thrusters and main propulsion system located on the propulsion module received a major overhaul after being integrated with each other. Piloted flights of the T series began in June 1980 with *Soyuz T-2* and concluded with *Soyuz T-15* flown in March 1986.

Soyuz Flights (1967–2010)

Vehicle Type*	1967–1970	1971–1975	1976–1980	1981–1985	1986–1990	1991–1995	1996–2000	2001–2005	2006–2010[a]	Total
Soyuz	9	11	19	2	–	–	–	–	–	41
Soyuz T	–	–	1	11	1	–	–	–	–	13
Soyuz TM	–	–	–	–	10	11	9	3	–	33
Soyuz TMA	–	–	–	–	–	–	–	7	11	18
Combined	9	11	20	13	11	11	9	10	11	105

*The Soyuz, Soyuz T, Soyuz TM, and Soyuz TMA spacecraft were nonreusable and replaced after each flight.

[a]Flights through April 2010.

Yet another improved design, Soyuz TM (Transport Modification), replaced the T series in mid-1986. The TM series was equipped with the more modern Kurs rendezvous-and-docking system necessary for trips to *Mir*. The orbital module was changed to include a new window to aid during docking. The propulsion module received another overhaul, and a new, lighter parachute system improved the landing capability of the spacecraft. The first Soyuz TM piloted flight to *Mir* occurred in February 1987. In December 1990 a Soyuz TM carried the first Japanese cosmonaut, Toyohiro Akiyama, to *Mir*.

The Soyuz TMA was introduced in late 2002 to accommodate new NASA requirements for flights to the *ISS*, among them its use as a lifeboat for resident crews. The TMA series was equipped with a larger seating capacity for taller passengers, received changes to its automated landing system, and was retrofitted with newer computer displays and control panels.

The TMA series proved to be a safe and reliable ferry and service spacecraft when it was called on to rescue the *ISS* following the Space Shuttle *Columbia* accident in February 2003. When NASA grounded its aging Shuttle fleet after the tragedy, Soyuz TMA and Progress spacecraft helped keep the wavering *ISS* program on track.

Despite its early mishaps, the Soyuz spacecraft served as a reliable vehicle for more than three decades. In all its utilitarian incarnations, from uncrewed Zonds to Progress cargo ships employed for *ISS* missions, Soyuz formed the backbone of the Soviet/Russian space program—and even inspired the Chinese crewed *Shenzhou* spacecraft launched in 2003. The robust Soyuz design and strong safety record were the hallmarks of the program. From April 1967 to April 2010, a total of 105 piloted Soyuz flights were flown with visits to eight space stations in low Earth orbit.

Lou Varricchio

See also: Interkosmos, Rocket and Space Corporation Energia

Bibliography

Rex Hall and David Shayler, *Soyuz: A Universal Spacecraft* (2003).

Dennis Newkirk, *Almanac of Soviet Manned Space Flight* (1990).

Asif A. Siddiqi, *Challenge to Apollo: The Soviet Union and the Space Race, 1945–1974* (2000).

Space Shuttle

Space Shuttle, the main component of the U.S. NASA Space Transportation System, includes the piloted orbiter vehicle, a giant external propellant tank, and two solid rocket boosters. The Space Shuttle became the first piloted vehicle to be launched to space and returned for repeated flights.

The concept of a space shuttle first emerged during the mid-1960s, when Project Apollo was passing its peak in funding. This raised the question of what NASA would do next. George Mueller, director of the manned-spaceflight program, set his sights on a large space station that was to fly to orbit atop a Saturn V. He also wanted low cost for the frequent spaceflights he believed it would attract. Existing launch vehicles, such as the Saturn I-B, looked unpromising in this respect. Mueller therefore decided that a reusable space shuttle, as an all-new design, could lower the logistical price tag that he sought. His space station concept failed to win support, for there was little demand for it beyond NASA. But during 1969 the new Richard Nixon presidential administration carried through a high-level planning effort.

NASA Administrator Thomas Paine, flushed by the success of the Apollo lunar landings, accepted Mueller's plans and took them several steps further. Paine proposed going to Mars. En route to that destination, Paine proposed to sow the Earth-Moon system thickly with space stations and to follow with a much larger space base. Space station modules would be adapted for the Mars voyage, with nuclear engines pushing the Mars ship on its way. To make these things possible, NASA would need a space shuttle for routine flight to orbit.

Paine had come to Washington as a liberal Democrat—and the White House was full of Republicans. Nixon particularly took the view that having reached the Moon, NASA had no need to go farther. Hence within 18 months NASA sustained three important setbacks. The first came from the Office of Management and Budget (OMB), which drew up the annual federal budget that the president sent to Congress. In November 1969 the OMB chopped Paine's request by $1 billion. The Mars plan was shelved along with the space base. Paine responded by falling back on Mueller's earlier plans, as he called for a shuttle/station. NASA officials described it as a single interrelated program that focused on the station, with a shuttle providing low-cost logistics. This approach nearly terminated both. Critics in Congress attacked NASA harshly during 1970, asserting that a shuttle/station was really the opening gambit in a program that still aimed at Mars. Such a program would be quite costly, and the shuttle/station survived by the narrowest margins in congressional votes.

A second setback to NASA happened when Paine left the agency in September 1970, and George Low, his former deputy, responded by changing NASA's approach. He knew that the station needed a shuttle, but a shuttle might find plenty to do even without a station. A shuttle had been planned as part of a program of human spaceflight. Low now argued that it could earn its way by launching and servicing the nation's satellites.

For this argument to hold, NASA needed support from the U.S. Air Force (USAF). Yet the USAF did not need NASA. The USAF had its own launch vehicles, including the powerful Titan III. USAF officials therefore insisted that NASA redesign a shuttle to meet military requirements—and to foot the entire bill with no financial help from the Pentagon.

NASA had envisioned a shuttle of modest size, carrying payloads of as little as 25,000 lb. The USAF was launching heavy reconnaissance satellites and insisted on a payload capability of 65,000 lb. Such a shuttle would be far heavier and therefore more costly, but NASA had no choice. In January 1971 agency officials agreed to design a shuttle to meet USAF demands.

This surrender actually pointed to victory, for the Space Shuttle now could take form as a national launch system, capable of serving civilian and military requirements. With this, NASA stilled the doubts of Congress. But the OMB was still to be heard from, and in May 1971 it dealt NASA a third setback. Its budget examiners declared that they intended to hold NASA to strict financial limits during future years. The agency had hoped to spend as much as $2 billion per year to develop the Shuttle, but the OMB cut this in half.

New NASA Administrator James Fletcher responded with a sweeping program of design studies that sought to cut this cost while retaining full 65,000 lb capability. Initial designs had called for a two-stage vehicle, both stages being winged and piloted while carrying propellants within their fuselages. To shrink the second stage (the orbiter), thereby cutting its development cost, its propellants went into a big external tank that would be discarded during each flight. The first stage was planned initially as an enormous airplane with 12 rocket engines. In final form, this stage took shape as a simple pair of solid-propellant rocket boosters.

The new designs carried higher costs per flight, but they met OMB's requirements. OMB staffers nonetheless took the view that because NASA had done so well in cutting development costs, it might do even better if squeezed further. By the end of 1971 Fletcher believed he could seek no more than a mini-shuttle, too small to serve the needs of the USAF.

Yet his position within the White House was stronger than he thought. George Shultz, director of OMB, worked closely with Nixon and shared a view that human spaceflight was something that the nation should continue to pursue. With this basic policy in place, Shultz was not about to quibble over the modest cost savings that might result by building a smaller shuttle. With Nixon's assent, Shultz gave support to Fletcher's request for funding that would allow NASA to build the full-size Shuttle that it wanted.

Low cost-per-flight remained a prime goal. Each Shuttle orbiter was to achieve this by being turned around for reflight in only two weeks. At the outset, there were several reasons to believe this might be feasible. The X-15, a piloted rocket plane with a liquid-propellant engine, had already demonstrated turnaround times of as few as six days, with individual X-15 craft flying as often as three times in a month.

Pan American World Airways had taken the lead in developing onboard fault detection. In the Shuttle, this capability promised to ease the problems of diagnosing faulty components that needed replacement. Existing rocket engines, built for one-time use, had nevertheless shown high reserves of endurance. One J-2 engine, built for Apollo, had demonstrated 103 starts and 6.5 hours of test-stand operation, without overhaul. This encouraged thoughts of a reusable rocket with a long life.

The Space Shuttle was to be the world's first reusable launch vehicle. This imposed particular demands in the areas of thermal protection and propulsion. Lockheed made a major contribution with its reusable tiles, which served as the principal form of

Mission specialist Peter J. K. Wisoff, bottom, wearing an Extravehicular Mobility Unit, works with the antenna on the European Retrievable Carrier, while payload commander G. David Low, on the Remote Manipulator System robot arm, hovers above during STS-57 in June 1993. (Courtesy NASA/Johnson Space Center)

thermal protection. New carbon composites also protected the nose and the wing leading edges, which faced particularly intense heat during reentry.

Rocketdyne made a similar contribution with its reusable Shuttle main engine. It worked at operating pressures as high as 7,211 psi, and its turbopumps were particularly demanding. The main fuel pump generated more than 61,000 hp, more than had driven the ocean liner *Titanic*. Yet this pump was small enough to fit on a dining room table.

In other areas the solid rocket boosters were the largest solid motors ever built. Fuel cells, for onboard power, achieved higher power, lighter weight, and much longer life than those of Apollo. Redundancy in onboard electronics, including computers, assured reliability. A manipulator arm, imported from Canada, extended an astronaut's reach.

The Shuttle's hydraulic systems followed conventional design as they moved control surfaces, steered the solid and liquid rocket engines, and operated the landing gear and

Space Shuttle Flights (1981–2010)

Orbiter	1981–1985	1986–1990	1991–1995	1996–2000	2001–2005	2006–2010[a]	Total
*Columbia**	6	4	8	8	2	–	28
*Challenger***	9	1	–	–	–	–	10
Discovery	6	5	10	7	3	7	38
Atlantis	2	5	8	7	4	5	31
Endeavour	–	–	9	6	4	5	24
Combined	23	15	35	28	13	17	131

**Columbia* was destroyed on reentry 1 February 2003.
***Challenger* was destroyed 72 seconds after launch on 28 January 1986.
[a]Flights through April 2010.

brakes. They drew power from a new source: a suite of onboard units that used hydrazine as the propellant. Hydrazine, burned with nitrogen tetroxide, also provided thrust from auxiliary rockets that maneuvered the Shuttle and controlled its attitude when in orbit.

At the start, NASA promised the OMB that the Shuttle launch cost would be $10.45 million per flight, in 1971 dollars. This estimate rested on a planned schedule of 60 flights per year. The first piloted Space Shuttle flight occurred on 12 April 1981. Commanded by veteran astronaut John Young and pilot Robert Crippen, the Shuttle orbiter *Columbia* circled Earth 36 times during 54 hours before returning to Earth. NASA conducted only four additional Shuttle flights by the end of 1982, but its managers expected to gain experience and to move gradually toward more ambitious flight rates.

A 60-per-year flight schedule proved to be out of reach. Still, by 1985, with four orbiters in the Shuttle fleet (*Columbia*, *Challenger*, *Discovery*, and *Atlantis*), NASA's program managers were ready to project 24 flights per year by 1990. For 1986 the number of projected flights was 15. But on 28 January 1986, the Shuttle *Challenger* exploded in midair 72 seconds after launch and was destroyed. The immediate fault lay with the solid rocket boosters and was fixed with a redesign. Yet there was a broader and more significant issue: pressed by demands of the flight rate, managers had made decisions that had compromised safety.

In August 1986 President Ronald Reagan issued a new space policy that ended attempts to use the Shuttle as a general-purpose launch vehicle. Instead it was to fly only a few times per year and was to be reserved for missions that required its specialized capabilities, such as carrying astronauts to refurbish a satellite that was already in orbit.

During the 1970s Shuttle managers had sought to attract payloads by quoting the lowest launch costs. However, experience showed that the Shuttle in fact was the most costly launch vehicle in the world. By 1996, a decade after the *Challenger* accident, the cost per flight was $550 million in 2004 dollars.

Because it carried astronauts, the Shuttle showed as early as 1984 that it could do more than simply launch satellites. On its 11th flight, the Shuttle conducted a retrieval and repair of the *Solar Maximum* spacecraft, making it ready for further service. Later that year another mission retrieved two communications satellites that had been stranded in low orbit. Subsequent retrievals were more deliberate. In January 1990 the Shuttle recovered the *Long Duration Exposure Facility*, launched in 1984. A 1996 mission retrieved a Japanese satellite that had reached orbit the previous year.

Shuttle support of the *Hubble Space Telescope* (*HST*) became a highlight of such missions. Following launch in 1990, this orbiting telescope proved to have a main mirror with a faulty shape, which prevented it from focusing properly. A Shuttle mission in 1993 installed "eyeglasses" that corrected its vision, allowing this telescope to achieve its full promise. Subsequent Shuttle missions in 1997, 1999, and 2002 installed new instruments and replaced aging components, repeatedly renewing *HST* for continued life.

Many Shuttle flights worked to extend the roles of astronauts. The German firm Entwicklungsring Nord (ERNO) had developed *Spacelab*, a small laboratory that the Shuttle could carry in its payload bay. The ninth Shuttle mission was also the first *Spacelab* mission, in 1983. Three more *Spacelab* flights followed during 1985; then, following the loss of *Challenger*, the program stood down for a hiatus of nearly six years. But *Spacelab* flew again in 1991 and then made 10 more flights before the program concluded in 1998.

In 1991, communism fell within the Soviet Union. One year later, agreements at the highest levels of government opened the door to a merger of the U.S. and Russian space programs. As cooperation flourished between these nations, the Shuttle took on new roles. In 1994 it carried its first cosmonaut, Sergei Krikalev. One year later the Shuttle initiated a succession of visits to the Russian space station *Mir*. The first such visit picked up astronaut Norman Thagard, who had flown to *Mir* aboard a Russian launch vehicle. Flights to *Mir* continued until 1998, with Americans living aboard this station for months at a time.

In this fashion the Shuttle helped Russia bypass the need to pursue its own reusable vehicle. This nation indeed had done that, building a large expendable launch vehicle called Energia and arranging for it to carry a reusable orbiter called Buran. Energia flew in May 1987, carrying a payload with a record weight of 100 metric tons, while Buran followed with an unpiloted flight in November 1988. These flights were the only ones of their kind. Energia never flew again, while Buran, which closely resembled the Shuttle orbiter, wound up in Moscow's Gorky Park, not far from a Ferris wheel.

The year 1998, which saw the end of both *Spacelab* and Shuttle flights to *Mir*, featured the beginning of construction of their mutual successor: the *International Space Station* (*ISS*). The first Shuttle flight in support of its assembly took place late that year. Then, beginning in 2000, the Shuttle became virtually an element of the *ISS*. Of 15 consecutive flights, 14 supported this station, either by bringing equipment or by ferrying members of its crew.

The loss of the Shuttle *Columbia* in February 2003, due to damage to its thermal protection, dealt an important setback to both programs. The *Columbia* Accident Investigation Board (CAIB) report, released that August, raised a clear likelihood that additional Shuttles would be destroyed in future accidents. The aftermath of *Challenger* brought a stand-down of nearly three years before a return to flight. The aftermath of *Columbia* brought a stand-down of 28 months before an attempted return to flight. This stand-down was extended after a flight of *Discovery* in July 2005 showed that NASA had not solved the problem of ensuring the safety of this thermal

RICHARD HARRISON TRULY (1937–)

Richard Truly, a U.S. astronaut and Navy vice admiral, played a vital role in the Space Shuttle program as a pilot and administrator. In 1965 he was selected to the U.S. Air Force Manned Orbiting Laboratory (MOL) program. After MOL's cancellation in 1969, Truly joined NASA as an astronaut. He piloted Space Transportation System (STS)-2 in 1981, the first time a Shuttle was flown into space a second time. In 1983 he commanded STS-8, the first Shuttle flight to launch and land at night. Following the 1986 *Challenger* accident, Truly led the rebuilding of the Shuttle program. He was NASA administrator from 1986 to 1992 and the first astronaut to lead the agency.

Tim Chamberlin

protection. The long suspension in Shuttle flights forced NASA to rely on Russian Soyuz vehicles to taxi *ISS* crews to and from the station. The delay also hindered construction of the station, interrupting plans to reach a "core complete" configuration.

In other respects, the Shuttle proved to be a mixed blessing. This was particularly true in the field of expendable launch vehicles. The *Spacelab* program was part of a package deal whereby Germany assisted France in developing a new and powerful set of rockets called Ariane. Ariane was conceived as a challenge to the Shuttle, which NASA believed would replace all but the smallest of existing expendable launchers. In pursuing Ariane, the French took the view that the Shuttle would fail to meet its goals, that the world would continue to need expendables, and that France intended to be ready when customers came to its door.

In viewing the Shuttle as offering all things to all people, NASA was not about to continue to build expendables. It shut down their production and proceeded to fly off its inventory. Hence, following the loss of *Challenger*, the only launch vehicles certified for NASA's use were two Deltas and three Atlas-Centaurs. Ariane by then had been flying since 1979. It experienced its own setback when a launch failed in May 1986, but overall the French had been prescient. While NASA worked to resume production of its Atlas and Delta expendables, the French racked up new orders for Ariane and took a commanding position as a leader in the launch industry.

Delays within the Shuttle program, both during development and during the long post-*Challenger* stand-down, also inflicted delays on some of the programs that the Shuttle was to serve. This was particularly true of *Galileo*. This mission to Jupiter was initially slated for the 26th Shuttle flight in 1982. Amid the Shuttle program's slippages, it went to the seventh planned flight, still in 1982. Further delays then put off its launch to 1986, the year of *Challenger*.

Galileo had been slated to fly with the Centaur upper stage, which burned liquid hydrogen and liquid oxygen. Those propellants gave this stage enough power to propel *Galileo* on a direct trajectory to Jupiter that could avoid time-consuming swingbys of the inner planets. In addition, NASA was convinced that this stage was safe enough to fly on the Shuttle with astronauts.

But following the loss of *Challenger*, the agency's senior managers had second thoughts. In May Administrator James Fletcher canceled the Shuttle-capable Centaur on grounds of safety. As a replacement NASA fell back on the existing Inertial Upper Stage, which was already in use. But it burned solid propellant rather than hydrogen and oxygen, and it lacked the capability of the Centaur. Consequently, NASA abandoned plans for a direct flight to Jupiter. Instead the new mission plan called for *Galileo* to follow a roundabout route that enabled the spacecraft to gain energy through several inner-planet flybys. Launched in 1989, this spacecraft reached Jupiter in 1995.

Meanwhile the advent of President Reagan's Strategic Defense Initiative in 1983 raised the prospect that the USAF would need a great deal of new space launch capability. As early as February 1984, two years before the loss of *Challenger*, Defense Secretary Caspar Weinberger declared that total reliance on the Shuttle "represents an unacceptable national security risk." In June the USAF decided that beginning in 1988, it would remove 10 payloads from the Shuttle and fly them on improved versions of the Atlas or Titan III. It then arranged to upgrade the latter and to produce a new version, the Titan IV, which could lift 39,000 lb to low Earth orbit.

With the USAF losing interest in the Shuttle and with commercial satellites out of reach as payloads, interest burgeoned in potential post-Shuttle replacements. For a time the USAF took the lead in seeking to build the National Aero-Space Plane, which was designed as a single-stage craft that could fly to orbit using advanced airbreathing engines known as scramjets. These proved to lack the necessary performance, while the flight vehicle, the X-30, proved to produce more drag than expected. New and lightweight materials failed to meet researchers' goals, and after 1990 the program was gradually abandoned.

NASA and Lockheed Martin also pursued their own Shuttle replacement. This was to be a commercial single-stage rocket called VentureStar, which was to use its own advanced materials for light weight. As a prelude NASA sought to build the X-33, which was to demonstrate feasibility. It did no such thing, as its propellant tank proved to lie beyond the state of the art, suffering structural failure in tests. The abandonment of the X-33 in 2001 showed that the near-term future of spaceflight would lie with conventional expendables. In 2004 NASA initiated plans to build a successor to the Shuttle called the Crew Exploration Vehicle (CEV) as part of President George W. Bush's Vision for Space Exploration. The CEV was later called Orion under the space agency's Constellation program, but President Barack Obama proposed its cancellation in his Fiscal Year 2011 budget. Obama changed course two months later in April 2010 with a plan for a scaled-back version of Orion to be used as a lifeboat for the *ISS*.

The Shuttle no longer serves the USAF; it carried its last military payload in 1992. But the close ties between the Shuttle and the *ISS* show that decades after George Mueller envisioned a shuttle/station, this combined program finally came into existence.

T. A. Heppenheimer

See also: Buran, Disasters in Human Spaceflight

Bibliography

David M. Harland, *The Story of the Space Shuttle* (2004).

T. A. Heppenheimer, *Development of the Space Shuttle, 1972–1981* (2002).

———, *The Space Shuttle Decision, 1965–1972* (2002).

Dennis Jenkins, *Space Shuttle. The History of the National Space Transportation System: The First 100 Missions* (2001).

Space Station Freedom

Space Station Freedom was NASA's space station program from 1984 to 1993. President Ronald Reagan embraced the idea of a space station program with international participation and in 1984 directed NASA to build an orbiting outpost that would be permanently occupied within a decade by a crew of eight astronauts. When NASA first proposed the Space Shuttle, its primary purpose was to help build a space station, but budgetary constraints required that the station be delayed. Reagan's decision put the plan back on track. NASA envisioned the station as a jumping-off point for a renewal of lunar exploration and as a step toward future planetary missions, but financial constraints forced this ambitious plan to be put on hold. The station would instead become a scientific platform in low Earth orbit.

The first design selected for the space station was Grumman's Power Tower. This design included a 120 m keel set vertically for stability, a cluster of pressurized modules at either end to facilitate simultaneous observation of Earth and the sky, and a horizontal truss in the middle with solar panels. NASA replaced this design in 1985 in favor of the Dual Keel concept by Lockheed and McDonnell-Douglas. This had two 150 m vertical trusses linked top and bottom by 45 m spars for increased structural strength. Like the Power Tower, it would perform terrestrial and astronomical work, but the instruments would be remotely operated from a single cluster of modules on the horizontal truss halfway up that carried the solar panels. Within months of this revision, the enormous dual keel had been dropped from the configuration because of projected costs. This left the module cluster on the truss with the solar panels.

To give the project a sense of identity, in 1988 Reagan, reflecting his hostility toward what he had dubbed "the evil empire" of the Soviet Union, named it Space Station Freedom. That same year, an intergovernmental agreement was signed by NASA, the European Space Agency, the Canadian Space Agency, and Japan to firmly establish international participation in the project.

Congressional budget reviews led to a succession of redesigns to cut costs. In 1990 Congress gave NASA 90 days to create a design that could be constructed within five years using the Space Shuttle and upgraded for permanent habitation. To reduce the number of flights, the truss was shortened by 30 percent, the modules were shortened by 40 percent to ensure that they could be delivered fully outfitted, and the crew was cut from eight to four astronauts.

Daniel Goldin, named NASA administrator in April 1992, instigated another redesign to further slash costs. When Reagan began the project, it had a budget of $8 billion, but after numerous redesigns (intended to save money), it had consumed $11 billion without placing any hardware in orbit. In March 1993 President William

Clinton took office with a mission to reduce the national deficit. Clinton ordered Goldin to provide a range of more cost-effective plans for the space station. Three options were provided. NASA invited Russia to participate in the project by the time Clinton made his decision. The provocative name Freedom was dropped, initially in favor of Alpha, reflecting the designation of the option that Clinton selected, but this was changed to the *International Space Station*.

David M. Harland

See also: European Space Agency, National Aeronautics and Space Administration

Bibliography

John E. Catchpole, *The International Space Station: Building for the Future* (2008).
David M. Harland and John E. Catchpole, *Creating the International Space Station* (2002).

Space Transportation System. *See* Space Shuttle.

Spacelab

Spacelab, a laboratory research facility carried on the Space Shuttle, resulted from a U.S. presidential mandate to increase international participation in the space program after Apollo. *Spacelab* was a cooperative effort between NASA and the European Space Agency (ESA), conceived as a short mission laboratory. It became the first international space effort of its kind.

Germany was the primary national sponsor of the project for ESA and, through a prime contract to German corporation Entwicklungsring Nord (ERNO), provided the *Spacelab* module and ground support equipment and gave initial flight technical support. NASA provided technical assistance through the Marshall Space Flight Center and the necessary support for integration into the Space Transportation System (STS).

ERNO's design, delivered in 1982, included a pressurized laboratory module with a shirtsleeve environment and an open pallet for instruments needing access to space. The total cost of the project was estimated at $750 million. Two modules, five pallets, or combinations of modules and pallets fit in the Space Shuttle cargo bay and held several tons of experiments. NASA and ESA split experiment access equitably by weight, unless someone purchased the *Spacelab* module to fly experiments, such as Germany during missions D-1 and D-2 (STS-61A and STS-55) and the United States and Japan during mission SL-J (STS-47). This significantly improved flight opportunities for international researchers.

Spacelab's laboratory module provided interchangeable, standardized racks, data, power, and coolant for experiments. Missions were multidisciplinary or could be dedicated to a single discipline such as life science, physical science, astronomy, astrophysics, or Earth observation. *Spacelab* gave researchers the opportunity to fly experiments on multiple flights, ensured flight-tested hardware and techniques, and guaranteed crew monitoring of inflight systems for both module and pallet.

NASA created two new astronaut positions for these science-heavy payloads: mission specialist and payload specialist. Mission specialists were astronauts with

Spacelab Mission Chronology (Part 1 of 4)

Year	Mission	Spacelab Payload	Mission Highlights
1981	Space Transportation System (STS)-2	Office of Space and Terrestrial Application (OSTA)-1	First flight to calibrate Earth remote sensing instruments for atmospheric, oceanographic, and environmental studies.
1982	STS-3	Office of Space Science-1	First broad measure of spacecraft environment. Discovered Shuttle glow.
1983	STS-7	OSTA-2	Showed that crystal growth in microgravity was possible.
1983	STS-9	Spacelab-1	Study of spectral variability from galactic X-ray sources (binary system/neutron star and black hole). Confirmation of Marangoni convection effect in space.
1984	STS-41D	Office of Applications and Space Technology-1	Demonstration of large lightweight solar arrays for use on space stations—largest structure extended in space for the time.
1984	STS-41G	OSTA-3	Reflight of remote sensing instruments from OSTA-1. First verification that biomass burning in remote locations had broad effects on atmospheric chemistry.
1985	STS-51B	Spacelab-3	Two triglycine sulfate (TGS) crystals were grown, and for the first time scientists could compare ground and microgravity growth.
1985	STS-51F	Spacelab-2	Found that plant seeds can germinate and seedlings can grow in the space environment.
1985	STS-61A	Spacelab-D1	Developed float-zone crystal growth technique. Produced crystal morphologies by vapor crystal growth unseen previously.
1985	STS-61B	Experimental Assembly of Structures in Extravehicular Activity/ Assembly Concept for Construction of Erectable Space Structures	Tested assembly methods for large structures in microgravity.
1986	STS-61C	Microgravity Science Laboratory (MSL)-2	Successful test of an acoustic levitator for materials sample processing.
1990	STS-35	Astronomical Observatory (Astro)-1	First spaceborne observation of stars during Earth night.

(Continued)

Spacelab Mission Chronology (Part 2 of 4)

Year	Mission	Spacelab Payload	Mission Highlights
1991	STS-40	Spacelab Life Sciences 1 (SLS-1)	Established process for diminished immune response to microgravity.
1992	STS-42	International Microgravity Laboratory (IML)-1	Great success with a solution crystal growth of a pyroelectric detector material (TGS), with uniform faces and structure never obtained on Earth.
1992	STS-45	Atmospheric Laboratory for Applications and Science (ATLAS)-1	First detection of deuterium atoms in the mesosphere. Observations of a layer of sulfuric acid and water aerosol, from Mt. Pinatubo's 1991 eruption.
1992	STS-47	Spacelab-J	Successful test of acoustic levitator furnace, and several materials processing methods.
1992	STS-50	United States Microgravity Laboratory (USML)-1	Discovered the effects of wall-contact on striations and twinning in crystal growth. Zeolite crystals, important as catalysts in the petroleum industry, grew 10 percent to 50 percent larger and free of gravity defects.
1992	STS-52	United States Microgravity Payload (USMP)-1	First experiment system capable of being controlled from the ground by the principal investigator in real time (Lambda Point Experiment). First time ambient Shuttle acceleration data was available in real time.
1993	STS-56	ATLAS-2	Atmospheric Trace Molecular Spectroscopy made first chemical measurement of arctic stratospheric chemistry at sunrise and made detailed measurements of mesospheric chemistry.
1993	STS-55	Spacelab-D2	Confirmed results of and refined many experiments flown on STS-61A/Spacelab-D1. Significantly added to theoretical understanding of Marangoni (surface energy driven) convection and critical point phenomena.
1993	STS-58	SLS-2	Showed that metabolic changes in blood cells are due to microgravity induced changes in the lipid-phospolipid structure of their cell membrane.

Spacelab Mission Chronology (Part 3 of 4)

Year	Mission	Spacelab Payload	Mission Highlights
1994	STS-59	Space Radar Laboratory (SRL)-1	First space radar to use three frequency bands at once, increasing data accuracy.
1994	STS-62	USMP-2	First photographs of dendrite growth in microgravity, which helped refine existing theory.
1994	STS-65	IML-2	Found microstructural developmental alterations never before seen in tungsten-heavy alloys. First successful attempt to electromagnetically levitate a sample in microgravity.
1994	STS-64	LIDAR In-space Technology Experiment (LITE)	First flight of LITE optical radar to study atmosphere.
1994	STS-66	ATLAS-3	Continued observations made on ATLAS-1 and 2, allowing multi-year, dual hemisphere, and cross-seasonal comparisons of mesospheric and stratospheric chemistry data for the first time.
1994	STS-68	SRL-2	Reflight of SRL-1 to obtain Earth observations data of the same areas at different seasons.
1995	STS-67	Astro-2	Continued work begun on Astro-1.
1995	STS-71	Spacelab-Mir	Bird eggs developed normally in microgravity. Cereal plant seeds sprout, grow, and reproduce in microgravity.
1995	STS-73	USML-2	First use of ground-to-air digital television.
1996	STS-75	USMP-3	Determined that growth rate and tip radius variances from theory in dendrites during growth in microgravity are due to specimen size, not microgravity.
1996	STS-78	Life and Microgravity Spacelab	Vertebrate (fish) reproduced fertile offspring in microgravity, no developmental anomalies in either generation.
1997	STS-83	MSL-1	Mission cut short by transport problems; reflown on STS-94 as MSL-1R.
1997	STS-87	USMP-4	Obtained most precise temperature measurement ever in space: one-billionth of a degree Kelvin, in a liquid helium analog for very tiny computer chips.

(Continued)

Spacelab Mission Chronology (Part 4 of 4)

Year	Mission	Spacelab Payload	Mission Highlights
1997	STS-94	MSL-1R (reflight)	Established successful protocol for uptake of radionuclides in plants in microgravity.
1998	STS-90	Neurolab	Found structural changes in the brain areas related to vestibular sensitivity in vertebrates. Crewmembers showed a greater dependency on visual cues in microgravity.

a strong science background, while payload specialists were researchers who trained with astronauts for one mission.

Originally planned for a 10-year lifespan, between 1982 and 1998 *Spacelab* science hardware flew 36 missions. While its first two flights (STS-2 and STS-3) tested hardware components, the other flights were dedicated science missions with nearly 1,000 science experiments. These included the Astronomical Observatory missions, Atmospheric Laboratory for Applications and Science (ATLAS) Earth observation flights, four dedicated life sciences flights, and many physical science missions. NASA discontinued *Spacelab* after STS-90, but is still using hardware developed for and tested by *Spacelab* for support and research on the *International Space Station.*

Karen Murphy

See also: European Space Agency, Germany, Microgravity Materials Research

Bibliography

Walter Froehlich, *Spacelab: An International Short-Stay Orbiting Laboratory* (1983).
Douglas R. Lord, *Spacelab, an International Success Story* (1987).
Marshall Space Flight Center Spacelab. http://history.msfc.nasa.gov/chronology/index.html.

Tourism

Tourism in space has been given serious thought by entrepreneurs since the pioneering spaceflights of NASA's Apollo project. The concept of opening space travel to civilians who are willing to pay for a trip beyond Earth's atmosphere was introduced as early as 1967 when hotel entrepreneur Barron Hilton, in a talk given to the American Astronautical Society, proposed an orbiting hotel, to be followed by a hotel on the Moon. Although the presentation did not address the technical challenges of constructing a hotel in space, Hilton did have students at Cornell University's School of Hotel Management study the feasibility of the idea.

Hilton and Pan American World Airways stirred up public interest in space travel by paying to have their names appear on a space station and spaceship in the popular movie *2001: A Space Odyssey*, released in 1968. Pan American followed this publicity stunt later that year by taking reservations for flights to the Moon. More than 93,000

American multimillionaire Dennis Tito, the world's first paying space tourist, appears after his landing in the Central Asian steppes northeast of Arkalyk, Kazakhstan, in May 2001 after a 10-day vacation on the International Space Station. (Courtesy AP/ Wide World Photos)

interested passengers got their names on the air carrier's First Moon Flights Club waiting list, but never had any real chance of flying into space as tourists, because the market was still in its infancy.

By the mid-1980s new studies on tourism in space began to emerge. In 1986 Patrick Collins, cofounder of the Space Future website, and David Ashford, founder of Bristol Spaceplanes, made a presentation to the International Astronautical Federation (IAF) that predicted how the market would evolve. This included an initial pioneer phase, where a handful of tourists would pay as much as $1 million for a trip, to a mature, mass-market phase when trips could be obtained for about $2,000. Papers on potential vehicles for tourist trips into space and designs for orbiting space hotels also were presented at the IAF conference of 1987 and 1989. In 1986, Society Expeditions Space Travel Company, Inc., became the first company to advertise brief trips to low Earth orbit aboard a single-stage reusable vehicle called Phoenix. It collected several down payments on the $150,000 trip, but Phoenix was never completed and the venture fell through.

Following the collapse of the Soviet Union, Russia's ailing economy forced its space agency to look for new sources of funding. In 1990 the Tokyo Broadcasting System in Japan reputedly paid Russia between $12 and $14 million to fly reporter Toyohiro Akiyama to *Mir*. One year later, a British consortium paid for chemist Helen Sharman's trip to the Russian space station. These flights set a precedent by opening seats on Russian spacecraft to anyone who could afford to pay the multimillion-dollar price tag, including tourists.

In 1993, the Japanese Rocket Society began a market study on tourism in space. Encouraged by initial findings, the society formed the Transportation Research Committee to research designs on a passenger launch vehicle called Kankoh-Maru. On 18 May 1996 space enthusiast Peter Diamandis announced that his newly formed X-Prize Foundation was offering $10 million to the first private team that could build a reusable launch vehicle (RLV) that could carry three people to the edge of space, return, and launch again within two weeks. Although the prize would not cover the cost of developing an RLV, it generated public interest in space tourism—enough to attract more than 20 teams between 1996 and 2004.

In the mid-1990s NASA teamed with the Space Transportation Association (STA) to conduct a market study on tourism in space. This was the first time that NASA had officially expressed interest in the subject. In 1997 the first International Symposium on Space Tourism was held in Bremen, Germany. Also that year, *Apollo 11* astronaut Edwin "Buzz" Aldrin helped form the Space Tourism Society, a nonprofit organization dedicated to promoting tourist spaceflights. In 1998 Diamandis formed Space Adventures, Ltd., a company offering a variety of space-related tourist packages, such as rides on Russian MiG jets and spaceflight training. In June 1999 the STA hosted a conference in Washington to discuss the results of its joint study conducted with NASA.

Tourism in space began in earnest in 2000 when MirCorp, an Amsterdam-based firm developing uses for the Russian *Mir* space station, selected U.S. businessman Dennis Tito out of six candidates to fly to *Mir* as a tourist. Before a flight could be arranged, Russia's space agency Rosaviacosmos decided to deorbit the deteriorating station. In May 2000 Tito signed a new contract with Space Adventures for a trip to the *International Space Station* (*ISS*). Rosaviacosmos and launch provider Rocket and Space Corporation Energia agreed to supply transportation. NASA and the European Space Agency, the largest *ISS* partner, were initially hesitant to allow a civilian aboard the *ISS*, citing safety concerns, but finally gave approval. On 28 April 2001 Tito took a Russian Soyuz to the *ISS* for a historic 10-day vacation. One year later South African Internet entrepreneur Mark Shuttleworth followed Tito's example. Tito and Shuttleworth paid $20 million each to take their space vacations. U.S. millionaires

Space Tourists, 2001–2009

Name	Citizenship	Launch	Landing	Mission
Dennis Tito	United States	28 Apr. 2001	6 May 2001	*Soyuz TM-32*
Mark Shuttleworth	South Africa/ United Kingdom	25 Apr. 2002	5 May 2002	*Soyuz TM-34*
Gregory Olsen	United States	1 Oct. 2005	11 Oct. 2005	*Soyuz TMA-7*
Anousheh Ansari	Iran/United States	19 Sept. 2006	29 Sept. 2006	*Soyuz TMA-9*
Charles Simonyi	Hungary/United States	7 Apr. 2007	21 Apr. 2007	*Soyuz TMA-10*
Sheikh Muszaphar Shukor	Malaysia	10 Oct. 2007	21 Oct. 2007	*Soyuz TMA-11*
Richard Garriott	Britain/United States	12 Oct. 2008	23 Oct. 2008	*Soyuz TMA-13*
Charles Simonyi	Hungary/United States	26 Mar. 2009	8 Apr. 2009	*Soyuz TMA-14*
Guy Laliberté	Canada	30 Sept. 2009	11 Oct. 2009	*Soyuz TMA-16*

Gregory Olsen and Anousheh Ansari each paid for trips to the *ISS* in 2005 and 2006, respectively. Ansari was the first woman to fly in space as a tourist. Hungarian-born computer software executive Charles Simonyi became the first space tourist to visit the *ISS* twice, in 2007 and 2009. Other paying tourists who stayed on the *ISS* include Sheikh Muszaphar Shukor of Malaysia (2007); British-American Richard Garriott (2008), the son of Apollo-era astronaut Owen Garriott; and Canadian Guy Laliberté (2009).

As the new millennium began, a number of commercial companies were ready for a tourist market in space to take off. Bigelow Aerospace, a company started by Robert Bigelow, owner of the Budget Suites of America hotel chain, signed a technology transfer agreement with NASA in 2002 to work with TransHab, an inflatable habitation structure designed by NASA. The company began incorporating TransHab technology into future commercial space habitat designs. In March 2003 the Tate Gallery in the United Kingdom announced that British Extraterrestrial Architecture Laboratory Architects won its competition, Tate in Space, to design an art gallery that could be docked to the *ISS* or float freely, in an effort to capitalize on tourism in space. In September 2003 the California-based Space Island Group announced an agreement with Spacehab, Inc., the company that designed the space laboratory module for the Space Shuttle, to develop designs for habitation, research, and manufacturing modules for a future commercial space station.

By 2004 the prospect of more tourist flights in space reached another milestone. X-Prize competitor Mojave Aerospace Ventures of Mojave, California, successfully launched on 21 June 2004 *SpaceShipOne*, which became the first civilian spacecraft to leave the atmosphere, reaching a record-breaking altitude of 328,491 ft. Mojave Aerospace Ventures matched the flight on 29 September and 4 October to win the X-prize. British airline mogul and billionaire Richard Branson founded Virgin Galactic in 2004 to develop a fleet of commercial spacecraft for tourist flights into space based on the *SpaceShipOne* design. Branson, with Scaled Composites founder Burt Rutan, unveiled the *SpaceShipTwo* design in January 2008 and conducted the vehicle's first "captive carry" test flight in March 2010. Commercial suborbital flights are planned to follow, at $200,000 per ticket.

In December 2004 President George W. Bush signed into law the Commercial Space Launch Amendments Act, which set the regulatory standards for the suborbital space tourism industry. One year later, the Office of the Associate Administrator for Commercial Space Transportation issued proposed rules for crew qualifications and training, informed consent for passengers, and medical requirements.

By 2006 new commercial spaceports in California, New Mexico, Oklahoma, and Texas emerged as a result of business ventures to fly tourists into space. The Republic of Singapore and the United Arab Emirates also announced plans to build spaceports in their countries. Several companies in addition to Virgin Galactic had indicated plans to build vehicles capable of taking tourists to suborbital space and beyond, such as U.S.-based Armadillo Aerospace, Blue Origin, Rocketplane, Space Adventures, TGV Rockets, and XCOR Aerospace.

Bigelow Aerospace successfully flight tested and launched into orbit the first privately funded space habitat, *Genesis 1*, on a Russian Dnepr rocket in July 2006. The company launched a second space station module, *Genesis 2*, in June 2007, with improved communications systems and a multi-tank inflation system. Both experimental habitats were designed as precursors for bigger orbital platforms that would accommodate future space tourists.

Anne Simmons

See also: International Astronauts and Guest Flights, Space Stations

Bibliography

Lou Dobbs, *Space: The Next Business Frontier* (2001).

Federal Aviation Administration, *2004 U.S. Commercial Space Transportation Developments and Concepts: Vehicles, Technologies, and Spaceports* (2004).

John Spencer, *Space Tourism: Do You Want To Go?* (2004).

Voskhod

Voskhod was a Soviet Union multi-seated spacecraft derived from the Vostok program. Under pressure from Soviet Premier Nikita Khrushchev for continued space spectaculars to outdo the United States, Sergei Korolev's OKB-1 (Experimental Design Bureau) decided to modify the Vostok spacecraft to equal or surpass some aspects of the U.S. Project Gemini.

On 13 April 1964 Soviet leaders approved plans for Voskhod missions. Two Voskhod spacecraft were built for piloted missions. The first, called 3KV, was designed for a three-person crew, to exceed the two-person Gemini crew. The second, given the designation 3KD, was designed for two cosmonauts and equipped for extravehicular activity (EVA) so as to perform a spacewalk before Gemini. Korolev personally called the EVA project Vykhod ("Exit") because of the term's use in earlier studies.

The Voskhod design was similar to Vostok's in that both contained a reentry vehicle, or capsule, and an equipment module. Unlike Vostok, a new two-canopy parachute

The crew of Voskhod 1, from left to right, cosmonauts Vladimir Komarov, Boris Yegorov, and Konstantin Feoktistov. The mission, launched on 12 October 1964, was the first multi-crewed human spaceflight. (Courtesy Peter Gorin/NASA)

KONSTANTIN PETROVICH FEOKTISTOV (1926–2009)

Konstantin Feoktistov, a Soviet Union spacecraft designer and cosmonaut, was the first civilian to travel in space. He was one of the principal designers of the Vostok, Voskhod, Soyuz, and other piloted spacecraft of OKB-1 (later RSC Energia). Feoktistov served as a crewmember of *Voskhod 1*, the first three-person mission to fly in space, which launched on 12 October 1964. He was among the initiators of the Soviet civilian space station program. From 1970 to 1990, Feoktistov served as a key manager for the design and development of the Salyut and *Mir* space stations.

Peter A. Gorin

system with a soft landing rocket engine ensured that the Voskhod crew could land while inside the reentry vehicle. Three suspended seats replaced the single ejection seat originally equipped inside the Vostok capsule. The Voskhod spacecraft was equipped with a backup solid propellant braking engine.

The Voskhod spacecraft was launched by a more powerful version of the R-7 rocket. In the three-seated 3KV, the crew cabin was so crowded that cosmonauts were unable to wear space suits. In contrast, two cosmonauts riding in the 3KD had to wear space suits designed for EVAs. The life-support system for their space suits replaced the third seat. The 3KD had an inflatable airlock attached to one of the hatches. The Voskhod spacecraft was about 5 m long and 2.43 m in diameter.

A desire to launch the Voskhod spacecraft before the two-person Gemini led Soviet designers to abandon an emergency escape system, which was not ready in time. This made flights more dangerous because any rocket failure in the first 50 seconds of launch would be fatal for the cosmonauts. Only one uncrewed test flight was conducted before piloted missions.

Cosmonauts Vladimir M. Komarov (commander), Konstantin P. Feoktistov (flight engineer), and Boris B. Yegorov (physician) were launched on *Voskhod 1*, the first spacecraft to put more than one person into orbit, on 12 October 1964. Feoktistov was one of the leading designers of the Vostok and Voskhod spacecraft. During the mission, Soviet Premier Nikita Khrushchev, who had pressed for the program, was removed from office and replaced by Leonid Brezhnev. *Voskhod 1* landed safely after a mission lasting 24 hours and 17 minutes.

Voskhod 2 was launched on 18 March 1965 with Pavel I. Belyayev (commander) and Alexei A. Leonov (researcher) on board. Leonov performed the world's first EVA in space, which lasted about 20 minutes. He was forced to reduce the air pressure in his space suit to get back into the airlock and was dangerously fatigued after the task. The crew landed 386 km off course in an uninhabited, snow-covered forest in the Soviet Union and spent two days in the wilderness before being evacuated. Despite the landing error, the mission was hailed as a huge success.

The piloted Voskhod flights were the pinnacle of the early Soviet human spaceflight program. In 1966 Voskhod was canceled to free resources for the development of the

Soyuz spacecraft. This marked the end of the Soviet lead in the space race and its long list of firsts in human spaceflight.

Peter A. Gorin

See also: Korolev, Sergei; Rocket and Space Corporation Energia

Bibliography

Rex Hall and David Shayler, *The Rocket Men: Vostok and Voskhod, The First Soviet Manned Spaceflights* (2001).
Dennis Newkirk, *Almanac of Soviet Manned Space Flight* (1990).
Asif A. Siddiqi, *Challenge to Apollo: The Soviet Union and the Space Race, 1945–1974* (2000).

Vostok

Vostok was the world's first piloted spacecraft. Sergei Korolev's OKB-1 (Experimental Design Bureau), the Soviet Union's top missile development organization, began work on the spacecraft (originally named Object OD-2 satellite) in early 1957. Future cosmonaut Konstantin Feoktistov and OKB-1's department of space systems, which was directed by space pioneer Mikhail Tikhonravov, created Vostok's initial design, while Oleg G. Ivanovsky became the spacecraft's lead designer. In May 1958 the Soviet government approved Vostok after major theoretical research had been completed. Not until 1960 did the spacecraft receive its Vostok designation.

Three Vostok spacecraft models were developed: a flight test prototype, a photoreconnaissance satellite, and a piloted satellite. All three had the same bus, but different onboard systems. Along with OKB-1, 122 other research-and-development organizations were involved in the Vostok program as subcontractors.

Vostok consisted of two major structural parts made of aluminum: a spherical reentry vehicle, or capsule, and a non-pressurized equipment module. The reentry vehicle was pressurized and housed pilot, life-support, communications, navigation, and landing systems. The reentry vehicle for uncrewed missions was equipped with a self-destructive device. Feoktistov proposed a spherical shape for the reentry vehicle because at that time a sphere was better developed from an aerodynamic standpoint than any other suitable shape. The reentry vehicle had three round hatches, 1 m in diameter. One hatch was used for a pilot's entry and ejection, the second covered the parachutes, and the third was for auxiliary access. The pilot also had access to a porthole and an optical orientation device. The reentry vehicle's external diameter was 2.3 m and total internal volume was 5.2 m^3. The reentry vehicle contained enough vital supplies for 10 days. It was covered by heat-absorbing material of a variable thickness. An intentionally shifted center of gravity ensured a proper orientation of the reentry vehicle during the descent phase.

The equipment module consisted of two truncated cones connected by their larger bases. At one end, the equipment module was connected to the reentry vehicle by four metal bands, and at the other it had a liquid propellant braking engine. The equipment module housed systems for orientation, temperature control, telemetry, flight control communications, and propellant tanks. This included tanks of compressed nitrogen used for orientation thrusters, which were arranged in a collar around the reentry vehicle's attachment point. Dry cell batteries powered the spacecraft and were located in the equipment module. All electrical connections between the reentry vehicle and the equipment module were assembled in a single cable mast. The equipment module was 2.25 m long and 2.43 m in diameter.

The total length of the spacecraft in orbit was 4.41 m (without antennae). The spacecraft was launched from the Baikonur Cosmodrome in Kazakhstan by three stage variants of the R-7 ballistic missile, 8K72 and 8K72K, which also were publicly known as Vostok rockets.

Vostok was not maneuverable but was capable of changing its orientation. Cosmonauts were trained to operate the spacecraft manually. However, some psychologists at the time expressed doubt in the ability of pilots to control the spacecraft in the vacuum of space. The prime mode of operation was either automatic or by radio commands from the flight control center. After deceleration by the braking engine, the reentry vehicle separated from the equipment module and descended in the atmosphere on a ballistic trajectory. As a standard procedure, the cosmonaut was ejected from the reentry vehicle at an altitude of 7 km and landed with a separate parachute. From the outset the pilot's ejection was assumed safer than landing in the reentry vehicle, because the designers could not guarantee a soft landing on the ground. Landing inside the reentry vehicle was considered only during an emergency.

A cosmonaut wore a specially designed SK-1 space suit and was seated in an ejection seat in a reclined position. In the event of a landing in water, the space suit was designed to keep the cosmonaut afloat. The space suit was covered by a bright orange overcoat, had an inflating collar, and was equipped with an emergency kit that included an inflatable raft. The ejection seat also served as an emergency escape system during launch.

OKB-1 tried to work around apparent limitations to the Vostok design. The possible failure of the spacecraft's only braking engine could have stranded a cosmonaut in space. To avoid this scenario, the spacecraft's orbit was purposely kept low to ensure a natural deceleration and descent into the atmosphere. However, the rocket's velocity control was not perfected at that time, causing dangerous variations in orbital altitude. Engineers could not guarantee whether the ejection seat would work in the event of a rocket explosion on the launch pad.

Although Vostok was the next logical step after Sputnik, by 1960 it was at the center of a politically motivated competition between the Soviet Union and the United States during the Cold War. Soviet Premier Nikita Khrushchev used Sputnik and Vostok as a way to demonstrate technological superiority over the United States. To promote Khrushchev's aggressive foreign policy, Korolev agreed to skip a Vostok suborbital flight and go directly to an orbital mission. According to a government decree, the first piloted launch was planned for December 1960. However, that date was delayed as seven Vostok uncrewed test flights worked through various problems. On 15 May 1960 a reentry vehicle test failed after an

YURI ALEKSEYEVICH GAGARIN (1934–1968)

(Courtesy Library of Congress)

Yuri Gagarin, a Soviet Union cosmonaut and Air Force colonel, became the first human to travel in space on 12 April 1961, aboard *Vostok 1*. His historic flight lasted one orbit and demonstrated Soviet technological supremacy during the height of the Cold War. From 1961 to 1968, he attended the Zhukovsky Air Force Academy and was a commander at the Cosmonaut Training Center (TsPK) in Star City, Russia. He died after his MiG-15 crashed during a training flight on 27 March 1968. Gagarin's popularity as a national hero resulted in the training center being named for him.

Peter A. Gorin

orientation error caused the spacecraft to raise its orbit instead of lower it. On 28 July 1960 two dogs were killed inside a Vostok spacecraft when it exploded during launch. A third flight less than a month later successfully placed into orbit two more dogs, Belka and Strelka, where they spent a day in space. This flight returned safely and was the first Soviet satellite landing. On 1 December 1960 a fourth test flight carried two dogs into space; however, the reentry vehicle was destroyed after a navigation error. That same month, another flight carrying two dogs failed to reach orbit but landed safely in Siberia. The last two test flights in March 1961 served as dress rehearsals for a human spaceflight. Both carried a dog and a mannequin into orbit and landed safely.

On 12 April 1961 Yuri Gagarin successfully piloted *Vostok 1* into orbit and became the first human in space. The mission lasted 108 minutes during one orbit. Several potentially dangerous situations occurred during the flight: the orbit apogee was higher than normal, the cosmonaut reported delayed separation of the reentry vehicle from the equipment module, and the reentry vehicle missed its landing zone. Nevertheless, Gagarin safely ejected from the Vostok capsule and parachuted to the ground. The mission was a great propaganda success, trumping NASA's first suborbital Mercury flight. The event made Gagarin an overnight international celebrity and was so prestigious that it became celebrated in the Soviet Union as Cosmonautics Day. The American public was shocked by the news, and newly elected President John Kennedy responded a month later by declaring an acceleration of U.S. space exploration goals, including landing an American on the Moon by the end of the decade.

Gherman Titov piloted *Vostok 2* on 6–7 August 1961. Although the cosmonaut experienced motion sickness due to weightlessness, the first time this occurred in orbit, the mission demonstrated the human ability to live and work in space. The mission lasted 25 hours and 11 minutes.

Vostok 3 and *Vostok 4* represented the first joint space mission. Andrian Nikolayev piloted *Vostok 3* into space on 11 August 1962, while Pavel Popovich rode *Vostok 4* into orbit one day later. The trajectories of both spacecraft brought the cosmonauts within 6.4 km of each other during the mission. *Vostok 3* lasted three days, 22 hours, and 22 minutes, while *Vostok 4* lasted two days, 22 hours, and 57 minutes.

Another joint mission was launched with *Vostok 5* and *6*. Valeri Bykovsky piloted *Vostok 5* into space on 14 June 1963. Two days later Valentina Tereshkova became the first woman to fly into space aboard *Vostok 6*. Both spacecraft flew within 4.8 km of each other during the mission before returning to Earth. *Vostok 5* lasted four days, 23 hours, and 6 minutes, the longest Vostok flight, which was not surpassed by another Soviet flight until *Soyuz 9* in June 1970. Bykovsky's endurance record was overshadowed by Tereshkova's flight, which lasted two days, 22 hours, and 50 minutes, because its popularity had worldwide impact and represented another significant boost to Soviet prestige. *Vostok 6* was the last Vostok flight and culminated a lengthy list of Soviet achievements in space.

After the completion of the Vostok program, its design became a basis for the Voskhod multi-seated spacecraft. There were also unrealized plans to develop a maneuverable Vostok-ZH spacecraft for an orbital assembly of a piloted circumlunar spacecraft from separate blocks. Vostok spacecraft were also modified by OKB-1 into the Zenit reconnaissance satellites.

Peter A. Gorin

See also: Animals in Space, Cosmonauts

Bibliography

Rex Hall and David Shayler, *The Rocket Men: Vostok and Voskhod, The First Soviet Manned Spaceflights* (2001).

Dennis Newkirk, *Almanac of Soviet Manned Space Flight* (1990).

Asif A. Siddiqi, *Challenge to Apollo: The Soviet Union and the Space Race, 1945–1974* (2000).

Zond. *See* Soviet Manned Lunar Program.

MICROGRAVITY SCIENCE

Microgravity science became an important feature of human spaceflight programs as they evolved to long-duration missions to understand the space environment's effects on physical materials and biological organisms.

Due to strong interest in microgravity effects on humans, the first microgravity experiments were focused on animal (and then human) biology, starting with the dog Laika carried on *Sputnik 2*, and chimpanzees carried on suborbital Mercury precursor flights. Along with human biomedical research conducted on human spaceflight missions, both the United States and the Soviet Union developed dedicated biological satellites. The U.S. Biosatellite program consisted of three satellites carrying animals and plants, recovered by midair capture, similar to the Corona reconnaissance film return capsules. Flown in 1966, 1967, and 1969, the Biosatellite experiments showed that plant roots formed abnormally in space, and that young dividing cells were more susceptible to radiation than older cells. The Soviet Union's first dedicated biological satellite was *Kosmos 110*, which flew with two dogs in 1966. The Bion program consisted of 11 Kosmos flights from 1973 to 1996, carrying experiments from the Soviet Union and Russia, the United States, China, and European nations, which discovered loss of muscle and bone mass and changes to endocrine systems.

The advent of U.S. and Soviet space stations provided many more opportunities for microgravity experimentation. *Skylab* included experiments with spider webs, rice growth, and some fluid and crystallization experiments. The Salyut stations included fluid transfer experiments and the start of an extensive program of crystallization and materials processing experiments that grew the first zeolites and protein crystals in space. *Salyut 6* and *7* included furnaces, called Kristall and Korund, to heat metals and induce crystal growth.

The Space Shuttle program included the European Space Agency–built *Spacelab*, which fit in the Shuttle payload bay. From 1982 to 1998 *Spacelab* provided capabilities for both internal and external experiments. It flew four dedicated life-sciences missions and many physical-science experiments. The Shuttle also provided capabilities for Getaway Special and Hitchhiker experiments that fit into the Shuttle on a space-available basis. These experiment programs, along with the Space Experiment Module carriers, were combined into the Space Shuttle Small Payloads Project, which flew starting in the early 1990s. Together these capabilities

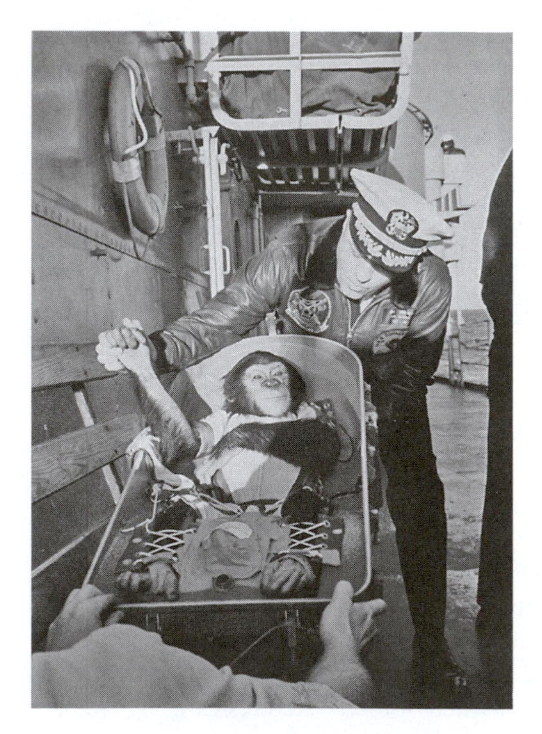

Chimpanzee Ham is brought aboard a recovery ship after his flight in space from Cape Canaveral, Florida, in January 1961. (Courtesy NASA/Kennedy Space Center)

enabled hundreds of experiments. By the 1990s these included tests of cryogenic fluid transfer, capillary thermal control systems, and thin-film semiconductor processing.

Materials science also utilized free-flying orbital platforms that tested pallets of materials for how they behaved in free space. This information was then used to select better materials for spacecraft and space instrumentation design. These systems included the *Long Duration Experiment Facility*, the Mir Environmental Effects Payload, and the Materials International Space Station Experiment, which exposed materials to the space environment on the outside of the *International Space Station* (*ISS*).

Provision of long-duration microgravity experiment capabilities was among the primary justifications for the *Mir* and the *ISS* space stations. *Mir*'s *Kristall* module contained five furnaces for microgravity materials processing experiments, including semiconductor and crystal growth experiments. The *ISS* provided the Microgravity Experiment Glovebox, which enabled fluid experiments, such as research on colloids (fluids with particle suspensions) in magnetic fields, fluid mixing without convection, and others. The lack of convection in microgravity environments allowed measurement of mass diffusion coefficients with great precision, which was useful for many Earth-based processes. Other experiments, such as flame experiments, were crucial for long-term safety in space stations.

Up to the early 2000s, microgravity experimentation had primarily been accomplished through government funding. In the late 1970s and 1980s, there was great interest, partly spurred by NASA to utilize the Space Shuttle and the *ISS* in private funding of microgravity experiments. However, space experiments proved expensive and too long to perform in comparison to ground-based research. With few exceptions, corporations funded few experiments in comparison to governments. Private industry also attempted to provide experimentation capabilities, such as the abortive Industrial Space Facility (ISF), which would enable the government to lease space on a free-flying commercial platform. The ISF foundered in the late 1980s because NASA supported its own Space Station Freedom program. The *Spacehab* module had more success, as a pressurized module smaller than *Spacelab* that NASA could rent to provide facilities in the Space Shuttle payload bay for astronaut-tended experiments. NASA used it on board 22 Shuttle flights starting in 1993. The *Spacehab* research double module was lost with the *Columbia* mission in February 2003.

Stephen B. Johnson

See also: Animals in Space, Biomedical Science and Technology, Space Stations, *Spacelab*

Bibliography

Commission on Physical Sciences, Mathematics, and Applications, Space Studies Board, *Microgravity Research in Support of Technologies for the Human Exploration and Development of Space and Planetary Bodies* (2000).

Kenneth Souza et al., *Life into Space: Space Life Sciences Experiments, NASA Ames Research Center, Kennedy Space Center, 1991–1998* (2000).

Robert Zimmerman, *Leaving Earth: Space Stations, Rival Superpowers, and the Quest for Interplanetary Travel* (2003).

Bion Mission Chronology

Bion Program			
Designation	**Launch Date/ Duration**	**Test Subjects**	**Milestones**
Kosmos 605 (*Bion 1*)	31 October 1973 22 days	Turtles, rats, insects, fungi	Study of weightlessness effects on cellular level
Kosmos 690 (*Bion 2*)	22 October 1974 21 days	Turtles, rats, insects, fungi	Research of radiation effects and protection from radiation
Kosmos 782 (*Bion 3*)	25 November 1975 20 days	Turtles, rats, insects, plants	Artificial gravity experiment. Landed earlier than planned due to weather conditions.
Kosmos 936 (*Bion 4*)	3 August 1977 19 days	Rats, insects, plants	Missed the landing zone but was found later. Test subjects survived.
Kosmos 1129 (*Bion 5*)	25 September 1979 19 days	Rats, insects, quail eggs, plants, mold	Study of weightlessness effects on reproductive cycles in mammals, birds, insects
Kosmos 1514 (*Bion 6*)	14 December 1983 5 days	Monkeys, rats, fish, insects, plants	
Kosmos 1667 (*Bion 7*)	10 July 1985 7 days	Monkeys, rats, fish, newts, insects, plants	
Kosmos 1887 (*Bion 8*)	29 September 1987 13 days	Monkeys, rats, fish, frogs, insects, plants	Missed the landing zone but was found later. Most test subjects survived.
Kosmos 2044 (*Bion 9*)	15 September 1989 14 days	Monkeys, rats, fish, newts, insects, plants	Temperature control system failed at the end of the mission. Most test subjects survived.
Kosmos 2229 (*Bion 10*)	29 December 1992 12 days	Monkeys, rats, frogs, insects, plants	Landed three days earlier than planned due to temperature control failure. Test subjects survived.
Bion 11	24 December 1996 15 days	Monkeys, rats, frogs, insects, plants	One of the monkeys died during the postflight examination.

Bion

Bion, the Soviet and Russian biological research satellite program, improved on Soviet suborbital flights of dogs in the 1950s. The world's first biosatellite, *Sputnik 2*, was launched in 1957 with the dog Laika on board. Animals also flew aboard Vostok and Zond test flights during the 1960s. The first dedicated biological mission to fly in space was the 22-day flight of two dogs in a converted Voskhod spacecraft (*Kosmos 110*) in 1966, which prompted the beginning of the Bion program by the Moscow-based Institute of Biomedical Problems of the Russian Academy of Science.

The Bion satellite, developed in the early 1970s by TsSKB Progress in Samara, Soviet Union, was derived from the Vostok spacecraft. The satellite was 6 m long and 2.4 m in diameter. It consisted of an aggregate module, a spherical reentry vehicle, and an additional power supply container. The interior layout of the reentry vehicle was designed to sustain living test subjects and safely return them to Earth. The test subjects consisted of various flora and fauna samples and large animals, such as monkeys (two per flight) with surgically implanted electrodes for monitoring brain activities. To ensure safe recovery of the test subjects at a landing site, a special mobile laboratory complex was developed from inflatable shelters with artificial environments to support up to 30 people in extreme weather conditions.

Bion satellites were launched from Plesetsk, Soviet Union, to low Earth orbit by a Soyuz variant of the R-7 rocket. Normal flight durations were 14–22 days. Eleven Bion missions were flown from 1973 to 1996. The primary role of Bion missions was to study the effect of microgravity on living organisms in support of long-duration human spaceflights. The missions identified several hazards, such as loss of muscle mass, calcium erosion in bones, and unusual changes in endocrine glands. Severe abnormalities were also observed in insects' reproductive cycles. Bion experiments contributed to a better understanding of the negative effects associated with spaceflight and allowed development of new methods for maintaining the health of cosmonauts during and after missions. Some early Bion experiments also studied the effects of high radiation levels in combination with weightlessness (*Kosmos 690*) and artificial gravity created by a centrifuge inside the reentry vehicle (*Kosmos 782*).

The Bion program was unique in terms of its unusually broad international participation. Starting in 1975, 11 countries, including China, the United States, and members of the European Space Agency, contributed to experiments on board Bion satellites. In 1997 NASA refused to sponsor further Bion experiments with primates following the unexpected death of a monkey flown during *Bion 11*. Funding shortfalls forced Russia to indefinitely postpone the *Bion 12* mission, originally scheduled for 1998.

Peter A. Gorin

See also: Animals in Space, Sputnik, Voshkod, Vostok

Bibliography

Kristen E. Edwards, "The U.S.–Soviet/Russian *Cosmos Biosatellite* Program," *Quest* 7, no. 3 (1999): 20–35.

Nicholas L. Johnson and David M. Rodvold, *Europe and Asia in Space, 1993–1994* (1995).

Biosatellite Program. *See* United States Biosatellite Program.

Cosmos Biosatellites. *See* Bion.

Industrial Space Facility

Industrial Space Facility was a concept that sprang from the fertile mind of U.S. space engineer Max Faget (1923–2004), who designed the Mercury spacecraft and held a patent on the design concept for the Space Shuttle.

In the early 1980s, at the beginning of the Shuttle era, many start-up space companies emerged in the expectation that the Shuttle would reduce launch prices and create new industry. One of them was Space Industries, Inc. (SII), established by Faget in 1982. Its primary product was the Industrial Space Facility (ISF), a mobile home–sized laboratory module that the Space Shuttle would deploy into orbit. Inside the ISF, a variety of experiments and commercial production facilities would operate inside its pressurized volume. As originally conceived, the ISF would accommodate up to 11,000 kg of experiments and manufacturing hardware, and its solar arrays were to produce 20 kW of electrical power. The ISF attitude would be maintained by deploying a 30.5 m long boom from the bottom to provide gravity gradient stabilization.

The ISF's operational concept was to fly autonomously in a 300 km high orbit, without people onboard, while the experiments and production facilities operated automatically. The Space Shuttle would make regular servicing flights to the ISF and dock with it, and astronauts would then enter the pressurized volume to service and resupply experiments and to gather products for return to Earth. One of the primary customers of the ISF was to be the McDonnell-Douglas electrophoresis operations in space (EOS) system. EOS would produce extremely pure pharmaceutical products in the microgravity environment. Unfortunately for EOS and the ISF, advances in biotechnology led to the capability to produce the same kinds of ultra-pure products on Earth at a fraction of the cost. The EOS project soon died and with it ISF's hopes.

In an attempt to salvage the ISF, SII tried to market it to NASA as a test bed for space station technologies and operations. At first ISF was well received by NASA, but by the late 1980s it was perceived as a potential threat to NASA's Space Station Freedom project. This, plus the fact that no firm commercial customers could be identified for the ISF, led SII to postpone the ISF project in 1989, which effectively killed the project.

Michael Engle

See also: Space Shuttle, Space Station Freedom, *Spacelab*

Bibliography

Wolf Bailey, "The Industrial Space Facility: The Next Logical Step?" *Quest* 9, no. 1 (2001): 49–55.

David M. Harland, *The Story of the Space Shuttle* (2004).

Dennis R. Jenkins, *Space Shuttle. The History of the National Space Transportation System: The First 100 Missions* (2001).

Microgravity Biology Experiments

Microgravity biology experiments study the effects of the space environment on life. After World War II the need to determine the viability of living organisms in the space environment had emerged as a high priority for the United States and the Soviet Union. This was due largely to each nation's growing technological capabilities and competing political desires to transport humans at high altitudes and into space.

Because it was unknown whether life forms could actually survive during spaceflight, many high-altitude balloon experiments were developed in the 1940s–50s to study the biological response of microbes, insects, rodents, cats, dogs, and monkeys. These spacefaring biological pathfinders were flown to altitudes around 30 km for up to 28 hours. During this same period suborbital rocketry and associated life science missions were being developed, tested, and used to study similar organisms at much higher altitudes. Although some of the early balloons and rockets were plagued with problems that caused the loss of live specimens, many were successful. In general, organisms that were recovered from nominal missions with properly functioning life-support systems were found to have survived the flights without noticeable ill effects. In some cases a few animals, such as mice, displayed behaviors that suggested evidence of microgravity adaptation. The results from these experiments provided valuable information that helped support the first successful human missions into space.

Coinciding with NASA's formation in 1958 was the creation of the Lovelace Committee (named after its chair Randolph Lovelace II) to bring together the leading experts working in space life sciences to address problems associated with humans and spaceflight. NASA established a Bioscience Advisory Committee, which in turn led to the formation of NASA's Office of Life Sciences, established on 1 March 1960 to oversee agency biological and medical programs. Upon the recommendations of Soviet space leaders Sergei Korolev and Mstislav Keldysh, the Russian Institute for Biomedical Problems (IMBP) was established in Moscow on 26 October 1963. In 1965 Anatoliy Blagonravov and other Soviet experts worked closely with NASA Associate Administrator Hugh Dryden and U.S. scientists to prepare *Foundations of Space Biology and Medicine*, a collection of writings on human ecology in space. The books were published in 1975 and ultimately led to wider international cooperation regarding space life science research.

Following the first human spaceflight, the field of space life sciences during the 1960s embarked on a new focus: enabling humans to go to the Moon. As a result, initial life-support and life-science hardware and tools aboard piloted missions focused mainly on optimizing human performance on orbit. Because crew time to operate experiments and handle biological specimens in space was limited, specialized habitats were designed to automatically foster the growth of organisms in space. These experiments were repeated in labs on the ground to help ensure a full comparative understanding of the effects of spaceflight.

By the mid-1960s improvements in habitat design allowed the United States and the Soviet Union to develop and fly autonomous, uncrewed research laboratories specifically designed to study biological processes in microgravity. NASA activated its Biosatellite Program in 1963 and tested the effects of microgravity on a monkey, flies, and other organisms. In 1966 the Soviet Union began its highly successful uncrewed biological research satellite program called Bion, where the growth of specimens ranging from microbes to monkeys was monitored to investigate mechanisms for physiological, biochemical, and behavioral changes associated with the space environment. Following the 1971 signing of the U.S.-Soviet Science and Applications Agreement, the two nations worked cooperatively in joint studies in space biology and medicine from 1975 to 1990 on seven Bion missions. During these experiments NASA and the IMBP played major roles in research design, experiment integration, and data interpretation.

Several crewed orbital platforms were constructed to prove that humans could live and work in space for many months, including the Soviet Union's Salyut space station program (1971–85) and *Mir* (1986–2001), and NASA's Skylab program (1973–74), all of which allowed researchers to characterize a variety of poorly understood physiological changes in cosmonauts and astronauts during long-duration missions. Early medical experiments in space involved careful characterization of astronaut and cosmonaut metabolic, cardiovascular, muscle, and bone changes and the use of tools, such as the ergometer to determine its possible use as a countermeasure to the effects of weightlessness. The results showed that humans and other organisms learned to adapt to microgravity.

With the advent of the Space Shuttle program in 1981, reusable laboratories, such as Spacelab, fostered cooperation among scientists and engineers in Asia, Europe, and the United States. Studies conducted during the Shuttle-Mir program (1994–98) focused on how microgravity affected the growth and development of microbes, plants, and small animals. To determine what biological changes were happening and why, clever experiments were chosen to support Shuttle missions. Comparative studies by NASA researchers cultivated organisms in a spectrum of gravity, including microgravity in space, hypergravity produced in centrifuges on Earth, and in normal Earth gravity. This was done to understand changes in organism behavior and structure. In addition to varying gravity levels, researchers monitored or altered radiation doses.

The construction of the *International Space Station*, begun in 1998, allowed biological research with tools that simulated artificial gravity in support of long-duration human exploration, such as return missions to the Moon and trips to Mars.

Jon Rask

See also: Aerospace Medicine, Animals in Space, Closed Loop Ecological Systems, *Skylab*, *Spacelab*

Bibliography

Melvin Calvin and Oleg Gazenko, *Foundations of Space Biology and Medicine*, 3 vols. (1975).

Kristen Edwards, "The U.S.-Soviet/Russian Cosmos Biosatellite Program," *Quest* 7, no. 3 (1999): 20–35.

Emily Morey-Holton, "The Impact of Gravity on Life," in *Evolution on Planet Earth: The Impact of the Physical Environment*, ed. Lynn Rothschild and Adrian M. Lister (2003).

Kenneth Souza et al., *Life into Space: Space Life Sciences Experiments, NASA Ames Research Center, Kennedy Space Center, 1991–1998* (2000).

Microgravity Fluids Research

Microgravity fluids research studies substances that flow under an applied force in microgravity environments where gravitational force is 1 percent or less of Earth's normal gravity. These substances include liquids, gases, or even solids, such as soil, that can liquefy during earthquakes. Removing gravitational force—and with it the effects of gravity-driven buoyancy, sedimentation, and convection—causes fluids to behave and flow differently when controlled by other secondary forces. Microgravity poses many challenges to engineers who build systems for crew-rated space vehicles, but it also offers opportunities for unique fundamental fluids research.

The first fluids experiments in space were part "gee-whiz" curiosity, where crewmembers were encouraged to play with liquids to see what would happen, and part practicality, testing theories and models developed in ground-based laboratories. Astronaut Stuart Roosa conducted two fluids experiments during *Apollo 14*'s return from the Moon: one where he heated liquids inside a container, proving that there would be convective motion caused by surface tension; and a liquid transfer demonstration, where he used a scale-model system to pump liquid from one baffled tank to another, proving that the system would work efficiently. The Skylab program was expanded to include testing of vapor cloud formation, liquid films, and water drop formation and movement. During the Apollo-Soyuz Test Project in July 1975, a joint mission between the United States and Soviet Union, attempts were made to improve electrophoresis techniques for possible manufacturing of vaccines and other products in space.

Cosmonauts performed fluids research on Salyut space stations from the late 1970s to the mid-1980s. On *Salyut 5*, a method to transfer liquids using special capillary pumps without electricity was tested. During *Salyut 6* and *Salyut 7*, crews investigated crystallization in and out of solution and the effects of convection currents in fluids.

Microgravity fluids research did not fully mature until 1983, when the *Spacelab* module first flew aboard the Space Shuttle. Developed by the European Space Agency (ESA), *Spacelab* featured a pressurized laboratory where astronauts could work in shirtsleeves. The module was lined with standardized payload racks, each equipped with power couplings, ventilation, and computer access. *Spacelab* made it possible for investigators to fly more complex research payloads: furnaces for heating a range

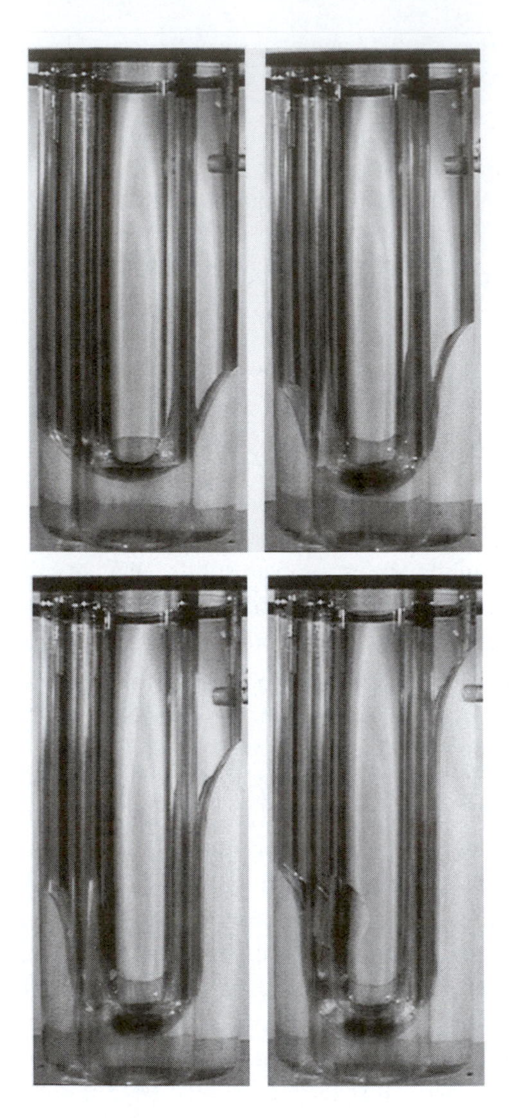

The Interface Configuration Experiment was flown during STS-73 in 1995, as part of the second U.S. Microgravity Laboratory (USML-2). Over time the photos show a change in the shape of the interface between a liquid and a gas in a sealed, slightly asymmetrical container. Under the force of Earth's gravity, the interface would remain nearly flat, but in microgravity, the interface shape and location changes significantly in the container, resulting in major shifts of liquid arising from small asymmetries in the container shape. (Courtesy NASA/Marshall Space Flight Center)

of solutions, metals, and alloys; freezers for preserving the results; and a glovebox where astronauts could conduct hands-on experiments involving messy or potentially hazardous materials.

Special equipment installed inside *Mir* space station's *Kristall* module was used to conduct advanced fluids experiments during the 1990s. For example, crews took digital photographs of phase-change phenomena at critical points of gas and liquid systems, were able to control accurately the temperatures of fluids during tests, and investigated liquid phase sintering (to cause to become a coherent mass by heating without melting) in metallic alloys.

Based on lessons learned from Skylab, Space Shuttle missions, and the Soviet and Russian space station programs, the first fluids investigation on the *International Space Station* (*ISS*) initiated in June 2001 studied colloids—a complex fluid composed of particles suspended in liquid that can be used to make paints, foams, and aerosols. As the colloidal samples aged during several weeks in orbit, they yielded surprising results, such as demixing and separation into a liquid and vapor, something that had not been seen in ground-based laboratories. The Microgravity Science Glovebox, provided by ESA and installed in 2002, was the first rack designed to handle a wide variety of fluids research. Its enclosed work volume was used to induce bubble movement in heated samples, demonstrate convection caused by concentration and temperature gradients in miscible (able to be mixed without separation of two phases) fluids, and create colloidal suspensions containing particles that can be altered by magnetic fields. The investigations were conducted to help researchers create new and improved materials.

Anne Simmons

See also: *International Space Station*, *Mir*, Salyut, *Spacelab*

Bibliography

Commission on Physical Sciences, Mathematics, and Applications, Space Studies Board, *Microgravity Research in Support of Technologies for the Human Exploration and Development of Space and Planetary Bodies* (2000).

David M. Harland, *The Story of Space Station Mir* (2005).

Fred J. Kohl et al., "The NASA Microgravity Fluid Physics Program—Knowledge for Use on Earth and Future Space Missions" (2002).

Robert Zimmerman, *Leaving Earth: Space Stations, Rival Superpowers, and the Quest for Interplanetary Travel* (2003).

Microgravity Materials Research

Microgravity materials research investigates the formation, structure, and behavior of metals and alloys, glasses and ceramics, polymers, electronic materials, and biomaterials. The results are used to design new and improved materials tailored to a specific application. Materials serve an important role in spaceflight, providing valuable protection against radiation, extreme pressures, acceleration forces, and impact events. Conversely, spaceflight has played an important role in the advancement of materials research as the removal of gravity produces novel effects, such as extremely regular crystal growth.

Early observations on *Apollo 14* in 1971 sparked the first true microgravity (1 percent or less of normal Earth gravity) experiments. Using a heater, sealed composite samples were melted. The resolidified samples showed that the particles were not distributed according to density, as would happen if processed on Earth. The resulting structure could only happen in space.

Several experiments on *Skylab* revolved around composite metals and solidification. *Skylab* crews completed 32 hours of materials experiments from May 1973 to February 1974. The results of organic and inorganic crystallization experiments sparked particular interest, as researchers discovered that the crystals were more uniform in structure and larger than Earth-grown counterparts were.

These early experiments set the tone for microgravity materials programs worldwide. The materials used for spacecraft shielding, systems, and space suits emerged from ground-based laboratories and testing facilities. Space, on the other hand, would become the playground of fundamental research—investigations into surface tension, thermally induced convection, phase change—in anticipation that a better understanding of basic phenomena would lead to better materials processing.

By the mid-1970s several countries, including three with a particular interest in microgravity materials research (Germany, Japan, and the United States), initiated sounding-rocket programs. Launched on a parabolic trajectory, sounding rockets provide several minutes of microgravity conditions as they fall to Earth. This method proved adequate to conduct basic dendritic and crystal growth experiments. Easy to launch and relatively affordable, sounding rockets became a popular research and education tool.

The United States and the Soviet Union conducted joint materials research experiments during the Apollo-Soyuz Test Project in July 1975. Metals were heated, melted, and cooled in the Apollo docking module's electric furnace—one of many experiments carried over from *Skylab*. A year later, cosmonauts on board the military space station *Salyut 5* (*Almaz 3*) completed crystal growth experiments, using a Kristall furnace, and tried to produce dibenzyl and toluene alloys.

During the 1980s and early 1990s, the Soviet Union conducted about 500 materials processing experiments on *Salyut 6* and *7* and the *Mir* space station. This led to the first zeolite (open, crystalline structures that can be used as absorbents and filters) and protein crystals grown in space. *Salyut 6* was equipped with two Kristall furnaces for work on semiconductors, and a Splav-01 alloy furnace could heat metals to 990°C. Because of their robust power requirements, the furnaces were not used simultaneously. In addition to the Kristall furnaces, *Salyut 7* was retrofitted with a Korund furnace about three months after launch in July 1982. The unit was capable of heating samples to 1,270°C and could be controlled from the ground without a crew on board. *Mir*'s *Kristall* module, added to the station in June 1990, contained five separate furnaces for manufacturing semiconductor and crystal growth samples.

Simple crystal growth experiments became the mainstay of materials research during the early years of the U.S. Space Shuttle program because they could be conducted in Dewar flasks and bioreactors small enough to store in the crew compartment. The introduction of Spacelab, a reusable laboratory developed by the European Space Agency with Germany as the primary sponsor, made it possible to use fully the orbiter's cargo bay for research. Spacelab featured a pressurized module where the crew could comfortably conduct experiments.

In April 1984 the crew of Space Transportation System-41C placed the *Long Duration Exposure Facility* (LDEF) into orbit where it remained until January 1990, exposing its contents—10,000 samples of materials used for shielding and optical, electrical, environmental, and other spacecraft systems—to the harsh environment of space. This was the first time that materials slated for use in spacecraft were tested in situ. The results identified reliable materials, eliminated several for use on long-duration missions, and validated predictive models used at ground-based test facilities. LDEF would inspire other similar experiments, including the Mir Environmental Effects Payload, which flew on *Mir* in 1996, and the Materials International Space Station Experiment, a series of sample exposure pallets first attached to the outside of the *International Space Station* (*ISS*) in 2001.

The *ISS* was designed as a microgravity research laboratory. Shortly after it became permanently crewed in 2001, experiments focused on protein crystal growth, zeolites, and colloids (particles that self-assemble to form superlattices used in paints, foams, and aerosols), which require several weeks on-orbit to fully form and evolve. The Microgravity Science Glovebox, installed in 2002, allowed microgravity materials research to investigate processing methods that would help remove flaws, such as pores formed by bubbles that can weaken materials. The *ISS* also served as a test bed for space-based materials processing techniques that could be used during future long-duration missions to the Moon or other planets.

Anne Simmons

See also: *International Space Station*, *Spacelab*

Bibliography

Donald C. Gillies, *New Directions in NASA's Materials Science Program* (2001).
Dieter Langbein, *Material Research in Microgravity* (1984).
Liya L. Regel, "Research Experiences on Materials Science in Space aboard *Salyut* and *Mir*," *Space Station Freedom Utilization Conference* (1992).
Günther Seibert et al., *A World without Gravity* (2001).

Shuttle Small Payloads Project

The Shuttle Small Payloads Project (SSPP) was created and managed by Goddard Space Flight Center (GSFC) in the mid-1980s to support development, integration, and mission operations of low-cost science and technology payloads for the Space Shuttle. From the beginning of the Shuttle program, it became clear that unused volume in the Shuttle payload bay could be utilized by so-called payloads of opportunity. GSFC developed a new class of low-cost, quick-turnaround Shuttle carriers to fly science and technology experiments using standard orbiter and carrier interfaces. The intent was to manifest payloads on a space-available basis, with little or no impact to Shuttle integration operations—the ship-and-shoot concept.

The first of these payload carriers was the Get Away Special (GAS), a tertiary-class payload system designed to allow experiments to fly within, or be ejected from, canisters with volumes up to 5 cu ft. GAS experiments could be passive or provided with on/off commanding from the Shuttle crew. GAS payloads could be mounted individually to the side of the Shuttle payload bay or in sets of up to 12 on a mechanical truss (bridge) that spanned the width of the payload bay. The first GAS payload flew on Space Transportation System-2 (STS) in 1981. A total of 167 GAS canisters were flown from 1981 to 2003, which included seven cross-bay bridges.

Hitchhiker Missions
Discoveries and Accomplishments (Part 1 of 13)

Payload Name	Mission, Launch Date	Experiments (Customers)	Research Disciplines	Experiment Accomplishments and Discoveries
HH-G1 (Hitchhiker-Goddard)	STS-61C (Space Transportation System) 12 Jan 1986	CPL (Capillary Pumped Loop) (Goddard Space Flight Center—GSFC)	Thermal engineering, heatpipe technology	Characterized microgravity operation of multi-power capillary pumped-loop heat transfer system
		PACS (Particle Analysis Cameras for Shuttle) (United States Air Force—USAF)	Orbital contamination	Observed far-field emissions from stellar groupings, Earth-limb airglow, and metropolitan areas; provided on-orbit contaminant data to help assess impact on remote sensing from Shuttle
		SEECM (Shuttle Environmental Effect on Coated Mirror) (NASA) (Perkin-Elmer—PE)	Optics, contamination	Demonstrated improved reflectivity of precontaminated mirrors due to atomic oxygen
BBXRT (Broadband X-ray Telescope)	STS-35 2 Dec 1990	BBXRT (GSFC)	X-ray astronomy	First focusing X-ray telescope operating over a broad energy range (0.3–12 keV) with moderate energy resolution; resolved Iron K-line in two binary stars, detected evidence of line broadening in New General Catalog (NGC)-4151, and studied cooling flow in clusters; obtained data on 82 X-ray sources from 157 observations over 185,000 seconds.
STP-1 (Space Test Payload)	STS-39 28 Apr 1991	ALFE (Advanced Liquid Feed Experiment) (McDonnell Douglas Aerospace Corporation—MDAC) (Air Force Research Laboratory—AFRL)	Propellant management	Characterized technology and techniques required for on-orbit propellant management, with application to advanced spacecraft feed systems

Hitchhiker Missions
Discoveries and Accomplishments (Part 2 of 13)

Payload Name	Mission, Launch Date	Experiments (Customers)	Research Disciplines	Experiment Accomplishments and Discoveries
		APM (Ascent Particle Monitor) (USAF)	Ascent particulates	Measured particulates in Shuttle payload bay during ascent
		DSE (Data Systems Experiment) (GSFC)	Computer engineering	Tested advanced data management concepts
		SKIRT (Spacecraft Kinetic Infrared Test) (GSFC/USAF)	Infrared (IR) glow science, sensor technology	First IR measurements to identify molecular species of Shuttle glow; demonstrated use of cryogenic IR sensors
		UVLIM (Ultraviolet Limb Imaging) (Naval Research Laboratory—NRL/USAF)	Atmospheric science	Collected UV observations of atmosphere at Earth limb
ASP (Attitude Sensor Package)	STS-52 22 Oct 1992	ASP (University—U. Trieste)	Sensor technology, attitude determination	Tested three independent new sensors for application on future spacecraft
GCP (Shuttle Glow Experiment Cryogenic Heat Pipe Payload)	STS-53 2 Dec 1992	CryoHP (Cryogenic Heat Pipe) (GSFC/USAF)	Thermal engineering, heatpipe technology	First flight demonstration of cryogenic-oxygen heatpipe technology; first flight data of operation below 100K
		GLO-1 (Shuttle Glow Experiment) (U. Arizona)	IR glow science, atmospheric physics	First simultaneous optical detection of magnesium and its ion in a common volume of the ionosphere, helping in the identification of active chemical pathways; observed up/down motion of Earth's diurnal electric field, providing data on dawn/dusk asymmetry in ionospheric metal ion distribution

(Continued)

Hitchhiker Missions
Discoveries and Accomplishments (Part 3 of 13)

Payload Name	Mission, Launch Date	Experiments (Customers)	Research Disciplines	Experiment Accomplishments and Discoveries
DXS (Diffuse X-ray Spectrometer)	STS-54 13 Jan 1993	DXS (U. Wisconsin)	X-ray astronomy	Obtained first-ever spectra of the diffuse soft X-ray background in the energy band 0.15–0.284 keV (42–84 Å), the first direct evidence that a bubble of million-degree gas surrounds the solar system
SHOOT (Superfluid Helium On-Orbit Transfer)	STS-57 21 Jun 1993	SHOOT (GSFC)	Cryogenic fluid transfer technology	First flight demonstration of cryogenic fluid transfer in space—liquid helium at 720 liters (l)/hr; characterized operation of cryogenic technology components in a microgravity environment
COB (Capillary Pumped Loop Orbital Debris Radar Calibration Spheres University of Bremen Satellite)	STS-60 3 Feb 1994	CAPL (Capillary Pumped Loop) (GSFC)	Thermal engineering, heatpipe technology	First microgravity demonstration of advanced capillary pumped-loop thermal control system
		ODERACS (Orbital Debris Radar Calibration Spheres) (Johnson Space Center—JSC/USAF)	Orbital debris, radar tracking	(Ejected) Allowed ground-based detection of small objects in orbit to help calibrate radar systems used to monitor orbital debris; calibrated the principle polarization response of ground-based radar and the data processing system
		BremSat (U. of Bremen Satellite) (U. Bremen)	Orbital contamination	(Ejected) Characterized on-orbit atomic oxygen environment and electric charges produced by micrometeoroids
OAST-2 (Office of Aeronautics and Space Technology)	STS-62 4 Mar 1994	ECT (Emulsion Chamber Technology) (Marshall Space Flight Center—MSFC)	Emulsion chamber technology	Characterized space radiation for spacecraft and human shielding applications

Hitchhiker Missions
Discoveries and Accomplishments (Part 4 of 13)

Payload Name	Mission, Launch Date	Experiments (Customers)	Research Disciplines	Experiment Accomplishments and Discoveries
		CryoTP (Cryogenic Two-Phase) (GSFC/USAF)	Thermal engineering, heatpipe technology	First flight demonstration of cryogenic nitrogen heat-pipe technology and 120K phase-change thermal storage device for electronics applications
		SAMPIE (Solar Array Module Plasma Interaction Experiment) (Lewis Research Center—LeRC)	Solar array technology, plasma science	First retrievable high-voltage space plasma interaction experiment; first flight characterization of *ISS* and advanced photovoltaic cells; measured effects of oxygen/nitrogen plasma interaction on shuttle
		TES-1/2 (Thermal Energy Storage) (LeRC)	Thermal energy storage	Obtained first data on long-duration microgravity behavior of thermal storage fluoride salts, for solar dynamic power applications; confirmed predicted salt void behavior
		EISG/SKIRT (Experimental Investigation of Spacecraft Glow)/(Spacecraft Kinetic Infrared Test) (GSFC/JSC)	IR glow science	Measured effects of temperature on spacecraft glow from oxygen/nitrogen plasma interaction; confirmed that Shuttle IR glow phenomenon peaks strongly in ram direction, is zero in anti-ram, and is suppressed by venting nitrogen
ROMPS (Robot Operated Processing System)	STS-64 9 Sep 1994	ROMPS (GSFC)	Materials processing, robotics	Demonstrated commercial methods of rapid thermal processing of 100 thin-film semiconductor materials in microgravity; demonstrated robot control using capaciflector proximity sensor
CGP/ODERACS-2 (Cryogenic System Experiment Shuttle Glow Experiment Payload)	STS-63 3 Feb 1995	CSE (Cryogenic System Experiment) (Jet Propulsion Laboratory—JPL/Hughes)	Cryogenics	Validated and characterized on-orbit performance of two technologies that comprise a hybrid cryogenic system, to support design of future cryogenic systems for NASA and military spaceflight

(Continued)

Hitchhiker Missions
Discoveries and Accomplishments (Part 5 of 13)

Payload Name	Mission, Launch Date	Experiments (Customers)	Research Disciplines	Experiment Accomplishments and Discoveries
		GLO-2 (U. Arizona)	IR glow science, atmospheric physics	Collected data on ionospheric metal ion clouds, allowing better understanding of ionospheric electric fields
		ICBC (IMAX Cargo Bay Camera) (JSC/Image Maximum—IMAX)	Mission photography	Photographed *Mir* rendezvous operations for motion picture *Mission to Mir*, which premiered at the Smithsonian National Air and Space Museum on 21 May 1997
		ODERACS 2 (JSC/USAF)	Orbital debris, radar tracking	(Ejected) Verified the optical data analysis process leading to debris piece sizes; calibrated the orthogonal polarization response of ground-based radar relative to the principle polarization response
IEH-1 (International Extreme Ultraviolet Hitchhiker)	STS-69 7 Sep 1995	UVSTAR (Ultraviolet Spectrograph Telescope for Astronomical Research) (U. Arizona/European Space Agency—ESA)	Ultraviolet astronomy	Acquired stellar UV spectra, demonstrating use of UVSTAR instrument for Shuttle and space station
		SEH (Solar Extreme Ultraviolet Hitchhiker) (U. Southern California)	Solar astronomy, solar physics	Measured difficult-to-measure absolute solar extreme ultraviolet (EUV) flux, important as calibration data for other instruments
		GLO-3 (U. Arizona)	IR glow science, atmospheric physics	Observed ionospheric metal ion cloud build-up as function of time-of-day, providing insight into evolution of metal ion upwelling by Earth's electric fields
		CONCAP-IV/03 (Consortium Complex Autonomous Payload) (U. Alabama-Huntsville)	Materials science	Characterized organic thin-film growth for electro-optics applications; obtained diffusion-controlled transport; confirmed thin-film growth in microgravity was very robust and uniform, allowing production of

Hitchhiker Missions
Discoveries and Accomplishments (Part 6 of 13)

Payload Name	Mission, Launch Date	Experiments (Customers)	Research Disciplines	Experiment Accomplishments and Discoveries
				single-crystalline films; contributed to film growth modeling
CAPL-2	STS-69 7 Sep 1995	CAPL-2 (GSFC)	Thermal engineering, heatpipe technology	Flight demonstration of advanced capillary pumped-loop thermal control system incorporating modified starter pump
GPP (Shuttle Glow Experiment Photogrammetric Appendage Structural Dynamics Experiment Payload)	STS-74 11 Nov 1995	GLO-4 (U. Arizona)	IR glow science, atmospheric physics	Provided data to further understand altitude-dependent metal-ion density in Earth's electric field
		PASDE (Photogrammetric Appendage Structural Dynamics Experiment) (Langley Research Center—LaRC)	Structural dynamics, photogrammetry	Characterized *Mir* solar array dynamics; established low-cost, passive photogrammetry for application to the *International Space Station* (*ISS*) appendage structural response and verification
SLA-1 (Shuttle Laser Altimeter)	STS-72 11 Jan 1996	SLA-1 (GSFC)	Laser altimetry, tracking technology	First substantive altimetry of Earth from space, collecting 2.7 million laser measurements of land, ocean, and clouds; first flight demonstration of second-generation, diode-pumped Neodymium-doped Yttrium Aluminum Garnet (Nd:YAG) lasers; demonstrated surface lidar methods and dramatic improvements in laser altimetry data processing

(Continued)

Hitchhiker Missions
Discoveries and Accomplishments (Part 7 of 13)

Payload Name	Mission, Launch Date	Experiments (Customers)	Research Disciplines	Experiment Accomplishments and Discoveries
TEAMS (Technology Experiments Advancing Missions in Space)	STS-77 16 May 1996	GANE (Global Positioning System Attitude and Navigation Experiment) (JSC)	Navigation, attitude determination	Demonstrated use of Global Positioning System (GPS) for on-orbit attitude determination with application to *ISS*
		LMTE (Liquid Metal Thermal Experiment) (USAF)	Thermal engineering	First flight demonstration of liquid-metal heat-pipe operation, with 550°C temperature gradient
		PAMS (*Passive Aerodynamically Stabilized Magnetically Damped Satellite*) (GSFC)	Spacecraft stabilization technology	(Ejected) Tested a passive aerodynamic stabilization and magnetic damping system; verified proof-of-concept of propellant-free, aerodynamic stabilization for satellites
		VTRE (Vented Tank Resupply Experiment) (LeRC)	Tanking/refueling technology	First flight demonstration of an autonomous fluid transfer system; verified ability of vane propellant management devices to separate liquid and gas in microgravity
CryoFD (Cryogenic Flexible Diode)	STS-83 4 Apr 1997	CryoFD (GSFC/USAF)	Thermal engineering	Demonstrated the first U.S.-made loop heat pipe, and the highest capacity cryogenic (70–150K) oxygen heat pipes ever developed, with applications for space-based IR sensors
CryoFD-R (Cryogenic Flexible Diode Reflight)	STS-94 1 Jul 1997	CryoFD (GSFC/USAF)	Thermal engineering	Obtained reflight data for operation of cryogenic flexible-diode heat pipe in microgravity; verified operation of composite wick design
IEH-2	STS-85 17 Jul 1997	UVSTAR (U. Arizona/ESA)	UV astronomy	Collected data on H-Lyman intensity and water dissociation rate for Comet Hale-Bopp; obtained first evidence of

Hitchhiker Missions
Discoveries and Accomplishments (Part 8 of 13)

Payload Name	Mission, Launch Date	Experiments (Customers)	Research Disciplines	Experiment Accomplishments and Discoveries
				high-velocity stellar winds in two unpredicted ionization stages for three subdwarf O-stars
		SEH (U. Southern California)	Solar astronomy, solar physics	Provided solar flux data, such as that required for interpretation of Jovian EUV/ Far Ultraviolet—FUV data obtained by UVSTAR
		GLO-5/6 (U. Arizona)	IR glow science, atmospheric physics	Extended altitude range for measurements of ionospheric metal ions, helping to characterize the time-of-day versus latitude model of Earth's electric field
		DATA-CHASER (Distribution and Automation Technology Advancement-Colorado Hitchhiker and Student Experiment of Solar Radiation) (U. Colorado)	Solar astronomy, data systems technology	Demonstrated integrated advanced data system tools and technologies for improving space payload operations; established interactive human/automated payload control, distributed to remote users; measured the full-disk solar UV and soft X-ray irradiance, and imaged the Sun in Lyman-Alpha, allowing correlation of solar activity with radiation flux and associated Lyman-Alpha fluxes with individual active regions
TAS-1 (Technology Applications and Science)	STS-85 17 Jul 1997	CFE (Cryogenic On-Orbit Long-Life Active Refrigerator Flight Experiment) (USAF)	Thermal engineering	Demonstrated Joule-Thomson (J-T) cycle cryocooler operation, designed to provide two stages of cooling, for future space applications
		CVX (Critical Viscosity of Xenon) (LeRC)	Fluid viscometry	Most accurate measurement of critical exponent for xenon viscosity; first

(Continued)

Hitchhiker Missions
Discoveries and Accomplishments (Part 9 of 13)

Payload Name	Mission, Launch Date	Experiments (Customers)	Research Disciplines	Experiment Accomplishments and Discoveries
				measurement of viscoelasticity near the critical point of xenon
		ISIR (Infrared Spectral Imaging Radiometer) (GSFC)	Remote sensing, atmospheric science	Obtained high-resolution thermal IR imagery of clouds at 8.5 μm, with application for cloud particle discrimination; demonstrated feasibility of uncooled IR detector for remote sensing
		SLA-2 (GSFC)	Laser altimetry	Performed laser ranging from all major Earth surface types, allowing derivation of near-global sampling of land-cover elevation; extended SLA-1 dense-grid topography data to 57 degrees
		SOLCON/SOVA-1 (Solar Constant/Solar Constant and Variability) (Royal Meteorological Institute—RMI—Belgium)	Solar physics	Collected solar irradiance data required to better understand global climate change
		TPF (Two-Phase Flow) (GSFC)	Thermal engineering	First flight demonstration of capillary vapor-flow activation device; confirmed theories of multi-evaporator operation for thermal control systems
LHP—NaSBE (Loop Heat Pipe—Sodium-Sulfur Battery Experiment)	STS-87 19 Nov 1997	LHP (Texas A&M U., Dynatherm)	Thermal engineering	Demonstrated and characterized loop heat-pipe technology for spacecraft thermal management applications
		NaSBE (USAF/NRL)	Battery technology	Demonstrated microgravity operation of a battery with liquid electrodes and solid electrolyte; characterized microgravity

Hitchhiker Missions
Discoveries and Accomplishments (Part 10 of 13)

Payload Name	Mission, Launch Date	Experiments (Customers)	Research Disciplines	Experiment Accomplishments and Discoveries
				effects on mass transport and reactions at solid-electrolyte interface
SOLSE-1 (Shuttle Ozone Limb Sounding Experiment)	STS-87 19 Nov 1997	SOLSE (GSFC)	Remote sensing, atmospheric science	First UV spectrometry of scattered radiation from Earth's limb; solved problems of tangent height registration and scattered light; demonstrated use of limb scattering to derive high-resolution ozone profile, 15–50 km; verified viewing orientation for ozone retrieval and Charge Coupled Device-array technology for UV imaging
		LORE (Limb Ozone Retrieval Experiment) (GSFC)	Remote sensing, atmospheric science	Demonstrated that vertical profiles of ozone can be measured with high resolution using sunlight scattered in the atmospheric limb; extended SOLSE's knowledge of the limb down to 10 km above Earth's surface
IEH-3	STS-95 29 Oct 1998	UVSTAR (U. Arizona/ESA)	UV astronomy	Obtained first 0.1-nm resolution spectrum for 50–125 nm region; obtained first spectrally resolved images of Jovian system in 500–1250Å range, including Io plasma torus; performed line identification, flux calibration, classical atmospheric model comparison, and flux-variability measurements; obtained FUV spectra of several galactic and stellar sources
		SEH (U. Southern California)	Solar astronomy	Produced excellent full-disk, absolute solar EUV flux data for flux incident on Jovian system to help understand the EUV dynamics of the Io plasma torus; collected data from a major solar flare, yielding insight into spectral and energy changes in solar EUV and

(Continued)

Hitchhiker Missions
Discoveries and Accomplishments (Part 11 of 13)

Payload Name	Mission, Launch Date	Experiments (Customers)	Research Disciplines	Experiment Accomplishments and Discoveries
				soft X rays; (in conjunction with UVSTAR) demonstrated tight coupling between solar EUV input and planetary system output
		SOLCON-2 (RMI-Belgium)	Solar physics	Measured total solar irradiance; provided data to help calibrate and verify effects of aging on long-term solar radiometers
		STARLITE (Spectrograph Telescope for Astronomical Research) (U. Arizona)	IR imaging	Demonstrated on-orbit performance of a composite-structure telescope and UV imaging spectrometer in the space environment
		PANSAT (*Petite Amateur Navy Satellite*) (Naval Postgraduate School/Space Test Payload—NPS/STP)	Radio communications	(Ejected) Provided global spread-spectrum signals for amateur radio; provided students with experience in communications satellite design, development, integration, test, and mission ops
CryoTSU (Cryogenic Thermal Storage Unit)	STS-95 29 Oct 1998	CryoTSU (GSFC/USAF)	Thermal engineering	First flight demonstration of a cryogenic capillary pumped loop and a 60K phase-change thermal storage device with a superconducting bolometer; demonstrated 3000J of energy storage
MightySat/SAC A (Mighty Satellite/ Satellite de Aplicaciones Cientifico)	STS-88 4 Dec 1998	*SAC A* (Comision Nacional de Actividades Espaciales—CONAE, Argentina)	Remote sensing, solar cell technology	(Ejected) First flight demonstration of remote sensing for whale tracking; characterized new solar cell technology
		MightySat 1 (USAF/AFRL)	Solar cells, microelectronics, composites	(Ejected) Flight demonstration of advanced composite structure, microelectronics and packaging; characterized advanced solar cells; demonstrated shape-memory alloy

Hitchhiker Missions
Discoveries and Accomplishments (Part 12 of 13)

Payload Name	Mission, Launch Date	Experiments (Customers)	Research Disciplines	Experiment Accomplishments and Discoveries
				release device with many times less shock than conventional pyrotechnic devices
STARSHINE (Student-Tracked Atmospheric Research Satellite for Heuristic International Networking Experiment)	STS-96 17 May 1999	*STARSHINE 1* (NRL/ Starshine)	Orbital physics, student outreach	(Ejected) Facilitated understanding of relationship between upper atmosphere and solar activity; allowed determination of effects of solar EUV radiation on satellite orbital decay; enabled teacher/student participation in small satellite development and on-orbit tracking
HEAT (Hitchhiker Experiments Advancing Technology)	STS-105 10 Aug 2001	*SimpleSat* (Simple Satellite) (GSFC)	GPS attitude control/pointing	(Ejected) No contact with satellite post-deploy
		ACE (Advanced Carrier Electronics) (GSFC)	Flight avionics	Successfully demonstrated new modular carrier avionics subsystem
MACH-1 (Multiple Application Customized Hitchhiker)	STS-108 29 Nov 2001	CAPL-3 (GSFC/NRL)	Thermal engineering	Demonstrated multiple evaporator pumped-loop thermal control system, with reliable startup and continuous operation, for wide range of operating conditions; demonstrated 50 percent heat load sharing; established higher system confidence for use in future satellite thermal control systems
		PSRD (Prototype Synchrotron Radiation Detector) (JSC, Eidgenössische Technische Hochschule—ETH—Zurich)	High-energy astrophysics	Measured cosmic background radiation; tested and calibrated SRD technology before its application on a three-year *ISS* mission; characterized instrument's response to both *ISS* and undocked space environments

(Continued)

Hitchhiker Missions
Discoveries and Accomplishments (Part 13 of 13)

Payload Name	Mission, Launch Date	Experiments (Customers)	Research Disciplines	Experiment Accomplishments and Discoveries
		STARSHINE 2 (NRL/ Starshine)	Orbital physics, student outreach	(Ejected) Allowed more accurate predictions of the orbital decay of other satellites due to solar activity and UV radiation
FREESTAR (Fast-Reaction Experiments Enabling Science, Technology, and Research)	STS-107 16 Jan 2003	MEIDEX (Mediterranean Israeli Dust Experiment) (Israeli Space Agency—ISA, Tel Aviv U.)	Atmospheric science	Proved transient luminous events emit in near-IR, and correlation with thunderstorms; discovered synchronicity of lightning activity; demonstrated determination of desert aerosol height distribution from space via multispectral data; showed effects of desert dust on cloud precipitation production and terrestrial temperatures; proved theory that forest fire smoke inhibits cloud development and rain
		CVX-2 (Glenn Research Center—GRC)	Fluid viscometry	First measurement of shear thinning in a simple atomic fluid, enabling enhanced thinning techniques for products that use liquids (oil, plastics)
		SOLCON-3 (RMI, Belgium)	Solar physics	Collected solar irradiance data required to better understand global climate change
		LPT/CANDOS (Low-Power Transceiver/Communications and Navigation Demonstration on Shuttle) (GSFC)	Internet Protocol (IP) communications technology, navigation	First flight demo of Internet Protocol ops via NASA Communications Network /Tracking and Data Relay Satellite System (NASCOM/ TDRSS), software-defined radio, and on-board navigation via GPS Enhanced On-board Navigation System (GEONS)
		SOLSE-2/LORE (GSFC)	Remote sensing, atmospheric science	Measured limb-scattered radiance to allow determination of high-resolution profile of stratospheric ozone

In 1984 GSFC developed the Hitchhiker (HH) carrier system to enhance small-payload mission capabilities and to allow more complex interaction with Shuttle and crew. Standard mechanical and electrical interfaces to both Shuttle and payload experiment sides of the carrier were baselined to maintain the ship-and-shoot concept as much as practicable. Like GAS, HH offered payload bay side-mounted and cross-bay bridge configurations. However, HH offered a variety of instrument mounting configurations and ground-based commanding and telemetry from a GSFC control center. There were even self-contained ejection systems to deploy small satellites. GSFC HH personnel designed, developed, tested, and integrated the HH carrier systems and supported crew training and mission operations. HH missions were developed at relatively low cost and short times compared to other Shuttle payloads, utilizing reusable systems and small teams of typically less than two dozen engineers, managers, and technicians. The first HH payload, HH-G1, flew on STS-61C in January 1986. From 1986 to 2003, a total of 69 instruments and 28 payloads were flown on 25 Shuttle missions, the last on *Columbia*'s final flight, STS-107. During its 17-year history, HH instruments included astronomical telescopes, remote sensing cameras, robotic systems, and materials science demonstrations, in addition to nine ejected satellites.

Since HH was based on many of the same carrier components as the GAS system, both carriers were eventually managed by one project office, the SSPP Office. This enabled synergies between the two, resulting in lower operational costs of approximately $13 million per year, covering about three HHs and the GAS and Space Experiment Module (SEM) projects.

SEM was designed to host 10 simple experiments contained in modules mounted within a GAS canister and was targeted for elementary to high school students. Following the *Columbia* accident, SEM Satchels were designed to allow 20 sample vials to be delivered to the *International Space Station* (*ISS*) via Russian Proton resupply missions. During this short-lived program, 14 SEM canisters were flown inside the payload bay, and three in-cabin SEM Satchels flew on the *ISS*.

These three SSPP carrier systems—HH, GAS, and SEM—proved to be valuable resources for experimenters interested in flying low-cost instruments to orbit. The SSPP payloads also facilitated proof-of-concept testing during the Shuttle's sortie missions, before incorporating technologies into more complex or longer duration missions. SSPP provided Shuttle mission opportunities for hundreds of experimenters, from preschool students to the international science community. HH customers included all the NASA mission directorates, Department of Defense, industry, and academia. It also hosted experiments from international customers, such as the European Space Agency and the Argentina and Israel space agencies.

Michael R. Wright

See also: *International Space Station*, Space Shuttle

Bibliography

Shuttle Small Payloads Project. http://library01.gsfc.nasa.gov/host/hitchhiker/index14.html.

Smithsonian Astrophysics Observatory (SAO). http://www.cfa.harvard.edu/sao/index.html.

Michael R. Wright, "Shuttle Small Payloads: How to Fly Shuttle in the 'Faster, Cheaper, Better' World," IEEE Aerospace Applications Conference (February 1996).

United States Biosatellite Program

The United States Biosatellite Program was used in the 1960s to examine the effects of microgravity on living organisms launched into space. Begun in 1962 NASA's Ames Research Center managed the U.S. Biosatellite Program, and the General Electric Company was contracted to build satellites for three missions.

By 1963 NASA approved fewer than 20 experiments out of more than 170 proposals for the program, which would test the effects of weightlessness on plants, insects, and a monkey. Each biosatellite contained a capsule equipped with a life-support system that used an oxygen-nitrogen atmosphere and a thermal control system. The biosatellites were to be recovered in midair by a military cargo plane using a cable that grabbed the satellite's parachute during freefall after reentry through the atmosphere. The spacecraft could also be rescued at sea, if necessary.

Biosatellite 1 was launched on 14 December 1966 from Cape Kennedy, Florida, on a Thrust-Augmented Delta (TAD) rocket with 13 experiments on board. During its scheduled three-day mission, bacteria, bread mold, flour beetles, plants, wasp larvae, and fruit flies were purposely exposed to varying levels of gamma radiation, while pepper plants, wheat seedlings, frog eggs, and amoebas were protected in a separate chamber. A retro-rocket failure stranded the biosatellite in low Earth orbit for two months, and the spacecraft was eventually lost during an unplanned reentry.

Biosatellite 2 was launched on 7 September 1967 on a TAD and carried the same payload as the first mission. The biosatellite's capsule was recovered successfully in midair by a C-130 after NASA ended the mission early because of communications problems. Experiments conducted during the 45-hour spaceflight provided significant amounts of data and showed that in microgravity, plant root systems grew abnormally; that young, rapidly dividing cells were the most susceptible to radiation; and that animal cells had adapted to weightlessness better than plant cells.

Biosatellite 3 was launched on 29 June 1969 on a Long Tank TAD N booster with Bonny, a 14 lb male pigtail monkey, on board. NASA ended the scheduled 30-day spaceflight early after Bonny became ill. The reentry capsule containing the sick monkey was recovered from the Pacific Ocean after nearly nine days in space. Bonny died less than one day later from a heart attack. The mission, which yielded limited in-flight data, was the last flight of the U.S. Biosatellite Program. NASA decided to terminate the program to redirect funds for the crewed *Skylab* space station, which offered a better platform to conduct biological research.

Tim Chamberlin

See also: Aerospace Medicine, Animals in Space

Bibliography

Gregory P. Kennedy, "U.S. Biosatellites," in *Magill's Survey of Science, Space Exploration Series*, ed. Frank N. Magill, vol. 1 (1989).

John A. Pitts, *The Human Factor: Biomedicine in the Manned Space Program to 1980* (1985).

SPACEFLIGHT PASSENGERS

Spaceflight passengers include all humans and animals that have traveled beyond an altitude of 62 miles, the distance where scientists consider the Earth's atmosphere ends and space begins. As passengers, humans have accepted the risks associated with spaceflight, an inherently dangerous activity that has killed men and women during flight testing and actual missions. Humans have been subjected to abnormal stresses associated with microgravity, such as space sickness, and have been trained to deal with such hazards to perform their duties.

Animals were flown in space as early as the late 1940s during high-altitude rocket flights. Animals have served as test subjects during spaceflights to help determine the physical limits associated with the effects of microgravity. Primates and dogs have been the public's most memorable subjects from early flights, especially the Soviet dog Laika, the first animal to orbit Earth in 1957. Laika became a heroic symbol in the Soviet Union, but her death in orbit caused an uproar among animal rights activists, particularly in the United Kingdom. Nearly as famous were the chimpanzees, Enos and Ham, sent into space by NASA to test the Mercury capsules. Since then, dozens of species of animals have served as passengers, including mice, spiders, and frogs, many of them for university research and elementary and high school projects.

More than 500 humans (as of 2010) have traveled to space since the historic flight of cosmonaut Yuri Gagarin in April 1961. Astronauts (United States) and cosmonauts (Soviet Union/Russia) comprise the majority of humans who have traveled into space. The first astronauts were chosen in April 1959 and the first cosmonauts chosen in February 1960. Those initial selections were limited to military test pilots, but as both groups grew in numbers and their roles became more diverse, academic qualifications became just as important as piloting skills. During the height of the space race of the 1960s, the United States and the Soviet Union enlisted aeronautical engineers, scientists, and physicians to fly in space. A geologist was one of the 12 astronauts who walked on the Moon.

By the late 1970s spaceflight passengers from outside the United States and the Soviet Union became much more common. The Soviet Interkosmos program allowed scientists from countries of the former communist bloc to fly in space. The decision to open the cosmonaut corps to other countries was most likely a political response to NASA's decision to fly European scientists on the Space Shuttle during the 1980s as part of the Spacelab program.

Vostok 6 in June 1963, which carried sport parachutist Valentina Tereshkova, the first female to fly in space, was considered a stunt to upstage the American human spaceflight effort. While NASA had prequalified women as possible astronaut candidates during the Mercury project, they were not admitted into the astronaut corps until 1978. NASA hoped to rekindle interest and support in the U.S. space program when it opened its astronaut corps to teachers in August 1984, selecting two women. Unfortunately, the highly publicized Teacher in Space Project turned into a public relations nightmare for the space agency when social studies teacher Sharon Christa McAuliffe

The dog Laika, a passenger on Sputnik 2, *became the first living creature in orbit in 1957. She died from heat and stress a few days into the mission. (Courtesy AP/Wide World Photos)*

was to become the first citizen passenger to fly in space but was killed when the Space Shuttle *Challenger* exploded after takeoff in January 1986.

The deaths of astronauts and cosmonauts in space or on the ground frequently became national tragedies. While cosmonauts killed in training were kept secret, after being put in orbit, the Soviet Union could not hide the death of Soviet cosmonauts. The loss of four cosmonauts in the early Soyuz flights (*Soyuz 1*, 1967; *Soyuz 11*, 1971) led to national expressions of mourning. The death of Yuri Gagarin in a 1968 training flight caused a similar response. In the United States, disasters such as the 1967 *Apollo 204* fire, the *Challenger* tragedy, and the loss of *Columbia* in 2003 had sobering effects on the public's view of human spaceflight, as some questioned whether NASA placed astronauts at too much risk. Congressional investigations ensued in each case, but in the end, Congress and Administration accepted the risks, and flights ultimately resumed after major efforts to improve safety.

While most people who have flown in space work for various space agencies, in 1990 the Japanese Tokyo Broadcasting System paid $12–14 million to fly reporter Toyohiro Akiyama on *Soyuz TM11*. Having flown numerous "cosmonaut researchers" from several nations in the Interkosmos program, Soviet leaders realized they could earn hard cash by flying paying customers. Helen Sharman from the United Kingdom (UK) flew on *Soyuz TM12* the following year, paid by a consortium in the UK. The first individual to pay his own way as a true space tourist was U.S. businessman Dennis Tito, who flew in April 2001 to the *International Space Station* on *Soyuz TM32*.

But it was not without struggle, as NASA and the European Space Agency publicly criticized the Russian Space Agency's decision, calling the trip an unnecessary safety risk. Purely private (no government funding or vehicles used) suborbital flights began in 2004 when Scaled Composites successfully flew *SpaceShipOne* to 100 km altitude.

Tim Chamberlin and Stephen B. Johnson

See also: Economics of Human Spaceflight, National Aeronautics and Space Administration, Politics of Prestige, Soviet Union

Bibliography

Tim Furniss and David J. Shayler, *Praxis Manned Spaceflight Log 1961–2006* (2007).

Green Peyton, *Fifty Years of Aerospace Medicine, 1918–1968* (1968).

Asif A. Siddiqi, *Challenge to Apollo: The Soviet Union and the Space Race, 1945–1974* (2000).

Animals in Space

Animals in space have historically been used to test human physiological and psychological response to spacecraft environments and effects of microgravity. More than 40 species have been employed, typically mammals, insects, or water dwellers. Scientists have selected species based on environmental constraints and research goals.

Primates and dogs tested basic survivability in the 1950s and early 1960s. They tolerated confinement and immobilization, performed tasks on cue, and exhibited identifiable emotional responses. Later satellite and space station research examined system- and molecular-level physiology. Scientists studied frog otoliths because they are similar to that of humans, and they observed crickets because their neurons are easily isolated. Rats and Drosophila (fruit flies) have been the most popular subjects, given the quantity of data, strains, and cell lines bred for specific traits, plus the commonality of knowledge about them worldwide.

The United States first tested animals in conditions similar to spaceflight (low air pressure, controlled oxygen environment, and cold temperatures) while developing B-17 bombers in the 1930s. Germany and the United States did so while studying explosive decompression aboard pressurized aircraft during the same decade. During World War II, German jet and rocket studies employed mammals to study rapid ascent and abrupt acceleration and deceleration. In the 1940s–50s, U.S. scientists observed animal reactions aboard balloons, jets, rocket sleds, centrifuges, and in acoustical facilities.

American and Soviet researchers first launched animals by rocket in 1948 and 1951, respectively, studying weightlessness and cosmic radiation. The Soviet dog Laika became the first living creature in orbit in 1957, and American primates helped qualify Mercury capsules for human use shortly afterward. Beginning in the late 1960s, on-orbit studies used small containable colonies, animals, embryos, and cells to study adaptation and long-term effects of microgravity and radiation on bone density, muscle atrophy, wound healing, and reproduction. By 2005 Australia and more than 20 countries in North America, Europe, and Asia had conducted animal experiments in space.

Maura Phillips Mackowski

See also: Bion, Mercury, Vostok

Bibliography

Charles A. Dempsey, *Fifty Years of Research on Man in Flight: The Air Force Aerospace Medical Research Laboratory* (1985).
Green Peyton, *Fifty Years of Aerospace Medicine, 1918–1968* (1968).
Asif A. Siddiqi, *Challenge to Apollo: The Soviet Union and the Space Race, 1945–1974* (2000).
Loyd S. Swenson Jr. et al., *This New Ocean: A History of Project Mercury* (1966).

Astronauts

Astronauts are educated and trained to pilot, navigate, or participate in the flight of a U.S. spacecraft. The term comes from the Greek language, with "astro" meaning star and "naut" meaning sailor, combining into "sailor of the stars." Astronaut became a household word soon after the newly formed NASA announced the start of its crewed space program, Project Mercury, in 1958. The chance to be the first human in space, and the great risk associated with trying to get there, made the astronaut a hero and cultural icon in the eyes of space enthusiasts around the world.

NASA created a committee comprised of engineers, flight surgeons, psychologists, and psychiatrists to interview and to select the agency's astronauts. The first group of astronauts (officially announced in April 1959) was comprised of seven military test pilots, who became famous as the Mercury Seven. They were Scott Carpenter, L. Gordon Cooper Jr., John Glenn Jr., Virgil "Gus" Grissom, Walter Schirra Jr., Alan Shepard Jr., and Donald "Deke" Slayton. NASA selected military test pilots because the stress of flying high-speed jet aircraft was believed to be similar to human spaceflight. At the time, it was unclear how the human body would react to the hazardous environment of space.

Each of the original seven astronauts, with the exception of Slayton, made his first flight as part of Project Mercury. Slayton was removed from flying status shortly after his selection when doctors revealed he had a heart condition. NASA offered him the position to coordinate astronaut activities and training, a job that involved participation in the astronaut selection process and mission assignments through the end of Project Apollo. Slayton's position became a permanent institution at NASA, while Slayton himself eventually made it to space as a crew member of the Apollo-Soyuz Test Project (ASTP) in 1975.

Most astronaut training and related activities were conducted at Johnson Space Center (JSC) in Houston, Texas, where NASA built special facilities, such as neutral buoyancy tanks to perfect extravehicular activity techniques, and full-scale mockups of spacecraft, including Space Shuttle simulators.

The Mercury Seven astronauts, whose selection was announced on 9 April 1959, are: front row, left to right, Walter H. Schirra, Jr., Donald K. "Deke" Slayton, John H. Glenn, Jr., and Scott Carpenter; back row, Alan B. Shepard, Jr., Virgil I. "Gus" Grissom, and L. Gordon Cooper, Jr. (Courtesy NASA/Johnson Space Center)

Compared to the Mercury Seven, the selection of the second group of astronauts was more focused on academic qualifications. Named in September 1962, nine astronauts comprised the second group, many of whom flew on missions during Project Gemini and Apollo. This included Neil Armstrong, the first human to walk on the Moon on 20 July 1969 during *Apollo 11*.

In October 1963 NASA selected 14 more astronauts. Edwin "Buzz" Aldrin Jr., who flew with Armstrong to the surface of the Moon, was among this group. In June 1965 the first group of six scientist-astronauts was selected. In anticipation of the planned lunar landings, NASA nearly doubled the size of its astronaut corps in April 1966 when it named 19 new astronauts. Some from this group flew on Apollo missions, while others had to wait many years to fly on the Space Shuttle. The fifth group was the last chosen for flights to the Moon.

Beginning in October 1967 NASA selected 11 scientist-astronauts to work on Apollo Applications Programs—human spaceflight missions that would incorporate Apollo technology for uses other than flying to the Moon. This group, along with the seven additional astronauts who were transferred to NASA in August 1969 from the U.S. Air Force's canceled Manned Orbital Laboratory program, had to wait until the Space Shuttle became operational before getting a chance to fly in space, after Congress reduced the number of Apollo flights in 1972.

NASA waited nine years before it selected the first group of Space Shuttle astronauts. Because of the long delay between ASTP and the first Space Shuttle flight, many astronauts retired or moved to other positions, so NASA needed a greater number than previous selections. In January 1978 it named 35 new astronauts. This group included the first female astronauts (Anna Fisher, Shannon Lucid, Judith Resnik, Sally Ride, Margaret Rhea Seddon, and Kathryn Sullivan) and the first African American astronaut, Guion Bluford Jr. During the Space Shuttle era, NASA substantially changed its criteria for selection of astronauts. It required higher academic qualifications, such as a master's degree or doctorate, and a new team-oriented attitude. NASA also applied a minority-favorable hiring policy.

From 1980 to 2000 NASA selected 10 new groups of astronauts. During that time the space agency assigned astronauts one of three designations: pilot, mission specialist, or payload specialist. Payload specialist referred to individuals selected and trained by commercial or research organizations for flights containing unique payloads.

In 2004 a new group of 11 astronaut candidates (including three educator-astronauts) and three international candidates was selected. In 2009 a new group of nine astronaut candidates and five international candidates was selected.

Pablo de Leon

See also: Apollo, Apollo-Soyuz Test Project, Astronaut Training, Extravehicular Activity, Gemini, *International Space Station*, Mercury, Shuttle-Mir, *Skylab*, Space Shuttle

Bibliography

NASA, *Astronaut Fact Book* (1992).

NASA JSC Astronaut Office. http://www.nasa.gov/centers/johnson/home/index.html.

Alan B. Shepard and Donald K. Slayton, *Moon Shot: The Inside Story of America's Race to the Moon* (1994).

Donald K. Slayton and Michael Cassutt, *Deke! An Autobiography* (1994).

Cosmonauts

Cosmonauts, like their U.S. counterparts, have been taught to pilot, navigate, or participate in the flight of Soviet and Russian spacecraft. Unlike astronauts, many cosmonauts remained anonymous until the waning years of the Soviet Union. The highly secretive posture that characterized the Soviet space program during the Cold War kept many cosmonauts from enjoying the recognition and prestige given to astronauts. While some cosmonauts were considered heroes in the former Soviet Union, their identities were often revealed only after they flew in space, so many did not receive recognition for years, if at all.

In July 1959 OKB-1 (Experimental Design Bureau), experts in aerospace medicine from the Soviet Academy of Sciences, and representatives of several scientific research institutions met to formulate the criteria for cosmonaut selection and training. Soviet Air Force pilots were chosen as the best possible candidates for their aviation and navigation skills and underwent tests at the Central Military Scientific Aviation Hospital (TsVNIAG) in Moscow. After passing this medical screening, the State Interdepartmental Commission, also called the Credential Committee or Mandate Commission, approved or disqualified candidates based on political, personal, and moral beliefs. This included an examination by the Soviet Komitet Gosudarstvennoi Bezopasnosti, State Security Committee (KGB) of all aspiring cosmonauts.

By June 1960 20 men had been selected to represent the first group of cosmonauts, six of whom were considered front runners for the first piloted human spaceflight (*Vostok 1*): Yuri Gagarin, Anatoliy Kartashov, Andrian Nikolayev, Pavel Popovich, German Titov, and Valentin Varlamov. The first group of cosmonauts underwent physical training at the Cosmonaut Training Center, about 40 km northeast of Moscow, and flight training at OKB-1 in nearby Kaliningrad. Gagarin was the first human to fly in space on 12 April 1961, followed by Titov during *Vostok 2* on 6 August, Nikolayev during *Vostok 3* on 11 August 1962, Popovich during *Vostok 4* one day later, and Valeri Bykovsky during *Vostok 5* on 14 June 1963. Despite the success of these flights, the Soviet human spaceflight program suffered setbacks from 1961 to 1963 that alarmed General Nikolai Kamanin, director of cosmonaut training. Six cosmonauts were dismissed for medical or disciplinary reasons, including Kartashov and Varlamov, and a flash fire during an isolation chamber test killed cosmonaut Valentin Bondarenko. Of the original 20 selected from the first group, only 12 flew in space, during which one cosmonaut was killed—Vladimir Komarov—after reentry on board *Soyuz 1* on 23 April 1967.

Five women represented the second group of cosmonauts, including Valentina Tereshkova, the first woman to fly in space during *Vostok 6* on 16 June 1963. Their selection in 1962 was prompted in part by Chief Designer Sergei Korolev's aim to recruit women in addition to engineers and physicians to bolster the overall development of the Soviet human spaceflight program. Their participation held significant propaganda value because NASA never officially recognized women as astronauts at this time, even though several U.S. female pilots trained for the job outside NASA.

A third group consisting of 15 men was added in 1963. The class included engineers for the first time, diversifying the skill set of cosmonaut teams. Colonel Georgi Beregovoi joined the ranks in 1964, despite being older than the stipulated age limit of 35, and brought experience to the relatively young cadre of space pilots. Konstantin Feoktistov, one of the top engineers at OKB-1 who played an integral role in the design of the Vostok and Voskhod spacecraft, was given special consideration when approved to be a cosmonaut in 1964. He was the first civilian to travel in space during *Voskhod 1* and spurred other civilian engineers to seek cosmonaut duty. Twenty-three more cosmonauts—12 pilots, 9 engineers, and 2 doctors—were selected in 1965 in anticipation of an increased rate of spaceflights on the new Soyuz (*7K-OK*).

Following Korolev's death in January 1966, OKB-1 was renamed the Central Construction Bureau of Experimental Machine Building (TsKBEM) and formed its own group of eight cosmonauts, publicly referred to as testers, a few months later. The group underwent medical screening at TsVNIAG to be eligible for Soyuz spaceflights. Beginning in March 1967 the Air Force relinquished control of medical tests to the civilian-led Institute for Biomedical Problems (IMBP) in Moscow. The change resulted in more design bureaus and research institutions forming their own cosmonaut groups separate from the Air Force. This included Vladimir Chelomei's OKB-52, the Soviet Academy of Sciences, the Central Research Institute for Machine Building

(TsNIIMash), and others. The cosmonauts from these groups provided a majority of the engineering assignments on early Soyuz flights into the mid-1970s and later.

Twelve Air Force officers were selected to be cosmonauts in May 1967, including three scientists from the Science Research Institute for Air Defense Forces. Nine more cosmonauts were added in April 1970, but the loss of the *Soyuz 11* crew in June 1971 during a reentry accident grounded piloted spaceflights for more than two years and slowed the number of new cosmonaut selections. While many physicians had been recruited to train as cosmonauts, only a small number actually flew in space. This included Valeri Polyakov, selected as a cosmonaut from the IMBP in 1972, who eventually stayed on *Mir* for a record 437 days during 1994–95.

Cosmonauts Alexei Leonov and Valeri Kubasov trained at Johnson Space Center (JSC) and Kennedy Space Center during multiple visits from 1973 to 1975 as part of the Apollo-Soyuz Test Project. From 1976 to 1978, 11 Soviet Air Force pilots were chosen as cosmonauts for the Buran shuttle program. It was during this time that the Soviet Union also initiated its Interkosmos program and added 16 international research cosmonauts, plus two more in 1979, for nine flights to Salyut space stations through 1981. Countries represented by these cosmonauts included Bulgaria, Cuba, the former Czechoslovakia, the former German Democratic Republic (East Germany), Hungary, Mongolia, Poland, Romania, and Vietnam.

Throughout the 1980s at least 32 civilian pilots and engineers from various design bureaus were approved as cosmonauts to fly on *Salyut 6* and *Salyut 7* or *Mir*. Some of the more notable cosmonauts from this group included pilot Svetlana Savitskaya (chosen in 1980), the second Soviet woman to fly in space aboard *Soyuz T 7* and *Soyuz T 12*, and engineer Sergei Krikalev (chosen in 1985). Krikalev spent more time in space than any other human during two missions to *Mir*, two Space Shuttle flights, and two trips to the *International Space Station* (*ISS*).

By 1986 the Soviet human spaceflight program had been downsized to focus on *Mir*. No more flights to Salyuts were scheduled, and the Buran program had made little progress, so many cosmonauts either retired or assumed other space-related jobs. From 1987 to 1991, with the Soviet Union facing financial and political turmoil, only 15 new cosmonauts were recruited from the Air Force. Some cosmonauts selected for Buran were transferred to space station operations once it became clear the program would be canceled after the Soviet Union's collapse.

In 1992 the United States and Russia agreed to participate in the Shuttle-Mir program, a cooperative venture that allowed cosmonauts to fly on the Space Shuttle and astronauts to work on *Mir*. Five Russian cosmonauts trained at JSC from 1992 to 1998, where they became familiar with Shuttle systems, including the Shuttle Remote Manipulator System, extravehicular activity guidelines, and emergency egress operations. They also learned to overcome language barriers. Cosmonauts started training for flights to the *ISS* in 1996 at facilities in Canada, Europe, Russia, and the United States. After a five-year layoff, two new cosmonauts from the Russian Air Force were selected from 1996 to 1997, and 14 more from 1997 to 2003. From 1990 to 2003, Russia's Mandate Commission had approved at least 26 civilian candidates as cosmonauts from various design bureaus and agencies. This included six Soviet journalists in 1990, none of whom flew.

From 2003 to 2010, only one new group of cosmonauts was selected. It included five Russian Air Force officers and two Russian civilians.

Tim Chamberlin

See also: Almaz, Apollo-Soyuz Test Project, Extravehicular Activity, *International Space Station*, *Mir*, Salyut, Shuttle-Mir, Soyuz, Voskhod, Vostok

Bibliography

Rex D. Hall and David J. Shayler, *The Rocket Men: Vostok and Voskhod, the First Soviet Manned Spaceflights* (2001).

Rex D. Hall et al., *Russia's Cosmonauts: Inside the Yuri Gagarin Training Center* (2005).

Asif A. Siddiqi, *Challenge to Apollo: The Soviet Union and the Space Race, 1945–1974* (2000).

Disasters in Human Spaceflight

Disasters in human spaceflight and accidents have marred the history of human space exploration since the 1960s, resulting in the deaths of astronauts and cosmonauts while training and during spaceflight missions.

The Space Shuttle Challenger *breaks apart on 28 January 1986, a few seconds after a ruptured O-ring in the right solid rocket booster caused an explosion. (Courtesy NASA/Kennedy Space Center)*

The Soviet Union's human spaceflight program suffered its first fatality during a training exercise on 23 March 1961. Cosmonaut Valentin Bondarenko died as a result of a fire that broke out inside the pure-oxygen environment of a pressurized chamber during an altitude test. Ted Freeman was the first American astronaut killed after a midair mishap on 31 October 1964. He died when his T-38 jet crashed on approach to Ellington Air Force Base, Houston, Texas, after his cockpit canopy was struck and shattered by a snow goose. Less than two years later astronauts Elliott See and Charles Bassett, the prime crew for *Gemini 9*, were killed when their T-38 jet crashed into the McDonnell-Douglas factory in St. Louis, Missouri, on 28 February 1966. Poor weather and pilot error were cited as contributing to the accident.

Three more astronauts were killed on Launch Pad 34 at Kennedy Space Center (KSC), Florida, on 27 January 1967. The crew of the Apollo-Saturn 204 (AS-204) mission, Commander Virgil "Gus" Grissom with pilots Edward White and Roger Chaffee, entered their command module for a launch rehearsal with mission controllers. The capsule was sealed and pressurized with 100 percent oxygen. Soon after, controllers heard astronauts' cries of fire in the spacecraft. Within seconds, the crew was dead. An investigation concluded the fire started when a wiring short sparked, spreading flames quickly in the pure-oxygen environment. The astronauts died from asphyxia due to inhalation of toxic gases. A NASA-appointed investigative board made many recommendations after a review of the accident that led to major design and engineering modifications to the Apollo spacecraft. The resulting changes, including a two-gas pressurization system and simplified emergency egress procedures, were crucial to the safety and success of later flights. The AS-204 mission was eventually designated *Apollo 1* in honor of the crew.

The tragic loss of the first Apollo crew slowed the progress of the American lunar program and gave the Soviet Union an opportunity to regain some of the ground it lost after several setbacks and failures to its own space program. Any hope the Soviet Union had of gaining on the United States evaporated when tragedy struck its maiden Soyuz flight. *Soyuz 1* launched from the Baikonur Cosmodrome on 23 April 1967 with cosmonaut Vladimir Komarov on board. Major problems occurred almost immediately. A solar panel failed to deploy, leaving the spacecraft with only half power, followed by a failed attempt to maneuver the spacecraft. On the 13th orbit, mission control demanded that Komarov abort the mission. The vehicle successfully reentered the atmosphere, but a pressure sensor failure prevented the main parachute from deploying. The backup chute was deployed, but became entangled with the drag chute. As a result, the vehicle crashed into the ground at more than 200 miles per hour, killing Komarov instantly. The disaster grounded the Soyuz program for almost two years while engineers redesigned several systems on the craft, including the parachute system.

NASA mourned the loss of another astronaut on 5 October 1967. C. C. Williams was killed when the controls of the T-38 jet he was flying jammed and the jet went into a rolling dive near Tallahassee, Florida. Williams ejected, but died when his parachute failed to open.

The Soviet Union was dealt a huge psychological blow when cosmonaut Yuri Gagarin, the first man to fly in space, died in a plane crash at age 34. On

Human Spaceflight Disasters

Date of Mishap	Mission	What Happened	Country	Crew
27 January 1967	*Apollo 1*	Fire on launch pad killed crew during rehearsal.	USA	Virgil "Gus" Grissom, Ed White, Roger Chaffee
24 April 1967	*Soyuz 1*	Cosmonaut killed after parachute failure led to crash.	Soviet Union	Vladimir Komarov
29 June 1971	*Soyuz 11*	Depressurization of crew cabin killed all onboard.	Soviet Union	Viktor Patsayev, Georgi Dobrovolsky, Vladislav Volkov
28 January 1986	*Challenger* (Space Transportation System STS-25/ 51L)	Shuttle exploded 73 seconds after takeoff.	USA	Richard Scobee, Michael Smith, Judith Resnick, Ronald McNair, Ellison Onizuka, Gregory Jarvis, Christa McAuliffe
1 February 2003	*Columbia* (STS-107)	Shuttle broke apart 16 minutes before scheduled landing.	USA	Rick Husband, William McCool, Kalpana Chawla, David Brown, Michael Anderson, Laurel Clark, Ilan Ramon

27 March 1968, Gagarin was killed along with his instructor in the crash of a MiG-15 while on a routine training flight near Moscow. It is uncertain what caused the craft to lose control and crash, but weather conditions were poor, which likely contributed to the inability of Gagarin or the instructor to correct the situation prior to the crash. Gagarin, who was a national hero, attracted hundreds of thousands of mourners to his funeral.

The inherent risk of spaceflight led to the deaths of three more cosmonauts as they completed their *Soyuz 11* mission. Launched on 6 June 1971 and at the end of a record 23-day mission on board the first Soviet space station, *Salyut 1*, the crew (Commander Georgi Dobrovolsky and pilots Vladislav Volkov and Viktor Patsayev) fired explosive bolts to separate the reentry capsule from the rest of the spacecraft. The force of the separation accidentally opened a cabin pressure relief valve. Miles above Earth, precious air leaked out, causing rapid depressurization of the crew cabin and the death of all three cosmonauts, who were not wearing pressure suits. While the *Soyuz 11* capsule returned safely to Earth, recovery crews could not resuscitate the cosmonauts. As a result of this tragedy, designers introduced a protective pressure suit for crewmembers, which all Soyuz crews have since worn during launch, reentry, and docking activities. The Soviet Union did not return to *Salyut 1*, and it was more than two years before it attempted another human spaceflight mission.

The two greatest catastrophic accidents in human spaceflight involved the Space Shuttle *Challenger*, during Space Transportation System (STS)-51L in 1986, and the Space Shuttle *Columbia*, during mission STS-107 in 2003. *Challenger* lifted off from KSC on an unusually cold morning on 28 January 1986. The *Challenger* crew consisted of mission commander Francis R. Scobee; pilot Michael J. Smith; mission specialists Ronald E. McNair, Ellison S. Onizuka, and Judith A. Resnik; and payload specialists Gregory B. Jarvis and Christa McAuliffe, the first teacher selected to fly in space. At 73 seconds into the mission, the vehicle exploded, killing all seven crewmembers. The Rogers Commission investigating the accident blamed faulty O-ring seals situated between joints in the solid rocket boosters. The cold weather made the usually flexible rings stiff so they did not seal properly, allowing hot gases to burn through one of the boosters and into the external fuel tank, igniting the liquid oxygen and hydrogen. The disaster grounded the Shuttle fleet for more than two years and resulted in launch delays for several robotic spacecraft and satellites. After the accident, NASA enforced a new policy not to carry commercial satellites on the Shuttle and acknowledged the need for a mixed U.S. fleet of expendable and reusable launchers.

Columbia launched on 16 January 2003. The *Columbia* crew consisted of Commander Rick Husband; pilot Willie McCool; mission specialists Kalpana Chawla, Laurel Clark, Mike Anderson, and David Brown; and Israeli payload specialist Ilan Ramon. At 81 seconds into the launch, a piece of insulating foam on the external fuel tank broke loose, striking the orbiter's left wing. In subsequent discussions during the mission, Space Shuttle managers determined that the foam strike posed no threat to the crew or vehicle. On 1 February, shortly after reentry, Mission Control lost contact with the crew. Approximately 16 minutes from landing, *Columbia* had broken apart, killing all seven crewmembers and scattering debris across Texas and Louisiana. The Columbia Accident Investigation Board (CAIB) found that the foam strike created a hole on the leading edge of the orbiter's left wing, allowing hot gases to enter the wing and melt structural components—causing the eventual structural breakup of the vehicle. The board cited NASA's culture, safety program, and flawed decision making as factors contributing to the mishap. Spurred by the recommendations of the CAIB, NASA redesigned the external fuel tank, established an independent safety center, and implemented changes designed to address cultural and decision-making problems.

Lisa M. Reed

See also: Apollo, Soyuz, Space Shuttle

Bibliography

Columbia Accident Investigation Board (CAIB) Final Report, August 2003 (2003).
David J. Shayler, *Disasters and Accidents in Manned Spaceflight* (2000).
Asif A. Siddiqi, *Challenge to Apollo: The Soviet Union and the Space Race, 1945–1974* (2000).

Educator-astronaut

Educator-astronaut, an incarnation of U.S. Teacher in Space Project (TISP), is a teacher who has been selected to become a Space Shuttle mission specialist. President Ronald Reagan announced the start of TISP on 27 August 1984. NASA received more than 12,000 applications from teachers as part of the space agency's effort to rekindle the American public's interest in human spaceflight. Sharon Christa McAuliffe, a high school social studies teacher from New Hampshire, and Barbara Morgan, an elementary school teacher from Idaho, were among the applicants selected in July 1985. An apple on the mission patch designated TISP.

McAuliffe was scheduled to teach lessons during live television transmissions on board Space Shuttle *Challenger* flight 51-L. She was to become the first citizen passenger to fly in space as part of the crew. Shortly after takeoff on 28 January 1986, McAuliffe was killed after an explosion destroyed the *Challenger*. TISP was grounded along with the Shuttle fleet after the mishap. Morgan, McAuliffe's backup, remained part of NASA's Education Office as a Teacher in Space designee from March to July 1986. Other Teacher in Space finalists, termed space ambassadors, helped NASA develop additional space education programming for public schools during the years following the *Challenger* accident.

In 1992 TISP became the Teaching from Space Program (TFSP), reflecting a change in priority. TISP focused on the teacher as a civilian participant. TFSP focused more on educational activities taught by the entire Shuttle crew and broadcast on NASA television.

Morgan returned to NASA in 1998 as the first educator-astronaut. After two years of training, she was assigned to the Astronaut Office Space Station Operations Branch at Johnson Space Center and later became a prime communicator at Mission Control with on-orbit crews of the *International Space Station*. She eventually flew in space on board Space Shuttle *Endeavour* flight STS-118 in 2007 and retired from NASA a year later.

In 2002 NASA Administrator Sean O'Keefe announced changes leading to the development of the Educator Astronaut Program, including the selection of a new class of educator-astronauts to be fully trained as mission specialists. Students nominated more than 8,800 teachers to the Educator Astronaut Program, and more than 1,100 applications were completed. The astronaut candidate class of 2004 included the first three mission specialist-educators (MS-E): Joseph Acaba, an Earth science teacher from Inglewood, California; Richard Arnold II, a science and math teacher from Bucharest, Romania; and Dorothy Metcalf-Lindenburger, a science teacher from Vancouver, Washington. Their primary job was to connect missions with classroom activities and inspire the next generation of space explorers.

Melinda Marsh

See also: Space Shuttle

Bibliography

Colin Burgess, *Teacher in Space* (2000).
NASA's Educator Astronaut Program. http://education.ssc.nasa.gov/neap.asp.

International Astronauts and Guest Flights

International astronauts and guest flights were ushered into the human spaceflight programs of the Soviet Union and the United States in the late 1970s to broaden cooperative efforts in space science and strengthen international relations. The first citizen outside the Soviet Union and the United States to fly into space was Vladimir Remek, a cosmonaut from Czechoslovakia who flew in March 1978 aboard *Soyuz 28* to *Salyut 6* as part of the Interkosmos program. The Soviet Union created Interkosmos to promote collaborative efforts among countries of the former communist bloc and financed the training and launch costs of each of the foreign participants, called cosmonaut-researchers.

Critics of the Interkosmos program argued that it was a propaganda stunt orchestrated by the Soviet government in response to U.S. plans to fly European astronauts on Space Shuttle missions equipped with the *Spacelab* module. The U.S. human spaceflight program remained grounded during development of its Space Shuttle program, while the Interkosmos missions were flown. Interkosmos enabled cosmonaut-researchers to fly to space on nine missions from 1978 to 1981. Among the countries that participated after Remek's flight were Bulgaria, Cuba, the former German Democratic Republic (East Germany), Hungary, Mongolia, Poland, Romania, and Vietnam.

After 1981 the Soviet Union signed bilateral agreements with other countries to allow international astronauts to accompany cosmonauts on flights to Soviet space stations. This included India (Rakesh Sharma was launched aboard *Soyuz T-11* to *Salyut 7* on 3 April 1984), France (Jean-Loup Chretien was launched aboard *Soyuz T-6* to *Salyut 7* on 24 June 1982—Chretien also flew to *Mir* aboard *Soyuz TM-7* on 26 November 1988 where he stayed for 26 days), Syria (Mohammed Faris was

Interkosmos Program Flights*

Flight Name	Launch Date	Country Involved	International Crew Member
Soyuz 28	2 March 1978	Czechoslovakia	Vladimir Remek
Soyuz 30	27 June 1978	Poland	Miroslav Hermaszewski
Soyuz 31	26 August 1978	East Germany	Sigmund Jäehn
*Soyuz 33***	10 April 1979	Bulgaria	Georgi Ivanov
Soyuz 36	26 May 1980	Hungary	Bertalan Farkas
Soyuz 37	23 July 1980	Vietnam	Pham Tuân
Soyuz 38	18 September 1980	Cuba	Arnaldo Tamayo-Mendez
Soyuz 39	22 March 1981	Mongolia	Zhugderdemidiyn Gurragcha
Soyuz 40	14 May 1981	Romania	Dumitru Prunariu

*Interkosmos program flights docked with the *Salyut 6* space station.
**Mission was aborted after the Soyuz crew failed to dock with the space station.

launched aboard *Soyuz TM-3* to *Mir* on 22 July 1987), and Afghanistan (Abdul Ahad Mohmand was launched aboard *Soyuz TM-6* to *Mir* on 29 August 1988). Training for international cosmonauts took place at the Gagarin Cosmonaut Training Center (TsPK) in Star City, Russia.

In 1991 *Soyuz TM-13* was launched with the first cosmonaut from Kazakhstan, Toktar Aubakirov. Aubakirov was added at the last minute and flew partly as an effort by Russia to encourage the new independent Kazakhstani government to continue allowing launches from the Baikonur Cosmodrome. The first Austrian cosmonaut, Franz Viehboeck, also was a crewmember of *Soyuz TM-13*. The Austrian government paid Russia $7 million for his seat on the flight.

The European Space Agency selected in 1978 its first group of astronauts—Ulf Merbold (West Germany), Claude Nicollier (Switzerland), and Wubbo Ockels (the Netherlands)—to train as payload specialists on the first Space Shuttle flight with the European-built *Spacelab*. The three started training in Houston, Texas, in 1980 and were certified as NASA payload specialists in 1981. In 1983 Merbold was the first non-American to fly on board the Space Shuttle on Space Transportation System (STS)-9.

The first Canadian in space was payload specialist Marc Garneau, who operated the CANEX experiments during STS-41G in 1984. As of 2010, nine Canadians have flown in space.

In 1990 a Japanese television station paid Russia $12–14 million to fly reporter-cosmonaut Toyohiro Akiyama on *Soyuz TM-11* to *Mir* for seven days. This was the first commercially funded spaceflight. During the flight, Akiyama made daily television broadcasts.

Helen Sharman won a British contest and flew aboard *Soyuz TM-12* in May 1991 to *Mir*. She was the United Kingdom's first female citizen to fly into space as part of the commercially funded Project Juno.

The United States also opened its doors to international astronauts to foster cooperation with other countries. Saudi Prince Sultan Salman Abdul aziz Al-Saud served as a payload specialist on STS-51G in 1985. The same year, Rodolfo Neri-Vela from Mexico flew as part of the crew of STS-61B.

The first elected official to fly in space was Sen. E. J. "Jake" Garn (R-Utah) aboard STS-51D in 1985. Garn was chairperson of the Senate committee with oversight of NASA's budget. Rep. Bill Nelson (D-Florida) followed Garn aboard STS-61C in 1986, the last Shuttle flight before the *Challenger* accident.

The first teacher selected to fly in space, Christa McAuliffe, perished aboard *Challenger* in January 1986 along with the rest of the crew. Many international astronauts selected as payload specialists lost their chance to fly on the Shuttle as a result of the *Challenger* disaster. NASA became more cautious about involving non-career astronauts, especially international astronauts, on flights. Many years passed before NASA reconsidered sending non-U.S. astronauts to space on the Shuttle.

The Shuttle-Mir program, negotiated after the collapse of the Soviet Union, signified a new partnership between the United States and Russia. The program paved the way for cosmonauts to travel on the Space Shuttle and allowed astronauts to travel on Soyuz vehicles and live on *Mir*. In 1994 cosmonaut Sergei Krikalev became the first Russian cosmonaut to fly on the Space Shuttle. One year later Norman Thagard became the first American astronaut to fly on a Soyuz launch vehicle and stay on *Mir*.

The first Israeli astronaut, Ilan Ramon, was selected in 1997 as a payload specialist. He flew on STS-107 in 2003, a dedicated science-and-research mission. He died along with the rest of the crew when the Space Shuttle *Columbia* disintegrated during reentry, 16 minutes before the scheduled landing after a 15-day flight.

Pablo de Leon

See also: Interkosmos, *Mir*, Salyut, Shuttle-Mir, Soyuz, Space Shuttle, Tourism

Bibliography

Bryan Burrough, *Dragonfly: NASA and the Crisis aboard Mir* (1998).
Michael Cassutt, *Who's Who in Space* (1993).
Dennis Newkirk, *Almanac of Soviet Manned Space Flight* (1990).
David Shayler, *Disasters and Accidents in Manned Spaceflight* (2000).
Robert Zimmerman, *Leaving Earth: Space Stations, Rival Superpowers, and the Quest for Interplanetary Travel* (2003).

Mission Specialists. *See* Astronauts.

Payload Specialists. *See* Astronauts.

Spationaut. *See* International Astronauts and Guest Flights.

Taikonaut. *See* International Astronauts and Guest Flights.

Teacher in Space Project. *See* Educator-Astronaut.

Women Astronauts

Women astronauts have come from the former Soviet Union, followed by the United States, the United Kingdom, Canada, Japan, and France. While the United States was the second to place women in orbit, NASA hired more women to become astronauts than all other space agencies combined. Only women from the former Soviet Union, the United States, and France have qualified as pilots. Belgium selected a woman to train as an astronaut in 1992, but she did not fly.

The history of women in space reflects Cold War politics and late-twentieth-century changes in gender roles and expectations. Soviet sport parachutist Valentina Tereshkova became the first woman in space on 16 June 1963 on board *Vostok 6*. She had been one of five training for a mission that Soviet Premier Nikita Khrushchev had authorized to demonstrate the equality of women under communism. In contrast, NASA at the time chose only military jet test pilots. Because women were legally forbidden to fly for U.S. military services, they were prevented from qualifying as astronaut candidates. However, aeromedical specialist Randolph Lovelace II, U.S. Air Force Brigadier General Don Flickinger, and aviator Jacqueline Cochran privately evaluated a group of experienced female pilots between February 1960 and

Astronaut Eileen Collins became the first female Space Shuttle commander during STS-93 in July 1999. She is shown in the flight deck of Space Shuttle Columbia *during the mission. (Courtesy NASA/Johnson Space Center)*

July 1961 based on the physical and psychological standards established for Project Mercury astronauts. Thirteen passed these Mercury pretraining examinations. As a result, NASA Administrator James Webb named the first woman to complete the tests, Jerrie Cobb, as his special consultant, but NASA never selected a woman to train as an astronaut for the Mercury, Gemini, or Apollo projects.

The Soviet Union continued to use gender competitively. *Voskhod 5* was intended to be an all-female mission in 1966 with a space walk, but it was dropped for political and technical reasons. Aerobatic pilot Svetlana Savitskaya became the second woman to fly in space aboard *Soyuz T-7* in August 1982. Two women served as backups for that mission. Savitskaya flew again in July 1984, making the first space walk by a woman four months before American Kathryn Sullivan. Savitskaya was also the first woman aboard a space station, *Salyut 7*.

NASA's policy of selecting only men for its astronaut corps stood until 1978 when the space agency selected six women to join the class of Astronaut Group Eight. This included Anna Fisher, Shannon Lucid, Judith Resnik, Sally Ride, Margaret Rhea Seddon, and Sullivan. Ride became the first American woman to fly in space on 18 June 1983 aboard the Space Shuttle *Challenger* during mission Space Transportation System (STS)-7. Throughout the Space Shuttle era, the required credentials for female and male astronauts were the same. Most had doctorates in the sciences, engineering, or medicine, plus research experience. Others were scientists and engineers from industry, the military, or NASA, typically with graduate degrees. The two female NASA Shuttle pilots, through 2005, were military aviators.

VALENTINA VLADIMIROVNA TERESHKOVA (1937–)

Valentina Tereshkova, a Soviet Union cosmonaut, became the first woman to travel in space on 16 June 1963, aboard *Vostok 6*. Her historic flight served to enhance the prestige of the Soviet space program. The flight duration surpassed all U.S. spaceflights combined up to that point. She married *Vostok 3* pilot Andriyan G. Nikolaev on 3 November 1963 in the first state-hosted wedding in Soviet history. Tereshkova worked as a chairperson of the government-sponsored organization for international cultural exchange after graduation from the Zhukovsky Air Force Academy in 1969. She officially retired from the military and the cosmonaut corps in 1997 as an Air Force major-general.

Peter A. Gorin

Six women qualified for spaceflights outside the standard career path. Chemist Helen Sharman of Mars Confectionary, Ltd., won a British contest and flew aboard *Soyuz TM-12* in May 1991 to the Soviet space station *Mir* and carried out medical and agricultural research. Christa McAuliffe, a New Hampshire high school history teacher, won a NASA competition to become the first teacher in space, but was killed in the 1986 *Challenger* accident. Montana grade school teacher Barbara Morgan, McAuliffe's backup, became the first educator-astronaut in 1998. Dorothy Metcalf-Lindenburger, a science teacher from Vancouver, Washington, was one of three educator mission specialists hired in 2004. Two journalists, Japanese Ryoko Kikuchi and Soviet Svetlana Omelchenko became cosmonauts in 1989 and 1990, but the Soviet journalist-in-space program was canceled when the Soviet Union broke up. Kikuchi was a backup and did not fly.

Four women, all Americans, have died in space. Resnick and McAuliffe perished on the *Challenger*; astronauts Laurel Clark and Kalpana Chawla were killed during the *Columbia* reentry accident in 2003.

Several honors have been bestowed to women for their contributions to human spaceflight. The Congressional Space Medal of Honor was given to Lucid in 1996, for her 188-day stay on *Mir*, and to Resnick and McAuliffe posthumously in 2004 as *Challenger* crewmembers. Clark, Chawla, and others aboard *Columbia* received the Congressional Gold Medal of Honor posthumously in 2004. The former Soviet Union granted Tereshkova, Savitskaya, and cosmonaut Elena Kondakova Hero of the Soviet Union status. Sharman was selected an officer in the Order of the British Empire. French astronaut Claudie Haigneré and Shuttle pilot Eileen Collins received France's Legion of Honor.

In July 1999 Collins became the first female commander of a Shuttle mission, STS-93. That same month Haigneré became the first woman to qualify as a Soyuz return commander. Astronauts Susan Helms and Peggy Whitson were the only women to have served as flight engineers on the *International Space Station* (*ISS*). Whitson was also named the first *ISS* science officer in September 2002.

SALLY KRISTEN RIDE
(1951–)

(Courtesy National Archives)

Sally Ride, an astronaut, became the first U.S. woman to travel in space on 18 June 1983 aboard the Space Shuttle *Challenger* during Space Transportation System (STS)-7. She achieved celebrity status during the flight and played a pivotal role in progress for women. She served on the Presidential Commission on the Space Shuttle *Challenger* Accident in 1986 and filed the "Ride Report" in 1987, recommending future missions for NASA. She left the agency in 1987 and became a professor of physics at the University of California, San Diego, in 1989. She served on the *Columbia* Accident Investigation Board in 2003.

Tim Chamberlin

Many astronauts and cosmonauts who left active status resumed research careers; some chose a different role in government. Tereshkova, Savitskaya, and Kondakova were elected to the Soviet parliament. Haigneré was appointed France's Minister for Research and New Technologies in June 2002. About one-fifth of the Americans assumed NASA management jobs, some in a position to influence Moon and Mars mission planning.

Maura Phillips Mackowski

See also: *Mir*, Soyuz, Space Shuttle, Vostok

Bibliography

JSC Astronaut. http://www.jsc.nasa.gov/Bios/.
NASA, *Astronaut Flight Book* (2005).
Asif A. Siddiqi, *Challenge to Apollo: The Soviet Union and the Space Race* (2000).
Margaret A. Weitekamp, *Right Stuff, Wrong Sex: America's First Women in Space Program* (2004).

Human Spaceflight Missions by Country, through February 2011 (Part 1 of 51)

CHINA					
Shenzhou					
Mission	**Launch**	**Landing**	**Duration**	**Crew**	**Highlights**
Shenzhou 5	15 Oct 03	15 Oct 03	21h, 23m	Yang Liwei (P)	China's first piloted spaceflight
Shenzhou 6	12 Oct 05	17 Oct 05	4d, 19h, 32m	Fei Junlong (C), Nie Haisheng (O)	China's first two-person crew in space
Shenzhou 7	25 Sep 08	28 Sep 08	2d, 20h, 28m	Zhai Zhigang (C), Liu Buoming (O), Jing Haipen (O)	China's first three-person crew in space; Zhai performs first EVA by a taikonaut

SOVIET UNION / RUSSIA					
Vostok					
Mission	**Launch**	**Landing**	**Duration**	**Crew**	**Highlights**
Vostok 1	12 Apr 61	12 Apr 61	1h, 48m	Yuri Gagarin (P)	First human spaceflight
Vostok 2	06 Aug 61	7 Aug 61	1d, 1h, 18m	Gherman Titov (P)	First day long flight
Vostok 3	11 Aug 62	15 Aug 62	3d, 22h, 22m	Andrian Nikolayev (P)	Joint flight with *Vostok 4*
Vostok 4	12 Aug 62	15 Aug 62	2d, 22h, 57m	Pavel Popovich (P)	Joint flight with *Vostok 2*
Vostok 5	14 Jun 63	19 Jun 63	4d, 23h, 8m	Valeri Bykovsky (P)	Joint flight with *Vostok 6*
Vostok 6	16 Jun 63	19 Jun 63	2d, 22h, 51m	Valentina Tereshkova (P)	First woman to fly in space
Voskhod					
Mission	**Launch**	**Landing**	**Duration**	**Crew**	**Highlights**
Voskhod 1	12 Oct 64	13 Oct 64	1d, 0h, 17m	Vladimir Komarov (C), Konstantin Feoktistov (E), Boris Yegorov (X)	First three-person crew in space
Voskhod 2	18 Mar 65	19 Mar 65	1d, 2h, 2m	Pavel Belyayev (C) Alexei Leonov (R)	Leonov conducts first extra vehicular activity (EVA)

Human Spaceflight Missions by Country, through February 2011 (Part 2 of 51)

Soyuz				
Mission	**Launch, Landing**	**Duration**	**Crew**	**Highlights**
Soyuz 1	23 Apr 67, 24 Apr 67	1d, 2h, 48m	Vladimir Komarov (C)	Spacecraft's parachute failed to deploy properly, killing Komarov on landing
Soyuz 3	26 Oct 68, 30 Oct 68	3d, 22h, 51m	Georgi Beregovoi (C)	Rendezvous with uncrewed *Soyuz 2*; docking attempt failed
Soyuz 4	14 Jan 69, 17 Jan 69	2d, 23h, 21m	Vladimir Shatalov (C), Aleksei Yeliseyev (E) (dn), Yevgeni Khrunov (R) (dn)	Docks with *Soyuz 5*; first docking of two piloted spacecraft
Soyuz 5	15 Jan 69, 18 Jan 69	3d, 0h, 54m	Boris Volynov (C), Aleksei Yeliseyev (E) (up), Yevgeni Khrunov (R) (up)	Docks with *Soyuz 4*; Yeliseyev, Khrunov conduct EVA to *Soyuz 4*
Soyuz 6	11 Oct 69, 16 Oct 69	4d, 22h, 43m	Georgi Shonin (C), Valeri Kubasov (E)	Kubasov performs first welding of metals in space
Soyuz 7	12 Oct 69, 17 Oct 69	4d, 22h, 40m	Anatoli Filipchenko (C), Vladislav Volkov (E), Viktor Gorbatko (R)	Rendezvous with *Soyuz 6* and *Soyuz 8*; docking with *Soyuz 8* fails
Soyuz 8	13 Oct 69, 18 Oct 69	4d, 22h, 51m	Vladimir Shatalov (C), Aleksei Yeliseyev (E)	Rendezvous with *Soyuz 7* and *Soyuz 8*
Soyuz 9	1 Jun 70, 19 Jun 70	17d, 16h, 59m	Andriyan Nikolayev (C), Vitali Sevastyanov (E)	First human spaceflight launch conducted at night
Soyuz 10	22 Apr 71, 24 Apr 71	1d, 23h, 46m	Vladimir Shatalov (C), Aleksei Yeliseyev (E), Nikolai Rukavishnikov (R)	Docks with *Salyut 1*; could not enter station due to faulty hatch on *Soyuz*
Soyuz 11	6 Jun 71, 29 Jun 71	23d, 18h, 22m	Georgi Dobrovolsky (C), Vladislav Volkov (E), Viktor Patsayev (R)	First and only resident crew of *Salyut 1*; crew dies during reentry accident
Soyuz 12	27 Sep 73, 29 Sep 73	1d, 23h, 16m	Vasili Lazarev (C), Oleg Makarov (E)	Test flight of new *Soyuz* capsule
Soyuz 13	18 Dec 73, 26 Dec 73	7d, 20h, 56m	Pyotr Klimuk (C), Valentin Lebedev (E)	Crew uses Orion 2 astrophysical camera for imaging stars and Earth
Soyuz 14	3 Jul 74, 19 Jul 74	15d, 17h, 30m	Pavel Popovich (C), Yuri Artyukhin (E)	First and only resident crew of *Salyut 3*
Soyuz 15	26 Aug 74, 28 Aug 74	2d, 0h, 12m	Gennadi Sarafanov (C), Lev Demin (E)	Fails to dock with *Salyut 3*

(Continued)

Human Spaceflight Missions by Country, through February 2011 (Part 3 of 51)

Soyuz				
Mission	**Launch, Landing**	**Duration**	**Crew**	**Highlights**
Soyuz 16	2 Dec 74, 8 Dec 74	5d, 22h, 24m	Anatoli Filipchenko (C), Nikolai Rukavishnikov (C)	Test flight of a Soyuz spacecraft in preparation for *ASTP*
Soyuz 17	10 Jan 75, 9 Feb 75	29d, 13h, 20m	Aleksey Gubarev (C), Georgi Grechko (E)	First resident crew of *Salyut 4*
Soyuz 18A	5 Apr 75, 5 Apr 75	21m:27s	Vasili Lazarev (C), Oleg Makarov (E)	Launch to *Salyut 4* aborted as a result of stage one rocket failure
Soyuz 18B	24 May 75, 26 Jul 75	62d, 23h, 20m	Pyotr Klimuk (C), Vitali Sevastyanov (E)	Second resident crew of *Salyut 4*
Soyuz 19 Apollo-Soyuz Test Project (ASTP)	15 Jul 75, 21 Jul 75	5d, 22h, 31m	Alexei Leonov (C), Valeri Kubasov (E)	First joint mission between the United States and the Soviet Union
Soyuz 21	6 Jul 76, 24 Aug 76	49d, 6h, 24m	Boris Volynov (C), Vitali Zholobov (E)	First resident crew of *Salyut 5*; mission cut short after fumes build up in station
Soyuz 22	15 Sep 76, 23 Sep 76	7d, 21h, 52m	Valeri Bykovsky (C), Vladimir Aksyonov (E)	Earth observation mission with MKF 6 multispectral camera
Soyuz 23	14 Oct 76, 16 Oct 76	2d, 0h, 7m	Vyacheslav Zudov (C), Valeri Rozhdestvensky (E)	Fails to dock with *Salyut 5*; mission cut short due to docking system failure
Soyuz 24	7 Feb 77, 25 Feb 77	17d, 17h, 26m	Viktor Gorbatko (C), Yuri Glazkov (E)	Second resident crew of *Salyut 5*
Soyuz 25	9 Oct 77, 11 Oct 77	2d, 0h, 45m	Vladimir Kovalyonok (C), Valeri Ryumin (E)	Fails to dock with *Salyut 6*; mission cut short due to docking system failure
Soyuz 26	10 Dec 77, 16 Jan 78	37d, 10h, 6m	Yuri Romanenko (C) (up), Georgi Grechkon (E) (up), Vladimir Dzhanibekov (dn), Oleg Makarov (dn)	First resident crew of *Salyut 6* (Romanenko, Grechkon)
Soyuz 27	10 Jan 78, 16 Mar 78	64d, 22h, 53m	Vladimir Dzhanibekov (C) (up), Oleg Makarov (E) (up), Yuri Romanenko (dn), Georgi Grechkon (dn)	First guest crew of *Salyut 6*; first space station dual occupancy

Human Spaceflight Missions by Country, through February 2011 (Part 4 of 51)

Soyuz				
Mission	**Launch, Landing**	**Duration**	**Crew**	**Highlights**
Soyuz 28	2 Mar 78, 10 Mar 78	7d, 22h, 16m	Aleksey Gubarev (C), Vladimir Remek (R)	First Interkosmos mission (*Salyut 6*)
Soyuz 29	15 Jun 78, 3 Sep 78	79d, 15h, 24m	Vladimir Kovalyonok (C) (up), Aleksandr Ivanchenkov (E) (up), Vladimir Bykovsky (dn), Sigmund Jaehn (dn)	Second resident crew of *Salyut 6* (Kovalyonok, Ivanchenkov)
Soyuz 30	27 Jun 78, 5 Jul 78	7d, 22h, 3m	Pyotr Klimuk (C), Miroslav Hermaszewski (R)	Second Interkosmos mission (*Salyut 6*)
Soyuz 31	26 Aug 78, 2 Nov 78	67d, 20h, 13m	Vladimir Bykovsky (C) (up), Sigmund Jaehn (C) (up), Vladimir Kovalyonok (dn), Aleksandr Ivanchenkov (dn)	Third Interkosmos mission (*Salyut 6*)
Soyuz 32	25 Feb 79, 13 Jun 79	108d, 4h, 25m	Vladimir Lyakhov (C) (up), Valeri Ryumin (E) (up), uncrewed Soyuz (dn)	Third resident crew of *Salyut 6*
Soyuz 33	10 Apr 79, 12 Apr 79	1d, 23h, 1m	Nikolai Rukavishnikov (C), Georgi Ivanov (R)	Fourth Interkosmos mission; fails to dock with *Salyut 6*
Soyuz 34	6 Jun 79, 19 Aug 79	73d, 18h, 17m	Uncrewed *Soyuz* (up), Vladimir Lyakhov (dn), Valeri Ryumin (dn)	For use by *Soyuz 32* crew for return due to *Soyuz 33* mission failure
Soyuz 35	19 Apr 80, 3 Jun 80	55d, 1h, 28m	Leonid Popov (C) (up), Valeri Ryumin (E) (up), Valeri Kubasov (dn), Bertalan Farkas (dn)	Fourth resident crew of *Salyut 6* (Popov, Ryumin)
Soyuz 36	26 May 80, 31 Jul 80	65d, 20h, 54m	Valeri Kubasov (C) (up), Bertalan Farkas (R) (up), Viktor Gorbatko (dn), Pham Tuan (dn)	Fifth Interkosmos mission (*Salyut 6*)
Soyuz T 2	5 Jun 80, 9 Jun 80	3d, 22h, 20m	Yuri Malyshev (C), Vladimir Aksyonov (E)	Test flight of new Soyuz spacecraft; docks with *Salyut 6*
Soyuz 37	23 Jul 80, 11 Oct 80	79d, 15h, 17m	Viktor Gorbatko (C) (up), Pham Tuan (R) (up), Leonid Popov (dn), Valeri Ryumin (dn)	Sixth Interkosmos mission (*Salyut 6*)
Soyuz 38	18 Sep 80, 26 Sep 80	7d, 20h, 43m	Yuri Romanenko (C), Arnaldo Tamayo Mendez (R)	Seventh Interkosmos mission (*Salyut 6*)

(Continued)

Human Spaceflight Missions by Country, through February 2011 (Part 5 of 51)

Soyuz				
Mission	**Launch, Landing**	**Duration**	**Crew**	**Highlights**
Soyuz T 3	27 Nov 80, 10 Dec 80	12d, 19h, 8m	Leonid Kizim (C), Oleg Makarov (E), Gennadi Strekalov (E)	*Salyut 6* repair and upgrade mission
Soyuz T 4	12 Mar 81, 26 May 81	74d, 17h, 37m	Vladimir Kovalyonok (C), Viktor Savinykh (E)	Fifth resident crew of *Salyut 6*
Soyuz 39	22 Mar 81, 30 Mar 81	7d, 20h, 42m	Vladimir Dzhanibekov (C), Zhugderdemidiyn Gurragcha (R)	Eighth Interkosmos mission (*Salyut 6*)
Soyuz 40	14 May 81, 22 May 81	7d, 21h, 42m	Leonid Popov (C), Dumitru Prunariu (R)	Ninth Interkosmos mission (*Salyut 6*)
Soyuz T 5	13 May 82, 27 Aug 82	106d, 5h, 6m	Anatoli Berezovoi (C) (up), Valentin Lebedev (E) (up), Leonid Popov (dn), Aleksandr Serebrov (dn), Svetlana Savitskaya (dn)	First resident crew of *Salyut 7* (Berezovoi, Lebedev)
Soyuz T 6	24 Jun 82, 2 Jul 82	7d, 21h, 51m	Vladimir Dzhanibekov (C), Aleksandr Ivanchenkov (E), Jean Loup Chretien (R)	Joint mission between France and the Soviet Union (*Salyut 7*)
Soyuz T 7	19 Aug 82, 10 Dec 82	113d, 1h, 51m	Leonid Popov (C) (up), Aleksandr Serebrov (E) (up), Svetlana Savitskaya (R) (up), Anatoli Berezovoi (dn), Valentin Lebedev (dn)	Savitskaya becomes second woman to fly in space (*Salyut 7*)
Soyuz T 8	20 Apr 83, 22 Apr 83	2d, 0h, 18m	Vladimir Titov (C), Gennadi Strekalov (E), Aleksandr Serebrov (E)	Fails to dock with *Salyut 7*; mission cut short due to docking system failure
Soyuz T 9	27 Jun 83, 23 Nov 83	149d, 10h, 46m	Vladimir Lyakhov (C), Aleksandr Aleksandrov (E)	Crew docks with and enters *Salyut 7*; station's second resident crew
Soyuz T 10A	26 Sep 83, 26 Sep 83	05m:13s	Vladimir Titov (C), Gennadi Strekalov (E)	Launch pad fire forces first use of emergency escape tower
Soyuz T 10B	8 Feb 84, 11 Apr 84	62:22h, 41m	Leonid Kizim (C) (up), Vladimir Solovyov (E) (up), Oleg Atkov (X) (up), Yuri Malyshev (dn), Gennadi Strekalov (dn), Rakesh Sharma (dn)	Third resident crew of *Salyut 7* (Kizim, Solovyov, Atkov)

Human Spaceflight Missions by Country, through February 2011 (Part 6 of 51)

Soyuz				
Mission	**Launch, Landing**	**Duration**	**Crew**	**Highlights**
Soyuz T 11	3 Apr 84, 2 Oct 84	181d, 21h, 49m	Yuri Malyshev (C) (up), Gennadi Strekalov (E) (up), Rakesh Sharma (R) (up), Leonid Kizim (dn), Vladimir Solovyov (dn), Oleg Atkov (dn)	Joint mission between India and Soviet Union (*Salyut 7*)
Soyuz T 12	17 Jul 84, 29 Jul 84	11d, 19h, 15m	Vladimir Dzhanibekov (C), Svetlana Savitskaya (E), Igor Volk (R)	Savitskaya becomes the first woman to perform a spacewalk (*Salyut 7*)
Soyuz T 13	6 Jun 85, 26 Sep 85	112d, 3h, 12m	Vladimir Dzhanibekov (C), Viktor Savinykh (E) (up), Georgi Grechko (dn)	Fourth resident crew of *Salyut 7* (Dzhanibekov, Savinykh)
Soyuz T 14	17 Sep 85, 21 Nov 85	64d, 21h, 52m	Vladimir Vasyutin (C), Aleksandr Volkov (E), Georgi Grechko (E) (up), Viktor Savinykh (dn)	Partial exchange of *Salyut 7*'s fourth resident crew
Soyuz T 15	13 Mar 86, 16 Jul 86	125d, 1m	Leonid Kizim (C), Vladimir Solovyov (E)	*Mir* 1 crew; transfer to *Salyut 7* from 5 May to 26 June
Soyuz TM 2	5 Feb 87, 30 Jul 87	174d, 3h, 26m	Yuri Romanenko (C) (up), Aleksandr Laveykin (E), Aleksandr Viktorenko (dn), Mohammed Faris (dn)	*Mir* 2 crew (Romanenko, Laveykin); Laveykin departs early due to medical condition
Soyuz TM 3	22 Jul 87, 29 Dec 87	160d, 7h, 17m	Aleksandr Viktorenko (C) (up), Aleksandr Aleksandrov (E), Mohammed Faris (R) (up), Yuri Romanenko (dn), Anatoli Levchenko (dn)	*Mir* visit; Aleksandrov replaces Laveykin as part of *Mir* 2 crew.
Soyuz TM 4	21 Dec 87, 17 Jun 88	178d, 22h, 54m	Vladimir Titov (C) (up), Musa Manarov (E) (up), Anatoli Levchenko (R) (up), Anatoliy Solovyov (dn), Viktor Savinykh (dn), Aleksandr Panayotov Aleksandrov (dn)	*Mir* 3 crew (Titov, Manarov); repair of the TTM telescope

(Continued)

Human Spaceflight Missions by Country, through February 2011 (Part 7 of 51)

Soyuz				
Mission	**Launch, Landing**	**Duration**	**Crew**	**Highlights**
Soyuz TM 5	7 Jun 88, 7 Sep 88	91d, 10h, 46m	Anatoliy Solovyov (C) (up), Viktor Savinykh (E) (up), Aleksandr Aleksandrov (R) (up), Vladimir Lyakhov (dn), Abdul Mohmand (dn)	*Mir* visit; *Soyuz* crew works on science experiments with *Mir* 3 crew
Soyuz TM 6	29 Aug 88, 21 Dec 88	114d, 5h, 34m	Vladimir Lyakhov (C) (up), Valeri Polyakov (X) (up), Abdul Mohmand (R) (up), Vladimir Titov (dn), Musa Manarov (dn), Jean Loup Chretien (dn)	*Mir* visit; Polyakov joins Titov and Manarov as member of *Mir* 3 crew
Soyuz TM 7	26 Nov 88, 27 Apr 89	151d, 11h, 8m	Aleksandr Volkov (C), Sergei Krikalev (E), Jean Loup Chretien (R) (up), Valeri Polyakov (dn)	*Mir* 4 crew (Volkov, Krikalev, Polyakov); Chretien spacewalk is first by international cosmonaut
Soyuz TM 8	5 Sep 89, 19 Feb 90	166d, 6h, 58m	Aleksandr Viktorenko (C), Aleksandr Serebrov (E)	*Mir* 5 crew; perform five EVAs to test SPK maneuvering unit
Soyuz TM 9	11 Feb 90, 9 Aug 90	179d, 1h, 18m	Anatoliy Solovyov (C), Aleksandr Balandin (E)	*Mir* 6 crew
Soyuz TM 10	1 Aug 90, 10 Dec 90	130d, 20h, 36m	Gennadi Manakov (C), Gennadi Strekalov (E), Toyohiro Akiyama (dn)	*Mir* 7 crew (Manakov, Strekalov)
Soyuz TM 11	2 Dec 90, 26 May 91	175d, 1h, 51m	Viktor Afanasyev (C), Musa Manarov (E), Toyohiro Akiyama (R) (up), Helen Sharman (dn)	*Mir* 8 crew (Afanasyev, Manarov); Akiyama (Japanese journalist) makes daily television broadcasts
Soyuz TM 12	18 May 91, 10 Oct 91	144d, 15h, 22m	Anatoli Artsebarsky (C), Sergei Krikalev (E) (up), Helen Sharman (R) (up), Toktar Aubakirov (dn), Franz Viehboeck (dn)	*Mir* 9 crew (Artsebarsky, Krikalev); Sharman from United Kingdom
Soyuz TM 13	2 Oct 91, 25 Mar 92	175d, 2h, 52m	Aleksandr Volkov (C), Toktar Aubakirov (R) (up), Franz Viehboeck (R) (up), Sergei Krikalev (dn), Klaus Dietrich Flade (dn)	*Mir* 10 crew (Volkov joins Krikalev); first Soyuz mission with two researchers

Human Spaceflight Missions by Country, through February 2011 (Part 8 of 51)

Soyuz				
Mission	**Launch, Landing**	**Duration**	**Crew**	**Highlights**
Soyuz TM 14	17 Mar 92, 10 Aug 92	145d, 14h, 11m	Aleksandr Viktorenko (C), Aleksandr Kaleri (E), Klaus Dietrich Flade (R) (up), Michel Tognini (dn)	*Mir* 11 crew (Viktorenko, Kaleri); Flade from Germany; first Russian mission after breakup of Soviet Union
Soyuz TM 15	27 Jul 92, 1 Feb 93	188d, 21h, 41m	Anatoliy Solovyov (C), Sergei Avdeyev (E), Michel Tognini (R) (up)	*Mir* 12 crew (Solovyov, Avdeyev); Togini from France
Soyuz TM 16	24 Jan 93, 22 Jul 93	179d, 0h, 44m	Gennadi Manakov (C), Aleksandr Poleshchuk (E), Jean Pierre Haignere (dn)	*Mir* 13 crew (Manakov, Poleshchuk)
Soyuz TM 17	1 Jul 93, 14 Jan 94	196d, 17h, 45m	Vasili Tsibliyev (C), Aleksandr Serebrov (E), Jean Pierre Haignere (R) (up)	*Mir* 14 crew (Tsibliyev, Serebrov); Haignere from France; *TM 17* strikes *Mir*'s Kristall module twice during departure
Soyuz TM 18	8 Jan 94, 9 Jul 94	182d, 0h, 27m	Viktor Afanasyev (C), Yuri Usachyov, (E), Valeri Polyakov (X) (up)	*Mir* 15 crew; Polyakov performs series of medical experiments during 437-day stay on Mir
Soyuz TM 19	1 Jul 94, 4 Nov 94	125d, 22h, 54m	Yuri Malenchenko (C), Talgat Musabayev (E), Ulf Merbold (dn)	*Mir* 16 crew (Malenchenko and Musabayev join Polyakov)
Soyuz TM 20	4 Oct 94, 22 Mar 95	169d, 5h, 22m	Aleksandr Viktorenko (C), Yelena Kondakova (E), Ulf Merbold (R) (up), Valeri Polyakov (dn)	*Mir* 17 crew (Viktorenko and Kondakova join Polyakov); Merbold from Germany
Soyuz TM 21	14 Mar 95, 11 Sep 95	181d, 0h, 41m	Vladimir Dezhurov (C) (up), Gennadi Strekalov (E) (up), Norman Thagard (R) (up), Anatoliy Solovyov (dn), Nikolai Budarin (dn)	*Mir* 18 crew (Dezhurov, Strekalov, Thagard); *Mir* 19 crew (Solovyov, Budarin) launched on STS 71
Soyuz TM 22	3 Sep 95, 29 Feb 96	179d, 1h, 42m	Yuri Gidzenko (C), Sergei Avdeyev (E), Thomas Reiter (R)	*Mir* 20 crew; Reiter becomes first German to perform an EVA
Soyuz TM 23	21 Feb 96, 2 Sep 96	193d, 19h, 8m	Yuri Onufrienko (C), Yuri Usachyov (E), Claudie Andre Deshays (dn)	*Mir* 21 crew (Onufrienko, Usachyov); crew joined by Shannon Lucid on 24 Mar 96 from STS 76

(Continued)

Human Spaceflight Missions by Country, through February 2011 (Part 9 of 51)

Soyuz				
Mission	**Launch, Landing**	**Duration**	**Crew**	**Highlights**
Soyuz TM 24	17 Aug 96, 2 Mar 97	196d, 17h, 26m	Valeri Korzun (C), Aleksandr Kaleri (E), Claudie Andre Deshays (R) (up), Reinhold Ewald (dn)	*Mir* 22 crew (Korzun, Kaleri join Shannon Lucid); Andre Deshays from France
Soyuz TM 25	10 Feb 97, 14 Aug 97	184d, 22h, 8m	Vasili Tsibliyev (C), Aleksander Lazutkin (E), Reinhold Ewald (R) (up)	*Mir* 23 crew (Tsibliyev and Lazutkin join Jerry Linenger); Ewald from Germany
Soyuz TM 26	5 Aug 97, 19 Feb 98	197d, 17h, 35m	Anatoliy Solovyov (C), Pavel Vinogradov (E), Leopold Eyharts (dn)	*Mir* 24 crew (Solovyov and Vinogradov join Michael Foale)
Soyuz TM 27	29 Jan 98, 25 Aug 98	207d, 12h, 51m	Talgat Musabayev (C), Nikolai Budarin (E), Leopold Eyharts (E) (up), Yuri Baturin (R) (dn)	*Mir* 25 crew (Musabayev and Budarin join Andy Thomas); Eyharts from France
Soyuz TM 28	13 Aug 98, 28 Feb 99	198d, 16h, 31m	Gennadi Padalka (C), Sergei Avdeyev (E) (up), Yuri Baturin (R) (up), Ivan Bella (dn)	*Mir* 26 crew (Padalka, Avdeyev); Baturin becomes first Russian politician to fly in space
Soyuz TM 29	20 Feb 99, 28 Aug 99	188d, 20h, 16m	Viktor Afanasyev (C), Jean Pierre Haignere (R), Ivan Bella (R) (up), Sergei Avdeyev (dn)	*Mir* 27 crew (Afanasyev and Haignere join Avdeyev)
Soyuz TM 30	4 Apr 00, 16 Jun 00	72d, 19h, 42m	Sergei Zalyetin (C), Aleksandr Kalen (E)	*Mir* 28 crew (Last resident *Mir* crew)
Soyuz TM 31	31 Oct 00, 6 May 01	186d, 21h, 49m	Yuri Gidzenko (C) (up), William Shepherd (E) (up), Sergei Krikalev (E) (up), Talgat Musabayev (dn), Yuri Baturin (dn), Dennis Tito (dn)	*ISS* 1 crew (Shepherd, Gidzenko, Krikalev); crew returned on STS 102
Soyuz TM 32	28 Apr 01, 31 Oct 01	185d, 21h, 23m	Talgat Musabayev (C) (up), Yuri Baturin (E) (up), Dennis Tito (T) (up), Victor Afanasyev (dn), Konstantin Kozeyev (dn), Claudie Haignere (dn)	*ISS* visit; Tito, an American, becomes first space tourist; *Soyuz TM 32* becomes *ISS* escape vehicle
Soyuz TM 33	21 Oct 01, 5 May 02	195d, 18h, 52m	Victor Afanasyev (C) (up), Konstantin Kozeyev (E) (up), Claudie Haignere (E) (up), Yuri Gidzenko (dn), Roberto Vittori (dn), Mark Shuttleworth (dn)	*ISS* visit; *TM 33* new lifeboat

Human Spaceflight Missions by Country, through February 2011 (Part 10 of 51)

Soyuz				
Mission	**Launch, Landing**	**Duration**	**Crew**	**Highlights**
Soyuz TM 34	25 Apr 02, 10 Nov 02	198d, 17h, 38m	Yuri Gidzenko (C) (up), Roberto Vittori (E) (up), Mark Shuttleworth (T) (up), Sergei Zalyotin (dn), Frank De Winne (dn), Yuri Lonchakov (dn)	*ISS* visit; Shuttleworth becomes first South African in space and second space tourist; *TM 34* new lifeboat
Soyuz TMA 1	30 Oct 02, 4 May 03	185d, 22h, 56m	Sergei Zalyotin (C) (up), Frank De Winne (E) (up), Yuri Lonchakov (E) (up), Nikolai Budarin (dn), Kenneth Bowersox (dn), Donald Pettit (dn)	*ISS* visit; *TMA 1* new lifeboat
Soyuz TMA 2	26 Apr 03, 27 Oct 03	183d, 22h, 47m	Yuri Malenchenko (C), Edward Lu (E), Pedro Duque (dn)	*ISS* 7 crew (Malenchenko, Lu); Lu is first American to launch and land on a Soyuz spacecraft
Soyuz TMA 3	18 Oct 03, 30 Apr 04	194d, 18h, 33m	Aleksandr Kaleri (C), Michael Foale (E), Pedro Duque (E) (up), Andre Kuipers (dn)	*ISS* 8 crew (Kaleri, Foale and Duque)
Soyuz TMA 4	19 Apr 04, 24 Oct 04	187d, 21h, 16m	Gennadi Padalka (C), E. Michael Fincke (E), Andre Kuipers (E) (up), Yuri Shargin (dn)	*ISS* 9 crew (Padalka, Fincke and Kuipers)
Soyuz TMA 5	14 Oct 04, 24 Apr 05	192d, 19h, 2m	Saliszan Sharipov (C), Leroy Chiao (E), Yuri Shargin (E) (up), Roberto Vittori (dn)	*ISS* 10 crew (Sharipov, Chiao and Shargin)
Soyuz TMA 6	15 Apr 05, 11 Oct 05	179d, 0h, 23m	Sergei Krikalev (C), John Phillips (E), Roberto Vittori (E) (up), Gregory Olsen (dn)	*ISS* 11 crew (Krikalev, Phillips and Vittori)
Soyuz TMA 7	1 Oct 05, 8 Apr 06	189d, 19h, 53m	Valeri Tokarev (C), William McArthur Jr. (E), Gregory Olsen (T) (up), Marcos Pontes (dn)	*ISS* 12 crew (Tokarev, McArthur and Olsen)
Soyuz TMA 8	30 Mar 06, 29 Sept 06	182d, 22h, 43m	Pavel Vinogradov (C), Jeffrey Williams (E), Marcos Pontes (T) (up), Anousheh Ansari (dn)	*ISS* 13 crew (Vinogradov, Williams and Pontes)

(Continued)

Human Spaceflight Missions by Country, through February 2011 (Part 11 of 51)

Soyuz				
Mission	**Launch, Landing**	**Duration**	**Crew**	**Highlights**
Soyuz TMA 9	18 Sep 06, 21 Apr 07	215d, 8h, 22m	Mikhail Tyurin (C), Michael Lopez Alegria (E), Anousheh Ansari (T) (up), Charles Simonyi (dn)	*ISS* 14 crew (Tyurin, Lopez Alegria and Ansari)
Soyuz TMA 10	7 Apr 07, 21 Oct 07	196d, 17h, 5m	Oleg Kotov (C), Fyodor Yurchikhin (E), Charles Simonyi (T) (up), Sheikh Muszaphar Shukor (dn)	*ISS* 15 crew (Kotov, Yurchikhin and Simonyi)
Soyuz TMA 11	10 Oct 07, 19 Apr 08	191d, 19h, 7m	Yuri Malenchenko (C), Peggy Whitson (E), Sheikh Muszaphar Shukor (T) (up), Soyeon Yi (dn)	*ISS* 16 crew (Malenchenko, Whitson and Shukor)
Soyuz TMA 12	8 Apr 08, 24 Oct 08	198d, 16h, 20m	Sergei Volkov (C), Oleg Kononenko (E), Soyeon Yi (T) (up), Richard Garriott (dn)	*ISS* 17 crew (Volkov, Kononenko and Yi)
Soyuz TMA 13	12 Oct 08, 8 Apr 09	178d, 15m	Yuri Lonchakov (C), Edward Michael Fincke (E), Richard Garriott (T) (up), Charles Simonyi (dn)	*ISS* 18 crew (Lonchakov, Fincke and Garriott)
Soyuz TMA 14	26 Mar 09, 11 Oct 09	198d, 16h, 42m	Gennadi Pedalka (C), Michael Barratt (E), Charles Simonyi (T) (up), Guy Laliberte (dn)	*ISS* 19 crew (Pedalka, Barratt and Simonyi)
Soyuz TMA 15	27 May 09, 1 Dec 09	187d, 20h, 42m	Roman Romanenko (C), Frank De Winne (E), Robert Thirsk (E)	*ISS* 20 crew (Romanenko, De Winne and Thirsk) join *ISS* 19 crew and expand overall *ISS* resident occupancy to six persons
Soyuz TMA 16	30 Sep 09, 18 Mar 10	169d, 4h, 10m	Maksim Surayev (C), Jeffrey Williams (E), Guy Laliberte (T) (up)	*ISS* 21 crew (Surayev, Williams and Laliberte) join *ISS* 20 crew
Soyuz TMA 17	20 Dec 09	117d, 8h, 28m	Oleg Kotov (C), Soichi Noguchi (E), Timothy Creamer (E)	*ISS* 22 crew (Kotov, Noguchi, and Creamer) join *ISS* 21 crew
Soyuz TMA 18	2 Apr 10		Aleksandr Skvortsov (C), Mikhail Korniyenko (E), Tracy Caldwell-Dyson (E)	*ISS* 23 crew (Skvortsov, Korniyenko, Caldwell-Dyson) join *ISS* 22 crew

Human Spaceflight Missions by Country, through February 2011 (Part 12 of 51)

Salyut					
Name	**Launch**	**Reentry**	**Days in orbit**	**Type**	**Highlights**
Salyut 1	19 Apr 71	11 Oct 71	175	Civilian space station	First space station; resident crew: *Soyuz 11*
Salyut 2 (*Almaz 1* or *OPS 1*)	3 Apr 73		13	Almaz military space station	Station lost on 14 Apr 73 after depressurization
Salyut 3 (*Almaz 2* or *OPS 2*)	25 Jun 74	24 Jan 75	213	Almaz military space station	Resident crew: *Soyuz 14*
Salyut 4	26 Dec 74	3 Feb 77	770	Civilian space station	Resident crews: *Soyuz 17* and *18B*
Salyut 5 (*Almaz 3* or *OPS 3*)	22 Jun 76	8 Aug 77	412	Almaz military space station	Resident crews: *Soyuz 21* and *24*
Salyut 6	29 Sep 77	28 Jul 82	1,763	Civilian space station	Resident crews: *Soyuz 26*, *29*, *32*, *35*, and *T 4*
Salyut 7	19 Apr 82	7 Feb 91	3,216	Civilian space station	Resident crews: *Soyuz T 5*, *T 9*, *T 10B*, *T 13*, and *T 14*

Mir
Mir was launched on 20 February 1986 and reentered the atmosphere on 23 March 2001. The station was in orbit for 5,510 days.

Expedition #, Launch Vehicle	**Launch, Landing**	**Docking, Undocking**	**Crew**	**Highlights**
Expedition 1, SL 4	13 Mar 86, 16 July 86	15 Mar 86 on *Soyuz T 15*, 16 Jul 86 on *Soyuz T 15*	Leonid Kizim (C), Vladimir Solovyov (E)	First resident crew of *Mir*; flew to *Salyut 7* on 5 May, returning to *Mir* on 26 June
Expedition 2, SL 4	5 Feb 87, 29 Jul 87	7 Feb 87 on *Soyuz TM 2*, 29 Jul 87 on *Soyuz TM 3*	Yuri Romanenko (C), Aleksander Laveykin (E), Aleksander Aleksandrov (E)	Aleksandrov (arrives on *Soyuz TM 3*) replaces Laveykin; Kvant 1 module docks with *Mir* on 9 April
Expedition 3, SL 4	21 Dec 87, 17 Jun 88	23 Dec 87 on *Soyuz TM 4*, 17 Jun 88 on *Soyuz TM 6*	Vladimir Titov (C), Musa Manarov (E), Valeri Polyakov (X) (up)	Polyakov arrived on *Soyuz TM 6* to become third member of Expedition 3
Expedition 4, SL 4	26 Nov 88, 27 Apr 89	28 Nov 88 on *Soyuz TM 7*, 27 Apr 89 on *Soyuz TM 7*	Aleksandr Volkov (C), Sergei Krikalev (E), Valeri Polyakov (X) (dn)	Volkov and Krikalev spend 151 days on *Mir*; Polyakov ends 240-day stay on *Mir*

(Continued)

Human Spaceflight Missions by Country, through February 2011 (Part 13 of 51)

Mir				
Expedition #, Launch Vehicle	**Launch, Landing**	**Docking, Undocking**	**Crew**	**Highlights**
Expedition 5, SL 4	5 Sep 89, 19 Feb 90	7 Sep 89 on *Soyuz TM 8*, 19 Feb 90 on *Soyuz TM 8*	Aleksandr Viktorenko (C), Aleksandr Serebrov (E)	Kvant 2 module docks with *Mir* on 2 December; Viktorenko and Serebrov perform five EVAs
Expedition 6, SL 4	11 Feb 90, 9 Aug 90	13 Feb 90 on *Soyuz TM 9*, 9 Aug 90 on *Soyuz TM 9*	Anatoliy Solovyov (C), Aleksandr Balandin (E)	Kristall module docks with *Mir* on 10 June; Solovyov and Balandin perform two EVAs
Expedition 7, SL 4	1 Aug 90, 10 Dec 90	3 Aug 90 on *Soyuz TM 10*, 10 Dec 90 on *Soyuz TM 10*	Gennadi Manakov (C), Gennadi Strekalov (E)	Manakov and Strekalov spend 130 days on *Mir*; perform one EVA
Expedition 8, SL 4	2 Dec 90, 26 May 91	4 Dec 90 on *Soyuz TM 11*, 26 May 91 on *Soyuz TM 11*	Viktor Afanasyev (C), Musa Manarov (E)	Afanasyev and Manarov spend 175 days on *Mir*; perform four EVAs
Expedition 9, SL 4	18 May 91, 10 Oct 91	20 May 91 on *Soyuz TM 12*, 10 Oct 91 on *Soyuz TM 12*	Anatoli Artsebarsky (C), Sergei Krikalev (E) (up)	Artsebarsky and Krikalev perform six EVAs
Expedition 10, SL 4	2 Oct 91, 25 Mar 92	4 Oct 91 on *Soyuz TM 13*, 25 Mar 92 on *Soyuz TM 13*	Aleksandr Volkov (C), Sergei Krikalev (E) (dn)	Volkov and Krikalev perform one EVA; Krikalev ends 311-day stay on *Mir*
Expedition 11, SL 4	17 Mar 92, 10 Aug 92	19 Mar 92 on *Soyuz TM 14*, 10 Aug 92 on *Soyuz TM 14*	Aleksandr Viktorenko (C), Aleksandr Kaleri (E)	First resident crew following the breakup of the Soviet Union
Expedition 12, SL 4	27 Jul 92, 1 Feb 93	29 July 92 on *Soyuz TM 15*, 1 Feb 93 on *Soyuz TM 15*	Anatoliy Solovyov (C), Sergei Avdeyev (E)	Solovyov and Avdeyev spend 188 days on *Mir*; perform four EVAs

Human Spaceflight Missions by Country, through February 2011 (Part 14 of 51)

Mir				
Expedition #, Launch Vehicle	**Launch, Landing**	**Docking, Undocking**	**Crew**	**Highlights**
Expedition 13, SL 4	24 Jan 93, 22 Jul 93	26 Jan 93 on *Soyuz TM 16*, 22 Jul 93 on *Soyuz TM 16*	Gennadi Manakov (C), Aleksandr Poleshchuk (E)	Manakov and Poleshchuk spend 179 days on *Mir*; perform two EVAs
Expedition 14, SL 4	1 Jul 93, 14 Jan 94	3 Jul 93 on *Soyuz TM 17*, 14 Jan 94 on *Soyuz TM 17*	Vasili Tsibliyev (C), Aleksandr Serebrov (E)	Tsibliyev and Serebrov spend 196 days on *Mir*; perform five EVAs; install Rapana truss
Expedition 15, SL 4	8 Jan 94, 9 Jul 94	10 Jan 94 on *Soyuz TM 18*, 9 Jul 94 on *Soyuz TM 18*	Viktor Afanasyev (C), Yuri Usachyov (E), Valeri Polyakov (X) (up)	Crew conducts studies of Earth's magnetosphere; Polyakov begins record 438-day stay on *Mir*
Expedition 16, SL 4	1 Jul 94, 4 Nov 94	3 Jul 94 on *Soyuz TM 19*, 4 Nov 94 on *Soyuz TM 19*	Yuri Malenchenko (C), Talgat Musabayev (E), Valeri Polyakov (X)	First successful manual docking of a Progress supply vehicle by Malenchenko
Expedition 17, SL 4	4 Oct 94, 22 Mar 95	6 Oct 94 on *Soyuz TM 20*, 22 Mar 95 on *Soyuz TM 20*	Aleksandr Viktorenko (C), Yelena Kondakova (E), Valeri Polyakov (dn)	Polyakov ends 438-day stay aboard *Mir*.
Expedition 18, SL 4	14 Mar 95, 7 Jul 95	16 Mar 95 on *Soyuz TM 21*, 4 Jul 95 on STS 71	Vladimir Dezhurov (C), Gennadi Strekalov (E), Norman Thagard (R)	Thagard first American on *Mir*; Spektr module docks with *Mir* on 1 June
Expedition 19, *Atlantis* (STS 71)	27 Jun 95, 11 Sep 95	29 Jun 95 on STS 71, 11 Sep 95 on *Soyuz TM 21*	Anatoly Solovyev (C), Nikolai Budarin (E)	First Space Shuttle docking with *Mir*
Expedition 20, SL 4	3 Sep 95, 29 Feb 96	5 Sep 95 on *Soyuz TM 22*, 29 Feb 96 on *Soyuz TM 22*	Yuri Gidzenko (C), Sergei Avdeyev (E), Thomas Reiter (R)	Germany's Reiter spends 179 days on *Mir*, longer than any other international cosmonaut

(Continued)

Human Spaceflight Missions by Country, through February 2011 (Part 15 of 51)

Mir				
Expedition #, Launch Vehicle	**Launch, Landing**	**Docking, Undocking**	**Crew**	**Highlights**
Expedition 21, SL 4	21 Feb 96, 2 Sep 96	23 Feb 96 on *Soyuz TM 23*, 2 Sep 96 on *Soyuz TM 23*	Yuri Onufrienko (C), Yuri Usachyov (E), Shannon Lucid (R)	Lucid boards *Mir* on 24 March (STS 76); Priroda module docks with *Mir* on 26 April
Expedition 22, SL 4	17 Aug 96, 2 Mar 97	19 Aug 96 on *Soyuz TM 24*, 2 Mar 97 on *Soyuz TM 24*	Valeri Korzun (C), Aleksandr Kaleri (E), Shannon Lucid (R), John Blaha (R), Jerry Linenger (R)	Blaha boards *Mir* on 19 September (STS 79); Linenger boards *Mir* on 15 January (STS 81)
Expedition 23, SL 4	10 Feb 97, 14 Aug 97	12 Feb 97 on *Soyuz TM 25*, 14 Aug 97 on *Soyuz TM 25*	Vasili Tsibliyev (C), Aleksander Lazutkin (E), Jerry Linenger (R), Michael Foale (R)	Fire fills *Mir* with smoke on 23 February; Foale boards *Mir* on 17 May (STS 84); Progress M 34 collides with Spektr module on 25 June
Expedition 24, SL 4	5 Aug 97, 19 Feb 98	7 Aug 97 on *Soyuz TM 26*, 19 Feb 98 on *Soyuz TM 26*	Anatoliy Solovyov (C), Pavel Vinogradov (E), Michael Foale (R), David Wolf (R), Andrew Thomas (R)	Wolf boards *Mir* on 27 September (STS 86); Thomas boards *Mir* on 24 January (STS 89)
Expedition 25, SL 4	29 Jan 98, 25 Aug 98	31 Jan 98 on *Soyuz TM 27*, 25 Aug 98 on *Soyuz TM 27*	Talgat Musabayev (C), Nikolai Budarin (E), Andrew Thomas (R)	Musabayev and Budarin perform five EVAs; repair damaged solar panel on Spektr module
Expedition 26, SL 4	13 Aug 98, 28 Feb 99	15 Aug 98 on *Soyuz TM 28*, 28 Feb 99 on *Soyuz TM 28*	Gennadi Padalka (C), Sergei Avdeyev (E) (up)	Avdeyev begins 379-day stay on *Mir*
Expedition 27, SL 4	20 Feb 99, 28 Aug 99	22 Feb 99 on *Soyuz TM 29*, 28 Aug 99 on *Soyuz TM 29*	Viktor Afanasyev (C), Sergei Avdeyev (E) (dn), Jean Pierre Haignere (R)	France's Haignere breaks international endurance record set by Thomas Reiter with 188-day stay on *Mir*

Human Spaceflight Missions by Country, through February 2011 (Part 16 of 51)

Mir				
Expedition #, Launch Vehicle	**Launch, Landing**	**Docking, Undocking**	**Crew**	**Highlights**
Expedition 28, SL 4	4 Apr 00, 16 Jun 00	6 Apr 00 on *Soyuz TM 30*, 16 Jun 00 on *Soyuz TM 30*	Sergei Zalyetin (C), Aleksandr Kalen (E)	Last Mir resident crew fixes small air leak; spend 72 days on station

UNITED STATES					
Mercury					
Mission	**Launch**	**Landing**	**Duration**	**Crew**	**Highlights**
Mercury Redstone 3 (*Freedom 7*)	5 May 61	5 May 61	15m:22s	Alan Shepard Jr. (P)	First American to fly in space
Mercury Redstone 4 (*Liberty Bell 7*)	21 Jul 61	21 Jul 61	15m:37s	Virgil Grissom (P)	Spacecraft sank after splashdown
Mercury Atlas 6 (*Friendship 7*)	20 Feb 62	20 Feb 62	04h, 55m	John Glenn Jr. (P)	First American to orbit Earth
Mercury Atlas 7 (*Aurora 7*)	24 May 62	24 May 62	04h, 56m	Scott Carpenter Jr. (P)	Duplicate flight of Mercury Atlas 6
Mercury Atlas 8 (*Sigma 7*)	3 Oct 62	3 Oct 62	09h, 13m	Walter Schirra Jr. (P)	Six orbit engineering test flight
Mercury Atlas 9 (*Faith 7*)	15 May 63	16 May 63	1d, 10h, 20m	L. Gordon Cooper Jr. (P)	Evaluates effects of weightlessness in space
X 15					
Mission	**Launch**	**Landing**	**Duration**	**Crew**	**Highlights**
Flight 90	19 Jul 63	19 Jul 63	11m, 24s	Joseph Walker (P)	First rocket powered aircraft to reach space
Flight 91	22 Aug 63	22 Aug 63	11m, 9s	Joseph Walker (P)	Suborbital flight reaching altitude of 354,200 ft

(Continued)

Human Spaceflight Missions by Country, through February 2011 (Part 17 of 51)

Gemini					
Mission	**Launch**	**Landing**	**Duration**	**Crew**	**Highlights**
Gemini 3 (Molly Brown)	23 Mar 65	23 Mar 65	04h, 52m	Virgil Grissom (C), John Young (P)	First American spaceflight with two-person crew
Gemini 4	3 Jun 65	7 Jun 65	4d, 1h, 56m	James McDivitt (C), Edward White II (P)	White performs first American EVA
Gemini 5	21 Aug 65	29 Aug 65	7d, 22h, 55m	Gordon Cooper (C), Charles Conrad Jr. (P)	First use of fuel cells for electric power
Gemini 7	4 Dec 65	18 Dec 65	13d, 18h, 35m	Frank Borman (C), James Lovell Jr. (P)	Joint rendezvous mission with *Gemini 6 A*
Gemini 6 A	15 Dec 65	16 Dec 65	1d, 1h, 51m	Walter Schirra Jr. (C), Thomas Stafford (P)	First U.S. piloted rendezvous (*Gemini 7*)
Gemini 8	16 Mar 66	17 Mar 66	10h, 41m	Neil Armstrong (C), David Scott (P)	First piloted space docking (with uncrewed Agena)
Gemini 9 A	3 Jun 66	6 Jun 66	3d, 0h, 21m	Thomas Stafford (C), Eugene Cernan (P)	Cernan performs laborious, two-hour EVA
Gemini 10	18 Jul 66	21 Jul 66	2d, 22h, 47m	John Young (C), Michael Collins (P)	First use of Agena Target Vehicle propulsion systems
Gemini 11	12 Sep 66	15 Sep 66	2d, 23h, 17m	Charles Conrad Jr. (C), Richard Gordon Jr. (P)	Gemini capsule reaches record altitude, 739.2 m
Gemini 12	11 Nov 66	15 Nov 66	3d, 22h, 35m	James Lovell Jr. (C), Edwin Aldrin Jr. (P)	Aldrin refines space-walk techniques during EVA

Apollo					
Mission	**Launch**	**Landing**	**Duration**	**Crew**	**Highlights**
Apollo 7	11 Oct 68	22 Oct 68	10d, 20h, 9m	Walter Schirra Jr. (C), Donn Eisele (CP), Walter Cunningham (LP)	First American three-person crew in space; first mission with live telecasts in orbit

Human Spaceflight Missions by Country, through February 2011 (Part 18 of 51)

Apollo					
Mission	**Launch**	**Landing**	**Duration**	**Crew**	**Highlights**
Apollo 8	21 Dec 68	27 Dec 68	6d, 3h, 1m	Frank Borman (C), James Lovell Jr. (CP), William Anders (LP)	First piloted spaceflight to enter lunar orbit; first views of the far side of the Moon
Apollo 9	3 Mar 69	13 Mar 69	10d, 1h, 1m	James McDivitt (C), David Scott (CP), Russell Schweickart (LP)	Successful demonstration of Lunar Module
Apollo 10	18 May 69	26 May 69	8d, 0h, 3m	Thomas Stafford (C), John Young (CP), Eugene Cernan (LP)	Lunar Module approaches within 5.5 m of Moon; test run for *Apollo 11*
Apollo 11	16 Jul 69	24 Jul 69	8d, 3h, 19m	Neil Armstrong (C), Michael Collins (CP), Edwin Aldrin Jr. (LP)	Armstrong and Aldrin are first humans to walk on the Moon (20 July 1969)
Apollo 12	14 Nov 69	24 Nov 69	10d, 4h, 36m	Charles Conrad Jr. (C), Richard Gordon Jr. (CP), Alan Bean (LP)	Conrad and Bean perform two lunar EVAs; bring back items from Surveyor 3
Apollo 13	11 Apr 70	17 Apr 70	5d, 22h, 55m	James Lovell Jr. (C), John Swigart Jr. (CP), Fred Haise Jr. (LP)	Moon landing aborted after oxygen tank explodes on third day of mission
Apollo 14	31 Jan 71	9 Feb 71	9d, 0h, 2m	Alan Shepard Jr. (C), Stuart Roosa (CP), Edgar Mitchell (LP)	Shepard and Mitchell perform two lunar EVAs
Apollo 15	26 Jul 71	7 Aug 71	12d, 7h, 12m	David Scott (C), Alfred Worden (CP), James Irwin (LP)	Scott and Irwin use lunar rover during three EVAs; first telecast of lunar liftoff
Apollo 16	16 Apr 72	27 Apr 72	11d, 1h, 51m	John Young (C), Thomas Mattingly II (CP), Charles Duke (LP)	Young and Duke perform three lunar EVAs totaling 20h, 17m
Apollo 17	7 Dec 72	19 Dec 72	12d, 13h, 52m	Eugene Cernan (C), Ronald Evans (CP), Harrison Schmitt (LP)	Cernan and Schmitt perform three lunar EVAs; damage and repair rover

(Continued)

Human Spaceflight Missions by Country, through February 2011 (Part 19 of 51)

Apollo					
Mission	**Launch**	**Landing**	**Duration**	**Crew**	**Highlights**
Apollo-Soyuz Test Project	15 Jul 75	24 Jul 75	9d, 1h, 28m	Thomas Stafford (C), Vance Brand (CP), Donald Slayton (DP)	Joint flight with *Soyuz 19*; first U.S. Soviet rendezvous of spacecraft

Skylab
Skylab was launched on 14 May 1973 and reentered the atmosphere on 11 July 1979. The station was in orbit for 2,249 days.

Mission	**Launch**	**Landing**	**Duration**	**Crew**	**Highlights**
Skylab 2	25 May 73	22 Jun 73	28d, 0h, 50m	Charles Conrad Jr. (C), Joseph Kerwin (SP), Paul Weitz (P)	First resident crew; repair damage to station sustained during launch; conduct 392 hours of experiments
Skylab 3	28 Jul 73	25 Sep 73	59d, 11h, 10m	Alan Bean (C), Owen Garriott (SP), Jack Lousma (P)	Second resident crew; conduct 1,081 hours of experiments
Skylab 4	16 Nov 73	8 Feb 74	84d, 1h, 15m	Gerald Carr (C), Edward Gibson (SP), William Pogue (P)	Last resident crew; conduct observations of Comet Kohoutek among other experiments

Space Shuttle

Flight #, Space Transportation System #, Orbiter	**Launch, Landing**	**Duration**	**Crew**	**Highlights**
1, STS 1, *Columbia*	12 Apr 81, 14 Apr 81	2d, 6h, 22m	John Young (C), Robert Crippen (P)	First winged spacecraft to orbit Earth; orbiter loses 16 tiles, 148 sustain damage
2, STS 2, *Columbia*	12 Nov 81, 14 Nov 81	2d, 6h, 14m	Joe Engle (C), Richard Truly (P)	*Columbia* is first piloted spacecraft to fly twice; Failure of fuel cell forces early end to mission
3, STS 3, *Columbia*	22 Mar 82, 30 Mar 82	8d, 0h, 6m	Jack Lousma (C), Charles Gordon Fullerton (P)	External tank no longer painted white; payload includes OSS 1

Human Spaceflight Missions by Country, through February 2011 (Part 20 of 51)

Space Shuttle				
Flight #, Space Transportation System #, Orbiter	**Launch, Landing**	**Duration**	**Crew**	**Highlights**
4, STS 4, *Columbia*	27 Jun 82, 4 Jul 82	7d, 1h, 11m	Thomas Mattingly (C), Henry Hartsfield (P)	SRBs lost due to parachute failure; classified Department of Defense (DoD) payload; first concrete runway landing
5, STS 5, *Columbia*	11 Nov 82, 16 Nov 82	5d, 2h, 15m	Vance Brand (C), Robert Overmeyer (P), Joseph Allen (MS1), William Lenoir (MS2)	First scheduled EVA canceled after space suit malfunction; two satellites deployed
6, STS 6, *Challenger*	4 Apr 83, 9 Apr 83	5d, 0h, 25m	Paul Weitz (C), Karol Bobko (P), Donald Peterson (MS1), Story Musgrave (MS2)	First *Challenger* flight; first shuttle EVA by Peterson and Musgrave; *TDRS 1* deployed
7, STS 7, *Challenger*	18 Jun 83, 24 Jun 83	6d, 2h, 25m	Robert Crippen (C), Frederick Hauck (P), Sally Ride (MS1), John Fabian (MS2), Norman Thagard (MS3)	Ride becomes first American woman in space; two satellites deployed
8, STS 8, *Challenger*	30 Aug 83, 5 Sep 83	6d, 1h, 10m	Richard Truly (C), Daniel Brandenstein (P), Dale Gardner (MS1), Guion Bluford (MS2), William Thornton (MS3)	Bluford becomes first African American in space; first Shuttle launch and landing at night
9, STS 9, *Columbia*	28 Nov 83, 8 Dec 83	10d, 7h, 48m	John Young (C), Brewster Shaw (P), Owen Garriot (MS1), Robert Parker (MS2), Bryon Lichtenberg (PS1), Ulf Merbold (PS2)	Merbold becomes first European Space Agency astronaut to participate in a U.S. space shuttle flight; *Spacelab 1* aboard
10, STS 11 (41 B*), *Challenger*	3 Feb 84, 11 Feb 84	7d, 23h, 17m	Vance Brand (C), Robert Gibson (P), Bruce McCandless (MS1), Robert Stewart (MS2), Ronald McNair (MS3)	First untethered EVAs by McCandless and Stewart using MMU; first landing at Kennedy Space Center
* NASA assigned manifest numbers beginning with the 10th Shuttle flight and ending with the 25th flight. The first digit represented the fiscal year; the second digit represented the launch site (Kennedy Space Center); the letter represented the sequence of launches.				

(Continued)

Human Spaceflight Missions by Country, through February 2011 (Part 21 of 51)

Space Shuttle				
Flight #, Space Transportation System #, Orbiter	**Launch, Landing**	**Duration**	**Crew**	**Highlights**
11, STS 13 (41 C), *Challenger*	6 Apr 84, 13 Apr 84	6d, 23h, 41m	Robert Crippen (C), Francis Scobee (P), George Nelson (MS1), James van Hoften (MS2), Terry Hart (MS3)	Repair of Solar Maximum satellite during EVA; crew deploys *Long Duration Exposure Facility*
12, STS 16 (41 D), *Discovery*	30 Aug 84, 5 Sep 84	6d, 0h, 57m	Henry Hartsfield (C), Michael Coats (P), Judith Resnik (MS1), Steven Hawley (MS2), Richard Mullane (MS3), Charles Walker (PS1)	First flight of *Discovery*; Walker is first commercial payload specialist; crew deploys three satellites
13, STS 17 (41 G), *Challenger*	5 Oct 84, 13 Oct 84	8d, 5h, 25m	Robert Crippen (C), Jon McBride (P), Kathryn Sullivan (MS1), Sally Ride (MS2), David Leestma (MS3), Marc Garneau (PS1), Paul Scully Power (PS2)	Garneau becomes first Canadian in space; Sullivan becomes first U.S. woman to perform an EVA
14, STS 19 (51 A), *Discovery*	8 Nov 84, 16 Nov 84	7d, 23h, 46m	Frederick Hauck (C), David Walker (P), Anna Fisher (MS1), Dale Gardner (MS2), Joseph Allen (MS3)	Allen and Gardner retrieve two malfunctioning satellites (*Palapa B2*, *Westar 6*) during EVA
15, STS 20 (51 C), *Discovery*	24 Jan 85, 27 Jan 85	3d, 1h, 34m	Thomas Mattingly (C), Loren Shriver (P), Ellison Onizuka (MS1), James Buchli (MS2), Gary Payton (PS1)	First dedicated DoD mission; classified military payload
16, STS 23 (51 D), *Discovery*	12 Apr 85, 19 Apr 85	6d, 23h, 56m	Karol Bobko (C), Donald Williams (P), Margaret Rhea Seddon (MS1), Jeffrey Hoffman (MS2), David Griggs (MS3), Charles Walker (PS1), Sen. Jake Garn (PS2)	Garn becomes first U.S. senator in space; crew deploys two satellites

Human Spaceflight Missions by Country, through February 2011 (Part 22 of 51)

Space Shuttle				
Flight #, Space Transportation System #, Orbiter	**Launch, Landing**	**Duration**	**Crew**	**Highlights**
17, STS 24 (51 B), *Challenger*	29 Apr 85, 6 May 85	7d, 10m	Robert Overmeyer (C), Frederick Gregory (P), Don Lind (MS1), Norman Thagard (MS2), William Thornton (MS3), Lodewijk van den Berg (PS1), Taylor Wang (PS2)	*Spacelab 3* aboard; observation of effects of weightlessness on two monkeys and 24 rodents
18, STS 25 (51 G), *Discovery*	17 Jun 85, 24 Jun 85	7d, 1h, 40m	Daniel Brandenstein (C), John Creighton (P), Shannon Lucid (MS1), John Fabian (MS2), Steven Nagel (MS3), Patrick Baudry (PS1), Salman Abdul aziz Al Saud (PS2)	Salman Al Saud becomes first Arab in space; crew deploys three satellites
19, STS 26 (51 F), *Challenger*	29 Jul 85, 6 Aug 85	7d, 22h, 46m	Gordon Fullerton (C), Roy Bridges (P), Story Musgrave (MS1), Anthony England (MS2), Karl Henize (MS3), Loren Acton (PS1), John David Bartoe (PS2)	Shutdown of one main engine during launch results in successful abort to orbit; *Spacelab 2* aboard
20, STS 27 (51 I), *Discovery*	27 Aug 85, 3 Sep 85	7d, 2h, 18m	Joe Engle (C), Richard Covey (P), James van Hoften (MS1), John Lounge (MS2), William Fisher (MS3)	Fisher and van Hoften perform EVA to repair satellite (*Leasat 3*); crew deploys three satellites
21, STS 28 (51 J), *Atlantis*	3 Oct 85, 7 Oct 85	4d, 1h, 46m	Karol Bobko (C), Ronald Grabe (P), David Hilmers (MS1), Robert Stewart (MS2), William Pailes (PS1)	First flight of *Atlantis*; Second dedicated DoD classified mission

(Continued)

Human Spaceflight Missions by Country, through February 2011 (Part 23 of 51)

Space Shuttle				
Flight #, Space Transportation System #, Orbiter	**Launch, Landing**	**Duration**	**Crew**	**Highlights**
22, STS 30 (61 A), *Challenger*	30 Oct 85, 6 Nov 85	7d, 46m	Henry Hartsfield (C), Steven Nagel (P), James Buchli (MS1), Guion Bluford (MS2), Bonnie Dunbar (MS3), Reinhard Furrer (PS1), Ernst Messerschmid (PS2), Wubbo Ockels (PS3)	Largest crew to fly in space; dedicated German *Spacelab D1* aboard with 75 numbered experiments
23, STS 31 (61 B), *Atlantis*	27 Nov 85, 3 Dec 85	6d, 21h, 6m	Brewster Shaw (C), Bryan O'Connor (P), Mary Cleave (MS1), Sherwood Spring (MS2), Jerry Ross (MS3), Rudolfo Neri Vela (PS1), Charles Walker (PS2)	Neri becomes first Mexican astronaut in space; crew deploys three satellites
24, STS 32 (61 C), *Columbia*	12 Jan 86, 18 Jan 86	6d, 2h, 5m	Robert Gibson (C), Charles Bolden (P), Franklin Chang Diaz (MS1), Steven Hawley (MS2), George Nelson (MS3), Robert Cenker (PS1), Rep. Bill Nelson (PS2)	Launch scrubbed six times; Nelson becomes first U.S. congressman in space; crew deploys one satellite
25, STS 33 (51 L), *Challenger*	28 Jan 86	01m, 13s	Francis Scobee (C), Michael Smith (P), Judith Resnik (MS1), Ellison Onizuka (MS2), Ronald McNair (MS3), Gregory Jarvis (PS1), Christa McAuliffe (PS2)	First Shuttle launch from Pad 39B; explosion 73 seconds after liftoff results in loss of orbiter and crew
26, STS 26R, *Discovery*	29 Sep 88, 3 Oct 88	4d, 1h, 1m	Frederick Hauck (C), Richard Covey (P), John Lounge (MS1), George Nelson (MS2), David Hilmers (MS3)	Return to flight mission; crew deploys *TDRS 3*

Human Spaceflight Missions by Country, through February 2011 (Part 24 of 51)

Space Shuttle				
Flight #, Space Transportation System #, Orbiter	**Launch, Landing**	**Duration**	**Crew**	**Highlights**
27, STS 27R, *Atlantis*	2 Dec 88, 6 Dec 88	4d, 9h, 6m	Robert Gibson (C), Guy Gardner (P), Richard Mullane (MS1), Jerry Ross (MS2), William Shepherd (MS3)	Third dedicated DoD mission; classified military payload
28, STS 29R, *Discovery*	13 Mar 89, 18 Mar 89	4d, 23h, 40m	Michael Coats (C), John Blaha (P), James Bagian (MS1), James Buchli (MS2), Robert Springer (MS3)	Crew deploys *TDRS 4*
29, STS 30R, *Atlantis*	4 May 89, 8 May 89	4d, 0h, 58m	David Walker (C), Ronald Grabe (P), Norman Thagard (MS1), Mary Cleave (MS2), Mark Lee (MS3)	Deployment of *Magellan* spacecraft on trajectory toward Venus to conduct radar mapping mission
30, STS 28R, *Columbia*	8 Aug 89, 13 Aug 89	5d, 1h, 1m	Brewster Shaw (C), Richard Richards (P), James Adamson (MS1), David Leestma (MS2), Mark Brown (MS3)	Fourth dedicated DoD mission; classified military payload
31, STS 34, *Atlantis*	18 Oct 89, 23 Oct 89	4d, 23h, 40m	Donald Williams (C), Michael McCulley (P), Franklin Chang Diaz (MS1), Shannon Lucid (MS2), Ellen Baker (MS3)	Deployment of *Galileo* spacecraft on trajectory for six-year trip to Jupiter
32, STS 33R, *Discovery*	23 Nov 89, 28 Nov 89	5d, 0h, 8m	Frederick Gregory (C), John Blaha (P), Story Musgrave (MS1), Manley Carter (MS2), Kathryn Thornton (MS3)	Fifth dedicated DoD mission
33, STS 32R, *Columbia*	9 Jan 90, 20 Jan 90	10d, 21h, 2m	Daniel Brandenstein (C), James Wetherbee (P), Bonnie Dunbar (MS1), G. David Low (MS2), Marsha Ivins (MS3)	Retrieval of *Long Duration Exposure Facility* using remote manipulator system

(Continued)

Human Spaceflight Missions by Country, through February 2011 (Part 25 of 51)

Space Shuttle				
Flight #, Space Transportation System #, Orbiter	**Launch, Landing**	**Duration**	**Crew**	**Highlights**
34, STS 36, *Atlantis*	28 Feb 90, 4 Mar 90	4d, 10h, 19m	John Creighton (C), John Casper (P), Richard Mullane (MS1), David Hilmers (MS2), Pierre Thuot (MS3)	Sixth dedicated DoD mission; classified military payload
35, STS 31R, *Discovery*	24 Apr 90, 29 Apr 90	5d, 1h, 17m	Loren Shriver (C), Charles Bolden (P), Steven Hawley (MS1), Bruce McCandless (MS2), Kathryn Sullivan (MS3)	Deployment of *Hubble Space Telescope* at altitude of 380 miles
36, STS 41, *Discovery*	6 Oct 90, 10 Oct 90	4d, 2h, 11m	Richard Richards (C), Robert Cabana (P), William Shepherd (MS1), Bruce Melnick (MS2), Thomas Akers (MS3)	Deployment of ESA-built *Ulysses* spacecraft to study polar regions of the Sun
37, STS 38, *Atlantis*	15 Nov 90, 20 Nov 90	4d, 21h, 55m	Richard Covey (C), Frank Culbertson (P), Robert Springer (MS1), Carl Meade (MS2), Charles Gemar (MS3)	Seventh dedicated DoD mission; classified military payload
38, STS 35, *Columbia*	2 Dec 90, 11 Dec 90	8d, 23h, 6m	Vance Brand (C), Guy Gardner (P), Jeffrey Hoffman (MS1), John Lounge (MS2), Robert Parker (MS3), Samuel Durrance (PS1), Ronald Parise (PS2)	Astro 1 observatory enables stellar observations with four onboard telescopes
39, STS 37, *Atlantis*	5 Apr 91, 11 Apr 91	5d, 23h, 34m	Steven Nagel (C), Kenneth Cameron (P), Jerry Ross (MS1), Jerome Apt (MS2), Linda Godwin (MS3)	Deployment of *Gamma Ray Observatory* (*GRO*); Ross, Apt perform EVA to fix *GRO* high gain antenna

Human Spaceflight Missions by Country, through February 2011 (Part 26 of 51)

Space Shuttle				
Flight #, Space Transportation System #, Orbiter	**Launch, Landing**	**Duration**	**Crew**	**Highlights**
40, STS 39, *Discovery*	28 Apr 91, 6 May 91	8d, 7h, 23m	Michael Coats (C), Blaine Hammond (P), Guion Bluford (MS1), Gregory Harbaugh (MS2), Richard Hieb (MS3), Donald McMonagle (MS4), Charles Veach (MS5)	Eighth dedicated DoD mission; release of Multi Purpose Release Canister
41, STS 40, *Columbia*	5 Jun 91, 14 Jun 91	9d, 2h, 15m	Bryan O'Connor (C), Sidney Gutierrez (P), Margaret Rhea Seddon (MS1), James Bagian (MS2), Tamara Jernigan (MS3), Drew Gaffney (PS1), Millie Hughes Fulford (PS2)	Spacelab Life Sciences 1 on board; crew conducts experiments on 30 rodents and small jellyfish
42, STS 43, *Atlantis*	2 Aug 91, 11 Aug 91	8d, 21h, 22m	John Blaha (C), Michael Baker (P), Shannon Lucid (MS1), James Adamson (MS2), G. David Low (MS3)	Deployment of *TDRS 5*; Lucid becomes first woman to make three spaceflights
43, STS 48, *Discovery*	12 Sep 91, 18 Sep 91	5d, 8h, 28m	John Creighton (C), Kenneth Reightler (P), James Buchli (MS1), Charles Gemar (MS2), Mark Brown (MS3)	Deployment of Upper Atmosphere Research Satellite
44, STS 44, *Atlantis*	24 Nov 91, 1 Dec 91	6d, 22h, 52m	Frederick Gregory (C), Terence Henricks (P), Story Musgrave (MS1), Mario Runco (MS2), James Voss (MS3), Thomas Hennen (PS1)	Ninth dedicated DoD mission; mission cut short due to inertial measurement unit failure
45, STS 42, *Discovery*	22 Jan 92, 30 Jan 92	8d, 1h, 16m	Ronald Grabe (C), Stephen Oswald (P), Norman Thagard (MS1), David Hilmers (MS2), William Readdy (MS3), Roberta Bondar (PS1), Ulf Merbold (PS2)	*Spacelab IML 1* on board

(Continued)

Human Spaceflight Missions by Country, through February 2011 (Part 27 of 51)

Space Shuttle				
Flight #, Space Transportation System #, Orbiter	**Launch, Landing**	**Duration**	**Crew**	**Highlights**
46, STS 45, *Atlantis*	24 Mar 92, 2 Apr 92	8d, 22h, 10m	Charles Bolden (C), Brian Duffy (P), Kathryn Sullivan (MS1), David Leestma (MS2), Michael Foale (MS3), Byron Lichtenberg (PS1), Dirk Frimout (PS2)	First flight of Atmospheric Lab for Applications and Science (ATLAS 1)
47, STS 49, *Endeavour*	7 May 92, 16 May 92	8d, 21h, 19m	Daniel Brandenstein (C), Kevin Chilton (P), Pierre Thuot (MS1), Kathryn Thornton (MS2), Richard Hieb (MS3), Thomas Akers (MS4), Bruce Melnick (MS5)	First *Endeavour* flight; four EVAs by crew, including recovery and redeployment of *Intelsat 6*
48, STS 50, *Columbia*	25 Jun 92, 9 Jul 92	13d, 19h, 31m	Richard Richards (C), Kenneth Bowersox (P), Bonnie Dunbar (PC), Ellen Baker (MS2), Carl Meade (MS3), Lawrence DeLucas (PS1), Eugene Trinh (PS2)	U.S. Microgravity Laboratory 1 makes first flight; mission sets new Shuttle duration record
49, STS 46, *Atlantis*	31 Jul 92, 8 Aug 92	7d, 23h, 16m	Loren Shriver (C), Andrew Allen (P), Jeffrey Hoffman (PC), Franklin Chang Diaz (MS2), Claude Nicollier (MS3), Marsha Ivins (MS4), Franco Malerba (PS1)	Deployment of ESA *European Retrievable Carrier* and *Tethered Satellite System 1*
50, STS 47, *Endeavour*	12 Sep 92, 20 Sep 92	7d, 22h, 31m	Robert Gibson (C), Curtis Brown (P), Mark Lee (PC), Jerome Apt (MS2), N. Jan Davis (MS3), Mae Jemison (MS4), Mamoru Mohri (PS1)	First African-American woman in space (Jemison); first married couple in space (Lee, Davis)

Human Spaceflight Missions by Country, through February 2011 (Part 28 of 51)

Space Shuttle				
Flight #, Space Transportation System #, Orbiter	**Launch, Landing**	**Duration**	**Crew**	**Highlights**
51, STS 52, *Columbia*	22 Oct 92, 1 Nov 92	9d, 20h, 57m	James Wetherbee (C), Michael Baker (P), Charles Veach (MS1), William Shepherd (MS2), Tamara Jernigan (MS3), Steven MacLean (PS1)	Deployment of *LAGEOS 2*; U.S. Microgravity Payload 1 on board
52, STS 53, *Discovery*	2 Dec 92, 9 Dec 92	7d, 7h, 21m	David Walker (C), Robert Cabana (P), Guion Bluford (MS1), James Voss (MS2), Michael Clifford (MS3)	Tenth and final dedicated DoD mission
53, STS 54, *Endeavour*	13 Jan 93, 19 Jan 93	5d, 23h, 39m	John Casper (C), Donald McMonagle (P), Mario Runco (MS1), Gregory Harbaugh (MS2), Susan Helms (MS3)	Deployment of *TDRS 6*; Runco and Harbaugh perform EVA tasks to improve spacewalk techniques
54, STS 56, *Discovery*	8 Apr 93, 17 Apr 93	9d, 6h, 9m	Kenneth Cameron (C), Stephen Oswald (P), Michael Foale (MS1), Kenneth Cockrell (MS2), Ellen Ochoa (MS3)	*ATLAS 2* on board
55, STS 55, *Columbia*	26 Apr 93, 6 May 93	9d, 23h, 41m	Steven Nagel (C), Terrence Henricks (P), Jerry Ross (MS1), Charles Precourt (MS2), Bernard Harris (MS3), Ulrich Walter (PS1), Hans Schlegel (PS2)	Dedicated German *Spacelab* D2 on board with 88 numbered experiments
56, STS 57, *Endeavour*	21 Jun 93, 1 Jul 93	9d, 23h, 46m	Ronald Grabe (C), Brian Duffy (P), G. David Low (MS1), Nancy Sherlock (MS2), Peter Wisoff (MS3), Janice Voss (MS4)	*Spacehab 1* on board with 22 experiments

(Continued)

Human Spaceflight Missions by Country, through February 2011 (Part 29 of 51)

Space Shuttle				
Flight #, Space Transportation System #, Orbiter	**Launch, Landing**	**Duration**	**Crew**	**Highlights**
57, STS 51, *Discovery*	12 Sep 93, 22 Sep 93	9d, 20h, 12m	Frank Culbertson (C), William Readdy (P), James Newman (MS1), Daniel Bursch (MS2), Carl Walz (MS3)	First landing at night at Kennedy Space Center
58, STS 58, *Columbia*	18 Oct 93, 1 Nov 93	14d, 0h, 14m	John Blaha (C), Richard Searfoss (P), Margaret Rhea Seddon (MS1), William McArthur Jr. (MS2), David Wolf (MS3), Shannon Lucid (MS4), Martin Fettman (PS1)	*Spacelab* Life Sciences 2 on board; mission sets new Shuttle duration record by 4h, 45m
59, STS 61, *Endeavour*	2 Dec 93, 13 Dec 93	10d, 19h, 59m	Richard Covey (C), Kenneth Bowersox (P), Kathryn Thornton (MS1), Claude Nicollier (MS2), Jeffrey Hoffman (MS3), Story Musgrave (MS4), Thomas Akers (MS5)	First *Hubble Space Telescope* servicing mission includes five EVAs totaling 35h, 28m
60, STS 60, *Discovery*	3 Feb 94, 11 Feb 94	8d, 7h, 10m	Charles Bolden Jr. (C), Kenneth Reightler Jr. (P), Jan Davis (MS1), Ronald Sega (MS2), Franklin Chang Diaz (MS3), Sergei Krikalev (MS4)	First Russian on Shuttle mission (Krikalev); Spacehab 2 on board with 12 experiments
61, STS 62, *Columbia*	4 Mar 94, 18 Mar 94	13d, 23h, 18m	John Casper (C), Andrew Allen (P), Pierre Thuot (MS1), Charles Gemar (MS2), Marsha Ivins (MS3)	U.S. Microgravity Payload 2 on board
62, STS 59, *Endeavour*	9 Apr 94, 20 Apr 94	11d, 5h, 50m	Sidney Gutierrez (C), Kevin Chilton (P), Jay Apt (MS1), Michael Clifford (MS2), Linda Godwin (PC), Thomas Jones (MS4)	Use of Space Radar Laboratory 1 to map 20 percent of Earth's surface; testing of new thermal protection tiles

Human Spaceflight Missions by Country, through February 2011 (Part 30 of 51)

Space Shuttle				
Flight #, Space Transportation System #, Orbiter	**Launch, Landing**	**Duration**	**Crew**	**Highlights**
63, STS 65, *Columbia*	8 Jul 94, 23 Jul 94	14d, 17h, 56m	Robert Cabana (C), James Halsell Jr. (P), Richard Hieb (MS1), Carl Walz (MS2), Leroy Chiao (MS3), Donald Thomas (MS4), Chiaki Mukai (PD1)	Spacelab IML 2 on board; mission sets new Shuttle duration record by 17h, 42m
64, STS 64, *Discovery*	9 Sep 94, 20 Sep 94	10d, 22h, 51m	Richard Richards (CO), Blaine Hammond Jr. (P), Jerry Linenger (MS1), Susan Helms (MS2), Carl Meade (MS3), Mark Lee (MS4)	First untethered EVA in more than 10 years; Lee and Meade test new EVA backpack
65, STS 68, *Endeavour*	30 Sep 94, 11 Oct 94	11d, 5h, 47m	Michael Baker (C), Terrence Wilcutt (P), Thomas Jones (PC), Steven Smith (MS1), Daniel Bursch (MS2), Peter Wisoff (MS3)	Use of Space Radar Laboratory 2 to image volcanic eruption in Russia
66, STS 66, *Atlantis*	3 Nov 94, 14 Nov 94	10d, 22h, 35m	Donald McMonagle (C), Curtis Brown Jr. (P), Ellen Ochoa (PC), Joseph Tanner (MS2), Jean Francois Clervoy (MS3), Scott Parazynski (MS4)	First flight of CRISTA SPAS 01; *ATLAS 3* on board
67, STS 63, *Discovery*	3 Feb 95, 11 Feb 95	8d, 6h, 30m	James Wetherbee (C), Eileen Collins (P), Bernard Harris Jr. (PC), Michael Foale (MS1), Janice Voss (MS2), Vladimir Titov (MS3)	First female Shuttle pilot (Collins); rendezvous with *Mir*; *Spacehab* 3 on board with 20 experiments
68, STS 67, *Endeavour*	2 Mar 95, 18 Mar 95	16d, 15h, 10m	Stephen Oswald (C), William Gregory (P), Tamara Jernigan (PC), John Grunsfeld (MS1), Wendy Lawrence (MS2), Samuel Durrance (PS1), Ronald Parise (PS2)	ASTRO 2 observatory on board; mission sets new Shuttle duration record by 1d, 22h, and 40m

(Continued)

Human Spaceflight Missions by Country, through February 2011 (Part 31 of 51)

Space Shuttle				
Flight #, Space Transportation System #, Orbiter	**Launch, Landing**	**Duration**	**Crew**	**Highlights**
69, STS 71, *Atlantis*	27 Jun 95, 7 Jul 95	9d, 19h, 23m	Robert Gibson (C), Charles Precourt (P), Ellen Baker (PC), Gregory Harbaugh (MS2), Bonnie Dunbar (MS3), Anatoliy Solovyov (*Mir* 19) (up), Nikolai Budarin (*Mir* 19) (up), Vladimir Dezhurov (*Mir* 18) (dn), Gennadi Strekalov (*Mir* 18) (dn), Norman Thagard (*Mir* 18) (dn)	First shuttle docking with *Mir*; mission is 100th U.S. human spaceflight from Kennedy Space Center; returning crew equals largest in history (STS 30)
70, STS 70, *Discovery*	13 Jul 95, 22 Jul 95	8d, 22h, 21m	Terrence Henricks (C), Kevin Kregel (P), Nancy Sherlock (MS1), Donald Thomas (MS2), Mary Ellen Weber (MS3)	Deployment of TDRS G (final TDRS mission)
71, STS 69, *Endeavour*	7 Sep 95, 18 Sep 95	10d, 20h, 30m	David Walker (C), Kenneth Cockrell (P), James Voss (PC), James Newman (MS2), Michael Gernhardt (MS3)	First mission where two satellites are deployed and retrieved (*Spartan 201 03*, *WSF 2*)
72, STS 73, *Columbia*	20 Oct 95, 5 Nov 95	15d, 21h, 53m	Kenneth Bowersox (C), Kent Rominger (P), Kathryn Thornton (PC), Catherine Coleman (MS1), Michael Lopez Alegria (MS2), Fred Leslie (PS1), Albert Sacco Jr. (PS2)	Launch scrubbed six times; Spacelab USML 2 on board
73, STS 74, *Atlantis*	12 Nov 95, 20 Nov 95	8d, 4h, 32m	Kenneth Cameron (C), James Halsell Jr. (P), Chris Hadfield (MS1), Jerry Ross (MS2), William McArthur Jr. (MS3)	Second *Mir* docking; delivers new, Russian-built Shuttle docking module and solar arrays

Human Spaceflight Missions by Country, through February 2011 (Part 32 of 51)

Space Shuttle				
Flight #, Space Transportation System #, Orbiter	**Launch, Landing**	**Duration**	**Crew**	**Highlights**
74, STS 72, *Endeavour*	11 Jan 96, 20 Jan 96	8d, 22h, 2m	Brian Duffy (C), Brent Jett Jr. (P), Leroy Chiao (MS1), Winston Scott (MS2), Koichi Wakata (MS3), Daniel Barry (MS4)	Retrieval of Japanese Space Flyer Unit
75, STS 75, *Columbia*	22 Feb 96, 9 Mar 96	15d, 17h, 41m	Andrew Allen (C), Scott Horowitz (P), Franklin Chang Diaz (PC), Jeffrey Hoffman (MS1), Maurizio Cheli (MS2), Claude Nicollier (MS3), Umberto Guidoni (PS1)	Satellite *TSS 1R* lost after tether snaps just short of full deployment
76, STS 76, *Atlantis*	22 Mar 96, 31 Mar 96	9d, 5h, 17m	Kevin Chilton (C), Richard Searfoss (P), Ronald Sega (MS1), Michael Clifford (MS2), Linda Godwin (MS3), Shannon Lucid (MS4) (up)	Third *Mir* docking; Lucid joins *Mir* Expedition 21
77, STS 77, *Endeavour*	19 May 96, 29 May 96	10d, 40m	John Casper (C), Curtis Brown (P), Andrew Thomas (MS1), Daniel Bursch (MS2), Mario Runco Jr. (MS3), Marc Garneau (MS4)	*Spacehab* 4 on board
78, STS 78, *Columbia*	20 Jun 96, 7 Jul 96	16d, 21h, 49m	Terrence Henricks (C), Kevin Kregel (P), Richard Linnehan (MS1), Susan Helms (PC), Charles Brady Jr. (MS3), Jean Jacques Favier (PS1), Robert Thirsk (PS2)	First live video from flight deck of ascent and reentry; mission sets new Shuttle duration record by 6h, 39m

(Continued)

Human Spaceflight Missions by Country, through February 2011 (Part 33 of 51)

Space Shuttle				
Flight #, Space Transportation System #, Orbiter	**Launch, Landing**	**Duration**	**Crew**	**Highlights**
79, STS 79, *Atlantis*	16 Sep 96, 26 Sep 96	10d, 3h, 20m	William Readdy (C), Terrence Wilcutt (P), Jay Apt (MS1), Thomas Akers (MS2), Carl Walz (MS3), John Blaha (MS4) (up), Shannon Lucid (MS4) (dn)	Fourth *Mir* docking; Blaha joins *Mir* Expedition 22
80, STS 80, *Columbia*	19 Nov 96, 7 Dec 96	17d, 15h, 54m	Kenneth Cockrell (C), Kent Rominger (P), Tamara Jernigan (PC), Thomas Jones (MS2), Story Musgrave (MS3)	Longest Shuttle flight to date; sets new Shuttle duration record by 17h, 55m
81, STS 81, *Atlantis*	12 Jan 97, 22 Jan 97	10d, 4h, 56m	Michael Baker (C), Brent Jett (P), Peter Wisoff (MS1), John Grunsfeld (MS2), Marsha Ivins (MS3), Jerry Linenger (MS4) (up), John Blaha (MS4) (dn)	Fifth *Mir* docking; Linenger joins *Mir* Expedition 23
82, STS 82, *Discovery*	11 Feb 97, 21 Feb 97	9d, 23h, 38m	Kenneth Bowersox (C), Scott Horowitz (P), Joseph Tanner (MS1), Steven Hawley (MS2), Gregory Harbaugh (MS3), Mark Lee (PC), Steven Smith (MS5)	Second *Hubble Space Telescope* servicing mission extends observatory's wavelength range into the near infrared
83, STS 83, *Columbia*	4 Apr 97, 8 Apr 97	3d, 23h, 14m	James Halsell (C), Susan Still (P), Janice Voss (PC), Michael Gernhardt (MS2), Donald Thomas (MS3), Roger Crouch (PS1), Greg Linteris (PS2)	Microgravity Science Lab 1 on board; mission cut short due to fuel cell problem; reflown on STS 94.

Human Spaceflight Missions by Country, through February 2011 (Part 34 of 51)

Space Shuttle				
Flight #, Space Transportation System #, Orbiter	**Launch, Landing**	**Duration**	**Crew**	**Highlights**
84, STS 84, *Atlantis*	15 May 97, 24 May 97	9d, 5h, 21m	Charles Precourt (C), Eileen Collins (P), Jean Francois Clervoy (PC), Carlos Noriega (MS2), Edward Lu (MS3), Yelena Kondakova (MS4), Michael Foale (MS5) (up), Jerry Linenger (MS5) (dn)	Sixth *Mir* docking; Foale joins *Mir* Expedition 23
85, STS 94, *Columbia*	1 Jul 97, 17 Jul 97	15d, 16h, 46m	James Halsell (C), Susan Still (P), Janice Voss (PC), Michael Gernhardt (MS2), Donald Thomas (MS3), Roger Crouch (PS1), Greg Linteris (PS2)	First reflight of same vehicle, crew, and payload (MSL 1) due to failure of fuel cell on STS 83
86, STS 85, *Discovery*	7 Aug 97, 19 Aug 97	11d, 20h, 28m	Curtis Brown (C), Kent Rominger (P), Jan Davis (PC), Robert Curbeam (MS2), Stephen Robinson (MS3), Bjarni Tryggvason (PS1)	Second flight of *CRISTA SPAS 02* (first on STS 66)
87, STS 86, *Atlantis*	25 Sep 97, 6 Oct 97	10d, 19h, 22m	James Wetherbee (C), Michael Bloomfield (P), Vladimir Titov (MS1), Scott Parazynski (MS2), Jean Loup Chretien (MS3), Wendy Lawrence (MS4), David Wolf (MS5) (up), Michael Foale (MS5) (dn)	Seventh *Mir* docking; Wolf joins *Mir* Expedition 24; first joint U.S.-Russian EVA during a Shuttle mission (Titov, Parazynski)
88, STS 87, *Columbia*	19 Nov 97, 5 Dec 97	15d, 16h, 35m	Kevin Kregel (C), Steve Lindsey (P), Kalpana Chawla (MS1), Winston Scott (MS2), Takao Doi (MS3), Leonid Kadenyuk (PS1)	U.S. Microgravity Payload 4 on board; Doi becomes first Japanese citizen to perform an EVA

(Continued)

Human Spaceflight Missions by Country, through February 2011 (Part 35 of 51)

Space Shuttle				
Flight #, Space Transportation System #, Orbiter	**Launch, Landing**	**Duration**	**Crew**	**Highlights**
89, STS 89, *Endeavour*	23 Jan 98, 31 Jan 98	8d, 19h, 48m	Terrence Wilcutt (C), Joe Edwards (P), James Reilly (MS1), Michael Anderson (MS2), Bonnie Dunbar (PC), Salizhan Sharipov (MS4), Andrew Thomas (MS5) (up), David Wolf (MS5) (dn)	Eighth *Mir* docking; Thomas joins *Mir* Expedition 24 crew (last astronaut to live on *Mir*)
90, STS 90, *Columbia*	17 Apr 98, 3 May 98	15d, 21h, 51m	Richard Searfoss (C), Scott Altman (P), Richard Linnehan (PC), Kathryn Hire (MS2), Dafydd Williams (MS3), Jay Buckey (PS1), James Pawelczyk (PS2)	Neurolab carried on board with 26 numbered experiments
91, STS 91, *Discovery*	2 Jun 98, 12 Jun 98	9d, 19h, 55m	Charles Precourt (C), Dominic Gorie (P), Franklin Chang Diaz (MS1), Wendy Lawrence (MS2), Janet Kavandi (MS3), Valeri Ryumin (MS4), Andrew Thomas (MS4) (dn)	Ninth *Mir* docking; last Shuttle visit to *Mir*
92, STS 95, *Discovery*	29 Oct 98, 7 Nov 98	8d, 21h, 45m	Curtis Brown (C), Steve Lindsey (P), Stephen Robinson (PC), Scott Parazynski (MS2), Pedro Duque (MS3), Chiaki Mukai (PS1), John Glenn Jr. (PS2)	Glenn returns to space 36 years, 8 months, and nine days after his first flight (Mercury Atlas 6)
93, STS 88, *Endeavour*	4 Dec 98, 16 Dec 98	11d, 19h, 19m	Robert Cabana (C), Frederick Sturckow (P), Jerry Ross (MS1), Nancy Sherlock (MS2), James Newman (MS3), Sergei Krikalev (MS4)	Integration of *Unity* and *Zarya* modules

Human Spaceflight Missions by Country, through February 2011 (Part 36 of 51)

Space Shuttle				
Flight #, Space Transportation System #, Orbiter	**Launch, Landing**	**Duration**	**Crew**	**Highlights**
94, STS 96, *Discovery*	27 May 99, 6 Jun 99	9d, 19h, 14m	Kent Rominger (C), Rick Husband (P), Tamara Jernigan (MS1), Ellen Ochoa (MS2), Daniel Barry (MS3), Julie Payette (MS4), Valeri Tokarev (MS5)	First flight to dock with *ISS*
95, STS 93, *Columbia*	23 Jul 99, 28 Jul 99	4d, 22h, 50m	Eileen Collins (C), Jeffrey Ashby (P), Catherine Coleman (MS1), Steven Hawley (MS2), Michel Tognini (MS3)	Collins becomes first female Shuttle commander; deployment of *Chandra X-ray Observatory*
96, STS 103, *Discovery*	20 Dec 99, 28 Dec 99	7d, 23h, 12m	Curtis Brown Jr. (C), Scott Kelly (P), Steven Smith (PC), Jean Francois Clervoy (MS2), John Grunsfeld (MS3), Michael Foale (MS4), Claude Nicollier (MS5),	Third *Hubble Space Telescope* servicing mission equips observatory with six new gyroscopes
97, STS 99, *Endeavour*	11 Feb 00, 22 Feb 00	11d, 5h, 40m	Kevin Kregel (C), Dominic Gorie (P), Gerhard Thiele (MS1), Janet Kavandi (MS2), Janice Voss (MS3), Mamoru Mohri (MS4)	Shuttle Radar Topography Mission on board
98, STS 101, *Atlantis*	19 May 00, 29 May 00	9d, 20h, 10m	James Halsell Jr. (C), Scott Horowitz (P), Mary Ellen Weber (MS), Jeffrey Williams (MS), James Voss (MS), Susan Helms (MS), Yuri Usachyov (MS)	Transfer of more than 3,300 pounds of equipment to *ISS*

(Continued)

Human Spaceflight Missions by Country, through February 2011 (Part 37 of 51)

Space Shuttle				
Flight #, Space Transportation System #, Orbiter	**Launch, Landing**	**Duration**	**Crew**	**Highlights**
99, STS 106, *Atlantis*	8 Sep 00, 20 Sep 00	11d, 19h, 12m	Terrence Wilcutt (C), Scott Altman (P), Daniel Burbank (MS1), Edward Tsang Lu (MS2), Richard Mastracchio (MS3), Yuri Malenchenko (MS4), Boris Morukov (MS5)	Transfer of more than 6,000 pounds of equipment to *ISS*
100, STS 92, *Discovery*	11 Oct 00, 24 Oct 00	12d, 21h, 44m	Brian Duffy (C), Pamela Melroy (P), Koichi Wakata (MS1), Leroy Chiao (MS2), Peter Wisoff (MS3), Michael Lopez Alegria (MS4), William McArthur Jr. (MS5)	Installation of Zenith Z1 Truss and third pressurized mating adapter
101, STS 97, *Endeavour*	1 Dec 00, 11 Dec 00	10d, 19h, 58m	Brent Jett (C), Michael Bloomfield (P), Joseph Tanner (MS1), Carlos Noriega (MS2), Marc Garneau (MS3)	Installation of first set of solar arrays during three EVAs
102, STS 98, *Atlantis*	7 Feb 01, 20 Feb 01	12d, 21h, 21m	Kenneth Cockrell (C), Mark Polansky (P), Robert Curbeam (MS1), Marsha Ivins (MS2), Thomas Jones (MS3)	Installation of U.S. Laboratory *Destiny*
103, STS 102, *Discovery*	8 Mar 01, 21 Mar 01	12d, 19h, 51m	James Wetherbee (C), James Kelly (P), Andrew Thomas (MS1), Paul Richards (MS2), James Voss (Exp 2) (up), Susan Helms (Exp 2) (up), Yuri Usachyov (Exp 2) (up), Sergei Krikalev (Exp 1) (dn), William Shepherd (Exp 1) (dn), Yuri Gidzenko (Exp 1) (dn)	First flight of multipurpose logistics module *Leonardo*

Human Spaceflight Missions by Country, through February 2011 (Part 38 of 51)

Space Shuttle				
Flight #, Space Transportation System #, Orbiter	**Launch, Landing**	**Duration**	**Crew**	**Highlights**
104, STS 100, *Endeavour*	19 Apr 01, 1 May 01	11d, 21h, 31m	Kent Rominger (C), Jeffrey Ashby (P), Chris Hadfield (MS1), John Phillips (MS2), Scott Parazynski (MS3), Umberto Guidoni (MS4), Yuri Lonchakov (MS5)	Installation of Canadarm2; first flight of multipurpose logistics module *Raffaello*
105, STS 104, *Atlantis*	12 Jul 01, 25 Jul 01	12d, 18h, 37m	Steve Lindsey (C), Charles Hobaugh (P), Michael Gernhardt (MS1), Janet Kavandi (MS2), James Reilly (MS3)	Installation of joint airlock module (*Quest*) to *Unity* node
106, STS 105, *Discovery*	10 Aug 01, 22 Aug 01	11d, 21h, 14m	Scott Horowitz (C), Frederick Sturckow (P), Patrick Forrester (MS1), Daniel Barry (MS2), Frank Culbertson (Exp 3) (up), Vladimir Dezhurov (Exp 3) (up), Mikhail Tyurin (Exp 3) (up), James Voss (MS3) (Exp 2), Susan Helms (Exp 2) (dn), Yuri Usachyov (Exp 2) (dn)	Second flight of multipurpose logistics module Leonardo
107, STS 108, *Endeavour*	5 Dec 01, 17 Dec 01	11d, 19h, 37m	Dominic Gorie (C), Mark Kelly (P), Linda Godwin (MS1), Daniel Tani (MS2), Yuri Onufrienko (Exp 4) (up), Daniel Bursch (Exp 4) (up), Carl Walz (Exp 4) (up), Frank Culbertson (Exp 3) (dn), Vladimir Dezhurov (Exp 3) (dn), Mikhail Tyurin (Exp 3) (dn)	*ISS* visit; second flight of multipurpose logistics module *Raffaello*; crew carries U.S. flags found at 9/11 attack sites on board *ISS*

(Continued)

Human Spaceflight Missions by Country, through February 2011 (Part 39 of 51)

Space Shuttle				
Flight #, Space Transportation System #, Orbiter	**Launch, Landing**	**Duration**	**Crew**	**Highlights**
108, STS 109, *Columbia*	1 Mar 02, 12 Mar 02	10d, 22h, 11m	Scott Altman (C), Duane Carey (P), John Grunsfeld (MS1), Nancy Sherlock (MS2), Richard Linnehan (MS3), James Newman (MS4), Michael Massimino (MS5)	Fourth *Hubble Space Telescope* servicing mission equips observatory with Advanced Camera for Surveys
109, STS 110, *Atlantis*	8 Apr 02, 19 Apr 02	10d, 19h, 44m	Michael Bloomfield (C), Stephen Frick (P), Rex Walheim (MS1), Ellen Ochoa, (MS2) Lee Morin (MS3), Jerry Ross (MS4), Steven Smith (MS5)	Installation of Center Integrated Truss Assembly S0 and Mobile Transporter
110, STS 111, *Endeavour*	5 Jun 02, 19 Jun 02	13d, 20h, 36m	Kenneth Cockrell (C), Paul Lockhart (P), Franklin Chang Diaz (MS1), Philippe Perrin, (MS2), Valeri Korzun (Exp 5) (up), Sergei Treschev (Exp 5) (up), Peggy Whitson (Exp 5) (up), Yuri Onufrienko (Exp 4) (dn), Carl Walz (Exp 4) (dn), Daniel Bursch (Exp 4) (dn)	*ISS* visit; third flight of multipurpose logistics module *Leonardo*; transfer of Mobile Remote Service Base System onto Mobile Transporter
111, STS 112, *Atlantis*	7 Oct 02, 18 Oct 02	10d, 19h, 59m	Jeffrey Ashby (C), Pamela Melroy (P), David Wolf (MS1), Sandra Magnus (MS2), Piers Sellers (MS3), Fyodor Yurchikhin (MS4)	Installation of Integrated Truss Assembly S1 and Crew Equipment Translation Aid Cart A

Human Spaceflight Missions by Country, through February 2011 (Part 40 of 51)

Space Shuttle				
Flight #, Space Transportation System #, Orbiter	**Launch, Landing**	**Duration**	**Crew**	**Highlights**
112, STS 113, *Endeavour*	24 Nov 02, 7 Dec 02	13d, 18h, 49m	James Wetherbee (C), Paul Lockhart (P), Michael Lopez Alegria (MS1), John Herrington (MS2), Kenneth Bowersox (Exp 6) (up), Nikolai Budarin (Exp 6) (up), Donald Pettit (Exp 6) (up), Valeri Korzun (Exp 5) (dn), Sergei Treschev (Exp 5) (dn), Peggy Whitson Exp 5) (dn)	Installation of Integrated Truss Assembly P1
113, STS 107, *Columbia*	16 Jan 03, 1 Feb 03	15d, 22h, 20m	Rick Husband (C), William McCool (P), David Brown (MS1), Kalpana Chawla (MS2), Michael Anderson (PC), Laurel Clark (MS4), Ilan Ramon (PS1)	Ramon first Israeli in space; foam debris damages orbiter's left wing during launch; results in vehicle breakup and loss of crew during reentry
114, STS 114, *Discovery*	26 Jul 05, 09 Aug 05	13d, 21h, 32m	Eileen Collins (C), James Kelly (P), Soichi Noguchi (MS1), Stephen Robinson (MS2), Andrew Thomas (MS3), Wendy Lawrence (MS4), Charles Camarda (MS5)	Return to flight mission; *ISS* visit; third flight of multipurpose logistics module *Raffaello*; first EVA to repair damage to Shuttle tiles
115, STS 121, *Discovery*	4 Jul 06, 17 Jul 06	12d, 18h, 36m	Steven Lindsey (C), Mark Kelly (P), Michael Fossum (MS1), Lisa Nowak (MS2), Piers Sellers (MS3), Stephanie Wilson (MS4), Thomas Reiter (E) up	Fourth flight of multipurpose logistics module *Leonardo;* Reiter transfers to *ISS* as part of *ISS* Expedition 13

(Continued)

Human Spaceflight Missions by Country, through February 2011 (Part 41 of 51)

Space Shuttle				
Flight #, Space Transportation System #, Orbiter	**Launch, Landing**	**Duration**	**Crew**	**Highlights**
116, STS 115, *Atlantis*	9 Sep 06, 21 Sep 06	11d, 19h, 6m	Brent Jett (C), Chris Ferguson (P), Joe Tanner (MS1), Dan Burbank (MS2), Steve MacLean (MS3), Heidemarie Stefanyshyn-Piper (MS4)	*ISS* visit; testing of new equipment and techniques for repairing Space Shuttle heat shields
117, STS 116, *Discovery*	10 Dec 06, 22 Dec 06	12d, 20h, 44m	Mark Polansky (C), William Oefelein (P), Robert Curbeam (MS1), Joan Higginbotham (MS2), Nicholas Patrick (MS3), Arne Christer Fuglesang (MS4), Sunita Williams (E) (up), Thomas Reiter (dn)	Installation of new P5 truss segment; Williams transfers to *ISS* as part of *ISS* Expedition 14
118, STS 117, *Atlantis*	8 Jun 07, 22 Jun 07	13d, 20h, 11m	Frederick Sturckow (C), Lee Archambault (P), Patrick Forrester (MS1), Steven Swanson (MS2), John Olivas (MS3), James Reilly (MS4), Clayton Anderson (E) up, Sunita Williams (E) (dn)	Installation of S3/S4 truss segment; Anderson transfers to *ISS* as part of *ISS* Expeditions 15 and 16
119, STS 118, *Endeavour*	8 Aug 07, 21 Aug 07	12d, 17h, 56m	Scott Kelly (C), Charles Hobaugh (P), Tracy Caldwell (MS1), Richard Mastracchio (MS2), Dafydd Rhys Williams (MS3), Barbara Morgan (MS4), Benjamin Drew (MS5)	*ISS* assembly flight; Morgan is first educator astronaut to fly in space

Human Spaceflight Missions by Country, through February 2011 (Part 42 of 51)

Space Shuttle				
Flight #, Space Transportation System #, Orbiter	**Launch, Landing**	**Duration**	**Crew**	**Highlights**
120, STS 120, *Discovery*	23 Oct 07, 7 Nov 07	15d, 2h, 23m	Pamela Melroy (C), George Zamka (P), Scott Parazynski (MS1), Stephanie Wilson (MS2), Douglas Wheelock (MS3), Paolo Nespoli (MS4), Daniel Tani (E) (up), Clayton Anderson (dn)	Successful repair and deployment of solar array; Tani transfers to *ISS* as part of *ISS* Expedition 16
121, STS 122, *Atlantis*	7 Feb 08, 20 Feb 08	12d, 18h, 22m	Stephen Frick (C), Alan Poindexter (P), Stanley Love (MS1), Leland Melvin (MS2), Rex Walheim (MS3), Hans Schlegel (MS4), Leopold Eyharts (E) (up), Daniel Tani (E) (dn)	Installation of Columbus module; Eyharts transfers to *ISS* as part of *ISS* Expedition 16
122, STS 123, *Endeavour*	11 Mar 08, 27 Mar 08	15d, 18h, 11m	Dominic Gorie (C), Gregory Johnson (P), Robert Behnken (MS1), Michael Foreman (MS2), Takao Doi (MS3), Richard Linnehan (MS4), Garrett Reisman (E) (up), Leopold Eyharts (E) (dn)	Assembly of Special Purpose Dexterous Manipulator (SPDM), named Dextre; Reisman transfers to *ISS* as part of *ISS* Expedition 17
123, STS 124, *Discovery*	31 May 08, 14 June 08	13d, 18h, 13m	Mark Kelly (C), Kenneth Ham (P), Karen Nyberg (MS1), Ronald Garan (MS2), Michael Fossum (MS3), Akihiko Hoshide (MS4), Gregory Chamitoff (E) (up), Garrett Reisman (E) (dn)	Assembly of Japanese *Kibo* module; Chamitoff transfers to *ISS* as part of *ISS* Expeditions 17 and 18

(Continued)

Human Spaceflight Missions by Country, through February 2011 (Part 43 of 51)

Space Shuttle				
Flight #, Space Transportation System #, Orbiter	**Launch, Landing**	**Duration**	**Crew**	**Highlights**
124, STS 126, *Endeavour*	14 Nov 08, 30 Nov 08	15d, 20h, 30m	Christopher Ferguson (C), Eric Boe (P), Heidemarie Stefanyshyn-Piper (CS1), Donald Pettit (CS2), Stephen Bowen (CS3), Robert Kimbrough (CS4); Sandra Magnus (E) (up), Gregory Chamitoff (E) (dn)	Fifth flight of multipurpose logistics module *Leonardo;* upgrades enable larger crews to reside on station; Magnus transfers to *ISS* as part of *ISS* Expedition 18
125, STS 119, *Discovery*	15 Mar 09, 28 Mar 09	12d, 19h, 30m	Lee Archambault (C), Dominic Antonelli (P), John Phillips (MS1), Steven Swanson (MS2), Joseph Acaba (MS3), Richard Arnold (MS4), Koichi Wakata (E) (up), Sandra Magnus (E) (dn)	Crew completes construction of the *ISS* Integrated Truss Structure; Wakata transfers to *ISS* as part of *ISS* Expeditions 18, 19 and 20
126, STS 125, *Atlantis*	11 May 09, 24 May 09	12d, 21h, 38m	Scott Altman (C), Gregory Johnson (P), Michael Good (MS1), Katherine McArthur (MS2), John Grunsfeld (MS3), Michael Massimino (MS4), Andrew Feustel (MS5)	Fifth *Hubble Space Telescope* servicing mission to replace gyroscopes and batteries, and perform upgrades to equipment
127, STS 127, *Endeavour*	15 Jul 09, 31 Jul 09	15d, 16h, 45m	Mark Polansky (C), Douglas Hurley (P), David Wolf (MS1), Julie Payette (MS2), Christopher Cassidy (MS3), Thomas Marshburn (MS4), Timothy Kopra (E) (up), Koichi Wakata (E) (dn)	Crew performs work on Japanese Exposed Facility and Japanese *Kibo* module; Kopra transfers to *ISS* as part of *ISS* Expedition 20

Human Spaceflight Missions by Country, through February 2011 (Part 44 of 51)

Space Shuttle				
Flight #, Space Transportation System #, Orbiter	**Launch, Landing**	**Duration**	**Crew**	**Highlights**
128, STS 128, *Discovery*	29 Aug 09, 12 Sep 09	13d, 20h, 54m	Frederick Sturckow (C), Kevin Ford (P), Patrick Forrester (MS1), John Olivas (MS2), José Hernández (MS3), Arne Fuglesang (MS4), Nicole Stott (E) (up), Timothy Kopra (E) (dn)	Sixth flight of multipurpose logistics module *Leonardo*; Stott transfers to *ISS* as part of *ISS* Expeditions 20 and 21
129, STS 129, *Atlantis*	16 Nov 09, 27 Nov 09	10d, 19h, 16m	Charles Hobaugh (C), Barry Wilmore (P), Michael Foreman (MS1), Robert Satcher (MS2), Randolph Bresnik (MS3), Leland Melvin (MS4), Nicole Stott (E) (dn)	*ISS* assembly flight; First flight of an EXPRESS Logistics Carrier
130, STS 130, *Endeavour*	8 Feb 10, 22 Feb 10	13d, 18h, 6m	George Zamka (C), Terry Virts (P), Kathryn Hire (MS1), Stephen Robinson (MS2), Robert Behnken (MS3), Nicholas Patrick (MS4)	Crew adds *Tranquility* node to *ISS* along with observatory module *Cupola*.
131, STS 131, *Discovery*	5 Apr 10, 20 April 10	15d, 2h, 47m	Alan Poindexter (C), James Dutton (P), Dorothy Metcalf-Lindenburger (MS1), Stephanie Wilson (MS2), Richard Mastracchio (MS3), Naoko Yamazaki (MS4), Clayton Anderson (MS5)	*ISS* assembly flight; seventh flight of multipurpose logistics module *Leonardo*.
132, STS 132, *Atlantis*	14 May 10, 26 May 10	11d, 18h, 28m	Kenneth Ham (C), Dominic Antonelli (P), Michael Good (MS1), Piers Sellers (MS2), Stephen Bowen (MS3), Garrett Reisman (MS4)	*ISS* assembly flight; last scheduled flight of Shuttle *Atlantis*

(Continued)

Human Spaceflight Missions by Country, through February 2011 (Part 45 of 51)

Space Shuttle				
Flight #, Space Transportation System #, Orbiter	**Launch, Landing**	**Duration**	**Crew**	**Highlights**
133, STS 133, *Discovery*	29 October 10 (*)		Mark Kelly (C), Gregory Johnson (P), Edward Fincke (MS1), Gregory Chamitoff (MS2), Andrew Feustel (MS3), Roberto Vittori (MS4)	*ISS* assembly flight; last scheduled flight of Shuttle *Discovery*
134, STS 134, *Endeavour*	28 February 11 (*)		Steven Lindsey (C), Eric Boe (P), Benjamin Drew (MS1), Michael Barratt (MS2), Timothy Kopra (MS3), Nicole Stott (MS4)	*ISS* assembly flight; last scheduled flight of Shuttle *Endeavour* and last planned mission of the Space Shuttle program
(*) Projected launch date				

Shuttle *Mir*
STS 60 marked the beginning of the Shuttle-Mir program, a joint effort between the United States and Russia that brought astronauts and cosmonauts together in space for the first time since the Apollo-Soyuz Test Project in 1975.

Mission, Orbiter/ Vehicle, Docking #	**Launch, Landing**	**Docking, Undocking**	**Crew**	**Highlights**
STS 60, *Discovery*, None	3 Feb 94, 11 Feb 94	None	Charles Bolden Jr. (C), Kenneth Reightler Jr. (P), Jan Davis (MS1), Ronald Sega (MS2), Franklin Chang Diaz (MS3), Sergei Krikalev (MS4)	Krikalev becomes first cosmonaut on a Space Shuttle flight
STS 63, *Discovery*, near *Mir*	3 Feb 95, 11 Feb 95	None	James Wetherbee (C), Eileen Collins (P), Bernard Harris Jr. (PC), Michael Foale (MS1), Janice Voss (MS2), Vladimir Titov (MS3)	First Shuttle rendezvous with *Mir*
Soyuz TM 21, SL 4, only Soyuz docking	14 Mar 95, 11 Sep 95	16 Mar 95, 11 Sep 95	Vladimir Dezhurov (C) (up), Gennadi Strekalov (E) (up), Norman Thagard (R) (up), Anatoliy Solovyov (dn), Nikolai Budarin (dn)	Thagard, the first American to live on board *Mir*, stays for 115 days

Human Spaceflight Missions by Country, through February 2011 (Part 46 of 51)

Shuttle *Mir*				
Mission, Orbiter/ Vehicle, Docking #	**Launch, Landing**	**Docking, Undocking**	**Crew**	**Highlights**
STS 71, *Atlantis*, Shuttle docking 1	7 Sep 95, 18 Sep 95	29 Jun 95, 5 Jul 95	Robert Gibson (C), Charles Precourt (P), Ellen Baker (PC), Gregory Harbaugh (MS2), Bonnie Dunbar (MS3), Anatoliy Solovyov (Mir 19) (up), Nikolai Budarin (Mir 19) (up), Vladimir Dezhurov (Mir 18) (down), Gennadi Strekalov (Mir 18) (down), Norman Thagard (Mir 18) (down)	Solovyov and Budarin stay on board *Mir*; both returned on 11 September 1995 on board *Soyuz TM 21*
STS 74, *Atlantis*, Shuttle docking 2	12 Nov 95, 20 Nov 95	15 Nov 95, 18 Nov 95	Kenneth Cameron (C), James Halsell Jr. (P), Chris Hadfield (MS1), Jerry Ross (MS2), William McArthur Jr. (MS3)	Hadfield becomes first Canadian mission specialist on board a Space Shuttle
STS 76, *Atlantis*, Shuttle docking 3	22 Mar 96, 31 Mar 96	24 Mar 96, 29 Mar 96	Kevin Chilton (C), Richard Searfoss (P), Ronald Sega (MS1), Michael Clifford (MS2), Linda Godwin (MS3), Shannon Lucid (MS4) (up)	Lucid stays on board *Mir*, the second American to live on the station, for 179 days
STS 79, *Atlantis*, Shuttle docking 4	16 Sep 96, 26 Sep 96	19 Sep 96, 23 Sept 96	William Readdy (C), Terrence Wilcutt (P), Jay Apt (MS1), Thomas Akers (MS2), Carl Walz (MS3), John Blaha (MS4) (up), Shannon Lucid (MS4) (down)	Blaha stays on board *Mir*, the third American to live on the station, for 118 days
STS 81, *Atlantis*, Shuttle docking 5	12 Jan 97, 22 Jan 97	15 Jan 97, 20 Jan 97	Michael Baker (C), Brent Jett (P), Peter Wisoff (MS1), John Grunsfeld (MS2), Marsha Ivins (MS3), Jerry Linenger (MS4) (up), John Blaha (MS4) (down)	Linenger stays on board *Mir*, the fourth American to live on the station, for 123 days

(Continued)

Human Spaceflight Missions by Country, through February 2011 (Part 47 of 51)

Shuttle *Mir*				
Mission, Orbiter/ Vehicle, Docking #	**Launch, Landing**	**Docking, Undocking**	**Crew**	**Highlights**
STS 84, *Atlantis*, Shuttle docking 6	15 May 97, 24 May 97	17 May 97, 22 May 97	Charles Precourt (C), Eileen Collins (P), Jean Francois Clervoy (PC), Carlos Noriega (MS2), Edward Lu (MS3), Yelena Kondakova (MS4), Michael Foale (MS5) (up), Jerry Linenger (MS5) (down)	Foale stays on board *Mir*, the fifth American to live on the station, for 134 days
STS 86, *Atlantis*, Shuttle docking 7	25 Sep 97, 6 Oct 97	27 Sep 97, 3 Oct 97	James Wetherbee (C), Michael Bloomfield (P), Vladimir Titov (MS1), Scott Parazynski (MS2), Jean Loup Chretien (MS3), Wendy Lawrence (MS4), David Wolf (MS5) (up), Michael Foale (MS5) (down)	Wolf stays on board *Mir*, the sixth American to live on the station, for 119 days
STS 89, *Endeavour*, Shuttle docking 8	23 Jan 98, 31 Jan 98	24 Jan 98, 29 Jan 98	Terrence Wilcutt (C), Joe Edwards (P), James Reilly (MS1), Michael Anderson (MS2), Bonnie Dunbar (PC), Salizhan Sharipov (MS4), Andrew Thomas (MS5) (up), David Wolf (MS5) (down)	Thomas stays on board *Mir*, the seventh and last American to live on the station, for 130 days
STS 91, *Discovery*, Shuttle docking 9	2 Jun 98, 12 Jun 98	4 Jun 98, 8 Jun 98	Charles Precourt (C), Dominic Gorie (P), Franklin Chang Diaz (MS1), Wendy Lawrence (MS2), Janet Kavandi (MS3), Valeri Ryumin (MS4), Andrew Thomas (MS4) (down)	*Discovery* was the last Space Shuttle to dock with *Mir*

Human Spaceflight Missions by Country, through February 2011 (Part 48 of 51)

International Space Station					
Expedition #, Launch Vehicle	**Launch, Landing**	**Docking, Undocking**	**Duration**	**Crew**	**Highlights**
Expedition 1, *SL 4*	31 Oct 00, 21 Mar 01	2 Nov 00 on *Soyuz TM 31*, 18 Mar 01 on STS 102	140d, 23h	William Shepherd (C), Yuri Gidzenko (E), Sergei Krikalev (E)	First resident crew of *ISS*; experiments include MACE II, CEO, PCG EGN Dewar, and SEEDS
Expedition 2, *Discovery*	8 Mar 01, 22 Aug 01	10 Mar 01 on STS 102, 20 Aug 01 on STS 105	167d, 6h	Yury Usachev (C), James Voss (E), Susan Helms (E)	Usachev and Voss perform first EVA from *ISS*; conduct 18 separate science experiments
Expedition 3, *Discovery*	10 Aug 01, 17 Dec 01	12 Aug 01 on STS 105, 15 Dec 01 on STS 108	128d, 20h	Frank Culbertson (C), Vladimir Dezhurov (E), Mikhail Tyurin (E)	First use of Russian-built airlock and docking port Pirs; crew conducts 19 science experiments
Expedition 4, *Endeavour*	5 Dec 01, 19 Jun 02	7 Dec 01 on STS 108, 15 Jun 02 on STS 11	195d, 19m	Yury Onufrienko (C), Daniel Bursch (E), Carl E. Walz (E)	Bursch and Walz break U.S. spaceflight endurance record; crew conducts 27 science experiments
Expedition 5, *Endeavour*	5 Jun 02, 7 Dec 02	7 Jun 02 on STS 111, 2 Dec 02 on STS 113	184d, 22h	Valery Korzun (C), Peggy Whitson (E), Sergei Treschev (E)	Whitson becomes first *ISS* science officer; crew conducts 29 science experiments
Expedition 6, *Endeavour*	23 Nov 02, 3 May 03	25 Nov 02 on STS 113, 3 May 03 on *Soyuz TMA 1*	161d, 1h	Kenneth Bowersox (C), Donald Pettit (E), Nikolai Budarin (E)	Pettit serves as second *ISS* science officer; crew conducts 19 science experiments
Expedition 7, *SL 4*	25 Apr 03, 27 Oct 03	28 Apr 03 on *Soyuz TMA 2*, 27 Oct 03 on *Soyuz TMA 2*	184d, 22h	Yuri Malenchenko (C), Ed Lu (E)	Lu serves as third *ISS* science officer; crew conducts 18 science experiments
Expedition 8, *SL 4*	18 Oct 03, 29 Apr 04	20 Oct 03 on *Soyuz TMA 3*, 29 Apr 04 on *Soyuz TMA 3*	194d, 18h	Michael Foale (C), Alexander Kaleri (E), Pedro Duque (E)	Crew conducts 27 science experiments
Expedition 9, *SL 4*	18 Apr 04, 23 Oct 04	21 Apr 04 on *Soyuz TMA 4*, 23 Oct 04 on *Soyuz TMA 4*	187d, 21h	Gennady Padalka (C), Mike Fincke (E), André Kuipers (E)	Fincke serves as fourth *ISS* science officer; crew conducts 21 science experiments

(Continued)

Human Spaceflight Missions by Country, through February 2011 (Part 49 of 51)

International Space Station					
Expedition #, Launch Vehicle	**Launch, Landing**	**Docking, Undocking**	**Duration**	**Crew**	**Highlights**
Expedition 10, *SL 4*	13 Oct 04, 24 Apr 05	15 Oct 04 on *Soyuz TMA 5*, 24 Apr 05 on *Soyuz TMA 5*	192d, 19h	Leroy Chiao (C), Salizhan Sharipov (E), Yuri Shargin (E)	Crew conducts 22 science experiments; Sharipov tosses Russian nanosatellite into orbit during EVA
Expedition 11, *SL 4*	14 Apr 05, 10 Oct 05	16 Apr 05 on *Soyuz TMA 6*, 10 Oct 05 on *Soyuz TMA 6*	179d, 23h	Sergei Krikalev (C), John Phillips (E), Roberto Vittori (E)	Crew photograph *Discovery* orbiter (STS 114) thermal protection system; conduct 19 science experiments
Expedition 12, *SL 4*	30 Sep 05, 8 Apr 06	3 Oct 05 on *Soyuz TMA 7*, 8 Apr 06 on *Soyuz TMA 7*	189d, 19h	William McArthur (C), Valery Tokarev (E), Gregory Olsen (T)	Crew conduct two EVAs; deploy old Russian space suit (SuitSat) with radio transmitter
Expedition 13, *SL4*	29 Mar 06, 28 Sep 06	31 Mar 06 on *Soyuz TMA 8*, 28 Sept 06 on *Soyuz TMA 8*	182d, 22h	Pavel Vinogradov (C), Jeffrey Williams (E), Marcos Pontes (T), Thomas Reiter (E)	Crew conduct three EVAs; Pontes departs with Expedition 12; Reiter arrives during STS 121
Expedition 14, *SL4*	18 Sep 06, 21 Apr 07	20 Sep 06 on *Soyuz TMA 9*, 21 Apr 07 on *Soyuz TMA 9*	215d, 8h	Michael Lopez Alegria (C), Mikhail Tyurin (E), Anousheh Ansari (T), Thomas Reiter (E), Sunita Williams (E),	Reiter remains with Expedition 14 after Expedition 13 departs with Ansari; Williams arrives on STS 116; Reiter departs on STS 116
Expedition 15, *SL4*	7 Apr 07, 21 Oct 07	9 Apr 07 on *Soyuz TMA 10*, 21 Oct 07 on *Soyuz TMA 10*	196d, 17h	Fyodor Yurchikhin (C), Oleg Kotov (E), Charles Simonyi (T), Sunita Williams (E), Clayton Anderson (E)	Williams remains with Expedition 15 after Expedition 14 departs with Simonyi; Anderson arrives on STS 117; Williams departs on STS 117
Expedition 16, *SL4*	10 Oct 07, 19 Apr 08	12 Oct 07 on *Soyuz TMA 11*, 19 Apr 08 on *Soyuz TMA 11*	191d, 19h	Peggy Whitson (C), Yuri Malenchenko (E), Sheikh Muszaphar Shukor (T) , Daniel Tani (E), Léopold Eyharts (E), Garrett Reisman (E)	Shukor departs with Expedition 15; Tani arrives on STS 120; Eyharts arrives on STS 122; Reisman arrives on STS 123

Human Spaceflight Missions by Country, through February 2011 (Part 50 of 51)

International Space Station					
Expedition #, Launch Vehicle	**Launch, Landing**	**Docking, Undocking**	**Duration**	**Crew**	**Highlights**
Expedition 17, *SL4*	8 Apr 08, 24 Oct 08	10 Apr 08 on *Soyuz TMA 12*, 24 Oct 08 on *Soyuz TMA 12*	198d, 16h, 20m	Sergei Volkov (C), Oleg Kononenko (E), So-yeon Yi (T), Garrett Reisman (E), Gregory Chamitoff	Reisman remains with Expedition 17 after Expedition 16 departs with Yi; Chamitoff arrives on STS 124; Reisman departs on STS 124
Expedition 18, *SL4*	12 Oct 08, 8 Apr 09	14 Oct 08 on *Soyuz TMA 13*, 8 Apr 09 on *Soyuz TMA 13*	178d, 15m	Michael Fincke (C), Yury Lonchakov (E), Richard Garriott (T), Gregory Chamitoff (E), Sandra Magnus (E), Koichi Wakata (E),	Chamitoff remains with Expedition 18 after Expedition 17 departs with Garriott; Magnus arrives on STS 126; Chamitoff departs on STS 126; Wakata arrives on STS 119
Expedition 19, *SL4*	26 Mar 09, 11 Oct 09	28 Mar 09 on *Soyuz TMA 14*, 11 Oct 09 on *Soyuz TMA 14*	198d, 16h, 42m	Gennady Padalka (C), Michael Barratt (E), Charles Simonyi (T), Koichi Wakata (E)	Wakata remains with Expedition 19 after Expedition 18 departs with Simonyi
Expedition 20, *SL4*	27 May 09, 1 Dec 09	29 May 09 on *Soyuz TMA 15*, 1 Dec 09 on *Soyuz TMA 15*	187d, 20h, 42m	Gennady Padalka (C), Michael Barratt (E), Koichi Wakata (E), Roman Romanenko (E), Frank De Winne (E), Robert Thirsk (E), Tim Kopra (E), Nicole P. Stott (E)	Padalka, Barratt and Wakata remain on *ISS* to join Expedition 20; Kopra arrives on STS 127; Wakata departs on STS 127; Stott arrives on STS 128; Kopra departs on STS 128
Expedition 21, *SL4*	30 Sep 09, 18 Mar 10	2 Oct 09 on *Soyuz TMA 16*, 18 Mar 10 on *Soyuz TMA 16*	169d, 4h, 10m	Frank De Winne (C), Robert Thirsk (E), Roman Romanenko (E), Nicole Stott (E), Jeffrey Williams (C), Maxim Suraev (E), Guy Laliberté (T)	De Winne, Thirsk and Romanenko remain on *ISS* to join Expedition 21; Laliberté departs on *Soyuz TMA 14*; Stott departs on STS 129
Expedition 22, *SL4*	20 Dec 09	22 Dec 09 on *Soyuz TMA 17*, 2 June 10 on *Soyuz TMA 17*	163d, 5h, 32m	Jeffrey Williams (C), Maxim Suraev (E), Oleg Kotov (E), Soichi Noguchi (E), Timothy Creamer (E)	Addition of Tranquility node, which housed the observatory module (Cupola—Dome)

(Continued)

Human Spaceflight Missions by Country, through February 2011 (Part 51 of 51)

International Space Station					
Expedition #, Launch Vehicle	**Launch, Landing**	**Docking, Undocking**	**Duration**	**Crew**	**Highlights**
Expedition 23, *SL4*	2 Apr 10	4 Apr 10 on *Soyuz TMA 18*		Oleg Kotov (C), Soichi Noguchi (E), Timothy Creamer (E), Alexander Skvortsov (E), Tracy Caldwell Dyson (E), Mikhail Kornienko (E)	

Ansari X Prize					
Mission	**Launch**	**Landing**	**Duration**	**Crew**	**Highlights**
Space-ShipOne, Flight 15 (private)	21 Jun 04	21 Jun 04	24m, 5s	Michael Melville (P)	First privately funded spaceflight; reaches suborbital altitude of 328,491 ft
Space-ShipOne, Flight 16 (private)	29 Sep 04	29 Sep 04	24m, 11s	Michael Melville (P)	Suborbital spaceflight reaches altitude of 337,569 ft; first of two flights in bid to win Ansari X Prize

Ansari X Prize					
Mission	**Launch**	**Landing**	**Duration**	**Crew**	**Highlights**
Space-ShipOne, Flight 17 (private)	4 Oct 04	4 Oct 04	23m, 56s	Brian Binnie (P)	Suborbital spaceflight reaches 367,442 ft; wins $10 million prize

Key to abbreviations: C: Commander; CP: Command Module Pilot; DP: Docking Module Pilot; E: Flight Engineer; LP: Lunar Module Pilot; P: Pilot; PC: Payload Commander; PS: Payload Specialist; MS: Mission Specialist; O: Operator; R: Researcher; SP: Science Pilot; T: Tourist; X: Physician; d: days; h: hours; m: minutes; s: seconds; dn: joins crew/rides launch vehicle returning to Earth; up: joins crew/rides launch vehicle to station. Unless noted, launch and landing dates for spaceflights were based on Universal Time.

Abbreviations

Corp.	Corporation
Ltd.	Limited

Units

Å	Angstrom
arcsec	arcsecond
arcmin	arcminute
AU	Astronomical Unit
bar	bar
bps	bits per second
C	centigrade
cm	centimeter
eV	electron-Volt
ft	feet
G	gauss
Gbps	gigabits per second
GHz	gigahertz
h	hour
Hz	Hertz
in	inch
J	Joule
K	Kelvin
keV	kilo electron-Volt
kg	kilogram
km	kilometer
l	liter
lb	pounds
m	meter
mb	millibar
mm	millimeters
Mbps	megabits per second
MHz	megahertz
MeV	mega electron-Volt
min	minutes

mph	miles per hour
μm	micrometer, micron
nm	nanometer
Oe	Oersteds
R_E	Earth radii
rpm	revolutions per minute
μs	microsecond
T	Tesla
V	Volts
W	Watts
We	Watts of electricity

metric ton 1,000 kilograms

Index

Bold page numbers indicate the main article about the topic.

St. Louis Community College
at Meramec
LIBRARY